Computational Methods for Microstructure-Property Relationships

Somnath Ghosh • Dennis Dimiduk
Editors

Computational Methods for Microstructure-Property Relationships

Springer

Editors
Somnath Ghosh
Ohio State University
Dept. Mechanical Engineering
W. 19. Ave. 201
43210 Columbus Ohio
USA
ghosh.5@osu.edu

Dennis Dimiduk
Wright-Patterson Air Force Base
Materials & Manufacturing Directorate
Air Force Research Lab.
45433-7702 Dayton Ohio
USA
dennis.dimiduk@wpafb.af.mil

ISBN 978-1-4419-0642-7 e-ISBN 978-1-4419-0643-4
DOI 10.1007/978-1-4419-0643-4
Springer New York Dordrecht Heidelberg London

Library of Congress Control Number: 2010935949

Printed on acid-free paper

Springer is part of Springer Science+Business Media (www.springer.com)

This book is dedicated to all those individuals whose discontent with the present state of knowledge and vision for the future make the research wheels turn

Preface

Design of and with materials plays an intrinsic role in today's challenging world of high performance structural components and applications. They constitute an integral part of comprehensive structural design, given the opportunity offered by optimal materials design for structural performance and life enhancement. These opportunities impose high demands on effective modeling and simulation methodologies to establish quantitative relations between the material microstructure and physical properties at different length scales. The rapid advances in computer and computational sciences enable sophisticated simulations that unravel the underpinnings of complex material microstructure on behavior. In concert with outstanding advances in experimental methods, these computational tools are increasingly able to enhance the fundamental understanding of microstructure–property relations, thus enabling materials and process design for improved performance and life.

The field of computational materials modeling transcends traditional disciplinary boundaries between mechanics, materials science, physics and chemistry, mathematics and computer science. In addition, it is creating a true synergy between experiments and modeling in terms of incorporation of physics, calibration, and validation. The results of these unified efforts at various scales are yielding unprecedented levels of rigor and accuracy in predictions of complicated phenomena that have previously eluded scientists and engineers. The role of microstructure on physical properties and performance is emerging as a quantitative discipline with broad and direct implications on material design.

This book "Computational Methods for Microstructure–Property Relationships" is an attempt to capture this rapid advancement at a period of time, with a glimpse of what is yet to come in this very dynamic emergent field of science and technology. It introduces state-of-the-art advances in computational modeling as well as experimental approaches for materials structure–property relations. Representing a body of collected works by well-known professionals in the field, it covers topics ranging from materials modeling principles with a multiscale perspective to materials design. It presents the current state of knowledge for a wide collection of research areas related to materials assessment in structures–materials interactions. The collection aims at establishing the necessity of a robust integrated computational mechanics and computational materials science framework, together with an experimental validation protocol, that treats heterogeneous materials at microstructural and continuum scales. Selectively encompassing both computational mechanics and computational materials science disciplines, it offers an analysis of current techniques and selected topics important to industry researchers, such as deformation, creep,. and fatigue of primarily metallic materials. It emphasizes modeling at continuum and heterogeneous microstructural scales, e.g. crystalline or grain scales, validated with experimental observations. This book in intended for researchers in academia, industry, and government laboratories to understand the issues and challenges involved in predicting performance and failure in materials, with a focus on the engineering structure–materials interaction. Researchers, engineers, and professionals involved with predicting the performance and failure of materials are expected to find this book a valuable reference.

This book is topically divided into four essential parts. The first part deals with 3D image-based materials structure data collection and representation and microstructure builders for mechanical response simulations.

Chapter "Serial Sectioning Methods for Generating 3D Characterization Data of Grain-and Precipitate-Scale Microstructures" introduces serial-sectioning methods for 3D characterization of grain- and precipitate-scale microstructures. It focuses on the use of serial-sectioning methods and associated instrumentation as a means for collecting microstructural, crystallographic, and chemical data. Chapter "Digital Representation of Materials Grain Structure" discusses the state-of-the-art methods in the field of microstructure representation with focus on the following: data collection, feature identification, mesh generation, quantitative descriptors, and synthetic structure generation. Chapter "Multi-Scale Characterization and Domain Partitioning for Multi-Scale Analysis of Heterogeneous Materials" discusses the development of a multiscale characterization methodology leading to a microstructural morphology-based domain partitioning method for materials having nonuniform heterogeneous microstructure. The set of methods is intended to provide a concurrent multiscale analysis model with the initial computational domain that delineates regions of statistical homogeneity and heterogeneity. The method is intended as a preprocessor to multiscale analysis of mechanical behavior and damage of heterogeneous materials. Chapter "Coupling Microstructure Characterization with Microstructure Evolution" discusses the synergistic coupling of quantitative microstructure characterization via experimental imaging techniques, with computer simulations of microstructural evolution using the phase-field method. Having experimental images as inputs, the chapter describes uses of the phase-field method at different length scales to explore mechanisms of microstructural evolution, extract material parameters, conduct physics-based repairs of experimentally reconstructed microstructures, and evolve the microstructure for different time, temperature, stress, etc. regimes.

Part of this volume is devoted to materials constitutive laws and kinematical approaches, together with their coupling of material structure to responses via simulation codes. These are presented in next six chapters. Chapter "Representation of Materials Constitutive Responses in Finite Element Based Design Codes" surveys

FEM-based tools for simulating materials behavior and reviews the material models available in commercial codes. Chapter "Accounting for Microstructure in Large Deformation Models of Polycrystalline Metallic Materials" analyzes the influence of microstructure on large-strain mechanical behavior for metallic polycrystalline materials. Results of a macroscale continuum internal state variable-based model for tantalum are compared with those from a multiscale polycrystalline plasticity approach having explicit representation of the polycrystalline aggregate. In Chapter "Dislocation Mediated Continuum Plasticity: Case Studies on Modeling Scale Dependence, Scale-Invariance, and Directionality of Sharp Yield-Point", the authors discuss a field dislocation dynamics theory to account for the emergence of inhomogeneous dislocation distributions at mesoscopic length scales, as well as their coupling to initial and boundary conditions and consequences on mechanical behavior. Size effects and scale-invariant intermittency are interpreted through field dislocation dynamics. Anisotropy of strain hardening induced by the emergence of internal stress fields is also reviewed in this chapter. Chapter "Dislocation-Mediated Time-Dependent Deformation in Crystalline Solids" shows a methodology for incorporating the effects of slip gradients associated with intra-grain deformation heterogeneity in crystal-plasticity-based finite element simulation. The treatise quantifies the orientation dependence of the misorientation field in the polycrystalline microstructure and introduces a modification of the kinematic decomposition that accommodates distortions arising from the presence of a static dislocation distribution. Chapter "Modeling Heterogeneous Intra-Grain Deformations Using Finite Element Formulations" is a review of two crystal plasticity-based methodologies for the prediction of microstructure–property relations in polycrystalline aggregates. These include a mean-field, second-order viscoplastic self-consistent method and a Fast Fourier Transform-based full-field method. Numerical examples demonstrate that models like the FFT-based formulation can explicitly account for interaction between individual grains. Finally, chapter 11 discusses multiscale modeling of plastic deformation and strength in crystalline materials with emphasis on models and experiments below the grain level. Specifically, the chapter deals with experimental advances and theoretical models for characterizing dislocations at the subgrain level.

The third part introduces computational mechanics for time dependency of materials with links to fracture mechanics and multi-time scaling methods for fatigue in next three chapters. Chapter "Stochastic Upscaling for Inelastic Material Behavior from Limited Experimental Data" develops time-dependent plastic deformation and creep models for crystalline solids using dislocation-level mechanics. The theory uses microstructural information to develop broad quantitative mechanistic relationships that match the observed phenomenology. The discussion includes mobility-controlled systems, where dislocations move through the crystal under stress and interaction of dislocations with discrete obstacles for a range of alloys. Chapter "DDSim: Framework for Multiscale Structural Prognosis" introduces a prototype hierarchical computational simulation system called damage and durability simulator (DDSim) for prognosis of fatigue life of airframe components. While this prototype focuses on fatigue cracking, the framework can be extended to other modes of damage. Chapter "Modeling Fatigue Crack Nucleation Using Crystal Plasticity Finite Element Simulations and Multi-Time Scaling" addresses two important aspects of predicting fatigue crack nucleation in polycrystalline alloys under dwell cyclic loading. The first is a microstructure-sensitive criterion for dwell fatigue crack initiation in polycrystalline titanium alloys, while the second part of this chapter discusses a wavelet transformation-based multi-time scaling (WATMUS) algorithm for accelerated crystal plasticity finite element simulations. The WATMUS algorithm significantly enhances the computational efficiency for fatigue life prediction.

Finally, the fourth part of this book deals with some additional emerging topics in the next three chapters. Chapter "Challenges Below the Grain Scale and Multiscale Models" examines selected experimental methods at different length scales that are important tools in building models for location specific design. While special experimental techniques are needed to probe the material at finer scales to assess the local behaviors, testing methods at all scales are discussed to demonstrate the breadth of experimental capability available at each scale of the material. A stochastic upscaling approach for strain-hardening plastic materials from limited experimental data based on random matrix theory is introduced in chapter "Emerging Methods for Matching Simulation and Experimental Scales". The uncertainty characterized by constitutive tangential matrices can be construed as a reflection, on the coarse scales, of fluctuations of the fine scale features from which constitutive matrices are constructed. Finally, chapter "Simulation-Assisted Design and Accelerated Insertion of Materials" introduces some emerging concepts for robust design of materials and challenges for the synthesis of modeling and simulation and materials design. The distinction between materials design and multiscale modeling is elucidated in this chapter with emphasis on top-down requirements on material structure and performance to meet product requirements.

The editors note that this work would not have been possible without continued financial and technical support from their employers, namely The Ohio State University and the Air Force Research Laboratory, Materials and Manufacturing Directorate. They also gratefully acknowledge the research support from various sponsoring agencies, viz. the Defence Advanced Research Projects Agency (Program Director: Dr. Leo Christodoulou), The Air Force Office of Scientific Research (Directors: Drs. Lyle Schwartz and Tom Russell; Program Directors: Drs. Craig Hartley and David Stargel), The Army Research Office (Program Director: Dr. Bruce Lamattina), and the Office of Naval Research (Program Director: Dr. Julie Christodoulou).

In closing, the editors would like to extend their sincere thanks and appreciation to all the contributing authors of this volume for embracing our template vision and providing excellent state-of-the-art articles on different topics in the general field. They are also thankful to the Springer editorial staff, particularly Alex Greene and Andrew Leigh, for their tremendous support with the production of this book. Somnath Ghosh expresses his love and deep appreciation to his wife Chandreyee, son Anirban, and mother Lalita for their constant encouragement and support throughout this project. Dennis Dimiduk offers his deepest appreciation to his wife Lisa whose love and support made this work possible. He also extends thanks to the current and past members of the advanced metals team at AFRL who helped to form aspects of the vision represented in this book.

Columbus, Ohio
Dayton, Ohio

Somnath Ghosh
Dennis M. Dimiduk
April 2010

Contents

Contributors

A. Acharya Department of Civil and Environmental Engineering, Carnegie Mellon University, Pittsburgh, PA 15213, USA, acharyaamit@cmu.edu

F.L. Addessio Theoretical Division, Los Alamos National Laboratory, Los Alamos, NM 87545, USA

Masoud Anahid Department of Mechanical Engineering, The Ohio State University, W496 Scott Laboratory, 201 West 19th Avenue, Columbus, OH 43210, USA

D. Backman Worcester Polytechnic Institute, Mechanical Engineering, Worcester, MA, USA

David F. Bahr School of Mechanical and Materials Engineering, Washington State University, Pullman, WA, USA

A.J. Beaudoin Department of Mechanical Sciences and Engineering, University of Illinois at Urbana Champaign, Urbana, IL 61801, USA

R. Brenner Laboratoire des Proprietés Mécaniques et Thermodynamiques des Matériaux, Université Paris XIII, Av J.-B.Clement, 93430 Villetaneuse, France

Robert A Brockman Universal Energy Systems, Dayton, USA

C.A. Bronkhorst, Theoretical Division, Los Alamos National Laboratory, Los Alamos, NM 87545, USA, cabronk@lanl.gov

P. Ponte Castañeda Department of Mechanical Engineering and Applied Mechanics, University of Pennsylvania, Philadelphia, PA 19104, USA

O. Castelnau Laboratoire des Proprietés Mécaniques et Thermodynamiques des Matériaux, Université Paris XIII, Av J.-B.Clement, 93430 Villetaneuse, France

E.K. Cerreta Materials Science and Technology Division, Los Alamos National Laboratory, Los Alamos, NM 87545, USA

Pritam Chakraborty Department of Mechanical Engineering, The Ohio State University, W496 Scott Laboratory, 201 West 19th Avenue, Columbus, OH 43210, USA

Yoon Suk Choi Universal Energy Systems, Dayton, USA,
Yoon-Suk.Choi@wpafb.af.mil

Yuwen Cui Department of Materials Science and Engineering, The Ohio State University, Columbus, OH 43210, USA

Glenn Daehn Department of Materials Science and Engineering, The Ohio State University, Columbus, OH 43210, USA

Sonjoy Das Massachusetts Institute of Technology, Cambridge, MA 02139,
sonjoy@mit.edu
and
Department of Civil Engineering, University of Southern California,
Los Angeles, CA, USA

Paul Dawson Sibley School of Mechanical and Aerospace Engineering, Cornell University, Ithaca, NY 14853, USA, prd5@cornell.edu

Dennis M Dimiduk Materials and Manufacturing Directorate, Air Force Research Laboratory, Wright-Patterson Air Force Base, OH, USA,
dennis.dimiduk@wpafb.af.mil

John Emery Solid Mechanics Division, Sandia National Laboratories, Albuquerque, NM 87113, USA, jmemery@sandia.gov

Claude Fressengeas LPMM, Universite Paul Verlaine - Metz/CNRS Ile du Saulcy, 57045 Metz Cedex 01, France, claude.fressengeas@univ-metz.fr

Jobie Gerken ANSYS, Inc., Canonsburg, PA 15317, USA

Roger Ghanem Department of Civil Engineering, University of Southern California, Los Angeles, CA 90089, USA, ghanem@usc.edu

Somnath Ghosh Department of Mechanical Engineering, The Ohio State University, W496 Scott Laboratory, 201 West 19th Avenue, Columbus, OH 43210, USA, ghosh.5@osu.edu

G. T. Gray III Materials Science and Technology Division, Los Alamos National Laboratory, Los Alamos, NM 87545, USA

Michael A. Groeber Materials and Manufacturing Directorate, Air Force Research Laboratory, Wright-Patterson Air Force Base, OH, USA,
Michael.Groeber@wpafb.af.mil

E. N. Harstad Engineering Sciences Division, Sandia National Laboratories, Albuquerque, NM 87185, USA

Anthony R. Ingraffea Cornell Fracture Group, 643 Rhodes Hall, Cornell University, Ithaca, NY 14853, USA, ari1@cornell.edu

Deepu S. Joseph Department of Mechanical Engineering, The Ohio State University, W496 Scott Laboratory, 201 West 19th Avenue, Columbus, OH 43210, USA

R.A. Lebensohn Materials Science and Technology Division, Los Alamos National Laboratory, MS G755, Los Alamos, NM 87545, USA, lebenso@lanl.gov

Ning Ma Corporate Strategic Research, ExxonMobil Research & Engineering Company, 1545 Route 22 East, Rm LB248, Annandale, NJ 08801, USA

P. J. Maudlin Theoretical Division, Los Alamos National Laboratory, Los Alamos, NM 87545, USA

Tito Marin Department of Industrial Engineering, University of Parma, Parma, Italy

D.L. McDowell Georgia Institute of Technology, GWW School of Mechanical Engineering, Atlanta, GA, USA, david.mcdowell@me.gatech.edu

Michael Mills Department of Materials Science and Engineering, The Ohio State University, Columbus, OH 43210, USA, mills.108@osu.edu

Andrew H. Rosenberger Materials & Manufacturing Directorate, Air Force Research Laboratory, Wright-Patterson Air Force Base, OH, USA, Andrew.Rosenberger@wpafb.af.mil

Chen Shen GE Global Research, 1 Research Circle, Niskayuna, NY 12309, USA

Michael D. Uchic Materials and Manufacturing Directorate, Air Force Research Laboratory, Wright-Patterson Air Force Base, OH, USA, Michael.Uchic@wpafb.af.mil

Yunzhi Wang Department of Materials Science and Engineering, The Ohio State University, Columbus, OH 43210, USA, wang.363@osu.edu

Hussein M. Zbib School of Mechanical and Materials Engineering, Washington State University, Pullman, WA, USA, zbib@wsu.edu

Ning Zhou Department of Materials Science and Engineering, The Ohio State University, 2041 College Road, Columbus, OH 43210, USA

Microstructure–Property–Design Relationships in the Simulation Era: An Introduction

Dennis M. Dimiduk

Abstract Computational methods are affecting a paradigm change for using microstructure–property relationships within materials and structures engineering. This chapter examines the emergent use of quantitative computational tools for microstructure–property–design relationships, primarily for structural alloys. Three major phases are described as a historical "serial paradigm," current "integrated computational materials engineering" and, future "virtual materials systems" emerging from advances in multiscale materials modeling. The latter two phases bring unique demands for integrating microstructure representations, constitutive descriptions, numerical codes, and experimental methods. Importantly, these approaches are forcing a fundamental restructuring of materials data for structural engineering wherein data centers on a hierarchy of model parameterizations and validations, rather than the current application-specific design limits. Examining aspects of current research on microstructure-sensitive design tools for single-crystal turbine blades provides an accessible glimpse into future computational tools and their data requirements. Finally, brief descriptions set context and interrelationships for the remaining chapters of the book.

1 Microstructure–Property–Design Relationships and Structural Materials Engineering

Present-day advancements in microstructure–property relationships are coming about via computational methods. The efforts largely recognize that microstructure–property relationships evolve over a wide range of scales and that both technical and computational advances must occur for adequate representations of these

---

D.M. Dimiduk (✉)
Materials and Manufacturing Directorate, Air Force Research Laboratory, Wright-Patterson Air Force Base, OH, USA
e-mail: dennis.dimiduk@wpafb.af.mil

S. Ghosh and D. Dimiduk (eds.), *Computational Methods for Microstructure-Property Relationships*, DOI 10.1007/978-1-4419-0643-4_1,

relationships within predictive tools. However, many of these efforts fall short of full recognition that engineered materials are systems. What is needed is a computational methodology and framework for systems engineering of materials and the sciences that support such an approach. The systems engineering of materials within a simulation environment will provide advances to both the materials utilization and representations for usefully advancing quantitative microstructure–property relationships. To better understand what is needed from the computational framework, it is useful to briefly examine materials in present-day engineering.

About 100 years ago, a defining aspect of materials science and engineering (MSE) had its origin in the first microscopy studies of materials structure; yet, nearly a half-century would pass before their impact evolved into the MSE discipline (Smith 1988; Cahn 2001; Olson 1997). With the study of microstructure (including defect structure or statistical extrema), the materials engineer gained an important tool by which processes and properties are controlled. Microstructure–properties science was born and has expanded ever since. MSE now recognizes four major disciplines of practice that identify the unique character of the MSE field: processing–structure–properties–design, irrespective of the material type, class, or design application (see Fig. 1). However, unlike other mainstream engineering

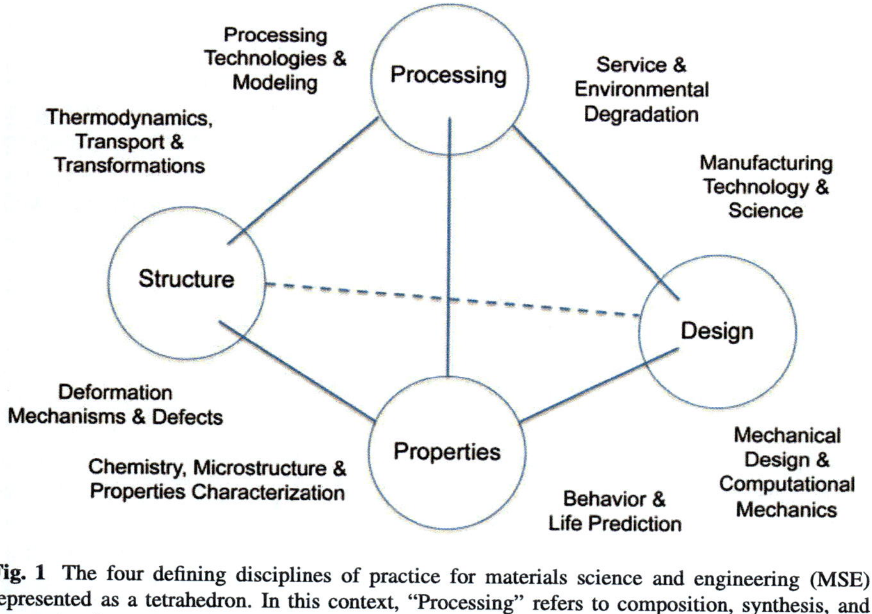

Fig. 1 The four defining disciplines of practice for materials science and engineering (MSE) represented as a tetrahedron. In this context, "Processing" refers to composition, synthesis, and processing in general. "Structure" refers to all aspects of microstructure, including both intrinsic and extrinsic defects from the atomic to macroscopic scales. "Properties" and "Design" refer to materials performance and behavior and, to engineering design rather than materials design, respectively. Selected examples of the types of studies and activities that tend to link the major disciplines are shown about the periphery of the figure

disciplines (e.g., civil, chemical, electrical, and mechanical) that largely came into existence as the quantitative frameworks for them emerged, the same cannot be said for MSE.

More broadly speaking, structural alloys tend to be defined from two different perspectives. Materials producers (and patent law) associate them with compositions of matter and the prescribed synthesis and process paths by which they are formed into engineered products. Alternatively, from a structures-design engineering perspective, materials are viewed as contextual databases containing representative measured values of property bounds, including statistical minima, as functions of selected variables, such as temperature or state of stress. These are often represented within models. The contextual aspect of those databases usually relates to specific application products and manufacturing processes. Additionally, the structures engineer also attributes the businesses and practices that make materials available as products to the materials engineering discipline. Historically, the metallurgist, ceramist or chemical engineer and, more recently the materials-scientist or engineer carried out the onerous task of melding these perspectives into unified activities and practices for safe and affordable structures. Within this engineering reality, the notion of microstructure–property relationships is only implicit at best. While both materials and process engineers and part designers recognize that structural materials have significant microstructural variations, there are few quantitative tools and standards that permit integration of that knowledge into the broad engineering process, especially in a predictive manner. Consequently, outside of the MSE community, processing–properties–design relationships (not including structure) are generally recognized via the "allowables" (i.e., distributions of property values including statistically defined minima) for using a material for a given design. For most of today's products, one typically defines an application and then seeks to define and document a processing specification through specific suppliers by which a selected composition of matter will reproducibly lead to properties (performance) for that application. Material microstructure descriptors, such as grain size (ASTM 2002), are only used as specifications of the material for process assurance.

For higher value engineered structures (e.g., a gas turbine engine disk), a database of processing–properties relationships typically develops in which the data are reduced to phenomenological constitutive laws that are linked to the application design process via finite element method simulations of the part configuration. With few exceptions, these constitutive relationships are assumed to hold over volumes of material that are essentially on the scale of the part (meter scale), even though every metallurgist or materials scientist understands that heterogeneities or defects that affect properties exist over many length scales from the full part to the atomic level. Clearly, there is a disconnection between MSE, and the broader engineering community as the notion of processing–structure–properties–design essentially does not exist beyond MSE.

This disconnection is a rational result of the fact that even after a century of development the quantitative links between processes and structure and, structure and properties, are insufficiently advanced to permit direct systems-oriented optimization of materials and products (in "Simulation-Assisted Design and Accelerated Insertion of Materials" by McDowell and Backman). There is simply too much complexity associated with the kinetics of processes to quantitatively define the resulting microstructures within the equally complex hierarchy of length and timescales of the applications. For electronic materials, the length and timescales may be extremely small fractions of seconds and nanometer dimensions, while for structural composites they may be at the scale of the components and system dimensions (meters) over timeframes of years.

Fortunately, current advances in computing capabilities and MSE tools bring opportunities for not only expanding the quantitative basis for processing–structure–property–design relationships within simulation environments but also for redefining aspects of MSE within those simulation environments. In so doing, MSE becomes a quantitative engineering discipline for structural materials and several aspects of its relationship to other engineering disciplines will be redefined. Full recognition of this opportunity stems from considering aspects of the use of computer modeling and simulations along the evolutionary path of MSE.

2 Computational Materials Science for Microstructure

Computers and simulation were available essentially since the origins of MSE as a recognized discipline. Several phases of their use in MSE are linked to growth in computational capacity and databases. In the 1950s and 1960s, the computer was commonly used to model specific phenomena, usually within a mean-field, especially where numerical solutions to differential equations were necessary. From a materials engineering perspective, perhaps the best example of this is the computer calculation of phase diagrams or the "CalPhad" method that was well developed by the end of the 1960s (Kaufman and Bernstein 1970). During that period foundations were built for materials-oriented computer simulations that last to this day (see additional diverse examples such as computing diffraction contrast of transmission electron microscopy images (Head et al. 1973) and plasticity analysis for metal deformation (Mandel 1973; Kocks 1987) to name but two others. Importantly, even though the foundational sciences were known more than 40 years ago, neither the computational capacity nor the necessary databases were sufficiently developed for the CalPhad method to have significant engineering impact at that time for alloy or process development. Only about 10 years ago did the method begin to add value to engineered products and the practices of MSE. Today, after more than a decade of sustained development investments for engineering, CalPhad techniques are becoming a part of standard industrial methods (NMAB 2008; Backman et al. 2006).

A second phase in the maturation of materials computational methods occurred during the 1970s and 1980s through research in process modeling. Simulation codes evolved that are still in use today (ProCAST http, DEFORM http). These codes, based on continuum fields and state variables without treatments of microstructure, are essential to design engineering of aerospace and other industrial parts and components. Also during this period, methods for solving a range of materials challenges from the electronic structure of materials to techniques for plasticity and stress analysis continued to advance (Hafner 2000; McDowell 2000). Methods for simulating plasticity under crystallographic constraint within the finite-element method gave new insights into behavior at the mesoscopic scale, including strain localization during crystal slip (Asaro 1983).

One could say that during the late 1970s through mid 1980s, computational materials science (CMS) came into its own as a discipline of study. Here, the term CMS refers to the activities of a widespread community of investigators that are developing simulation tools to represent unit mechanisms exhibited by materials. These include such techniques as electronic structure methods for selected material properties and thermodynamic quantities (Hafner 2000; van de Walle et al. 2002; Liu et al. 2006); empirical atomistic methods that offer insight into understanding dislocation core structures, surfaces and grain boundaries (Daw and Baskes 1984; Vitek 1985; Tschopp et al. 2008); dislocation dynamics methods (Devincre et al. 2001; Ghoniem et al. 2000); phase field methods and, many others. A good compendium of such methods may be found in the work edited by Yip (2005).

However, the majority of the CMS-based advances in understanding mechanisms of materials behavior had little or no impact on materials engineering. While the quantitative nature of simulated results improved, too frequently they lacked comprehensive context or sufficient accuracy for use in engineering design. The few applications of simulation-based methods to industrial-world microstructure–properties engineering tended to use simulation results to provide qualitative insight into existing engineering processes (see for example Dimiduk 1998); however, there were notable exceptions (Shercliff and Ashby 1990). There are numerous reasons for this, but obvious among them was insufficient computational capacity together with the integrated data available during that time.

Throughout the 1980s and 1990s, many materials simulation efforts were performed in relative isolation within the MSE, physics, mechanics, and chemistry communities having few linkages to engineering techniques or design tools. Unlike other engineering disciplines, one might suppose that the role of simulations within MSE was viewed as only interesting or important for understanding qualitative behavior trends since so little community-wide work was carried out to establish standard techniques and methods as foundations for industrial practice. Consequently, simulation-centric materials engineering methods continued to evolve in a piece-wise fashion within proprietary corporate communities. Throughout this period, there were few efforts outside of the process-modeling discipline that attempted to integrate mechanistic or heuristic knowledge within simulations to understand the microstructure–property relationships in engineered products, as a standard methodology of practice. Although some researchers recognized that the quantitative aspects of microstructure–property relationships were underdeveloped (Cedar 2000), CMS was often characterized as simply "applied quantum mechanics" (Bernholc 1999). During this period, CMS was essentially a "cottage industry" of models and modelers of and to itself (Dimiduk et al. 2004a).

During the mid and late 1990s, a few industry, government and academic leaders began to see the limiting aspects of this state of CMS (see for example Olson 1997; Christodoulou DARPA-AIM http; Fraser CAMM http). These leaders recognized that CMS approaches to materials modeling typically originated from the "bottom up" of the length and time scales and that such approaches rarely made an impact on the practices or efficiencies of design engineering, especially for structural materials. Further, there was recognition in the MSE community that significant computing capability was becoming sufficiently widespread that new approaches to simulation-based materials engineering should be attempted from the "top down." As a result, two notable new initiatives in computational-based materials engineering were initiated in the first year of the new millennium (NMAB 2008).

3 Integrated Computational Materials Engineering

3.1 Materials Readiness and the Evolving Microstructure–Properties–Design Paradigm

To best understand the uniqueness of the integrated computational materials engineering (ICME) approach and its impact on the practices associated with microstructure–properties–design relationships, it is useful to first understand the concepts of materials engineering readiness. Materials development and process engineering involves significant open-ended risk and cost. To manage and mitigate that risk, the MSE community adopted various frameworks for assessing readiness along the pathway toward product application. These frameworks are similar to ones used for other engineering but are tailored (especially within major manufacturing companies) to materials and processes disciplines. Figure 2 illustrates the highest-level structured "stage-gate" process that exists within most materials and processes practice. Typically, ten levels of readiness are defined and the progression of application-specific technologies through these levels occurs within well-defined engineering templates. These templates demand specific test data, cost assessments, manufacturing source qualification, etc. that gain fidelity *and scope* at each stage of development. This serially staged paradigm of materials and processes technology maturity to some degree reflects the learning curve that innately exists for anything new. Unfortunately, the expanding scope required at each step is a key limiter to this paradigm that adds significant risk, quite often cost, and certainly time.

Reviews of case studies of materials development that follow a serial paradigm have shown that it leads to serious challenges for materials development and limits the opportunities for coupling materials and process advancements within mainstream engineering design practice (NMAB 2008; NMAB 2004; Lipsitt et al. 2001; Dimiduk 2001; Dimiduk et al. 2003). The serial paradigm leads to what has been called the "valley-of-death" for new materials and processes. That valley exists for several reasons including funding gaps, long time requirements for experimental

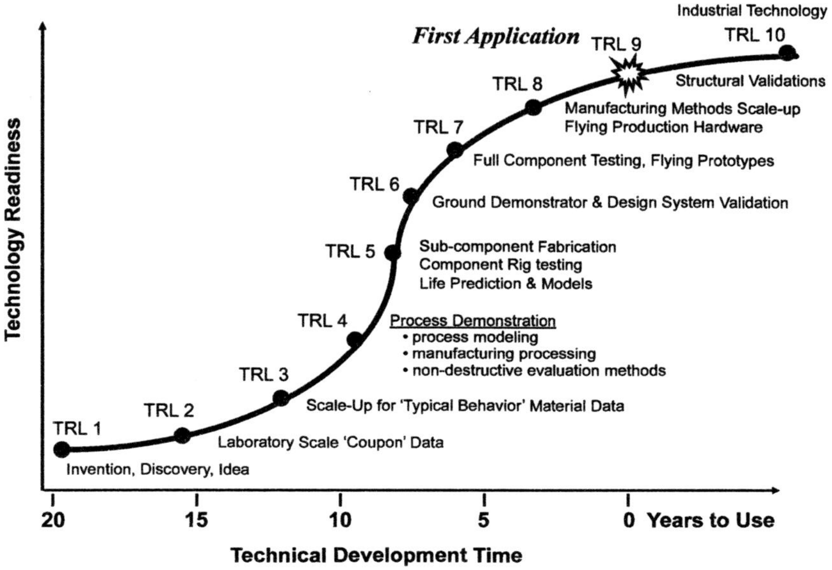

Fig. 2 General technology readiness levels (TRL) for materials. The ten stages of materials readiness are adopted from the broader engineering readiness metrics used for products and systems. Historically, achieving the transition from TRL 3 to TRL 5 is the most difficult step. The reason for this is that technical risks typically remain high at TRL 3; however, the financial outlays required to mitigate them also grow much more substantially at this stage by comparison to the lower levels. Better materials and processes simulation tools are needed throughout, but especially for risk mitigation through the TRL 3–5 maturity levels. For aerospace materials, evolutionary advances (such as modified alloy compositions within established applications) are known to require 7–12 years to reach first use. For more challenging completely new materials, such introducing ceramic composites or TiAl alloys in turbine engines, the time span for achieving fist-use readiness exceeded 26 and 36 years, respectively

or empirical iterations and, what may loosely be called a "point contact" interface between present-day design engineering and materials engineering.

To further illustrate this point of contact, Fig. 3 schematically depicts the broad engineering procedural steps that may be used to select the geometric configuration of a manufactured aerospace metal component. The figure also shows selected materials and processes procedural steps that are taken to assure appropriate microstructure–property relationships are maintained in the final product. Inspection of the figure reveals that the primary interface between the design process and the materials development process lies in the steps needed to assure that validated constitutive descriptions (or minima curves and allowables) are available for the design optimization procedures. Thus, within this schematic depiction, the interface between the communities is a point contact. This point of contact includes not only the constitutive laws that reside within component design codes, but also their empirical validation against databases that must sufficiently encompass the variations of microstructure–property relationships judged to be important to the specific

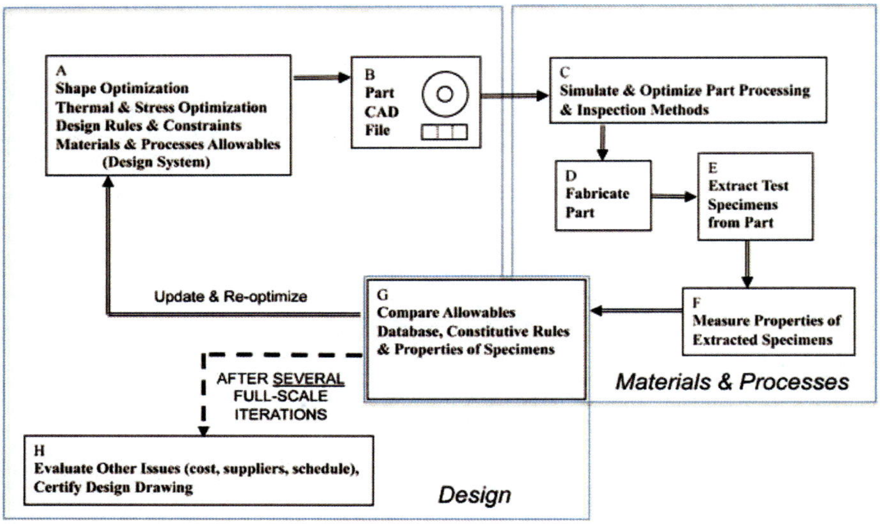

Fig. 3 Schematic representation of activities within today's experiment-intensive processing–properties–design serial paradigm for materials engineering. The methodology has no explicit consideration of microstructure. Microstructural effects are only implicitly considered when extracting specimens and as selected specifications for parts. Microstructural effects/variation is represented through expensive, time-consuming testing and multiple full-scale process and test iterations are usually required

design. Given that the allowables databases are produced from application-specific, full-scale development hardware, this serial approach inevitably leads to a conservative estimation of material performance and does so through a costly process.

Since part-specific and feature-specific microstructures and properties cannot be accounted for within the design system, the observed "worst-case" uncertainties are assigned to all parts at all locations (Christodoulou and Larsen 2004). Consequently, microstructure–property relationships are specified and controlled in the context of their application databases alone, usually via testing of full-scale prototype parts. However, further advancement in the design process demands a less conservative and more realistic, probabilistic approach (McClung et al. 2008; Millwater and Osborn 2006). That new demand is driving the MSE community toward developing predictive tools for location-specific properties that can be used within probabilistic design tools.

Herein lies one major hurdle for microstructure–properties sciences and materials development in general. As long as materials behavior can only be indirectly defined within the very specific contexts of their applications, via extensive testing of samples excised from full-scale prototypes that may not even directly capture the design or materials-limiting features of interest, materials development will always entail long development times and high costs. That fundamental limitation in the procedure for obtaining and representing materials performance data presently places the whole of MSE into a unique domain that is outside of those of the other engineering disciplines. The time scales, cost structures and design tools are simply mismatched, while the risk is high. Today one develops empirical knowledge of materials response to chemistry and process iterations within the stage-gated templates described previously, such that learned practitioners of the engineering disciplines can support design judgments. Those judgments inevitably entail reasonable assurances to business managers that the financial investments in scale-up and advances in technology readiness are affordable within business plans and product timing. For the future, materials development needs to be achieved via a new materials-to-design paradigm. Essential to that paradigm is that the materials readiness structure (readiness templates) be re-cast to maximize the scope of readiness information at the earliest stages; then, to expand only their fidelity with added development investments and time. Fortunately, efforts toward building these are well underway.

3.2 Accelerated Insertion of Materials, Virtual Aluminum Castings, and the ICME Paradigm

Today, the computational tools that facilitate quantitative support for the development and investment judgments required for materials scale-up are just emerging. Examples of these exist within the ICME demonstration efforts that occurred during this decade (NMAB 2008). Essentially, the underlying concept behind the efforts is that having simulation tools for all aspects of new product and materials development will reduce development time while lowering costs and risks. Two notable examples of the ICME paradigm will now be discussed.

Within the aerospace sector, the Accelerated Insertion of Materials (AIM) program was sponsored by the US Defense Advanced Research Projects Agency (DARPA) and the United States Air Force, to examine and restructure the paradigms for metal and organic-composite materials development (NMAB 2008; Backman et al. 2006; Dimiduk et al. 2003, 2004a, b). Similarly, within the automotive sector, the Virtual Aluminum Castings (VAC) program was sponsored by Ford Motor Company (NMAB 2008; Allison et al. 2006). In the specific sense of microstructure–property relationships, the efforts showed that representing the microstructural aspects of materials (especially including kinetics and mechanical behavior), via models that are integrated within design-engineering optimization protocols and software, yields dividends to the product development cycle. Importantly, the case studies showed that even elementary theory and empirical models have a substantial positive impact on the design engineering process when fully integrated within a computational environment (NMAB 2008; Dimiduk et al. 2003). Somehow, that important payoff to engineering was missed by most of the CMS and MSE research communities and to this day is not developed as an integrated materials engineering standard practice. Also, when viewed from the perspectives of these demonstrations, there is now a clear justification for expanding the fidelity

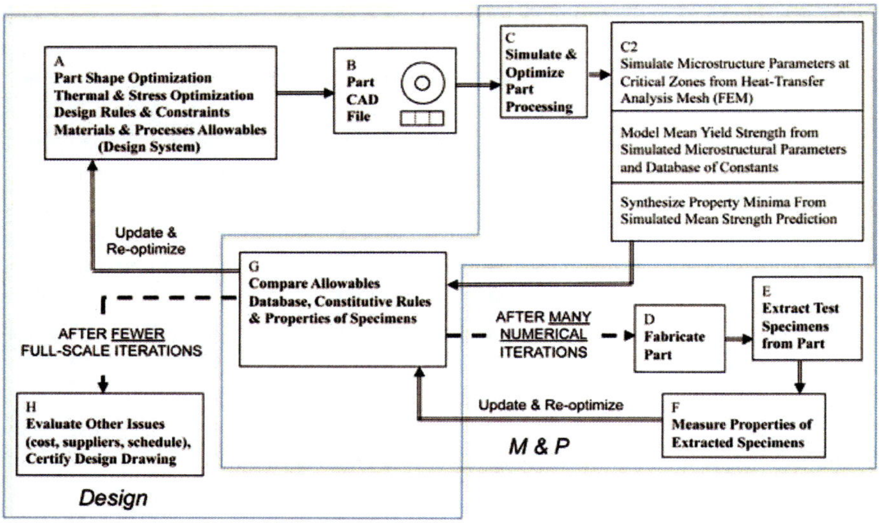

Fig. 4 Schematic representation of activities within ICME paradigm for materials engineering (see text for explanation). Methodology explicitly includes microstructural-based design via microstructure evolution within process models and, mechanical property models being applied to various regions of designed part. Including microstructure–property relationships via simulations means that the domains of design and materials engineering overlap much more significantly within ICME than within the historical paradigm for materials engineering. The ICME paradigm includes the early cases of explicitly using processing–structure–properties–design within closed-loop engineering frameworks

of microstructure–property representations and predictive capabilities and, also a somewhat general template for both focusing those developments and then integrating them into the product value stream as they occur.

The materials and processes development paradigm has changed with the evolution of ICME. Figure 4 shows a similar schematic as the one previously described in Fig. 3, but with modifications that reflect broad procedural changes brought about via the ICME approach as it was applied in the AIM program. Two aspects of the new procedure are noteworthy. First, as shown by the expanded activities associated with step "C2" (in the upper right-hand side of the figure), specific simulation tools focused on microstructure–property relationships enter into the development paradigm. Second, utilizing such tools fundamentally changes the experimental activity that currently takes place to empirically assure the manufactured products perform in the desired fashion. Rather than many full-scale synthesis and processing trials followed by sectioning and testing, many of the results of such efforts are now anticipated via simulations. Having models, even in empirical form, integrated with the design process permits iteration and optimization via design tools and minimizes the time-consuming and expensive procedures associated with full-scale prototype product development. Thus, the overlap between engineering design and MSE fields of practice has expanded. That expansion is the direct result of using simulation tools to provide a more quantitative and structured description of the microstructure–property relationships of materials. A widespread acceptance by a peer group of engineers, systematic reductions in the types, cost and quantity of data needed and, the predictive nature or capabilities of the microstructure–properties relationship tools used within such a paradigm, are all direct measures of the quantitative advance of the field. Future advances in computational methods for microstructure–property relationships should be evaluated by those metrics.

3.3 The Evolving Needs for Materials Data

Another important aspect of the ICME paradigm for materials not explicitly shown in Fig. 4 was a significant aspect of the both AIM and VAC feasibility demonstrations. That aspect pertained to the development of models and the nature of experimental data. Within the historical processing–properties–design paradigm for materials, critical design data exists almost entirely in the form of measured mechanical properties obtained from production-scale hardware – again, having little explicit connection to microstructure. However, the ICME paradigm changes the structure and types of data that are essential to design. *Under the ICME approach, data must be associated with models and supported simulation codes*. Also, specific types of data are collected for the primary purpose of validating codes. That data often extends outside of the ranges typically associated with prototype parts and may be associated with certain pedigree-type materials and microstructures. Fortunately, just as simulation tools and models are becoming more advanced, test and evaluation procedures are becoming automated and miniaturized. Critical data can increasingly be measured from small-scale samples prepared to validate kinetics or mechanical behavior domains for models.

The last paragraph discusses points that are nontrivial and merit further comment. For example, the nature of data intrinsic to an expanding ICME paradigm is data associated with simulation tools and their validation. Those tools by their inherent architectures and operative material models define the data required for their use. In this respect, the ICME paradigm is in its infancy and aspects of data taxonomy and efficiencies must be developed for the purposes of supporting simulations. However, even from the initial case studies just described, the ICME paradigm suggests a different view of data and materials informatics than the one described by recent reports on the subject (Cebon and Ashby 2006; Arnold 2006). Those reports essentially describe higher-fidelity extensions of the classical MSE-design paradigm – a paradigm constrained by the empirical development of handbook materials allowables. Within that paradigm, the role of the computer is "passive" in that it primarily facilitates the organization of greater quantities of information. When data sets are sufficiently large and too complex for typical human interrogation, this paradigm may not exclude cases of the "blind discovery" of new relational knowledge (data mining) in a more "active" mode.

However, computational tools and simulation environments are beginning to synthesize data that may be fused with conventional empirical measurements (Liu et al. 2006; van de Walle et al. 2002). The practice is likely to spread far beyond its present use within alloy thermodynamics. Yet, there is little readiness for this within the old processing–properties–design paradigm and the practice is limited even within the current ICME paradigm. The MSE and design communities have a formidable task ahead of them to define appropriate data architectures and a taxonomy that will not only permit full "active" utilization of materials simulations in the design process but also maintain efficient certifiable engineering practices throughout the new simulation era. Within an emergent paradigm called here "virtual materials systems" that taxonomy and the actual data are facets of the substantially expanded and quantitative nature of microstructure–property relationships. Finally, the new ICME paradigm suggests that the materials allowables view of data will change to more effectively utilize the active power of materials simulations for "synthesizing" data and providing quantitative insights into materials response.

3.4 ICME: Lessons Learned

There is value to considering lessons learned from the initial case studies of AIM and VAC. The recent report by the US National Materials Advisory Board discusses some of these lessons (NMAB 2008), but a selected three global aspects are highlighted here. First, for the longer term, the contrasting primary attributes of engineering design and MSE must be bridged. For engineering design, those attributes include a simulation-centric community of practitioners, education structures that convey such practices, well developed and supported simulation tools that are integrated with heuristic data and, the expectation that many rapid-time-frame simulations will be carried out as a routine part of the design process. Conversely, the primary attributes of MSE in this regard currently include long lead times for experiment results within a data focused community of practitioners, an educational system that is just now grappling with an appropriate treatment of ICME and its tools, relatively few established and supported simulation codes that are still too separated from heuristic data and, a general expectation that when simulations are done they will commonly be characterized by relatively few large-scale simulations performed in a supercomputing environment. As aspects of the previous discussion and portions of this book support, the gap between these communities exists in no small part because of the still underdeveloped quantitative sciences and standards associated with materials microstructure–properties kinetics and mechanical behavior.

A second lesson contained in ICME is that the engineering design paradigm needs to evolve to explicitly include material heterogeneity within engineered parts (read microstructure–property–design relationships). In present day design practice, those aspects of heterogeneity not broadly included in databases or represented in analytical and simulation tools tend to be captured via heuristic rules that constrain the design process. For example, heterogeneities within materials lead to a variation in the performance for identically designed parts and populations of those parts perform differently. Consequently, that variability in parts often leads to the costly replacements of part populations based upon time in service, rather than conditional replacement tied to specific part behavior (Christodoulou and Larsen 2004). In the longer term the development of simulation tools must strive to mitigate the need for heuristic rules by integrating sensor measurements of the service history and environment into materials response models. In that way, the design and user communities would also gain tools to assess variations in the part lives that result from variations in their application environments. Conversely, it is important for the ICME and CMS communities of practice to recognize that in most cases of engineering design, there is an incomplete understanding of details of the use or operational environment even with sensor measurements. Thus, heuristic rules will always be a part of the design system to varying degrees and both design and materials must strive for robustness.

A third key lesson focuses on the notion that all engineering design proceeds from *representations* of the desired properties, behavior phenomena, part-geometry, design constraints and, the materials from which parts are constructed. The materials engineer should ask of every item of interest "how should this item be modeled and represented in the optimization framework for design?" The whole of this book focuses on selected aspects of microstructure–property representations. However, engineering design demands representations for many additional aspects of the product value-stream that could interact effectively within simulation-based materials properties tools, including product cost (in "Simulation-Assisted Design and Accelerated Insertion of Materials" by McDowell and Backman). Thus, the nature of the representation used in simulation for a selected attribute or property is one clear measure of present-day understanding, relative importance and tractability of the attribute within the design and simulation environment.

The examples of ICME to date suggest that simulation-based approaches to microstructure–properties relationships can be effectively used in the design process to add value to engineered products. Thus, it is reasonable to expect that the fidelity of those tools for the representations of microstructure–property relationships will also grow. Consequently, it is useful to consider where the expansions of microstructure–property science may lead within the modeling and simulation era.

4 Multiscale Materials Modeling, Materials Systems Simulation Science, and Virtual Materials Systems

Present-day advancements in microstructure–property relationships are coming about via the techniques of multiscale materials modeling, especially concurrent multiscale modeling. Those efforts largely recognize that microstructure–property relationships evolve over a wide range of scales and that advances must occur for adequate representations of these within predictive tools. However, even most of those efforts fall short of full recognition that engineered materials are systems and, engineering demands standards of practice. What is needed for MSE and CMS is a systems approach to materials simulations and the sciences that supports such an approach. The systems engineering of materials within a simulation environment will provide the usefully structured advances to both materials utilization and the tools for quantitatively representing microstructure–property relationships.

Given the context of materials engineering discussed previously, it is useful to peer into the future of microstructure–properties science and engineering. This book captures one view of that future through a selected look at a few of the advanced techniques in the field as well as some of the pacing state-of-the-art capabilities and challenges. The set of techniques is drawn from the editors' viewpoint that microstructure–properties relationships science is headed toward the development of "virtual materials systems" in every sense of the term. That is, just as the biological sciences are slowly evolving toward computer-based representations of systems (such as humans for example) that somewhat virtually function in the same ways as their real-world counterparts, so too MSE should strive to supply computer-based systems representations of materials that mimic the real-world behavior at all scales (Wikipedia Virtual Human http). This view of the future demands a full embrace of simulation tools as an integrating theme and, in some respects, a defining aspect of the quantitative microstructure–property sciences. The view also requires that research embrace the notion that materials rarely perform outside of a systems context (for example, a turbine engine airfoil system, or an automotive engine valve system, etc.).

4.1 Microstructure–Property Representation and Simulation

The engineering objective is to represent microstructure–property relationships together with engineered part designs within virtual materials systems. Quantitative predictions of part performance are obtained via numerous statistical instantiations of the material microstructure and the resultant simulated responses of those structures for a current part configuration. To achieve such simulation environments, one must recognize the fact that there are only four primary domains of freedom for simulations that collectively determine the quality and fidelity of the resultant predictions. Figure 5 depicts those domains for representing each aspect of materials microstructure–property relationships in a computational environment. Aspects of concurrent multiscale materials modeling strive to expand these four domains of materials representation by having the structure representation and perhaps even the constitutive description(s) evolve in an adaptive fashion as heterogeneities (such as local deformation or micro-cracking) evolve out of the initial representation. Simulation science involves quantitative management of error metrics, clear descriptions of failure criteria and, an intimate knowledge of the computing environment employed for the simulation set.

As depicted in Fig. 5, there must be a multiscale representation of the engineered structure and its microstructure that includes coarse-scale domains of the part,

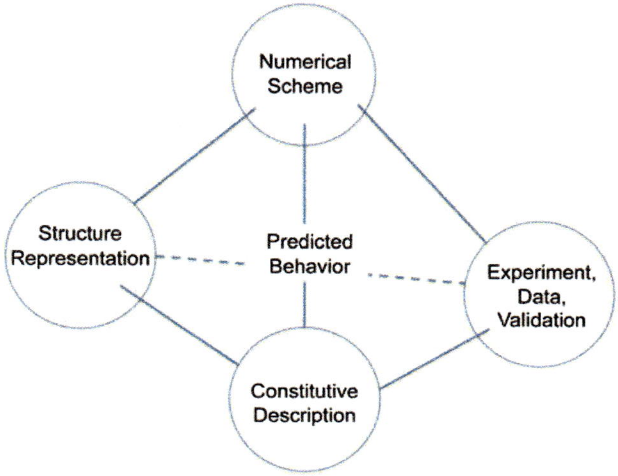

Fig. 5 For any property of interest in design, there exists a multiscale hierarchy of microstructural effects that must be represented within simulation codes. However, as this figure depicts, within the computational environment there are only four broad domains of freedom for representing all of those aspects of the material. Recognition of these four domains provides a means for assigning each aspect of the material to the simulation environment and, by doing so, clearly identifies the coarse-graining inherent to the selected technique

extrinsic defect structures, intrinsic microstructure, their statistics at various scales and, even the smallest-scale aspects that affect chemical kinetics. That representation must dovetail with the constitutive descriptions of the system energetics and evolution. For example, the constitutive descriptions may involve pseudopotential formulations for electronic interactions at one lower length-scale, empirical atomic interaction potentials at another scale, mean-field thermally activated process models at a still larger scale, as well as the myriad mechanical behavior descriptions that are captured in present-day property models and design codes. However, to select the most appropriate modeling and simulation development pathway, it is not enough to know the constitutive relationships and structure representation alone; one must also know the context of the system, or design requirements that are to be simulated. Both of these in turn must align with all aspects of computational tractability of the simulation methods, represented in the figure by the domain of numerical schemes. For example, today there is little ability to represent thermally activated processes within parametric dislocation dynamics simulations, making the present form of that numerical environment a poor choice for studying the creep behavior of materials. Similarly, as some of the chapters in this book reflect, there is a growing ability to use explicit grain-level representations of microstructure within continuum constitutive descriptions of flow to examine deformation localization and instabilities for a variety of materials. However, going still further, as these explicit microstructural methods emerge so too must the constitutive descriptions evolve since the present ones tend to coarse-grain at an inappropriate scale. Evolving methods for concurrent multiscale simulations will eventually permit localization and time-dependent failure initiation when structure representations and constitutive rules are tailored for those methods of solution.

Finally, the aspects of the material that are not represented within the previously described domains of structure representation, constitutive laws or numerical schemes, must be brought to the simulation via measured quantities or empirical calibration parameters. Obviously, there should be recognition that experiments are as much a part of multiscale materials modeling as the simulations themselves. Consequently, there are new quantitative tools emerging for approaching those experimental challenges (Zhao 2006; Uchic et al. 2006; in "Emerging Methods for Matching Simulation and Experimental Scales" by Rosenburger).

4.2 Single-Crystal Turbine Blades: An Emerging Case Study

The design and manufacture of turbine engine airfoils is a multibillion dollar-per-year industry. The materials and designs used for the single-crystal high-pressure turbine airfoils represent a limiting aspect of these ubiquitous engines. Since the efficiencies of the engines depend upon the maximum temperature of the gas path, there is a sustained need to find materials and designs that permit continued gains.

4.2.1 A Prototype Challenge

In recent years, the operating temperatures of turbine engines have risen beyond the melting point of the Ni-superalloy single-crystal materials used to make the hot section airfoils. Clever designs and manufacturing methods that permit cooling air to flow through the interior of the airfoil, coupled with complex zirconia-based coating systems on the exterior portions exposed to the combustion gases led to such high-performance capabilities (Reed 2006). They have also resulted in complex states of time-dependent stress during service and are an interesting example of applications where dimensional constraint imposed by aero-thermal design interacts with materials at dimensional scales comparable to microstructural dimensions. The continued evolution of these highly engineered hybrid material systems demands advances in design methodology, which today resides principally with anisotropic elasticity and homogeneous descriptions of material response (Meric et al. 1991; Arakere and Swanson 2002; Harrison et al. 2004). What is needed for turbine blade design is a computationally tractable, higher-fidelity design system that permits a better analysis of the spatial–temporal stress state and damage accumulation in a representative environment.

Figure 6 shows that the aerodynamic and cooling geometry design features of cooled airfoils (wall thickness, cooling channels, ribs, etc.) are on the scale of the primary material microstructure. Thus any variation of the material may lead to variations in airfoil behavior from region-to-region and from airfoil-to-airfoil, simply as a result of those variations occurring at differing locations relative to the designed geometric features. How does one best use computational methods,

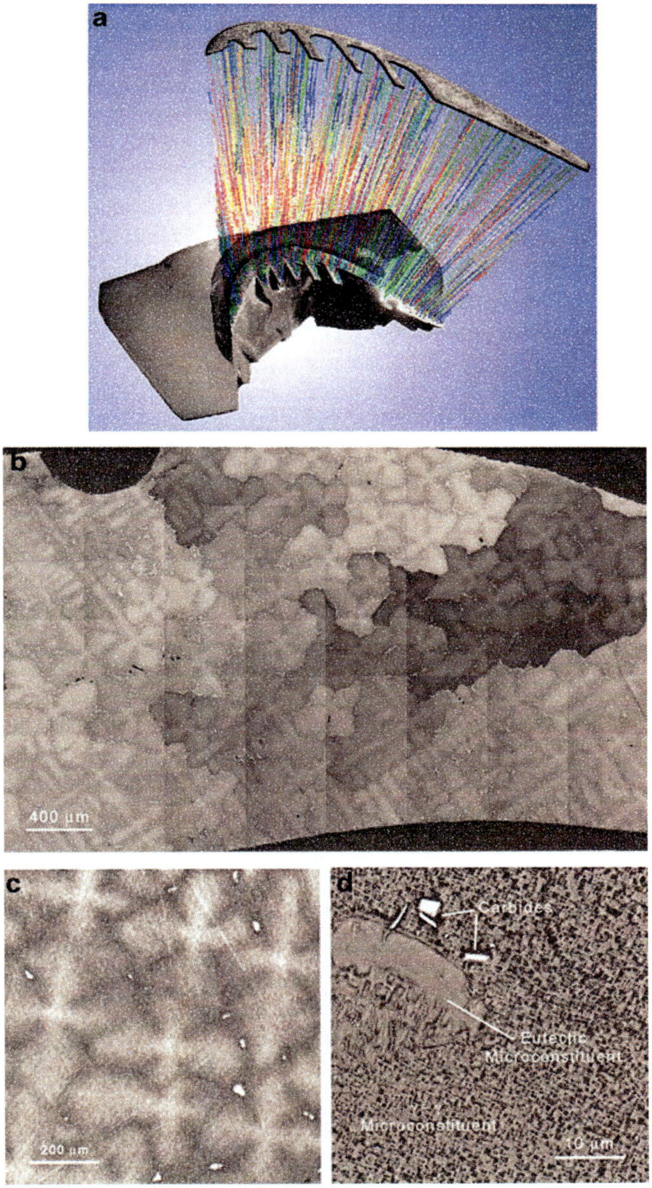

Fig. 6 (**a**) Dendritic microstructure representation for a cast single crystal Ni-base superalloy turbine blade produced from serial sectioning and optical metallography. Colored lines map dendrite cores from base to top. Top of figure shows a metallographic section revealing dendrite cores. Note that blade has been "filet" cut to reveal dendrite core locations relative to cooling channels of the airfoil. (**b**) Backscattered electron image of blade cross-section showing crystal orientation contrast associated with low-angle misoriented grains. (**c**) Optical micrograph of etched cross section showing dendrites and white eutectic particles. (**d**) Backscattered electron images of etched cross section showing the finer microconstituents of a typical superalloy blade. Images provided by M. Groeber

especially for microstructure–property relationships, to permit such assessments within the design process? How might an airfoil designer assess the probability of the weakest-link or life-limiting microstructural feature occurring at the geometric feature that most limits the design? To begin to answer these questions, one must examine not only the nature of the material microstructure–property relationships, but also their representation in design simulation codes, as previously suggested by Figure 5. One must devise representations of the material's structure and its response to time-dependent loading states, all within some context of numerical frameworks that may be used to perform design simulations.

Figure 6 also shows examples of microstructural variables that can be important to performance variations across a population of turbine airfoils of a constant design and manufacturing process. These arise from both the complexity of the superalloys themselves and the methods of their manufacture (Pollock and Tin 2006). Figure 6b–e shows examples of misoriented or low-angle grain boundaries, the dendritic microstructure that results from chemical segregation during casting solidification and, a mixture of eutectic microconstituent, carbides and pores, respectively. Freckle grains may lead to locally high stresses since these are local polycrystalline regions. The low-angle boundaries are a generally accepted feature of the otherwise single-crystal materials; however, the superalloys exhibit severe crystal orientation sensitivity to their creep behavior (MacKay and Maier 1982). Thus, design should be able to assess the stress states relative to these features when the airfoils are configured. While much of the dendritic structure is annealed away during heat treatment, the homogenization is never complete and studies show that internal stresses develop at the scale of the dendrite spacing (Epishin et al. 2004). Pore and eutectic microconstituents (including carbides) tend to be locally soft or hard relative to the matrix, thus concentrating strain under load and leading to fatigue crack initiation (Yi et al. 2007; Liu et al. 2008). Thus, each of these features affect internal stresses but is not taken into account within current design methods. In fact, even when design practice extends beyond treatments of the single-crystal superalloy as an elastic solid, these microstructural features are not directly included, perhaps accounting for some of the variability between performance and design.

4.2.2 Deficiencies in the Processing–Properties–Design Paradigm

The standard practice for including creep or fatigue response into a design falls back on the previously described processing–properties-design paradigm for materials engineering. The material is represented by a database of design minima curves from testing and basic feature configurations derived from experience. Nondestructive inspection methods are used to assure that cast blade crystal orientations are within the bounds set by the design curves and to selectively inspect for other defects. To establish those limitations, one might produce cast bars (having net sections much larger than the airfoils themselves) from which test specimens are prepared to evaluate creep and fatigue properties at a macroscopic scale. While a single primary crystallographic orientation and the secondary dendrite arm spacing may be evaluated for those bars, efforts rarely track/control other aspects of microstructure. Thus, the processing–properties–design paradigm employed in this case implicitly assumes that: (1) the cast microstructure of those specimens represents the same microstructures found in the turbine blade configurations, (2) the stochastic variations of properties within populations of tested specimens encompasses the property variations occurring within turbine blades and, (3) perhaps most importantly, that each of the microstructural details controlling properties is a homogeneous or equal-likelihood-of-occurrence entity over the configuration of the turbine blade. Within this paradigm, the "local continuum" (local) approximation is implicitly invoked well above the scale of key microstructural features, simply by the choice of specimens and scales used for determining material behavior.

Nowhere within the processing–properties–design paradigm does one explicitly assure that the correlation lengths (de-correlation lengths) for the stochastic variations of microstructural features are exceeded, even though the local approximation cannot hold below such correlation lengths. Said differently, for any statistically varying aspect (descriptor) of microstructure, there is a minimum volume of material or number of cases of that aspect of the material that must be included in the tested volume, so that statistical fluctuations have no significant influence on the behavior of the volume. For sample sizes that equal or exceed the correlation length, any random sample is expected to behave similarly and independently of any other sample (they are de-correlated). Further, nowhere during these procedures, except perhaps by de-rating material performance capability (design minima) does one tailor the design process for the fact that some microstructure attributes cannot be homogenized. As examples, at the scale of the turbine blade feature sizes, freckle defects, and low-angle grain structures occur having only one to three features through the wall thickness. These are not present in sufficient numbers to be statistically or homogeneously represented over the airfoil. Coarse features such as these do not have a correlation length within the context of the turbine blade and, may not have one even at the larger scale of the tested specimens.

The notion of establishing the de-correlation length for controlling microstructural features is an important one for setting foundations of microstructure–property relationship simulations. Such correlation lengths exist in a hierarchical way across microstructural scales and descriptors. That notion is pervasive and may lead to better foundations from which to build quantitative MSE tools and techniques. For example, chapters of this book suggest that fundamental scientific questions remain open regarding the viability of establishing representative volume elements (RVE) for evolving path-dependent plastic properties (e.g., see "Representation of Materials Constitutive Responses in Finite Element-Based Design Codes" by Choi and Brockman, "Accounting for Microstructure in Large Deformation Models of Polycrystalline Metallic Materials" by Bronkhorst et al.). At the scale of dislocations and substructure evolution, no quantitative theory exists and empirical approaches have not developed much beyond the scalar "dislocation density" and associated hypotheses. At the scales of grain structure, basic questions centered on establishing the correlation length for grain–grain interaction effects in a 3d elastic-viscoplastic zone or for 3d plastic front propagation, remain relatively unaddressed except for highly idealized cases (in "Representation of Materials Constitutive

Responses in Finite Element-Based Design Codes" by Choi and Brockman, Simonovski et al. 2004). From a multiscale science and physics perspective such absences make the prospects for accurate predictive simulations of failure properties, such as crack initiation, rather remote since the kinematical driving forces would not be known even if the atomic processes could be adequately represented. For such properties, one engineering challenge is to establish a protocol, using accessible methods for microstructure RVE construction, to assess the validity and inaccuracies of property simulations and to gain insights into the weakest volume elements that inevitably control mechanical behavior. Consequently, aspects of material property variability and design minima are perhaps reflections of the underdeveloped materials–design interface in that there are no sufficient methods to treat unsolved materials science issues within engineering design. These may also illustrate a growth area needed within microstructure–properties based materials engineering and computational methods for microstructure–property relationship.

4.2.3 A Look Forward

Current research is exploring the use of concepts developed by Ghosh et al. (2001, 2007, 2008; Swaminathan et al. 2006; Swaminathan and Ghosh 2006) for 2d simulations of long-fiber composites, to build a framework for microstructure–property–design simulations of turbine blades (Groeber et al. 2009). At the scale of the entire blade, many analysis iterations are needed and a substantial number of volumetric analysis nodes are demanded simply from the spatial variation and complexity of the blade features. Therefore, the structure representations and analysis methods must be computationally quick or fast acting.

Following Figs. 5 and 6, there are at least four steps to representing turbine airfoils at this scale. First, one needs to define a structure representation having sufficient fidelity to represent both the engineering design geometry and the microstructural features too large or heterogeneous to be represented within a single local continuum entity. One may also choose to represent distinct defect structures as identifiable features at this scale. Second, validated constitutive descriptions are needed for the property response of selected interest. These flow rules are assigned to the discrete continua of the structure representation. However, a key open aspect of these flow rules is the level of finer-scale microstructure and/or failure criteria that they represent (MacLachlan et al. 2001; Harrison et al. 2004; Ma et al. 2008; Choi et al. 2009). Usually, an anisotropic elastic-viscoplastic yield function or state-variable model, with or without crystallography and/or a damage model, would be the highest level of complexity that could be carried at this scale. Third, steps one and two need to be established within a numerical framework that is self-consistent with those selections. Within Ghosh's scheme, a concurrent adaptive finite element method is preferred since such methods permit natural strain or damage localization during the strain evolution and couples them to lower length-scale aspects of the microstructure. Finally, a formal engineering protocol requires that a parameterization and validation testing methodology be established at the same dimensional scale as the microstructural discreteness selected for the blade representation. For this example, that engineering requirement implies isolating, sectioning and testing various feature and specimen sizes from actual airfoils, rather than from separate test bars as current procedures employ.

The first three steps just described constitute an adaptation of what is termed a "Level 0" (L0) or part domain analysis within Ghosh's scheme. However, one still needs to rigorously tie these to lower scale microstructural features and micromechanisms of behavior as deformation and damage evolve during analysis. To achieve this, Ghosh's method defines a "Level 1" (L1) analysis domain at a lower scale that is a homogenized material point, for two reasons. Within the L1 domain, a statistically equivalent representative volume element (SERVE) may be defined from microstructure characterization and descriptor set development coupled with standard asymptotic homogenization methods (Swaminathan et al. 2006; Swaminathan and Ghosh 2006). These SERVEs are used to numerically compute the anisotropic yield functions used for L0 analysis and may be iterated to include the intrinsic statistical variations of microstructure. In addition, as an L0 domain simulation run evolves, the coupling of far-field loads to geometric and coarse microstructural features begins to localize against pre-selected criteria. The concurrent adaptive numerical scheme permits a new L1 SERVE analysis domain to be inserted, as a periodic domain that introduces a higher level of microstructural fidelity that interacts with the now localizing (stress) fields. Methods for establishing these domains and selected criteria for tracking localization have been previously described and are treated by Ghosh, et al. (Valiveti and Ghosh 2007, in "Multiscale Characterization and Domain Partitioning for Multiscale Analysis of Heterogeneous Materials").

For turbine blade analysis, establishing the L1 analysis tools presents several challenges. By definition, the L1 representation can only comprise microstructural features whose correlation length is smaller than the SERVE for that domain. That is, the microstructural features must be amenable to computational homogenization at that scale. In this example, low-angle grain structures and freckle grains would not qualify; however, the dendritic structure may.

Characterization methods are needed for the dendritic microstructure to ascertain how many dendrite features are sufficient to establish the internal dendritic stress state. As Fig. 7 shows, Shade (Shade 2008) has made some progress in this regard by showing that differences in flow stress between dendrite cores and interdendritic regions can be directly measured. Also, the L1 representation must carry most aspects of the γ–γ' microconstituent and any variations of it that may occur at the scale of the dendrites. What remains unclear in this example of the L1 domain is how much of the interdendritic microconstituents (eutectics, carbides, and pores) can be or should be represented at the L1 scale. Hence, a challenge for the materials engineer developing computational methods for microstructure–property relationships is to establish not only the microstructural de-correlation lengths, testing methods that correspond to those lengths and rigorous statistical representations of the microstructure variability, but also to establish all of these within standardized protocols that are consistent with simulation capabilities.

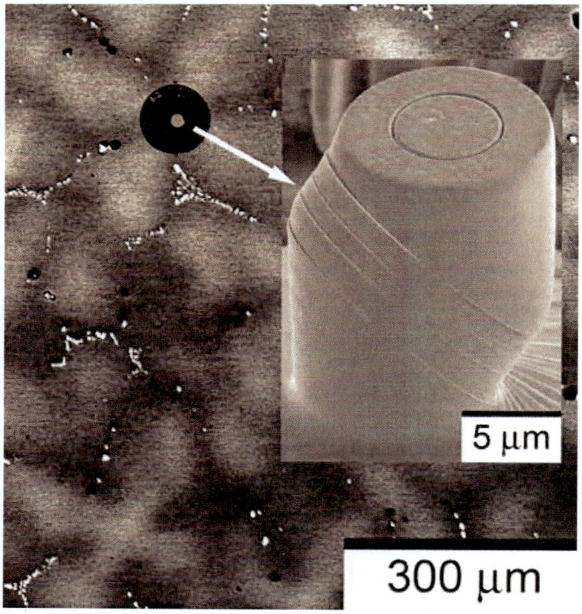

Fig. 7 Microcrystal compression sample machined from dendrite core structure of cast single crystal superalloy and tested (Shade 2008)

As Ghosh et al. describe, a quantitative partitioning of the computational domain will inevitably lead to identifiable microstructural features that cannot or should not be homogenized within a selected L1 domain (in "Multiscale Characterization and Domain Partitioning for Multiscale Analysis of Heterogeneous Materials" by Ghosh). For the turbine blade example, one may anticipate that the eutectic microconstituent or pores that exceed some size dimension would fall into that realm. These are defined as Level 2 (L2) features. Any such features need to be represented within their own micromechanics frameworks that account for plastic processes and damage accumulation within those entities. Even coarse-scale extrinsic defects may fall into a similar L2 domain since they cannot be homogenized via asymptotics. For the intrinsic microstructural variation, as simulations proceed at the L0 and L1 levels, again localization is expected that will exceed the bounds of pre-selected failure criteria established within failure or damage models. Once that occurs, the adaptive scheme inserts an L2 domain into localization fields and permits a still higher fidelity representation of behavior to evolve in simulations. For the turbine blade material example, one may envisage that L2 computational methods consist of crystal plasticity models, non-local formulations of damage, or crack initiation models of various forms. Clearly, the concurrent multiscale adaptive scheme described here can be explicitly tied to microstructural–dependent properties and heterogeneous materials, to the extent that its various parts can be build in a computationally viable way.

4.3 *Advanced Engineering Design: A Virtual Materials Systems Paradigm*

Previously, the notion of virtual materials systems was introduced as a parallel to virtual biological systems. One may expect that as virtual materials systems become closer to reality, the interface between design engineering and MSE will further dissolve. Figure 4 described what is becoming current-day practice within the ICME paradigm for materials, which already contains broad overlap between design and materials. Here, that view is contrasted with a futuristic view partially described in previous reports (Dimiduk et al. 2004a, b). The view outlined the potential of virtual materials systems or virtual processing–microstructure–property–design relationships to affect the engineering design practices, especially for developing an unknown or new material. Figure 8 describes a design environment wherein the usual suite of engineering tools for shape and product performance in a mechanical engineering sense (boxes A and B) are integrally coupled to a comparable suite of design tools for microstructure–property relationships. Boxes E, F, and G describe microstructure sensitive representations of the part. From this one may synthesize probability of part population behavior, then link these to system fleet probability of behavior via other probabilistic tools.

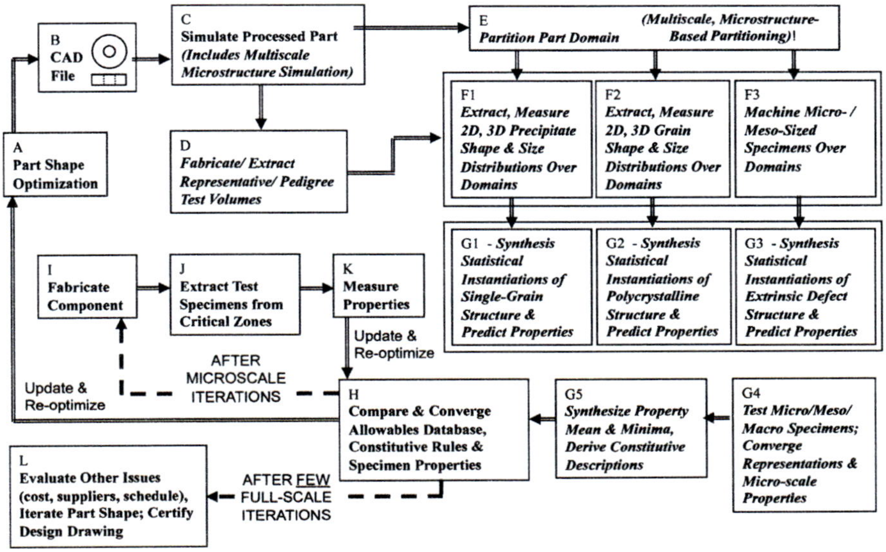

Fig. 8 Schematic depiction of the microstructure–property aspects of the virtual materials systems paradigm. This paradigm places significant emphasis on building and using a validated representation of the material, via a suite of integrated small-scale experimental and simulation methodologies, as an integral part of the system design process. Within such an approach, there are no real boundaries between engineering design and materials engineering

One may envisage an environment (Fig. 8) for which process modeling includes spatial-temporal simulations of microstructure evolution, both at the level of primary constituent kinematics (grain, fiber or primary matrix constituent level) and at the lower levels of microconstituent and defect chemical and kinetics behavior (box C). Utilizing those tools results in a virtual description of the part domain. The figure shows that the process modeling procedures result in two key attributes for the remaining design system. First, as suggested by box D, experiments would be initiated that are defined from the results of process simulations and are specifically focused on evaluating critical microstructures or pedigreed materials for bounding models.

In parallel to the experimental activity, box E suggests that the preliminary designed part may be partitioned into microstructural simulation domains. These can be defined from both the spatial-temporal variations of continuum state-variable fields and from expectations of extrinsic defect influences on behavior. For each of those domains, SERVE must be constructed to manage the intrinsic evolution of both the grain or primary-constituent kinematics, and the lower-scale, single-grain or microconstituent level kinetics under service loads and environmental conditions. Construction of the RVE suite involves both small-scale experimental measurements and a more-substantial set of simulation-based activities that includes building synthetic statistical instantiations of microstructures that include imposed extrinsic defect structures. Those activities are shown in boxes F1-G4 in Fig. 8. The activities depicted in box G5 represent the use of the small-scale SERVE suite within simulation frameworks that derive a statistically relevant set of material responses to include probabilities of performance. These result in larger-scale constitutive descriptions with damage mechanisms and "materials allowables" for part design. The synthetically derived materials behavior descriptions are then reconciled and adjusted using information from historical databases for similar material behavior (box H).

From this point, the whole set of procedures may be iterated and updated toward some optimization criteria, until there is a converged quantitative view of the expected microstructure–property relationships over all domains of a part, consistent with the desired design performance criteria. Only after such reconciliation of design goals would one have to prepare full-scale test articles for full certification of the part. Clearly, within such a future paradigm, there is little separation between MSE and design engineering. For such a paradigm, *design with materials* becomes symbiotically fused with *design of materials*. This is done in such a way that simulated responses are fused with heuristically known responses resulting in both cost and performance risk reductions.

For such a long-term view of microstructure–property–design engineering as depicted in Fig. 8 to become a reality, it is important that virtual materials systems build from technologies that are viable today, but remain extensible into the representations of tomorrow. As previously discussed, one area that already poses a present-day challenge, for which today's decisions will impact the broader longer-term evolution of the field, is data types and management. For example, the last 5 years have seen step-wise growth in the techniques for both characterizing microstructures in 3d and simulating their behavior. Those successes brought about what some have a called a "data tsunami." Both models and experiments now overwhelm data management, storage and, most importantly analysis capabilities. The very existence of such data, together with the interests in mining such data, calls for extensible data structures that carry the data pedigree throughout and, for automated, unsupervised analysis tools especially for microstructural analysis.

5 The Present Book

No book on microstructure–property sciences and techniques can be comprehensive, nor was it the goal of the editors of this work to cover the topic in a comprehensive way. Rather, the purpose of this book is to provide insights into selected aspects of microstructure–property science and provide views of what is in the realm of the possible when a computation and simulation centric perspective is adopted. By doing this, the editors believe that a vision for the future of the field can be shown, specific advances in the field conveyed and, gaps in the computational methods highlighted.

The structure of the book follows to great degree from the introductory context discussed previously. The view is that the broad goal of attaining virtual materials systems provides the guiding principles and, that the four domains of multiscale materials modeling (Fig. 5) together with microstructure–property science provide the more detailed structure. Thus, the book consists of four parts. Part I describes selected methods for attaining virtual materials structure and directly tying that information to the computational domain, beginning with methods for experimentally determining 3d microstructure. Part II of the book shifts the focus onto virtual material response within the simulation environment, which is directly tied to the constitutive descriptions and kinematical frameworks selected for simulation. In Part III, the book describes selected numerical techniques and simulation frameworks for treating some aspects of engineering challenges associated with time-dependent material behavior at microstructural scales.

While the first three parts of this work develop many key aspects of microstructure–property science from a computational perspective, there is still a great deal missing. Within Part IV of the book, three selected and current broader-interest topics in the MSE community are presented. This part includes a chapter that describes the multiscale framework for mechanical behavior testing that is evolving in parallel to the ICME paradigm and, a view of computational stochastic methods for describing effects across length scales. The final chapter returns to a broader look at the design and MSE fields and provides another perspective on the challenges of bringing microstructure–property information into the systems engineering optimization domain.

The editors of this work operate from the strong belief that there is a quiet revolution under way within microstructure–property science that is being driven by the continuing advances in computing capabilities. That revolution is allowing designers to explicitly include microstructure heterogeneity and location- or feature-specific behavior directly within the design process. Ultimately, this revolution will lead to the availability of virtual materials systems, to more portable materials that are no longer so closely tied to application-specific processes and descriptions and, to transformed approaches to materials data and data structures. While much of this book describes the advances from the perspectives of metals technologies, there is ample reason to believe that the structuring of the challenges and aspects of the techniques are general and applicable to broad classes of materials beyond metals. Nonetheless, even for metals many of the techniques require significant further developments to realize engineering gains. Our hope is that many will be swayed by the art of the desirable that is conveyed herein and find ways to transform that into the art of the possible while continuing to advance the field.

Acknowledgements There are many incremental contributors to the viewpoints expressed both in this introduction and the throughout remainder of the book. Among those contributors, the author gratefully acknowledges important and formulating discussions with Profs. H.L. Fraser, S. Ghosh and J.C. Williams; and with Drs. R.E. Dutton, J.P. Simmons, C. Woodward, Dr. C. Hartley, M.G. Mendiratta, T.A. Parthasarathy, J.M. Larsen, L. Christodoulou, S. Wax, D. Backman, H.A. Lipsitt, M.J. Blackburn and Mr. J. Schirra. We also gratefully acknowledge financial support from the Air Force Office of Scientific Research, under the direction of Dr. C. Hartley, and the Defense Advanced Research Projects Agency, especially during early periods of this effort.

References

Allison J, Li M, Wolverton C and Su X (2006) Virtual aluminum castings: an industrial application of ICME. JOM 58(11):28–35.

Arakere NK and Swanson G (2002) Effect of crystal orientation on fatigue failure of single crystal nickel base turbine blade superalloys. J Eng Gas Turb Power, ASME 124:161–176.

Arnold SM (2006) Paradigm shift in data content and informatics infrastructure required for generalized constitutive modeling of materials behavior. MRS Bull 31:1013–1021.

Asaro RJ (1983) Crystal plasticity. J Appl Mech 50:921–934.

ASTM Std E1181-02 (2002) Standard test methods for characterizing duplex grain sizes. ASTM International, West Conshohocken, PA.

Backman DG, Wei DY, Whitis DD, Buczek MB, Finnigan PM and Gao D (2006) ICME at GE: accelerating the insertion of new materials and processes. JOM 58(11):36–41.

Bernholc J (1999) Computational materials science: the era of applied quantum mechanics. Phys Today 52:30–35.

Cahn RW (2001) The coming of materials science. Elsevier Science, Ltd., Oxford, UK.

Cebon D and Ashby MF (2006) Engineering materials informatics. MRS Bull 31:1004–1012.

Cedar G (2000) Materials science needs and is getting quantitative methods. Phys Today 53:75–76.

Choi Y-S, Wen Y-H, Parthasarathy TA, Woodward C and Dimiduk DM (2009) A new microstructure-sensitive crystallographic constitutive model for creep of Ni-base single-crystal blade alloys. TMS Annual Meeting.

Christodoulou L and Larsen JM (2004) Using materials prognosis to maximize the utilization potential of complex mechanical systems. JOM 55:15–19.

Christodoulou L Defense Advanced Research Projects Agency (DARPA) DARPA-AIM. http://www.darpa.mil/dso/thrusts/matdev/aim/index.html.

Daw M and Baskes M (1984) Embedded-atom method: derivation and application to impurities, surfaces, and other defects in metals. Phys Rev B 29:6443–6453.

DEFORM (2009) http://www.deform.com

Devincre B, Kubin L, Lemarch C and Madec R (2001) Mesoscopic simulations of plastic deformation. Mater Sci Eng A 309–310:211–219.

Dimiduk DM, Martin PL and Dutton R (2003) Accelerated insertion of materials: the challenges of gamma alloys are really not unique. In: Kim Y-W, Clemens H and Rosenberger, A (eds) Gamma titanium aluminides, TMS, Warrendale, PA, 15–22.

Dimiduk DM, Parathasarathy TA, Rao SI, Choi Y-S and Uchic MD (2004a) Predicting the microstructure-dependent mechanical performance of materials for early-stage design. In: Ghosh S, Castro JM and Lee JK (eds) Materials processing and design: modeling, simulation, and applications, NUMIFORM 2004, AIP CP712, American Institute of Physics, Springer-Verlag, New York, 1705.

Dimiduk DM, Parthasarathy TA, Rao SI, Choi Y-S, Uchic MD, Woodward C and Simmons JP (2004b) Structural alloy performance prediction for accelerated use: evolving computational materials science & multiscale modeling. In: Ghoniem NM (ed) Conference proceedings of the second international conference on multiscale materials modeling, University of California Los Angeles, Los Angeles, CA, 9–11.

Dimiduk DM (2001) Gamma titanium-aluminide technology within the advanced propulsion community. In: Waltrup PF (ed) 15th International Symposium on Air Breathing Engines, Bangalore, India, 3–7 September, 2001, International Society for Air Breathing Engines (ISOBE) and American Institute of Aeronautics and Astronautics (AIAA), Kansas City, MO, paper #1026.

Dimiduk DM (1998) Systems engineering of gamma titanium aluminides: impact of fundamentals on development strategy. Intermetallics 6:613–621.

Epishin A, Link T, Brückner U, Fedelich B and Portella P (2004) Effects of segregation in nickel-base superalloys: dendritic stresses. In: Green KA, Pollock TM, Harada H, Howson TE, Reed RC, Schirra JJ and Walston S (eds) Superalloys 2004, Tenth International Symposium, TMS, Warrendale, PA, 537–543.

Fraser HL Center for Accelerated Maturation of Materials (CAMM). http://www.camm.ohio-state.edu/index.html

Ghoniem NM, Tong SH and Sun LZ (2000) Parametric dislocation dynamics: a thermodynamics-based approach to investigations of mesoscopic plastic deformation. Phys Rev B 61:913–927

Ghosh S, Lee K and Raghavan P (2001) A multi-level computational model for multi-scale damage analysis in composite and porous materials. Int J Solids Struct 38:2335–2385.

Ghosh S, Bai J and Raghavan P (2007) Concurrent multi-level model for damage evolution in microstructurally debonding composites. Mech Mater 39:241–266.

Ghosh S, Dakshinamurthy V, Hu C and Bai J (2008) Multi-scale characterization and modeling of ductile failure in cast aluminum alloys. Int J Comp Meth Eng Sci Mech 9:1–18.

Groeber M, Dimiduk DM, Uchic MD and Woodward C (2009) Integration of 3D structure information for a Ni-base superalloy into computational models for behavior prediction. TMS Annual Meeting.

Hafner J (2000) Atomic-scale computational materials science. Acta Mater 48:71–92.

Harrison GF, Tranter PH, Shepherd DP and Ward T (2004) Application of multi-scale modeling aeroengine component life assessment. Mater Sci Eng A 365:247–256.

Head AK, Humble P, Clarebrough LM, Morton AJ and Forwood CT (1973) Computed electron micrographs and defect identification. North Holland, Amsterdam.

Kaufman L and Bernstein H (1970) Computer calculation of phase diagrams with special reference to refractory metals. In: Margrave JL (ed) Refractory materials: a series of monographs, Academic, New York.

Kocks UF (1987) Constitutive behavior based on crystal plasticity. In: Miller AK (ed) Unified constitutive equations for plastic deformation and creep of engineering alloys, Elsevier, London, 1–88.

Lipsitt HA, Blackburn MJ and Dimiduk DM (2001) High–temperature structural applications. In: Westbrook JH and Fleischer RL (eds) Intermetallic compounds principles and practice, Vol. 3, Wiley, New York, 471–499.

Liu L, Husseine NS, Torbet CJ, Kumah DP, Clarke R, Pollock TM and Jones J W (2008) In situ imaging of high cycle fatigue crack growth in single crystal nickel-base superalloys by synchrotron X-radiation. J Eng Mater Technol 130:021008–1–6.

Liu Z-K, Chen L-Q and Rajan K (2006) Linking length scales via materials informatics. JOM 58(11):42–50.

Ma A, Dye D and Reed RC (2008) A model for the creep deformation behaviour of single-crystals superalloy CMSX-4. Acta Mater 56:1657–1670.

MacKay RA and Maier RD (1982) The influence of orientation on the stress rupture properties of nickel base superalloy single crystals. Metall Trans 13A:1747–1754.

MacLachlan DW, Wright LW, Gunturi S and Knowles DM (2001) Constitutive modeling of anisotropic creep deformation in single crystal blade alloys SRR99 and CMSX-4. Int J Plast 17:441–467.

Mandel J (1973) Equations constitutives et directeurs dans les milieux plastiques et viscoplastiques. Int J Sol Struct 9:725–740.

McClung RC et al. (2008) Turbine Rotor Material Design, Phase 2 Final Report, Southwest Research Institute, University of Texas at San Antonio, Mustard Seed Software, GE Aviation, Honeywell, Pratt & Whitney, Rolls-Royce Corporation, FAA Grant 99-G-016, Federal Aviation Administration, Washington, DC.

McDowell DL (2000) Modeling and experiments in plasticity. Int J Solid Struct 37:293–309.

Meric L, Pourbanne P and Cailletaud G (1991) Single crystal modeling for structural calculations: part 1-model presentation. J Eng Mater Tech 113:162–170.

Millwater HR and Osborn RW (2006) Probabilistic sensitivities for fatigue analysis on turbine engine disks. Int J Rotat Mach 28487:1–12.

National Materials Advisory Board (NMAB) (2004) Committee on Accelerating Technology Transition: Accelerating technology transition: bridging the valley of death for materials and processes in defense systems. National Academies Press, Washington, DC.

National Materials Advisory Board (NMAB) (2008) Committee on Integrated Computational Materials Engineering: integrated computational materials engineering a transformational discipline for improved competitiveness and national security. National Academies Press, Washington, DC.

Olson GB (1997) Computational design of hierarchically structured materials. Science 277:1237–1242.

Pollock TM and Tin S (2006) Nickel-based superalloys for advanced turbine engines: chemistry, microstructure, and properties. J Prop Power 22:361–374.

ProCAST (2009) http://www.esi-group.com/products/casting/procast

Reed RC (2006) The superalloys: fundamentals and applications, Cambridge University Press, Cambridge, UK.

Shade PA (2008) Small scale mechanical testing techniques and application to evaluate a single crystal nickel superalloy. Ph.D. Thesis, The Ohio State University, 68–109.

Shercliff HR and Ashby MF (1990) A process model for age hardening of Al alloys—part I. the model. Acta Metall 38:1789–1802.

Simonovski I, Kovac M and Cizelj L (2004) Estimating the correlation length of inhomogeneities in a polycrystalline material. Mater Sci Eng A 381:273–280.

Smith CS (1988) A history of metallography. MIT Press, Cambridge, MA.

Swaminathan S and Ghosh S (2006) Statistically equivalent representative volume elements for composite microstructures, part II: with damage. J Compos Mater 7:605–621.

Swaminathan S, Ghosh S and Pagano NJ (2006) Statistically equivalent representative volume elements for composite microstructures, part I: without damage. J Compos Mater 7:583–604.

Tschopp MA, Spearot DE and McDowell DL (2008) Influence of grain boundary structure on dislocation nucleation in FCC metals. In: Hirth JP (ed) Dislocations in solids, Vol 14, Elsevier, Oxford, UK, 42–139.

Uchic MD, Dimiduk DM, Wheeler R, Shade PA and Fraser HL (2006) Application of micro-sample testing to study fundamental aspects of plastic flow. Scripta Mater 54:759–764.

Valiveti DM and Ghosh S (2007) Morphology based domain partitioning of multi-phase materials: a preprocessor for multi-scale modeling. Int J Num Meth Eng 69:1717–1754.

van de Walle A, Asta M and Ceder G (2002) The alloy theoretic automated toolkit: a user guide. CALPHAD 26:539–553.

Vitek V (1985) Effect of dislocation core structure on the plastic properties of metallic materials. In: Dislocations and properties of real materials, The Institute of Metals, London.

Wikipedia Virtual Human (2009) http://en.wikipedia.org/wiki/Virtual_Physiological_Human, 31 March 2006.

Yi JZ, Torbet CJ, Feng Q, Pollock TM and Jones JW (2007) Ultrasonic fatigue of single crystal Ni-base superalloy at 1000°C. Mater Sci Eng A 443:142–149.

Yip S (2005) Handbook of materials modeling, Springer, Heidelberg.

Zhao JC (2006) Combinatorial approaches as effective tools in the study of phase diagrams and composition–structure–property relationships. Prog Mater Sci 51:557–631.

Serial Sectioning Methods for Generating 3D Characterization Data of Grain- and Precipitate-Scale Microstructures

Michael D. Uchic

Abstract This chapter provides an overview of the current state-of-the-art for experimental collection of microstructural data of grain assemblages and other features of similar scale in three dimensions (3D). The chapter focuses on the use of serial sectioning methods and associated instrumentation, as this is the most widely available and accessible technique for collecting such data for the foreseeable future. Specifically, the chapter describes the serial sectioning methodology in detail, focusing in particular on automated systems that can be used for such experiments, highlights possibilities for including crystallographic and chemical data, provides a concise discussion of the post-experiment handling of the data, and identifies current shortcomings and future development needs for this field.

1 Introduction

In the previous chapter, the concept of integrated computational materials engineering (ICME) via microstructurally informed, multiscale simulations was introduced. For this type of endeavor, it is incumbent that the required microstructural information be on hand as either input or validation for these simulations to properly account for microstructural dependencies. Today, this information is most commonly found in the form of mean values for selected features, e.g., average grain size, average precipitate size or spacing, or in more advanced models, distributions of these microstructural descriptors are required.

To provide as complete and unbiased description of microstructure as possible, the field of materials characterization is gradually developing and adopting methods that provide quantitative microstructural information in three-dimensions (3D). The desire for 3D microstructural data is relatively straightforward. Primarily, it is

M.D. Uchic (✉)
Materials and Manufacturing Directorate, Air Force Research Laboratory, Wright-Patterson Air Force Base, OH, USA
e-mail: Michael.Uchic@wpafb.af.mil

S. Ghosh and D. Dimiduk (eds.), *Computational Methods for Microstructure-Property Relationships*, DOI 10.1007/978-1-4419-0643-4_2,

because 3D data provides access to some very important geometric and topological quantities that cannot be determined a priori by classical stereological methods that utilize only 2D images (DeHoff 1983). These quantities include assessing the true size, shape, distribution of both individual features and that of their local neighborhoods, determining the connectivity between features or networks, and counting of the number of features per unit volume (De Hoff 1983; Wolfsdorf et al. 1997).

Experimental methods that enable 3D characterization have undergone dramatic improvements in the past decade, due in large part to advances in both computing power and visualization and analysis software that have been enabling factors for both the collection and interpretation of these massive data sets. The 3D data collection process requires significantly more effort compared to conventional 2D analysis, which has spurred the development of fully automated instruments that are capable of collecting such information (Alkemper and Voorhees 2001; Spowart et al. 2003), as well as software programs that take in the raw data stack and provide as output reconstructions and analysis of the microstructural features in 3D [see for example, IMOD (Kremer et al. 1996)]. The diverse size range of microstructural features has resulted in the development of a suite of instruments to address the collection of 3D data at various size scales. This ranges from counting individual atoms in nanometer-sized needles (Miller and Forbes 2009) to interrogating features within manufactured components (MA Groeber, DM Dimiduk, MD Uchic, C Woodward 2009, unpublished research) – a difference of 7–9 orders of magnitude in scale – and cube of this value for volumetric coverage! The state-of-the-art for the field of 3D materials characterization has been the focus of recent collections of papers in a number of materials journals (Spanos 2006; Uchic 2006; Thornton and Poulsen 2008), and has also been the topic of a number of symposia at materials society meetings, for example, the 3D Materials Science symposia I to VI at the TMS national meetings.

As an aside, a similar renaissance in 3D characterization methodologies has already occurred in the biological and medical sciences, with instruments that are more suited to either sectioning soft matter, or in some cases making use of instrumentation that cannot be directly applied to opaque materials such as confocal laser microscopy. Nevertheless, the significant overlap in problems of data handling, data segmentation and feature extraction, 3D visualization, and surface meshing has accelerated the maturation of this methodology for the structural materials community.

This chapter focuses on one aspect of microstructural characterization with respect to the ICME field, which is to discuss the methodologies that can be used to quantify the 3D microstructure associated with grain ensembles or other features that are of similar scale such as second-phases, dendrites, precipitates, dispersoids, and voids. These are ubiquitous features found in most structural alloys, and these features as a whole range in size from multiple millimeters to tens-of-nanometers in scale.

There are two main experimental pathways to collect information over this size range. The first is the use of X-rays, which are nondestructive and therefore allow for time-dependent studies that examine microstructural changes due to thermal or mechanical input, i.e., 4D experiments (Juul Jensen et al. 2006). There are a number of different techniques that can be used to provide image contrast in X-ray tomography experiments (Ice 2004). The most common method obtains information by reconstructing a suite of transmission (absorption) images taken at various projections. This technique is very sensitive to differences in atomic number and density, so that microstructural features which are quite different in these characteristics – such as porosity relative to the matrix – can be readily detected as shown in Fig. 1a. Other methods utilize diffraction contrast and either ray tracing methods (Schmidt et al. 2004; Juul Jensen et al. 2006; Ludwig et al. 2009) or other spatial localization methods (Larson et al. 2002; Ice 2006) to define features such as individual grains from grain aggregates. These diffraction-contrast methods have been greatly advanced in the past few years, and for selected techniques have been demonstrated to rapidly produce 3D characterization data of grain ensembles as shown in Fig. 1b. The primary disadvantage of these experiments is that they require the use of very high-intensity X-rays to produce data that has acceptable signal-to-noise levels, such as those produced by synchrotron sources (Ice 2004). This requirement severely restricts the general availability and applicability of these methods until there is a revolutionary change in the ability to produce high brilliance X-rays in a laboratory setting.

The other method to acquire 3D characterization data at the macro-to-microscale is through serial sectioning experiments. Serial sectioning is much more accessible experimental methodology compared to synchrotron-based tomography, but this methodology has a significant disadvantage that the sample volume is inevitably consumed during the data collection process, which precludes any re-examination or re-use of the material after analysis. In spite of this drawback, serial sectioning

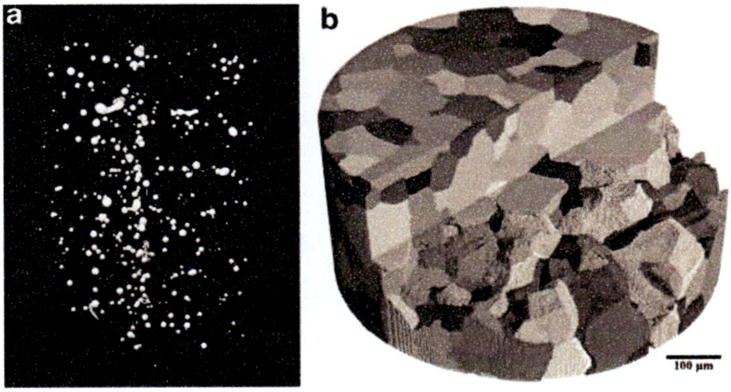

Fig. 1 Examples of microstructural data that can be obtained with synchrotron X-ray methods. **(a)** 3D reconstruction of the porosity in a cast single-crystal nickel base superalloy, CMSX-10, using transmission (absorption) X-ray tomography (Link et al. 2006). The dimensions of the reconstructed volume are $500 \times 500 \times 800\,\mu m$. Figure is used with permission from Elsevier. **(b)** 3D reconstruction of the 3D grain structure of a tensile sample of β-21 titanium alloy (Ludwig et al. 2009). The reconstruction contains 1,008 grains, and was collected using X-ray differential contrast tomography. Figure is used with permission from the American Institute of Physics

experiments are becoming an increasingly common procedure to characterize microstructure in 3D, especially in the past decade with respect to the development and usage of automated instruments to perform of this task. This chapter endeavors to provide an overview of this technology, and discuss the state-of-the-art with regards to characterizing grain and precipitate scale microstructural features in structural materials.

2 Serial Sectioning

For opaque materials, serial sectioning has been the most widely used method to acquire raw 3D characterization data at the macro-to-microscale, and in fact the first application of this methodology to examine the microstructure of structural metals was published over 90 years ago (Forsman 1918). Tomographic serial sectioning experiments are conceptually simple, being composed of two steps that are iteratively repeated until completion of the experiment. The first is to prepare a nominally flat surface, which can be accomplished by a variety of methods – a noninclusive list includes cutting, polishing, ablating, etching, and sputtering – where ideally a constant depth of material removal has occurred between each section. The second step is to collect two-dimensional (2D) characterization data after each section has been prepared, although data could also be collected continually during material removal depending on the particular sectioning method that is employed. After collection of the series of 2D data files, computer software programs are used to construct a 3D array of the characterization data that can be subsequently rendered as an image or analyzed for morphological or topological parameters.

The 2D characterization data collected during a serial sectioning experiment can be comprised of number of different types and/or quantities of information. For example, in the particular case of characterizing grain microstructures, this could consist of using optical microscopy to image the structure of etched grain boundaries, as well as using an SEM to collect electron backscatter diffraction maps on key sections to characterize the average grain orientation, which was recently demonstrated by Spanos, Lewis, Rowenhorst and co-workers at the Naval Research Laboratory (Lewis et al. 2006; Spanos et al. 2008), as shown in Fig. 2. From a practical perspective, the main criteria for determining whether to incorporate a particular image or data map into a serial sectioning experiment is whether the microstructural feature or features of interest can be readily classified from this information, especially via unsupervised computer segmentation processes. In the most commonly performed experiment, the characterization data consists of a single 2D image per section (Mangan et al. 1997; Kral and Spanos 1999; Lund and Voorhees 2002; Holzer et al. 2006). Other examples include multiple images that highlight different aspects of the microstructure (Jorgensen et al. 2009), crystallographic (Wall et al. 2001) or chemical maps (Kotula et al. 2006; Schaffer et al. 2007), or conceivably any other 2D spatial measurement that is of interest (such as local measurements of resistivity or elastic modulus, etc.). The process of sectioning and

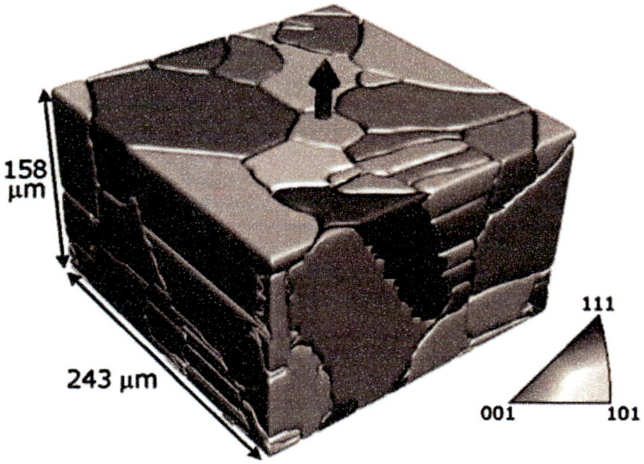

Fig. 2 3D reconstruction of the austenite phase in a commercial austenitic stainless steel alloy AL-6XN (Lewis and Geltmacher 2006). The data set was produced via manual serial sectioning that incorporated collection of both optical images and EBSD maps. The volume contains 138 grains, and the *arrow* represents the normal of the serial sectioning plane. Although not readily visible because of the gray-scale coloring of this printing, the color of each grain corresponds to the crystallographic orientation relative to the *arrow*, which was determined by EBSD. Figure is adapted with permission from Elsevier

data collection is repeated until the desired sample volume has been interrogated, or perhaps more realistically for manual implementations of this methodology, the motivation to continue collecting the data falls below a critical value.

One of the key aspects in the design of a serial sectioning experiment is to determine the minimal spatial resolution required by the subsequent microstructural analysis. For example, a serial sectioning study that will quantify aspects of feature shape such as surface area will require a much greater spatial resolution than one that is simply counting the number of features per unit volume. A rule-of-thumb is that one should strive for a minimum of ten sections per feature, although this is simply an ad hoc estimate. A better approach is to perform a critical examination of the effect that spatial resolution has on the accuracy or bias of any of the quantitative measurements-of-interest on simple test objects prior to initiating the experiment (Wojnar et al. 2004). Ideally, one would like to section at the finest possible step size and also collect high-resolution 2D data to generate the highest fidelity 3D data structures as possible. In practice, this goal is tempered by a number of factors. First, the precision of the sectioning technique should be assessed. The typical serial sectioning experiment employs a section thickness where the variability between sections is a small fraction of the total section thickness ($<10\%$), as historically most studies assume a constant section thickness and do not take this variation into account when reconstructing the data. Second, the spatial resolution of the 2D characterization technique should be considered as well.

For example, an experiment that utilizes X-ray spectroscopy via electron-beam irradiation [i.e., energy-dispersive spectroscopy (EDS) or wavelength-dispersive spectroscopy (WDS)] will have a much larger interaction volume and consequently a much poorer spatial resolution compared to an experiment that uses secondary electron imaging. Thus, the spatial resolution of each characterization technique that is employed should set a minimum bound, to prevent collecting "empty magnification" in either the 2D imaging plane or the sectioning depth. Additional feasibility issues include the proportional increase in time needed to complete a serial sectioning experiment as the spatial resolution is increased in the sectioning direction or the imaging plane, as well as the concomitant increase in computational resources for data handling and storage.

Like any experimental methodology, serial sectioning has both advantages and disadvantages. One advantage is that both sectioning equipment and 2D characterization instruments are commonly found in most materials laboratories, therefore, manual implementation of this methodology can readily initiated, with the only significant "cost" being that of instrument and personnel time to perform this repetitious experiment. A wide variety of materials characterization methods are optimized for analysis of planar surfaces, such as light microscopy, scanning electron or ion microscopy, scanning probe microscopy or its derivatives, surface analysis techniques (EDS, XPS, Auger, SIMS, RBS, etc.), so that a wide range of materials characteristics can be obtained. The type of data that is collected during a serial sectioning experiment can have a profound impact on the ease of identifying and classifying microstructural features, and so this selection should be carefully considered in the design of the experiment. The primary disadvantage associated with this technique is the destruction of the sample, which for some applications is unacceptable. Another potential but much-less common issue is that the volume that is analyzed is always adjacent to or at the free-surface, which might affect the characterization measurement in an undesirable way.

Manual demonstrations of this experimental methodology can be found periodically throughout the 1960s to the early 2000s, and a review of these studies can be found in the papers by Kral and co-authors (Kral et al. 2000, 2004). While serial sectioning experiments can be performed in this manner, the repetitive nature of this experiment is ideally suited for automation using instruments that are designed specifically for this application. Automation clearly reduces the tedium associated with the experiment (DeHoff 1983; Kammer and Voorhees 2008), thereby providing significant gains in terms of the amount of data that can be collected. In addition, there are other potential benefits to instrument automation, such as reductions in data variability via machine inspection and metrology. For example, pattern recognition methods can be incorporated to increase the accuracy and precision of the serial section thickness (Groeber et al. 2006), or image analysis methods can be used to adjust instrument settings so that the intensity histogram or image sharpness remains unchanged throughout the experiment.

At present, there are only three devices that are capable of automatically acquiring 3D grain-level data in structural materials via serial sectioning. The next section describes these devices in some detail. Two of the devices – the Alkemper and Voorhees micromiller and Robo-Met.3D – utilize optical microscopy as the sole characterization method. These devices have been constructed and are suitable for characterization of the microstructure of millimeter scale volumes with micron-level precision. The third device, the focused ion beam–scanning electron microscope (FIB-SEM), is not specifically designed for serial sectioning experiments, but has been adapted for this purpose through the use of machine control software scripts. This device has approximately 1–2 orders of magnitude improvement in imaging resolution as well as sectioning fidelity compared to the other serial sectioning instruments, so that micron and submicron features can be accurately characterized. Also, both crystallographic and chemical data can be collected as part of the serial sectioning experiment using commercially available detectors that can be installed on the FIB-SEM. However, this instrument cannot currently characterize larger volumes like the other two devices.

3 Automated Serial Sectioning Instrumentation

3.1 Alkemper–Voorhees Micromiller

This serial sectioning instrument was developed at Northwestern University by Alkemper and Voorhees (A&V) around the year 2000 (Alkemper and Voorhees 2001), to augment previous efforts by the Voorhees group (Wolfsdorf et al. 1997) to quantify materials microstructure in 3D using microtome milling, i.e., physical cutting with a rotating diamond knife. Unlike biological microtomy studies in which the thin section prepared by the cutting process is of interest, here the microtome blade is used as an end-mill to prepare an optical-quality surface in soft ductile metals and alloys that do not react adversely with the diamond blade, such as Pb, Sn, Al and Cu alloys (Kammer and Voorhees 2008). Images of this tomographic instrument are shown in Fig. 3.

At the heart of the system is a commercial microtome that is outfitted to a rotary micro-milling attachment. The microtome performs the sectioning operation by moving a sample underneath the rotating micro-milling head using a linear stage. The key improvements by A&V were to incorporate an optical microscope with a digital camera, an etching/washing/drying station, and a linear variable differential transformer (LVDT) sensor within the commercial micromiller system. The cleaning, etching, and drying station was located next to the cutting head to remove machining chips after sectioning, and also to reveal microstructural features that could be selectively attacked via chemical etching. The linear stage then translates the sample underneath the optical microscope where an image of the freshly prepared surface can be captured using a digital camera. These first two modifications eliminated the need to remove the sample from the micromiller to collect an image of the serial section surface, which resulted in significant reductions in the time needed to complete one cycle of the experiment, as well as

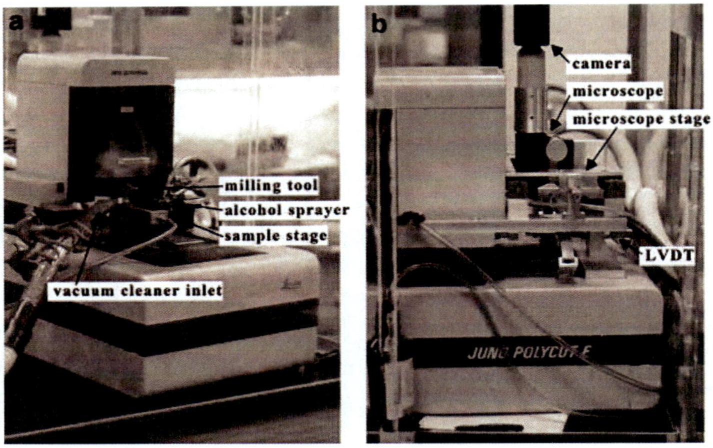

Fig. 3 The automated serial sectioning device developed by Alkemper and Voorhees (Alkemper and Voorhees 2001). (**a**) Image of the commercial Reichert-Jung Polycut E micromiller system, with modifications by A&V for cleaning the sample after the micromilling operation. (**b**) View of the microscope attachment and LVDT installed by A&V to perform automated serial sectioning. Figure is adapted from (Alkemper and Voorhees 2001) with permission from Wiley

minimizing positioning errors due to removal and subsequent re-mounting of the sample that affected the accuracy of both the cutting and imaging operations. The third modification of incorporating the LDVT allowed for an independent measurement of the spatial position of the sample when placed underneath the optical microscope. As a result, the absolute lateral position of the sample was quantitatively known for each slice, and this information was used to correct for in-plane translational errors in each of the 2D images that comprised the 3D data stack without resorting to image matching/correlation methods, which is useful in preventing alignment errors associated with using only internal features for this task (Russ 2002).

With any serial sectioning experimental methodology, one needs to quantify both the accuracy and precision of the sectioning process, as these variations lead to both systematic and random distortions in the sectioning direction of the 3D data stack. In particular, this issue is vitally important for instruments that do not monitor or measure this quantity during the experiment. The A&V micromiller utilizes a precision mechanical feed to advance the depth of the milling head over an adjustable range of 1–20 μm, but this device as constructed does not provide closed-loop control over the sectioning depth. Alkemper and Voorhees solved this problem by developing a custom calibration standard, and subsequent analysis of the 3D reconstruction of the calibration standard allowed A&V to determine that both machine errors as well as thermal expansion associated with heating of the system during operation were shown to affect the accuracy of the sectioning thickness (Alkemper and Voorhees 2001). Furthermore, they used the custom standard to quantify the average section thickness during steady-state operation, as well as identified a minimum warm-up time for instrument operation prior to which there were significant variations in the sectioning thickness.

Examples of some of the data sets that have been collected with this device are shown in Fig. 4, which have been used to systematically quantify the dendrite coarsening process in simple binary alloy systems. The micro-milling process is fast, and optical cameras can also quickly acquire image data. As a result, this device can provide upward of 20 sections per hour, resulting in 3D data sets that are comprised of hundreds of images that are prepared in less than a day. It is clear from Fig. 4 that this device is useful for the 3D characterization of materials at the millimeter size-scale,

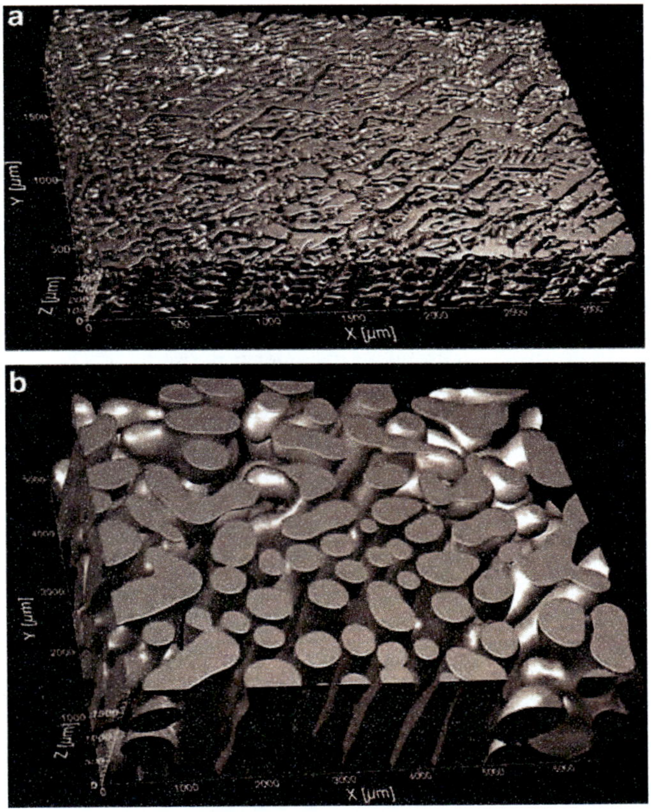

Fig. 4 Application of the A&V micromilling serial sectioning device to study dendrite coarsening processes (Kammer et al. 2006) (**a**) 3D reconstruction of the dendrite structure of a Pb-Sn alloy after a 3-min coarsening time. The solid corresponds to the β-Sn solid-solution dendrites, while the voids in the reconstruction correspond to the Pb-Sn eutectic phase. (**b**) 3D reconstruction of an Al-Cu alloy after a 3-week coarsening experiment. The solid corresponds to Al dendrites, while the voids in the reconstruction correspond to the Al-Cu eutectic phase. Note that this reconstructed volume is approximately 6 × 5 × 43 mm in dimension. Figure is adapted with permission from Elsevier

as long as two conditions are met: one, the materials are compatible with diamond blade sectioning and two, the microstructural features of interest can be selectively identified from machined surfaces using only optical images and chemical etching.

3.2 *RoboMet.3D*

Another serial sectioning device, RoboMet.3D, was developed by Spowart and Mullens (Spowart et al. 2003; Spowart 2006), and an image of this system is shown in Fig. 5. This device is conceptually similar to the A&V micromiller in that there are three stations to the system – sectioning, etching/washing/drying, and optical imaging. However, the RoboMet.3D can be used to examine a much broader range of materials compared to the A&V micromiller because it uses a precision mechanical polishing system for material removal rather than micromilling. Mechanical polishing is the most commonly used method to prepare materials for metallographic analysis, and therefore the sectioning system is well suited to examining many structural materials. Each of the stations is a physically separate unit on the RoboMet.3D, and therefore a six-axis robot arm is employed to move the sample between the various stations, and also holds the sample while it is being washed, etched, and dried with forced air. Notably, the commercially available optical microscope that is used on these systems is a fully automated device in its own right, being able to perform tasks such as focusing, contrast adjustments, and the capture of large-area montages of the serial section surface without human intervention.

The sample-of-interest is mounted on a custom holder that minimizes rotational movement of the sample between polishing and imaging operations, but small lateral

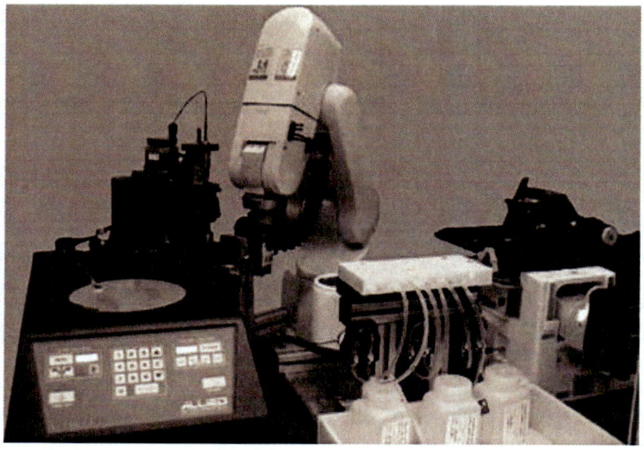

Fig. 5 The automated serial sectioning device RoboMet3.D (Spowart et al. 2003; Spowart 2006). From left-to-right in the image are the precision metallographic polisher, six-axis robot, etching/washing/drying station, and a motorized inverted optical microscope. Figure is used with permission from Elsevier

translations of the sample that occur between consecutive imaging operations are not measured – these must be removed via image analysis methods as will be discussed later in this chapter. Also, RoboMet.3D does not use closed-loop control over the material removal process, but rather section-to-section consistency is maintained by keeping common variables in the polishing process fixed, such as the time of polishing, applied load, wheel speed, in addition to using fresh diamond lapping films as the polishing media. With this protocol, RoboMet.3D has been demonstrated to achieve very good control over material recession rates, with a repeatability of $\pm 0.03\,\mu m$ for a section thickness of $0.8\,\mu m$ (Spowart 2006).

The device can prepare sections ranging from ~ 0.1 to $10\,\mu m$ in thickness and complete the sectioning cycle up to 20 times per hour (Spowart 2006), where the cycle time is mostly dependent on the sectioning depth, the etching time needed to resolve the feature-of-interest, and the on the quantity, resolution, and number of images that are acquired per section. Figure 6 shows representative data from the RoboMet.3D that highlights the good volumetric coverage that can be obtained with this device, as well as the diversity of materials that can be examined with this instrument. Like the A&V micromiller, this device can be successfully employed if two conditions are met: one, that a single polishing step provides both adequate material removal rates and sufficient metallographic surface quality and two, the microstructural features of interest can be identified using optical imaging methods.

3.3 Focused Ion Beam–Scanning Electron Microscopes

For smaller-scale grain and precipitate structures – those that are approximately $10\,\mu m$ in scale or smaller – FIB-SEM are well suited to characterize these features in 3D via serial sectioning (Dunn and Hull 1999; Inkson et al. 2001; Uchic et al. 2006, 2007). FIB columns are able to focus highly energetic ions (typically Ga^{+}) to small spot sizes that are on the order of 5–20 nm. The interaction of these energetic ions with a target results in localized material removed via ion sputtering interactions (Orloff et al. 2003). FIB microscopes are well suited to perform serial sectioning via cross-section milling with extremely fine resolution, and at the extreme can provide average serial section thickness of approximately 10–15 nm using closed-loop control measures (Bansal et al. 2006; Holzer et al. 2006). This value is at least one order of magnitude finer than the section thickness that can be nominally achieved with traditional mechanical removal methods such as metallographic polishing. Using the appropriate software control scripts, a typical serial sectioning experiment will usually encompass material volumes that are larger than $1{,}000\,\mu m^3$ with voxel dimensions approaching tens-of-nanometers. This combination of spatial coverage and resolution cannot be achieved with any other tomographic instrument.

FIB-SEM microscopes have other advantages relative to the task of serial sectioning. Cross-section ion milling is an almost universally applicable method for

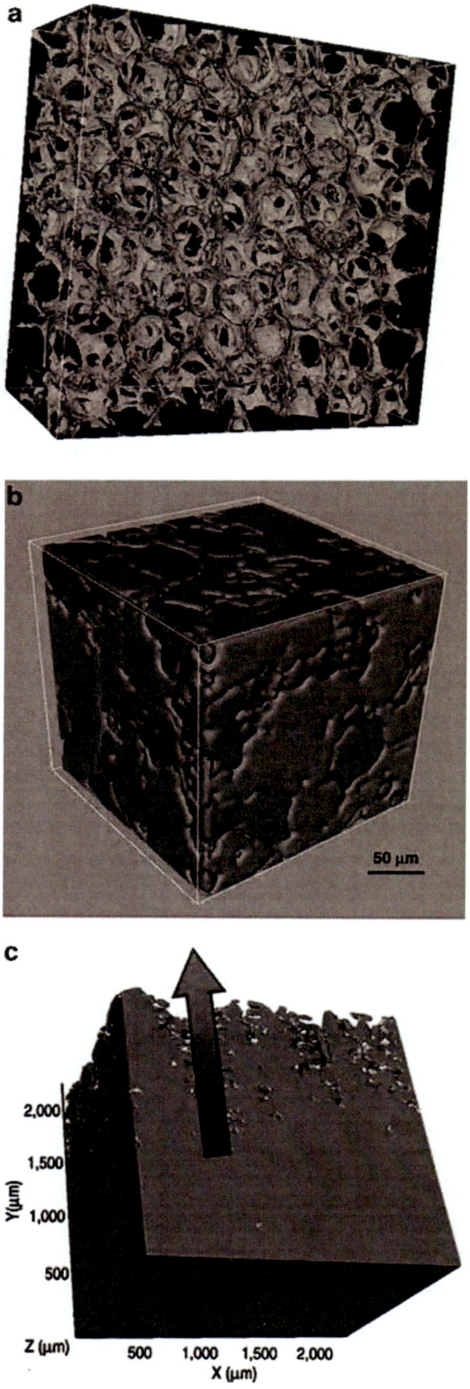

Fig. 6 Examples of 3D data sets that have been produced using RoboMet3.D. (**a**) Iso-surface rendering of a structural carbon foam (Maruyama et al. 2006). Dimensions of the 3D reconstruction are $1{,}526 \times 1{,}526 \times 776\,\mu m$, and the average serial section thickness is $3.5\,\mu m$. Figure is used with permission from Elsevier. (**b**) 3D reconstruction of a powder-compacted Fe-Cu alloy, which utilized an average serial section thickness of $1.2\,\mu m$ (Spowart 2006). Figure is used with permission from Elsevier. (**c**) 3D reconstruction of the mushy zone during directional solidification of a commercial cast single crystal Ni-base superalloy René N4 (Madison et al. 2008), which utilized an average serial section thickness of $2.2\,\mu m$. Figure is used with permission from Springer

preparing planar surfaces, and has been successfully applied to metallic alloys, ceramics, polymers, electronic materials and biological materials, although for some systems both low beam currents and sample cooling are required to prevent alteration of the starting microstructure. In comparison with most polishing or cutting methods, ion sputtering is a relatively low damage process that not only preserves the details of hard-to-prepare microstructures like those composed of both soft and hard phases, or brittle materials that contain significant porosity, but the depth of the damage layer is small enough to permit the usage of surface-damage sensitive techniques like electron backscattered diffraction (EBSD) for selected metallic alloys (Groeber et al. 2006; Konrad et al. 2006; Zaefferer et al. 2008).

One other significant advantage of FIB-SEM microscopes is ability to incorporate imaging and surface analysis methodologies that can greatly mitigate the difficulty in classifying various microstructural features like grains and precipitates. These characterization methods include high-resolution backscattered electron (BSE) images that exhibit atomic-number contrast to differentiate between multiple phases, ion-induced secondary electron (ISE) images that often exhibit channeling contrast which can differentiate individual grains in polycrystalline materials (Orloff et al. 2003), EBSD mapping for local crystallographic orientation measurements, and EDS (Kotula et al. 2006), WDS, or secondary ion mass spectroscopy (SIMS) (Dunn and Hull 1999) for local chemical spectra mapping. The information limits for these methods are often of the same order of magnitude as the FIB sectioning capabilities, thus a properly outfitted microscope can provide the user tremendous flexibility in selecting which types of structural, chemical, or crystallographic information are important for their particular characterization study. For example, if only structural information is required, then image data may suffice. Chemical or crystallographic analysis can be included to help with the identification of particular phases and grain orientations, respectively.

The procedure for performing an automated FIB-SEM serial sectioning experiment is roughly similar to the methods that use bulk sectioning processes. First the sample volume-of-interest is initially prepared, as FIB serial sectioning experiments have some specific sample geometries that can improve the quality of data that is collected (Uchic et al. 2006, 2007). Next, software control scripts are used to move the sample between sectioning and characterization steps. In their most basic form, these scripts perform the primary functions of cross-section ion milling of the sample and collection of electron images, as shown in Fig. 7. For these experiments, the sample does not need to move if it is placed at a position where both electron and ion columns can simultaneously image the same region of the sample, which greatly simplifies both the experimental setup and the complexity of the control scripts required to automate of the experiment. Like the other serial sectioning instruments discussed previously, the use of machine control scripts allows for a more consistent serial slice thickness, reduces the time needed per acquisition cycle, and enables the microscopes to run unattended for periods of about 1–2 days. More advanced control scripts incorporate image recognition procedures to minimize the effects of electrical, thermal, or mechanical drift, as well as to incorporate a wider range of data signals like ISE imaging or EBSD mapping, which requires accurate sample

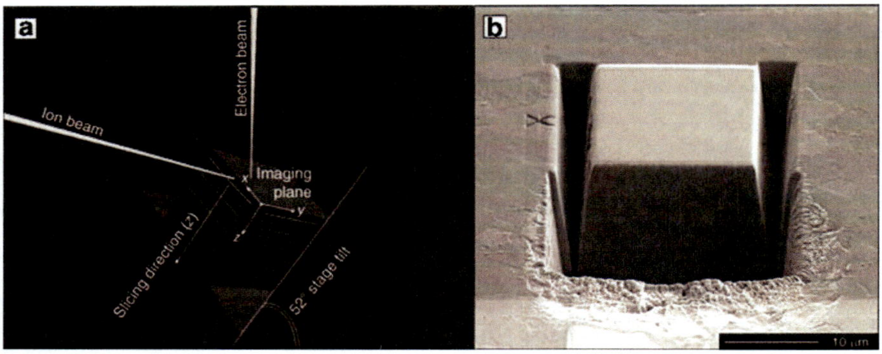

Fig. 7 The standard sample geometry and its orientation within a FIB-SEM microscope for a typical serial sectioning experiment (Holzer et al. 2006; Uchic et al. 2007). (**a**) Schematic of the experimental set-up for an FIB-SEM serial sectioning experiment where cross-section ion milling is used to controllably remove material at the micro- to nanoscale, and electron imaging is used to characterize the freshly prepared surface. If only electron imaging or EDS is used to characterize the sample surface, then the sample does not need to be moved during the tomographic experiment. (**b**) SEM image of a sample volume prior to sectioning. The trenches that surround the sample volume allow the electron beam to image the serial-sectioned surface, and also help prevent sputtered and re-deposited material from obscuring the surface-of-interest. Figure is used with permission from the Materials Research Society

repositioning after complex five-axis stage movements (Groeber et al. 2006). Note that for the experiments that use image recognition methods, these microscopes provide closed-loop control over sectioning thickness (unlike the two previously described serial sectioning instruments), thus enabling the user to simply select the serial sectioning depth and eliminating the need for calibration runs.

Commercial FIB-SEM instrumentation is installed in many laboratories worldwide, and therefore this system is the most widely used of the three automated serial sectioning systems discussed in this chapter. Figure. 8 shows representative examples of 3D data that have been collected with these instruments, which highlights the diversity of size scales that can be examined (Fig. 8a), the ability to characterize precipitate- and multiple-phase microstructures at the microscale (Figs. 8b, c), and the ability to examine grain-level microstructures that includes the automated collection of orientation information (Fig. 8d). This figure demonstrates that FIB-SEM microscopes epitomize a new breed of multimodal serial sectioning instruments, as they are currently capable of high-fidelity characterization of the morphology, crystallography, and chemistry of micron and submicron size features in 3D. One final note is that these experiments can require significant instrument time to complete an experiment depending on the size of the volume that is examined and the type of data that is collected. Although some experiments require only a few hours – like those where the milling step takes a couple of minutes to execute, and a single electron image is collected for each section – others that include chemical or crystallographic maps, or those that attempt to interrogate volumes that have submillimeter dimensions may require multiple days.

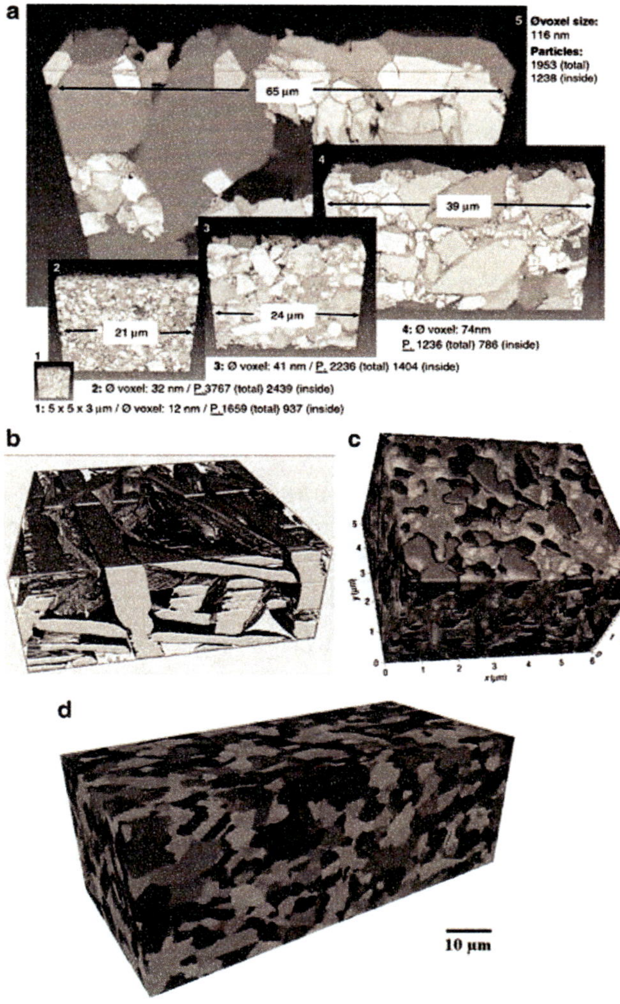

Fig. 8 Representative examples of 3D reconstructions that have been collected using a dual beam FIB-SEM. (**a**) Five reconstructions of portland cement agglomerants where the average particle size is different in each reconstruction, ranging from 0.68 to 14.2 µm (Holzer et al. 2006). The voxel edge length as well as the number of particles identified in each volume is listed next to each reconstruction. Figure is used with permission from Wiley. (**b**) Reconstruction of α-laths from a Ti-6242 colony (Simmons et al. 2009). Figure is used with permission from the IOP. (**c**) Anode of a solid-oxide fuel cell, which is composed of three phases: Ni (*light gray*), pores (*dark gray*), and yttria-stabilized zirconia (translucent phase) (Wilson et al. 2006). Figure is used with permission from the Nature Publishing Group. (**d**) A 3D reconstruction of the grain structure from a powder-metallurgy Ni-based superalloy, IN100. The volume shown has dimension of 96 × 36 × 46 µm with a voxel edge length of 250 nm, and EBSD data was used to form this reconstruction, which provided information on both the morphology and crystallography of the grain structure (Groeber et al. 2008). Figure is used with permission from Elsevier

4 Data Processing and Segmentation

After completion of the serial sectioning experiment using either manual or automated methods, the stack of 2D data files need to be combined and processed in such a way so that the microstructural features that are within the 3D data stack can be classified. Said differently, for each object of interest in the data, i.e., every grain, precipitate, dendrite, void, and so on, one has to identify all of the voxels that are associated with that object. This procedure is referred to as data segmentation. Once all of the objects and their voxels are identified and classified, the process of quantifying microstructural characteristics using computational methods can be readily performed; this topic is the subject of other chapters in this book, and therefore will not be covered here.

While the concepts of segmentation and classification is straightforward – as humans we perform this task during every waking moment – in practice this can be the toughest and often rate-limiting step to transforming the serial sectioning data into a useful form. The difficulty with data segmentation is utterly dependent on the type, quantity, and quality of data from the serial sectioning experiment, as well as the complexity of the microstructure of the material being examined. This is especially true for serial sectioning experiments that only collect image data, because of the general difficulty in using semi- or fully automated computer segmentation algorithms to defining objects from visually complex images. Note that unsupervised computerized segmentation processes are a necessary component of tomographic experiments because of the sheer volume of information that is contained within the data sets, thus eliminating the possibility of human-assisted segmentation except when only a handful of features require classification.

A necessary step (typically performed before image segmentation) is to ensure that the spatial registry of each 2D data file that collectively comprises the 3D data stack is accurate. Each data file may have to be translated, stretched, rotated, or possibly all three in combination to account for the fact that the sample may have been in different physical locations when the 2D characterization data was collected, to correct for systematic or random distortions that may have also occurred during data collection (for example, image foreshortening), or to overlay data that may have been acquired at different spatial resolutions. The need for data alignment and registration is extremely common for most tomographic experiments, although manually performed experiments generally have more section-to-section variability unless specific measures such as dedicated sample fixtures are employed. One solution is to have an independent record of the spatial position of the sample relative to the 2D characterization device during data collection, like the LVDT data used in the A&V micromilling device. However, most tomographic instruments do not have the capability to provide this information at the present time.

Another method to perform data alignment is through the use of fiducial markers that are within or on the outside of the sample volume. A commonly used method is the use of multiple hardness indentations (Kral and Spanos 1999). As long as the indentations are relatively large so that their shape is not grossly changed between consecutive sections, these markers not only provide an independent reference to adjust for any in-plane affine transformation of the data, but also the relative change of the diameter of the indent from section-to-section provides a local gage of the sectioning depth that can be used to correct for the position of the data along the sectioning direction within the 3D stack. Other fiducial marker strategies have also been proposed, involving machining patterns directly onto the side of the sample (Spanos et al. 2008) or onto a chip that is glued to the sample (Wall et al. 2001) that also provide both in-plane and depth removal information.

A third option is to simply use the internal features that are present in the data for registration, but this method can introduce systematic errors unless the microstructure is relatively isotropic (Russ 2002). There are a number of signal processing algorithms that can be used for spatial alignment. Cross-correlation or other convolution methods are very commonly used if the data that is being registered is very similar from slice to slice, for example, image data that has approximately the same intensity and brightness variations throughout the data stack (Gulsoy et al. 2008). There are newer techniques such as mutual information (Pluim et al. 2003) that can also be used for this purpose, which is especially helpful in registering different forms of data, such as combining images with chemical or crystallographic maps (Gulsoy et al. 2008).

After correcting for spatial registry, another common operation is to use interpolation methods to adjust the x-y resolution of the 2D data files to match the average sectioning depth, to produce cubic voxels for the 3D array. Finally, one usually selects a subregion from the 3D array, to define a volume that eliminate areas that have poor data quality or other artifacts, and if needed, to minimize the size of the data volume to prevent problems with computer memory allocation.

The next step after data alignment is data segmentation, which Gonzales and Woods define for an image as "subdividing an image into its constituent regions or objects" (Gonzales and Woods 2002). For example, let us consider a two-phase microstructure that consists of a matrix with precipitates, as shown in Fig. 9. If one can obtain images of this structure that display a large contrast difference between

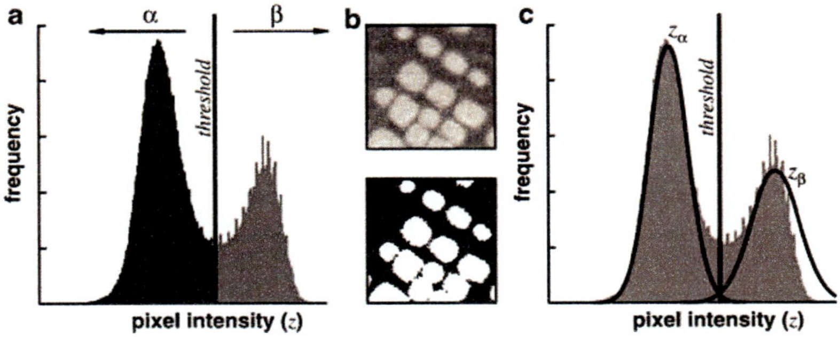

Fig. 9 (**a**) Intensity histogram for the ISE image shown in the upper frame of (**b**), which is of γ' precipitates from a Ni-based superalloy. For this histogram, the threshold value that correctly separates the pixels associated with γ' from the matrix can be readily drawn by eye, or determined analytically by fitting the intensity histogram to two Gaussian peaks (**c**). The application of this threshold can be seen in the lower frame of (**b**). Figure is used with permission from the IOP (Simmons et al. 2009)

the two phases and the contrast/brightness levels are maintained throughout the experiment, data segmentation might consist only of a single threshold operation for the entire 3D stack. Unfortunately, the complexity found in most microstructures and/or limitations of the experimental data often makes life much more difficult! The combined subject of signal processing of images and its relationship to image segmentation is far too complex to cover even in brief detail in this chapter, as there are textbooks written to examine this subject (Gonzales and Woods 2002; Russ 2002). Rather, the reader is encouraged to review these resources, and this discussion will restrict its comments to the following two recommendations to help ameliorate this process.

The first recommendation is in spite of the fact that the algorithms for performing image segmentation are becoming ever more sophisticated, especially for problems encountered in materials characterization [see for example (Jorgensen et al. 2009; Simmons et al. 2009)], there is no substitution for optimizing the "quality" of image data (Russ 2002). In particular, the experimentalist should strive to find an imaging condition or perhaps multiple imaging conditions that enable simple signal processing operations to readily threshold the features-of-interest. A recent example of this strategy is highlighted in the paper by Wilson et al. (2009), who utilized energy-biased low-kV backscattered images to readily define the various phases observed in solid-oxide fuel cell cathodes.

The second recommendation is to consider incorporating crystallographic or chemical maps rather than only image data, even though the amount of time needed to complete the experiment is usually significantly greater when these techniques are included. For the particular problem of characterizing grain structures, one method that is very attractive is to collect orientation information during the tomographic experiment, such as EBSD maps. The commercial manufacturers that supply such instrumentation have already developed a segmentation and analysis methodology to convert the Kikuchi band pattern generated by the interaction of the electron beam with the sample into a crystallographic orientation. Thus, the "difficult" part of image segmentation has already been solved, and more importantly, the singular characteristic that defines a grain – common crystallographic orientation – is an inherent part of the data collected during the experiment. The studies of Groeber (Groeber et al. 2006, 2008) and Zaefferer (Konrad et al. 2006; Zaefferer et al. 2008) have demonstrated that grain structures can be readily defined using this methodology. Similar comments hold as well for chemical spectra image data, especially when used in conjunction with automated phase analysis software, such as that developed by Kotula et al. (2003).

5 Summary Comments: Future Developments and Needs

The serial sectioning methodologies and new experimental instrumentation shown in this chapter demonstrate a direct pathway to collecting and providing 3D data for both grain and precipitate structures. However, these achievements have only partially completed the anticipated ICME need for fully autonomous systems that can provide a complete description of microstructural distributions from the micro-to-macroscale. The necessity for autonomous systems is driven by both the inherent statistical nature of microstructural arrangements and the accelerated pace of twenty-first century research programs.

For serial sectioning devices that operate at the millimeter scale, current automated instrumentation allows one to readily acquire structural information of grain-level microstructures via optical imaging techniques and simple mechanical removal methods, but collection of other basic information such as crystallographic or chemical information requires manual intervention. At the microscale, there is a greater quantity of spectral information that can be currently integrated into automated FIB-SEM data collection processes, via either electron or ion imaging and their derivatives, or chemical or crystallographic mapping.

This chapter concludes with a list of future instrumentation and other advancements from the author's perspective that should make for significant improvements in performing robust serial sectioning experiments, and ultimately make a positive impact in the field of ICME:

- Automated serial sectioning instrumentation that provides both high-spatial resolution (nanometer-to-submicrometer voxels) and multimodal data collection that is also capable of interrogating millimeter scale and larger volumes. One example of such instrumentation would be to integrate fine-scale macroscale sectioning methods with an SEM outfitted with the current state-of-the-art EBSD and silicon-drift-detector EDS systems.
- Extending mechanical or other macroscale sectioning methods like ultrafast laser-ablation (Echlin and Pollock 2008) to enable automated characterization of coarse microstructural features within full-scale engineering components.
- Further incorporation of machine inspection and metrology tools to improve both the repeatability and accuracy of the sectioning process and the task of image stack registration.
- Improved multimodal data collection. For example, the ability to simultaneously collect both chemical and crystallographic data for automated phase analysis has already been commercialized for 2D data collection using electron microscopes (e.g., the Pegasus and Trident systems by EDAX, Inc, or the INCASynergy system by Oxford Instruments). However, the capability to perform similar data acquisition outside a high-vacuum environment remains to be demonstrated.
- Continued improvement in data segmentation algorithms.
- Real-time analysis and feedback to the characterization subsystems with respect to feature segmentation and classification during data acquisition. To the author's knowledge, almost every tomographic experiment is performed asynchronously, that is, the data collection process is performed independently of data segmentation and classification. However, real-time interaction between the analysis software and characterization instruments would be especially useful for destructive techniques such as serial sectioning. One can envision that this capability would help ensure that all features within the analysis area would be positively

identified before proceeding to the next section. Also, experimental acquisition times should be significantly reduced if the tomographic instrument collected information only at the location and frequency with which it is needed.

References

Alkemper J, Voorhees PW (2001) Quantitative serial sectioning analysis. J Microsc 201:388–394

Bansal RK, Kubis A, Hull R, Fitz-Gerald JM (2006) High-resolution three-dimensional reconstruction: a combined scanning electron microscope and focused ion-beam approach. J Vac Sci Technol B 24(2):554–561

DeHoff RT (1983) Quantitiatve serial sectioning analysis: preview. J Microsc 131:259–263

Dunn DN, Hull R (1999) Reconstruction of three-dimensional chemistry and geometry using focused ion beam microscopy. Appl Phys Lett 75:3414–3416

Echlin M, Pollock T (2008) Femtosecond Laser Serial Sectioning: A New Tomographic Technique. WCCM8/ECCOMAS

Forsman O (1918) Undersökning av rymdstrukturen hos ett kolstå av hypereutectoid sammansättning. Jernkontorets Ann 102:1–30

Gonzales RC, Woods RE (2002) Digital Image Processing, 2nd edn. Prentice Hall, Upper Saddle River, NJ

Groeber MA, Haley BK, Uchic MD, Dimiduk DM, Ghosh S (2006) 3D reconstruction and characterization of polycrystalline microstructures using a FIB-SEM system. Mater Char 57:259–273

Groeber MA, Ghosh S, Uchic MD, Dimiduk DM (2008) A framework for automated analysis and simulation of 3D polycrystalline microstructures. Part I: statistical characterization. Acta Mater 56:1257–1273

Gulsoy EB, Simmons JP, De Graef M (2008) Application of joint histogram and mutual information to registration and data fusion problems in serial sectioning microstructure studies. Scripta Mater 60:381–384

Holzer L, Muench B, Wegmann M, Gasser P, Flatt R (2006) FIB-nanotomography of particulate systems—Part I: particle shape and topology of interfaces. J Am Ceram Soc 89:2577–2585

Ice GE (2004) X-ray microtomography. In: Vander Voort GF (ed) ASM Handbook, Vol. 9, Metallography and Microstructure, pp. 461–464. ASM International, Materials Park, OH

Ice GE, Pang JWL, Barabash RI, Puzrev Y (2006) Characterization of three-dimensional crystallographic distributions using polychromatic X-ray microdiffraction. Scritpa Mater 55:57–62

Inkson BJ, Mulvihill M, Möbus G (2001) 3D determination of grain shape in a FeAl-based nanocomposite by 3D FIB tomography. Scripta Mater 45:753–758

Jorgensen PS, Hansen KV, Larsen R, Bowen JR (2009) A framework for automated segmentation in three dimensions of microstructural tomography data. Ultramicroscopy. doi: 10.1016/j.ultramic.2009.11.013

Juul Jensen D, Lauridsen EM, Margulies L, Poulsen HF, Schmidt S, Sorensen HO, Vaughan GBM (2006) X-ray microscopy in four dimensions. Mater Tod 9:18–25

Kammer D, Mendoza R, Voorhees PW (2006) Cylindrical domain formation in topologically complex structures. Scripta Mater 55:17–22

Kammer D, Voorhees PW (2008) Serial sectioning and phase-field simulations. MRS Bull 33:603–610

Kotula PG, Keenan MR, Michael JR (2003) Automated analysis of SEM X-ray spectral images: a powerful new microanalysis tool. Microsc Microanal 9:1–17

Kotula PG, Keenan MR, Michael JR (2006) Tomographic spectral imaging with multivariate statistical analysis: comprehensive 3D microanalysis. Microsc Microanal 12(1):36–48

Konrad J, Zaefferer S, Raabe D (2006) Investigation of orientation gradients around a hard Laves particle in a warm-rolled Fe_3Al-based alloy using a 3D EBSD-FIB technique. Acta Mater 54:1369–1380

Kral M, Spanos G (1999) Three-dimensional analysis of proeutectoid cementite precipitates. Acta Mater 47:711–724

Kral MV, Mangan MA, Rosenberg RO, Spanos G (2000) Three-dimensional analysis of microstructures. Mater Charact 45:17–23

Kral MV, Ice GE, Miller MK, Uchic MD, Rosenberg RO (2004) Three dimensional microscopy. In: Vander Voort GF (ed) ASM Handbook, Vol. 9, Metallography and Microstructure. ASM International, Materials Park, OH

Kremer JR, Mastronarde DN, McIntosh JR (1996) Computer visualization of three-dimensional image data using IMOD. J Struct Biol 116:71–76

Larson BC, Yang W, Ice GE, Budai JD, Tischler JZ (2002) Three-dimensional X-ray structural microscopy with submicrometre resolution. Nature 415:887–890

Lewis AC, Bingert JF, Rowenhorst DJ, Gupta A, Geltmacher AB, Spanos G (2006) Two- and three-dimensional microstructural characterization of a super-austenitic stainless steel. Mater Sci Eng A 418:11–18

Lewis AC, Geltmacher AB (2006) Image-based modeling of the response of experimental 3D microstructures to mechanical loading. Scripta Mater 55:81–85

Link T, Zabler S, Epishin A, Haibel A, Bansal M, Thibault X (2006) Synchrotron tomography of porosity in single-crystal nickel base superalloys. Mat Sci Eng A 425:47–54

Ludwig W, Reischig P, King A, Herbig M, Lauridsen EM, Johnson G, Marrow TJ, Buffiere JY (2009) Three-dimensional grain mapping by x-ray diffraction contrast tomography and the use of Friedel pairs in diffraction data analysis. Rev Sci Instrum 80:033905

Lund AC, Voorhees PW (2002) The effect of elastic stress on microstructural development: the three-dimensional microstructure of a γ-γ' alloy. Acta Mater 50:2585–2598

Madison J, Spowart JE, Rowenhorst DJ, Pollock TM (2008) The three-dimensional reconstruction of the dendrite structure at the solid-liquid interface of a Ni-based single crystal. JOM 60(7): 26–30

Mangan MA, Lauren PD, Shiflet GJ (1997) Three-dimensional reconstruction of Widmanstätten plates in Fe-123Mn-08C. J Microsc 188:36–41

Maruyama B, Spowart JE, Hooper DJ, Mullins HM, Druma AM, Druma C, Alam MK (2006) A new technique for obtaining three-dimensional structures in pitch-based carbon foams. Scripta Mater 54:1709–1713

Miller MK, Forbes RG (2009) Atom probe tomography. Mater Charact 60:461–469

Orloff J, Utlaut M, Swanson L (2003) High Resolution Focused Ion Beams: FIB and Its Applications. Kluwer Academic/Plenum, New York

Pluim JPW, Maintz JBA, Viergever MA (2003) Mutual information based registration of medical images: a survey. IEEE Trans Med Imaging 22:986–1004

Russ JC (2002) The Image Processing Handbook, 4th edn. CRC Press, Boca Raton, FL

Schaffer M, Wagner J, Schaffer B, Schmied M, Mulders H (2007) Automated three-dimensional X-ray analysis using a dual-beam FIB. Ultramicroscopy 107:587–597

Schmidt S, Nielsen SF, Gundlach C, Margulies L, Huang X, Juul Jensen D (2004) Watching the growth of bulk grains during recrystallization of deformed metals. Science 305:229–232

Simmons JP, Chuang P, Comer M, Spowart JE, Uchic MD, De Graef M (2009) Application and further development of advanced image processing algorithms for automated analysis of serial section image data. Mod Sim Mater Sci Eng 17:025002–0250024

Spanos G (2006) Foreword: scripta materialia viewpoint set on 3D characterization and analysis of materials. Scripta Mater 55:3

Spanos G, Rowenhorst DJ, Lewis AC, Geltmacher AB (2008) Combining serial sectioning, EBSD analysis, and image-based finite element modeling. MRS Bull 33:597–602

Spowart JE (2006) Automated serial sectioning for 3-D analysis of microstructures. Scripta Mater 55:5–10

Spowart JE, Mullens HM, Puchala BT (2003) Collecting and analyzing microstructures in three dimensions: a fully automated approach. JOM 55:35–37

Thornton K, Poulsen HF (2008) Three-dimensional materials science: an intersection of three-dimensional reconstructions and simulations. MRS Bull 33:587–595

Uchic MD (2006) 3D microstructural characterization: methods, analysis, and applications. JOM 58:24

Uchic MD, Groeber MA, Dimiduk DM, Simmons JP (2006) 3D microstructural characterization of nickel superalloys via serial-sectioning using a dual beam FIB-SEM. Scripta Mater 55:23–28

Uchic MD, Holzer L, Inkson BJ, Principe EL, Munroe P (2007) Three-dimensional microstructural characterization using focused ion beam tomography. MRS Bull 32:408–416

Wall MA, Schwartz AJ, Nguyen L (2001) A high-resolution serial sectioning specimen preparation technique for application to electron backscatter diffraction. Ultramicroscopy 88:73–83

Wilson JR, Kobsiriphat W, Mendoza R, Chen HY, Hiller JM, Miller DJ, Thornton K, Voorhees PW, Adler SB, Barnett SA (2006) Three-dimensional reconstruction of a solid-oxide fuel-cell anode. Nat Mater 5:541–544

Wilson JR, Duong AT, Gameiro M, Chen HY, Thornton K, Mumm DR, Barnett SA (2009) Quantitative three-dimensional microstructure of a solid oxide fuel cell cathode. Electrochem Commun 11:1052–1056

Wojnar L, Kurzydlowski JK, Szala J (2004) Quantitative image analysis. In: Vander Voort GF (ed) ASM Handbook, Vol. 9, Metallography and Microstructure. ASM International, Materials Park, OH

Wolfsdorf TL, Bender WH, Voorhees PW (1997) The morphology of high volume fraction solid-liquid mixtures: an application of microstructural tomography. Acta Mater 45:2279–2295

Zaefferer S, Wright SI, Raabe D (2008) Three-dimensional orientation microscopy in a focused ion beam-scanning electron microscope: a new dimension of microstructural characterization. Metall Mater Trans A 39A:374–389

Digital Representation of Materials Grain Structure

Michael A. Groeber

Abstract Recent initiatives to accelerate the insertion of materials and link the materials design and systems design processes have called for the advancement of microstructure–property relationships. To achieve these goals, the development of digital microstructure models in conjunction with computational methods for simulating material response is a necessity. There have been significant advancements in the collection and representation of microstructure, which coupled with computational power increases, has yielded microstructure models with increasing complexity and accuracy. It is the emphasis of this chapter to discuss the state-of-the-art methods and current limitations in the field of microstructure representation. Specific focus will be paid to the areas of: experimental data collection, feature identification, mesh generation, quantitative characterization, and synthetic structure generation. In presenting the status of the field, the key links to other fields that must be developed will also be addressed wherever possible.

1 Introduction

This book is motivated by the practical assessment that many questions remain to be answered regarding the effect of microstructure on materials response. Attempts to answer these questions have produced microstructure–property relationships with varying levels of detail and empiricism. The empiricism of such relationships has limited the integration of materials design in the systems design process. Instead, the systems design community tends to use databases developed through extensive experimental testing to select existing materials with a set of required performance properties. The last chapter of this book outlines the prospect of designing new materials to a desired set of properties and doing so in an accelerated fashion. One

M.A. Groeber (✉)
Materials and Manufacturing Directorate, Air Force Research Laboratory,
Wright-Patterson Air Force Base, OH, USA
e-mail: Michael.Groeber@wpafb.af.mil

S. Ghosh and D. Dimiduk (eds.), *Computational Methods for Microstructure-Property Relationships*, DOI 10.1007/978-1-4419-0643-4_3,

factor limiting the advancement of microstructure–property relationships to allow for accelerated materials design is the disjointed treatment of microstructure by the systems design, materials processing and materials development communities. The current "industry standard" in the systems design community is to treat microstructure as a set of notes or ASTM specifications on a part drawing. As a result, the processing community uses these notes/specifications as guidelines and quality control metrics when producing materials. This approach to microstructure description is inconsistent with that used in most computational microstructure–property models. This chapter attempts to address these inconsistencies in microstructure description and outlines a process for digital representation of microstructure, which allows for the more accurate inclusion of microstructure in property models.

It has become common practice to generate microstructure models for property prediction simulations with limited amounts of microstructural information included. Typically, simple geometric shapes or tessellations are used to represent microstructural features with little attention paid to accuracy beyond average values. Often though, the use of simple average quantities such as "grain size" is likely to be inadequate in some microstructure–property relationships; instead one may need to consider the possibility that the full three-dimensional (3D) microstructure is important. The complexity and interplay of features in the 3D structure dictates that calculations by hand are generally impractical. Fortunately, computational power has provided a new pathway for the materials scientist to investigate microstructure–property relationships through microstructure modeling. However, computational modeling of materials requires the ability to generate digital microstructures in which the most relevant features are sufficiently described. There are multiple ways to tackle the challenge of generating 3D digital representations of microstructure. The obvious approach is to explicitly represent structure by collecting sample volumes of microstructure experimentally, whether it involves reconstructing direct images from serial sectioning or by reconstructing diffraction information from various 3D tomography techniques. A second, less direct approach is to represent structure statistically by developing tools to generate synthetic structures with statistics equivalent to some desired set, likely obtained from some experimental observation technique(s). Both of these approaches will be discussed in detail in this chapter.

With regard to the explicit representation of structure, this chapter focuses on the specific topics of experimental interrogation techniques, data processing, and mesh generation. Experimental techniques to characterize microstructure in 3D have undergone dramatic improvements in the past decade, and there now exists a host of methodologies that are capable of collecting 3D microstructural information that range from counting individual atoms (nm) to imaging macro-scale volumes (mm). However, it is important to note that there are still gaps that exist in structure collection; for example, dislocation structures cannot be readily imaged in 3D. The state-of-the-art for this field has been reviewed recently in a Viewpoint Set for *Scripta Materialia* (Spanos 2006) and thus will not be presented here. Currently, there are two main experimental interrogation pathways to collect information about microstructural features in 3D. Serial sectioning experiments are more commonplace but consume the sample as part of the experiment, while X-ray methods are nondestructive but typically require the use of high-intensity X-ray sources. For the X-ray methods, there are a handful of groups worldwide that are working toward spatially resolved crystallographic analysis of grain structures in 3D using high-intensity X-ray systems (Schmidt et al. 2004; Lauridsen et al. 2006; Budai et al. 2004, 2008; Lienert et al. 2007). These methods have a significant advantage in that the sample remains intact after analysis, allowing for the possibility of time-dependent studies of microstructural changes due to thermal or mechanical input. Nonetheless, these experiments require the high brilliance of a synchrotron source that puts significant restrictions on the general applicability of the methods. Therefore, this chapter will skew its focus toward the processing and analysis of grain-level data using more universally accessible serial sectioning experiments. Electron backscattered diffraction (EBSD) maps are often coupled with the sectioning process as an integral component of the characterization method. The incorporation of crystallographic maps enables a straightforward approach to define and segment the individual grains or precipitates that compose aggregate assemblies, and also allow for orientation-based data analysis. Most of the data-processing steps presented here are tailored for the EBSD-based section data, but all are generic in their end goal of reconstructing and segmenting features with limited experimental artifacts. Finally, the reconstructed and processed data must be translated into a meshed structure for simulation. Various methods for surface representation and mesh generation will be presented with focus paid to the complexity, quality and physical accuracy of their resultant meshes.

The quantification of microstructure and generation of equivalent synthetic structures will be discussed in the context of the statistical representation of materials structure. The motivation for such an approach is the ability to generate many different representative microstructures that match statistically the measurements on the material of interest to a pre-determined degree of accuracy. The ability to generate synthetic structures is meant to limit the need for abundant data collection as well as supplement when direct 3D information is unavailable. The experimental methods discussed in the section on explicit representation of structure provide direct 3D information. Three-dimensional (3D) characterization methods are then required to quantify the structure so as to serve as the input information for the synthetic structure generation process. In contrast to the use of direct 3D statistics, the prospect of inferring 3D statistical descriptors from 2D observations is also a topic of interest. These stereological approaches are necessary to develop due to the fact that 3D experimental techniques remain unavailable to a large portion of the materials community.

The development of computational materials models with specific focus on the accurate incorporation of microstructure is the ultimate goal of this chapter. Two approaches for representing structure, explicit and statistical, will be addressed. The major steps in each of these processes will be described. The explicit approach requires experimental investigation, data processing, and mesh generation. The statistical approach consists of statistical quantification of experimental observations and synthetic structure generation also requiring subsequent mesh generation.

2 Challenges and Previous Work

2.1 Characterization

Characterizing microstructure has been classically limited to two-dimensional (2D) measurements and simplistic extrapolations to three dimensions (3D). Many microstructural features can be estimated with reasonable accuracy in this manner, but arguably far more are inaccurate, if calculable at all. Limitations in these techniques have fueled a drive toward direct 3D data collection. Early attempts to collect direct 3D data (Hillert 1962; Hopkins and Kraft 1965; Rhines et al. 1976) were somewhat successful, but only recently have experimental and computational tools advanced to make 3D data collection readily feasible. Additionally, many of the early 3D experiments provided only visualizations and not quantitative microstructural descriptors. Equally important in the collection of microstructure are the current limits on the automation of microstructure feature identification and segmentation. Direct 3D data has proven to be exhaustive and tedious to reconstruct and segment, which has greatly hindered its applicability. Even 2D measurements have been slowed by the common inability to distinguish features of microstructure. Some microstructures display the ability to easily segment features using only either optical or scanning electron microscope (SEM) imaging, but this is not always the case, especially as microstructures grow in complexity. The development of orientation imaging microscopy (OIM) via EBSD has cleared a path to automated segmentation of grains and phases over restricted scales, but does increase data collection times. Finally, and arguably most important, the quantitative description of microstructure has, in some cases, been left incomplete and inconsistent. Often only average values are measured neglecting the distributions of features, leaving the microstructure incompletely described. The distributions of parameters are certainly more arduous to measure and this is usually the cause for their absence in many characterizations. However, recent developments in data collection and feature identification have allowed for these measurements to be made relatively easily for many microstructures. Additionally, parameters are generally measured independently, resulting in the inability to correlate relationships between parameters. This is a major concern when representing microstructure, because correlations may describe the tendency of features to cluster, which may have significant affects on properties. Inconsistencies in microstructure description may arise due to a number of descriptors that quantify the same microstructural feature or parameter, but vary in their meanings and dimensionalities [e.g., mean linear intercept, equivalent sphere diameter (ESD), true grain volume, and mean width]. The various descriptors, when scaled to be comparable, can deviate significantly, causing confusion in which to use.

2.2 Modeling

The representation of microstructure has traditionally been significantly limited by computational power, which has led to the common practice of grossly simplifying microstructure in many computational models. It is quite a standard approach to represent microstructural features as simple geometrical shapes, often with no size distribution, because "size" is rarely a part of model physics. Even in cases where attempts are made to match quantified microstructural parameters; there are generally not more than one or two parameters represented and rarely a correlation between multiple parameters. For some properties (i.e., texture or yield stress), it is possible that detailed morphological structure is unnecessary. However, other properties (i.e., fracture or strain hardening) are likely to require not only initial detailed structure, but evolving structure as well. The assumptions made in the representation of microstructure may significantly hinder the ability of physical models to predict material response. That is, a model which is developed from physical understanding of a process may be applied to a nonphysical microstructure, and as a result the simulation is inherently constrained and biased. Further, when deviations between simulations and experiments are encountered, they become difficult to classify as errors in microstructure representation or errors in the physical model itself. The inability to decipher the cause of error forces incorporation of scaling constants and other relatively empirical factors to match results. Additionally, when explicit 3D structures are available, the inability to mesh the structures with a consistent quality has limited the ability to analyze the structures. The mesh generation issue remains a key concern for computation materials modeling. Finally, determination of the necessary detail to be incorporated into a simulated volume has proven complicated. Identification of the so-called representative volume element (RVE), which specifies/bounds the extent to which the structure needs to be described to accurately simulate the underlying structure, has become an entire area of study. The combination of detailed structure quantification, accurate digital representation, robust meshing techniques, properly developed constitutive laws and significant computational power is required to determine an RVE for a given property. Given that the RVE is likely to change for various properties and the significance of a simulation is strongly linked to the use of an RVE, it becomes immediately clear that the aforementioned tools are key in the development of computational materials modeling.

3 Explicit Representation of Structure

The most straightforward and likely most accurate method for representation of 3D microstructure is to explicitly translate an experimentally collected volume into a computational model. The direct inclusion of experimentally obtained microstructural information requires few or even no assumptions about the microstructure itself; although the data collection technique may require some prior knowledge of the microstructure. There are many experimental techniques that are capable of yielding information about the 3D grain structure of a material. In general, these techniques tend to fall in one of the two categories: X-ray based methods and serial sectioning methods. Both of these general categories have advantages over the other, as well as inherent limitations. X-ray based techniques have the major advantage of being nondestructive, but are often limited by resolution and cost/availability. Serial sectioning, given the many tools capable of undertaking it, provides adequate resolution across a significantly wide range of length scales (i.e., hundreds of nanometers to tens of millimeters). However, the process is destructive and can be time-consuming and erratic. This chapter will only briefly mention some aspects about the collection processes themselves; rather the processing, analysis, and application of the resultant data will be the main focus. Further technical details can be found earlier in this book as well as elsewhere for the various X-ray (Schmidt et al. 2004; Lauridsen et al. 2006; Budai et al. 2004, 2008; Lienert et al. 2007) and serial sectioning (Rowenhorst, Groeber, Raabe, Voorhies) techniques.

Serial sectioning experiments are comprised of two main tasks – sectioning and data collection – which are repeated until the desired volume of material has been interrogated. The sectioning process involves the removal of a known volume slice of material, usually determined primarily by the size-scale of the microstructural features that are to be examined. A typical rule-of-thumb is to acquire a *minimum* of ten sections per average feature if feature shape is to be determined. The section thickness is often a compromise between the desire for high-fidelity data and the constraints of both personnel and instrument time to collect the data sets. After sectioning, the surface-of-interest is characterized through standard imaging methods or other mapping methods, such as EBSD, electron dispersive spectroscopy (EDS), etc. The resultant data obtained from a serial sectioning experiment is thus a series of parallel images that likely contain some alignment errors, require some form of feature segmentation, and may need some smoothing to account for the digital (or pixilated) images of the microstructure.

X-ray-based experiments involve exposing the sample of interest with an incident beam of X-rays and measuring various aspects of the interaction of the X-rays with the sample. The fluorescence, absorption and diffraction scattering are examples of measurable results. The group of techniques that provide a description of the position and topology of the internal features of the sample are generally referred to as tomographic techniques. Within the broad class of tomography, there are multiple methods for obtaining contrast between the features of interest; for example, phase contrast tomography. In addition to obtaining topological/morphological information, if diffraction scattering is analyzed, the crystallographic information of the features can be measured as well. The limit on resolution and size of the data collected is closely linked to the intensity of the incident X-ray beam. The resultant data obtained from the X-ray experiment is a set of 3D positions with associated values of absorption (or other measures) and possibly coupled with crystallographic information. Similar to serial sectioning, there are usually artifacts in the X-ray data as well, many of which require filtering to allow for feature segmentation.

The important point to consider is that the data collection technique, while it may certainly dictate/demand some specific processing substeps, does not greatly change the major post-collection steps necessary to represent the experimental data as a 3D model of microstructure. Every 3D dataset requires a thorough application of a series of tools to align consecutive images, segment features of interest, and

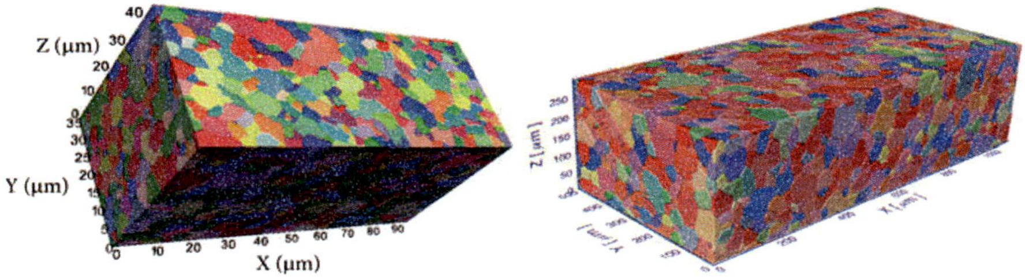

Fig. 1 Sample reconstructions of (*left*) a nickel-base superalloy obtained through FIB-based sectioning and (*right*) a titanium alloy obtained by manual mechanical polishing

represent feature boundaries. Each of these major issues will be discussed with more focus toward serial section data. Figure 1 shows examples of some reconstructed 3D volumes obtained by serial sectioning.

3.1 Reconstruction and Feature Identification

3.1.1 Image Alignment and Stacking

It is often the case that misalignment between sections occurs during the sectioning experiment and can cause difficulties in the subsequent analysis of the dataset. These misalignments can be minimized through the use of fiducial marks coupled with manual and automated alignment tools during the data collection process, but can rarely be totally avoided. This section will address techniques used to adjust sections after the collection process has completed. First, the more general alignment solution for image data is discussed and then the more unique condition involving specialized data, such as EBSD, is presented.

The most straightforward registration procedures involve applying simple translations to sections to improve section-to-section alignment. In the general case, this can be done with images and image processing techniques such as least-square difference fitting or image convolution. In these procedures, images are translated (generally at multiples of the pixel size) in the x and y directions of the image until either a minimum difference or maximum product between pixels in consecutive images is obtained. Subpixel alignment can also be obtained by moving images fractions of the pixel size and interpolating the values on the new grid. Rotational alignment can also be achieved through these image processing procedures if interpolation is used to generate a grid coincident with the reference image. It should be mentioned that for these techniques to function properly, consecutive images should have consistent conditions (i.e., brightness and contrast). Often, this requirement necessitates a certain amount of image processing prior to the alignment procedures.

In many cases, the registration of separate serial sections requires more than simple translations to gain proper alignment of the image stack. This is most prevalent when combining data collected by different methods, such as micrographs from a light optical microscope and EBSD scans from an SEM, where there may be different coordinate systems, orientations, and scaling between pixels. More complex alignment procedures are also needed when the sectioning experiment lacks a fixed reference frame. If the sections are obtained by sectioning slabs through a sample rather than polishing away material, the slabs, when mounted separately, may be rotated, translated and tilted relative to one another. Alignment of these images begins by identifying a number of point features (examples would include the centers of grains, triple junctions or sample edges). One then must choose a reference frame for to which images are aligned, such as an optical image (here by noted as X', Y'). The points (X, Y) in a corresponding EBSD scan, for example, should then be brought into coincidence with the optical image reference frame. The transformation from the EBSD image to the reference optical image is given by:

$$\begin{bmatrix} X' \\ Y' \\ 1 \end{bmatrix} = T \begin{bmatrix} X \\ Y \\ 1 \end{bmatrix}, \tag{1}$$

where T is a two-dimensional transformation matrix of the form:

$$T = \begin{bmatrix} t_{11} & t_{21} & t_{31} \\ t_{12} & t_{22} & t_{32} \\ 0 & 0 & 1 \end{bmatrix}.$$

For an affine transform, where image rotation and separate translation, scaling and skewing are all allowed for the two directions, T is solved for all values without constraint, using the pseudo inverse matrix (a least squares fit to a system of linear equations).

However, when allowing skewing of the images, it is not possible to independently determine the rigid rotation of the two coordinate frames, which is necessary to correct the measurement of the crystallographic orientations between each EBSD section. Therefore, a limited transformation matrix can be used that only allows independent scaling in the x-direction and y-direction, S, translations in the x-direction and y-direction, P, and finally an image rotation (of angle θ) normal to the sectioning plane, R. Thus

$$T = RPS, \tag{2}$$

where:

$$R = \begin{bmatrix} \cos(\theta) & -\sin(\theta) & 0 \\ \sin(\theta) & \cos(\theta) & 0 \\ 0 & 0 & 1 \end{bmatrix}; P = \begin{bmatrix} 1 & 0 & P_x \\ 0 & 1 & P_y \\ 0 & 0 & 1 \end{bmatrix}; S = \begin{bmatrix} S_x & 0 & 0 \\ 0 & S_y & 0 \\ 0 & 0 & 1 \end{bmatrix}$$

The terms in the transformation matrix are then determined by a least-squares optimization. As mentioned, the decoupling of the matrix T is only required to correct crystallographic data associated with EBSD and is not necessary to do for data without such information. Further, the z-components of the matrices can be included to correct tilt errors introduced by remounting or nonparallel polishing.

As mentioned, alignment errors can be corrected in image space by processes such as least-square difference fitting, image convolutions, and Fourier transforms (Gulsoy et al. 2009; Simmons et al. 2009). However, these methods may not utilize all of the data in an image equally and image conditions can vary significantly between sections. Varying image conditions can create problems identifying features as the same feature in consecutive images. EBSD or other special data types enable the use of all data points equally and generally provide a more invariant parameter to follow between sections. This is generally true because while contrast in electron images (i.e., secondary or backscatter) does have some dependence on the structure of the individual microstructural constituents, there is not a unique probing of the structure like is the case with EBSD or chemical mapping, for example. In the case of EBSD, the misorientation between a data point and the corresponding data point on a neighboring section can be calculated for all points in a given section. A parameter (Ψ) can then be created to define the amount of misalignment between the two sections. The definition of Ψ is given by:

$$\Psi(k) = \sum_{j=0}^{y_{\max}} \sum_{i=0}^{x_{\max}} \psi(i, j, k), \tag{3}$$

where $x_{\max}$ and $y_{\max}$ are the total number of data-points in the corresponding directions and an example of $\psi(i, j, k)$ is given by:

$$\psi(i, j, k) = \begin{cases} M & [P(i, j, k), P(i, j, k-1)] \\ 0 & \text{otherwise}, \end{cases},$$

where $M\ [P(i, j, k), P(i, j, k-1)]$ is the misorientation between points i, j, k and $i, j, k{-}1$. The calculation of misorientation is discussed further in the following section on feature segmentation. The section k can be translated, by multiples of the EBSD (or other data-type) step-size, in the x and y direction until a position with minimum Ψ is located. The parameter $\psi(i, j, k)$ can be adjusted to handle criterion tailored to other data types. Subpixel alignment can also be obtained connecting triple points on consecutive sections and minimizing their alignment (Rollett and Manohar 2004).

3.1.2 Feature Segmentation and Clean-Up

Segmentation of individual features is necessary to allow for the measurement of each separately. Additionally, identifying and separating microstructural constituents provides the ability to investigate features removed from their surroundings. Grain segmentation is greatly aided by the quantitative orientation information provided by the EBSD maps, as local orientation information is the most direct means to group voxels which reduces issues of image contrast, thresholding and feature identification. There have been other, image-based techniques shown to be applicable to feature segmentation, but few have been utilized to segment features in 3D without significant user intervention (Rowenhorst, DeGraef). Feature segmentation using EBSD allows for the complete automation of the segmentation process. During segmentation, grains are identified as groups of voxels that share a similar orientation. Algorithms for identifying these groups of voxels have been presented previously (Groeber 2007; Bhandari et al. 2007; Ghosh et al. 2008). Commercial analysis packages (TSL, HKL) likely use a similar approach to identifying grains. The major steps of the algorithm are outlined in this section.

The initiation of each identified grain is the selection of a seed voxel. Generally, it is a useful idea to select a voxel that has been deemed to be of "good" quality. Quality is defined by the EBSD data collection software and refers to the sharpness of the pattern, which is often correlated with the confidence in the assigned orientation. Usually, the highest quality data tends to lie within the center of a grain, and these voxels serve as a reliable point to begin grain segmentation. A grain is then defined as the set of voxels contiguous to and with the same orientation as the seed voxel. The requirement of the voxels to have the same orientation (within a defined tolerance) is based on the fact that all regions within a grain should share a similar orientation. A list of voxels assigned to the grain is created, which initially contains only the seed voxel. The voxels that neighbor the seed voxel are checked to determine whether they have a similar orientation. The misorientation is measured to evaluate the orientation difference between the seed voxel and each of its neighbors. The value of misorientation is given by the following equation:

$$\theta = \min \left| \cos^{-1} \left[\frac{tr(O_c g_A g_B^{-1} O_c) - 1}{2} \right] \right|, \tag{4}$$

where O_c is the crystal symmetry operator and g_A and g_B are the rotation matrices of voxel A and B and are given by:

$$g_i = \begin{pmatrix} \cos\varphi_1 \cos\varphi_2 - \sin\varphi_1 \sin\varphi_2 \cos\Phi & \sin\varphi_1 \cos\varphi_2 + \cos\varphi_1 \sin\varphi_2 \cos\Phi & \sin\varphi_2 \sin\Phi \\ -\cos\varphi_1 \sin\varphi_2 - \sin\varphi_1 \cos\varphi_2 \cos\Phi & -\sin\varphi_1 \sin\varphi_2 + \cos\varphi_1 \cos\varphi_2 \cos\Phi & \cos\varphi_2 \sin\Phi \\ \sin\varphi_1 \sin\Phi & -\cos\varphi_1 \sin\Phi & \cos\Phi \end{pmatrix},$$

where $(\varphi_1, \Phi, \varphi_2)$ are the Bunge Euler angles of the voxel. If the misorientation is less than the defined tolerance (i.e., $\sim 5°$), then that voxel is added to the list of voxels of the grain being segmented. Each voxel on the list undergoes the same process of checking its neighboring voxels until no new voxels are added to the list. After the list is complete, a new seed point is generated and the voxel assignment is repeated for the next grain. An option during segmentation is to terminate the process when no unassigned voxels remain above a data quality tolerance. This ensures that no grains can be formed that include all low quality voxels. "Clean-up" routines can be implemented to handle low quality points either before or after assignment. Some examples of clean-up routines will be discussed later.

The process of grouping voxels is not unique to data with orientation information. Similar algorithms can be used for image data or chemical data, where voxels of similar grayscale or chemistry are grouped together. In all three cases, it may be important to consider the possibility of gradients in the data. Rarely is the interface between neighboring features perfectly sharp in the collected data. The grouping criterion may allow for additive deviation from the original seed voxel if only immediate neighbors are checked. Possible solutions include comparing the new voxel to both its immediate neighbors and the original seed voxel or comparing the new voxel to an updated average value of all the previously grouped voxels.

Clean-up routines are important to handle low-quality data as well as treat features that may be nonphysical or difficult for the simulation tools to handle. Clean-up should ideally be performed in 3D, to provide the most information during the clean-up process. In any experimental technique, there will be data points that are indexed incorrectly, due either to sample preparation or to collection error. These points will either be left unassigned during the feature segmentation process or they will be identified as a feature of their own. The latter case results in a large number of extremely small features that will be difficult for the simulation to handle. A minimum feature size criterion can be used to filter any extremely small features. A feature which contains less than a defined number voxels can be dissolved and its voxels reassigned to neighboring features. The motivation to remove extremely small grains is the inability to accurately characterize features made of so few voxels and the trouble obtaining results in the simulation at those regions. The minimum size should be carefully defined and not selected arbitrarily. One important factor in selecting the minimum size is the number of voxels needed to generate a reasonable description of feature shape. Another factor in selecting the minimum size is the fraction of features that will be removed by the filter. Generally, it is undesirable to remove a large percentage of features with any one filter. If more than a few percent of the features are removed due to the minimum size criteria, it may be an indication that the resolution or quality of the data is insufficient.

Any remaining voxels, whether unassigned during or dissolved after the segmentation process, should still be assigned to neighboring features to create a fully dense structure. There are multiple options for deciding how to assign these remaining points, each of which has advantages and disadvantages. The remaining low-quality points can be assigned to the feature with which they share the most surface area, to the feature which owns the highest quality neighbor of the unassigned voxel, or to the largest feature which they neighbor. In addition to the assignment of problem voxels, there are a variety of additional data processing procedures that can be applied to the 3D EBSD data, many of which are simple extensions of the 2D filters supplied in commercial EBSD analysis programs. For example, the average orientation of each grain can be calculated and assigned to all the voxels that constitute the grain. The average orientation is the orientation that minimizes the total misorientation with all the voxels in the grain. The average orientation can be solved for numerically as shown by Barton and Dawson (2001). Generally, an adequate initial estimate of the average orientation can be obtained by transforming the orientation of each voxel into a single fundamental zone and finding the center of mass of the resultant point cloud in orientation space. Additionally, the dataset can be scanned for grains are that are fully contained within another grain and as a result have only one neighbor. These grains could potentially be small subgrains that are misoriented only slightly larger than the misorientation tolerance. Data sets may also contain grains with special boundaries (i.e., twin boundaries) that can be identified and omitted from certain analyses if desired. Grains sharing these special misorientations can be merged together to leave only general grain boundaries in the data set. These processing possibilities are provided as selected examples and not meant to comprise a full list of the options available to further process 3D EBSD data.

3.2 Feature Surface Representation and Mesh Generation

The following section will discuss a set of mesh generation techniques. Each technique will be critiqued with focus on its surface representation/structure, mesh quality, difficulty of implementation, inherent physicality, and user bias. It is not the author's intent to identify a "best" technique, but rather to highlight the strengths and weaknesses of each technique. All of these techniques are designed to accept as input, a discrete voxel-based structure that has been segmented and labeled as unique features. It is not necessary for the discrete data to be on a cubic grid, although the grid should be orthogonal to produce the best results in most cases. Additionally, the data need not be obtained experimentally; a synthetic structure builder can be used to generate the voxel-based structure, which will be discussed later in this chapter. Figure 2 shows a grain with different surface representations to illustrate the effect of the techniques discussed here.

Fig. 2 Surface representations of a grain using (*left*) voxel-based meshing (*middle*) marching cubes without smoothing as commonly used in direct image-based meshing and (*right*) marching cubes coupled with line tension-based smoothing

3.2.1 Voxel-Based Mesh

Many, if not all, experimental methods collect data on a regular, discrete gird. The resultant data can then be treated as either a set of pixels in 2D or a set of voxels in 3D. Pixels and voxels provide a useful construction for automated mesh generation without the need for a separate surface representation step. The pixels or voxels can be directly imported into a finite element (FE) or fast Fourier transform (FFT) analysis code as brick elements or integration points, respectively. Another advantage of the voxel-based mesh is that the voxels (or brick elements) have good quality metrics. Two common measures of a mesh's quality are the distribution of dihedral angles of the elements and the size distribution of the elements. Voxels, even rectangular voxels, have all dihedral angles equal to 90°, which is nearly optimal for FE analysis. In general, all dihedral angles between 20° and 160° are acceptable for most analyses. The size distribution of a voxel-based mesh is a delta function at the grid spacing of the experimental data. Thus, there is no distribution in size of the voxels, unless some nonhomogenous decimation process has been employed. With a single element size, there is no concern of tiny elements much smaller than the average element size. Tiny elements are problematic when simulating the structure using FE.

However, the voxel-based mesh is not without issues to consider. Voxels create two major concerns that must be discussed to determine the suitability of this technique. First, the voxel-based construction produces a surface structure that is aliased or "stair-stepped". The stepped nature of the surface is directly linked to the facts that the elements have all dihedral angles equal to 90° and no surface representation step is used. The edges and corners of elements on the surface create sharp discontinuities between features and can give rise to mesh instabilities. These sharp features can behave as stress concentrations during deformation simulations, leading to artificial stress and strain localizations. Second, the lack of distribution in the size of elements often creates an abundant number of elements. If the microstructural features vary in size, it is practical to have elements that vary in size as well. Also, even if the microstructural features are of uniform size, the mesh size need not be uniform. The gradient in the property of interest can be used to grade the mesh size, where larger elements should be used where the gradient is small and smaller elements used where the gradient is large. In general, feature boundaries are areas of high gradient and the feature centers exhibit lower gradients. It is possible to decimate the voxel structure by combining neighboring voxels together in regions with small gradients in the property of interest, but not without some additional mesh compatibility steps that will not be discussed here.

The voxel-based mesh technique is certainly the simplest to implement of the techniques to be presented here. The mesh quality is also very high. The other techniques can also produce high quality meshes, but none have the inherent quality of the voxel-based mesh. However, the voxel-based mesh has arguably the worst surface structure due mainly to its complete lack of physical meaning. The "stair-stepped" structure is solely an artifact of the data collection process and not any true physical structure. The user bias is eliminated by simply using the experimental data, which at least confines the error to only the data collection itself. Due to the relative simplicity of implementation, many investigators have used the voxel-based construct for both FE and FFT simulations.

3.2.2 CAD-Based Surface Fitting

The voxel structure obtained from experimental data collection can be used as input into a surface-fitting algorithm. It is often practical to perform the surface fitting and subsequent surface reconstruction in a CAD environment. The goal of CAD-based surface fitting is to correct the "stair-stepped" nature of the voxel structure. The process consists of identifying the set of voxels that make up the boundary/interface between two microstructural features and fitting a polynomial or spline surface through them. After completing this step for every interface, each microstructural feature is defined as the region bounded by all of the fit surfaces corresponding to the interfaces between itself and its neighbors. The intricacies of this approach have been detailed previously (Bhandari et al. 2007; Ghosh et al. 2008) and are far too extensive for this chapter.

The advantage of this technique is the smooth surface representation that results from the adjustment of the polynomial surfaces or splines. However, there are difficulties encountered during the fitting and reconstruction/bounding process. First, the surfaces can be adjusted to be increasing more complex by increasing the order of the polynomial or spline. Increasing the order of the surface generally improves the fit to the voxels, but also creates issues such as self intersection and local degenerate solutions. A practical limit for the surface order has been shown to be around three for a typical engineered microstructure (Bhandari et al. 2007; Ghosh et al. 2008). Second, creating the region bounded by a set of fit surfaces is not trivial. Each surface is fit independently and there can be incompatibilities at the areas of their intersection. Neighboring surfaces may not actually intersect or may intersect at a point significantly away from the true feature surface. This is due in large part to the lower number of data points at the edges of the surfaces. Also, data at the edges and corners of features tends to have a higher propensity for error. These issues combined with the high order fit surfaces can lead to spurious events at the areas where surfaces intersect. In addition, each feature is bounded separately and thus overlaps and gaps can be artificially created between neighboring features. These generally small artifacts are often the result of locally perturbing fit surfaces to ensure their intersection. Artifact cleaning routines can correct the overlaps and gaps by overlaying the original voxel data (Bhandari et al. 2007; Ghosh et al. 2008), but not without significant user intervention or advanced CAD programming.

Following the bounding of each feature and artifact correction, a volume mesh must be generated. Each surface of the bounded region can be discretized, with local curvatures of the surface dictating the local density of points on the surface. The points on the surface serve as nodes for an FE mesh. Additional nodes are required in the interior of the feature and can be positioned with various goals, such as a graded size distribution toward the center of the feature, where gradients are often lower.

The volume node generation process becomes a balance between spacing nodes homogeneously to ensure the best possible dihedral angle distribution and grading the node density to yield a graded mesh structure. Once the surface and volume nodes are created, the nodes can be connected by a Delaunay triangulation process to generate a set of tetrahedrons that will be the elements for a FE analysis. The nature of the Delaunay process results in a near-optimal mesh for the given nodes. However, the tetrahedral elements generated by the Delaunay triangulation will likely still require some clean-up processes to ensure proper element quality. In the interest of brevity, these clean-uproutines will not be discussed here, but can be found elsewhere (Barry 1995; Amenzua et al. 1995; Parthasarathy and Kodiyalam 1991).

The CAD-based technique is likely the most time intensive technique to implement. The mesh quality can be very high, provided the clean-up tools are designed properly, but it does not have the inherent quality of the voxel-based construct. The largest benefit of the CAD-based mesh is the smooth surface structure, which is arguably better than any of the other techniques. The "stair-stepped" structure is generally completely eliminated. However, the process for smoothing the surfaces is not inherently physical. For example, the surface fitting criteria are generally defined to smooth while minimizing deviation from the experimental data and not to approach some physical property such as minimum surface area or specified angle of surface intersection. The user bias can also be significant during the artifact correction steps, where overlaps and gaps must be assigned to neighboring features. Bhandari et al. (2007) developed come physically based tools to correct these artifacts with limited bias.

3.2.3 Direct Image-Based Meshing

Direct image-based meshing is similar to the voxel-based approach, but with some attempt to smooth the surface structure. The central tool used in most direct image-based approaches is the marching cubes algorithm first developed by Lorensen and Cline (1987) and modified by others (refs). The marching cubes algorithm breaks a voxel-based structure into tetrahedrons based on local voxel neighborhoods. The tetrahedrons are generally high quality elements because of their tendency to be right tetrahedrons with dihedral angles of 45° and 90°. The surface structure is also generally improved over that of the voxel-based representation. The voxels on the surfaces of features are broken in a manner that removes many of the abrupt 90° angles that causes the "stair-stepped" surface structure discussed previously.

The marching cubes algorithm creates an extremely large number of elements, because each voxel is broken into multiple tetrahedrons. The tetrahedrons tend to be close to uniform in size and thus, they are far from efficient in meshing the microstructural features. The tetrahedrons can be decimated to improve the mesh efficiency by reducing the number of elements. The decimation of the tetrahedral mesh is actually more straightforward than the decimation of the voxel-based structure directly. The surface representation, while improved, is still far from the smooth surfaces generated by the CAD-based surface-fitting approach. Some commercial programs designed to generate direct image-based meshes have smoothing tools built in, but it is not known to this author what, if any, physical criteria are used. It is also possible to combine techniques by running a marching cubes algorithm on the original voxel-based structure before passing the semi-smoothed structure on to a CAD-based approach. The semi-smooth structure is likely to improve the CAD-based approach by supplying a smoother initial surface structure, which may allow for the use of lower order fit surfaces.

The direct image-based approach has nearly the same pros and cons as the voxel-based approach, albeit the surface representation is improved. The ease of implementation, high-quality metrics, and limited user bias are key advantages, but the relatively poor surface representation and the lack of physical smoothing criteria are drawbacks.

3.2.4 Surface Area/Line Tension-Based Smoothing Methods

The surface area/line tension smoothing technique is an attempt to combine the direct image-based method with a semi-physical smoothing process. The first step in this method is to run the marching cubes algorithm on the original voxel-based structure. After the voxel-based structure is converted into a tetrahedral mesh, a surface smoothing algorithm is applied to improve the surface representation. The difference between this smoothing algorithm and the CAD-based approach is that this algorithm has a physical driving force and the features are not treated independently. The physical driving force is related to the general belief that interfaces between features will tend to minimize their surface area in an attempt to minimize the energy of the structure. The smoothing process is presented in detail by Lee (ref) and will only be outlined here.

The smoothing process begins with the identification of all quadruple points, which are points where four features meet. Each quadruple point is connected to another quadruple point (with three of the same four features) and these connections make up the triple line network. The triple lines are not made by connecting the points directly with a line segment; they are connected by following edges of the tetrahedrons in the directions of the original voxel edges. The triple lines are initially "jagged" from following along edges of the original voxels, but will be smoothed by balancing the minimization of both line length and deviation from the experimental data. The smoothing process involves starting at one end of the line and progressively skipping edge segments on the triple lines and monitoring the deviation of the new line from the original "jagged" line. If a user-defined tolerance is exceeded, the current segment is fixed and a new segment is initiated. The process continues until the other end of the triple line is reached. After all the triple lines have been smoothed, the surfaces must also be smoothed. The surfaces are treated as a series of lines connecting triple lines across the surface. These lines are smoothed using the same process as the triple lines. The end points of all the line segments on the triple lines and surface lines are the surface nodes, similar to the nodes placed on the fit surfaces in the CAD-based method. Volume nodes are added and connected using the same ideas discussed in the CAD-based section.

The surface area/line tension smoothing approach is an attempt to smooth the data based on physical ideas, while still being true to the experimental data. The implementation is rather simple, especially when compared to the CAD-based approach. The quality of the elements is similar to the CAD-based approach, due mainly to the similar Delaunay triangulation method for element creation used in both methods. The physically based smoothing criterion makes for the most "realistic" surface representation of the methods presented here. The user bias, however, is still somewhat significant due to the user-defined tolerance for deviation from the experimental data.

4 Statistical Representation of Structure

Representing structure in a statistical sense has multiple benefits to the computational material scientist. First, some material properties show a significant amount of scatter with the apparent cause linked to specific local feature interactions. Thus, an individual experimental volume may not accurately represent the variation in locally critical neighborhoods. Statistically equivalent instantiations of the experimental volume may highlight the likelihood of the occurrence of such neighborhoods and permit simulation of their influence on behavior. Additionally, statistically quantifying structure can offer insight into the representative nature of a structure in relation to a specific property of interest. That is, the convergence of a property and some set of statistical descriptors can be linked to determine a RVE for that property. Further, statistical-based structure builders can be used when the 3D structure is not available. If the 3D statistics of a structure can be inferred from 2D observations, which will be discussed later, then a 3D structure can be created where one was not available. Finally, the statistical description of structure, if coupled with rapid and robust instantiation tools, can allow for the compression of data from large experimental datasets to a list of statistical descriptors used to generate structures when needed.

The following section will discuss the critical steps in generation of statistically based synthetic structures, specifically statistical quantification methods, structure generation processes, and metrics for the validation of equivalence.

4.1 Quantitative Description of Structure

4.1.1 Feature Size and Volume

Feature volume can be calculated simply by summing the number of voxels that are assigned to each feature. Each voxel has an associated volume given by $V_{\text{voxel}} = \delta\varepsilon^2$, where δ is the section thickness and ε is the pixel/step-size of the 2D image or EBSD map. The feature volume is calculated by $V_{\text{grain}} = N_v \cdot V_{\text{voxel}}$, where N_v is the number of voxels in the grain. In addition to measuring the true feature volume, the

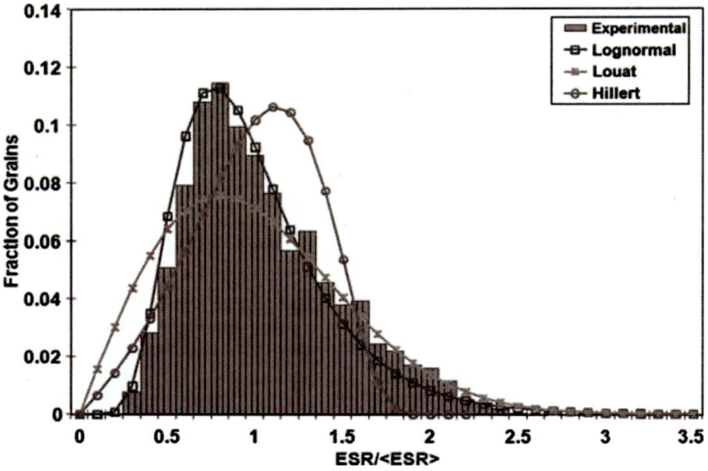

Fig. 3 Plot of equivalent sphere radius (ESR) for a nickel-base superalloy along with three fit theoretical distributions

equivalent sphere radius (ESR) can be calculated (Groeber 2007). The distribution of ESRs is useful for comparison with classical descriptions of feature size (or radius/diameter), which often involved some extrapolation from 2D. Additionally, various theoretical distributions have been developed to fit the ESR distribution (Zhang et al. 2004; Feltham 1957; Hillert 1965; Louat 1974). Figure 3 shows a sample distribution of ESR for a nickel-base superalloy along with fit theoretical distributions.

Another measure of feature size is the mean width of the feature. The mean width is given by:

$$L = \frac{1}{2\pi} \sum_{i=1}^{v} \varepsilon_i \beta_i, \tag{5}$$

where v is the number of edges on the feature, ε is the length of the edge, and β is the angle between the normals of the faces that meet at the edge. Mean width is a measure of the linear dimension of a feature and was shown by Hadwiger (1957) to exhibit the property of additivity. Mean width can be calculated analytically for all flat-faced polyhedra (MacPherson and Srolovitz 2007), which means that mean width can be easily computed for all grains after any of the surface mesh generation processes discussed previously.

4.1.2 Feature Shape

The irregular geometries that are typical of features in some materials make feature shape a difficult parameter to unambiguously describe. This is especially true due to the lack of general shape descriptors, where most descriptions of shape involve combining groups of size parameters to generate a unitless value (Russ 1986). Examples of possible shape descriptors include: length/width (aspect ratio), area/convex area (solidity), and length/fiber length (curl). A common practice is to fit an ellipsoid (or ellipse in 2D) to the feature (Saylor et al. 2004a; Brahme et al. 2006). A systematic method of generating ellipsoidal inclusions from voxel data obtained by serial-sectioning has also been developed (Li et al. 1999). In this method, the zeroth order moment (I_0), first-order moments (I_x, I_y, I_z), and second-order moments (I_{xx}, I_{yy}, I_{zz}) are first calculated for each feature by adding the contribution of each voxel belonging to an identified feature. The coordinates of the centroid for the best-fit ellipsoid are computed from the zeroth- and first-order moments as:

$$x_c = \frac{I_x}{I_0},\ y_c = \frac{I_y}{I_0},\ z_c = \frac{I_z}{I_0}. \tag{6}$$

Next, the principal directions corresponding to the principal axes of the ellipsoid are calculated from the eigenvalues and eigenvectors of the second-order moments $I_{ij}, i, j = 1, 2, 3$. The major axis ($2a$), minor axis ($2c$), and intermediate axis ($2b$) of the ellipsoidal grain are solved from the relations of the principal second moments of inertia as:

$$a = \left[\frac{A^4}{BC}\right]^{1/10}, b = \left[\frac{B^4}{AC}\right]^{1/10}, c = \left[\frac{C^4}{AB}\right]^{1/10}, \tag{7}$$

where A, B, and C are given by:

$$A = \left(\frac{15}{4\pi}\right)\left(\frac{I_1 + I_2 - I_3}{2}\right), B = \left(\frac{15}{4\pi}\right)\left(\frac{I_1 + I_3 - I_2}{2}\right),$$
$$C = \left(\frac{15}{4\pi}\right)\left(\frac{I_2 + I_3 - I_1}{2}\right).$$

In some cases, the ellipsoidal representation of features is an oversimplification that should be quantified. Correspondingly, an accuracy indicator of the ellipsoidal representation can be developed (Groeber 2007). The best-fit ellipsoid based on the moment analysis is scaled slightly to be of the same volume as the grain with the same aspect ratios and orientation. The fraction of the grain's voxels that lie within the best-fit ellipsoid is calculated. If the grain is perfectly ellipsoidal, then the value of this quantity would be very near to 1 (with the voxel size relative to the feature size controlling the nearness to 1). Decreasing values indicate more complex and likely concave shapes that are poorly represented by an ellipsoid. Figure 4 shows a sample plot of the distribution of grain aspect ratios for a nickel-base superalloy as well as the corresponding accuracy indicator plot.

In the case that the ellipsoidal representation is not sufficient, higher-order moments may need to be considered. However, there is more information in the second-order moments, beyond the aspect ratios, that is rarely used in microstructure description. For the second-order moments, there are three moment invariants.

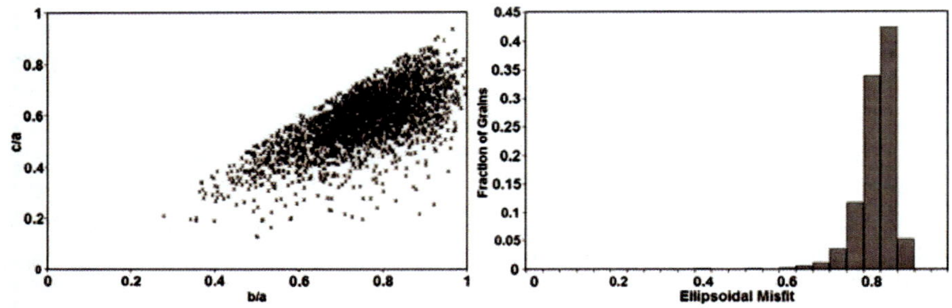

Fig. 4 Plot of (*left*) the distribution of grain aspect ratios for a nickel-base superalloy, obtained from a best-fit ellipsoid and (*right*) the accuracy indicator, termed ellipsoidal misfit, that describes the closeness of the fit ellisoids

The third moment invariant, where the first two are essentially the aspect ratios, offers more detail on the shape of a feature (MacSleyne et al. 2008). While it is true that the three invariants of the second-order moments are not sufficient in describing the true shape of the feature, there are techniques that can be used to incorporate all three invariants and improve on the current description of shape. For example, if a class of shapes is selected (i.e., superellipsoids or truncated octahedrons), the third moment invariant can be linked to parameters that fully define these shapes. In the case of the superellipsoid, the third invariant, known as Ω_3, can be linked to the exponent in the equation that defines a "superellipsoid" (MacSleyne et al. 2008):

$$\left(\frac{x}{a}\right)^n + \left(\frac{y}{b}\right)^n + \left(\frac{z}{c}\right)^n = 1, \tag{8}$$

where a, b, and c are the semi-axes of the superellipsoid. The relationship of Ω_3 to n is given by:

$$\Omega_3 = F^3[n], \tag{9}$$

where

$$F[n] = 20\frac{\Gamma\left[1+\frac{1}{n}\right]^3 \Gamma\left[\frac{5}{n}\right]}{\Gamma\left[\frac{3}{n}\right]\Gamma\left[1+\frac{3}{n}\right]^{5/3}}$$

and $\Gamma[x]$ is the complete Gamma function. Figure 5 shows the relationship of Ω_3 and n, along with some example feature shapes for selected values of n. It becomes obvious from the plot that utilizing Ω_3 enables the description of shape variation beyond just aspect ratio changes.

4.1.3 Number of Neighbors

In addition to the shape of features, the number of nearest neighbors is another example of a parameter that cannot be determined directly from 2D measurements

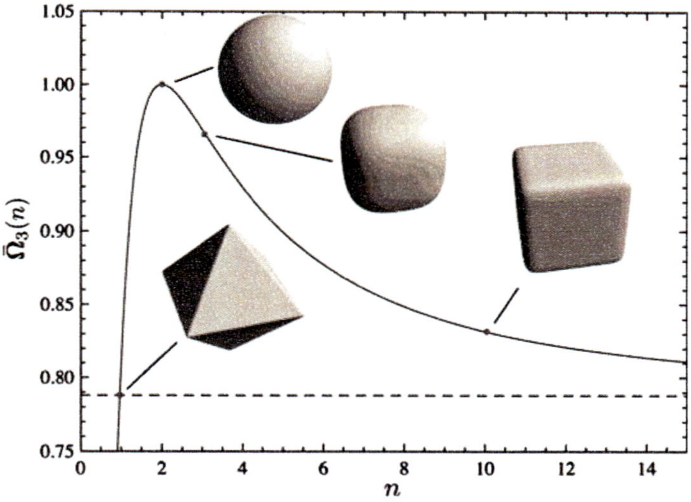

Fig. 5 Plot of the relationship between Ω_3 and n, where n is the exponent in the superellipsoid equation given by equation 8

but is easily determined in 3D. Here, the voxels of each feature are checked to see whether they neighbor another feature. If a neighboring voxel belongs to a different feature, the two features are neighbors. This process is exceedingly useful because it not only determines the number of neighbors, but it also identifies the feature connectivity in the structure, which allows for further automated investigation of any parameter involving neighbor interactions. Note that when checking neighboring voxels, only voxels that share a common face, not a common edge or corner, are considered. This requires the features share some actual area and avoids counting features that meet at only an edge or point. Figure 6 shows a sample plot of the number of neighbors distribution for a nickel-base superalloy.

4.1.4 Correlations between Parameters

The correlation or mutual relationship between parameters may be important because it provides additional information regarding the morphology and spatial arrangement of features. The feature connectivity coupled with the parameters of each individual feature allows for the relationships between parameters as well as the clustering of critical features to be studied. It is desirable to quantify the degree of correlation between two selected parameters. A quantity called the correlation ratio, η^2, can be used for this purpose (Kenney and Keeping 1947). The correlation ratio is a preferred metric in comparison with the correlation coefficient, r, because it can be used with nonlinear relationships. The correlation ratio is the square of the correlation coefficient (r^2) if the relationship is linear. If the relationship is

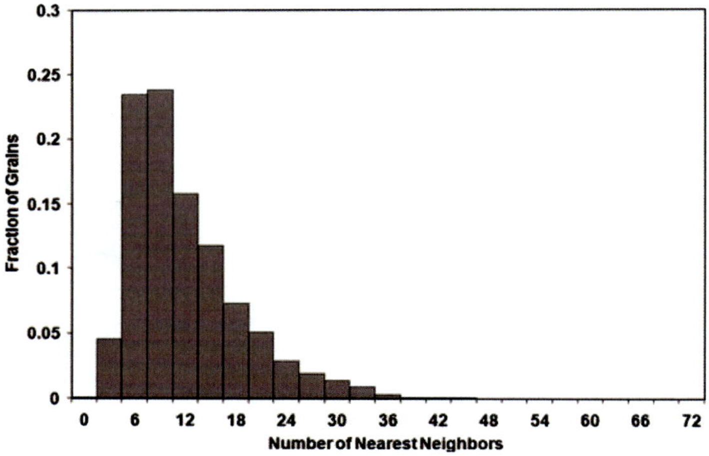

Fig. 6 Plot of the distribution of the number of neighbors for grains in a nickel-base superalloy

nonlinear, the magnitude of the correlation ratio is larger, but retains a value between 0 and 1. The formula for the correlation ratio is given by:

$$\eta^2 = \frac{\sum_{b=0}^{N_b} n_b \left(\bar{y}_b - \bar{y}\right)^2}{\sum_{i=0}^{N_g} \left(y_i - \bar{y}\right)^2}, \tag{10}$$

where N_g is the total number of features, y_i is the value of the ith feature, N_b is the number of bins, n_b is the number of observations in a given bin b and,

$$\bar{y}_b = \frac{\sum_{i=0}^{n_b} y_{bi}}{n_b}, \quad \bar{y} = \frac{\sum_{b=0}^{N_b} n_b \bar{y}_b}{\sum_{b=0}^{N_b} n_b},$$

where y_b is the value of the ith feature in bin b.

The correlation ratio can be thought of as the percent of the total variance of the dependent variable accounted for by the variance between groups of the independent variable. Seen in the equation is that if there is a large difference between the mean of the whole data set and the means of the individual bins, then the correlation ratio is high. Figure 7 shows a correlation plot relating feature size to a number of morphological parameters for a nickel-base superalloy. The grains of the superalloy were separated by their ESD and the averages of other morphological parameters were calculated for grains of similar ESD. The averages for each size bin were then normalized by the average of the corresponding morphological parameter for all sizes. It becomes clear which parameters are strongly correlated with size when observing the trends in Fig. 7. Parameters that deviate from a value of 1, such as number of neighbors and surface-to-volume ratio, are correlated with the grain diameter.

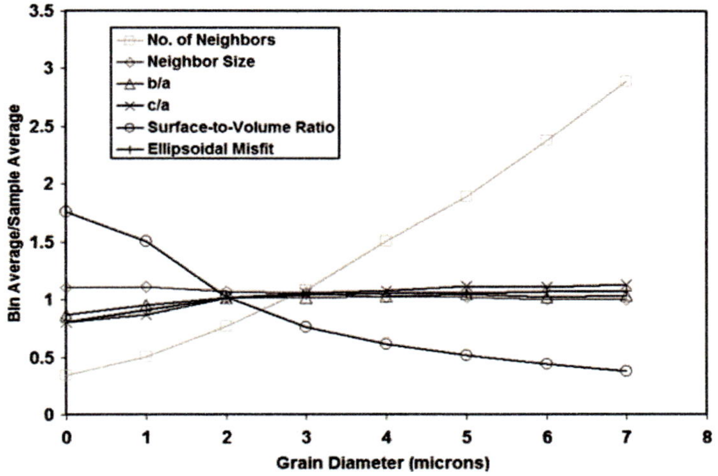

Fig. 7 Plot of the correlation between grain size (i.e., ESD = 2*ESR) and other morphological parameters. Bin average refers to the average of a parameter for all grains within a given size bin and sample average refers to the average of the parameter for all grains, regardless of size bin

4.1.5 Crystallographic Texture

Many of the traditional analysis of 3D EBSD data are directly analogous to the traditional two-dimensional counter analysis. For example, the volume fraction of a particular phase and the orientation distribution function (ODF), since the area fraction is exactly equal to the volume fraction for a random section through a material. However, a 3D analysis can often add information to these analyses, for example, only through a 3D analysis can the morphological texture of the grains be correlated with the crystallographic texture measured in the ODF.

Other measurements have 2D to 3D analogs that are related through stereological relationships that give a statistical equivalence of the properties. An example of this would be the distribution of misorientations across feature boundaries in a material. In the 2D section, one can easily measure the misorientation across the boundary, but the line-length of a given boundary is not directly related to that boundaries area. However, if enough boundaries are collected in 2D section with a similar misorientation, one can apply the stereological relation, $S_V = 4B_A/\pi$ where S_V is the surface area per unit volume, and B_A is the boundary length per unit area (Russ and DeHoff 1986). It should be noted a very large number of boundaries need to be present in the 2D section to obtain a statistically significant number of boundaries for each misorientation type for this analysis to be valid. In the case of a 3D reconstruction, the measurement of the misorientation distribution is relatively straightforward since, as discussed above, the feature's nearest neighbor misorientations and their boundary areas can be directly calculated from the 3D

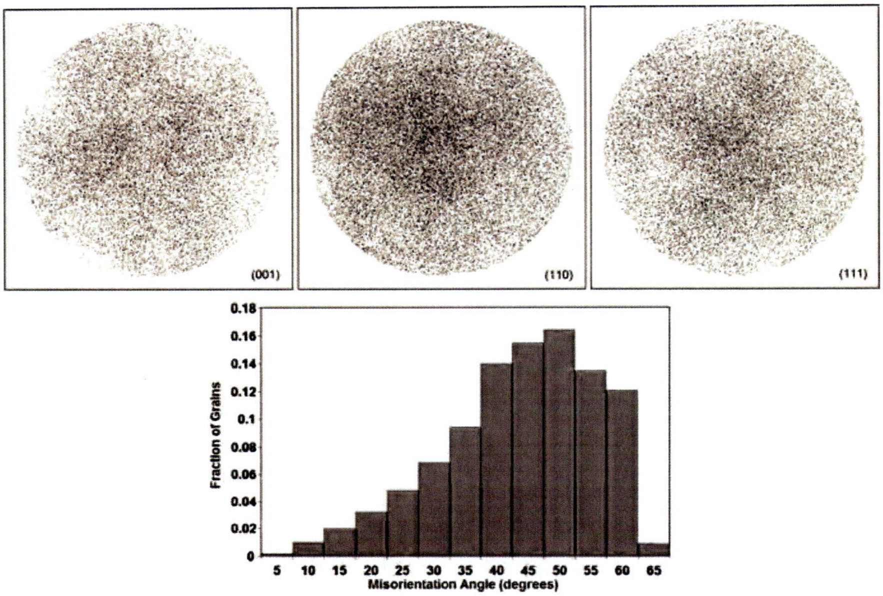

Fig. 8 Plot of (*top*) pole figures for a nickel-base superalloy and (*bottom*) the misorientation distribution of the same nickel-base superalloy. Note the misorientation angle reported here is the minimum of the set of equivalent misorientation angles and is often termed as the disorientation

reconstruction, thus the misorientation distribution function (MoDF) can be directly measured without assumption. Figure 8 shows a set of pole figures and the misorientation distribution for a nickel-base superalloy.

4.1.6 Interface Character Distribution

There are many analyses that can only be measured directly through 3D reconstruction of the material. In a very general sense, these analyses can be seen as the correlation of a specific 3D geometry with the crystallographic orientation. One of the most relevant applications of this is the determination of crystallographic interface orientations, which requires directly correlating the crystallographic orientation of an object with the local interface normal.

One of the most common 3D visualization techniques is to form a surface mesh of the object so that the interface of the object is described as a 3D mesh of discrete interconnected triangles. The conversion from a regularly gridded 3D array of data (such as a stack of images) is most often accomplished using a fast-marching cubes algorithm that converts the volumetric regular array data form to a surface mesh (Lorenson and Cline 1987). Often it is necessary to apply some degree of surface smoothing to remove pixel-like artifacts from the surface mesh, as discussed in the section on surface representation and mesh generation. Once the surface is described in terms of a set of triangles, the properties of these triangles can be used to quantify the interfaces of the 3D reconstruction, particularly the local interface normal. The normal $\hat{n}$ and the area A of each triangle in the surface mesh is given by:

$$\hat{n} = \frac{\vec{e}_1 \times \vec{e}_2}{\left|\vec{e}_1 \times \vec{e}_2\right|}; A = \left|\vec{e}_1 \times \vec{e}_2\right| / 2, \tag{11}$$

where $\vec{e}_1$ and $\vec{e}_2$ are the two edge vectors of the triangle.

One powerful construction that can be formed from this type of data is the interface normal distribution (IND) (Kammer et al. 2006). The IND is constructed by first placing the collection of all the interface normal vectors on to a unit sphere. The normals are then binned (weighted by the surface areas of the triangles) according to orientation, then normalized by a random distribution of orientations. Therefore, an orientation on the spherical histogram that has a strong intensity corresponds to a large surface area that shares that orientation. This spherical histogram is then projected to a 2D plot using a stereographic projection (typically along the $+z$ direction, but this is a matter of how one wants to represent the data), preserving the angular relationships between the orientations in the plot producing the IND plot. It should be noted that if the binning of the data occurred on the projected data space of the stereoplot, rather than on the unit sphere, the intensities of the histogram would be altered by the nonequal-area nature of the stereographic projection. Because the binning of the data occurs on the unit sphere, this artifact is avoided and a randomly distributed shape (like a sphere) would have a flat intensity. For INDs that contain peaks, the exact shape of peaks on the stereoplot would still be slightly altered by the projection especially close to the center poles and edges of the stereoplot. The stereographic projection has extremely large distortions for orientations that have a $-z$ component. To avoid this, often it is necessary to project each hemisphere separately, however this can be avoided if there is crystallographic symmetry, which allows the plot to be compressed to a stereographic triangle.

The inclusion of the local crystallographic orientation along with the local interface normal means that the interface normals can be expressed in the crystallographic coordinate system as well, creating a crystallographic interface normal distribution (CIND). The only variation in the construction of the CIND is that before the normals are binned on the unit sphere, they are rotated to the crystallographic coordinate system using the Euler angles of the object in question. There are two analyses that especially lend themselves to this type construction, the examination of individual objects and examining the overall crystal interface texture in a sample, which will be briefly reviewed.

Rowenhorst et al. (2006) used the CIND analysis to examine the facet planes on individual coarse martensite crystals, as shown in Fig. 9 (right). By identifying the peaks within the CIND, and fitting planes to the corresponding triangles the average crystallographic facet normal was determined. This article also introduced a unique visualization technique in which the interfaces in the 3D reconstruction of the martensite crystals were colored according to the local interface crystallographic normal (Fig. 9, middle). The coloring scheme is identical to that used to create an inverse pole figure (IPF) but unlike the IPF, where each point in the image is colored according the crystallographic direction that points along an arbitrary section plane

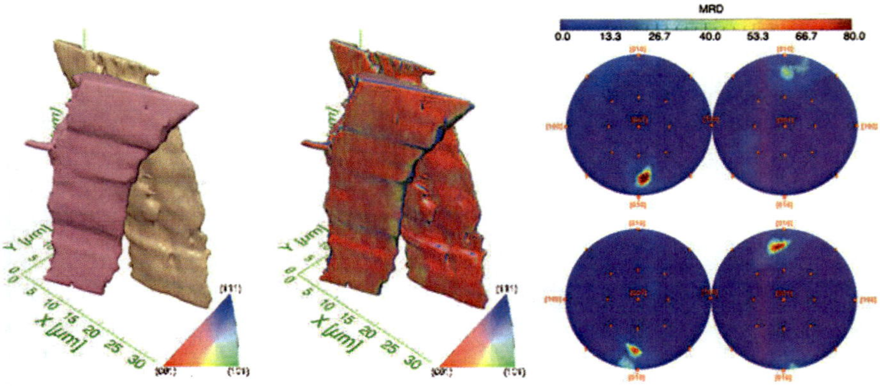

Fig. 9 Reconstruction of two coarse martensite crystals in HSLA-100 steel (*left*). Color indicates the crystallographic direction that is parallel with the z-axis. The difference in color indicates that the martensite crystals represent separate martensite variants (*middle*). Color indicates the crystallographic orientation of the local interface normal at each local patch of the interface. Note that while the crystals represent different variants, the crystal facets have similar crystallographic normals (*right*). Crystallographic interface normal distribution (CIND) plots of the two martensite crystals, where the upper and lower CINDs correspond, respectively, to the purple and yellow crystals in the left image. The peaks in the CIND indicate the average orientation of the facet planes of the crystals; peak intensities indicate Multiples of a Random Distribution (MRD)

direction (Fig. 9, left), here the interface was colored according to the crystallographic direction that is parallel with the surface normal. This visualization contains essentially the same exact information as the CIND construction with the added advantage of being able to link crystallographic orientation to specific morphological features, but unlike the CIND, the information presented is not quantitative.

Saylor et al. (2004b, c) significantly expanded on the CIND construction to include not only the interface distrpoibution of grain boundaries in polycrystalline MgO, but also to include the full five-parameter space of the grain boundaries, constructing the grain boundary character distribution (GBCD). By examining the full five parameter space, they were able to not only determine the interface texture for the system, but also the interface texture for particular grain misorientations. Since this initial study, the GBCD has been determined for many more materials systems including Al, Ni, Cu, Brass, and Ti-6–4 (Saylor et al. 2004b, c; Randle et al. 2005; Randle et al. 2008a, b).

4.1.7 N-Point Statistics

Two-point through N-point correlation functions are a useful set of descriptors that characterize the spatial arrangement of microstructural features. Generally, these correlation functions, most common of which is the two-point correlation function, are used to analyze spatial arrangement of particles or voids in materials such as: discontinuously reinforced composites, foams, geological samples, etc.

(Torquato 2001). However, correlation functions can also be utilized to describe the spatial arrangement of grains with certain microstuctural parameters. For example, the spatial arrangement of grains with a given orientation in a polycrystal could be quantified using correlation functions. Two-point correlation functions can be determined by systematically placing line segments in the structure and noting where the endpoints lie. Therefore, for a microstructure with two entities, be they phases, orientations, sizes, shapes, etc., there are exactly four two-point correlation functions: $\langle P_{11}\langle r\rangle\rangle$, $\langle P_{12}\langle r\rangle\rangle$, $\langle P_{21}\langle r\rangle\rangle$, $\langle P_{22}\langle r\rangle\rangle$. $\langle P_{nm}\langle r\rangle\rangle$ is the average probability that endpoints 1 and 2 of a randomly oriented line segment of length r lie in entity n and entity m, respectively (Tewari et al. 2006). Note that here the term entity is being used to mean a certain property and not a specific feature. Thus, for $\langle P_{11}\langle r\rangle\rangle$, it is not a requirement that both endpoints lie within the same particle/feature with entity 1, just that both endpoints lie within a particle with entity 1. Two-point correlation functions can also be directionally dependent and thus, the values of each correlation function can be calculated as a function of the orientation of the line segment within the sample. The extension of two-point correlation functions to n-point correlation functions simply involves placing an n-cornered polygon, rather than a line segment, in the sample and noting the position of the endpoints. The number of functions increases with n and with the number of entities in the sample.

N-point correlation functions could be used to describe the local neighborhood around a grain in a polycrystal. A specific property of a grain, its size, for example, could be chosen as entity 1 and all other sizes could be chosen as entity 2. Then line segments of varying lengths could be drawn from the centroids of all grains with size equal to (or near that of) entity 1. The resultant two-point correlation functions would describe the spatial arrangement of grains with size equal to (or near that of) entity 1. This same process could be carried out for the Schmid factor of a grain or its orientation, which would likely offer key insight into the clustering of critical orientations.

4.1.8 Limitations/Concerns when Using Statistical Descriptors

The statistical descriptors presented here certainly offer a great deal of information about the microstructure, but are by no means a complete set. Other classical morphological descriptors, such as the number of faces, edges, and corners of a feature have not been presented here, but are readily available in the 3D volume. Mean width was explained to be the integral of a feature's curvature and the IND analysis was shown to be the complete distribution of interface normals, but the curvatures of a grain's individual boundaries have not been correlated with the grain's size and shape or with the curvatures of neighboring boundaries. In general, the near-neighbor analyses have been limited to only contiguous neighbors and not next-nearest neighbors and beyond. These issues are not truly limitations of the descriptors themselves; rather they are an illustration of the abundant amount of information available in just one experimental volume. The limitation is actually in the ability to quantify every aspect of the structure, which has yet to be shown, is in fact necessary. It should be the goal of a concerted characterization-modeling effort to define to what extent microstructure needs to be quantified and represented for each property.

Some areas of the current analysis remain incomplete. The correlation between descriptors is one of the areas most lacking. Especially when three or more descriptors are related, it becomes difficult to quantify their dependencies on one another. This is most frequently the result of an insufficient number of features in the interrogated volume. This can be corrected by collecting larger volumes, but limitations of experimental capabilities and availabilities may be preventative. Additionally, the statistical analyses have not yet advanced to the point of accurately informing the experimentalist of the proper number of features to collect for each type of analysis. Another area that is currently underinvestigated, at least in the context of materials microstructure, is the analysis of rare events. The extreme-value statistics of the microstructure is not well understood and has not been well addressed in the quantification and comparison of structures. The standard distributions (i.e., lognormal, beta, weibull) used to describe most descriptors provide a "good" fit over much of the descriptor's range. However, when the extreme values are closely investigated, the experimental observations often appear to deviate systematically from the fit distribution in a manner that suggests more than simply sampling error. Until the extreme-value statistics is properly quantified and accounted for in digital microstructures, simulations of properties that are believed to be controlled by rare events or neighborhoods should be treated with some sense of skepticism.

4.2 Synthetic Structure Builders

The synthetic builders that will be discussed here consist of two major steps. First, the features to be placed in the synthetic volume are generated. These features are generated by sampling the size, shape, and morphological and crystallographic orientation distributions of the features observed by some experimental technique. Second, the features are placed in the volume with specific focus on the local neighborhoods created by neighboring features. The sampling procedure for generating representative features as well as the constraints used to place the features in the volume will be described here. Figure 10 shows some example synthetic volumes generated by the different techniques discussed in this section.

4.2.1 Representative Feature Generation

A representative feature generation process is responsible for the creation of a collection of idealized ellipsoidal (or alternative representation) features having distributions of size, shape, and morphological and crystallographic orientation equivalent to those observed in the experimental volume. In this representation, each grain is modeled as an ellipsoid as defined in the previous section on feature shape. The size corresponds to the volume of each ellipsoid, the shape corresponds to the

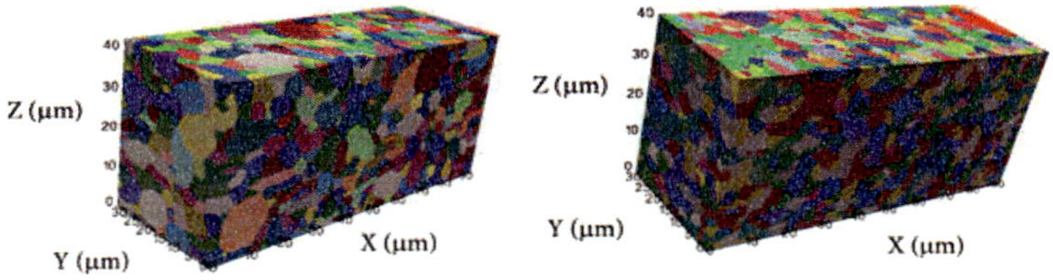

Fig. 10 Example synthetic structures produced by (*left*) the sequential grain placement algorithm using ellipsoidal grains and (*right*) the sequential grain placement algorithm using superellipsoidal grains. Note that the grains generally appear less "idealized" in the right image, where the superellipsoid representation described in Sect. 4.1.2 was used

aspect ratios of the principal axes ($b/a, c/a$ and c/b) and the morphological orientation corresponds to the orientation of the major principal axis ($a : a \geq b \geq c$) relative to the global coordinates. The first step in the process is to sample the experimental feature volume distribution, which is represented by the cumulative probability distribution function (CPDF) fit to the experimental data. Many investigations have shown the feature volume distribution to be best represented by a log-normal distribution (Zhang et al. 2004; Groeber et al. 2008a), whose CPDF is given by:

$$P(V) = \begin{cases} \dfrac{1}{2}\left[1 - \text{erf}\left(\dfrac{V_g^{\text{AVG}} - V}{V_g^{\text{STD}}\sqrt{2}}\right)\right] \to V < V_g^{\text{AVG}} \\ \\ \dfrac{1}{2}\left[1 + \text{erf}\left(\dfrac{V - V_g^{\text{AVG}}}{V_g^{\text{STD}}\sqrt{2}}\right)\right] \to V > V_g^{\text{AVG}} \end{cases}, \tag{12}$$

where V is the feature volume, $P(V)$ is its cumulative probability, which has 0 and 1 as its limits. The average feature volume (V_g^{AVG}) and the standard deviation (V_g^{STD}) are parameters that determine the precise shape of the distribution function. During volume assignment, a number within the limits of $P(V)$ is randomly generated and the corresponding volume, given by equation 1, is assigned. This assignment process is continued until the total volume of all features generated equals a threshold defined as 110% of the volume of the synthetic microstructural model. The additional volume is needed because some features may lie partially outside the domain of the microstructural model or overlap other features. This issue will be discussed further in the next subsection.

Subsequent to the volume assignment, feature shapes are assigned in conformity with CPDFs of the ellipsoid aspect ratios ($b/a, c/a$ and c/b) that have been established a priori from the experimental data. The corresponding CPDFs can be represented in terms of a beta distribution, with the form:

$$P(b/a) = \frac{\int_0^{b/a} t^{p-1}(1-t)^{q-1}\text{d}t}{B(p,q)}, \quad 0 \leq b/a \leq 1. \tag{13}$$

In equation 13, $B(p,q) = \int_0^1 t^{p-1}(1-t)^{q-1}\,dt$ is the beta function, p and q are the shape parameters, and $P(b/a)$ is the cumulative probability. The statistical analysis establishes the correlation between the shape and the size of each grain, represented by the aspect ratios and volume of each ellipsoid. To establish this correlation for the synthetic ellipsoidal features, each volume is converted to an ESD using the relation:

$$\mathrm{ESD} = 2\left(\frac{3}{4\pi}V\right)^{1/3}. \tag{14}$$

The correlation is determined by assigning aspect ratios to different volumetric bins that are represented by ranges of ESD values. The two aspect ratios that define an ellipsoid (b/a and c/a) each has a CPDF in each volume bin. The sampling of the b/a and c/a CPDFs is identical to that of the feature volume CPDF. This process ensures appropriate correlation between the shape and size distributions.

In addition to a correlation with volume, the aspect ratios b/a and c/a should also be mutually correlated. However, often there is an insufficient number of features in the experimental volume to determine this correlation table between the grain volume V, and both of the aspect ratios b/a and c/a. To overcome this shortcoming, individual correlation functions are generated between V and each of the aspect ratios separately. In this process, values of b/a and c/a are evaluated corresponding to randomly chosen $P(\frac{b}{a})$ and $P(\frac{c}{a})$ values from the $P(\frac{b}{a}) - (\frac{b}{a})$ and $P(\frac{c}{a}) - (\frac{c}{a})$ plots, respectively. The consequent aspect ratio $c/b = (c/a)/(b/a)$ is evaluated and its probability density $p(c/b) = \left[(c/b)^{p-1}(1-c/b)^{q-1}\right]/B(p,q), 0 \le c/b \le 1$ is ascertained from the experimentally observed distribution. The two individually generated aspect ratios are accepted with the same probability density of $p(c/b)$.

The third variable, the morphological orientation of each ellipsoidal feature, is defined by a set of rotations (θ, λ, ψ) needed to transform the global coordinates (X, Y, Z) onto the principal axes of the ellipsoid (A, B, C). The probability density function, $f(g)\Delta g = \Delta N_g/N$ is the probability of observing an orientation G in the interval $g \le G \le g + \Delta g$, where ΔN_g is the number of orientations between g and $g + \Delta g$ and N is the total number of experimentally observed ellipsoidal features. If $N^{(i)}$ is the number of observations in the i th orientation space element ranging from (θ, λ, ψ) and $(\theta + \Delta\theta, \lambda + \Delta\lambda, \psi + \Delta\psi)$, then the density of orientations can be expressed as $N^{(i)}/N$. To evaluate this density, the entire orientation space is defined as a finite cube with edge length π (180°) with the origin at one of its vertices. For the purpose of creating ranges in the orientation data, the orientation space is discretized into cubic bins of dimension $\pi/36$ or 5°. The morphological orientation density in each bin is calculated by dividing the number of orientations in the bin by the total number of orientations in the experimental data and normalizing by the size of the bin. Ellipsoidal orientations are created and assigned based on this probability density function.

In summary, the output of this process is a set of representative ellipsoidal features having statistically equivalent volume, aspect ratio, and morphological orientations as the experimental reference data. However, this process does not arrange the features in their appropriate spatial locations, which is the function of the next process. A final step in this procedure can be included either before or after the feature placement routine. This step is the assignment of crystallographic orientations to the generated features. The process of assigning crystallographic orientations is exactly identical to that of the morphological orientation assignment process previous described. Here, the only difference is that the rotations assigned are those to transform the global coordinate axes to the coordinate system of the crystal, rather than the principal axes of the grain.

4.2.2 Feature Placement

After generating a set of ellipsoids that is representative of the 3D features, the focus must be shifted to the placement of the ellipsoids in the volume. There are multiple issues to consider when packing the ellipsoids. The density of the objects represented by the ellipsoids is one of the largest factors in developing the packing algorithm. For example, ellipsoids representing particles of a low volume fraction phase will certainly be placed differently than ellipsoids representing grains in a fully dense polycrystalline material. In the fully dense grain example, care must be taken to pack the volume as densely as possible, but minimize overlap between ellipsoids to retain each ellipsoid's prescribed shape. In both cases, the local neighborhood of the ellipsoid (i.e., neighboring ellipsoids) must also be addressed during placement. The low volume fraction particles should be spaced equivalently to the experimental/reference data and the densely packed grains should neighbor grains of similar sizes and shapes as seen in the experimental/reference data.

Two inherently different, but viable options for ellipsoid packing will be discussed here. The first approach involves overpopulating the volume with a large number of ellipsoids. This approach is presented in greater detail elsewhere (Saylor et al. 2004a; Brahme et al. 2006). A large set of representative ellipsoids are placed into the model volume. The ellipsoids should have a total volume much larger than the volume to be filled and are allowed to overlap and extend outside of the volume. A simulated annealing procedure can then be employed to determine an "optimal" set of ellipsoids. An optimal set of ellipsoids would maximize space-filling, while minimizing overlap of ellipsoids. Saylor et al. (2004a) and Brahme et al. (2006) outline the development of a penalty function that promotes optimal space filling. It is easy to imagine the adjustment of this function to address the less dense packing of the distant particle case. At present, only the space-filling nature of the ellipsoids is addressed in this technique and not the local neighborhood of an ellipsoid. However, it is conceivable that penalty functions could be developed to encourage desired clustering or spacing of ellipsoids of given sizes or shapes. Once an active set of ellipsoids are selected by this method, a voxelized structure is generated using cellular automata (CA) where the centroid for each ellipsoid is a seed point in the simulation and voxels are added starting at this seed point until the entire structure is filled. The growth is constrained initially such that only those voxel locations that are located inside the ellipsoids are added. When each ellipsoid is completely filled, then the constraint is dropped and the remaining free volume is consumed.

A second approach to the ellipsoid packing problem is to sequentially place the ellipsoids while using statistical descriptors from the experimental/reference data as constraints. One such implementation of this approach is presented by Groeber et al. (2008b). Here, the set of representative ellipsoids should have a total volume much nearer to the model volume than the first approach. The ellipsoids are allowed to extend outside the volume, similar to the first approach, and thus the total volume should be some small amount (i.e., $\sim 10\%$) above the model volume. Each ellipsoid is randomly placed in a sequential fashion and checked against a number of constraints to determine whether its current position is acceptable. Constraints can include, but are not limited to: overlap limits, number of neighboring ellipsoids, and size distributions of neighboring ellipsoids. This approach generally yields optimal space-filling through the overlap limits and produces realistic neighborhoods by constraining placement to locations that improve the surroundings of previously placed grains. Although this technique presents some advantages over the previous, there are complications that arise as well. In any sequential process, there should be concerns of a failure to locate a suitable position, especially near the end of the process. Generally, the process is more efficient and successful when the largest grains are placed first, when there is sufficient room left for their placement. Additionally, the number of constraints greatly affects the feasibility of locating a suitable position and thus should be optimized.

4.3 *Measures of Goodness*

The "goodness" of the synthetic structure can be defined relative to two objectives, which may or may not be completely related. First, the statistical descriptors of the synthetic structure can be compared to those of the experimental structure. The similarity of these descriptors is certainly one measure of "goodness" and will be a focus of the following subsections. Second, the simulation results of the two structures can be compared. This definition of "goodness" is arguably a more practical one, in that the response of the material is often the overriding goal. However, depending on the property of interest, the two structures may be "equivalent" with respect to the statistics chosen, but still yield different simulation results. This is likely to be the case when the experimental structure is smaller than a RVE for the specific property or the statistical descriptors chosen are not directly linked to the response property. This section will not deal with the second definition; due mainly to a lack of simulation results, but it should be considered carefully in future structure–property relation investigations.

4.3.1 Size(s)

Measuring the "size" of a grain may appear to be a simple matter at first sight. Even in 2D sections, however, computing the circle-equivalent diameter yields a (slightly) different result than the average linear intercept (Underwood 1970). In 3D, one must

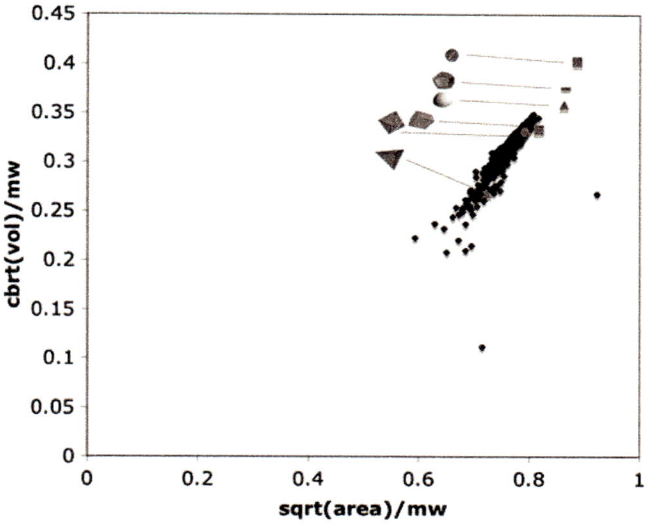

Fig. 11 Blasche diagram combining the properties that exhibit additivity in the three dimensions. The x-axis is the square root of surface area normalized by mean width and the y-axis is the volume normalized by mean width. The positions where some standard geometrical shapes reside on the plot are noted

be concerned with all three dimensions, of which measuring the volume and surface area of a grain is intuitively obvious. Less obvious is how best to measure the linear dimension of a grain, since there are so many possibilities (linear intercept, sphere-equivalent radius, etc.). The recent publication by MacPherson and Srolovitz (2007) on the theory of grain growth has, however, pointed out to the materials science community that "mean width" is not only a useful measure of integral curvature of objects such as grains, but that it also is unique in its property of additivity. Hadwiger (1957) showed that there is only one measure in each dimension that has the property of additivity, which means that the volume/area/mean width of the union of two overlapping objects is the sum of the separate quantities, minus the volume/area/mean width of the overlapping region. This suggests that the distributions of the three basic quantities (volume, area, and mean width) should be part of the validation of a digital microstructure. Moreover, ratios between pairs of these quantities, as shown in Fig. 11, also provide basic information on the shape of objects. Fitting to distributions of such ratios may also be part of the development of feature geometry in 3D models. The definition of mean width was presented in Sect. 4.1.1.

4.3.2 Shape(s)

A typical approach to quantifying the shape of grains is to fit an ellipsoid and report the aspect ratios (Groeber et al. 2008a; Saylor et al. 2004a). This approach, which was described in detail in Sect. 4.1.2, is useful in describing the distribution of the amount of elongation of the grains. However, aspect ratios are ambiguous in reference to many aspects of shape. For example, it is possible for an ellipsoid and a rectangular prism to have the same set of aspect ratios. The local curvatures of grain boundaries are often disregarded when a "simple" geometric feature is fit to represent a grain. It is this issue that makes shape one of the more complicated parameters to describe.

MacSleyne et al. (2008) have presented a method for distinguishing shapes by utilizing all three of the second-order moment invariants. The moment invariant technique creates a three-dimensional (τ_1, τ_2, Ω_3) moment invariant space to represent a grain's shape rather than the limited two-dimensional space defined by a pair of aspect ratios (τ_1 and τ_2). In the 3D moment invariant space, shapes with similar aspect ratios lie on the same arc, but are separated along the third dimension, Ω_3. This third dimension can potentially add extra information on the general class of shapes present in the microstructure. Additionally, combinations of the calculated moments can yield interesting insights into the types of shapes present in the structure. An example of a moment invariant analysis is shown in Fig. 12. The analysis provides the distribution of aspect ratios, which appears roughly equivalent for the two structures. However, when the value of Ω_3 is compared, there is a noticeable shift in the distribution between the two structures. Finally, the largest difference between the two structures can be seen in the comparison of the distribution of

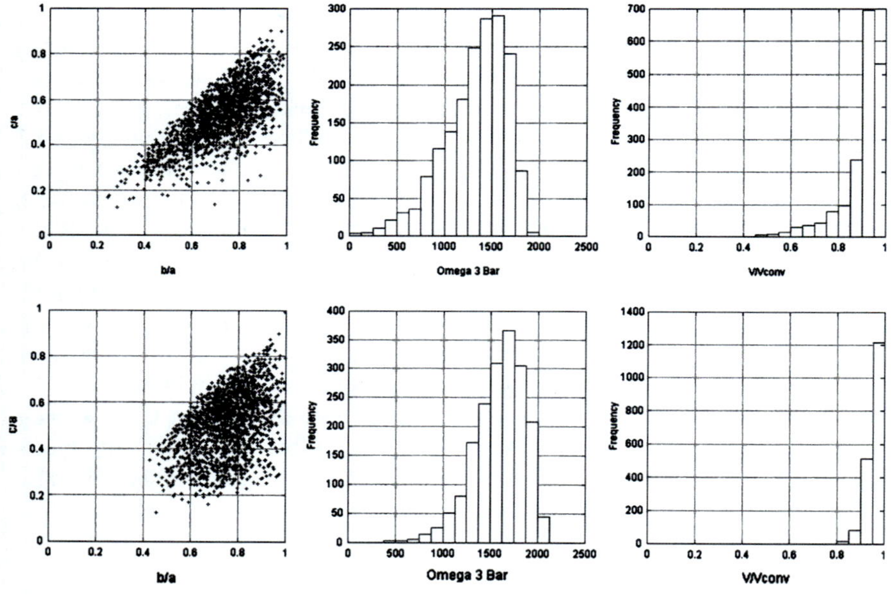

Fig. 12 Example of results from a moment invariant analysis. The upper set of plots is from an experimentally collected volume (Groeber et al. 2008a). The lower set of plots is from a synthetic microstructure generated with the goal of matching the experimental volume's statistics, using methods presented by Groeber et al. (2008a). The analysis highlights the need (and ability) to look past lower order descriptors like aspect ratios

V/V_{conv}. V_{conv} is the volume of the convex hull of the grain and V is the volume of the grain itself. The ratio of these two volumes is bounded by 0 and 1 and compares the relative concavity of the grain. It should be clear that the aspect ratio comparison alone does not accurately highlight many of the differences between the shapes in the two structures.

4.3.3 Neighborhood(s)

The local neighborhood of a grain can be a complicated aggregate of features that can be described by a number of different parameters. For example, the morphological descriptors of the neighboring features could be reported or their crystallographic relationship to the reference grain could be of more interest. Additionally, the approach to describing the local neighborhood of grain is likely to vary with the type of microstructure and data being investigated. A grain structure, which has been segmented, is likely to have a known connectivity of grains and contiguous neighbors can be characterized. In the case of low volume fraction second-phase particles, the nearest neighbors may not be known and a two-point statistics approach (Tewari et al. 2006; Torquato 2001) may be better suited.

For describing the morphology and connectivity of a grain's neighborhood, there are multiple distributions that can be created. First, the distribution of number of neighbors can be generated for all grains, as well as correlated with grain size by grouping grains of similar size. In addition to number of neighbors, the size distribution of the neighboring grains can also be considered. The size distribution of neighbors, when correlated with the size of the reference grain, offers insights into the tendency of grains of certain sizes to cluster together (i.e., Aboav–Wieve). The shapes of neighboring grains can be correlated with the reference grain's shape to quantify the clustering of similar shaped grains, which may evolve during recrystallization or deformation.

The crystallographic description of individual boundaries will be discussed in the next section, but there are other parameters that describe the crystallography of local grain aggregates to varying degrees. The MoDF can be calculated for the entire structure, which gives some insight into the local textures present in the material. However, the MoDF does not provide any knowledge of the spatial distribution of the misorientations in the MoDF. The known connectivity of the grains allows for the spatial description of the misorientations. For example, one could calculate the fraction of a grain's neighbors that have a critical misorientation value, be it high, low, or special. This approach could then be expanded to include secondary neighbors (i.e., neighbors of neighbors) and would ultimately offer a more local estimate of the clustering of grains with similar orientation. Two-point statistics can also be employed to describe distributions of orientations as well.

4.3.4 Boundary Character(s)

To generate a complete 3D microstructure, one must add grain (crystal lattice) orientations to the description. The current state-of-the-art is that the grain geometry is created first and then a set of orientations is optimized with respect to texture and grain boundary misorientation (Saylor et al. 2004a; Groeber et al. 2008b). The procedure relies on simulated annealing and is computationally straightforward on modern personal computers. This procedure has at least two significant limitations, however. The first is that it assumes that size and shape are uncorrelated with orientation. However, this is not always the case; Bozzolo et al. (2005) have demonstrated that in titanium that has been deformed and then recrystallized there are texture components that are more dominant in the small grains and vice versa. The second limitation is that it ignores the fact that grain boundary properties depend on the interface normal as well as the lattice misorientation across them. The full description of grain boundary character requires, in fact, five macroscopic parameters. Fitting orientations to include both texture and misorientation and interface normal distributions needs to be developed. Implementing such an algorithm in voxel-based representations requires some method to compute the local interface normal. Alternatively, interface normals are straightforward to compute in a surface or volumetric mesh representation of a microstructure, which has been discussed previously.

5 Inference of 3D Structure

It is often the case that the true 3D structure of a material is not available to include directly in a computational model. This can be attributed to the cost, availability, and complexity of experimental tools. As a result, it is still a reality for many to infer 3D structure from 2D observations. As previously mentioned, if 3D statistical descriptors can be inferred from 2D observations, the synthetic structure builders discussed in the previous section can be used to generate 3D structures for simulation. The following section will discuss some potential methods for inferring 3D statistics from 2D observations.

5.1 Link Between 2D and 3D Structure

The statistical reconstruction method described here is based on limited cross-sectional information from a given material; it is essential, however, that cross-sections are made on more than one sectioning plane, and preferably on three orthogonal planes. Statistical methods for reconstructing microstructures have been developed in a number of fields, especially for modeling geological materials (Fernandes 1996; Oren and Bakke 2002, 2003; Sundararaghavan and Zabaras 2005; Talukdar and Torsaeter 2002; Talukdar et al. 2002a, b, c). Saylor et al. proposed a method of constructing 3D models of polycrystalline materials based on the microstructural features observed in three orthogonal sections (Saylor et al. 2004a).

In this report, the microstructural features of interest include size and shape of grains, misorientation distribution, orientation distribution, and the relative placement of grains with respect to size. The procedure outlined by Saylor, along with adaptations for elongated grain shape noted by Brahme et al. (2006), is the basis for one of the procedures described here. Groeber (2007) also offered a methodology for inferring 3D structure from 2D measurements. In Groeber's study, the 2D sections were obtained by sectioning synthetically generated 3D structures, which provide a known set of 3D statistics to compare with the inferred statistics.

There is a substantial literature on the general stereological problem of reconstructing 3D microstructures based on limited section information. When treating microstructures as collections of general particles whose size, shape, and orientation are to be reconstructed (without regard to their packing), the problem is known to lack a solution (Cruz-Orive 1976a, b). However, for particles that are monodisperse (in size and shape), this problem is well known and has semianalytical solutions for which the names Cahn and Saltykov are well known in the materials literature (Cahn and Fullman 1956; Saltykov 1958). For a historical overview, see Underwood (1970). In contrast to these more general cases, polycrystalline grain structures have an added constraint since grains are not independent particles (i.e., low volume fraction) because they fill space. This constraint enhances the ability to accurately reconstruct a 3D distribution from 2D observations (Przystupa 1997). This section will attempt to address the previously less investigated problem of space-filling particles (i.e., grains).

5.2 *Probable Set Generation*

5.2.1 Monte Carlo Histogram Fitting

In this section, an example of generating a set of 3D (ellipsoidal) grains using statistical distributions calculated on three 2D, orthogonal, EBSD-based micrographs is given. The grain size distributions for the three orthogonal planes of a rolled aerospace aluminum alloy are shown in Fig. 13. Matching the statistics in the generated 3D structure to the measured 2D statistics is accomplished through a multistep process that includes: (1) generating representative ellipsoids (in terms of size and shape distributions), (2) placing those ellipsoids into a volume, (3) allowing that volume to be filled with voxels that are grouped as grains, and (4) modifying the grain structure by use of a (isotropic) Monte Carlo grain growth (Rollett and Manohar 2004). The first step in this process is the focus of this section, while the latter steps were previously discussed in the section on feature placement during synthetic structure generation.

The transition from 2D to 3D is accomplished by assuming the grain shape is that of an ellipsoid and considering that the observations on 2D sections are only a portion of the true size and shape of the actual grain. The probability distribution

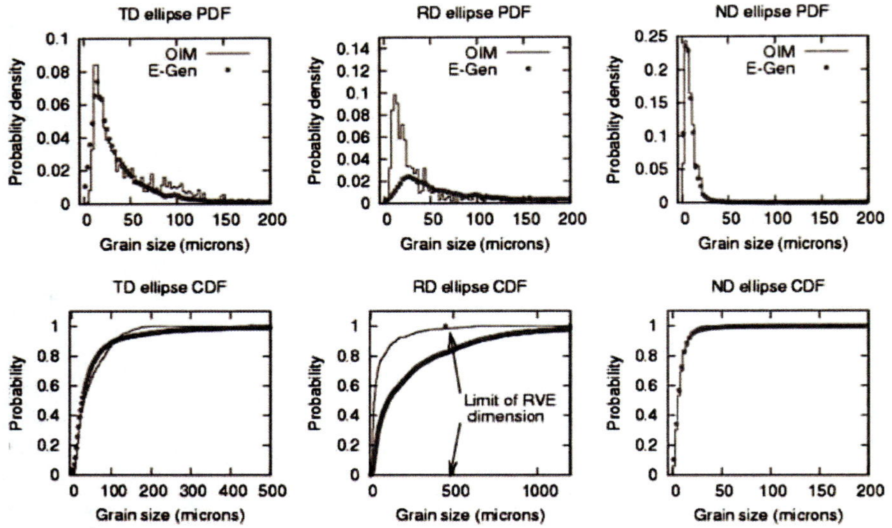

Fig. 13 The resulting (*top*) PDFs and (*bottom*) CDFs for the OIM input data and ellipsoids generated to represent the data for the ND, RD, and TD directions

functions (PDFs) of the ellipse dimensions obtained from the sliced ellipsoids are given as $f'(a)$, $f'(b)$, where a and b are the semiaxes of an ellipse, and the prime accent indicates that it is from a section. In this case, the ellipse dimensions are assumed to be independent, and that $a > b$. References to the cumulative distribution function, CDF, utilize the same notation but with a capital "F." The input data is used directly to create a CDF [i.e., F(TD)] of grain size where the range of the CDF is 0 to 1 and the domain is scaled in micrometers. The distribution for each direction is sampled and multiplied by a stereological constant to account for the fact that the CDF is generated from a 2D section. In this approach, the total number of ellipsoids generated is specified as input to the program and it proceeds to optimize the list such that they are a good match to the input data. The optimization is accomplished iteratively and the first step is to create an initial list of ellipsoids and then slice each of them many times and extract the 2D CDFs of grain size on each section. The RMS error between list and the input data is computed. The initial list is then modified by generating a new ellipsoid in the same manner as before and then randomly choosing an ellipsoid to replace from the list. The ellipsoid is replaced only if the new ellipsoid lowers the error of the system. The program completes when it has performed a number of user-specified iterations. The result of this method can be observed directly by comparing the PDFs and CDFs of the data and the simulated ellipsoids directly as can also be seen in Fig. 13. The distributions in both the ND and the TD directions are well matched to the input data. The discrepancy on the RD direction in this case is due to improper sampling of the grains in the RD direction (i.e., the majority of grains intersected the scan boundary).

5.2.2 Domain Constraint

The elemental assumption of this method is that the entire (and infinite) set of all possible ellipsoids can be bounded by observations of ellipses on experimentally collected, orthogonal 2D sections to leave a "most probable" set of ellipsoids. Factors such as the distributions of size, shape and orientation of the ellipses on the 2D sections are used to assign probabilities to groups of ellipsoids. This type of an approach is fundamentally different than the analytical developments made by Cruz-Orive (1976a, b) and DeHoff (1962) in that an exact solution is not the goal, rather a probable set is desired. An initial description of this technique is introduced by Groeber (2007).

To assign probabilities to groups of ellipsoids, the infinite domain of ellipsoids must be initially truncated and discretized. A five dimensional space is created to define the ellipsoids. Three dimensions correspond to the orientation of the principal axes of the ellipsoid and are inherently bounded by the finite dimensions of Euler space, which describe the orientation of the ellipsoid (i.e., its principal axes). The other two dimensions correspond to the two aspect ratios of the ellipsoid. The aspect ratio dimensions are not inherently bounded, but can be truncated by using the 2D observations to make assumptions about reasonable upper and lower bounds. By definition, the upper limit of the two aspect ratio dimensions is 1 (i.e., a sphere) and the lower limit can be set to be the smallest aspect ratio observed in the 2D sections. In practicality, the accuracy of the estimated lower limit will be directly related to the number of observations, and thus, it may be prudent to reduce the minimum observed value by an additional 25–50%. The volume of the ellipsoids is treated only as a distribution within each discrete bin in the 5D space, not as its own dimension. This is because volume only scales the dimensions of an ellipsoid and any resultant elliptic section through it, which has no effect on the following process.

Development of a probable set of ellipsoids is undertaken as an iterative process because of the inability to decouple the influences of ellipsoid shape (aspect ratios) and orientation on the resultant distribution of ellipses. The iterative process initiates by calculating a probable orientation distribution for the ellipsoids with an assumed uniform shape distribution. Then the shape distribution is updated using the calculated orientation distribution. Iteratively, each distribution is updated using the most recently calculated instance of the other distribution until a level of convergence is reached. Some details of the calculations are offered here as well as the process of extrapolation of the individual ellipses.

Calculation of a probable orientation distribution requires sectioning a large number of ellipsoids within each discrete orientation bin and observing their resultant elliptic sections. Each ellipsoid is assigned a set of aspect ratios in accordance with the shape distribution, which is initially uniform. A two-dimensional histogram of resultant ellipse orientation and resultant ellipse aspect ratio is created for all ellipsoids in each of the orientation bins. An example histogram, shown as contour plots, is shown in Fig. 14. The histograms for each orientation bin are then compared to the same histogram made from the actual observations on the experimental 2D sections. The simulated histograms are fit to the experimental histogram by a least squares method to determine the probability of each orientation bin.

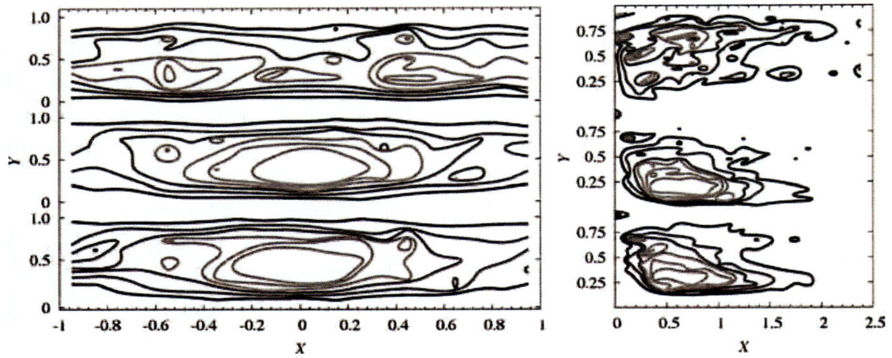

Fig. 14 Plot of (*left*) density of ellipse principal axis orientation vs aspect ratio and (*right*) density of ellipse normalized size vs aspect ratio. The top third of each plot shows ellipses on the plane normal to the z-direction, the middle third shows ellipses on the plane normal to the y-direction, and the bottom third shows ellipses on the plane normal to the x-direction. In the left plot, the x-axis refers to the x-component of a unit vector oriented along the major axis of the ellipse and the y-axis is the aspect ratio (b/a) of the ellipse. In the right plot, the x-axis is the ellipse area divided by the average ellipse area and the y-axis is the aspect ratio (b/a) of the ellipse

Upon calculating the orientation probability distribution, the shape distribution can be updated by sectioning a large number of ellipsoids within each discrete shape bin, represented by a set of aspect ratios. Each ellipsoid is assigned an orientation in accordance with the previously calculated orientation distribution. A two-dimensional histogram of resultant ellipse aspect ratio and resultant ellipse normalized size is constructed for all ellipsoids in each of the shape bins. The normalized size is the area of the resultant ellipse divided by the average resultant ellipse area. An example histogram is shown in Fig. 14. The histograms for each shape bin are then compared to the same histogram made from the actual observations on the experimental 2D sections. The simulated histograms are fit to the experimental histogram by a least squares method to determine the probability of each shape bin.

While sectioning the ellipsoids to determine a probable orientation and shape distribution, the distribution of fractional section size is constructed for each shape bin. The fractional section size is defined as the resultant ellipse area divided by the maximum possible resultant ellipse area for a given ellipsoid (i.e., when the ellipsoid is sectioned through the equatorial plane perpendicular to the minor axis). The fractional section size distribution is used to extrapolate the individual experimental ellipses. The shape of each experimental ellipse's parent ellipsoid is predicted by the probability of an ellipsoid of a given shape producing the experimental ellipse (i.e., its normalized size and aspect ratio). Once the parent ellipsoid's shape is assumed, the distribution of fractional section size for that shape can be used to convert the experimental section's area to the maximum possible section area for the parent ellipsoid. With an assumed shape and maximum possible section size, everything necessary to fully define the parent ellipsoid is available. This process is carried

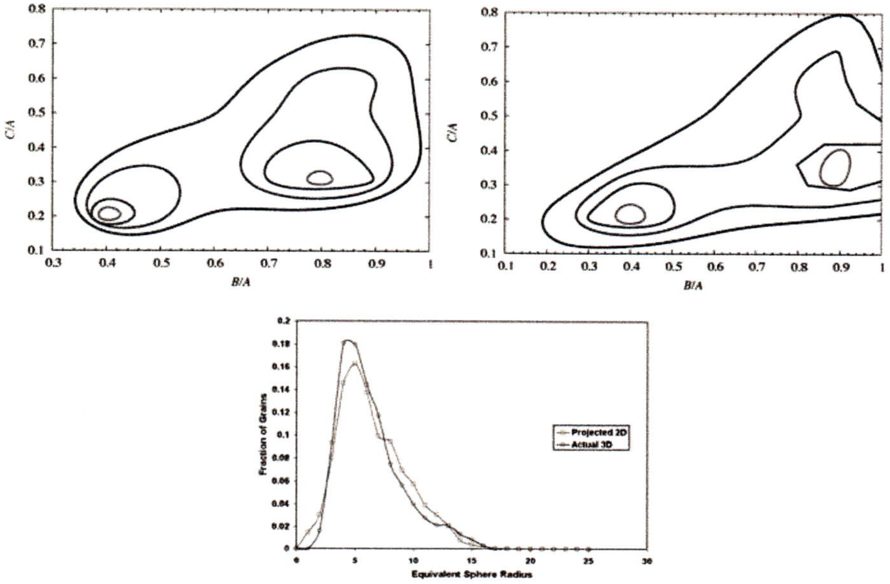

Fig. 15 Results of the Observation-Based Domain Constraint method. The upper right image is the true 3D shape distribution (ellipsoid aspect ratios). The upper left image is the "probable" shape distribution calculated by the method. The lower image is a comparison of the true 3D grain size distribution (equivalent sphere radius) and the calculated "probable" size distribution

out for each experimental ellipse, resulting in a set of "probable" ellipsoids, whose statistics can be used to generate synthetic 3D volumes. Figure 15 shows the results of the observation-based domain constraint process for a sample microstructure.

5.3 Limitations and Possibilities

Currently, there are a number of limitations that remain when inferring the true 3D microstructure of materials. First, the shape of the features being inferred has been limited to simple geometric shapes that can be easily described numerically, generally ellipsoids. This limitation need not be persistent, provided numerical descriptions of more complex shapes are developed. The moment invariant analysis mentioned previously may provide a key tool in developing this area.

Additionally, the linkage between neighborhoods of grains in 3D and grains in 2D is not immediately clear to this author. The number of neighbors, as well as their size, shape, etc, is likely to be a function of the packing of the grains, which may not be known from even several 2D sections. In principle, it would be possible to section many different, yet known, microstructures and attempt to develop a "look-up" table that correlates parameters such as: number of 2D neighbors to number of 3D neighbors.

6 Comments on Complex Microstructures

The microstructures investigated in the works presented in this chapter are relatively idealized. The nickel-base superalloy used in the work of Groeber et al. was chosen for its small feature size, propensity to yield high-quality data, and microstructural homogeneity. The aluminum alloy used in the work by Saylor et al. and the steel used in the work by Rowenhorst et al. both are single-phase materials with standard boundary structures and limited heterogeneities. These properties enabled the investigation of microstructure with comparatively little difficulty. All three microstructures certainly have inherent difficulties as well. The nickel-base superalloy contains a second phase that is difficult to distinguish from the matrix as well as twin grains that are often too small to sample properly with the tools used. The rolled aluminum alloy has a grain size too large for the desired experimental techniques (i.e., EBSD). The steel required the precipitation of a second phase to identify boundaries easily, which was met with some difficulty. However, the complications encountered do not approach the level of some heavily engineered, more topologically complex microstructures, such as beta-processed titanium with a basketweave structure. Techniques are being expanded to treat such microstructures, but are currently in the developmental stages.

It should not be a surprise that for the advancement of digital representation of grain-level microstructure to include more complex microstructures, there must be a coupled advancement in the ability to collect and quantify these microstructures. Additionally, clever techniques to homogenize or adaptively incorporate multiple microstructural scales will be an imperative to simulate microstructures that have critical features that exist of varying length scales. The utility of a microstructure representation can be limited greatly if the computational modeling community has no feasible method to simulate the structure. Again, this calls for a representation-modeling effort that defines the degree to which microstructure needs to be incorporated, both in the context of its effects on the property and on its ability to be simulated. For the example of beta-processed titanium, Venkatramani et al. (2008) have developed a homogenized grain representation that homogenizes the lamellar/lath structure of the titanium, which would be impossible to incorporate at the scale needed to include hundreds of prior beta grains. Thus, for properties that this homogenization is proper, the experimental techniques and representation tools can be tailored to identify and represent only the prior beta grains (and alpha colonies/variants) and omit the underlying lath-rib structure in an effort to increase simulation efficiency.

Finally, the synthetic structure generation process should likely attempt to parallel the natural process that produced the microstructure of interest. That is, the application of an increasing number of generic constraints on the placement of features in the synthetic volume is likely to inhibit the generation process. However, if the generation process follows the natural process, then many constraints may become obsolete and unnecessary. Two examples of this idea are the inclusion of twin grains in a polycrystal and the generation of a colony (or basketweave) structure in a titanium alloy. In the case of twin grains, it is difficult (if not impossible) to treat the twin and its parent grain as independent features and ensure that they will be placed in proper relation in the synthetic structure. Rather, it is potentially a better strategy to remove the twin grains, by merging them with their parent grains, and characterize the simplified structure. Then, a synthetic structure without twins can be generated to match the simplified structure. If the statistical description of the twin structure (i.e., fraction of grains with twins, twin plate thickness, number of twins per grain, etc.) is measured during the merging, then twins could be inserted into the simplified synthetic structure, which would simulate the natural process of twin nucleation in a parent grain. In the case of the colony structure of a titanium alloy, if the colonies themselves are treated as the features to be placed, it is a complicated process to place colonies into neighborhoods that have the proper orientation relationships (in accordance with the Burgers' relationship) and the correct topological structure (imposed by the prior beta grain boundaries). One option to circumvent this complication is to generate a synthetic structure that consists only of prior beta grains, with the statistics to match the experimental beta grain structure. The prior beta grains can be identified by grouping the colonies that came from the same beta grain, which is known through the Burgers' relationship. Then, the synthetic beta grains can be divided into colonies, where the orientations are already constrained by the beta orientation and the topology is already constrained by the beta grain boundaries. Similar to the twin example, the statistics of the "secondary" features (i.e., the twins or colonies) can be measured during grouping to locate the primary features (i.e., the parent grains or prior beta grains). Dividing the beta grains into colonies simulates the natural precipitation of the alpha phase in the prior beta grains. Linking the synthetic generation process to the natural process may simplify the representation of complex structures and it is also likely to enhance the physical significance of the generated structure.

7 Conclusions

This chapter has attempted to discuss the key areas necessary for the development of digital representations of materials microstructure at the grain level. Much of the detail in the various areas is presented elsewhere and has been noted wherever possible. It was the goal of the author to highlight the current "state-of-the-art" techniques and discuss their strengths and weaknesses. This field is still in a state of relative infancy and requires the cooperation of a number of other fields to properly evolve. The experimental community has elevated the ability and precision with which it can investigate microstructure in the last decade. It is important for communication with this community to tailor experiments to the needs of given representation requirements. After the generation of a microstructure representation, mesh generation remains a barrier to the simulation of models (without artifacts). The general meshing community has not been introduced to the needs of the materials modeling problem. Quantification of mesh error relative to the digital representation is one key metric, as well as the quality of elements required for a given simulation. Finally and arguably the most critical, the development of a connection between simulation results and statistical descriptors is imperative. Such an association is a necessary part of the pathway to the determination of a RVE for all materials properties. The improvement of constitutive relations requires knowledge of what descriptors influence properties and proper quantification of descriptors is key in defining their influence on properties. Digital representation of materials microstructure is an integral part in the determination of microstructure–property relationships, but cannot be treated as an independent step in the process. The full effect of developments presented here and those to come will only be realized when these collaborations have been cultivated.

Acknowledgments The author would like to acknowledge his collaborators, all of whom contributed through detailed discussions and in many cases developed some of the tools and techniques presented in this chapter. All of the sections in this chapter were heavily influenced by them and in some instances their own words and terminology were used. The works of Profs. Somnath Ghosh, Tony Rollett, and Marc DeGraef; as well as Drs. David Rowenhorst, Dennis Dimiduk, Mike Uchic, Sukbin Lee, Jeremiah MacSleyne, and Mrs. Yash Bhandari and Steve Sintay and many others have greatly advanced this field and inspired this author.

References

Amenzua E, Hormaza MV, Hernandez A, Ajurja MBG (1995) Adv Eng Softw 22:45–53
Barton NR, Dawson PR (2001) Metall Mater Trans 32A:1967–1975
Barry J (1995) SIAM J Sci Comput 16:1292–1307
Bhandari Y, Sarkar S, Groeber M, Uchic M, Dimiduk D, Ghosh S (2007) Comput Mater Sci 41:222–35
Brahme A, Alvi MH, Saylor D, Fridy J, Rollett AD (2006) Scripta Mater 55:75–80
Bozzolo N, Dewobroto N, Grosdidier T, Wagner F (2005) Mater Sci Eng A Struct Mater 397:346
Budai JD, Yang W, Larson BC, Tischler JZ, Liu W, Weiland H, Ice GE (2004) Mater Sci Forum 467–470:1373–1378
Budai JD, Liu W, Tischler JZ, Pan ZW, Norton DP, Larson BC, Yang W, Ice GE (2008) Thin Solid Films 576:8013–8021
Bullard JW, Garboczi EJ, Carter WC, Fuller ER (1995) Comput Mater Sci 4:103–116
Cahn JW, Fullman RL (1956) Trans Metall Soc AIME 206:610–612
Cruz-Orive LM (1976a) J Microsc 107:1–18
Cruz-Orive LM (1976b) J Microsc 107:235–253
DeHoff RT (1962) Trans Metall Soc AIME 224:474–486
Feltham P (1957) Acta Metall 5:97–105
Fernandes CP (1996) Phys Rev E 54:1734–1741
Ghosh S, Bhandari Y, Groeber M (2008) J Comput Aided Des 40(3):293–310
Groeber MA, Haley B, Uchic MD, Ghosh S (2004) In: Ghosh S, Castro J, Lee JK (Eds) Proceedings of NUMIFORM 2004. AIP Publishers, Melville, NY
Groeber MA (2007) Ph.D. Thesis, The Ohio State University
Groeber MA, Ghosh S, Uchic MD, Dimiduk DM (2008a) Acta Mater 56:1257–1273
Groeber MA, Ghosh S, Uchic MD, Dimiduk DM (2008b) Acta Mater 56:1274–1287
Gulsoy EB, Simmons JP, De Graef M 2009 Scripta Mater 60:381–384
Hadwiger H (1957) Vorlesungen über Inhalt, Oberfläche und Isoperimetrie. Springer, Berlin
Hillert M (1962) In: Zackay VF, Aaronson HI (Eds) The Decomposition of Austenite by Diffusional Processes. Interscience, New York

Hillert M (1965) Acta Metall 13:227–283
Hopkins RH, Kraft RW (1965) Trans AIME 233:1526–1532
Kammer D, Mendoza R, Voorhees PW (2006) Scripta Mater 55:17–22
Kenney JF, Keeping ES (1947) In: Mathematics of Statistics. Van Nostrand
Lauridsen EM, Schmidt S, Nielsen SF, Margulies L, Poulsen HF, Juul Jensen D (2006) Scripta Mater 55:51–56
Li M, Ghosh S, Richmond O, Weiland H, Rouns TN (1999) Mater Sci Eng A A265:153–173
Lienert U, Almer J, Jakobsen B, Pantleon W, Poulsen HF, Hennessey D, Xiao C, Suter RM (2007) Mater Sci Forum 539–543:2353–2358
Lorenson WE, Cline HE (1987) Comput Graph 21:163–169
Louat NP (1974) Acta Metall 22:721–724
MacPherson RD, Srolovitz DJ (2007) Nature 446:1053
MacSleyne J, Simmons JP, DeGraef M (2008) Model Sim Mater Sci Eng volume 16 045008
Oren PE, Bakke S (2002) Transport Porous Media 46:311–343
Oren PE, Bakke S (2003) J Pet Sci Eng 39:177–199
Parthasarathy VN, Kodiyalam S (1991) J Finite Elem Anal Des 9:309–320
Przystupa MA (1997) Scripta Mater 37:1701–1707
Randle V, Hu Y, Rohrer GS, Kim C-S (2005) Mater Sci Tech 21:1287–1292
Randle V, Rohrer GS, Hu Y (2008a) Scripta Mater 58:183–186
Randle V, Rohrer GS, Miller H, Coleman M, Owen G (2008b) Acta Mater 56:2363–2373
Rhines FN, Craig KR, Rousse DA (1976) Metall Trans A 7A:1729–1734
Rollett AD, Manohar P (2004) In: Raabe D (Ed) Continuum Scale Simulation of Engineering Materials. Wiley-VCH, Weinheim
Rowenhorst DJ, Gupta A, Feng CR, Spanos G (2006) Scripta Mater 55:11–16
Russ JC (1986) The Image Processing Handbook. CRC Press, West Palm Beach, FL
Russ JC, DeHoff RT (1986) Practical Stereology. Springer, Berlin
Saltykov SA (1958) Stereometric Metallography. Metallurgizdat, Moscow
Saylor DM, Morawiec A, Cherry KW, Rogan FH, Rohrer GS, Mahadevan S, Casasent D (2001) In: Gottstein G, Molodov DA (Eds) Proceedings of the First Joint International Conference on Grain Growth. Springer Verlag, Aachen
Saylor DM, Fridy J, El-Dasher BS, Jung KY, Rollett AD (2004a) Metall Mater Trans A 35A: 1969–1979
Saylor DM, El-Dasher BS, Adams BL, Rohrer GS (2004b) Metall Mater Trans 35A: 1981–1989
Saylor DM, El-Dasher BS, Rollett AD, Rohrer GS (2004c) Acta Mater 52:3649–3655
Schmidt S, Nielsen SF, Gundlach C, Margulies L, Huang X, Juul Jensen D (2004) Science 305:229–232
Simmons JP, Chuang P, Comer ML, Uchic M, Spowart JE, De Graef M (2009) Modell Simul Mater Sci Eng 17:025002
Spanos G (2006) Scripta Mater 55:3
Sundararaghavan V, Zabaras N (2005) Comput Mater Sci 32:223–239
Talukdar MS, Torsaeter O (2002) J Pet Sci Eng 33:265–282
Talukdar MS, Torsaeter O, Ioannidis MA (2002a) J Colloid Interface Sci 248:419–428
Talukdar MS, Torsaeter O, Ioannidis MA, Howard JJ (2002b) J Pet Sci Eng 35:1–21
Talukdar MS, Torsaeter O, Ioannidis MA, Howard JJ (2002c) Transport Porous Media 48:101–123
Tewari A, Spowart JE, Gokhale AM, Mishra RS, Miracle DB (2006) Mater Sci Eng A 428:80–90
Torquato S (2001) Random Heterogeneous Materials: Microstructure and Macroscopic Properties. Springer-Verlag, New York
Underwood E (1970) Quantitative Stereology. Addison-Wesley, New York
Venkatramani G, Kirane K, Ghosh S (2008) J Plasticity 28:428–454
Zaafarani N, Raabe D, Singh RN, Zaefferer S (2006) Acta Mater 54:1863–1876
Zhang C, Suzuki A, Ishimaru T, Enomoto M (2004) Metall Trans A 35A:1927–1932

Multiscale Characterization and Domain Partitioning for Multiscale Analysis of Heterogeneous Materials

Somnath Ghosh

Abstract This chapter discusses the development of a multiscale characterization methodology leading to microstructural morphology-based domain partitioning *MDP* methodology for materials with nonuniform heterogeneous microstructure. The comprehensive set of methods is intended to provide a concurrent multiscale analysis model with the initial computational domain that delineates regions of statistical homogeneity and inhomogeneity. The *MDP* methodology is intended as a preprocessor to multiscale analysis of mechanical behavior and damage of heterogeneous materials, e.g., cast aluminum alloys. It introduces a systematic three-step process that is based on geometric features of morphology. The first step simulates high-resolution microstructural information from low-resolution micrographs of the material and a limited number of high-resolution optical or scanning electron microscopy micrographs. The second step is quantitative characterization of the high-resolution images to create effective metrics that can relate microstructural descriptors to material behavior. The third step invokes a partitioning method to demarcate regions belonging to different length scales in a concurrent multiscale model. Partitioning criteria for domain partitioning are defined in terms of microstructural descriptors and their functions. The effectiveness of these metrics in differentiating microstructures of a 319-type cast aluminum alloy with different secondary dendrite arm spacings SDAS is demonstrated. The *MDP* method establishes intrinsic material length scales and consequently subdivides the computational domain for concurrently coupling macro- and micromechanical analyses in the multiscale model. Finally, a multiscale analysis of ductile fracture is conducted using a differentiated scale structure that has been laid out by the *MDP* algorithm. The chapter emphasizes the need for coupling multiscale characterization and domain decomposition with multiscale analysis of heterogeneous materials.

S. Ghosh, John B. Nordholt Professor (✉)
Department of Mechanical Engineering, The Ohio State University,
W496 Scott Laboratory, 201 West 19th Avenue, Columbus, OH 43210
e-mail: ghosh.5@osu.edu

S. Ghosh and D. Dimiduk (eds.), *Computational Methods for Microstructure-Property Relationships*, DOI 10.1007/978-1-4419-0643-4_4,

1 Introduction

Heterogeneous materials, such as alloy systems containing precipitates and defects, polymer, ceramic or metal matrix composites, functionally graded materials, are increasingly in use in the aerospace, automotive, electronics, defense, and other industries. Some are engineered at the microscale to possess optimal multifunctional properties such as low weight, high strength, superior energy absorption and dissipation, high impact and penetration resistance, superior crash-worthiness, better structural durability, etc. Many of these materials exhibit strong nonuniformities in the micro- and mesoscale morphology. Nonuniform distributions are observed in microstructures exhibiting clustering or preferred directionality, or those with irregularities in inclusion shapes and sizes. Furthermore, constituent material and interface properties can also contribute to the microstructural heterogeneity. As examples, micrographs of a silicon particulate reinforced aluminum alloy (DRA) and an epoxy matrix composite (PMC) consisting of graphite fibers are shown in Fig. 1. Processing methods, such as casting, powder metallurgy or resin transfer molding often result in strong morphological irregularities, e.g., in particulate spatial dispersion, phase shape or size, or even in the constituent material and interface properties.

Failure properties such as strain to failure, ductility, and toughness are highly sensitive to these variations. For instance, experimental studies on ductile failure in (Wang et al. 2003; Wang 2003; Boileau 2000; Poole and Dowdle 1998) have shown that morphology strongly affects microstructural damage nucleation due to particulate cracking and interfacial decohesion, as well as ductile damage growth by void growth and coalescence in the matrix. Argon et al. (1975) has shown that particles in clustered regions have a greater propensity toward cracking than those in regions of dilute concentration. This is caused by local stresses that increase rapidly with reduced distance between neighboring particles. Experimental work in (Caceres 1999; Caceres and Griffiths 1996) has demonstrated that larger and longer particles are prone to increased cracking and damage accumulation with higher dendrite

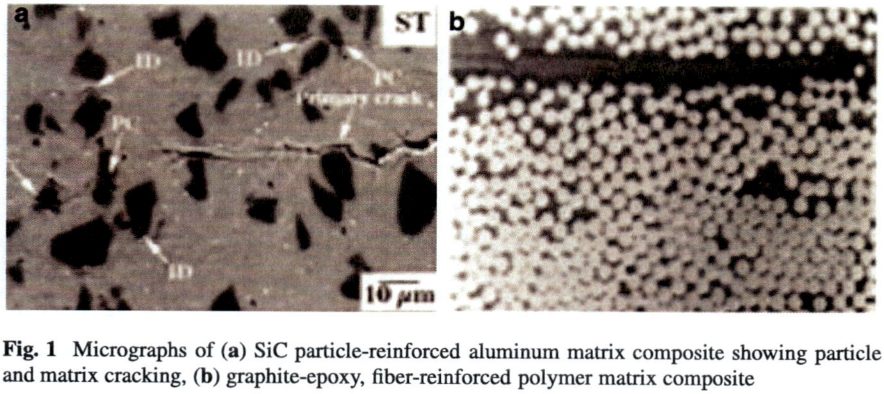

Fig. 1 Micrographs of (**a**) SiC particle-reinforced aluminum matrix composite showing particle and matrix cracking, (**b**) graphite-epoxy, fiber-reinforced polymer matrix composite

arm spacing. Consequently, special attention must be given to the microstructural morphology, when modeling these alloys for properties such as strain to failure, ductility, or fracture toughness.

Various computational models, e.g., (Christman et al. 1989; Gonzalez and Llorca 1996; Weissenbek et al. 1994), have been proposed for analyzing mechanical response and properties of multiphase materials using high-resolution finite element models. While most of them consider standard unit cell analysis, few efforts like (Weissenbek et al. 1994) have made extensions with creative boundary conditions to accommodate nonuniform effects. Predictive capabilities of these models are, however, limited for nonuniform microstructures, especially when failure properties are of interest. The models oversimplify the local morphology, in particular, the extremities of local morphological distributions that control damage initiation and growth. Recently some efforts have been made to model microstructural regions with larger number of spherical heterogeneities in three dimensions (Segurado and Llorca 2005; Boehm et al. 2004). A few studies, e.g., (Yang et al. 2000b; Shan and Gokhale 2004; Terada and Kikuchi 2000; Ghosh et al. 2000; Li et al. 1999a) have also focused on modeling realistic representations of nonuniform phase dispersions by combining digital image processing with microstructure modeling. The microstructure-based Voronoi cell finite element model or *VCFEM* (Ghosh et al. 2000; Li et al. 1999a; Li and Ghosh 2006; Hu and Ghosh 2008; Ghosh 2008) has shown significant promise in accurate and efficient analysis of large microstructural regions.

While advances in modern computing hardware and computational science have made analysis of microstructural regions with large number of heterogeneities possible, connecting local to overall structural failure is still far from maturation. A single-scale modeling scheme entails micromechanical analysis of the entire computational domain (structure) from initiation to failure, accounting for all morphological details. This is computationally prohibitive with current day computing facilities. Alternatively, effective multiscale modeling can offer significant relief from intense broad-based micromechanical computing through selective regions of microanalysis in an otherwise macroscopic computational domain. Multiscale modeling methods have currently gained considerable momentum for mechanical response and failure analysis of heterogeneous structures (Smit et al. 1998; Raghavan and Ghosh 2004a,b; Ghosh 2008; Ghosh et al. 2001; Fish and Shek 2000; Hao et al. 2004; Vemaganti and Oden 2001; Terada and Kikuchi 2000; Zohdi and Wriggers 1999; Chung and Tamma 1999; Xia et al. 2001). Large computational domains can be effectively handled in these techniques through different ways of information transfer between disparate scales. Two categories of methods have emerged in the multiscale analysis literature. The first category of hierarchical models passes information from lower to higher scales through homogenized constitutive material models and properties, or coarse graining (Jain and Ghosh 2008a,b; Ghosh et al. 2009). The second category introduces concurrent methods, which implement substructuring or embedding to delineate complementary regions of the computational domain corresponding to different resolutions or scales. Governing equations for different material scales are concurrently solved in a coupled manner in these models. Ghosh and coworkers have used special criteria to adaptively decompose computational domains and integrate aspects of both hierarchical and concurrent multiscale models in their analysis of heterogeneous domains (Raghavan et al. 2004; Ghosh 2008). Two-way coupling of scales are facilitated by preferential homogenization and localization in this method, which makes it suitable for problems involving localized damage and failure. Macroscopic analysis using bottom-up homogenization enhances the efficiency of computational analysis in regions of homogeneous deformation. On the other hand, top-down coupling of macroscopic and microscopic analyses is facilitated by cascading down to the microstructural levels at critical regions of localized failure. A schematic of the concurrent multilevel computational framework for multiscale modeling that is developed in (Ghosh et al. 2001; Raghavan and Ghosh 2004a,b; Raghavan et al. 2004; Ghosh 2008) is shown in Fig. 2. In these models, domain partitioning into macro- and micro-domains is adaptive and evolutionary. It is a continuous process that is based on the evolution of local stresses, strains and/or damage. Optimal domain partitioning can significantly enhance the efficiency of multiscale computational models by keeping the "zoomed-in" regions of micromechanical analysis to a minimum.

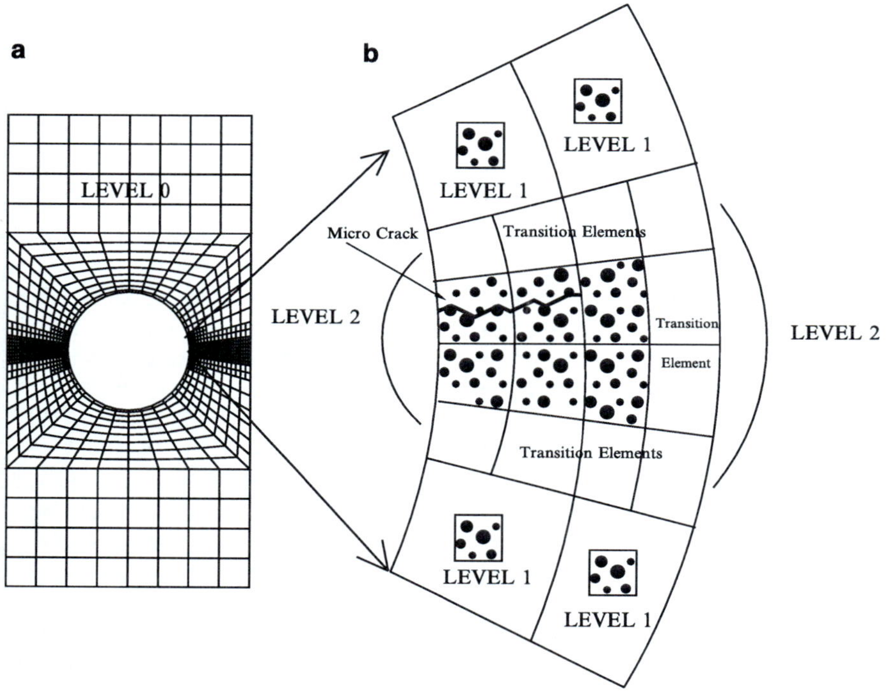

Fig. 2 Schematic of a coupled concurrent multilevel model showing: (**a**) *level-0* region of macroscopic continuum analysis with adaptive mesh refinement and zoom-in; and (**b**) blow-up of critical region containing *level-1* (swing region with RVE analysis) and *level-2* region (of pure micromechanical analysis)

A challenge in the implementation of multiscale modeling method for structures with nonuniform microstructure is the *a-priori* delineation of computational substructures. This should appropriately be based on information of morphological features and properties at the microstructural scale. Some multiscale models, e.g., in (Xia et al. 2001; Raghavan and Ghosh 2004a,b; Fish and Shek 2000; Vemaganti and Oden 2001; Terada and Kikuchi 2000; Zohdi and Wriggers 1999) have assumed periodic repetition of microstructural representative volume elements or RVEs over the entire computational domain. The RVEs themselves can contain reasonably large number of heterogeneities ($\sim$100). An underlying assumption in these models is that the microstructural morphologies themselves do not exhibit strong morphological gradients and homogenized representation of the local region will yield reasonably accurate macroscopic results. In concurrent multiscale modeling of materials with nonuniform microstructures, gradients or discontinuities in local morphological distributions require pockets of microstructural domains to be embedded in the otherwise homogenized macroscopic domain. It is therefore beneficial to have morphological information of the underlying microstructure at all points in the computational domain prior to analysis. A multiscale morphology-based domain partitioning (*MDP*) methodology has been developed in (Valiveti and Ghosh 2007; Ghosh et al. 2006) for multiphase materials. This serves as a preprocessor to multiscale analysis. The morphology-based domain partitioning *(MDP)* is intended for two reasons, viz.:

1. Determination of microstructural representative volume elements or RVEs that can be used in "bottom-up" homogenization of the computational domain.
2. Identification of regions where the morphology alone is sufficient to cause a breakdown in the homogenization assumption. For example, regions of dense clustering can cause the onset of dominant microstructural cracks or localization in the microstructure.

Embedded regions, requiring microstructural analysis, should then be coupled with complementary regions of homogenized macroscopic analysis.

A necessary requirement of the *MDP* method is that information of the microstructural morphology, at least with respect to important characterization functions, be available for all points of the computational domain. This can be a very challenging and time-consuming task, if the entire image has to be acquired by optical or scanning electron microscopy. A few methods have been suggested in the literature for dealing with this problem. The M-SLIP method of preparing a montage of a large number of high magnification microstructural images (nearly 400–500), followed by image compression has been proposed in (Shan and Gokhale 2004; Gokhale and Yang 1999). This method is effective for small domains where few images are necessary and the microstructural information is sufficient for evaluating point statistics (Tewari et al. 2004). However for large domains, this method of extracting microstructural images at each individual point can be exhaustive. Statistical image reconstruction techniques based on the n-point statistics have also been used in practice (Yeong and Torquato 1998; Rintoul and Torquato 1996; Manwart et al. 2000). These methods first generate characteristic functions, such as the lineal path function, of the morphology. The functions are subsequently used to regenerate the microstructure by a process called "Simulated Annealing" or *SA* (Yeong and Torquato 1998; Rintoul and Torquato 1996), for given area fraction and n-point statistics. While this method has the flexibility to use as many correlation functions as desired, it needs many iterations to evolve toward the expected microstructure. Also, the simulated annealing parameters should be appropriately chosen for monotonic convergence. This may limit its application in reconstructing large microstructural domains. A variant to the *SA*-based microstructure reconstruction has been proposed in (Kumar et al. 2006) for microstructures containing potential "hot spots" of high stress or strain localization. Other statistical methods of microstructure reconstruction include random generation of points in a large domain by representing the centroids of second-phase particles (Yang et al. 2000a). This is followed by replacing each of these points with a particle of a definite shape. This method has a low probability of accurately representing features of the actual microstructure due to the random generation process. Similar methods include the random sequential packing algorithm (*RSA*) introduced in (Cooper 1998) for simulating dispersions of regular shape particles and the Monte–Carlo technique discussed in (Everett and Chu 1992). All of these methods have had limited success with respect to convergence to the actual image. There are also few training based techniques such as the super-resolution method in (Freeman et al. 2000), where high frequency bands of sample images at high resolution are used as training sets to enhance the regular interpolated image. Important growth is happening in the recent days to the super-resolution techniques (Farsiu et al. 2004). It has become a fast growing field for enhancing the resolution of legacy video, to be viewed on modern high definition video screens. The resolution enhancements are typically a factor of 4 for a 2-D image because of Nyquist limitations.

The morphology-based domain partitioning or *MDP* method has been developed in (Valiveti and Ghosh 2007; Ghosh et al. 2006) for multiphase materials. It creates a morphology-based domain partitioning as a preprocessor to multiscale analysis of heterogeneous materials, e.g., cast aluminum alloys. The overall *MDP* process is founded upon three sequential building blocks.

1. The first step simulates necessary high-resolution microstructural information at all points of a computational domain from continuous low resolution images of the entire domain and few sample high-resolution images. The microstructural image representation and reconstruction method is discussed in Sect. 2.
2. The second step, discussed in Sect. 4, uses quantitative characterization of this high resolution microstructure using functions of the phase distribution, to create effective metrics that can relate microstructural features to the critical material properties. Effective parameters are identified for quantifying morphological characteristics based on micro-mechanical response analysis of various simulated micrographs. Such characterization is important in multiscale modeling for establishing length scale characteristics at different resolutions.
3. The third step invokes domain partitioning based on functions of the identified microstructural descriptors. The *MDP* process delineates regions corresponding to different length scales in a coupled concurrent multiscale model. Refinement

functions are defined in terms of microstructural characteristics and these are used to adaptively create multilevel domain partitioning. Representative volume elements in the microstructure are also identified in this step.

The efficacy of the *MDP* process is demonstrated for aluminum alloy microstructures. While the method described here is for 2D microstructures, extension to 3D is automatic, though it will involve additional methods for 3D data acquisition as well as for 3D image processing.

2 Reconstructing High-Resolution Microstructures from Low-Resolution Micrographs

A prerequisite for morphology-based partitioning of the computational domain as a preprocessor to multiscale modeling is information of high-resolution microstructure at all points of the domain. Since it is prohibitive to experimentally obtain contiguous high-resolution microscopic images at all points, it is desirable to simulate the local microstructure from high-resolution micrographs extracted from a few selected locations in the domain. The simulated micrographs should be accurate with respect to important morphological characteristics when compared with the actual micrograph. Image hallucination techniques have been developed in, e.g., (Baker and Kanade 2001) to generate high-resolution images from generic low-resolution images. These methods include extracting a plurality of primal sketch priors from training data. At the synthesis phase, the plurality of primal sketch priors are utilized to improve a low-resolution image by replacing one or more low-frequency primitives extracted from the low-resolution image with corresponding ones of the plurality of primal sketch priors. Hallucination has not generally been applied to material systems though. This section develops a different methodology for simulating such high-resolution microstructures at all locations in a low-resolution micrograph from high-resolution microstructural images at a few sample locations. The technique is applicable to optical microscopy or SEM-based micrographs of materials such as multi-phase alloys, metal, and polymer matrix composites, etc.

The examples considered in this chapter are mostly for a cast aluminum alloy W319, a higher silicon version of the AA319 alloy, having a nominal composition of Al-7%Si-3%Cu-0.4%Fe. It is a typical alloy used for various automotive components, such as engine blocks. A low-resolution scanning electron microscopy (SEM) image of its microstructure is shown in Fig. 3. The silicon particles are in the darker shades, while soft gray particles represent intermetallics. The silicon particles are pushed into the regions between secondary aluminum dendrites in the solidification process. Thus, their presence indicates the boundaries between two adjacent secondary aluminum dendrite arms. The distance between two arms is measured as secondary dendrite arm spacing (or SDAS). The W319 alloy examined here in cast with controlled colling rates to produce microstructures with three different SDAS values 23, 70, and 100 μm, respectively.

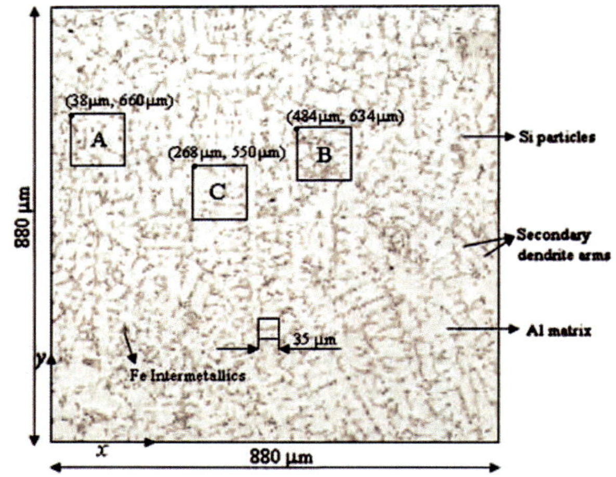

Fig. 3 Low magnification, low-resolution digital image of cast aluminum alloy W319, for which high resolution micrograph of a window C is desirable with available high-resolution micrographs at other locations A and B

The low-resolution micrograph in Fig. 3 does not provide adequate information required for microstructural characterization and modeling. The microstructure reconstruction process generates corresponding high-resolution images with clear delineation of the multiphase morphology. The digital micrograph of Fig. 3 can be resolved into a grid of pixels, with each pixel belonging to a certain level in the grayscale (white-black) hierarchy. For a region Ω_{mic} in the micrograph, the grayscale level of each pixel with centroid at (x, y) is represented by an integer valued indicator function $I^g(x, y)$. The indicator function is defined for a 8-bit monochrome grayscale image, as:

$$I^g(x, y) = \{p : 0 \leq p \leq 2^8 - 1; \forall \quad (1 \leq x \leq M); \ (1 \leq y \leq N) \in \Omega_{mic}\} \quad (1)$$

At each point of the micrograph, $I^g(x, y)$ may assume any integer value between 0 and 255.

2.1 Resolution Augmentation Problem

A magnified image of a small region of this micrograph is shown in Fig. 4a. In this chapter, magnification refers to the pixel size and hence a magnified image will have larger size of pixels with the same number of grayscale pixels as the original image. Resolution, on the other hand, corresponds to the number of pixels or pixel density in an image. Hence, a higher resolution image will have a higher pixel density with altered grayscale levels in regions of gradients. Thus, the number of pixels in the local image of Fig. 4a is the same as that in the original image window

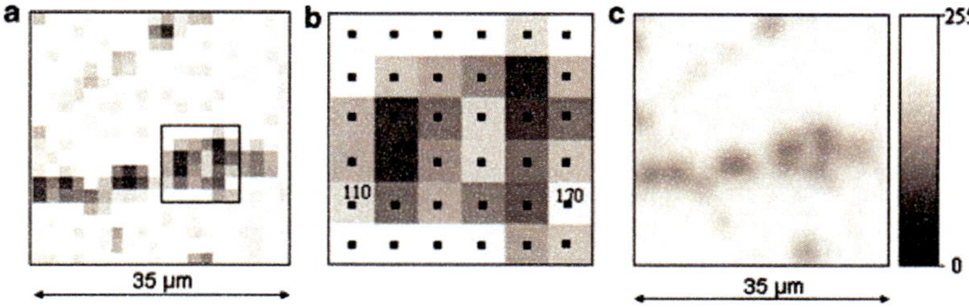

Fig. 4 High magnification 35 × 35 μm images of a region near C, shown in Fig. 3: (**a**) zoomed-in image showing larger pixels but with original resolution; (**b**) pixel representation of the square region marked in (**a**); (**c**) a higher resolution micrograph of (**a**) obtained by interpolation

of Fig. 3. Only the pixels in Fig. 4a are enlarged. Lower pixel densities in the low-resolution images are susceptible to a loss of information in the actual image. The microstructure reconstruction method is intended to replenish this lost information from data obtained from a few noncontiguous high-resolution images at different locations of the parent domain Ω_{mic}. Various augmentation methods exist in the literature. Polynomial interpolation methods for subpixel values in digital images have been developed in (Robert 1981), but they do not consider simultaneous grayscale variations in two orthogonal directions. Higher order interpolation by the B-spline kernel method (Unser et al. 1991) provides a more continuous representation of microstructural image. However, the interpolated micrographs are sometimes blurred as shown in Fig. 4c. Directional methods in (Jensen and Anastassiou 1995) interpolate along the edges of discontinuities rather than across them, which reduces the blurring effect. Computational efforts incurred in these methods are quite exhaustive in comparison with the improvements they provide over interpolation methods. Wavelet-based approaches have also been pursued for local interpolation (Prasad and Iyengar 1997). The high-resolution microstructure reconstruction developed in this work incorporates a wavelet-based interpolation of low-resolution images, which is followed by a gradient-based enhancement method (Valiveti and Ghosh 2007; Ghosh et al. 2006).

2.2 Wavelet-based Interpolation in the WIGE Algorithm

2.2.1 A Brief Discussion of Wavelet Basis Functions

Wavelet bases, discussed in (Chui 1992; Qian and Weiss 1993; Motard and Joseph 1994), are $L^2(\mathscr{R})$ and generally have compact support. Only the local coefficients in wavelet approximations are affected by abrupt changes in the solution. The construction of wavelet functions starts from a scaling or dilatation function $\phi(x)$ and a set of related coefficients $\{p(k)\}_{k\in\mathscr{Z}}$, which satisfy the two-scale relation

$$\phi(x) = \sum_k p(k)\phi(2x - k). \tag{2}$$

Translations of the scaling function $\phi(x-k)$ form an unconditional basis of a subspace $V_0 \subset L^2(\mathscr{R})$. Through a translation of ϕ by a factor of 2^n and dilation by a factor of $k \cdot 2^{-n}$ the unconditional basis is obtained for the subspace $V_n \subset L^2(\mathscr{R})$ as

$$\phi_{n,k}(x) = 2^{n/2}\phi(2^n x - k) \tag{3}$$

for a resolution level n. The scaling function ϕ is defined as orthonormal if translations at the same level of resolution satisfies the condition

$$\int_{-\infty}^{\infty} \phi_{n,k}(x)\phi_{n,l}(x)\mathrm{d}x = \delta_{k,l} \;\; \forall\, n,k,l \in \mathscr{Z}. \tag{4}$$

Consequently, the best approximation of a function $f(x)$ in the subspace V_n of $L^2(\mathscr{R})$ is expressed as the orthogonal projection of f on V_n as:

$$\mathscr{A}_n f(x) = \sum_k a_{n,k}\phi_{n,k}(x), \qquad \text{where } a_{n,k} = \int_{-\infty}^{\infty} f(x)\phi_{n,k}(x)\mathrm{d}x. \tag{5}$$

Approximation of $f(x)$ can be made at different resolution levels, and these approximations in subspaces $\cdots,\ V_{n-1},\ V_n,\ V_{n+1},\ \ldots,$ follow the relation

$$\begin{aligned}&\{0\} = V_{-\infty} \subset \cdots \subset V_{-1} \subset V_0 \subset V_1 \subset \cdots \subset V_{\infty} = L^2(\mathscr{R}), \text{ where}\\ &lim_{n\to-\infty} V_n = \bigcup V_n \text{ is dense in } L^2(\mathscr{R}) \text{ and } lim_{n\to-\infty} \cap_n V_n = \{0\}.\end{aligned} \tag{6}$$

In the multiresolution level transition, the information lost in the transition from level V_{n+1} to level V_n is characterized by an orthogonal complementary subspace W_n. A basis for the subspace W_n can be obtained is in the same manner as for scaling function, i.e., by dilating and translating the mother wavelet function

$$\psi(x) = \sum_k q(k)\psi(2x-k). \tag{7}$$

The subspaces spanned by the wavelet functions have the following essential properties:

(a) $V_{n+1} = V_n \oplus W_n \;\; \forall, i.e.,$ W_n is the orthogonal complement of V_n toV_{n+1}.
(b) For orthonormal bases, W_{n1} is orthogonal to W_{n2}.
(c) For orthonormal bases, $\oplus_{n=-\infty}^{\infty} W_n = L^2(\mathscr{R})$. (8)

An approximation of the function $f(x)$ at the n-th resolution level may be expressed as the orthogonal projection of f on W_n as

$$f \to \mathscr{D}_n f(x) = \sum_k b_{n,k}\psi_{n,k}(x), \qquad \text{where } b_{n,k} = \int_{-\infty}^{\infty} f(x)\psi_{n,k}(x)\mathrm{d}x. \tag{9}$$

Due to the orthonormality and multiresolution properties of wavelet basis functions, higher level approximate solutions can be generated from results of lower level solutions (see Chui 1992; Motard and Joseph 1994) by selective superposition of complementary solutions.

2.2.2 Wavelet Interpolated Indicator Functions

The localization property makes the wavelet basis a desirable representation tool for problems with localization and high solution gradients, or even singularities. Numerical experiments conducted in (Valiveti and Ghosh 2007) show that a wavelet-based interpolation with gradient-based enhancement or (*WIGE*) algorithm enjoys superior convergence characteristics over pure polynomial-based interpolation methods. This is mainly due to the better representation of local gradients with wavelets. Consider a polynomial and a wavelet-based reconstruction of a reference indicator function $I(x)$ shown in the Fig. 5. The function $I(x)$ is defined between points $x = \{-2.5, 2.5\}$ and has sharp discontinuities in the slope at $x = \{-1.5, -0.5, 0.5, 1.5\}$. With a polynomial representation, $I(x)$ is interpolated as:

$$I_{poly}(x) = \sum_{1 \leq p \leq n} C_p \, x^{p-1}, \tag{10}$$

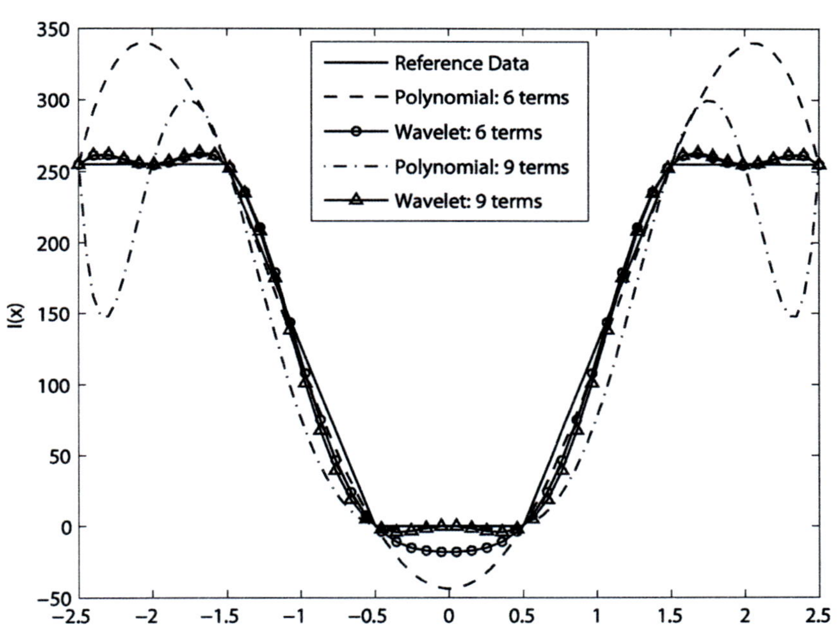

Fig. 5 Comparison of polynomial and wavelet interpolation in representing a reference indicator function with slope changes

where n represents the number of terms in the polynomial series, taken in the interpolation. The same indicator function may be expressed in terms of Gaussian wavelet functions as

$$I_{wvlt}(x) = \sum_{1 \leq q \leq n} C_q \, e^{-\frac{1}{2}\left(\frac{x-b_q}{a_q}\right)^2}. \tag{11}$$

Figure 5 shows a comparison of the reconstructed functions by the two basis functions. For a lower number of terms ($n = 6$), the polynomial-based interpolation shows large errors near the slope discontinuities. Increasing the number of terms does not show any marked improvement in the representation. Moreover, numerical instabilities are observed in the evaluation of coefficients with a higher number of polynomial terms. However, the wavelet-based representation captures the sharp changes in slope, without adversely affecting the overall function representation for the range of number of terms.

In the *WIGE* image reconstruction methodology (Valiveti and Ghosh 2007; Ghosh et al. 2006), the integer indicator function $I^g(x, y)$ in (1) is first interpolated in real space $\Re$ using a wavelet basis function. Let a window of the low-resolution image $\Omega_w^{lrsm} \in \Omega_{mic}$ encompass a $p \times q$ pixel grid. For a higher resolution image Ω_w^{hrsm}, the same window may be resolved into a $p' \times q'$ pixel grid, where $p' > p$ and $q' > q$. The grayscale level of each pixel in the $p \times q$ pixel grid corresponds to the value of the indicator function $I^g(x, y)$ at its centroid. The discrete form of $I^g(x, y)$ is thus represented by known values at a set of equispaced points in the low-resolution image window, as shown in Fig. 4b. Wavelet-based interpolation is used for estimating the indicator functions $I_{wvlt}^{g'}(x, y)(\in \hat{\Re})$ in the high-resolution $p' \times q'$ pixel grid, where $0 \leq \hat{\Re} \leq 255$ is a space of real numbers. Gaussian functions with continuous derivatives are popular wavelets bases (Everson et al. 1990) and can effectively represent sharp variations in characteristic features of an image. The Gaussian function is of the form:

$$\Phi_{m,n,k,l}(x, y) = e^{-\frac{1}{2}\left(\frac{x-b_n}{a_m}\right)^2} e^{-\frac{1}{2}\left(\frac{y-d_l}{c_k}\right)^2} \beta_{m,n,k,l}. \tag{12}$$

Here, subscripts (m, k) refer to the wavelet level in a multiresolution wavelet representation and (n, l) correspond to discrete translation of the bases in the x and y directions, respectively. The parameters (b_n, d_l) correspond to translation, while (a_m, c_k) are dilation parameters. The level (m, k) Gaussian wavelet interpolated indicator function $I_{wvlt}^{g'}(x, y)$ is expressed in terms of data obtained from a $p \times q$ pixel subregion of a low-resolution 2D image, as:

$$I_{wvlt}^{g'}(x, y) = \sum_{1 \leq n \leq p} \sum_{1 \leq l \leq q} \Phi_{m,n,k,l}(x, y) = \sum_{1 \leq n \leq p} \sum_{1 \leq l \leq q} e^{-\frac{1}{2}\left(\frac{x-b_n}{a_m}\right)^2} e^{-\frac{1}{2}\left(\frac{y-d_l}{c_k}\right)^2} \beta_{m,n,k,l}. \tag{13}$$

This yields a continuous image representation in terms of discrete $I^g(x, y)$ values of the low-resolution image in Ω_w^{lrsm}. The bases are constructed by translation from one pixel to the next in the $p \times q$ pixel subregion and the region is encoded with $p \times q$ Gaussian functions. The wavelet coefficients $\beta_{m,n,k,l}$ in (13) can be obtained by solving the matrix equation

$$\{\mathbf{I}^g\} = [\mathbf{F}]\{\mathbf{B}\}, \tag{14}$$

where $\{\mathbf{I}^g\}$, $[\mathbf{F}]$, and $\{\mathbf{B}\}$ are matrices of order $pq \times 1$, $pq \times pq$, and $pq \times 1$, respectively. The matrix $\{\mathbf{I}^g\}$ contains the values of the indicator functions from the available $p \times q$ pixel data. The matrix $[\mathbf{F}]$ contains the terms $e^{-\frac{1}{2}\left(\frac{x_i - b_n}{a_m}\right)^2} e^{-\frac{1}{2}\left(\frac{y_j - d_l}{c_k}\right)^2}$, $1 \le i \le p$ and $1 \le j \le q$ in (13), while the matrix $\{\mathbf{B}\}$ contains the unknown wavelet coefficients.

Very large values of p and q can lead to numerical instabilities in the solution of coefficients due to nearly linearly dependent columns in $[\mathbf{F}]$. Numerical studies conducted in (Valiveti and Ghosh 2007; Ghosh et al. 2006) have indicated that the values for which the system is stable are around $p = 6, q = 6$. Consequently, the low-resolution image is subdivided into basic building blocks, each containing a maximum of 6×6 pixels. The interpolated indicator function $I^{g'}_{wvlt}(x, y)$ for each block of the window containing a $p' \times q'$ pixel grid is constructed as a piecewise continuous wavelet function given in (13). A local coordinate system (x, y) is set up in each block, with the origin located at its centroid. Consequently, the matrix $[\mathbf{F}]$ in (14) will be identical for all 6×6 dimension (pixel) blocks. Hence, it needs to be computed only once irrespective of the size of the image to be reconstructed.

The interpolation method is tested on a low-resolution window marked **A**, in the W319 material micrograph of Fig. 3. The dimensions of the complete micrograph is $880 \times 880\,\mu$m, while that of an image window to be processed is $110 \times 110\,\mu$m. A magnified image of the window marked **A** consists of a $(p =)60 \times (q =)60$ pixel grid. Consequently, there are 10×10 blocks containing 6×6 pixel grids each, in this image window. A high-resolution SEM micrograph at the same location is shown in Fig. 6a, which has a $(p' =)480 \times (q' =)480$ pixel grid. This corresponds to a 64-fold increase in the pixel density from the low-resolution image. The reconstructed

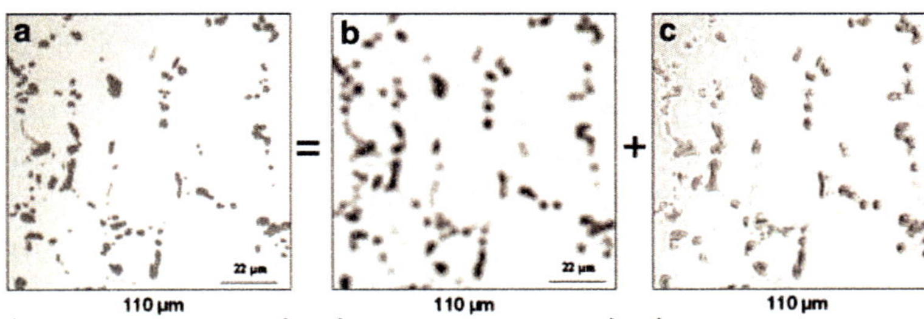

Fig. 6 High resolution micrographs at location A in Fig. 3: (**a**) actual high resolution micrograph; (**b**) micrograph obtained by wavelet based interpolation in the WIGE algorithm (**c**) difference micrograph between (**a**) and (**b**)

image will also have the same pixel density as the high-resolution image. The wavelet interpolated image on the 480 × 480 pixel grid is depicted in Fig. 6b. A pixel by pixel subtraction of $I^{g'}_{wvlt}(x, y)$ for image Fig. 6b from that of the high resolution image Fig. 6a is depicted in Fig. 6c. The difference micrograph clearly indicates that wavelet interpolation alone is not sufficient for reconstructing an accurate high-resolution microstructure. Subsequent image enhancement is essential.

2.3 *Gradient-based Probabilistic Enhancement of Interpolated Images in the WIGE Algorithm*

A probabilistic augmentation of the wavelet interpolated micrographs is developed in this section. A few high-resolution images at selected windows of the entire domain are chosen as "calibrating images" to generate correlation functions for the enhancement process. The method accounts for the location of these calibrating micrographs in relation to those being simulated in the overall domain.

The first step in this method is a pixel by pixel determination of the difference in the image indicator function values, between the high-resolution micrograph Ω^{hrsm}_{w} and the wavelet interpolated image Ω^{intm}_{w}. The difference indicator function in the higher pixel density grid is expressed as

$$I^{g'}_{diff}(x, y) = I^{g'}_{hrsm}(x, y) - I^{g'}_{wvlt}(x, y). \tag{15}$$

The corresponding difference micrograph Ω^{diff}_{w} for images in Figs. 6a,b is shown in Fig. 6c.

The augmentation methodology requires the creation of a correlation function between $I^{g'}_{diff}(x, y)$ and $I^{g'}_{wvlt}(x, y)$ that will predict the high-resolution image in Ω^{hrsm}_{w} from specific features of the interpolated image in Ω^{intm}_{w}. For an interpolated image, the pixel-wise grayscale indicator function (level) and its gradients in different directions are considered to be characteristic variables that adequately define the local material phase layout. At a point (x, y) in the pixel space, the value of $I^{g'}_{wvlt}(x, y)$ is an indicator of a given phase. In addition, gradients of $I^{g'}_{wvlt}(x, y)$ along opposite senses of two orthogonal directions, i.e., (x_+, x_-, y_+, y_-) are assumed represent the extent of a given phase. A discrete probability function that correlates these characteristic variables of the interpolated image Ω^{intm}_{w} and the indicator function value $I^{g'}_{diff}(x, y)$ of the difference image Ω^{diff}_{w} is first compiled at selected locations, where high-resolution calibrating micrographs are available. The functional form of this correlation is expressed in terms of the probability of occurrence of an indicator function and its gradients as

$$I^{g'}_{diff}(x, y) = P_{diff}\left(I^{g'}_{wvlt}(x, y), \frac{\partial I^{g'}_{wvlt}}{\partial x_+}, \frac{\partial I^{g'}_{wvlt}}{\partial x_-}, \frac{\partial I^{g'}_{wvlt}}{\partial y_-}, \frac{\partial I^{g'}_{wvlt}}{\partial y_+}\right). \tag{16}$$

Here, P_{diff} corresponds to the most probable value of the difference indicator function. The gradients are expressed as:

$$\frac{\partial I^{g'}_{wvlt}}{\partial x_\pm} \approx \frac{I^{g'}_{wvlt}(x \pm NG, y) - I^{g'}_{wvlt}(x, y)}{NG}$$
$$\frac{\partial I^{g'}_{wvlt}}{\partial y_\pm} \approx \frac{I^{g'}_{wvlt}(x, y \pm NG) - I^{g'}_{wvlt}(x, y)}{NG}, \qquad (17)$$

where NG (pixel offset value) is the number of pixels over which the gradient is approximated. The functional form of P_{diff} is not known a-priori. Hence, a *discrete probability table* is constructed from the calibrating micrographs Ω^{intm}_w and Ω^{diff}_w to construct this correlation map. A schematic of the probability table is shown in Fig. 7. The table is partitioned into discrete ranges or 'bins', based on ranges of $I^{g'}_{wvlt}$ and its gradients $\frac{\partial I^{g'}_{wvlt}}{\partial x_+}, \frac{\partial I^{g'}_{wvlt}}{\partial x_-}, \frac{\partial I^{g'}_{wvlt}}{\partial y_-}, \frac{\partial I^{g'}_{wvlt}}{\partial y_+}$ in Ω^{intm}_w in the four directions. The absolute values of gradients in each bin of the table are expressed as $|\ (I^{g'}_{wvlt})_{,x+}\ )\ |$, where $\left|(I^{g'}_{wvlt})_{,x+}\ )\right| = \left|\frac{\partial I^{g'}_{wvlt}}{\partial x_+}\right|$.

Each bin corresponds to a range of values for each of the five variables, and contains the values of the difference indicator function $I^{g'}_{diff}$ belonging to the image Ω^{diff}_w. The range of values to be assigned to each bin depends on the degree of variation of the variables. For example, the range $0 \leq I^{g'}_{wvlt} \leq 255$ can be divided into as high as 256 bins, with a single number in each bin, or as low as 2–3 bins. However with increasing number of bins, the number of $I^{g'}_{diff}$ entries in each bin will decrease and many of the bins may be empty for the calibrating micrographs considered. Data sparsity in any of the correlation bins renders the reliability of this probability table to be low. Numerical analysis-based convergence studies, discussed later in this chapter, have corroborated the sufficiency of a moderate number of divisions (∼10–15).

The range of divisions in the gradients should be such that they are able to distinguish between regions that belong to the interior and exterior of a given phase. A histogram of area fraction of the gradient $\frac{\partial I^{g'}_{wvlt}}{\partial x_+}$ for different pixel offset values NG in the image Fig. 6b is plotted in Fig. 8. Lower values of NG yield a better distribution of the gradients and hence a value of $NG = 1$ is used in this work. Additionally, the area fractions beyond the bounds $|\frac{\partial I^{g'}_{wvlt}}{\partial x_+}| \leq 8$ correspond to the second-phase particles in the microstructure. This conforms to the requirement that the algorithm should delineate regions within the second-phase particles from those outside. Consequently, the range of gradient values is separated into two groups: (a) $|\frac{\partial I^{g'}_{wvlt}}{\partial x_+}| \leq 8$ and (b) $|\frac{\partial I^{g'}_{wvlt}}{\partial x_+}| > 8$. The same division is applicable to gradients in the y direction too. Figure 7c shows the probability table with the discretized ranges of indicator function and its gradients. At a given pixel (x, y) in Ω^{intm}_w of Fig. 7a, $I^{g'}_{wvlt} = 105$, $\frac{\partial I^{g'}_{wvlt}}{\partial x_+} = 5.0$, $\frac{\partial I^{g'}_{wvlt}}{\partial x_-} = 4.0$, $\frac{\partial I^{g'}_{wvlt}}{\partial y_+} = 3.5$, *and* $\frac{\partial I^{g'}_{wvlt}}{\partial y_-} = 6.5$.

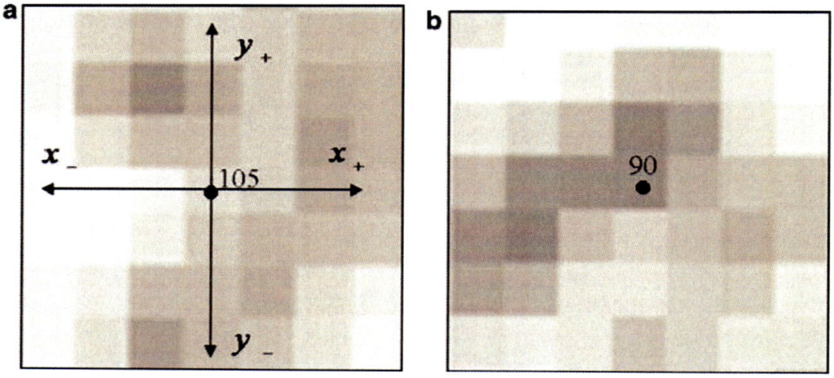

c

$I^{g'}_{wvlt}$	0-25	25-50	50-100	100-125	125-150	150-175	...	225-255
$\|(I^{g'}_{wvlt})_{,x+}\|$: 0-8 $\|(I^{g'}_{wvlt})_{,x-}\|$: 0-8 $\|(I^{g'}_{wvlt})_{,y+}\|$: 0-8 $\|(I^{g'}_{wvlt})_{,y-}\|$: 0-8				$I^{g'}_{diff}=90$				
$\|(I^{g'}_{wvlt})_{,x+}\|$: >8 $\|(I^{g'}_{wvlt})_{,x-}\|$: 0-8 $\|(I^{g'}_{wvlt})_{,y+}\|$: 0-8 $\|(I^{g'}_{wvlt})_{,y-}\|$: 0-8								
...								
$\|(I^{g'}_{wvlt})_{,x+}\|$: >8 $\|(I^{g'}_{wvlt})_{,x-}\|$: >8 $\|(I^{g'}_{wvlt})_{,y+}\|$: >8 $\|(I^{g'}_{wvlt})_{,y-}\|$: >8								

Fig. 7 Method of correlating interpolated and difference micrographs: (**a**) an interpolated region, (**b**) corresponding difference region, and (**c**) table with bins correlating the interpolated micrograph with the difference micrograph

The corresponding $I^{g'}_{diff} = 90$ is entered in the probability table Fig. 7c. The values of $I^{g'}_{diff}$ in the correlation bins vary from location to location for different samples. A histogram of the distribution of $I^{g'}_{diff}$ for a given bin corresponding to $\left(0 \leq I^{g'}_{wvlt} < 25,\ |\frac{\partial I^{g'}_{wvlt}}{\partial x_+}| \leq 8,\ |\frac{\partial I^{g'}_{wvlt}}{\partial x_-}| \leq 8,\ |\frac{\partial I^{g'}_{wvlt}}{\partial y_+}| \leq 8,\ |\frac{\partial I^{g'}_{wvlt}}{\partial y_-}| \leq 8\right)$ is shown in Fig. 9. Peaks in the histogram associate a high probability value of $I^{g'}_{diff}$ with a particular bin in the correlation table. This corresponds to the expected value of $I^{g'}_{diff} = P_{diff}\left(I^{g'}_{wvlt}, \frac{\partial I^{g'}_{wvlt}}{\partial x_+}, \cdots\right)$ that is selected for image enhancement of $I^{g'}_{wvlt}$ according to (15).

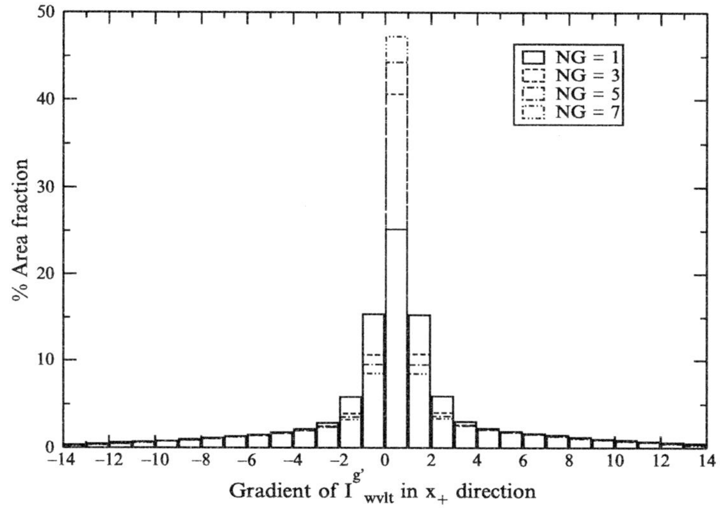

Fig. 8 Distribution histogram of the indicator function gradient $\frac{\partial I^{g'}_{wvlt}}{\partial x_+}$ in the interpolated image of Fig. 6b for various values of NG

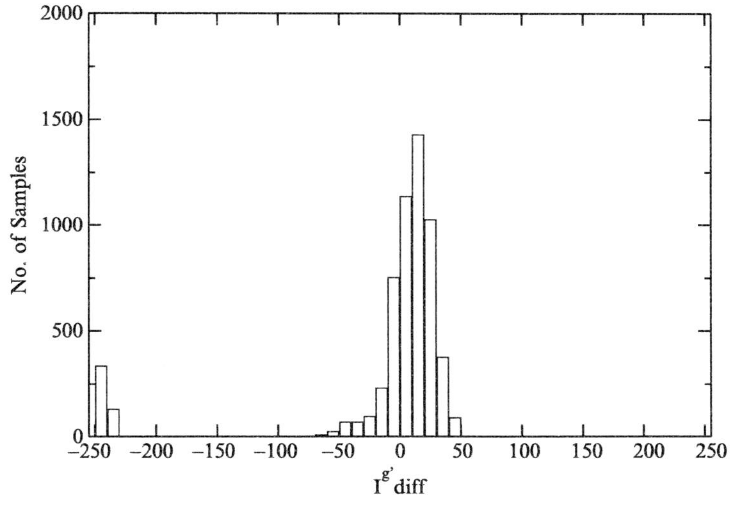

Fig. 9 A histogram of the distribution of $I^{g'}_{diff}$ in the difference image for a given bin corresponding to $\left(0 \leq I^{g'}_{wvlt} < 25, |\frac{\partial I^{g'}_{wvlt}}{\partial x_+}| \leq 8, |\frac{\partial I^{g'}_{wvlt}}{\partial x_-}| \leq 8, |\frac{\partial I^{g'}_{wvlt}}{\partial y_+}| \leq 8, |\frac{\partial I^{g'}_{wvlt}}{\partial y_-}| \leq 8\right)$

2.3.1 Accounting for Relative Locations of the Calibrating and Simulated Micrographs

For multiphase microstructures, the location of high-resolution calibrating micrographs in relation to the image being simulated is of considerable importance to the image augmentation process. A major assumption made is that if the calibrating micrographs contain the same constituent phases as the ones being simulated and if they are all produced by the same manufacturing process, the probability functions (P_{diff}) of local microstructural distributions will have a continuous variation across the micrographs, i.e., they are homogeneous. This similarity in the probability of local distributions is necessary for the calibration and augmentation processes to hold. For microregions with sharp contrast, the calibrating micrographs should belong to those regions that represent the essential features of the one being simulated. The effect of the proximity between calibrating and simulated images can be addressed by assigning distance-based weights to the expected values P_{diff} in the probability table. Micrographs closer to the simulated image will have a stronger influence than those farther away. The inverse dependence of a microstructure's correlation map on its spatial distance from each of the calibrating micrographs is represented by a 'shape function' type interpolation relation, commonly used in finite element analysis, i.e.:

$$P_{diff}(x, y) = \sum_{\alpha} N_{\alpha=1}^{\hat{\alpha}}(x, y) P_{diff}(x_\alpha, y_\alpha), \tag{18}$$

where $\hat{\alpha}$ is the total number of high-resolution calibration micrographs and N_α are the associated shape functions. When only two calibrating micrographs A and B are available as in Fig. 3, the most expected value of the enhancement $I_{diff}^{g'}(x, y)$ at a pixel in the simulated micro-image is obtained as

$$I_{diff}^{g'}(x, y) = P_{diff}(x, y) = \left(\frac{1-\xi}{2}\right) P_{diff}(x_A, y_A) + \left(\frac{1+\xi}{2}\right) P_{diff}(x_B, y_B). \tag{19}$$

Here, $\xi = \left(\frac{R_A - R_B}{R_{AB}}\right)$ and R_A and R_B are the distances of a pixel in the simulated image from the corresponding pixels in calibrating micrographs A and B, respectively, and R_{AB} is the distance between them. For microstructures containing a single predominant second-phase in the matrix, e.g., Si for cast aluminum alloys, the different locations e.g., **A** and **B** may have statistically equivalent expected values in the probability table of Fig. 7c. In this case, the effect of multiple locations in (18) will be minimal.

2.3.2 A Validation Test for the WIGE Algorithm

The effectiveness and convergence properties of the WIGE algorithm are tested by comparing characteristic metrics of the simulated microstructure with those for a real micrograph at the same location. The n-point statistics has been developed in

(Torquato 2002) as effective metrics for multiphase microstructure characterization. In this work, the 1-point, 2-point, and 3-point statistics are used for validation of the WIGE algorithm. For the low-resolution microstructural region of Fig. 3, high-resolution calibration micrographs are available for the windows at locations **A** and **B**. The WIGE algorithm is used to simulate a high-resolution image of the micrograph at the window **C**. A high-resolution SEM micrograph is available for this window **C** that can be used for validation. The 1-point probability function corresponds to the area fraction of the second phase particles in the micrograph. Its variation is plotted in Fig. 10 as a function of increasing number of divisions in the range of $I^{g'}_{wvlt}$, or bins in the probability table. The value at 0 bins corresponds to the micrograph with no enhancement. The simulated area fraction converges to the SEM image area fraction with about 10 discrete divisions or bins. The 2-point probability function is defined as the probability of finding two points at $r_1(x_1, y_1)$ and $r_2(x_2, y_2)$ (end-points of a line), separated by a distance $r = r_1 - r_2$, in the same phase in the microstructure, i.e.,

$$P_{ij}(r) = P\{I^b(x_1, y_1) = 1, I^b(x_2, y_2) = 1\}, \tag{20}$$

where $I^b = 1$ in the given phase and $I^b = 0$ otherwise. The % error in the 2-point probability function between the actual and simulated images is defined as

$$E_{2-point} = \frac{\sum_{r=0}^{r=L/2} |P_{ij}^{SEM} - P_{ij}^{sim}|}{\sum_{r=0}^{r=L/2} P_{ij}^{SEM}} \times 100\%. \tag{21}$$

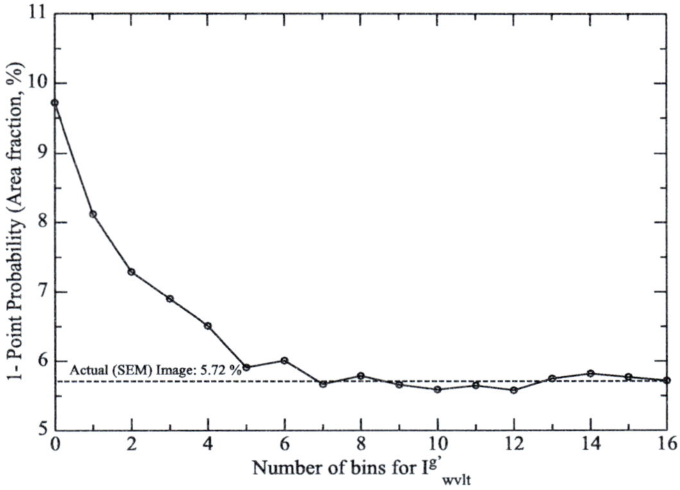

Fig. 10 Convergence of 1-point probability function with increasing number of divisions in the range of $I^{g'}_{wvlt}$ or bins for the simulated micrograph at region C of Fig. 3 by the WIGE algorithm

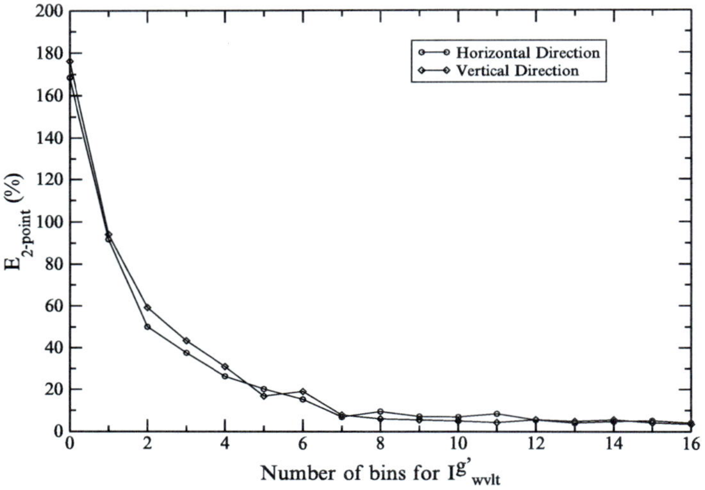

Fig. 11 Convergence of 2-point probability function with increasing number of divisions in the range of $I^{g'}_{wvlt}$ or bins for the simulated micrograph at region C of Fig. 3 by the WIGE algorithm

This error is evaluated along two orthogonal directions and plotted in Fig. 11. Once again, the convergence is fast and the error stabilizes to a near zero value for around 10 bins. Finally, the 3-point probability function is defined as the probability of finding three points at $r_1(x_1, y_1)$, $r_2(x_2, y_2)$ and $r_3(x_3, y_3)$ (vertices of a triangle) in the same phase, i.e.,

$$P_{ijk}(r) = P\{I^b(x_1, y_1) = 1, I^b(x_2, y_2) = 1, I^b(x_3, y_3) = 1\}. \qquad (22)$$

$P_{ijk}(r)$ is evaluated for three points at the vertices of an isosceles right triangle with interior angles 45°, 45°, 90°. The error in the 3-point probability function is defined in the same way as in (21) and is plotted in Fig. 12. The error in the 3-point probability function also stabilizes to near zero values for ~10 bins.

The lower order statistics provide information on phase dispersion and are relevant in domain partitioning. Higher order statistics like the 3-point probability function are important with respect to phase shapes that control the localization and damage behavior of the material. In conclusion, the convergence characteristics of the probability function enhanced WIGE algorithm are found to be quite satisfactory with respect to 1-point, 2-point, and 3-point correlation functions. Excellent agreement is seen in the WIGE simulated microstructural image and the corresponding actual micrograph shown in Fig. 13a,b. Thus, this method can be applied in a frame by frame sequence to all windows in the computational domain for obtaining high-resolution images. Corresponding to the resolution ratio between the high and low-resolution images, it can be concluded that a 64 fold scale up in the resolution is achieved by the WIGE algorithm in this study.

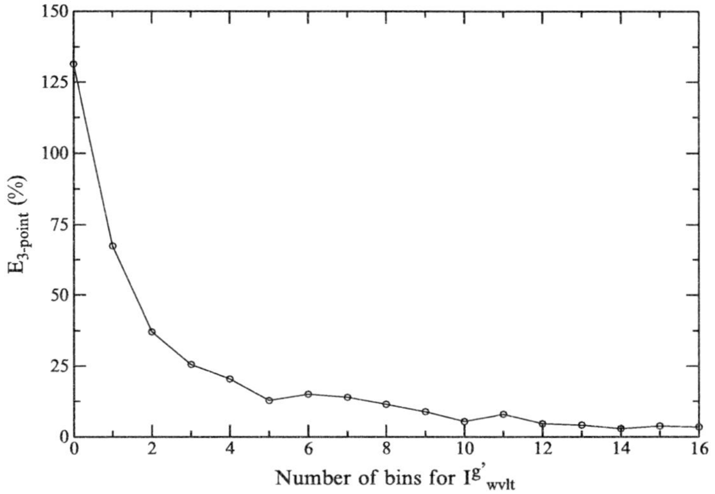

Fig. 12 Convergence of 3-point probability function with increasing number of divisions in the range of $I^{g'}_{wvlt}$ or bins for the simulated micrograph at region C of Fig. 3 by the WIGE algorithm

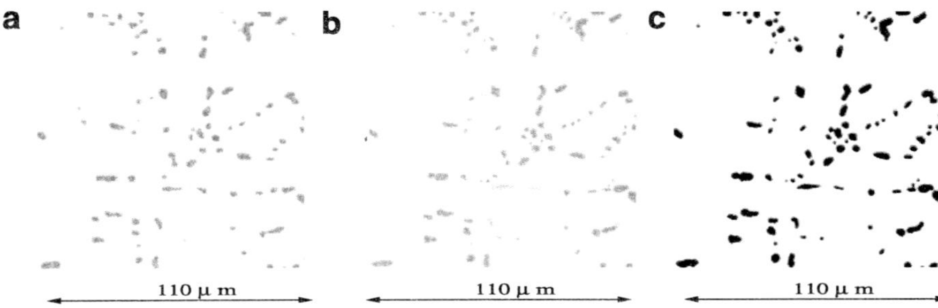

Fig. 13 High-resolution micrograph at location C of Fig. 3 by the WIGE algorithm: (**a**) simulated micrograph by using the correlation table, (**b**) the real-high resolution micrograph, and (**c**) binary high-resolution micrograph

3 Binary Image Processing for Noise Filtering

Prior to characterizing the simulated microstructure, it is necessary to process these images to eliminate noise and clearly delineate dominant phases. Hierarchy in the grayscale levels of digital images may be used for such image processing. During phase delineation, the indicator function values $I^g(x, y)$ of all pixels belonging to a given phase are assumed to fall within a narrow band of grayscale levels. Global thresholding is first conducted to enable phase delineation or segmentation in the micrographs. In global thresholding, $I^g(x, y)$ for the entire image is binarized with respect to a single threshold value. On the other hand, different values may be used in local thresholding based on the local variation of $I^g(x, y)$. The latter is necessary for those micrographs, where the same phase may have large differences in grayscale level representations at different locations. Global thresholding is deemed sufficient in this work, since the range of grayscale levels of each phase is assumed to have a narrow bandwidth. In a perfect image belonging to two distinct grayscale levels, global thresholding will yield a bimodal histogram of the percentage of pixels as a function of the grayscale levels. Two distinct peaks exist for such a bimodal histogram and the threshold value of $I^g(x, y)$ corresponds to the valley point between the peaks. However, in real images such as in Fig. 13, histograms are rarely bimodal. Various techniques have been suggested for evaluating the threshold value τ for histograms, in which distinct peaks are absent (Sahoo et al. 1988; Luthon et al. 2004). A simple technique is to evaluate τ from the shoulder region in the histogram, adjacent to the peak for the matrix phase that has a zero slope. The image can be binarized with respect to the indicator function value as:

$$\begin{aligned} I^b(x, y) &= 1 \quad \forall \quad 0 \le I^g(x, y) \le \tau \\ &= 0 \quad \forall \quad \tau < I^g(x, y) \le 2^\xi - 1. \end{aligned} \tag{23}$$

Heterogeneities, e.g., particles or voids, are consequently converted to a black image against a white matrix backdrop. For the W319 micrograph of Fig. 13b, a threshold value of $\tau = 225$ is obtained from the histogram in Fig. 14. The corresponding binary (black and white) image of the microstructure is shown in Fig. 13c. For multiphase microstructures with more than two phases, more than one threshold is necessary to separate different phases. The image processing algorithm in

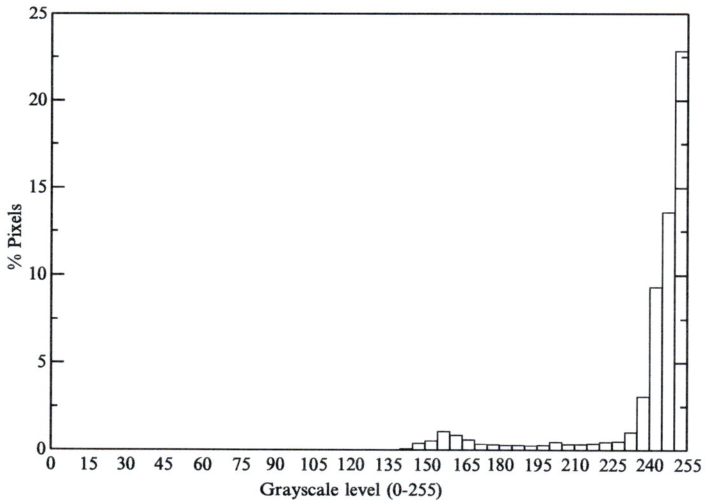

Fig. 14 Brightness histogram of high-resolution micrograph in Fig. 13b

this work does not make a distinction between Si particles and intermetallics in cast W319. Both phases are treated simply as inclusions in the matrix with a focus on their morphology.

Frequently, micrographs have significant noise due to tiny erroneous marks. The corresponding indicator functions $I^g(x, y)$ get transferred to the binary image indicators $I^b(x, y)$ based on their grayscale value. To prevent this, $I^g(x, y)$ is convoluted with a mean filter of mask size n pixels as shown in (Russ 1999). The process replaces each pixel at (x, y) with its respective local average grayscale level. The binary image also often contains tiny speckles due to thresholding. These speckles are unwanted noise and the binary image should be despeckled using a median filter on a kernel of mask size m pixels. The de-noising kernels help automation of the whole process without any user intervention. The binary domain represents a high-resolution computational domain Ω_{mic} necessary for microstructural characterization and analysis.

4 Functions for Microstructure Characterization

The simulated microstructure contains regions belonging to the matrix phase $\{\Xi_m\}$ and the N_c heterogeneities, represented as $\{\Xi_c^i : 1 \leq i \leq N_c\}$, i.e., $\Omega_{mic} = \Xi_m + \sum_{i=1}^{N_c} \Xi_c^i$. Characterization functions of microstructural parameters that have direct relevance to deformation and failure response of the material, e.g., those identified in (Wang et al. 2003; Wang 2003; Caceres et al. 1996), are developed. For instance, damage in cast W319 occurs by a combination of particle cracking, microcrack formation and growth in the matrix, and coalescence of microcracks. Particle cracking depends on size, aspect ratio, and clustering. Bigger particles with high aspect ratio or those within a cluster show a higher propensity toward cracking. Parameter descriptors and characterization functions are selected to quantify the size or shape of heterogeneities Ξ_c^i and their spatial distribution in Ω_{mic}.

4.1 Size Descriptors

Descriptors of area, perimeter, and longest diameter of heterogeneities are evaluated from binary image data $I^b(x, y)$ in the microstructural image Ω_{mic}, following their definitions.

1. *Area* (A^i) is measured in terms of the total number of pixels belonging to a heterogeneity Ξ_c^i:

$$A^i = \int_{\Xi_c^i} I_i^b(x, y)\, dx\, dy \tag{24}$$

The local area fraction $A_f^i = \frac{A^i}{A_{\Omega_{mic}}}$ is a more effective descriptor.

2. *Perimeter* (P^i) is measured by the number of pixel-edges in Ξ_c^i that interface with the matrix Ξ^m in a digital microstructure.
3. *Longest dimension* (d_{max}^i) is measured as the distance between the two farthest points in Ξ_c^i and is measured as:

$$d_{max}^i = Max\{|\bar{r}_{AB}| \quad \forall\ A(x_a, y_a), B(x_b, y_b) \in \Xi_c^i\} \tag{25}$$

4.2 Shape Descriptors

Shape descriptors, such as those prescribed in (Russ 1999; Seul et al. 2000), quantify the shape and surface irregularities of a heterogeneity Ξ_c^i. The following shape metrics are used for quantifying heterogeneous microstructures such as cast aluminum alloys.

1. *Roundness* (ρ^i) indicates how close the shape is to a circle. It is effective for arbitrary shapes for which the aspect ratio is not well defined, and is expressed as:

$$\rho^i = \frac{4A^i}{\pi (d_{max}^i)^2} \tag{26}$$

ρ^i varies from 1 for circular shapes to 0 for highly elongated phases.
2. *Edge smoothness* (χ^i) describes surface irregularities, e.g., sharp corners, even in the case of high overall roundness. Form factor is a metric that is defined in (Russ 1999) to delineate surface irregularities as:

$$ff^i = \frac{4\pi A^i}{(P^i)^2} \tag{27}$$

ff^i is sensitive to surface irregularities and varies from 1 for smooth surfaces to 0 for rough surfaces. It is also affected by the aspect ratio. To understand their effectiveness, ρ^i and ff^i of arbitrary shapes created in Fig. 15 are provided in Table 1. Although ff^i for shapes 8–11 captures the visible surface irregularities, it is very low for shapes 4–7, with smooth surfaces. The edge smoothness χ^i is consequently introduced to capture the surface irregularities by deemphasizing the aspect ratio.

$$\chi^i = \sqrt{ff^i \times \frac{\pi d_{max}^i}{(P^i)}} = \sqrt{\frac{4\pi^2 A^i d_{max}^i}{(P^i)^3}}. \tag{28}$$

In χ^i, the form factor is amplified by the effect of the largest dimension in the heterogeneity for better representation of surface irregularities. Also, the square root helps to create a better separation between different geometries, since the parameter is generally less than 1. χ closer to 0 indicates a large number of surface irregularities. However with the discrete pixel representation of boundaries in a digital image, χ can be closer to unity even for a perfect circle.

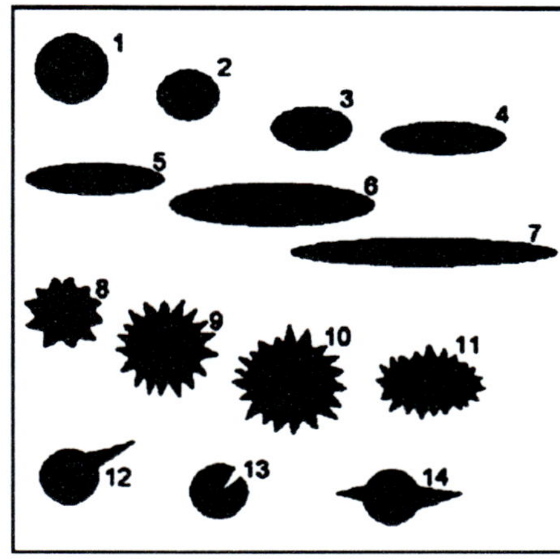

Fig. 15 Image with simulated heterogeneities for testing the shape description parameters

Table 1 Shape description parameters for image with simulated particles in Fig. 15

Particle No.	% Area Frtn.(A_f)	Roundness(ρ)	Form factor(ff)	Edge Smo.(χ)
1	1.25	0.9698	0.9909	0.6928
2	0.79	0.8431	0.8631	0.7069
3	0.87	0.5540	0.5559	0.7436
4	1.04	0.2776	0.2776	0.7115
5	1.08	0.2324	0.2325	0.6880
6	2.27	0.2179	0.2179	0.6796
7	1.87	0.1056	0.1073	0.5512
8	1.12	0.6848	0.7317	0.4318
9	1.89	0.6857	0.7193	0.2453
10	2.37	0.6605	0.7094	0.2675
11	1.58	0.5223	0.5495	0.3337
12	0.96	0.3471	0.3655	0.5806
13	0.75	0.9195	0.9274	0.4947
14	1.05	0.2508	0.2694	0.5749

4.3 *Spatial Distribution Descriptors*

Spatial distribution manifests the relative location of heterogeneities Ξ_c^i in the matrix Ξ_m. It can be quantified by spatial characterization functions that identify geometric properties such as isotropy, homogeneity, and clustering. A wide variety of spatial techniques exist in the literature. These include the Voronoi tessellation-based techniques in (Spitzig et al. 1985; Everett and Chu 1992; Li et al. 1999b; Ghosh et al. 1997a) for determining probability density functions, pair distribution functions for nearest-neighbor distances, local area fractions, etc., and the image-based characterization methods in (Yotte et al. 2004; Karnezis et al. 1998; Anson and Gruzleski 1999) for evaluating mean free path, nearest neighbor distance and for detecting clusters. The covariance function (Serra 1982) and other explicit descriptors are discussed here as morphological tools.

4.3.1 Covariance Function

The covariance function $K(h)$ is defined in (Serra 1982) as the Lebesgue measure of a deterministic compact set **X** in R^n that is eroded by **B** ={0,h}, a set of points at the ends of a vector **OH**. For microstructural images, **X** corresponds to the set of all the points that belong to the heterogeneities and **B** is a structural element consisting of two end pixels separated by a distance h and making an angle (α) with the reference axis. For $\alpha = 0°$ and $\alpha = 90°$ in the binary digital images, the function can be expressed as

$$K(h)_{(\alpha=0°)} = Mes(\mathbf{X} \ominus \mathbf{B}) = \int_{\mathbf{R^n}} I^b(x,y) \cdot I^b(x+h,y)\,\mathrm{d}x\,\mathrm{d}y$$

$$K(h)_{(\alpha=90°)} = \int_{\mathbf{R^n}} I^b(x,y) \cdot I^b(x,y+h)\,\mathrm{d}x\,\mathrm{d}y. \qquad (29)$$

where $I^b(x,y)$ (see (23)) is the binary indicator function of the image associated with the set **X**, and $\mathbf{X} \ominus \mathbf{B}$ indicates the erosion of set **X** by the element **B**. The set **X** in the binary microstructure is defined as $\mathbf{X} = \sum_{i=1}^{N_c} \Xi_c^i = \Omega_{mic} - \Xi_m$. The eroded set $\mathbf{X} \ominus \mathbf{B}$ may be expressed as $\mathbf{X} \cap \mathbf{X_h}$, where $\mathbf{X_h}$ is a translated set of **X**. For any point x,

$$x \in (\mathbf{X} \ominus \mathbf{B}) \quad \text{iff} \quad x, x+h \in \mathbf{X}. \qquad (30)$$

For $I^b(x,y) = 1$, $K(h)$ denotes the total number of events for which pixel points (x,y) and $(x+h,y)$ both belong to the second-phase particle region. Computationally, it is evaluated as the number of particle pixels that overlap when the image is translated by a distance h at an angle α to the reference direction, and overlaid on itself. The covariance function $K(h)$, normalized by the total number of pixels in the micrograph Fig. 13c, is plotted in Fig. 16 for $\alpha = 0°$ and $\alpha = 90°$. The plots capture the average properties at shorter translations ($\leq h$) as well as the behavior of the spatial distribution at larger translations. For smaller values of h, $K(h)$ corresponds to the intersection of a particle with its own translated image. Consequently, it decreases rapidly with h for decreasing self overlay. The small increase in $K(h)$ at higher values of h refers to the intersection with neighbors. Hence, the average nearest neighbor distance l_{nnd} of a micrograph corresponds to the smallest value of h at which $K(h)$ is a local minimum. The first local minima of $K(h)$ in Fig. 16 occur at 9.6μm for $\alpha = 0°$ and 8.5μm for $\alpha = 90°$. These correspond to the average nearest neighbor distances in the two orthogonal directions.

4.3.2 Cluster Index

Clustering manifests a high local density of heterogeneities that often leads to local stress concentrations under mechanical loading. The W319 micrograph in Fig. 3

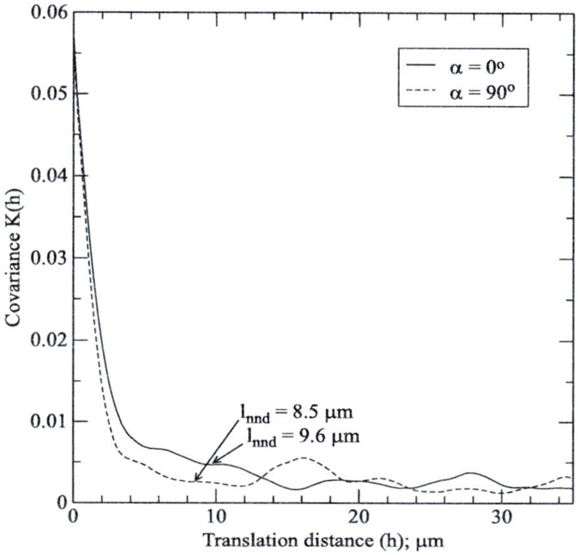

Fig. 16 Covariance(k(h)) plot for micrograph in Fig. 13c along two orthogonal directions

contains large regions of Al matrix surrounded by the silicon particles. This structure is formed by solidification, where the growing aluminum dendrites force the silicon particles into the spaces between dendritic arms. Particulate clustering in solidified aluminum alloy microstructures is thus related to the secondary dendrite arm spacing or SDAS, as well as the number of particles around a dendrite arm. The SDAS, measured as the average center to center distance of the dendrite arms, provides only a general idea about clustering (e.g., regions of higher SDAS have a higher degree of clustering) without any specific local information. For a better quantification of local clustering, two metrics, viz. the *Spacing Index (SI)* and the *Clustering Intensity (CI)* are introduced. These metrics quantify the size of the matrix that is free of second-phase particles and the number of particles concentrated in a particular region. These parameters are normalized with respect to a characteristic radius R_{ch}, defined as

$$R_{ch} = \sqrt{\frac{A_{image}}{\pi N}}, \tag{31}$$

where A_{image} is the image area and N is the total number of particles. R_{ch} signifies the interparticle distance for an ideal distribution of circular particles. *SI* is a measure of the dendrite arm size, which is estimated as the normalized radius of the biggest circle that can fit into the micrograph without intersecting any particles. However, as shown in Fig. 17, stray tiny particles in the matrix region can result in lower than reasonable values of the arm size.

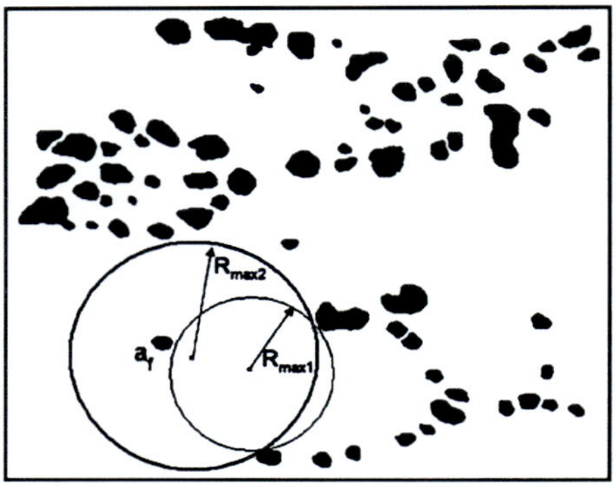

Fig. 17 Microstructure showing regions that have pockets of few second-phase particles in large matrix regions

To prevent this error, the spacing radius is evaluated beyond the first interfering particle to check whether the radius increases drastically (at least 25%). In that event, a new radius is used with an adjustment factor for enclosed particles, i.e.,

$$S_{ind} = \frac{R_{\max}(1 - a_f)}{R_{ch}}, \tag{32}$$

where a_f is the area fraction of the interfering particle. *CI*, on the other hand, quantifies the intensity of packing in a cluster. It is measured as the normalized difference between the maximum and minimum number of particles enclosed within a characteristic circle with radius R_{ch} (defined in (31)), i.e.,

$$CI = \frac{N^e_{\max}(x_{\max}, y_{\max}) - N^e_{\min}(x_{\min}, y_{\min})}{N_{\text{avg}}}, \tag{33}$$

where $N^e_{\max}$ and $N^e_{\min}$ are the maximum and minimum number of particles inside the characteristic circle at points $(x_{\max}, y_{\max})$ and $(x_{\min}, y_{\min})$, respectively, and N_{avg} is the average number of particles inside the characteristic circle over all points of the micrograph. Finally, the *Cluster Index* (κ), quantifying clustering in a microstructure, is defined as the product of spacing index *SI* and the clustering intensity *CI*, i.e.:

$$\kappa = SI * CI. \tag{34}$$

The effectiveness of κ in quantifying spatial distribution is demonstrated later.

4.3.3 Cluster Contour

Contour plots of parameters that represent local clustering are also helpful in identifying clusters. Such a contour plot can be generated using the characteristic radius R_{ch} as the field of influence of each heterogeneity. The total area of heterogeneities inside of each characteristic circle is measured as contour intensity (*COIN*) at a point. The cluster contour index ι is defined in terms of the contour intensity as:

$$\iota = 1 - \frac{Mean\ COIN}{Max\ COIN}. \tag{35}$$

The mean and maximum values of *COIN* are evaluated from all points of the micrograph. A contour index $\iota = 1.0$ denotes a cluster, while values closer to zero indicate uniform distribution. The contour index accounts for the area fraction of particles within a prescribed region while the cluster index considers the number of particles in this region. The microstructural descriptors can all be used to quantify the morphology of second-phase particles in a microstructure as discussed next.

4.4 Characterization of the W319 Microstructure

The tensile strength and ductility of the aluminum alloy W319 have been found to increase with decreasing SDAS in (Boileau 2000). The microstructure of the W319 alloy is characterized with respect to various size, shape, and distribution parameters prior to microstructural modeling to understand their effect on material behavior and failure response. High-resolution W319 micrographs at different SDAS values are characterized in this section to determine the sensitivity of the parameters and functions described in Sect. 4.

High-resolution, high-magnification micrographs of W319 with average SDAS values 23, 70, and 100μm are shown in Fig. 18a,b, and c respectively. The different size, shape, and clustering of second-phase particles are evident from these figures. The length scale of the high-resolution micrograph is important when comparing the microstructures of different SDAS. At lower length scales, e.g., with a micrograph size of ~100μm, the 23μm SDAS microstructure exhibits a clear delineation of particles, while the higher SDAS microstructures may not even contain any particles. At higher length scales of ~ 500μm, the resolution diminishes with a loss of feature clarity and hence a micrograph length scale of 220μm is adopted in this study. Size,

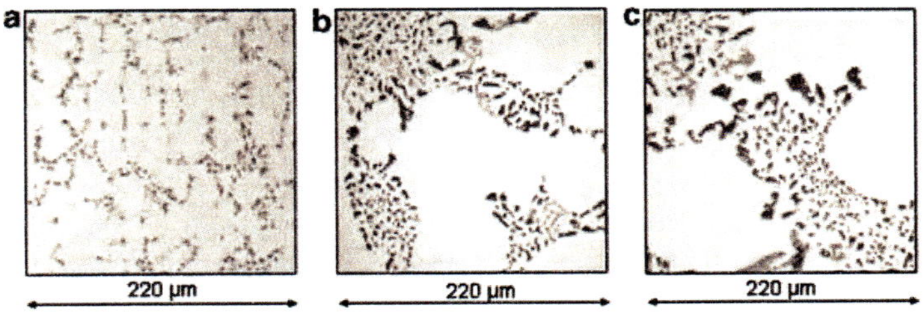

Fig. 18 W319 micrographs at various *SDAS* values. (**a**) *SDAS* = 23 μm, (**b**) *SDAS* = 70μm, (**c**) *SDAS* = 100μm

Table 2 Microstructure characterization parameters for the W319 alloys with different *SDAS* values, shown in Fig. 18

Parameter	23μm SDAS	70μm SDAS	100μm SDAS
Total A_f	6.90%	10.0%	11.0%
Min. Roundness(ρ)	0.285	0.120	0.145
Avg. Roundness(ρ)	0.720	0.476	0.486
Min. Edge smo.(χ)	0.369	0.139	0.281
Avg. Edge smo.(χ)	0.649	0.597	0.583
Cluster Index(κ)	14.92	19.12	23.35
Contour Index(ι)	0.81	0.83	0.84

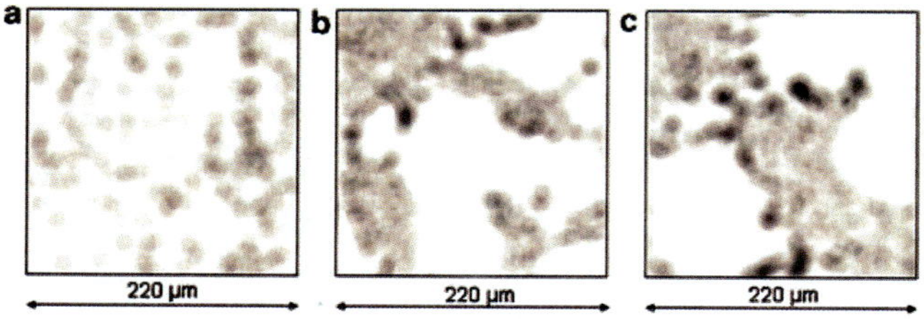

Fig. 19 Cluster contour plots of W319 micrographs shown in Fig. 18: **(a)** *SDAS* $= 23\mu$m. **(b)** *SDAS* $= 70\mu$m and **(c)** *SDAS* $= 100\mu$m

shape, and clustering parameters for the microstructures in Fig. 18 are tabulated in Table 2. The total area fraction of the combined silicon particles and intermetallics is found to increase with SDAS size. The decreasing roundness and edge smoothness with increasing SDAS, contributed by both silicon particles and intermetallics, capture the acicular particles in the higher SDAS material. The cluster index (κ) and contour index(ι) increase with SDAS, revealing higher particle density at higher SDAS. Cluster contour plots are shown in Fig. 19. These point to the higher variation of particle distribution with increasing SDAS value. The covariance plot of these micrographs at $\alpha = 0°$ in Fig. 20 shows that there is very little difference in the average nearest neighbor distance L_{nnd}, despite higher levels of clustering at higher SDAS.

4.5 Identification of Effective Spatial Distribution Descriptors

Multiscale characterization-based domain partitioning will require effective microstructure descriptors and characterization functions that can establish the relation between morphology and critical material response. Micromechanical damage analyses are conducted in this section for different simulated microstructures and the effectiveness of the spatial distribution functions is studied for their incorporation in domain partitioning criteria discussed in the subsequent sections.

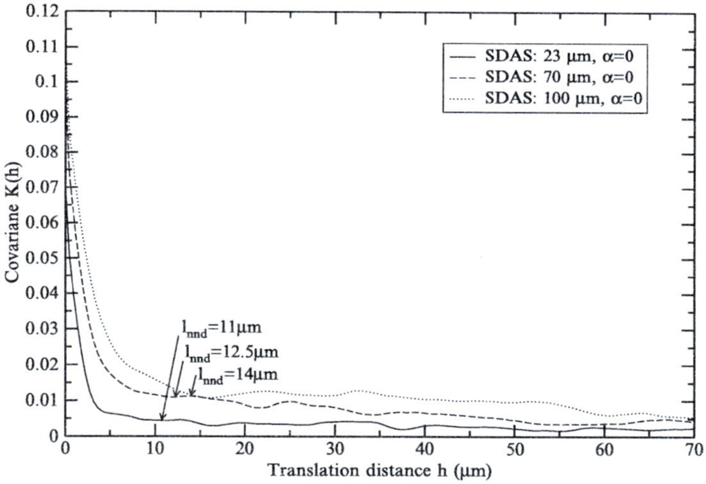

Fig. 20 Covariance function plots for W319 at various SDAS of Fig. 18

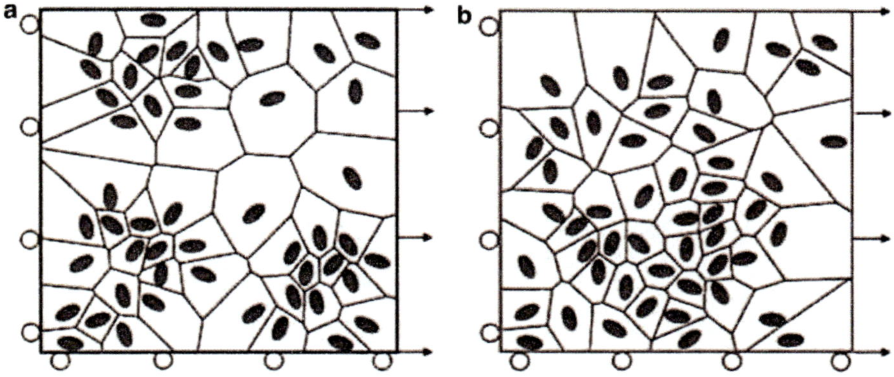

Fig. 21 VCFEM mesh showing the loading for simulated microstructures for identifying effective spatial distribution parameters; (**a**) with three small clusters, and (**b**) with one large cluster

Two micro-regions of 10% area fraction and containing 50 identical elliptical particles are simulated, as shown in Fig. 21a (three small clusters) and Fig. 21b (one large cluster).

The micromechanical analysis is performed with the Voronoi cell finite element model (*VCFEM*) (Ghosh et al. 2000; Li et al. 1999a; Li and Ghosh 2006; Hu and Ghosh 2008) for elastic-plastic deformation and damage by particle cracking. The particles are brittle with linear elastic material properties, while the matrix is assumed to be ductile and is modeled by J_2 plasticity theory with isotropic hardening. For each micrograph, the number of particles that cracked at 2% applied tensile

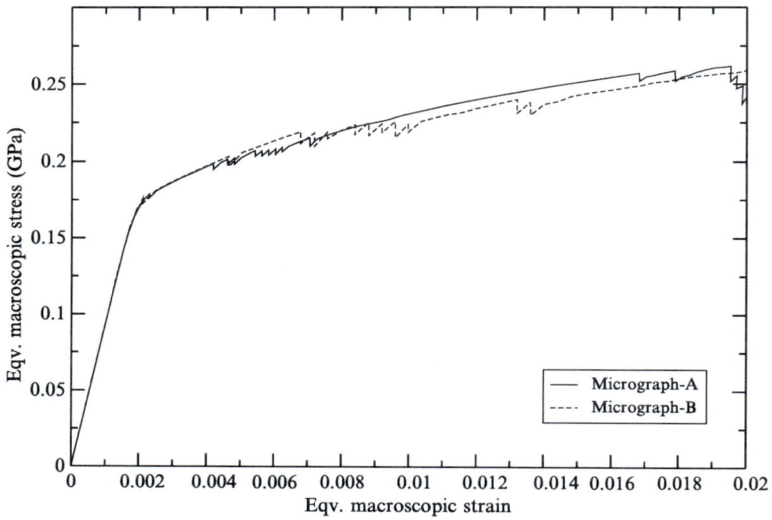

Fig. 22 Macroscopic stress-strain response for simulated micrographs of Fig. 21

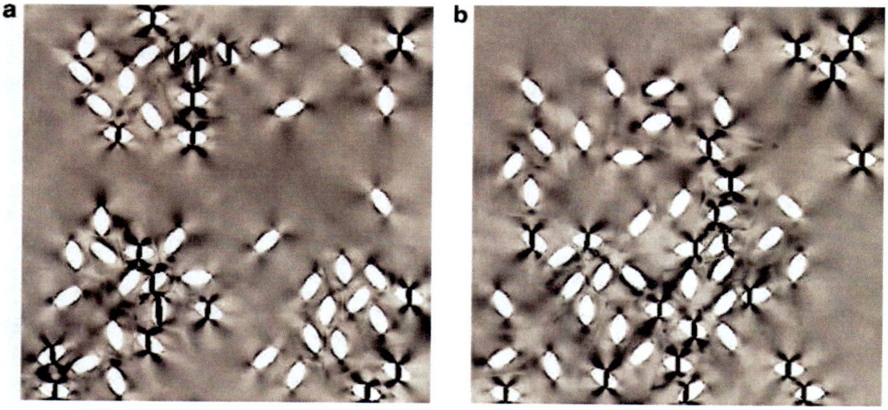

Fig. 23 Equivalent plastic strain and particle cracking for simulated microstructures of Fig. 21; (**a**) with three small clusters, and (**b**) with one large cluster

strain is considered as the measure of clustering. The volume-averaged stresses and strains are plotted in Fig. 22, wherein each drop corresponds to cracking of one or more particles. The particle cracking initiates earlier at lower values of strain ($\sim 0.2\%$) for micrograph A with three clusters. However, a higher number of cracked particles leading to a higher drop in the stress values is seen for micrograph B with a larger single cluster. The observations are further corroborated in the equivalent plastic strain contour plots of Fig. 23. In micrograph A, the cracking

Table 3 Spatial distribution parameters of two simulated micrographs in Fig. 21 and the number of simulation based cracked particles at 2% applied strain

Micrograph	Cluster Index(κ)	Contour Index(ι)	$\frac{1}{\hat{L}_{nnd}}$	No. of cracked particles
A	6.68	0.73	20.87	18
B	4.80	0.74	15.82	20

is predominantly contained within the cluster and does not percolate across the micrograph. However, a dominant path with a higher extent of particle cracking is observed in micrograph B, which causes the increased drop in stress carrying capacity. Table 3 shows a comparison of the cluster index (κ), contour index (ι) and an inverse nearest neighbor distance $\left(\frac{1}{\hat{L}_{nnd}}\right)$ with the number of cracked particles. The contour index is found to be the best indicator of the trend in the number of particles cracked, for many microregions simulated. Hence, the contour index (ι) is chosen as the spatial distribution descriptor in the morphology-based domain partitioning or **MDP** process to follow.

5 Domain Partitioning: A Preprocessor for Multiscale Modeling

An assumption made in the concurrent multilevel models of (Ghosh et al. 2001; Raghavan and Ghosh 2004b; Raghavan et al. 2004; Raghavan and Ghosh 2004a) is that the entire computational domain is initially homogenizable for macroscopic computations. However, many heterogeneous materials such as the W319 aluminum alloy consist of regions that display micro- and macro-length scale characteristics from morphological considerations alone (see Fig. 24). Local geometric features render some regions statistically inhomogenizable, i.e., statistically equivalent representative volume elements or SERVEs cannot be identified for these regions. Hence, in a true concurrent multiscale computational model, these regions of geometric nonhomogeneity should be identified prior to analysis and concurrently modeled at the microstructural length scales. Once the high-resolution microstructural features have been generated for all locations in the computational domain by the WIGE algorithm, the microstructural characterization functions and tools described in Sect. 4 can be used for delineating regions that necessitate different scale representation. The resulting computational domain is expressed as $\Omega_{comp} = (\cup_{i=1}^{N_{mac}} \Omega_{mac}^i) \bigcup (\cup_{i=1}^{N_{mic}} \Omega_{mic}^i)$, where the subscripts mac and mic correspond to regions that can and cannot be homogenized respectively. The objective of this section is to develop criteria that can enable the preanalysis partitioning of the computational domain into regions of homogeneity and inhomogeneity. Functions of the microstructure descriptors are developed to establish criteria for successive domain partitioning and refinement.

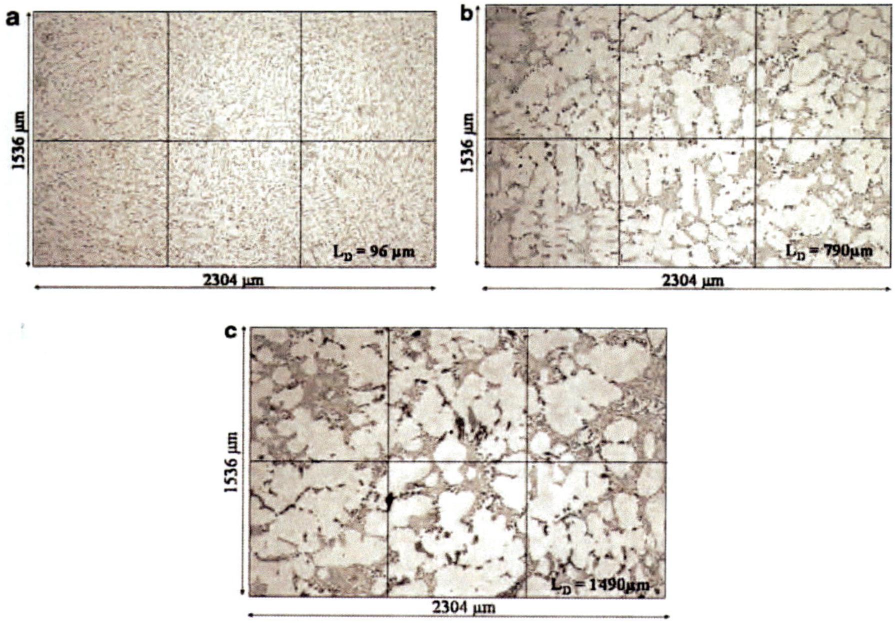

Fig. 24 Microstructural images of cast aluminum alloy W319 to be partitioned, (**a**) $SDAS = 23\mu$m, (**b**) $SDAS = 70\mu$m and (**c**) $SDAS = 100\mu$m

5.1 *Statistical Homogeneity and Homogeneous Length Scale (L_H)*

The n-point probability function S_n has been introduced in (Yeong and Torquato 1998), which for a statistically homogeneous media satisfies the condition

$$S_n(x_1, x_2, .., x_n) = S_n(x_1 + \bar{x}, x_2 + \bar{x}, ..., x_n + \bar{x}) = S_n(x_{12}, .., x_{1n}) \; \forall \, n \geq 1, \tag{36}$$

where $x_1, x_2, ...x_n$ are position vectors of n points in the medium, $\bar{x}$ corresponds to a fixed translation and $x_{ij} = x_j - x_i$. This implies that for a statistically homogeneous medium, S_n depends on the relative positions. The 1−point probability function S_1 (the volume or area fraction) is a constant everywhere, i.e., homogeneity can be assumed at regions, where S_1 does not vary significantly. A homogeneous length scale L_H in the material microstructure is established in (Spowart et al. 2001) from this consideration. L_H is the length scale above which the local variability in area fraction is smaller than a specified tolerance. It is evaluated in the following steps.

1. A large high-resolution microstructural domain of characteristic dimension L is divided into finite squares, each of size D.

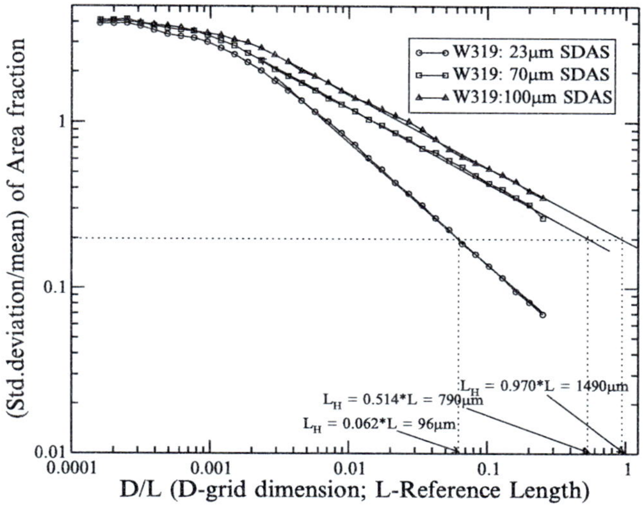

Fig. 25 Determination of the homogeneous length scale L_H for W319 with different *SDAS* values. The figure shows a linear fit in the log scale

2. The area fraction A_f of the heterogeneities in each square is evaluated. The ratio of standard deviation (σ_{A_f}) to the mean area fraction (μ_{A_f}) is defined as the coefficient of variation or *COV*. This corresponds to the variation of A_f between the squares.
3. The steps 2 and 3 are repeated for different sizes D.
4. For a Poisson distribution, the relation between the *COV* and the normalized square size $\frac{D}{L}$ is derived in (Spowart et al. 2001) as

$$COV(A_f) = \frac{\sigma_{A_f}}{\mu_{A_f}} = \left(\frac{\pi}{4A_f}\right)^{0.5} \left(\frac{D}{L}\right)^{-1}. \tag{37}$$

The corresponding *COV* varies linearly with the normalized square size $\frac{D}{L}$ in a logarithmic scale. Hence, the *COV* for the microstructural image is plotted as a function of $\frac{D}{L}$ on a logarithmic scale as shown in Fig. 25.
5. The intercept of the plot with the $\frac{D}{L}$ axis with a preset tolerance is evaluated. The corresponding size D is identified as the homogeneous length scale L_H. Below this threshold L_H, it is necessary to change from a homogeneous to a heterogeneous domain representation with explicit delineation of heterogeneities.

5.2 *Multiscale Domain Partitioning Criteria*

The **MDP** operation requires the following three ingredients:

- A high-resolution microstructure representation for the entire computational domain Ω_{comp}, at least with respect to key characteristic features;

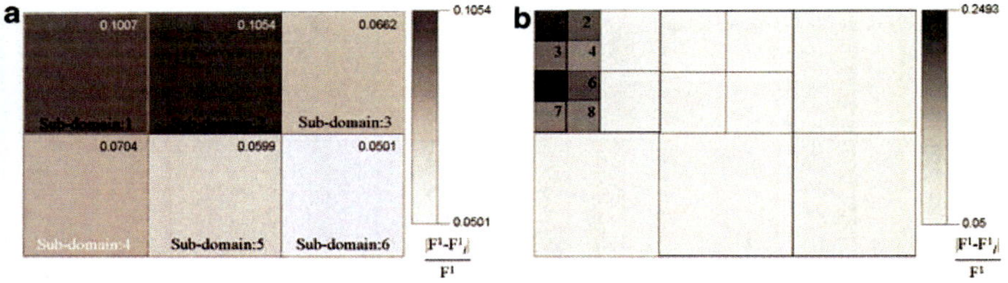

Fig. 26 Distribution of the partitioning function $\frac{|\mathbf{F^1} - \mathbf{F^1}_{(l)}|}{F^1}$ in the **MDP** process for W319 $SDAS = 23\mu$m: (**a**) before first cycle and (**b**) after the second cycle.

- The homogeneous length scale L_H; and
- Representative partitioning criteria in terms of key microstructural descriptors.

Since the extreme values of the microstructural morphology play important roles in the localization and failure behavior, descriptors that reflect these characteristics are considered important.

The method begins with a coarse discretization of Ω_{comp} into N_p^0 subdomains or partitions, as shown in Fig. 26a. A microstructural unit is defined as a high resolution, subhomogenization length scale, microstructural region Ω_{mic} of dimension $\xi * L_H$ where $\xi < 1$. The factor is chosen as $\xi = 0.5$ in this work. The i-th subdomain is assumed to be made up of M^i underlying microstructural units. Statistical functions representing the variation of a descriptor in the M^i microstructural units are evaluated for successive partitioning of the i-th subdomain. From the discussions in Sect. 4, the area fraction A_f, roundness ρ, edge smoothness χ, and contour index ι are microstructure descriptors that are used construct the refinement criteria functions. Two specific functions are introduced as described below:

1. $\mathbf{F^1}_i$: This function couples the size and distribution descriptors A_f and the contour intensity ι. It is constructed in terms of the mean parameters $\mu(A_f)$ and $\mu(\iota)$ for the M^i microstructural units within each subdomain i, and is expressed as

$$\mathbf{F^1}_i = \mu(A_f)\mu(\iota). \tag{38}$$

2. $\mathbf{F^2}_i$: A function that accounts for both shape and size parameters is defined as:

$$S_e = 1 + \sum_{k=1}^{N_c} \left[\frac{A_f^k}{A_f} (w_\rho(1 - \rho_k) + w_\chi(1 - \chi_k)) \right], \tag{39}$$

where A_f^k, ρ_k, and χ_k are the local area fraction, roundness, and edge smoothness of the k-th heterogeneity, respectively, and A_f is the overall area fraction in the microstructural region Ω_{mic}. N_c is the number of heterogeneities in Ω_{mic} and w_ρ, w_χ are assigned weights taken as $w_\rho = 0.5$, $w_\chi = 0.5$.

It should be noted that for microstructures where the aspect ratio or roughness are not pronounced, the value of S_e tends to 1.0. The contour index (ι) and the overall area fraction A_f are multiplied with S_e in the refinement function to capture spatial density of heterogeneities. The resulting function is written as:

$$\mathbf{F^2}_i = \mu(S_e \cdot A_f \cdot \iota). \tag{40}$$

The refinement functions $\mathbf{F^k}_i$; $\mathbf{k} = 1, 2$ are evaluated in each subdomain, together with those in each of its four divisions $\mathbf{F^k}_i(l), l = 1 \cdots 4)$. A subdomain ($i$) is partitioned only if the following criterion is attained for any of the four subregions.

$$\frac{|\mathbf{F^k}_i - \mathbf{F^k}_i(l)|}{\mathbf{F^k}_i} > C_{f1}, \quad \text{for any } l = 1 \cdots 4 \tag{41}$$

The prescribed tolerance is $C_{f1} = 0.10$ corresponding to 10% variation.

The successive partitioning process reduces the subdomain size locally, and may ultimately reach the homogeneous scale limit L_H. Once L_H is reached, only one additional step of further partitioning is possible. The level below L_H is not homogenizable and hence cannot be refined any further. A special criterion is required for this partitioning. Each of the subsequent partitions contains only one microstructural unit (M^{loc}) of dimension $0.5 * L_H$. It is not possible to evaluate the statistical functions $\mathbf{F^k}$, $k = 1, 2$ for a single (M^{loc}). The criterion is constructed in terms of the variation of average local area fraction A_f, an important descriptor that is present in both the functions $\mathbf{F^k}$, $k = 1, 2$. Partitioning below L_H is governed by the condition

$$\frac{|(A_f)_i - (A_f)_i(l)|}{(A_f)_i} > C_{f2}, \quad \text{for } l = 1 \cdots 4. \tag{42}$$

Any subdomain below the L_H threshold is characterized by significant variation in microstructure descriptor functions, e.g., the local area fraction. Consequently, those partitions for which the variation is really large are classified as nonhomogeneous and opened up for explicit microstructural representation in the multilevel model. The factor C_{f2} is taken as 0.75, corresponding to a 75% difference in the critical regions of the microstructure. The combined microstructure *simulation - characterization - partitioning* method delineates the hierarchy of scales in the computational model.

6 Numerical Execution of the MDP Method on the W319 Alloy

The morphology-based domain partitioning **MDP** methodology is applied to the microstructures of cast aluminum W319 alloy with respective SDAS values of 23, 70 and 100μm. The low-resolution computational micrographs Ω_{comp} of dimensions 2,304 × 1,536 μm for the alloys are shown in Fig. 24a, b, and c. The WIGE algorithm generates high-resolution images of all points in Ω_{comp} by constructing the correlation table like that in Fig. 7, from two $110 \times 110\,\mu$m high-resolution SEM image windows. For the *SDAS* $= 23\mu$m microstructure, the location of these two windows are shown as **A** and **B** in Fig. 3. Similar high resolution windows are also considered for the *SDAS* $= 70\mu$m and *SDAS* $= 100\mu$m microstructures. The logarithmic scale plot of COV vs. $\frac{D}{L}$ identifying L_H for the three *SDAS* is shown in Fig. 25. The reference dimension is taken as $L = 1,536\mu$m for these plots. The homogenization length scale L_H is calculated from an intercept tolerance value $COV = 0.2$ in the log–log plot of Fig. 25 for the three cases. L_H increases with SDAS and consequently larger regions need to be considered for accounting for their natural length scales in the multiscale modeling process.

The **MDP** process begins by dividing the computational domains Ω_{comp} for each of the three *SDAS* in Fig. 24 into 6 subdomains. This initial partition for *SDAS* $= 23\mu$m is shown in Fig. 24a. Successive partitioning progresses according to the refinement criteria in Sect. 5, until the subdomain size reaches L_H. The distribution of the characteristic functions $\frac{|\mathbf{F}^1{}_i - \mathbf{F}^1{}_i(l)|}{\mathbf{F}^1{}_i}$ for the first and second cycle domain partitioning in the microstructure with *SDAS* $= 23\mu$m are depicted in the contour plots of Fig. 26a and b, respectively. The corresponding partitions are shown in Fig. 27a and b. The characteristic functions in (41) for the first stage are reported in Table 4. For the microstructure with *SDAS* $= 70\mu$m, $L_H = 790\mu$m, and for *SDAS* $= 100\mu$m, $L_H = 1490\mu$m. Since the subdomain considered in the first

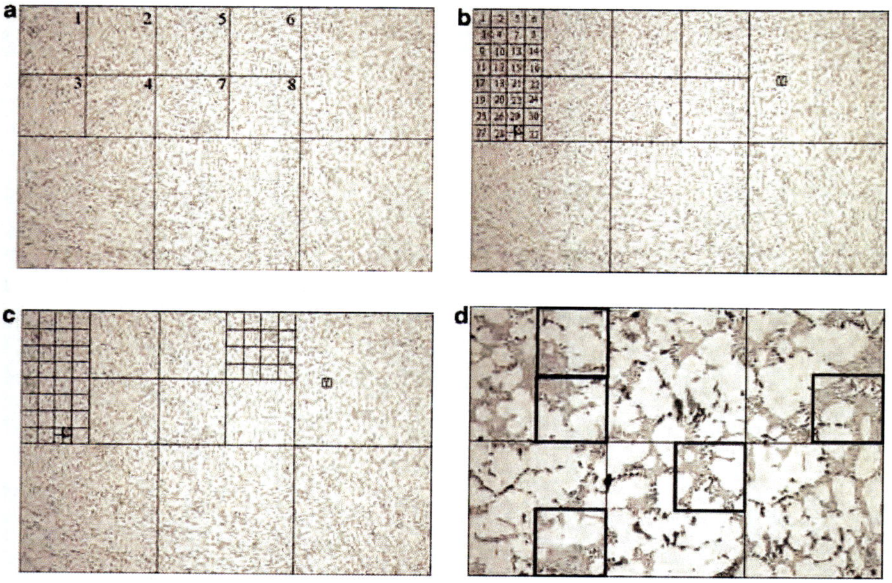

Fig. 27 Results of the **MDP** process for the different W319 microstructures: (**a**) partitioned domain after the first cycle for *SDAS* $= 23\mu$m, (**b**) final partitioned domain for *SDAS* $= 23\mu$m by the $\mathbf{F}^1$-based criterion, (**c**) final partitioned domain for *SDAS* $= 23\mu$m by the $\mathbf{F}^2$-based criterion, (**d**) regions of statistical inhomogeneity in the *SDAS* $= 100\mu$m microstructure

Table 4 Highest values of the refinement functions $\frac{|\mathbf{F^k}_i - \mathbf{F^k}_i(l)|}{\mathbf{F^k}_i}$ for the first cycle of domain partitioning in W319 ($SDAS = 23\mu$m) for subdomains 1–6 shown in Fig. 26a

Subdomain No.(i)	$\frac{\lvert\mathbf{F^1}_i - \mathbf{F^1}_i(l)\rvert}{\mathbf{F^1}_i}$	$\frac{\lvert\mathbf{F^2}_i - \mathbf{F^2}_i(l)\rvert}{\mathbf{F^2}_i}$
1	0.1007	0.1005
2	0.1054	0.1014
3	0.0662	0.0646
4	0.0704	0.0687
5	0.0599	0.0571
6	0.0501	0.0533

Table 5 Highest values of the refinement functions $\frac{|\mathbf{F^k}_i - \mathbf{F^k}_i(l)|}{\mathbf{F^k}_i}$ for the second cycle of domain partitioning in W319 ($SDAS = 23\mu$m) for subdomains 1–8 shown in Fig. 27a

Subdomain No.(i)	$\frac{\lvert\mathbf{F^1}_i - \mathbf{F^1}_i(l)\rvert}{\mathbf{F^1}_i}$	$\frac{\lvert\mathbf{F^2}_i - \mathbf{F^2}_i(l)\rvert}{\mathbf{F^2}_i}$
1	0.1264	0.1257
2	0.0267	0.0295
3	0.1862	0.1876
4	0.0170	0.0132
5	0.0664	0.0724
6	0.0982	0.1007
7	0.0577	0.0614
8	0.0833	0.0746

Table 6 Highest values of the refinement functions $\frac{|\mathbf{F^k}_i - \mathbf{F^k}_i(l)|}{\mathbf{F^k}_i}$ for the third cycle of domain partitioning in W319 ($SDAS = 23\mu$m)

Subdomain No.(i)	$\frac{\lvert\mathbf{F^1}_i - \mathbf{F^1}_i(l)\rvert}{\mathbf{F^1}_i}$	$\frac{\lvert\mathbf{F^2}_i - \mathbf{F^2}_i(l)\rvert}{\mathbf{F^2}_i}$
1	0.2273	0.2250
2	0.1683	0.1708
3	0.1504	0.1207
4	0.1140	0.1178
5	0.2493	0.2772
6	0.1717	0.1691
7	0.1227	0.1146
8	0.1223	0.1369
9		0.1143
10		0.1973
11		0.2102
12		0.2799

stage of their partitioning is smaller than their respective L_H, these two domains cannot be partitioned any further beyond the initial partitioning in this example.

After the first cycle, partitioning by $\mathbf{F^2}$ deviates from that by $\mathbf{F^1}$, as observed from the values in Table 5. The criterion using $\mathbf{F^1}$ results in 8 partitions, whereas that using $\mathbf{F^2}$ yields 12 partitions based on the $C_{f1} = 0.1$ cutoff value. The subsequent cycle values of $\frac{|\mathbf{F^k}_i - \mathbf{F^k}_i(l)|}{F^k_i}$ are calculated in Table 6 for different i's by the two criteria. The subdomain numbers i are labeled in Fig. 27a and b for partitioning by the function $\mathbf{F^1}$. The partitioning process continues until the size limit of $L_H = 96\mu$m is reached. The final partitioned computational domain for $SDAS = 23\mu$m with criteria based on functions $\mathbf{F^1}$ and $\mathbf{F^2}$ are shown in Fig. 27b and c, respectively.

Table 7 Comparison of microstructural characteristics of regions marked **X** and **Y** in Fig. 27b

Parameter	Micrograph **X**	Micrograph **Y**
No. of particles	26	14
Area fraction A_f	11.42%	4.31%
Least Roundness ρ	0.21	0.41
Least Edge Smoothness χ	0.37	0.56
Cluster Index κ	7.49	6.40
Contour Index ι	0.75	0.72

Partitioning with $\mathbf{F}^2$ leads to a higher number of subdomains. At this stage, (42) is used for delineating statistically homogeneous regions from inhomogeneous regions. Even with the difference in partitioning, the application of (42) yields the same inhomogeneous region in both cases as marked by the **X** in Fig. 27b and c. The size of the inhomogeneous domain is $\sim 48\mu$m. The microstructure characteristics of a typical inhomogeneous region **x** and a homogeneous region marked by **y** in Fig. 27b are shown in Table 7.

The partitioned domains for $SDAS = 70\mu$m and $SDAS = 100\mu$m microstructures, for which the homogenization length scales $L_H = 790$ μm and $L_H = 1490$ μm, respectively, are shown in Fig. 24b and c. The initial partitioning for these microstructures already brings the size of each partition below L_H, and hence no additional partitioning is conducted. However, the criterion of (42) is applied to each of the six subdivisions of the initial partitioning. For $SDAS = 70\mu$m no regions of statistical inhomogeneity are identified by this criterion. However, five regions are identified for $SDAS = 100\mu$m, as shown in Fig. 27d. The inhomogeneous regions identified by the **MDP** algorithm need to be modeled at the micromechanical level in the concurrent multiscale analyses and simulations.

7 Multiscale Analysis with the MDP Based Preprocessor

The computational domain, partitioned by the **MDP** algorithm, delineates homogeneous and morphologically inhomogeneous regions in a concurrent multilevel setting for multiscale analysis developed by the author in (Ghosh et al. 2001; Raghavan and Ghosh 2004b; Raghavan et al. 2004; Raghavan and Ghosh 2004a). Homogeneous regions are labeled as 'level-0' and are analyzed using macroscopic constitutive models, while morphologically inhomogeneous regions require explicit micromechanical analysis and are labeled as 'level-2'. Level-2 regions are characterized by significant departure from macroscopic uniformity, e.g., near a crack tip. They may be classified into two categories, viz.,

1. Regions of strong local nonhomogeneity that evolve from 'level-0' regions with high local gradients due to localized deformation and damage.
2. Microstructural regions that are inherently inhomogenizable even prior to deformation due to irregularities in the microstructural morphology.

The latter regions that require explicit micromechanical analyses are seeded by the **MDP** algorithm. The concurrent multiscale framework allows simultaneous analysis of complementary subdomains belonging to different length scales. A swing 'level-1' region has also been introduced in (Ghosh et al. 2001; Raghavan and Ghosh 2004b; Raghavan et al. 2004; Raghavan and Ghosh 2004a) to evaluate criteria for transformation of 'level-0' regions into 'level-2' regions. The location and extent of the three levels can change continuously with deformation and evolving damage in the domain.

The **MDP** partitioned computational domain with homogeneous and inhomogeneous regions are delineated in Fig. 27b,c. A finite element rendering of this partitioned domain for the multiscale modeling is shown in Fig. 28. The mesh structure is adaptively generated by successive refinement in the **MDP** algorithm.

The homogeneous 'level-0' region is discretized into 4-noded QUAD4 finite elements incorporating a homogenized continuum plasticity-damage (HCPD) material model for ductile fracture. The HCPD model, developed in (Ghosh et al. 2009), has the structure of the anisotropic Gurson–Tvergaard–Needleman (GTN) elastoplasticity model for porous ductile materials. It is assumed to be orthotropic in an evolving material principal coordinate system for the entire deformation history. The HCPD model parameters are calibrated from the homogenization results of micromechanical analysis of the microstructural *representative volume element* or RVE. The size and morphology of the RVE is important in determining the HCPD model parameters. Identification of a statistically equivalent RVE (SERVE) that locally represents the effective response of the microstructure in an average sense, has been conducted in (Ghosh et al. 2009; Swaminathan et al. 2006; Swaminathan and Ghosh 2006). The methods are briefly discussed below.

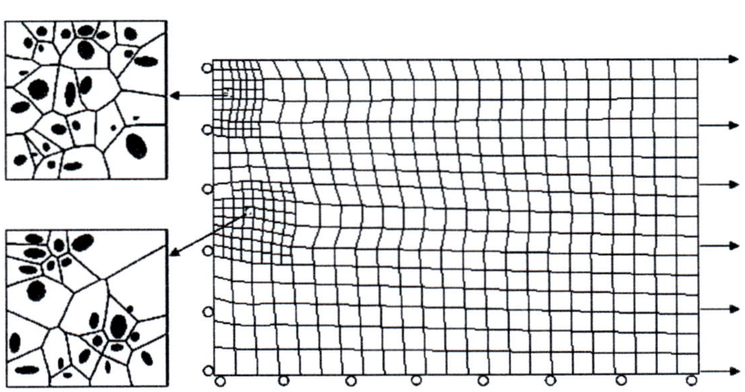

Fig. 28 Pre-deformation layout of the multilevel model for multiscale analysis of ductile fracture in W319 cast aluminum alloy. The **MDP** algorithm is responsible for designing the optimal combination of multilevel computational sub-domains

7.1 Identification of the RVE Size for Homogenization

The size of the microstructural representative volume element or RVE is an important parameter in the homogenization process for determining effective material properties. The concept of RVE was introduced by Hill (1963) as a microstructural subregion that is representative of the entire microstructure in an average sense. For microstructures with nonuniform dispersions, it is of interest to identify statistically equivalent RVEs or SERVEs that can be used for homogenization. Definition of the SERVE and methods of identification for nonuniform, pristine, and damaging composites have been discussed by Ghosh et al. (2006; 2006). A SERVE is identified as the smallest volume element of the microstructure exhibiting the following characteristics.

- Effective material properties for the SERVE should be representative of properties for the entire microstructure to within a prescribed tolerance.
- The SERVE identified should be independent of location from where it is extracted in the local microstructure.

Various statistical descriptors have been proposed to characterize and classify microstructures based on the spatial arrangement of heterogeneities. Pyrz (1994) has introduced the pair distribution function $g(r)$ and the marked correlation function $M(r)$ to characterize microstructures based on interinclusion distances. For observations within a finite window of area A, the pair distribution function $g(r)$ corresponds to the probability $g(r)\mathrm{d}r$ of finding an additional inclusion center between concentric rings of radii r and $r + \mathrm{d}r$, respectively. It characterizes the occurrence intensity of interinclusion distances and is expressed as (Ghosh et al. 1997a,b):

$$g(r) = \frac{1}{2\pi r}\frac{\mathrm{d}K(r)}{\mathrm{d}r}, \quad \text{where} \quad K(r) = \frac{A}{N^2}\sum_{k=1}^{N} I_k(r). \tag{43}$$

$K(r)$ is a second-order intensity function and $I_k(r)$ is defined as the number of additional centers of inclusions that lie within a circle of radius r about an arbitrarily chosen inclusion. With increasing r values, circles about particles that are near the edges of a finite-sized window may extend outside the observation window. Appropriate correction factors have been proposed in (Pyrz 1994; Ghosh et al. 1997a) to account for the edge effects in the evaluation of $K(r)$. A different method has been proposed in (Swaminathan et al. 2006) by repeating the microstructure periodically in both $x-$ and $y-$ directions for several period lengths. For a pure Poisson distribution, $K(r) = \pi r^2$, which corresponds to $g(r) = 1$. While $g(r)$ provides a univariate characterization in terms of geometry, the marked correlation function $M(r)$ results in a multivariate characterization of the microstructure. Every inclusion is marked by an appropriate descriptor to display the effect of a property variable on the geometrical arrangement of inclusions. The marked correlation function correlates any chosen field variable, e.g., stress, strain, or their dependent functions with the morphology of the microstructure. It is expressed as the ratio of a state variable dependent function $h(r)$ and the geometry-based pair distribution function $g(r)$ as:

$$M(r) = \frac{h(r)}{g(r)}, \tag{44}$$

where the function $h(r)$ is derived from the mark intensity function $H(r)$, i.e.,

$$h(r) = \frac{1}{2\pi r}\frac{\mathrm{d}H(r)}{\mathrm{d}r} \quad \text{and} \quad H(r) = \frac{1}{m^2}\frac{A}{N^2}\sum_{i}^{N}\sum_{k=1}^{k_i} m_i m_k(r). \tag{45}$$

Here, m_i is a 'mark' associated with the i-th inclusion and r is a measure of the radial distance of influence. A mark can be any chosen state variable in the microstructure that will influence the choice of an RVE for the problem considered. k_i is the number of inclusions, which have their centroids within a circle of radius r around the i-th inclusion, m is the mean of all the marks and N is the total number of inclusions. A declining value of $M(r)$ indicates reduced correlation between discrete entities in the microstructure. Thus, $M(r)$ provides a good estimate of the size of the RVE, which is interpreted as a region of influence in the microstructure. Various marks are experimented with, in the calculation of marked correlation functions for a real microstructure as shown in Fig. 29. They are:

1. Volume-averaged values of void volume fraction f_v,
2. Frobenius norm of tangent stiffness $\|\mathbf{E}^{tan}\|$,
3. Plastic work W_p over each Voronoi cell containing an inclusion.

These variables are postprocessed from solutions of computational micromechanical analyses by the locally enhanced VCFEM or LE-VCFEM developed in (Hu and Ghosh 2008). LE-VCFEM models ductile failure in the microstructure, which initiates with inclusion cracking and evolves with matrix cracking in the form of

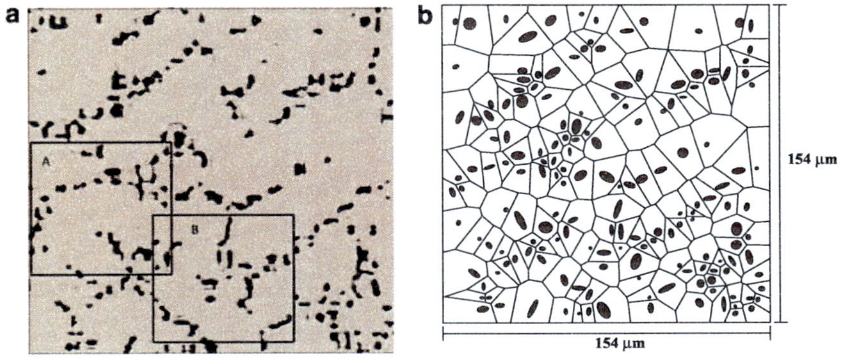

Fig. 29 (**a**) A micrograph of a cast aluminum alloy showing distribution of Si particles and intermetallics, (**b**) simulated microstructure, discretized into Voronoi cells by tessellation

void growth and coalescence. It is observed that the $M(r)$ functions, calculated with f_v and $\|\mathbf{E}^{\tan}\|$ as marks, vary with the loading conditions. On the other hand, $M(r)$ with W_p as the mark is independent of load conditions. Consequently, the plastic work-based $M_{wp}(r)$ is calculated for the scanning electron micrograph in 29(a, b) for three different loading conditions, viz. simple tension, bi-axial tension, and shear, respectively. The values of $M_{wp}(r)$ show very little dependence on the value of the overall strain state. The $M_{wp}(r)$ function for the different load conditions are plotted as functions of r in Fig. 30a. It stabilizes to near-unity values (to within a tolerance of $\sim 4\%$) at a radius of convergence r_p. For $r > r_p$, $M(r) \to 1$ and the local morphology ceases to have any significant influence on the state variables beyond this characteristic distance. The radius r_p corresponds to a local correlation length

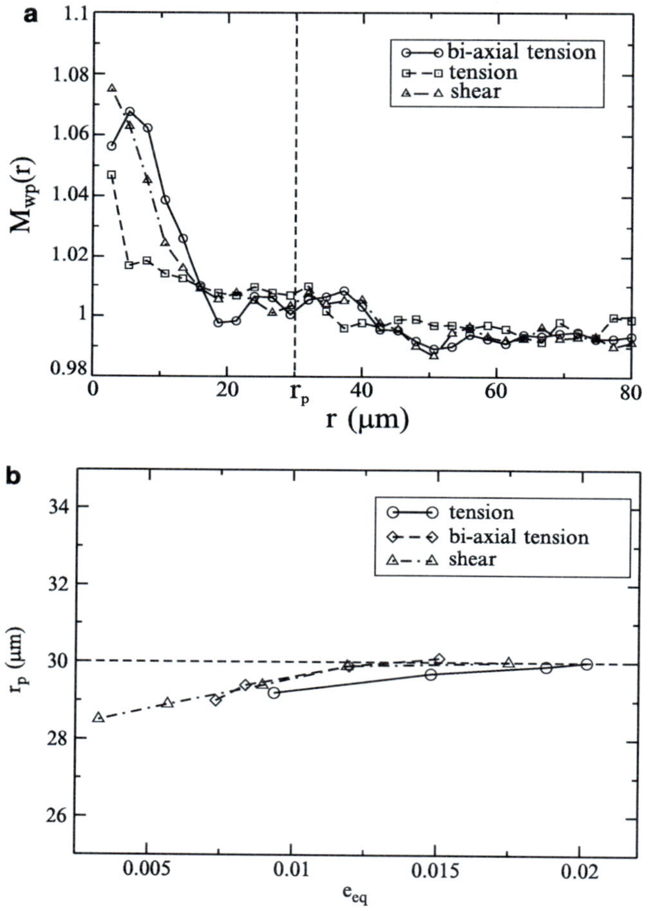

Fig. 30 **(a)** Plots of $M(r)$ for different loading conditions; **(b)** Evolution of r_p for different loading conditions

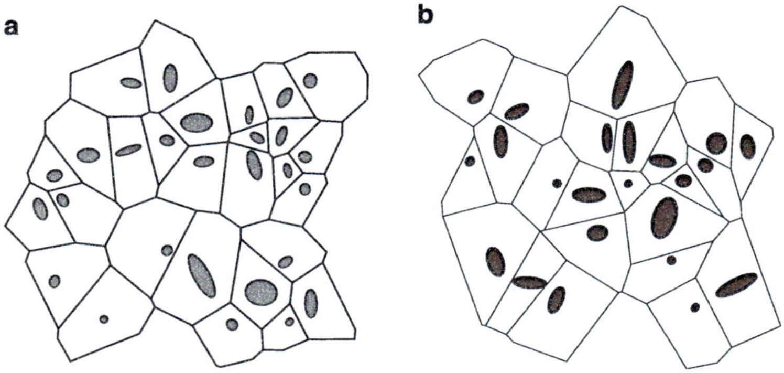

Fig. 31 (**a**) Window (A) with periodic boundary; (**b**) Window (B) with periodic boundary

that provides an estimate for the SERVE size. The correlation length r_p is plotted as a function of the equivalent strain e_{eq} for the different load conditions in Fig. 30b. The value of r_p does not change much with increasing strain and it converges to the same value 30 μm. Consequently, a window size of $\sim 2r_p$ or 60 μm is considered as the size of the SERVE for the microstructure in Fig. 29a,b.

Location independence of the 60 μm SERVE size is also verified by extracting RVEs from two locations A and B in the microstructure of Fig. 29a. The boundaries of the RVEs are created by periodically repeating the position of inclusions in the x and y directions, followed by tessellation. This is shown in Fig. 31. The local area fraction of inclusions in the windows A and B are 6.072% and 6.090%, respectively, in comparison with 6.078% for the entire microstructure. Homogenization of the results of micromechanical LE-VCFEM analyses is performed for these two RVEs, as well as for the whole microstructure of Fig. 29 under different loading conditions. Both intact and cracking inclusions are considered in the analyses. The homogenized stress-strain responses for the intact and cracking inclusions are plotted in Fig. 32a,b. The results for the two RVEs match well with those for the entire microstructure. This justifies the choice of r_p in determining the SERVE.

7.2 Level-1 and Level-2 Analysis with LE-VCFEM

Level-2 and level-1 subdomain analysis in the multiscale deformation and damage models, developed in (Ghosh et al. 2001; Raghavan and Ghosh 2004b; Raghavan et al. 2004; Raghavan and Ghosh 2004a; Ghosh 2008), requires explicit modeling of the microstructural response. The underlying microstructure in 'level-2' regions, together with microstructural analysis models, are shown in Fig. 28. Ductile heterogeneous materials can undergo catastrophic failure that initiates with particle fragmentation which evolves with void growth and coalescence in localized bands of intense plastic deformation and strain softening. The locally enhanced Voronoi Cell finite element model (LE-VCFEM) is developed in (Hu and Ghosh 2008) for

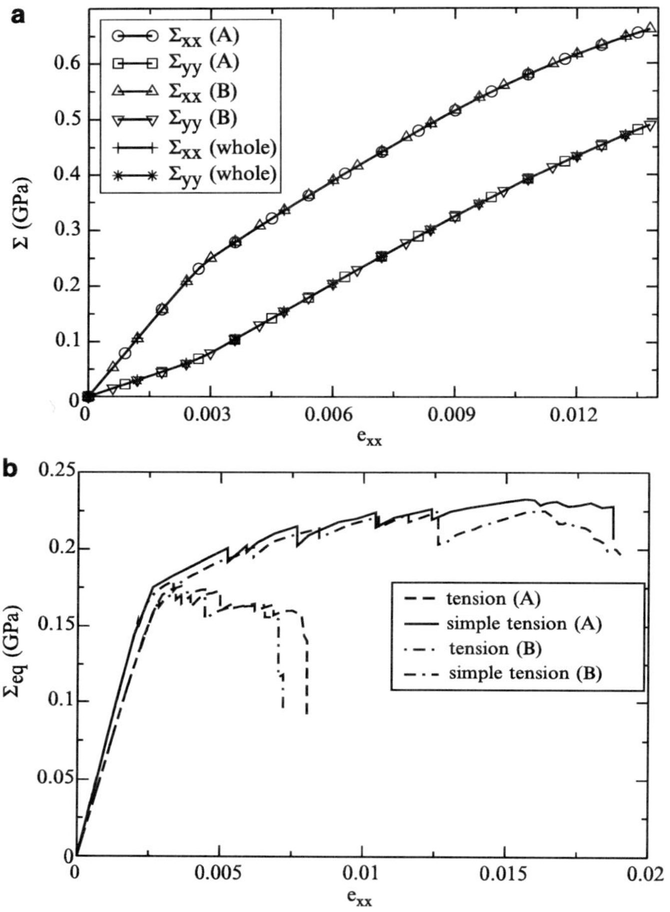

Fig. 32 Comparisons of macroscopic stress-strain response: (**a**) without inclusion cracking; and (**b**) with inclusion cracking

modeling the complex phenomenon of ductile failure in heterogeneous metals and alloys. In LE-VCFEM, finite deformation displacement elements are adaptively added to regions of localization in the otherwise assumed stress based hybrid Voronoi cell finite element to locally enhance modeling capabilities for ductile fracture. Adaptive h-refinement is used for the displacement elements to improve accuracy. Damage initiation by particle cracking is triggered by a Weibull model. The nonlocal Gurson-Tvergaard-Needleman model of porous plasticity is implemented in LE-VCFEM to model matrix cracking. An iterative strain update algorithm is used for the displacement elements.

Fig. 33 Stress (σ_{xx}) contours in the macro- and micro-elements of the multiscale model for an applied tensile strain of 0.35%. Cracked particles with the corresponding matrix cracks are also shown in the microstructures

7.3 Multiscale Analysis of Ductile Failure

The multiscale model for ductile fracture is subjected to uniaxial tension as shown in Fig. 28. For an applied tensile strain of 0.35% the macroscopic stress (Σ_{xx}) as well as the microstructural stress (σ_{xx}) distributions are shown in Fig. 33. Microstructures at the two 'level-2' regions show few particle cracking and associated matrix cracking due to void growth. With increasing strain, more 'level-0' elements exhibit large local gradients in plastic strains and void volume fraction. This eventually results in a topographical change to 'level-2' elements in this region of a dominant crack path. The multiscale model is thus capable of initiating dominant macroscopic ductile cracks from the microstructure by particle cracking, void growth, and coalescence in the macrostructure. This capability does not currently exist in most models of ductile failure.

8 Conclusions

This chapter discusses a microstructure morphology-based domain partitioning **MDP** methodology for materials with nonuniform heterogeneous microstructure. The comprehensive set of methods is intended to provide a concurrent multiscale analysis model with the initial computational domain that delineates regions of statistical homogeneity and inhomogeneity. The MDP methodology will act as a preprocessor to multiscale simulation of mechanical behavior and damage. The method is tested on an Aluminum alloy W319 with three different secondary dendrite arm spacing or SDAS.

The **MDP** methodology encompasses a three-step approach to achieve the overall goal. The methods and algorithms are based on geometric features of the morphology without any recourse to mechanical response. In the first step, high-resolution microstructural images are simulated from low-resolution optical or scanning electron micrographs and a limited set of high-resolution micrographs. It incorporates a wavelet interpolation of low-resolution images that is augmented by a grayscale gradient-based enhancement algorithm, termed WIGE algorithm. The algorithm can overcome the limitations of experimental acquisition of a large set of contiguous micrographs for creating a montage of images in any material domain. In experiments, perfect alignment of the microscope for nonoverlapping adjoining domains is a difficult task, aside from the time and expenses incurred in the acquisition process itself. The WIGE algorithm can aid significantly in this acquisition-reconstruction process and is discussed with application for W319 microstructure. Excellent convergence characteristics are observed for the reconstructed W319 micrographs with respect to 1-point, 2-point, and 3-point probability functions.

The second step involves the development of microstructure characterization tools that are able to identify morphological features of interest in the multiscale analysis. The tools incorporate parametric descriptors of size, shape, and spatial distributions that directly affect the mechanical and failure behavior of the material. The predictions of the characteristic functions of morphological descriptors are compared with the results of Voronoi cell finite element method or VCFEM simulations. The third and final step is the development of robust multiscale domain partitioning methods for delineating subdomains corresponding to different scales in concurrent multiscale analysis. The foremost task is the estimation of a homogeneous length scale or L_H, below which statistical inhomogeneity is strong to limit the use of homogenization. Following this, two criteria based on different functions of morphological descriptors are developed to govern domain partitioning. Successive domain partitioning continues according to these criteria till the L_H is reached. Subsequently, a different criterion is invoked to differentiate regions of statistical homogeneity from inhomogeneity. The latter corresponds to regions where explicit representation of the multiphase microstructure and micromechanical analysis is necessary.

The effectiveness of the **MDP** methodology as a preprocessor for multiscale analysis of cast aluminum alloy W319 at different *SDAS* is demonstrated satisfactorily in this chapter. The numerical analyses establish distinct requirements on the computational domains for the three SDAS microstructures based on their intrinsic length scales. Finally, a multiscale analysis of ductile fracture in the W319 microstructure is conducted using a differentiated scale structure that has been laid out by the **MDP** algorithm. The chapter emphasizes the need for coupling multiscale characterization and domain decomposition with multiscale analysis of heterogeneous materials. It is only through such coupling that the true origin of localization and failure can be probed and subsequently their growth can be tracked.

Acknowledgements This work has been supported by the National Science Foundation NSF Div Civil and Mechanical Systems Division through the GOALI grant No. CMS-0308666 (Program director: Dr. Clark cooper) and by the Army Research Office through grant No.DAAD19-02-1-0428 (Program Director: Dr. B. Lamattina). This sponsorship is gratefully acknowledged. Computer support by the Ohio Supercomputer Center through grant PAS813-2 is also gratefully acknowledged.

References

Anson, J.P. and Gruzleski, J.E.: The quantitative discrimination between shrinkage and gas microporosity in cast Aluminum alloys using spatial data analysis. Mater. Charac. **43**, 319–335 (1999)

Argon, A.S., Im, J. and Sofoglu, R.: Cavity formation from inclusions in ductile fracture. Metall. Mater. Trans. A. **6A**, 825–837 (1975)

Ghosh, S., Bai, J. and Paquet, D.: Homogenization based continuum plasticity-damage model for ductile failure of materials containing heterogeneities. J. Mech. Physics Solids **57**, 1017–1044 (2009)

Baker, S. and Kanade, T.: Super-resolution: reconstruction or recognition? Proc. 2001 IEEE-EURASIP Workshop on Nonlinear Signal and Image Processing, (2001)

Boehm, H.J., Han, W. and Eckslager, A.: Multi-inclusion unit cell studies of reinforcement stresses and particle failure in discontinuously reinforced ductile matrix composites. Comp. Model. Eng. Sci. **5**, 5–20 (2004)

Boileau, J.M.: The effect of solidification time on the mechanical properties of a cast 319 aluminum alloy. Ph.D. dissertation, Wayne State University (2000)

Caceres, C.H. and Griffiths, J.R.: Damage by the cracking of silicon particles in an Al-7Si-0.4Mg casting alloy. Acta Mater. **44**, 25–33 (1996)

Caceres, C.H., Griffiths, J.R. and Reiner, P.: The influence of microstructure on the bauschinger effect in an Al-Si-Mg casting alloy. Acta Mater. **44**, 15–23 (1996)

Caceres, C.H.: Particle cracking and the tensile ductility of a model Al-Si-Mg composite system. Aluminum Trans. **1**, 1–13 (1999)

Christman, T., Needleman, A. and Suresh, S.: An experimental and numerical study of deformation in metal-ceramic composites. Acta Metall. et Mater. **37**, 3029–3050 (1989)

Chui, C.K.: An introduction to wavelets. Academic (1992)

Chung, P.W. and Tamma, K.K.: Woven fabric composites: Developments in engineering bounds, homogenization and applications. Int. J. Numer. Meth. Eng. **45**, 1757–1790 (1999)

Cooper, D.W.: Random sequential packing simulation in three dimensions for spheres. Phys. Rev. A: **A38**, 522–524 (1998)

Everett, R.K. and Chu, J.H.: Modeling of non-uniform composite microstructures. J. Compos. Mater. **27**, 1128–1144 (1992)

Everson, R., Sirooich, L. and Sreenicasan, K.R.: Wavelet analysis of the turbulent jet. Phys. Lett. A. **145**, 314–322 (1990)

Farsiu, S., Robinson, D., Elad, M. and Milanfar, P.: Advances and challenges in super-resolution. Int. Jour. Imaging Syst. Tech., **14(2)**, 47–57 (2004)

Fish, J. and Shek, K.: Multiscale analysis of composite materials and structures. Comp. Sci. Tech. **60**, 2547–2556 (2000)

Freeman, W.T., Pasztor, E.C. and Carmichael, O.T.: Learning low-level vision. Int. J. Comp. Vision. **40**, 24–47 (2000)

Ghosh, S., Nowak, Z. and Lee, K.: Quantitative characterization and modeling of composite microstructures by Voronoi cells. Acta Mater. **45**, 2215–2234 (1997a)

Ghosh, S., Nowak, Z. and Lee, K.: Tessellation based computational methods in characterization and analysis of heterogeneous microstructures. Comp. Sci. Tech. **57**, 1187–1210 (1997b)

Ghosh, S., Ling, Y., Majumdar, B. and Kim, R.: Interfacial debonding analysis in multiple fiber reinforced composites. Mech. Mater. **32**, 561–591 (2000)

Ghosh, S., Lee, K. and Raghavan, P.: A multi-level computational model for multi-scale damage analysis in composite and porous materials. Int. J. Solids Struct. **38(14)**, 2335–2385 (2001)

Ghosh, S., Valiveti, D.M., Harris, S.H. and Boileau, J.: Microstructure characterization based domain partitioning as a pre-processor to multi-scale modeling of cast Aluminum alloys. Mod. Simul. Mater. Sci. Eng. **14**, 1363–1396 (2006)

Ghosh S.: Adaptive concurrent multi-level model for multi-scale analysis of composite materials including damage. In Kwon, Y., Allen, D.H. and Talreja, R. (eds.), Multiscale Modeling and Simulation of Composite Materials and Structures, pp 83–164, Springer (2008)

Gokhale, A.M. and Yang, S.: Application of image processing for simulation of mechanical response of multi-length scale microstructures of engineering alloys. Metall. Mater. Trans. **A30**, 2369–2381 (1999)

Gonzalez, C. and Llorca, J.: Prediction of the tensile stress-strain curve and ductility in Al/SiC composites. Scripta Metall. **35(1)**, 91–97 (1996)

Hill, R.: Elastic Properties of Reinforced Solids: Some Theoretical Principles. J. Mech. Phys. Solids **11**, 357–372 (1963)

Hu, C. and Ghosh, S.: Locally enhanced Voronoi cell finite element model (LE-VCFEM) for simulating evolving fracture in ductile microstructures containing inclusions, Int. J. Numer. Meth. Eng. **76(12)**, 1955–1992 (2008)

Hao, S., Liu, W.K., Moran, B., Vernerey, F. and Olson, G.B.: Multi-scale constitutive model and computational framework for the design of ultra-high strength, high toughness steels. Comp. Meth. Appl. Mech. Eng. **193(17–20)**, 1865–1908 (2004)

Jain, J.R. and Ghosh, S.: A 3D continuum damage mechanics model from micromechanical analysis of fiber reinforced composites with interfacial damage. ASME J. App. Mech. **75**, 031011-1-031011-15 (2008a)

Jain, J.R. and Ghosh, S.: Damage evolution in composites with a homogenization based continuum damage mechanics model. Inter. J. Damage Mech. (2008b) doi:10.1177/1056789508091563

Jensen, K. and Anastassiou, D.: Subpixel edge localization and the interpolation of still images. IEEE Trans. Image Proc. **4**, 285–295 (1995)

Karnezis, P.A., Durrant, G. and Cantor, B.: Characterization of reinforced distribution in cast Al-Alloy/SiC composites. Mater. Charac. **40**, 97–109 (1998)

Kumar, H., Briant, C.L., Curtin, W.A.: Using microstructure reconstruction to model mechanical behavior in complex microstructures. Mech. Mater. **38**, 818–832 (2006)

Lewalle, J.: Wavelet analysis of experimental data: some methods and the underlying physics. AIAA 94–2281, 25th AIM Fluid Dynamics Colorado Springs (1994)

Li, M., Ghosh, S. and Richmond, O.: An experimental-computational approach to the investigation of damage evolution in discontinuously reinforced aluminum matrix composites. Acta Mater. **47(12)**, 3515–3552 (1999a)

Li, M., Ghosh, S., Richmond, O., Weiland, H. and Rouns, T.N.: Three dimensional characterization and modeling of particle reinforced metal matrix composites part II: damage characterization. Mater. Sci. Eng. **A266**, 221–240 (1999b)

Li, S. and Ghosh, S.: Extended Voronoi cell finite element model for multiple cohesive crack propagation in brittle materials. Int. J. Numer. Meth. Eng. **65**, 1028–1067 (2006)

Luthon, F., Lievin, M. and Faux, F.: On the use of entropy power for threshold selection. Signal Proc. **84**, 1789–1804 (2004)

Manwart, C., Torquato, S. and Hilfer, R.: Stochastic reconstruction of sandstones. Physical Rev. E. **62**, 893–899 (2000)

Motard, R.L. and Joseph, B.: Wavelet applications in chemical engineering. Kluwer (1994)

Poole, W.J., Dowdle, E.J.; Experimental measurements of damage evolution in Al-Si eutectic alloys. Scripta Mater. **39**, 1281–1287 (1998)

Prasad L and Iyengar S.S.: Wavelet analysis with applications to image processing, CRC, 1997

Pyrz, R.: Quantitative description of the microstructure of composites: I. Morphology of unidirectional composite systems. Compos. Sci. Tech. **50**, 197–208 (1994)

Qian, S. and Weiss, J.: Wavelets and the numerical solution of boundary value problems. Appl. Math. Lett. **6(1)**, 47–52 (1993)

Raghavan, P. and Ghosh, S.: Adaptive multi-scale computational modeling of composite materials. Comp. Model. Eng. Sci. **5**, 151–170 (2004a)

Raghavan, P. and Ghosh, S.: Concurrent multi-scale analysis of elastic composites by a multi-level computational model. Comput. Meth. Appl. Mech. Eng. **5(2)**, 151–170 (2004b)

Raghavan, P., Li, S. and Ghosh, S.: Two scale response and damage modeling of composite materials. Fin. Elem. Anal. Des. **40(12)**, 1619–1640 (2004)

Rintoul, M.D. and Torquato, S.: Reconstruction of structure of dispersions. J. Colloid Int. Sci. **186**, 467–476 (1996)

Robert, R.K.: Cubic convolution interpolation for digital image processing. IEEE Trans. Acous. Speech Signal Proc. **ASSP-29**, 1153–1160 (1981)

Russ, J.C.: The Image Processing Handbook, 3rd Edition, CRC and IEEE, New York (1999)

Sahoo, P.K., Soltani, S., Wong, A.K.C.: A survey of thresholding techniques. Comp. Vision Graphics Image Proc. **41**, 233–260 (1988)

Segurado, J. and Llorca. J.: A computational micromechanics study of the effects of interface decohesion on the mechanical behavior of composites. Acta Mater. **53**, 4931–4942 (2005)

Serra, J.: Image Analysis and Mathematical Morphology, Academic (1982)

Seul, M., O'Gorman, L. and Sammon, M.J.: Practical Algorithms for Image Analysis, Cambridge University Press (2000)

Shan, Z. and Gokhale, A.M.: Digital image analysis and microstructure modeling tools for microstructure sensitive design of materials. Int. J. Plasticity **20**, 1347–1370 (2004)

Smit, R.J.M., Brekelmans, W.A.M. and Meijer, H.E.H.: Prediction of the mechanical behavior of nonlinear heterogeneous systems by multi-level finite element modeling. Comp. Meth. Appl. Mech. Eng. **155(1–2)**, 181–192 (1998)

Spitzig, W.A., Kelly, J.F. and Richmond, O.: Quantitative characterization of second-phase populations. Metallography **18**, 235–261 (1985)

Spowart, J.E., Mayurama, B. and Miracle, D.B.: Multiscale characterization of spatially heterogeneous systems: Implications for discontinuously reinforced metal-matrix composite microstructures. Mater. Sci. Eng. **A307**, 51–66 (2001)

Swaminathan, S., Ghosh, S. and Pagano, N.J.: Statistically equivalent representative volume elements for composite microstructures, Part I: Without damage. J. Compos. Mater. **7(40)**, 583–604 (2006)

Swaminathan, S. and Ghosh, S.: Statistically equivalent representative volume elements for composite microstructures, Part II: With damage. J. Compos. Mater. **7(40)**, 605–621 (2006)

Terada, K. and Kikuchi, N.: Simulation of the multi-scale convergence in computational homogenization approaches. Int. J. Solids Struc. **37**, 2285–2311 (2000)

Tewari, A., Gokhale, A.M., Spowart, J.E. and Miracle, D.B.: Quantitative characterization of spatial clustering in three-dimensional microstructures using two-point correlation functions. Acta Mater. **52**, 307–319 (2004)

Torquato, S.: Random Heterogeneous Materials: Microstructure and macroscopic properties, Springer, New York (2002)

Unser, M., Aldroubi, A. and Eden M.: Fast B-spline transforms for continuous image representation and interpolation. IEEE Trans. Pattern Anal. Mach. Int. **13**, 277–285 (1991)

Valiveti, D.M. and Ghosh, S.: Domain partitioning of multi-phase materials based on multi-scale characterizations: A preprocessor for multi-scale modeling. Int. J. Numer. Meth. Eng. **69(8)**, 1717–1754 (2007)

Vemaganti, K. and Oden, J.T.: Estimation of local modeling error and goal-oriented adaptive modeling of heterogeneous materials, Part II: A computational environment for adaptive modeling of heterogeneous elastic solids. Comput. Meth. Appl. Mech. Eng. **190**, 6089–6124 (2001)

Wang, Q.G., Caceres, C.H. and Griffiths, J.R.: Damage by eutectic particle cracking in Aluminum casting alloys A356/357. Metall. Mater. Trans. A. **34A**, 2901–2912 (2003)

Wang, Q.G.: Microstructural effects on the tensile and fracture behavior of Aluminum casting alloys A356/357. Metall. Mater. Trans. A. **34A**, 2887–2899 (2003)

Weissenbek, E., Boehm, H.J. and Rammerstoffer, F.G.: Micromechanical investigations of fiber arrangement effects in particle reinforced metal matrix composites. Comp. Mater. Sci. **3**, 263–278 (1994)

Xia, Z., Curtin, W.A. and Peters, P.W.M.: Multiscale modeling of failure in metal matrix composites. Acta Mater. **49**, 273–287 (2001)

Yang, N., Boselli, J. and Sinclair, I.: Simulation and quantitative assessment of homogeneous and inhomogeneous particle distributions in particulate metal matrix composites. J. Microscopy **201**, 189–200 (2000a)

Yang, S., Gokhale, A.M. and Shan, Z.: Utility of microstructure modeling for simulation of micro-mechanical response of composites containing non-uniformly distributed fibers. Acta Mater. **48**, 2307–2322 (2000b)

Yeong, C.L.Y. and Torquato, S.: Reconstructing random media. Physical Rev. E. **57**, 495–505 (1998)

Yotte, S., Riss, J., Breysse, D. and Ghosh, S.: PMMC cluster analysis. Comp. Model. Eng. Sci. **5**, 171–187 (2004)

Zohdi, T.I. and Wriggers, P.: A domain decomposition method for bodies with heterogeneous microstructure based on material regularization. Int. J. Solids Struct. **36**, 2507–2525 (1999)

Coupling Microstructure Characterization with Microstructure Evolution

Chen Shen, Ning Ma, Yuwen Cui, Ning Zhou, and Yunzhi Wang

Abstract Microstructure reconstruction in 3D and quantitative digital representation are enabling consideration of polycrystalline and multi-phase microstructures in mechanics codes in a realistic way. To take full advantage of these advances, we discuss in this chapter the synergy of coupling quantitative microstructure characterization by experimental imaging techniques with quantitative microstructural evolution modeling by image-based computer simulation techniques such as the phase field method. Specific attention will be paid to the fundamentals of the phase field method for microstructure representation and description of microstructure evolution, and procedures of using experimental images as model inputs. Through individual examples we show how to use the phase field method at different length scales to: explore mechanisms of microstructural evolution, extract important materials parameters, carry out physics-based repairs of experimentally reconstructed microstructures, and evolve existing microstructures or generate new microstructures to populate digital microstructural database for different time, temperature, stress, and other service conditions for mechanical property explorations.

1 Introduction

Material microstructures are defined by spatial arrangements of a large assembly of structural and chemical nonuniformities (imperfections or defects), such as dislocations, homophase and heterophase interfaces, concentration variation in multiphase alloys, and impurity segregation at structural defects. Since material properties are determined by the state of microstructure, it is essential to have an accurate representation of various microstructural features at appropriate length scales in any predictive models for material behaviors. Since most engineering

Y. Wang (✉)
Department of Materials Science and Engineering, The Ohio State University,
2041 College Road, Columbus, OH 43210, USA
e-mail: wang.363@osu.edu

S. Ghosh and D. Dimiduk (eds.), *Computational Methods for Microstructure-Property Relationships*, DOI 10.1007/978-1-4419-0643-4_5,

alloys are multicomponent, multiphase, and polycrystalline, their microstructures are extremely complex. The traditional microstructure models based on one-point correlation functions (such as volume fraction and average particle size) are neither sufficient to quantitatively define the microstructures nor adequate to allow for establishing a robust microstructure–property relationship. Image-based microstructure models such as Monte Carlo method, cellular automata method, and phase field method [see (Raabe 1998) for an overview] describe an evolving microstructure by tracking variation of microstructural state at each voxel (i.e., volumetric pixel) and hence provide complete information regarding the microstructure at a resolution defined by the size of the voxels. In this chapter, we discuss the synergy of coupling quantitative microstructure characterization by experimental imaging techniques with quantitative computer simulations of microstructural evolution using the phase field method. Utilizing various experimental images as direct inputs, we show detailed procedures on how to use the phase field method at different length scales to (1) explore mechanisms of microstructural evolution and extract important but difficult-to-measure material parameters, (2) carry our physics-based repairs of experimentally reconstructed microstructures, (3) evolve the microstructures and generate new microstructures to populate microstructural database for different time, temperature, stress, and other service conditions for mechanical property explorations. Fundamentals of phase field method for microstructure representation and description of microstructure evolution will first be introduced (Sect. 2), followed by discussions on model input parameters and procedures of using experimental images as initial microstructures (Sect. 3). These are the fundamental issues that have to be addressed for quantitative, materials-specific phase field simulations. Numerical algorithms are discussed in Sect. 4, followed by examples of applications in Sect. 5 and concluding remarks in Sect. 6.

2 Fundamentals of Phase Field Method

2.1 Description of Microstructure

In the phase field approach, chemical and structural nonuniformities in a multiphase and polycrystalline material are characterized by two types of continuum fields: conserved and nonconserved order parameters. Typical examples of the conserved order parameters include solute concentration, density and molar volume; while typical examples of the nonconserved order parameters include long-range order parameters for atomic ordering, inelastic displacement or inelastic strain (eigenstrain or transformation strain) for dislocations, martensitic particles and microcracks, and magnetization and polarization for ferromagnetic and ferroelectric transitions. These order parameters are well-defined physical properties. In cases where the choice of a well-defined physical quantity as the order parameter becomes difficult, such as in solidification and grain growth, phenomenological order parameters

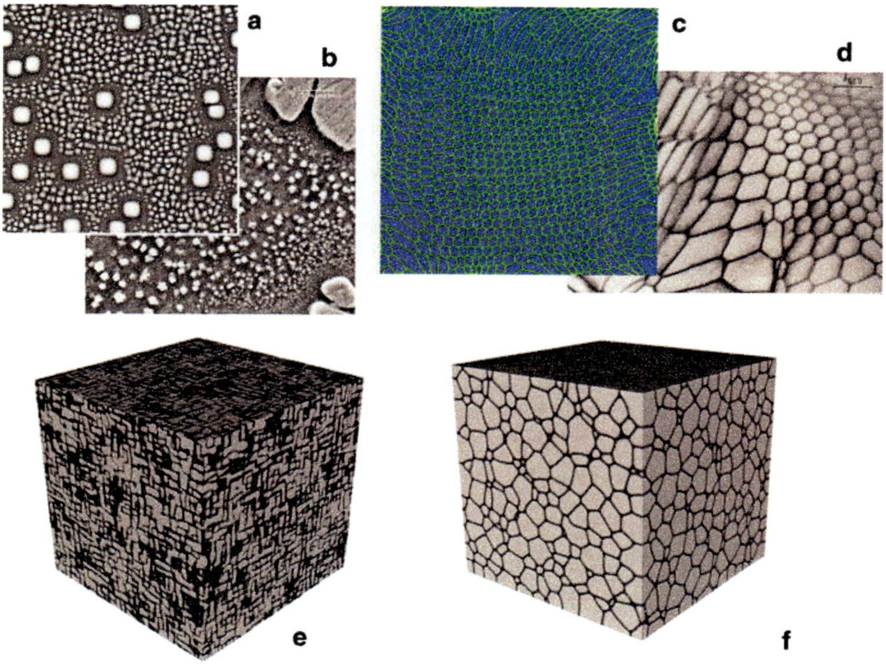

Fig. 1 Examples of microstructures predicted by phase-field simulations. (**a**) Bi-modal γ/γ' microstructure in a Ni-base superalloy for disk applications (courtesy of Y.H. Wen, UES Inc.) and (**b**) its comparison with experimental observation (courtesy of M.F. Henry, GE); (**c**) dislocation network formed on (111) plane in an FCC crystal and (**d**) its comparison with experimental observation [S. Amelinckx (1964) *The Direct Observation of Dislocations*]; (**e**) a γ/γ' microstructure in a Ni-base superalloy for air foil applications; (**f**) a polycrystalline microstructure

are introduced (Boettinger et al. 2002; Chen 2002). In the phase field method, these order parameters are defined as continuum fields (i.e., functions of position, $\mathbf{r}$, e.g., $\phi(\mathbf{r}, t)$) and are referred to as phase fields. Since a microstructure is an ever-evolving feature toward thermodynamic equilibrium, the descriptors ϕ generally include time t as the variable. A graphic plot of the order parameter fields produces images that are similar to the ones typically observed under a microscope (Fig. 1). Therefore, the field description of microstructures offers a natural link between microstructure characterization and microstructure modeling.

2.2 Governing Equations

Microstructure evolution in the phase field method, represented by time-evolution of the order parameter fields, is governed by total energy reduction along the steepest descent path. Since a field description of microstructure is mathematically a set of spatial coordinate dependent functions, the total energy is written as a *functional* (i.e., a function of functions),

$$E = E[\phi(\mathbf{r}, t)], \tag{1}$$

where the independent variable, ϕ, is itself a function of spatial coordinate. In the framework of thermodynamics, E is directly related to one of the free energies subject to given external constraints. If a system is under constant pressure and temperature, E shall represent the Gibbs free energy. The steepest descending direction of the total energy is given by "derivatives" of E with respect to each field variable and constitutes the thermodynamic driving force for the change of that variable at each location. Since E is a functional, the derivatives are regarded as variational derivatives in calculus of variations (any variation calculus textbook may be referred for mathematical background). While E is a scalar quantity its variational derivative

$$\delta E / \delta \phi(\mathbf{r}, t) \tag{2}$$

is a d-dimensional vector, where d is the total degrees of freedom of the system.

When a microstructure is represented by a conserved order parameter field, e.g., spatial distribution of solute concentration, the field variable can be defined as the mole fraction of the solute

$$\phi = X(\mathbf{r}, t). \tag{3}$$

For a multicomponent alloy there are n–1 composition fields, $\phi = \{X_i(\mathbf{r}, t), i = 1, ..., n-1\}$, where n is the total number of chemical species.

According to the so-called gradient thermodynamics (van der Waals 1893; Landau and Lifshitz 1935; Cahn and Hilliard 1958), when chemical or structural nonuniformities exist in a heterogeneous system the free energy of the system depends not only on local composition and structure but also on their spatial variations. Thus the total chemical free energy of a system of *nonuniform* concentration may be written formally as (Cahn and Hilliard 1958)

$$E^{\text{chem}} = E^{\text{chem}}[X(\mathbf{r}, t)] = \int f(X, \nabla X, \nabla^2 X, \ldots)\mathrm{d}\mathbf{r}. \tag{4}$$

With symmetry considerations and taking only the leading nonvanishing term in concentration gradient, (4) is reduced to

$$E^{\text{chem}} = \int V_m^{-1} \left[f_m(X) + \kappa \nabla X \cdot \nabla X \right] \mathrm{d}\mathbf{r} \tag{5}$$

where V_m is the molar volume and f_m is the molar free energy of a corresponding homogeneous material of composition X. The second term in the bracket takes into account the spatial variation (gradient) of X, where the gradient coefficient κ(in a unit of J·m^2·mol^{-1}) is related to the second derivatives of f in (4) [see (2.7) in (Cahn and Hilliard 1958)]. For brevity, we drop the independent variables $\mathbf{r}$ and t in X in (5) and hereafter.

If an existing spatial distribution of solute is not in thermodynamic equilibrium, its temporal evolution follows the Cahn–Hilliard generalized diffusion equation

$$\frac{1}{V_m}\frac{\partial X}{\partial t} = \nabla \cdot M \nabla V_m \frac{\delta E^{\text{chem}}}{\delta X} \tag{6}$$

where M is the chemical mobility (in a unit of $\text{J}^{-1}\cdot\text{mol}\cdot\text{m}^{-1}\cdot\text{s}^{-1}$). It is easy to verify that from (5) (assuming V_m is independent of spatial location)

$$V_m \frac{\delta E^{\text{chem}}}{\delta X} = \frac{\partial f_m}{\partial X} - 2\kappa \nabla^2 X. \tag{7}$$

Recall that $\Delta\mu = \partial f_m / \partial X$ is commonly regarded as the diffusion potential or exchange potential (a difference between the chemical potential of solute and solvent atoms), we write

$$\Delta\tilde{\mu} \equiv \Delta\mu - 2\kappa \nabla^2 X \tag{8}$$

as a generalized diffusion potential with the second term at the right-hand side as a gradient correction. With that, (6) can be rewritten as

$$\frac{1}{V_m}\frac{\partial X}{\partial t} = -\nabla \cdot \mathbf{J} \tag{9}$$

and

$$\mathbf{J} = -M \nabla (\Delta\tilde{\mu}). \tag{10}$$

Equations (9) and (10) give the conventional forms of diffusion equations, with $\mathbf{J}$ being the diffusion flux (in a unit of mol $\text{m}^{-2}\cdot\text{s}^{-1}$) defined in a laboratory frame of reference. Extension of the above governing equations into a multicomponent system is straightforward.

When a microstructure is characterized by structural nonuniformities, a set of nonconserved order parameters is used in phase field models. The number of order parameters is determined by the degrees of freedom of the microstructure. For example, three order parameters are needed for describing the Ni_3Al intermetallic (γ') phase to produce four antiphase domains in an $L1_2$ ordered crystal structure (Khachaturyan 1983; Wang et al. 1998; Poduri and Chen 1998), while the number of order parameters is equal to the number of crystallographic orientations of grains in a polycrystalline aggregate (Chen 1995).

The total energy functional formulation for a structurally nonuniform system is similar to that for a chemically non-uniform system presented above [(3)–(5)], with (3) being replaced formally by

$$\phi = \eta(\mathbf{r}, t) \tag{11}$$

to distinguish the structural (nonconserved) fields from the concentration (conserved) fields. One could also assign a subscript to η if there are multiple fields (see examples in Sect. 3). Nonconserved fields follow a different type of kinetic equation, known as time-dependent Ginzburg–Landau equation (Ginzburg and Landau 1950) or Allen–Cahn equation (Allen and Cahn 1979):

$$\frac{\partial \eta}{\partial t} = -L\frac{\delta E^{\text{chem}}}{\delta \eta} \tag{12}$$

where the unit of the kinetic coefficient L is $\text{J}^{-1}\cdot\text{m}^3 \cdot s^{-1}$.

2.3 Interface Property and Curvature

Note that the gradient term is a unique feature of the phase field method where the field representation of microstructures provides a unified description of multiple phase and polycrystalline microstructure with smooth transitions from one phase domain or grain to another. This is in contrast to conventional sharp-interface models where governing equations are solved separately for individual phase domains and grains bounded by explicit conditions along the moving interphase and grain boundaries. Physically, the spatial (nonlocal) interaction represented by the gradient term in gradient thermodynamics of nonuniform systems arises in a continuum-limit transition from its discrete counterparts (Hillert 1956; Lee and Aaronson 1980) or from microscopic theories in statistical mechanics [e.g., see (Langer 1969, 1971)]. It accounts for the change in atomic bonding from one location to its neighboring locations. In phase field models where the phase fields are physical order parameters, minimization of the total free energy determines the balance between the local free energy term that prefers an infinitely sharp interface and the gradient energy term that prefers an infinitely diffusion interface. Such a balance regulates the interface profile and gives interface energy, σ, and width, w, respectively, as

$$\sigma = 2V_m^{-1}\int_{\phi_1}^{\phi_2} \sqrt{\kappa \Delta f(\phi)}\mathrm{d}\phi \tag{13}$$

$$w = 2(\phi_2 - \phi_1)\sqrt{\kappa/\Delta f_{\max}} \tag{14}$$

where the quantities $\phi_1, \phi_2, \Delta f(\phi)$, and $\Delta f_{\max}$ are defined in Fig. 2.

In some phenomenological phase field models, typically those evolved from solidification models, the gradient term does not appear on every phase field introduced. For example, the gradient terms on concentration fields are usually neglected (Wheeler et al. 1992), or even with a particular purpose to remove the chemical contribution to interface properties (Kim et al. 1999). One benefit associated with the latter case is that the interface properties (energy and width) are not bounded by the physical constraint of the chemical free energy model [e.g., $\Delta f(\phi)$ in (13) and (14)] anymore and the models can be coarse-grained to much larger length scales (see Sect. 2.6 for details).

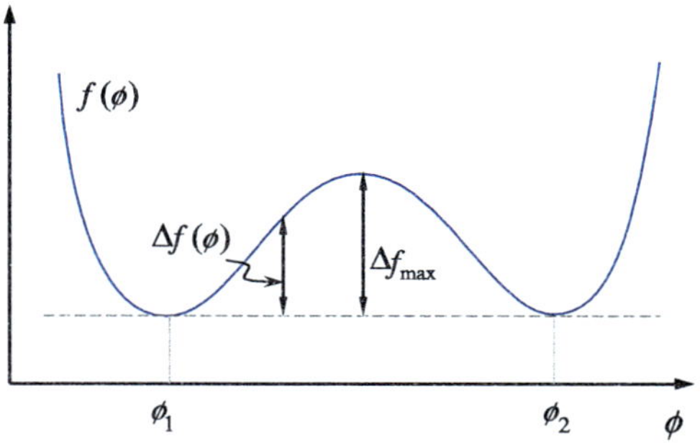

Fig. 2 Schematic drawing of the free energy density. The excess energy $\Delta f(\phi)$ between ϕ_1 and ϕ_2, which are the equilibrium phase-field values at the two sides of the interface, contributes to the interface energy

For both physically and phenomenologically defined phase fields, the gradient term provides another important feature in two- or three-dimensional space: the interface curvature. This can be seen more clearly if we make the expansion

$$\nabla^2\phi = \frac{\partial^2\phi}{\partial r^2} + \frac{d-1}{R}\frac{\partial\phi}{\partial r} \tag{15}$$

where R is the local radius of curvature of the interface and d is the dimensionality of the space. It can be seen (Langer 1992) that the second term at the right side contributes to local interface velocity, a term in proportion to $(d-1)/R$. Note that $\nabla^2\phi$ is proportional to the variational derivative of the gradient energy term, [e.g., see (5) and (7)]. Equation (15) implies that the gradient term contributes to the driving force from both the variation of phase field across (normal to) the interface, $\partial^2\phi/\partial r^2$, and the local radius of curvature, R. The interface curvature is the important driving force for grain growth and precipitate coarsening (Gibbs–Thomson effect).

2.4 Growth and Coarsening

The phase field kinetic equations (6) and (12) cover both growth and coarsening, with the driving force being the functional variation of the total free energy and the kinetic rate in proportional to the driving force. These equations can be reduced to sharp interface equations of domain (precipitate or grain) growth and coarsening [see, e.g., (Langer 1986)]. For a spherical domain, for example, they can be reduced, respectively to (Langer 1992):

$$\frac{\mathrm{d}R}{\mathrm{d}t} = \frac{L\kappa_\eta\Delta\eta}{\sigma}H - \frac{L\kappa_\eta(d-1)}{R} \tag{16}$$

$$\frac{\mathrm{d}R}{\mathrm{d}t} = \frac{D}{R}\left[\frac{\delta X}{\Delta X} - \frac{(d-1)d_0}{R}\right] \tag{17}$$

where σ is the interfacial energy, $\Delta\eta = \eta_\beta - \eta_\alpha$ is the difference between the values of the order parameter at the two equilibrium states, $D = V_m^2 M(\mathrm{d}^2 g_m/\mathrm{d}X^2)$ is the diffusivity, $d_0 = V_m^2\sigma(d-1)/(\mathrm{d}^2 g_m/\mathrm{d}X^2)/(\Delta X)^2$ is the correlation length, $\Delta X = X_\beta - X_\alpha$ is the miscibility gap, and δX is the concentration supersaturation. The growth rate, $\mathrm{d}R/\mathrm{d}t$, for a nonconserved field (16) reproduces a linear kinetic law under a constant external field, H, or a parabolic kinetic law for pure curvature driven kinetics (in grain growth for instance), depending on which of the two terms is dominant at the right-hand side of the equation. For conserved fields such as solute composition, the growth law can be parabolic, driven by supersaturation, δX, or cubic, driven by curvature, governed, respectively, by the two terms on the right-hand side of (17). The second term is what underlies the coarsening kinetics.

While the purely dissipative phase field kinetic equations [(6) and (12)] lead to monotonic decrease of total free energy with time, activation processes such as nucleation are often simulated in phase field models to account for microstructure formation via first-order phase transformations. Typically, this is accomplished by two means: the use of a Langevin force type of perturbative driving force in conjunction with the dissipative equations (Gunton et al. 1983), or an explicit introduction of nuclei into untransformed material volume based on evaluation of driving force for nucleation at local positions (Simmons et al. 2000). The former approach essentially is to mimic a physical process of nucleation via fluctuation-dissipation route and retains naturally physical fidelity of the process, while the latter focuses on the deterministic relation between thermodynamic driving force and nucleation activities and is a much more efficient and flexible means to account for nucleation during microstructure evolution in phase field simulations. Since nucleation is not a focus of the current chapter, readers are referred to a recent review in ASM Handbook Vol. 22. Modeling and Simulations: Processing of Metallic Materials (Shen and Wang 2010).

2.5 Long-Range Elastic Interactions

In solids a lot of microstructural constitutions carry stress, such as dislocations, coherent precipitates, cracks, and voids. The long-range elastic strain fields associated with these defects vary at the scale of the microstructure and dominate its evolution in most cases (Khachaturyan 1983; Wang et al. 1996; Johnson 1999; Shen and Wang 2005). The internal stress fields also react strongly with external load or residual stresses. The effects of stress fields on microstructural evolution range from dislocation substructure formation to precipitate morphology (e.g., equilibrium shape, habit planes) and spatial arrangement, to precipitate–dislocation interactions, to crack propagation, etc. In phase field models for solid state phase transformations and dislocations, the elasticity problems are treated in the framework of Eshelby (1957, 1959) using the general theory of phase field microelasticity by Khachaturyan and Shatlov (Khachaturyan 1983, 1966, 1967; Khachaturyan and Shatalov 1969) (hereafter *KS microelasticity theory* in short), where the total deformation is partitioned into elastic and inelastic ones, with the inelastic deformation characterized by an inelastic strain field

$$\varepsilon_{ij}^{T}(\mathbf{r}) = \sum_{p} \varepsilon_{ij}^{T}(\phi_p(\mathbf{r})) \tag{18}$$

which is a linear combination of contributions from each individual phase field, labeled by subscript, p, such as the stress-free transformation strain of a precipitate and eigenstrain of a dislocation loop. As a simplification, it may be approximated as

$$\varepsilon_{ij}^{T}(\mathbf{r}) = \sum_{p} \varepsilon_{ij}^{T0}(p)\phi_p(\mathbf{r}) \tag{19}$$

where $\varepsilon_{ij}^{T0}(p)$ is the coefficient of the linear term in a Taylor expansion of $\varepsilon_{ij}^{T}(\phi_p(\mathbf{r}))$ with respect to ϕ_p. In the case of concentration field ($\phi_p = X$), this is known as the Vagard's law. In the KS microelasticity theory, the elastic energy is formulated as a functional of the total strain (both elastic and inelastic) field and the elastic strain was relaxed instantaneously by minimizing the elastic energy with respect to the elastic displacement under a given inelastic strain through Green's function solution. This allows the elastic energy to be expressed in a close form as a function of the inelastic strain only:

$$\begin{aligned} E^{\mathrm{el}} = {} & \frac{1}{2}\int d\mathbf{r} C_{ijkl}\varepsilon_{ij}^{T}(\mathbf{r})\varepsilon_{kl}^{T}(\mathbf{r}) + \frac{V}{2}C_{ijkl}\bar{\varepsilon}_{ij}\bar{\varepsilon}_{kl} - \bar{\varepsilon}_{ij}\int d\mathbf{r} C_{ijkl}\varepsilon_{kl}^{T}(\mathbf{r}) \\ & -\frac{1}{2}⨍ \frac{d\mathbf{g}}{(2\pi)^3} n_i \tilde{\sigma}_{ij}^{T}(\mathbf{g})\Omega_{jk}(\mathbf{n})\tilde{\sigma}_{kl}^{T*}(\mathbf{g})n_l \end{aligned} \tag{20}$$

with $\sigma_{ij}^{T} = C_{ijkl}\varepsilon_{kl}^{T}$, $[\Omega^{-1}]_{jk} = n_i C_{ijkl} n_l$, $\mathbf{n} = \mathbf{g}/|\mathbf{g}|$, and $\mathbf{g}$ is a reciprocal space vector. $\tilde{\sigma}_{ij}^{T}(\mathbf{g}) = \int d\mathbf{r}\sigma_{ij}^{T}(\mathbf{r})\exp(-i\mathbf{g}\cdot\mathbf{r})$ is the Fourier transform of $\sigma_{ij}^{T}(\mathbf{r})$. The separation of the homogeneous strain $\bar{\varepsilon}_{ij}$ (the mean value of $\varepsilon_{ij}^{T}(\mathbf{r})$) gives flexibility of treating various boundary conditions of the elasticity problem (Khachaturyan 1983; Li and Chen 1997a; Wang et al. 2002).

Through (18)–(20), the elastic energy becomes a sole functional of the phase fields, ϕ_p, just as the chemical free energy given in (4). It is thus possible to compute the total driving forces as a combination of the chemical and elastic energies

$$\frac{\delta(E^{\mathrm{chem}} + E^{\mathrm{el}})}{\delta\phi_p} \tag{21}$$

which are the variational derivatives of the total energy with respect to the same set of phase fields $\{\phi_p\}$.

Recently, the KS microelasticity theory was extended to treat inhomogeneous (position-dependent) elastic modulus (Wang et al. 2002; Onuki 1989; Khachaturyan et al. 1995; Hu and Chen 2001), which extends further the applications to include cracks and voids (Wang et al. 2001), free surfaces (Wang et al. 2003), elasticity-induced rafting (Li and Chen 1997a, b; Zhou et al. 2008), etc. Therefore, the contributing phase fields to the inelastic strain in general can be many crystalline defects, including solute clusters, precipitates, dislocations, cracks, and voids. Their mutual elastic interactions are accounted for through the coupling in (18). This treatment provides phase field models with an ability of handling self-consistently the evolution of an arbitrary solid state microstructure consisting of various stress-carrying defects with interplay among the chemical free energy, interfacial energy, and elastic energy [for recent reviews, see (Shen and Wang 2005; Wang et al. 2005a)]. Because of the field description of defects, the complexity of computation is insensitive to the complexity of the defect configurations and morphologies. Essentially, the treatment is identical for defects of arbitrary types, shapes and spatial distributions (e.g., straight or curved, coplanar or noncoplanar dislocations).

2.6 Quantitative Phase Field Simulation and Length Scale

To evolve microstructures directly from experimental images and predict microstructural states at later times under a given set of processing conditions such as temperature and stress, it is essential to carry out quantitative phase field simulations at length scales consistent with the experimentally observed microstructures. Equations (13) and (14) show that interfaces in phase field models have finite equilibrium widths, unique chemical and structural variations within them (Fig. 2), and the associated interfacial energies. When applied at the natural (typically microscopic) length scales of a given defect (such as an individual dislocation, interface or nucleating precipitate), the phase field model has a unique advantage over the sharp-interface models in *predicting* the fundamental properties of the defect such as its equilibrium size and energy (Cahn and Hilliard 1958), and critical configuration and activation energy of a nucleus (Cahn and Hilliard 1959) rather than using them as inputs. One can refer to these phase field models as the *microscopic phase field* (MPF) models. As a matter of fact, early applications of the phase field method were to predict fundamental properties of extended defects, as demonstrated by Cahn and Hilliard who studied the equilibrium concentration variation across a coherent interface, the width of the interface and the corresponding interfacial energy (Cahn and Hilliard 1958), and the concentration profile and activation energy of a critical nucleus (Cahn and Hilliard 1959).

When applied at coarse-grained length scales (e.g., $\sim\mu$m), however, the phase field models yield interfacial widths far exceed their natural values. This is because the interfacial regions in a phase field model have to be "numerically" smooth (several grid size wide irrespective of the actual grid size) to ensure the accuracy in evaluation of the gradient terms. In these cases, the phase field models lose their intrinsic ability to *predict* the fundamental properties of interfaces, but retain their advantages over the sharp-interface models in treating complicated geometrical and topological changes of large defect ensembles during microstructural evolution (such as dendritic solidification, grain growth and domain coarsening, dislocation network formation and coarsening, and various phase transformations [for reviews see (Boettinger et al. 2002; Chen 2002; Shen and Wang 2005; Wang et al. 2005a, b; Wang and Chen 2000; Karma 2001)]. Their applications at different coarse-grained levels have offered many invaluable insights into the sequence of microstructural evolutions and mechanisms of pattern formation in many material systems during various material processes.

Coarse-grained phase field models can also provide quantitative rate information by matching phase field model parameters to the standard sharp-interface models at asymptotic limit of zero interface width (Wheeler et al. 1992), or more recently in the so-called thin-interface limit (Karma and Rappel 1998). Quantitative phase field simulations at coarse-grained mesoscale levels have since been carried out extensively with the new thin-interface analyses (Karma and Rappel 1998; 1996; Almgren 1999; McFadden et al. 2000; Elder et al. 2001). Some simple techniques (Shen et al. 2004a, b) based on physical arguments of equivalent driving forces were also developed to relax the restriction on the interface thickness. In addition, adaptive algorithms (Provatas et al. 1998; Feng et al. 2006) and new data structures (Gruber et al. 2006; Vedantam and Patnaik 2006) have been developed to increase the computational efficiency of quantitative phase field simulations.

One of the major obstacles for quantitative phase field simulations at coarse-grained length scales is the coupling between boundary width and boundary energy in the gradient thermodynamics. According to (13) and (14), for example, the boundary width and energy are determined by the interplay between the local free energy (in particular, Δf) and the gradient energy. If the local chemical free energy model is formulated based on available thermodynamic database such as the CALculation of PHAse Diagrams (CALPHAD) database, the values Δf of and κ in (13) and (14) are fixed and the phase field models predict equilibrium boundary widths that correspond to their natural values (typically of the order of nanometers depending on particle size and temperature). When applied at a coarse-grained length scale ($\sim\mu$m), however, the interfacial energies required to keep a diffuse interface at such a length scale will be several orders of magnitude higher than their physical values (Shen et al. 2004a). This will alter significantly the driving force for microstructure evolution.

To decouple boundary width from boundary energy in the phase field model so that it can be coarse-grained to arbitrary length scales without encountering unrealistically high interfacial energies, a multiphase-field model (Tiaden et al. 1998) was proposed. The thermodynamically consistent formulation of this model was developed by Kim, Kim and Suzuki (Kim et al. 1999) which is now known as the KKS model. In these formulations, the interface region is treated as a homogeneous mixture of the adjacent phases ($\alpha, \beta, \ldots$) characterized respectively by "phase compositions" (X_α, $X_\beta, \ldots$), and the diffusion potentials of each phase composition are required equal, i.e., $\partial f_\alpha/\partial X_\alpha = \partial f_\beta/\partial X_\beta = \ldots$, where $f_\alpha = f_\alpha(X_\alpha)$, $f_\beta = f_\beta(X_\beta)\ldots$, are the free energies of the respective phases. The latter constraint corresponds to a parallel-tangent construction and gives rise to a force (diffusion potential) balance condition among constituent phases in thermodynamics. Such a treatment,

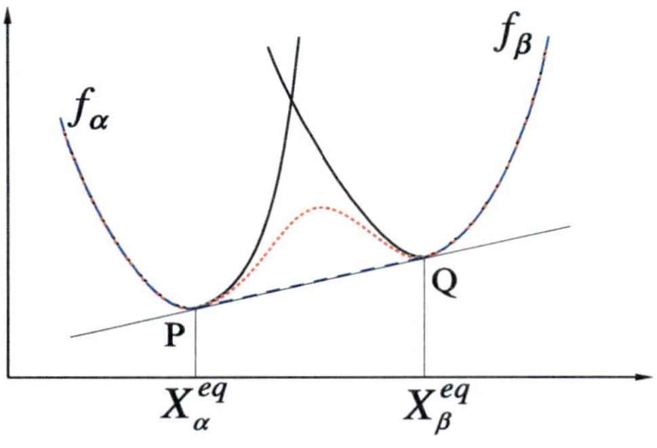

Fig. 3 Schematic drawing (after Kim 1999, #467) of free energy curves for: individual α and β phases (*solid curves*), WBM model (*dotted curve*), and KKS model (*dashed curve*). The chemical free energy contribution to interface energy is graphically represented by the area under the free energy curves and above the common tangent (PQ) between X_α^{eq} and X_β^{eq}. In the KKS model, the excess energy in the interface region is removed by making the free energy equal to that of a two-phase mixture (i.e., a straight line between P and Q)

combined with the absence of gradient terms on composition field, completely removed the contribution of composition profile across an interface to the interface energy (Fig. 3).

For grain growth, the order parameter is phenomenological and so is the local free energy. Thus, the physical constraint from Δf does not exist anymore and the model parameters can be determined fairly freely using desired grain boundary energy and mobility as inputs (Moelans et al. 2008).

2.7 Multicomponent Diffusion

The phase-field kinetic equation for conserved fields (6) can be extended directly to handle multicomponent diffusion. Since by definition the mole fractions of an n-component system are subject to the constraint

$$\sum_{i=1}^{n} X_i = 1. \tag{22}$$

There are only $n-1$ independent phase-field equations. By eliminating (any) one component, e.g., X_n, using (22), the molar free energy becomes

$$f_m = \sum_{i=1}^{n} \mu_i X_i = \sum_{i=1}^{n-1} \mu_i X_i + \mu_n X_n = \mu_n + \sum_{i=1}^{n-1} (\mu_i - \mu_n) X_i \tag{23}$$

where μ_i is the partial molar free energy (chemical potential) of the i-th component. The multicomponent diffusion can thus be described by the coupled $n-1$ equations:

$$\frac{1}{V_m}\frac{\partial X_i}{\partial t} = \nabla \cdot M_{ij} \nabla V_m \frac{\delta}{\delta X_j} \int f_m \mathrm{d}V = V_m \nabla \cdot \sum_{j=1}^{n-1} M_{ij} \nabla(\mu_j - \mu_n) \qquad (24)$$

for $i = 1, \ldots, n-1$. Here, we have dropped the gradient energy term as in the solidification models. The mobility coefficient in a multi-component equation is extended to a matrix form. Both f_m and M_{ij} are now functions of composition $(X_1, X_2, \ldots, X_n)$. Their values are typically obtained from assessed thermodynamic and mobility databases (Grafe et al. 2000; Zhu et al. 2002; Kobayashi et al. 2003; Zhu et al. 2004; Chen et al. 2004; Wu et al. 2008; Kitashima and Harada 2009).

2.8 Multiphase-Field Model

Steinbach et al. (Steinbach et al. 1996; Steinbach and Pezzolla 1999) generalized the phase field model (Wheeler et al. 1992) to treat multiple ($N > 2$) coexisting phases, each characterized by its local mole fraction, $\eta_\alpha(\mathbf{r}, t)$. By this definition, the phase fields, η_α, follow an additional constraint:

$$\sum_{\alpha=1}^{N} \eta_\alpha(\mathbf{r}, t) = 1 \qquad (25)$$

where η_α has a value of unity in the bulk phase α and zero in other phases. An intermediate value between 0 and 1 occurs only in the boundary regions (interphase interfaces and multiphase junctions) between the α phase and other phases. A general multibody phase-field equation for multiple phases was then decomposed into pairwise dual-interaction terms:

$$\frac{\partial \eta_\alpha}{\partial t} = -\sum_{\beta=1}^{\tilde{N}} \frac{L_{\alpha\beta}}{\tilde{N}} \left(\frac{\delta E^{\mathrm{chem}}}{\delta \eta_\alpha} - \frac{\delta E^{\mathrm{chem}}}{\delta \eta_\beta} \right) \qquad (26)$$

where $\tilde{N}$ is the number of phases (labeled by index β) adjacent to phase α with values η_β between (but not including) 0 and 1. The kinetic coefficient L is now taken in a form of pairwise coefficient between two joining phases. Equation (26) was derived (Steinbach and Pezzolla 1999) in a variational framework with the use of Lagrange multiplier to account for the interdependence among η_α (25) and resolved a previous issue of violation to interface stress balance (Steinbach et al. 1996).

Incorporation of compositional fields to the multiphase-field model was also developed in a solidification application (Tiaden et al. 1998) and later in the framework of Kim et al. (Kim et al. 2004) for modeling eutectic solidification of a ternary alloy. The methodology applies equally to solid-state phase transformations and grain growth (Wang et al. 2005a, b).

3 Model Input

To utilize experimental images as direct inputs, phase field models need to be tailored to specific alloy systems. This requires, in addition to the fundamental issues discussed in the preceding sections, matching a number of model parameters to material specific parameters. If one refers to (5), (6), and (12), these model parameters include the chemical free energy f_m, gradient coefficient κ, kinetic coefficients M and L, and molar volume V_m. To take into account elastic interactions one needs additional parameters such as elastic constants, lattice parameters, and lattice correspondence between the precipitate and matrix phases. The molar volume is usually assumed to be constant or a linear function of solute concentrations, with a typically value of $\sim 10^{-5} \mathrm{m}^3 \mathrm{mol}^{-1}$. The gradient coefficient κ can be determined by matching interfacial energy and width to experimental values in conjunction with the chemical free energy f_m, for example, according to the relations given in (13) and (14). This leaves two important independent model inputs: the chemical free energy and the kinetic coefficients (interface mobilities).

The free energy and chemical mobility of most alloy systems possess rather complex dependence on multiple field variables. For instance, a commercial alloy may have more than a dozen chemical elements and the Gibbs free energy and mobility matrix are both functions of the concentration of these elements, in addition to temperature and pressure. To present them in tractable forms for practical modeling, one generally needs (1) to choose appropriate base functions to represent these quantities, (2) to determine the coefficients in the base functions by fitting the free energy and mobilities to thermodynamic and mobility databases, or (3) to develop pseudobinary or pseudoternary databases for computational efficiency. All these tasks can be accomplished within the framework of the CALPHAD approach (Kaufman and Bernstein 1970; Saunders and Miodownik 1998).

3.1 CALPHAD Free Energy

The chemical free energy in the CALPHAD method is formulated in the framework of chemical thermodynamics and takes various forms according to the nature of crystal structure of a phase, such as a solution model for a disorder phase and a sublattice model for an ordered phase. For example, a regular solution model for a ternary alloy reads:

$$f(X_1, X_2, X_3) = \sum_{i=1}^{3} X_i G_i^0 + RT \sum_{i=1}^{3} X_i \ln X_i + \sum_{i=1}^{3} \sum_{j=i+1}^{3} X_i X_j L_{ij}^{00}, \quad (27)$$

where G_i^0 is the reference Gibbs energy for pure element i, L_{ij}^{00} the binary interaction parameters between element i and j. R and T are, respectively, gas constant and absolute temperature. This approach has been systematically matured in the past decade within CALPHAD, with successful extension to multicomponent commercial alloy systems and development of more sophisticated models taking into account structural transformations [e.g., the cluster variation model, see in (Saunders and Miodownik 1998)]. It is also able to provide information on atomic mobilities if taken in conjunction with a critical assessment of diffusion data (Andersson et al. 2002; Campbell et al. 2002). In the CALPHAD database, model parameters are optimized systematically by using available experimental data and more recently together with data from first principle calculations. As an example, the chemical free energy of a binary γ/γ' Ni–Al system using a four-sublattice model is given as (Ansara et al. 1997)

$$\begin{aligned} f(X,\eta_1,\eta_2,\eta_3) = {} & Xg_{\mathrm{Al}}^0 + (1-X)g_{\mathrm{Ni}}^0 + X(1-X)\sum_{i=0}^{3} L_i(2X-1)^i \\ & +4U_1X^2(\eta_1^2+\eta_2^2+\eta_3^2) \\ & +12U_4(1-2X)X^2(\eta_1^2+\eta_2^2+\eta_3^2) - 48U_4X^3\eta_1\eta_2\eta_3 \\ & +0.25RT\{X(1+\eta_1+\eta_2+\eta_3)\ln[X(1+\eta_1+\eta_2+\eta_3)] + \\ & [1-X(1+\eta_1+\eta_2+\eta_3)]\ln[1-X(1+\eta_1+\eta_2+\eta_3)] + \\ & X(1+\eta_1-\eta_2-\eta_3)\ln[X(1+\eta_1-\eta_2-\eta_3)] + \\ & [1-X(1+\eta_1-\eta_2-\eta_3)]\ln[1-X(1+\eta_1-\eta_2-\eta_3)] + \\ & X(1-\eta_1+\eta_2-\eta_3)\ln[X(1-\eta_1+\eta_2-\eta_3)] + \\ & [1-X(1-\eta_1+\eta_2-\eta_3)]\ln[1-X(1-\eta_1+\eta_2-\eta_3)] + \\ & X(1-\eta_1-\eta_2+\eta_3)\ln[X(1-\eta_1-\eta_2+\eta_3)] + \\ & [1-X(1-\eta_1-\eta_2+\eta_3)]\ln[1-X(1-\eta_1-\eta_2+\eta_3)]\} \end{aligned} \tag{28}$$

where η_1, η_2, η_3 are the site fractions. The model parameters g_{Al}^0, g_{Ni}^0, L_i, U_1, U_4 are obtained by using the CALPHAD technique. The free energy (28) has been applied directly in phase field modeling of microstructural evolution in Ni–Al (Zhu et al. 2002). It was further extended to multicomponent systems and utilized in phase field modeling by defining an individual set of order parameters for each element [see review in (Kitashima 2008)]. Another example is the chemical free energy for the α and β phases in an α/β Ti alloy, which can be expressed as

$$\begin{aligned} f^{\alpha,\beta} = {} & \sum_{i=\mathrm{Al,Ti,V}} X_i^0 f_i^{\alpha,\beta} + RT \sum_{i=\mathrm{Al,Ti,V}} X_i \ln X_i \\ & + \sum_{i=\mathrm{Al,Ti,V}} \sum_{j>1} \sum_{r=0}^{n} X_i X_j \left[{}^{r}L_{i,j}^{\alpha,\beta} (X_i - X_j)^r \right] \end{aligned} \tag{29}$$

where ${}^0f_i^{\alpha,\beta}$ is the Gibbs energy of species i in the α or β phase, and ${}^rL_i^{\alpha,\beta}$ is the interaction parameter between the species i and j in the α or β phase, the index (and exponent) r describes a regular solution model when its value is 0, which is the case for Ti–V solid solution having α phase structure, a subregular model when its value is 1, which is the case for Al–V solid solution in both α and β phase structures, and a sub-sub-regular model when its value is 2, which is the case for the Al–Ti solid solution in both α and β phase structures. These model parameters are currently available in the multicomponent thermodynamic databases for Ti-64 and other Ti alloys (http://www.thermocalc.com/Products/Databases/Descriptions/DBD_TTTI3.pdf;http://www.computherm.com/databases.html; Wang et al. 2005a, b; Zhang et al., 2005, 2010). Currently, there are no ternary interaction parameters available for Ti–Al–V system.

3.2 Pseudobinary and Pseudoternary Systems

A direct implementation of phase field code with a multicomponent database poses an extraordinary challenge because it needs a direct coupling of the phase field code with a complex multicomponent thermodynamic calculation engine in addition to solving multicomponent diffusion equations. Fortunately, similar to the expressions often quoted as the "equivalent" Al and Ni contents as γ' or γ former for commercial Ni-base superalloys (Gabb et al. 2000) or the "equivalent" Al and V contents for α or β stabilizer in commercial Ti-based alloys (Collings 1994), the situation becomes much simpler when the multicomponent alloys are treated as a pseudobinary Ni–Al (Zhang, unpublished work) or pseudoternary Ti–Al_x–V_y (Zhang et al. 2007) systems. Using commercial Ti-6wt%Al-4wt%V (Ti-64) alloy as an example, one may introduce two equivalent Al and V weight/mole fractions to account for the effects of alloying elements like C, N, O, Fe, H, and Si on the α and β volume fractions and the β transus temperature (Zhang et al. 2007), i.e.,

$$\mathrm{w(Al)}_x = \mathrm{w(Al)} + 2\mathrm{w(O)} + 5\mathrm{w(c)} + 5\mathrm{w(N)} \tag{30}$$

$$\mathrm{w(V)}_y = \mathrm{w(V)} + 1.5\mathrm{w(Fe)} + 25\mathrm{w(H)} + 0.5\mathrm{w(Si)}, \tag{31}$$

where $\mathrm{w(Al)}_x$ and $\mathrm{w(V)}_y$ are the equivalent Al and V weight fractions, and $\mathrm{w}(i)$ is the weight fraction of species i. Figure 4 shows a good agreement obtained between the calculated β volume fraction by using a pseudoternary Ti–Al–V database and the experimental data for the Ti-64 alloy.

3.3 CALPHAD Free Energy for Multiphase Systems

A combination of polynomial and CALPHAD approaches has become an effective means for treating multiphase systems of both structural and chemical

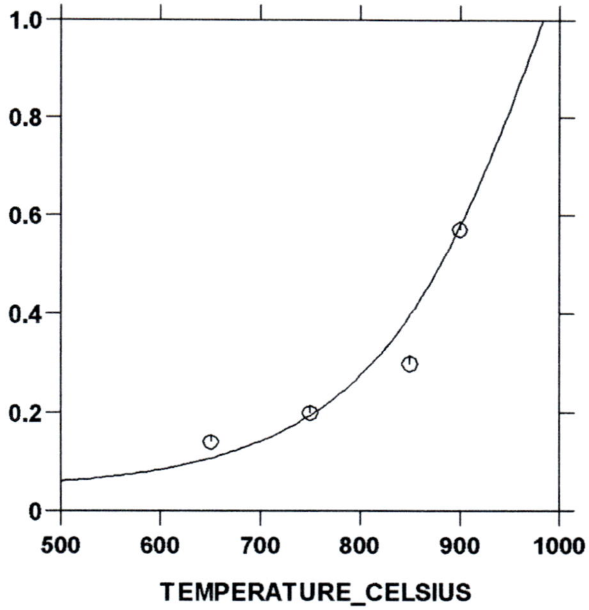

Fig. 4 Calculated β volume fraction for Ti-64 using the pseudo-ternary database (*solid line*) in comparison with experimental data (*open circles*)

nonuniformities. In this approach, the CALPHAD method provides the free energy of each individual phase, $f_p(p = 1, 2, \ldots, N)$, and simple polynomials of a set of nonconserved order parameters, η_p, are used to synthesize all phases in a single function:

$$f(\{X_i\}, \{\eta_p\}) = \sum_{p=1}^{N} \eta_p f_p(\{X_i\}) + \sum_{p=1}^{N} \sum_{q=p+1}^{N} \omega_{pq}\eta_p\eta_q. \tag{32}$$

This treatment simplifies the task for coupling different phases as well as multi-component compositions. It inherits the flexibility of adjusting interface properties through parameter ω_{pq} and the gradient coefficient κ from solidification models and, in the meantime, takes advantages of accurate material specific multicomponent free energies by the CALPHAD method. With this technique, the construction of the KKS model for scaling the interface width and thus the simulation length scale is also straightforward to implement [see, e.g., (Kim et al. 2004)]. An example of such a model applied for Ti-64 (Wang et al. 2005a, b) reads

$$f\left(X_{\mathrm{Al,Ti,V}}, \eta_1, \eta_1, \ldots, \eta_p\right) = h\left(\eta_0\right) f^{\alpha}\left(T, X^{\alpha}_{\mathrm{Al,Ti,V}}\right) + \left(1 - h\left(\eta_0\right)\right) f^{\beta}\left(T, X^{\beta}_{\mathrm{Al,Ti,V}}\right) + \omega \sum_{i \neq j}^{p} \left|\eta_i \eta_j\right| \tag{33}$$

where $h(\eta_0) = \eta_0 \left(6\eta_0^2 - 15\eta_0 + 10\right)$, η_0 is the order parameter for the matrix phase, and f^{α} and f^{β} are the chemical free energies of the α and β phases described by (29). A similar approach was also applied to Ni–Al (Zhu et al. 2004)

$$f(c,\eta_1,\eta_2,\eta_3,\eta_4) = h(\eta_1,\eta_2,\eta_3,\eta_4)f_{\gamma'} + [1 - h(\eta_1,\eta_2,\eta_3,\eta_4)]f_{\gamma} + \omega g(\eta_1,\eta_2,\eta_3,\eta_4), \tag{34}$$

where $f_{\gamma'}$ and f_{γ} are the chemical free energy of γ' and γ phases, respectively, $h = \sum_{i=1}^{4}[\eta_i^3(6\eta_i^2 - 15\eta_i + 10)]$ is an interpolation function with a value from 0 to 1, and a double-well function $g = \sum_{i=1}^{4}[\eta_i^2(1-\eta_i)^2] + \alpha\sum_{i=1}^{4}\sum_{j>i}^{4}\eta_i^2\eta_j^2$. Note in (34) that there are four phenomenological order parameters to represent the four antiphase domains at (1, 0, 0, 0), (0, 1, 0, 0), (0, 0, 1, 0), and (0, 0, 0, 1), instead of three physically defined site fractions (or long-range order parameters) in (28).

3.4 Free Energy for Grain Growth

As has been mentioned earlier, the order parameters are phenomenological and so is the local free energy in phase field models of grain growth. Thus, the exact form of the local free energy, f, is not important as long as it provides multiple degenerated minima corresponding to each grain orientation represented by η_i. A simple form that is commonly used reads

$$f\left(\eta_1,\eta_2,\cdots,\eta_p\right) = \frac{1}{2}\sum_{i=1}^{P}\left(-\eta_i^2 + \frac{1}{2}\eta_i^4\right) + \sum_{i=1}^{P}\sum_{j>i}^{P}\frac{b}{4}\eta_i^2\eta_j^2 \tag{35}$$

where the values of the phenomenological parameter b is determined by grain boundary energy and width. According to (13), (14), and (35), we have

$$\sigma \sim \sqrt{\frac{\kappa(b-1)}{4(b+1)}}, \tag{36}$$

$$w \sim \sqrt{\frac{4\kappa(b+1)}{b-1}}. \tag{37}$$

A general procedure on how to determine phase field model parameters for grain growth simulations from material's grain boundary properties can be found in (Moelans et al. 2008).

3.5 Chemical Mobility of Diffusion

The chemical mobility in multicomponent diffusion (24) may be given as

$$M_{ij} = \frac{1}{V_m}\sum_{l=1}^{n}(\delta_{il} - X_i)(\delta_{jl} - X_j)X_l M_l \tag{38}$$

where $M_l = M_l(\{X_i\}, \{\eta_p\})$ is the mobility of component l in a mixture of multiple phases characterized by $\{\eta_p\}$. It is in general dependent on both composition and long-range order parameters, for example (Chen et al. 2004),

$$M_l = M_l^{\alpha} + M_l^{\beta} - (M_l^{\alpha})^{\eta}(M_l^{\beta})^{1-\eta} \tag{39}$$

which is a combination of the atomic mobilities, M_l^{α} and M_l^{β}, of α and β phases that can be obtained from mobility databases. For example, the chemical mobility (defined in the laboratory reference frame with Ti as dependent species) for the α and β phases in a Ti alloy can be written as functions of the atomic mobility $M_i (i = \text{Al, V, Ti})$ that is stored in the mobility database (Chen et al. 2004; Wang et al. 2005a, b),

$$M_{ii}^{Ti} = V_m^{-1}\left\{(1 - X_i)X_i M_i - X_i\left[(1 - X_i)X_i(M_i - M_{Ti}) - X_i X_j(M_j - M_{Ti})\right]\right\} \tag{40}$$

$$M_{ij}^{Ti} = V_m^{-1}\left\{-X_i X_j M_j - X_j\left[(1 - X_i)X_i(M_i - M_{Ti}) - X_i X_j(M_j - M_{Ti})\right]\right\} \tag{41}$$

with $i, j = Al, V$, and $i \neq j$. In the framework of CALPHAD, the composition-dependent atomic mobility M_i is also described by the Redlich–Kister polynomial in the database (Jonsson 1994).

The mobility coefficient for the nonconserved order parameters, L, in (12) characterizes the contribution of interface kinetics. For solid-state phase transformations, since the interface motion is usually under diffusion-controlled limit, L can be determined at a vanishing kinetic coefficient condition [see, e.g., discussions in (Kim et al. 2004)].

3.6 Fast Diffusion Path (Boundary Diffusion)

At relatively low temperatures (below about 0.75–0.8T_m, where T_m is the melting temperature), boundary diffusion becomes important (Porter and Easterling 1981). By formulating an expression that distinguishes diffusion in grain or interphase boundary regions from that in bulk (Wang 2006), i. e.

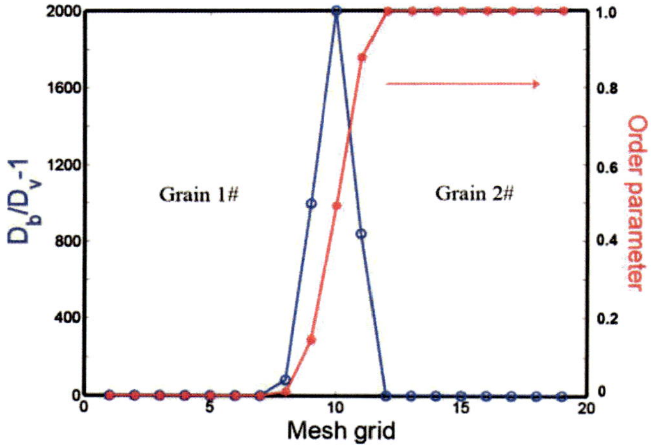

Fig. 5 Typical diffusivity variation within an α/α grain boundary region described by (42)

$$D_b = D_V\left(1 + a\sum_{i,j=1}^{n}\eta_i\eta_j\right) \tag{42}$$

where a is a temperature-dependent coefficient, as shown in Fig. 5, phase field models can describe boundary-bulk mixed diffusion.

3.7 Boundary Mobility

For purely curvature-driven grain growth or antiphase domain coarsening, the velocity of a moving grain boundary is determined by the mobility coefficient, L, for the order parameters in (12), the gradient coefficient in the local free energy, κ, and the curvature of the grain boundary (Allen and Cahn 1979; Fan and Chen 1997a)

$$v = L\kappa\left(\frac{1}{R_1} + \frac{1}{R_2}\right) \tag{43}$$

where R_1 and R_2 are the radii of curvature of the grain boundary. This relation holds when the grain boundary width is much smaller than its radii of curvature [e.g., $w/R \sim 1/4$ (Moelans et al. 2008)]. By comparing with the sharp-interface counterpart equation

$$v = M\sigma\left(\frac{1}{R_1} + \frac{1}{R_2}\right) \tag{44}$$

where M is grain boundary mobility and σ is grain boundary energy given by (12), one can obtain

$$M = \frac{L\kappa}{\sigma}. \tag{45}$$

Thus, mobility coefficient, L, in the phase field governing (12) is directly linked to boundary mobility.

3.8 *Input from Experiment Image as Initial Microstructure*

To use experimental scanning electron microscopy (SEM) or orientation imaging microscopy (OIM) images as starting microstructures for quantitative phase field modeling, the as-received images usually need to be processed. As illustrated in Fig. 6, the first step is to adjust the image contrast to outline the grain/phase boundaries while threshholding the image to remove unwanted details inside the grains, e.g., image contrast from non-equilibrium phases (such as secondary α) and fine concentration fluctuations with the α grains (Fig. 6a), and converting them into grayscale images. Ideally, the α phases can be separated in a grayscale image from the β background by threshholding the grayscale to binary image. Nevertheless, as SEM and OIM images generally carry rich information on composition, grain orientation, and structural defects, it is very rare for any image processing code to have all the grains and boundaries recognized automatically. One way to fill in the "missing pieces" in an experimental image is to use physics-based repairing procedures such as a phase-field like algorithm that will be discussed in Sect. 5.4. The other alternative is to use image processing software programs. An example is given in Fig. 6b. Although not being fully automated and thus time-consuming, the algorithm is simple and straightforward, i.e., brush-erasing an individual α grain in the grayscale image, saving the image as a new one and then comparing it with the old one. This enables the area that was occupied by the erased α grain to be recognized and assigned with a unique grain ID. Repeat this procedure until the last α grain is identified, and finally store all assignments in the two output files containing the α order parameter field and the α/β phase distribution, respectively (Fig. 6c). Both of these will serve as inputs to the phase field code. Figure 7 shows an OIM-like order parameter field output generated from the processed SEM image by this method. All the processes described above were coded in MATLAB (MATLAB® is a registered trademark of The Math Works, Inc.).

4 Numerical Algorithms

The basic phase field kinetic equations [(6) and (12)] are, respectively, fourth- and second-order partial differential equations when the gradient terms are taken into account. Using the forward Euler method for time integration and the central

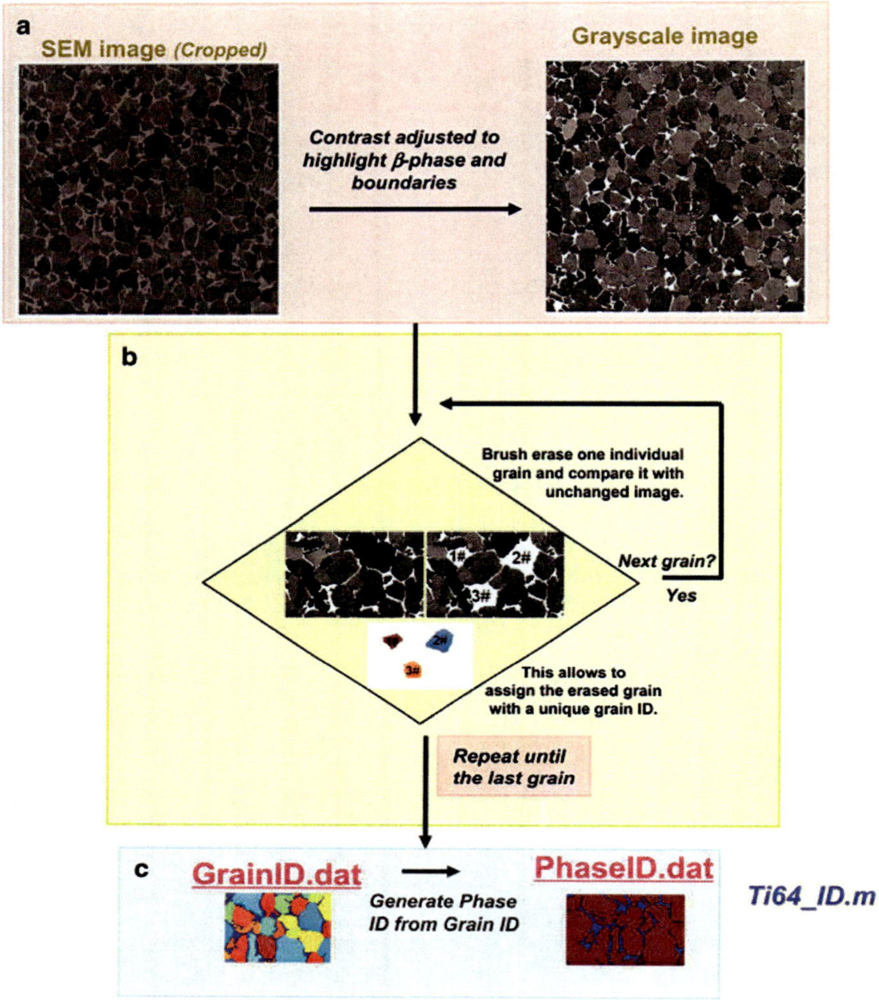

Fig. 6 Conversion of an SEM image to phase-field input. (**a**) Image processing of the as-received SEM image; (**b**) assigning grain IDs for the α phase; (**c**) identify phase IDs

difference approximation for space derivatives (FTCS) is the simplest numerical algorithm and works practically well. In applications associated with elasticity problems, the real space finite difference may be replaced by differentiation in the reciprocal (Fourier) space to accommodate the reciprocal-space formulation of the KS microelasticity equations [e.g., (20)].

The time integration can be improved with more advanced algorithms such as fourth-order Runge–Kutta method or fifth-order Runge–Kutta with adaptive time stepping. Stability of solution can benefit from the use of implicit methods including

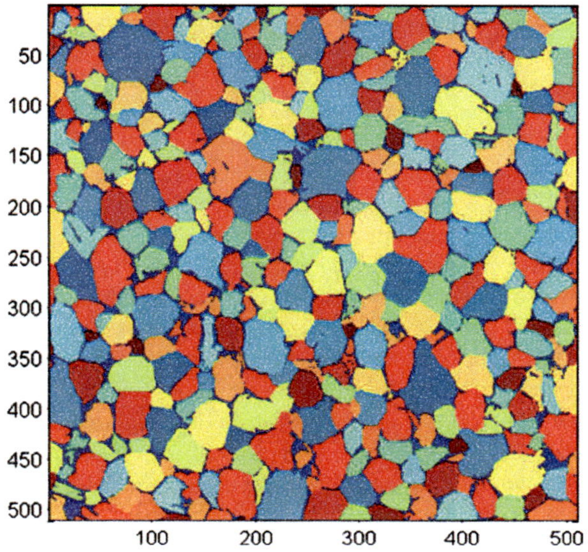

Fig. 7 OIM-like order parameter field generated from the processed SEM image, where β phase is in *dark blue* and its volume fraction is 8.7%

Cranck–Nicholson method, but at the cost of computation time spent on additional iterations within each time step. A reciprocal-space semi-implicit scheme introduced recently to phase field method (Chen and Shen 1998; Zhu et al. 1999) makes a good balance between the computational efficiency and stability.

On the spatial grid aspect, various adaptive algorithms have been developed in the past decade (Provatas et al. 1998; Braun 1997; Lan et al. 2002; Provatas and Greenwood 2005; Stogner et al. 2006). The spatial grid adaptation optimizes phase field computation efficiency based on the fact that the phase field evolution occurs significantly in the interface (boundary) regions and only moderately elsewhere (e.g., in matrix where diffusion occurs) or not at all (e.g., inside grains or structural domains). Using an adaptive grid size is essentially a strategy that accommodates two length scales, one associated with microstructure (precipitate, grain, or structural domain size) and the other associated with the interface (boundary) width, offering great efficiency when the two differ significantly.

Another strategy that works effectively when domain size is much greater than boundary width is to use efficient data structures (Gruber et al. 2006; Vedantam and Patnaik 2006). In a brute force phase field method of grain growth, (12) is solved for all order parameters at all grid nodes. Therefore, both the memory usage and the calculation time scales with system size and number of order parameters. This limits significantly the total number of grain orientations that can be handled in a simulation, which could cause unrealistically high grain coalescence events. To overcome this problem, Krill and Chen (Krill and Chen 2002) dynamically reassigned the spatial distribution of order parameters to reduce the number of order parameters that are required to avoid frequent domain coalescence. Such an approach is limited to isotropic grain growth. By examining the data structure of the order parameters in a phase field representation of a polycrystalline microstructure, one finds typically that many zero values that do not contribute to the solution of (12). This is because within a given grain only one order parameter is nonzero and several others carry nonzero values within the grain boundary regions (including junctions). To avoid solving the kinetic equation for those order parameters that have zero values and do not evolve, a new sparse data structure was developed for the phase field method (Gruber et al. 2006; Vedantam and Patnaik 2006) that stores and calculates only the nonzero order parameters. For example, if an order parameter is zero at a given grid node and the neighboring nodes, this order parameter will remain zero during the next time step. An update is needed for an order parameter at any node only if it is in the nonzero list of this node and its neighbors. Such a list contains items that hold both the value and the index of nonzero order parameters. The basic steps in the sparse phase field algorithm are, therefore, first to generate a list of unique nonzero order parameters in the neighborhood of each node and then to perform the update calculation for only those order parameters in the list. The list is dynamically maintained using the C++ vector class for efficiency. Although the sparse data distributes uniformly in space, its length and data-type vary, which causes difficulty in message passing between neighboring processors during parallel computation. This problem could be solved by utilizing advanced MPI_Datatype data passing protocol.

5 Examples of Application

5.1 Exploration of Mechanisms of Microstructural Evolution

The capability of producing realistic microstructures makes the phase field method a useful tool to explore mechanisms that drive a complex process in an experiment. For example, the motion of $1/2\,\langle 110\rangle$ dislocations in a γ/γ' microstructure in Ni-base superalloys can be controlled by multiple factors, including γ' particle size and shape, γ channel width, dislocation dissociation into Shockley partials, energies of various stacking faults in both γ and γ' phases, applied stress level and direction, coherency strain from γ/γ' interfaces, etc. To understand the contribution from individual factor and the resulting operation mechanisms, phase field modeling can be used in a postulate-and-verification fashion to identify the cause and effect, or as a comprehensive parametric study that maps out the dislocation behaviors in a multifactor space.

Figure 8a shows a typical observation of decorrelation of the motion of Shockley partials at the early stage of a creep test in a recent study on ME3/Rene104 alloy (Unocic et al. 2008). A representative microstructure configuration in phase field model is shown in Fig. 8b, which includes a periodic array of coarse (secondary) γ' particles separated by γ channels, randomly dispersed fine (tertiary) γ' particles

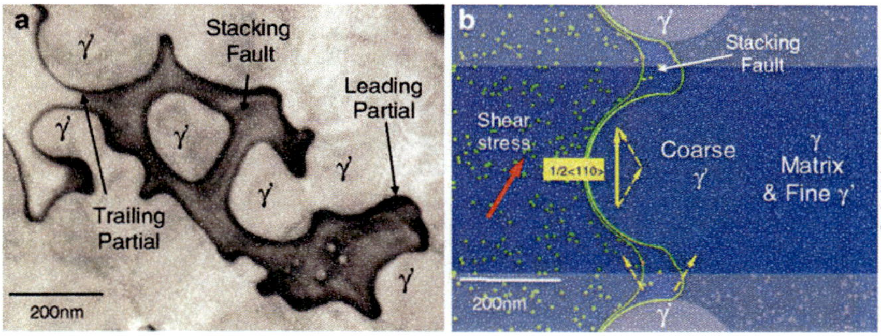

Fig. 8 (**a**) Experimental observation of de-correlation of Shockley partial dislocations in γ channel (courtesy of R Unocic and M J Mills). (**b**) Microstructure configuration in phase-field simulation

and a $1/2\langle 110\rangle$ screw dislocation initially placed on the left. A parametric simulation study that takes into account the applied stress direction and magnitude is shown in Fig. 9. The result indicates five distinctive regions, in which the dislocation (A) fills in the γ channel as a $1/2\langle 110\rangle$ type, (B) fills in the channel as two separate $1/6\langle 112\rangle$ Shockley partials, (C) shears γ' and creates planar fault of antiphase boundary, (D) decorrelates as only the leading Shockley partial goes through the channel, and (E) is arrested at the entrance of the γ channel. From the map it becomes clear that the decorrelation of Shockley partial dislocations is driven by the direction of the applied stress, which differentiates the resolved shear stress on the two Shockley partial Burgers vectors. The width of γ channel in the meantime acts as a threshold that determines whether a dislocation can pass through under a particular applied stress magnitude, according to Orowan critical stress (Hirth and Lothe 1982). The exhibited five regions is the result of the combination of the two factors.

In the above example, experimental characterization and image analysis provided essential inputs to the phase field modeling. The richer information such as precipitate morphology and size distribution, precipitate spatial location, and γ channel width variation can also be taken into account in the simulations. As a matter of fact, electron microscope images have been imported directly to the phase field model to construct microstructures in close resemblance to the real situation (Fig. 10). This allows a virtual in situ reproduction of the process in modeling under a similar experimental condition. Figure 11 shows a series of snapshots as a pair of two Shockley partials of identical Burger vector on two adjacent planes moves into a bimodal γ/γ' microstructure shown in Fig. 10 under an applied resolved shear stress of 600 MPa (Zhou et al. 2010a). Because thermally activated reordering process is accounted for in the model, the pair of Shockley partials can shear through easily the secondary γ' particles, leaving super-lattice extrinsic stacking fault (SESF) in them (indicated in Fig. 11b). In the meantime, decorrelation behavior similar to the

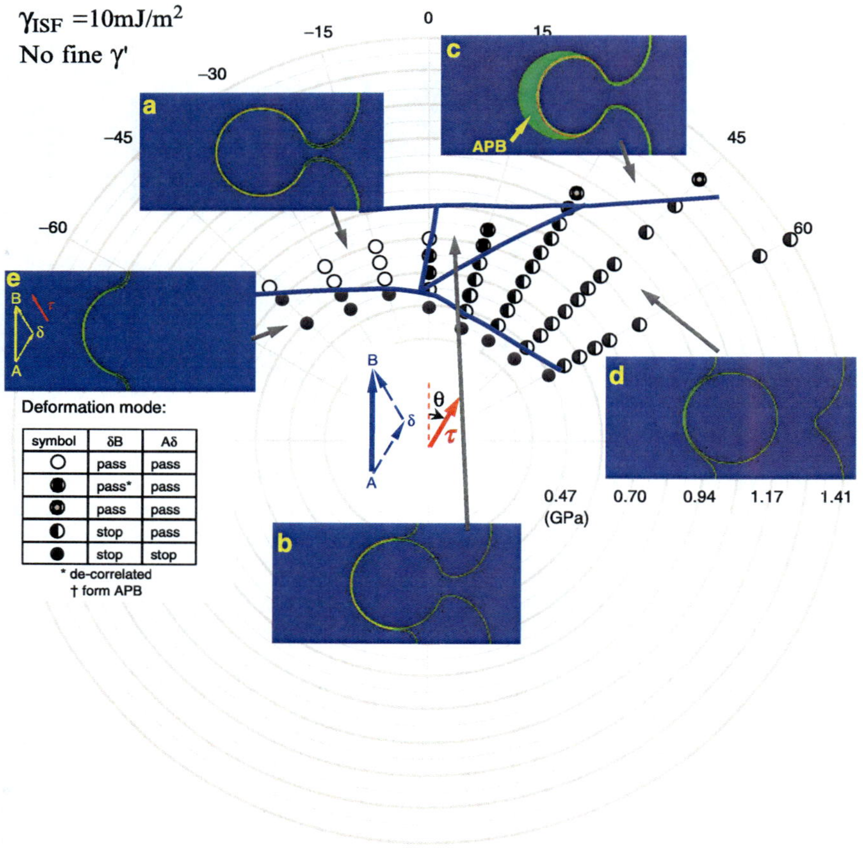

Fig. 9 Phase-field simulation of dislocation–precipitate interactions in a Ni-base superalloy shows five distinctive regions of behavior: (**a**) fills in the γ channel as a $1/2\langle 110\rangle$ type, (**b**) fills in the channel as two separate $1/6\langle 112\rangle$ Shockley partials, (**c**) shears γ' and creates planar fault of antiphase boundary, (**d**) de-correlates as only the leading Shockley partial passes through the channel, and (**e**) is arrested at the entrance of the γ channel

one shown in the previous example was also observed here. Formation of extrinsic stacking fault (ESF) and intrinsic stacking fault (ISF) is also indicated.

These parametric studies may shed light on the deformation mechanisms in superalloys under different stress and temperature conditions. The microstructure-sensitive micromechanisms discovered by the phase field simulations are being incorporated in microstructure-based crystal plasticity modeling.

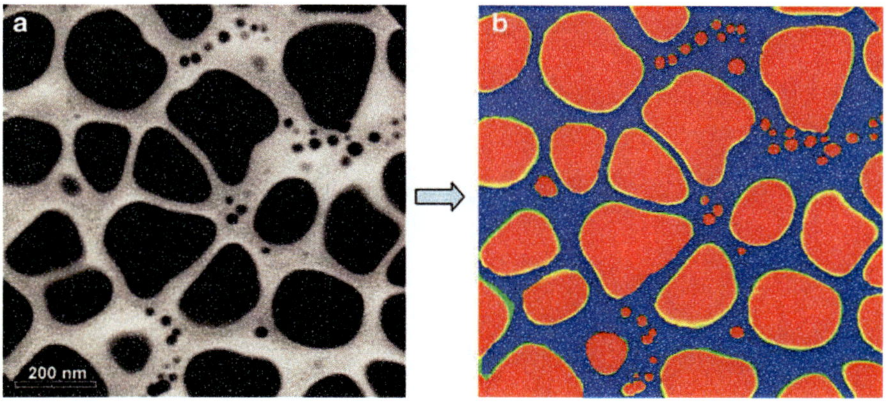

Fig. 10 (**a**) TEM images of a bi-modal γ' microstructure (courtesy of R Unocic and M J Mills). (**b**) The converted image as input microstructure in phase-field modeling

5.2 Extracting Materials Parameters by Evolving Experimental Images

Too often microstructure modeling is rendered qualitative because of the lack of materials parameters for model input. Typical examples of these parameters are interfacial energy, mobility, and atomic diffusivity along grain boundaries. Taking direct experimental images of a multiphase and polycrystalline microstructure as the initial input, running a "virtual experiment" through phase field simulations to evolve the microstructure to a later point in time, and comparing them to experimental observations allow for the extraction of these experimentally difficult-to-measure parameters. Here, we present one such example for Ti-64.

The microstructural evolution is described by temporal changes of both nonconserved order parameters, governed by (12), and Al and V concentrations, governed by (6). The free energy of the system is described by (33), wherein the model parameters of the chemical free energy in (29) and those of the atomic mobility parameters in (40) and (41) were taken from recent studies on phase equilibra and microstructural evolution in Ti-64 (Chen et al. 2004; Zhang et al. 2007). Some additional important material parameters that need to be obtained to perform the quantitative phase field modeling are the interphase interfacial energy and grain boundary energy and the phase field mobility L. These parameters are currently unknown and will be determined in the current example by parametric phase field simulations utilizing experimental images.

Following (13) and (14), and the general procedures described in (Moelans et al. 2008), different combinations of the phase field model parameters [e.g., the gradient energy coefficient, κ, the free energy hump between the equilibrium values of the order parameter, $\Delta f(\phi)$, and the boundary width, w, in (13) and (14)] allow for setting different specific energies for the α/α grain boundary and

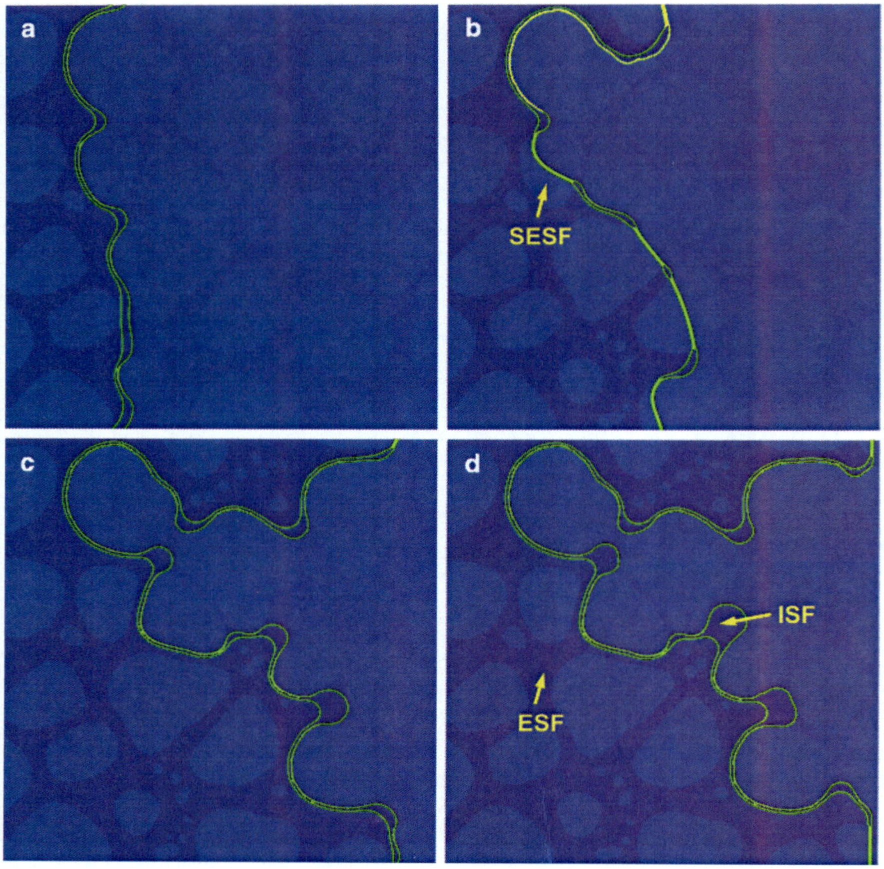

Fig. 11 **(a)**–**(d)** Snap shots of dislocation motion in γ/γ' microstructure converted from TEM image (Fig. 9a). Formation of super-lattice extrinsic stacking fault (SESF), extrinsic stacking fault (ESF), and intrinsic stacking fault (ISF) is also seen

the α/β interface. By considering both bulk and grain boundary diffusion (i.e., bulk-boundary mixed diffusion described in Sect. 3.6) and allowing adjustable interfacial energies for the α/α grain boundary and the α/β phase boundary, it is possible to run a "virtual phase field experiment" quantitatively from the SEM images obtained from experiment to obtain microstructures at later times that match experimental observations. This also allows one to study systematically the effects of high diffusivity paths along grain boundaries and the ratio of α/α grain boundary energy over α/β interfacial energy on α/β morphological evolution and grain growth kinetics. It is found that, while keeping all the other model parameters unchanged, changing the bulk-to-boundary diffusivity ratio [see (42)] and the interfacial energy ratio leads to significantly different morphologies of the α/β two-phase mixture and α grain growth kinetics. As a consequence, comparing the phase field "virtual

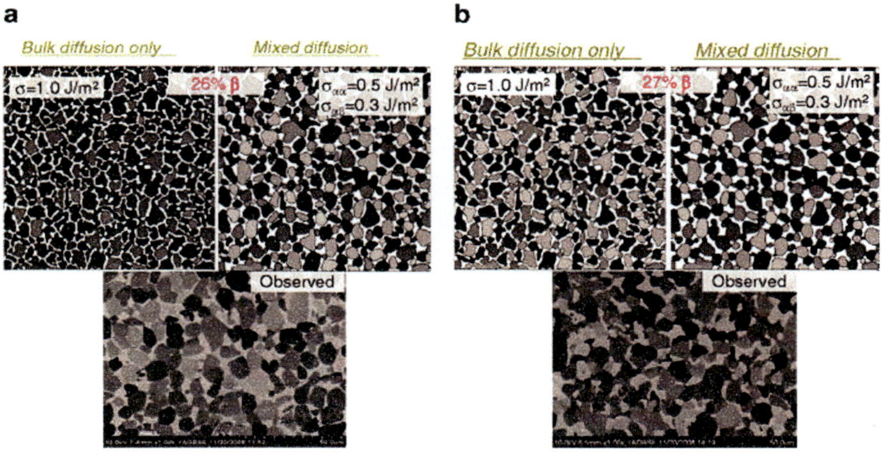

Fig. 12 Comparison between phase-field simulation predictions using different boundary energies and diffusivities (**top**) and experimental observation (**bottom**). In both simulations and experiment, the starting microstructure (shown in Fig. 6) was isothermally held at 800 °C for (**a**) 1 h and (**b**) 4 h. The volume fraction of the β phase has increased significantly during the isothermal holding (as compared to the starting microstructure)

experimental" results with the real experimental results allows one to find (e.g., by trial and error parametric study) materials parameters or their ratios that produce the best match between the two for a given heat treatment schedule. Examples are shown in Fig. 12. The simulations seem to indicate that α grain growth with β phase particles primarily located on α/α grain boundaries at 800°C is governed by a bulk-boundary mixed diffusion mechanism, and the consideration of fast diffusion along grain boundaries also alters the grain growth. Figure 13 shows that the power exponent deviates slightly from 3, being less than 3 for bulk diffusion only (Fig. 13a), but becomes larger than 3 for bulk-boundary mixed diffusion (Fig. 13b). More importantly, as shown by Fig. 14, the quantitative parametric phase field simulation results obtained at 800°C with $\gamma_{\alpha\alpha} = 0.5\,\text{Jm}^{-2}, \gamma_{\alpha\beta} = 0.3\text{Jm}^{-2}, D_b/D_v = 300$ and the product of grain boundary mobility and energy $m\sigma = 0.85\mu\text{m}^2\text{s}^{-1}$ seem to produce the best agreement with experimental observations (see the comparisons given in Fig. 14,a b). Although the values of these materials parameters can only be regarded as an estimate, they certainly provide useful information for other modeling and calculations.

5.3 *Texture Evolution During Grain Growth*

Grain growth is probably one of the simplest examples of microstructural evolution encountered in solids. However, the process is always complicated by the presence

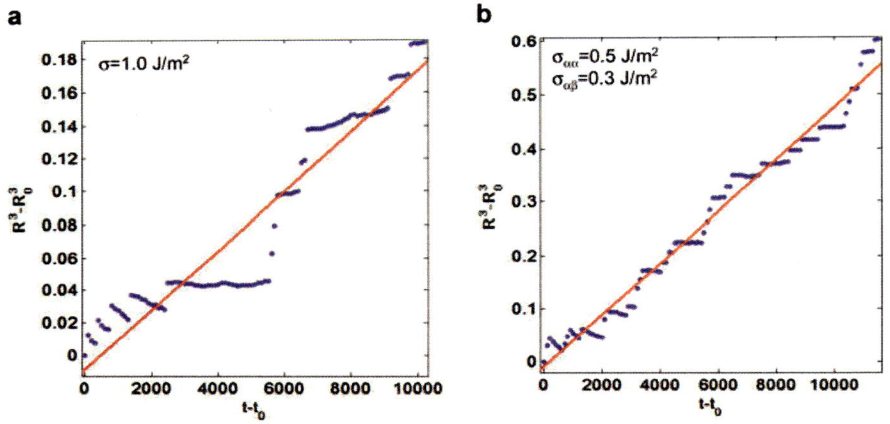

Fig. 13 Grain growth kinetics predicted under (**a**) bulk diffusion only and (**b**) bulk-boundary mixed diffusion for a two-phase α/β microstructure observed in Ti-64

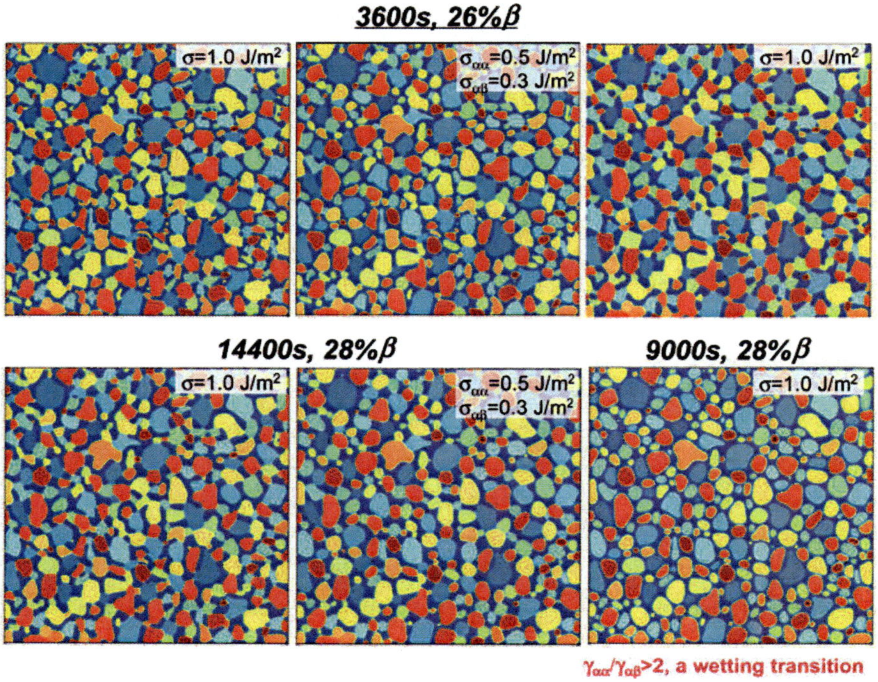

Fig. 14 Effect of α/α grain boundary and α/β interfacial energies on the morphology of the α/β two-phase mixture in Ti-64 using SEM image (shown in Fig. 6) as input. The ratio of grain boundary to bulk diffusivity used in the simulations is 300

of texture, second-phase particles, and impurity segregation at grain boundaries in commercial alloys. Phase field models of grain growth in both single-phase and two-phase alloy were first developed for systems of isotropic boundary properties (Chen 1995; Fan and Chen 1997a, b). The models were then extended to systems of anisotropic boundary properties and applied to study texture effect on grain growth (Kazaryan et al. 2002a; Ma et al. 2004). Effects of second phase particles and grain boundary segregation were also investigated by using the phase field methods (Fan et al. 1999; Cha et al. 2002; Ma et al. 2003; Moelans et al. 2005; Suwa et al. 2006; Gronhagen and Agren 2007; Kim and Park 2008).

There are two ways to describe a polycrystalline microstructure in phase field models of grain growth, one uses multiple orientation fields (Chen 1995) and the other uses a single orientation field (Warren et al. 2003). In both cases, the orientation fields are phenomenological order parameters introduced to distinguish grains from one orientation to another, although physically more rigorous order parameters could be introduced (Hoyt et al. 2001; Foiles and Hoyt 2006). These order parameters are nonconserved fields and their time-evolutions are governed by (12). In the following example, the multiorder parameter phase field model was used for its simplicity, and the focus is on incorporating texture information from OIM images in the phase field simulations.

In the multiorder parameter phase field model of grain growth, an arbitrary polycrystalline microstructure is described by a set of nonconserved order parameters, $\eta_1, \eta_2, \ldots, \eta_P$, with each of them representing a specific crystallographic orientation (shown by different colors in Fig. 15a). η_i has a value of 1 within the bulk of grains with orientation i, changes its value continuously from 1 to 0 (Fig. 15b) across the boundary regions, and remains 0 elsewhere. In Fig. 15c, the value of $\sum_{i,j\neq i}^{P} \eta_i^2 \eta_j^2$ is plotted, which is non-zero only within grain boundary regions and hence shows the grain boundary network. With the local free energy given by (35), the fundamental grain boundary properties (width, energy, and velocity) are described by (36), (37) and (43). If we assume that the kinetic coefficient L, the

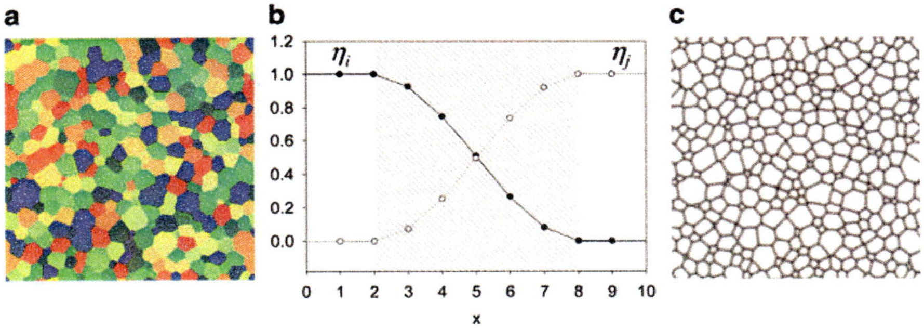

Fig. 15 Phase-field description of a polycrystalline microstructure. (**a**) Grain orientation map where each grain is represented by a unique order parameter with value of unity; (**b**) order parameter profiles across a grain boundary; (**c**) grain boundary network

gradient coefficient κ, and the parameter b are all constant, then (12) describes isotropic grain growth.

To describe texture evolution during grain growth and to take into account texture effect on grain growth kinetics, one needs to incorporate the dependence of grain boundary properties on misorientation and inclination into the phase field model. This was usually done by making L, κ, and b misorientation and inclination dependent under the constraint of constant grain boundary thickness (Moelans et al. 2008; Ma et al. 2006). The way of defining misorientation field is not unique. To be consistent with the field representation of a polycrystalline microstructure, the following function was used in the current example (Ma et al. 2004):

$$\theta(r) = \frac{\sum_{i,j\neq i}^{P} \eta_i^2 \eta_j^2 \theta_{ij}}{\sum_{i,j\neq i}^{P} \eta_i^2 \eta_j^2} \tag{46}$$

where θ_{ij} are precalculated misorientation angles between grains with orientations specified by η_i and η_j. Equation (46) assigns a constant misorientation angle to a narrow range of a grain boundary. Without losing generality, we assume that the energy anisotropy is characterized by a plateau for high-angle boundaries and by the Read–Shockley formula for small angle boundaries: (Read and Shockley 1950):

$$\sigma(\theta) = \begin{cases} \sigma_0 \frac{\theta}{\theta_m}\left(1 - \ln \frac{\theta}{\theta_m}\right) & \theta < \theta_m \\ \sigma_0 & \theta \geq \theta_m \end{cases} \tag{47}$$

where θ_m is the maximum angle at which the Read–Shockley equation still holds, and σ_0 is a constant. Correspondingly, the mobility anisotropy is characterized by (Huang et al. 2000):

$$L(\theta) = \begin{cases} L_0 \left(\frac{\theta}{\theta_m}\right)^5 & \theta < \theta_m \\ L_0 & \theta \geq \theta_m \end{cases}. \tag{48}$$

Note that the magnitude of θ_m determines the degree of anisotropy and it is assumed to be 20° following the experimental observation.

A grain boundary can be uniquely defined by misorientation between the two crystals involved and inclination of the grain boundary plane (Sutton and Balluffi 1995). The current example focuses on the implementation of misorientation. For those who are interested in inclination, the implementation in phase field grain growth model can be found in (Kazaryan et al. 2002b). If the two grains are defined as grain A and grain B, the misorientation angle of this grain boundary is then given by:

$$\theta = a\cos\left(\frac{Tr(\Delta g) - 1}{2}\right) = a\cos\left(\frac{\Delta g_{11} + \Delta g_{22} + \Delta g_{33} - 1}{2}\right) \tag{49}$$

with

$$\Delta g = g_A g_B^{-1} \tag{50}$$

where the orthogonal matrices g_A and g_B represent the orientations of grain A and B respectively, which can be specified by appropriate rigid body rotations with respect to the sample reference system.

Usually three Euler angles, $(\varphi_1, \Phi, \varphi_2)$, are used to describe the rotation of one crystal relative to another. The corresponding rotation operation is given by:

$$g = g^Z_{\varphi_2} g^X_{\Phi} g^Z_{\varphi_1} \tag{51}$$

where g^I_θ specifies a rotation with angle θ about axis I. Combining these equations, the misorientation of a grain boundary can be specified from the Euler angles of two neighbored grains.

In the case of cubic crystal, a grain boundary misorientation is invariant but Δg is different under the 48 symmetry operations. Therefore, the rotation angle defined by (49) is not unique. By convention, the least of these angles so obtained is the angle of misorientation. When the orientations in a polycrystalline system are totally random, the obtained misorientation distribution follows the so-called Mackenzie distribution (Mackenzie 1958).

Grain orientations are not always randomly distributed. Texture develops by preferred nucleation or growth primarily during deformation. The subsequent concurrent texture evolution and grain growth is a subject of intensive study in the literature. It should be noted that it is the misorientation rather than orientation that directly controls grain growth. Therefore, both texture intensity and its spatial distribution are important factors for grain growth (Ma et al. 2004).

Two types of β phase texture, $\{110\} < 211 >$ and $\{001\} < 110 >$, have been observed in forged Ti-64 alloys. In their experimental work, Semiatin et al. (2001) showed significant difference in grain growth kinetics between two texturally different but otherwise microstructurally identical samples after severe deformation. However, ideal parabolic growth law was also reported for the same alloy system under as cast condition (Semiatin et al. 1996). Apparently, different grain growth behavior could develop under various processing conditions even for the same alloy system. It is desirable to use directly OIM data that carry thermal mechanical history information of samples as inputs to the phase field model. Figure 16a shows an OIM micrograph obtained for β grains used in the simulations. Since β is a high temperature phase, a direct β phase OIM measurement is difficult. The β phase OIM micrograph shown in Fig. 16a was actually derived from the α grain orientations using a conversion technique developed by Glavicic and coworkers (Glavicic et al. 2003). A 512×512 uniform mesh was employed in the simulations. The OIM data were converted into an order parameter map with each order parameter corresponding to a unique crystallographic orientation within 5° tolerance. Model parameters b, κ, and L are precalculated based on (46)–(49), and stored in lookup tables. Figure 16b gives the microstructure just after the initial relaxation and Fig. 16c shows the evolved microstructure at later time. The corresponding grain growth kinetics is

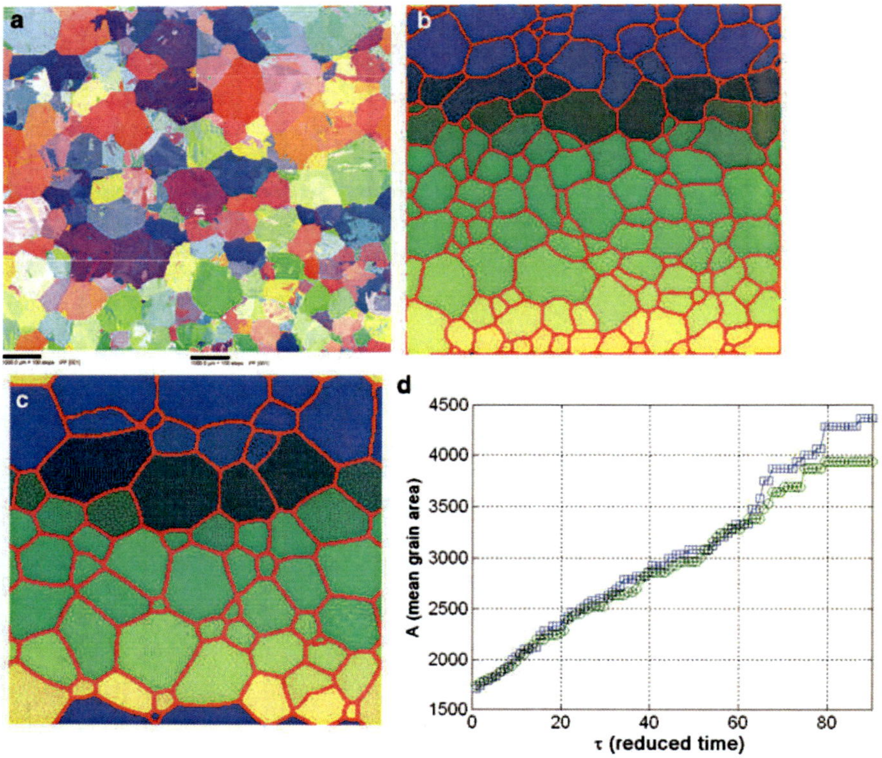

Fig. 16 Concurrent grain growth and texture evolution in Ti-64. (**a**) β grain orientation map obtained from experiment; (**b**) initial phase-field order parameter map; (**c**) evolved grain boundary network at later time; (**d**) grain growth kinetics with (*open circles*) and without (*open squares*) considering texture. Both area and time are in reduced units

plotted in Fig. 16d. For comparison, a parallel simulation was carried out with an assumption that all the grain boundaries are high angle boundary. In this example, the evolution of both orientation distribution and grain boundary network from the initial OIM image was obtained using the phase field method.

5.4 Physics-Based Repair of Experimental Microstructure Data Set

The direct usage of microstructure data sets obtained from reconstruction method based on FIB-SEM serial sectioning and imaging as model inputs is undermined by the noise from experimental uncertainty and misalignment. The resulted artifacts such as voids (missing pixels or voxels) or sharp corners could cause numerical instability to microstructure-based property computations. It is a nontrivial task to clean up and smooth the raw data using mathematical tools (Ghosh et al. 2008) and in the meantime to ensure that they evolve properly. The phase field method can take directly raw experimental data sets as inputs without numerical instabilities (Dobrich et al. 2004). It avoids the complicated mathematical geometry computation and requires no artificial criteria for alignment along grain boundary. Small voids and sharp corners are unstable (energetically unfavorable) defects and tend to be healed (annealed out) quickly during initial relaxation stages of a phase field simulation. Thus, the phase field method provides a natural way to repair raw experimental data sets. In addition, 3D microstructure reconstructions are costly and time-consuming and it would be more efficient and effective to utilize computer simulations to generate microstructural data at different time, temperature and stress from a given set of 3D experimentally reconstructed microstructure data.

As an example, the FIB-SEM data sets obtained by Ghosh et al. (2008) were used as input to the phase field simulations using the sparse data structure algorithm described in Sect. 5.4, with grains of relative disorientation with 4 degree grouped and indexed by a unique ID number. In the simulations, isotropic grain boundary properties (energy and mobility) were assumed and zero flux boundary condition was applied. The as-received grain boundary networks are shown in Figs. 17a and 18a, c, respectively, for the 2D and 3D data sets acquired. The 3D data set has been repaired using an FEM code developed by Ghosh et al. (Ghosh et al. 2008). In the unrepaired 2D data set, zigzags and broken grain boundary segments with missing pieces are seen clearly (Fig. 17a). Even in the FEM-repaired 3D dataset, roughness of grain boundary planes still exist (Fig. 18c). The 2D and 3D data sets after initial relaxation in the phase field simulations are plotted in Figs. 17b and 18b, respectively. Clearly, all the "defects" associated with the experimental data

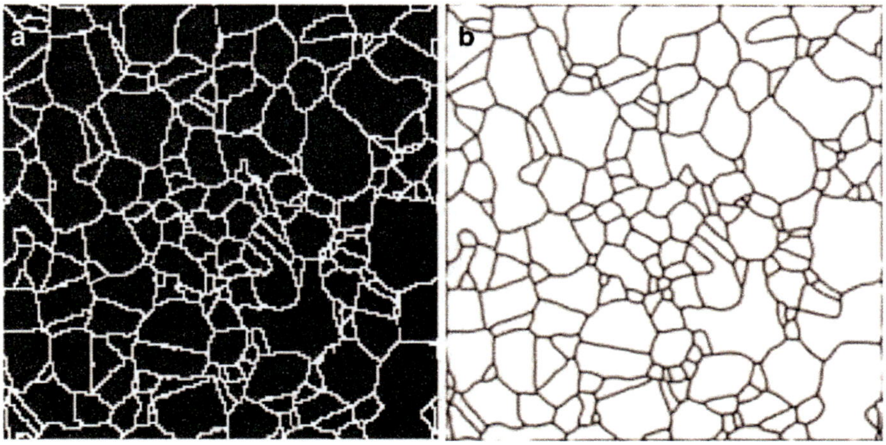

Fig. 17 (**a**) Direct OIM image from experiment; (**b**) relaxed image from phase-field simulation

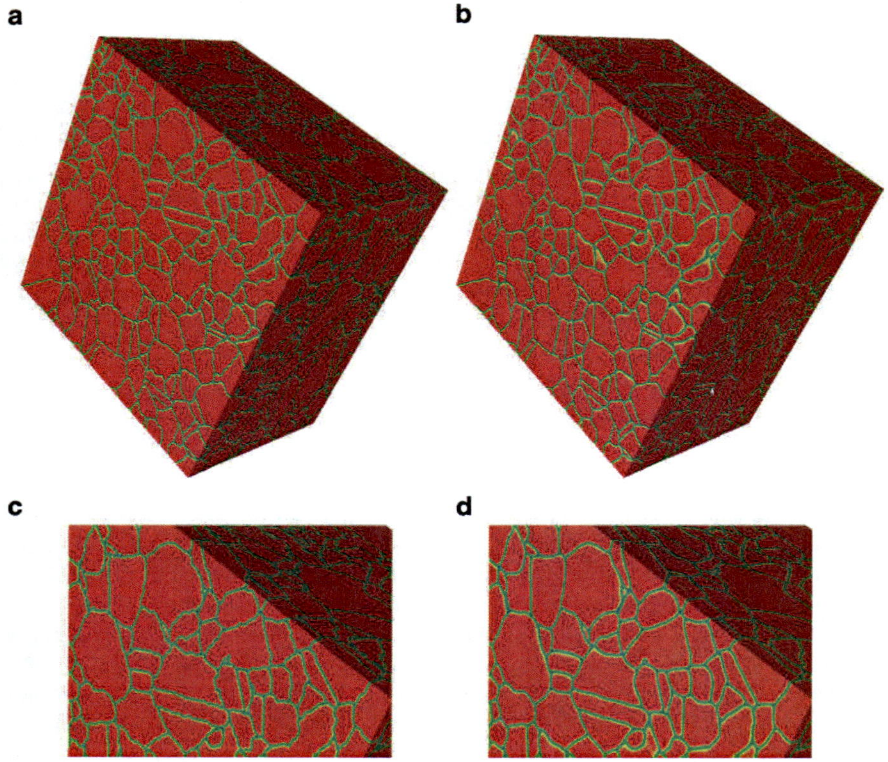

Fig. 18 (**a**) As received dataset; (**b**) relaxed image from phase-field simulation; (**c**) and (**d**) are enlarged figures of (**a**) and (**b**)

sets have disappeared and the shapes of the grains and the topology of the grain boundary networks have been preserved. Nevertheless, some of the broken grain boundary pieces also disappeared and leading to coalescence event. This can be prevented by using OIM image directly that contains grain orientation information to replace the grain boundary network data set.

Another example is the use of phase field simulations to repair tomographic characterization data set in polycrystalline Al–Sn by Krill and coworkers (Krill and Chen 2002; Dobrich et al. 2004; Krill et al. 2004). They used constrained phase field grain growth simulations to fill in the missing pieces in the measured 3D grain boundary network. In Al–Sn alloy, the position of grain boundary could be traced by X-ray due to Sn segregation. Figure 19a shows the tomographic reconstruction of Sn-rich grain boundaries. Due to the nonuniform Sn segregation, holes in grain boundary planes network are evident and the reconstructed grain structure contains a lot of missing information, as shown in Fig. 19b, which undermines quantitative grain size characterization. The identified grain structure (Fig. 19b) was mapped directly to a 3D phase field simulation cell with a unique order parameter assigned

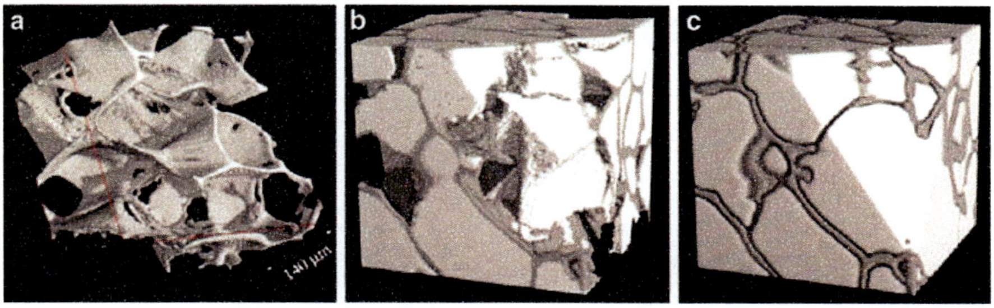

Fig. 19 (**a**) Tomographic reconstruction of Sn-rich grain boundaries; (**b**) identified grains from (**a**); (**c**) repaired grain structure using constrained phase-field method (Dobrich et al. 2004)

to each grain. The unidentified spaces were treated initially as liquid with no finite order parameters assigned. The liquid phase is unstable and will transform into solid quickly during the simulation. To preserve the geometry and position of the identified grain boundaries, their mobility was set to zero. With increasing simulation time, the nonzero order parameters grow and fill in all the spaces occupied originally by the liquid. The final repaired microstructure is shown in Fig. 19c. From these examples, one can see that the phase field method provides an unbiased, efficient, and reliable means to repair experimental data sets. With the development of sparse data structure in phase field models of grain growth models (see Sect. 5.4), large data sets containing tens of thousands of grains with unique IDs can be easily handled.

5.5 Generation of Digital Microstructures

There are increasing efforts in developing digital representation of microstructures and its inclusion in microstructure-based crystal plasticity models that predict mechanical behavior of advanced alloy systems (Searles et al. 2005; Groeber et al. 2007; Bhandari et al. 2007). Even though three-dimensional (3D) grain boundary networks and precipitate microstructures can be obtained experimentally by serial-sectioning technique [see, e.g., (Uchic 2006)] or diffraction techniques (see the example given in the proceeding section), they are costly and time-consuming process. The phase field method could be a useful tool in generating various digital microstructures in both 2D and 3D for different alloy systems under different processing condition for mechanical property explorations. To generate the 3D polycrystalline microstructure shown in Fig. 1d, for example, one can simply start the phase field simulation with a liquid phase and then add small crystals of random orientations or from any given orientation distribution using the existing explicit nucleation algorithm (Simmons et al. 2000). The size of the seed crystals should be above the critical size in the supercooled liquid phase. The seeds will then grow and impinge upon each other, and then grain growth will start, generating a 3D

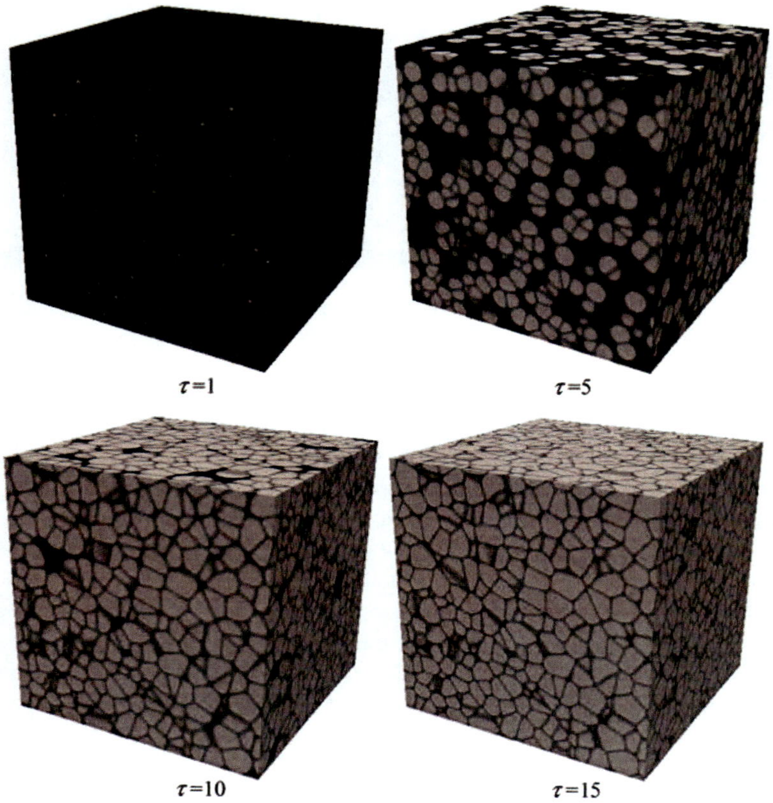

Fig. 20 Generation of a 3D polycrystalline microstructure using the phase-field method, starting from liquid with randomly placed seeding crystals. The growth and impingement of the seeding crystals produce an equi-axed polycrystalline microstructure. τ is reduced time

polycrystalline microstructure. This process is shown in Fig. 20. The system size used in this simulation is 256^3 and the system contains 3,000 different grain orientations. The different boundary thicknesses seen at reduced time $\tau = 15$ (Fig. 20d) is due to various projection angles of the 3D grain boundary planes on the computational unit cell surfaces.

Another advantage of using computer simulations to generate 3D microstructures is that one can evolve the microstructure under a given set of processing conditions to a variety of desired states for mechanical property modeling. For example, the two-phase γ/γ' microstructure shown in Fig. 1e was generated by phase field simulation of an aging process of a Ni-base superalloy with either positive or negative lattice misfit. The microstructure can be further evolved under a given service condition, e.g., under a uni-axial load at an elevated temperature (Fig. 21a, b) (Zhou et al. 2010b). The initial cuboidal γ' particle morphologies and the typical N-type and P-type rafting morphologies developed under the loading condition show

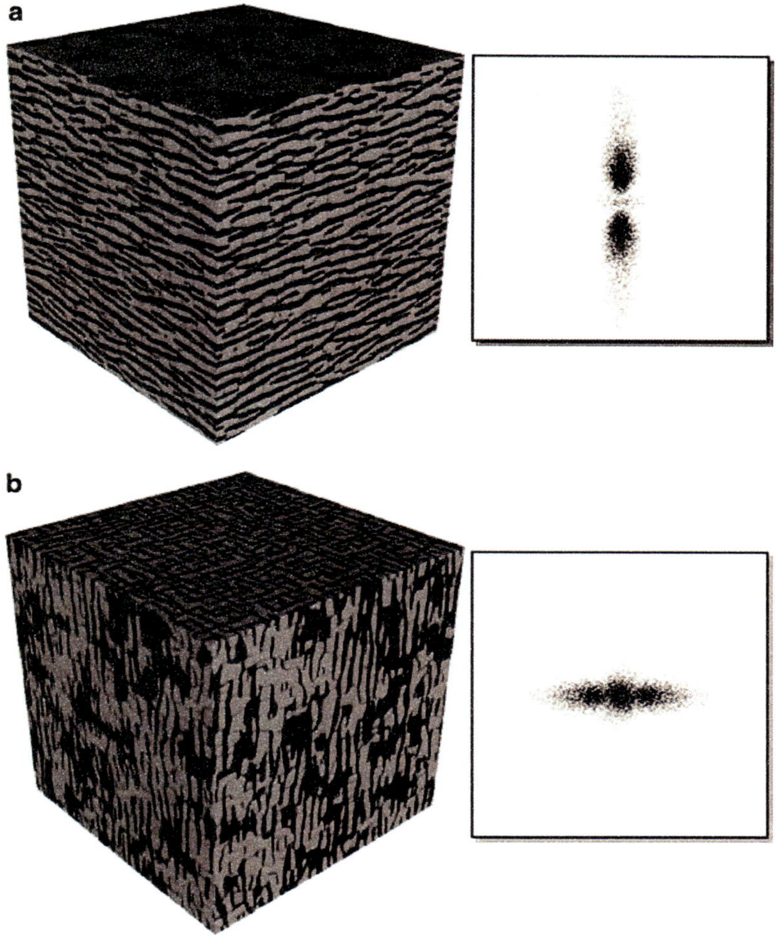

Fig. 21 Rafted microstructures developed from the simulated γ/γ' microstructures with $\pm0.3\%$ lattice misfit during isothermal aging at 1,300 K for 4.7 h (shown in Fig. 1e) after additional 5.6 h aging under 152 MPa tensile stress along [001], by assuming lattice misfit of (**a**) -0.3% and (**b**) $+0.3\%$ (Uchic 2006). The inset shows the Fourier transform (diffraction pattern) of the corresponding microstructure

remarkable resemblance to those observed in experiment (Fahrmann et al. 1999). The simulation linear dimension is 5μm, with grid size 20 nm. The simulation results were obtained by incorporating the KKS model (Kim et al. 1999) (see Sect. 2.6) that allows for treatment of solute diffusion at an increased length scale without altering artificially the driving forces for precipitate growth and coarsening. Accordingly, the chemical free energy for γ/γ' phases was chosen in the form of (34), with four phenomenologically defined order parameters to characterize the γ and γ' phases and the four types of antiphase domain in γ'. While the individual free

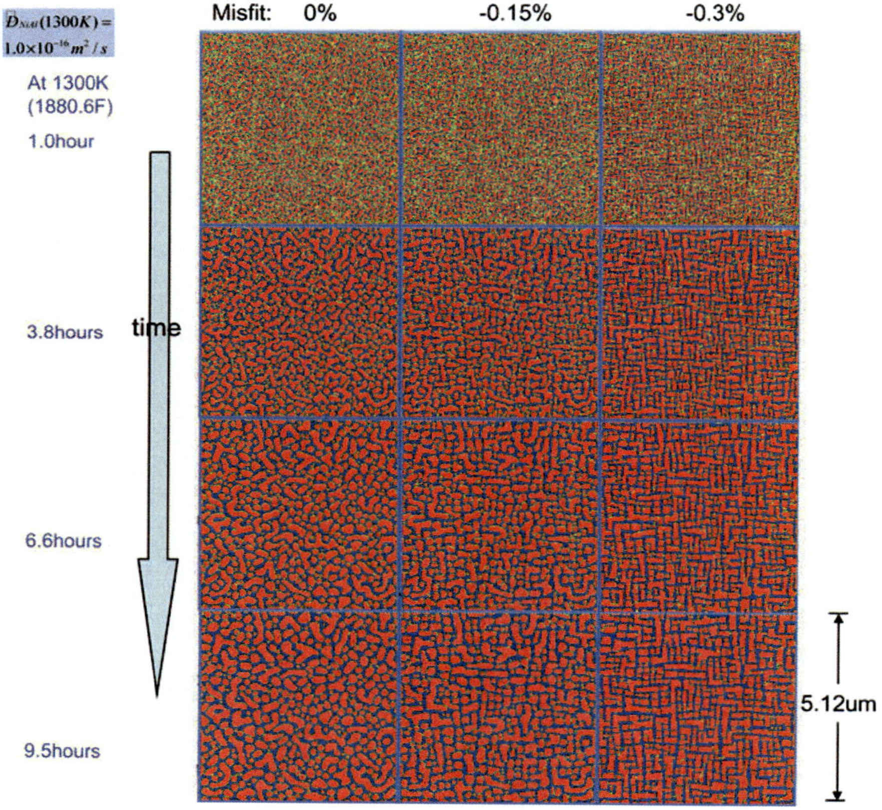

Fig. 22 γ/γ' microstructure in a pseudo-binary Ni-base alloy obtained under various material and processing conditions from phase-field simulations. γ' volume fraction is 60%

energies of γ and γ' phase could have been imported from CALPHAD database for specific alloys, only fitted parabolic polynomials were used in this example. Dislocation activities in the γ-channels under the external load and their interactions with the γ' particles were described by introducing a new set of nonconserved order parameters that characterize plastic strain fields associated with dislocations from each active slip systems (Zhou et al. 2010b).

Figures 22 and 23 show various γ/γ' microstructures generated by 2D phase field simulations for various Ni-base superalloys with different γ' volume fractions and lattice misfits at different aging times at 1,300 K. The same free energy model and material parameters used in the previous example were employed here. The microstructures are being used to explore the strength and creep deformation of the alloys.

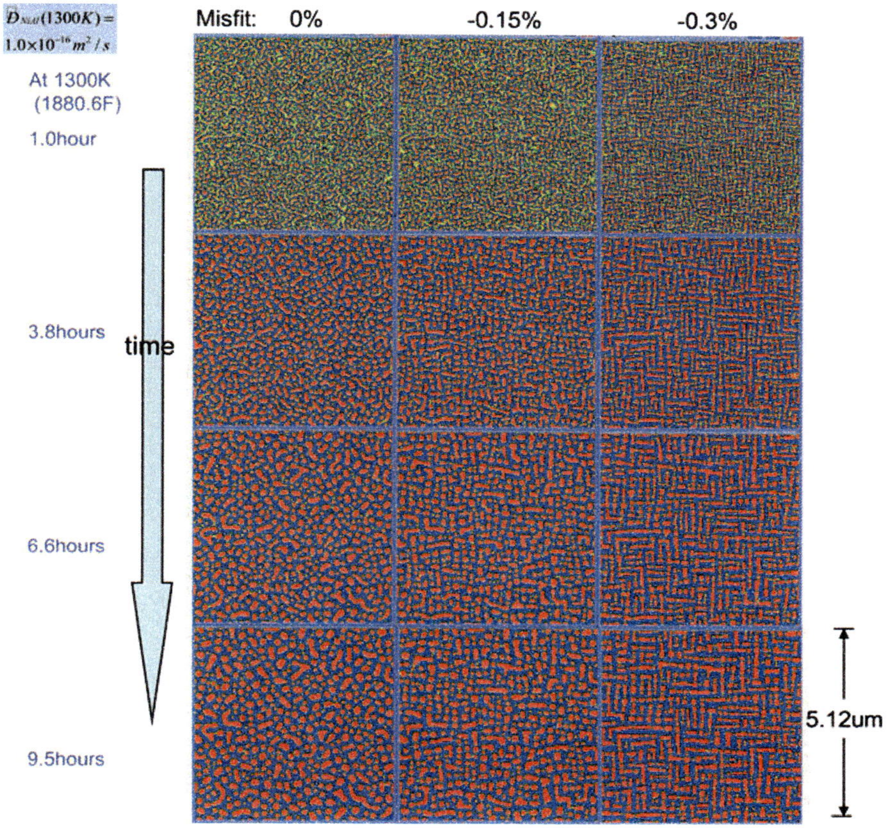

Fig. 23 γ/γ' microstructure in a pseudo-binary Ni-base alloy obtained under various material and processing conditions from phase-field simulations. γ' volume fraction is 40%

6 Summary

Fundamental of phase field method and critical issues regarding its applications to microstructural engineering and digital reconstruction of microstructures are reviewed. Some of the important issues in dealing with complex alloy systems including linking to material databases, using experimental images as initial microstructures, technical difficulties associated with length scales, and efficient numerical algorithms are addressed. Synergy of coupling material specific phase field simulations to advanced microstructure characterization techniques is demonstrated through various examples. When used at microscopic levels, the phase field models can be applied to understand and predict fundamental properties of extended defects such as interfaces and dislocations and micromechanisms of dislocation–precipitate interactions, using *ab initio* calculations as model inputs. Experimental characterizations play a key role in informing and focusing the phase field simulations in terms of precipitate morphology, dislocation configurations, and possible deformation mechanisms. When applied at various coarse-grained levels, the phase field models have the ability to produce various digital microstructures that contain a large assembly of both chemically and mechanically interacting defects and their complicated geometrical and topological changes during dynamic evolution. The method is shown as an effective and efficient tool to repair experimental microstructural data sets based on physical laws of microstructural evolution. In combination with experimental characterizations, phase field simulations are also shown to be able to extract useful information on difficulty-to-measure materials parameters, such as interfacial energy, diffusivity, and mobility.

Acknowledgments We gratefully acknowledge financial supports by the Office of Naval Research through the D 3D program (Grant No. N00014-05-1-0504), U.S. Air Force Office of Scientific Research through the Metals Affordability Initiative Program on Durable High Temperature Disks and the STW21 Program on Multi-Materials System with Adaptive Microstructures for Aerospace Applications (Grant No. FA9550-09-1-0014), and the National Science Foundation (Grant No. CMMI-0728069). The simulations were performed on supercomputers at the Arctic Region Supercomputing Center and the Ohio Supercomputing Center.

References

Allen SM, Cahn JW. A microscopic theory for antiphase boundary motion and its application to antiphase domain coarsening. Acta Metallurgica 1979;27:1085.

Almgren RF. Second-order phase field asymptotics for unequal conductivities. SIAM Journal on Applied Mathematics 1999;59:2086.

Andersson JO, Helander T, Hoglund L, Shi PF, Sundman B. THERMO-CALC & DICTRA, computational tools for materials science. CALPHAD 2002;26:273.

Ansara I, Dupin N, Lukas HL, Sundman B. Thermodynamic assessment of the Al-Ni system. Journal of Alloys and Compounds 1997;247:20.

Bhandari Y, Sarkar S, Groeber M, Uchic M, Dimiduk D, Ghosh S. 3D polycrystalline microstructure reconstruction from FIB generated serial sections for FE Analysis. Computational Materials Science 2007;41:222.

Boettinger WJ, Warren JA, Beckermann C, Karma A. Phase-field simulation of solidification. Annual Review of Materials Research 2002;32:163.

Braun RJ. Adaptive finite-difference computations of dendritic growth using a phase-field model. Modelling Simul. Materials Science and Engineering 1997;5:365.

Cahn JW, Hilliard JE. Free energy of a nonuniform system. I. Interfacial free energy. Journal of Chemical Physics 1958;28:258.

Cahn JW, Hilliard JE. Free energy of a nonuniform system. III. Nucleation in a two-component incompressible fluid. Journal of Chemical Physics 1959;31:688.

Campbell CE, Boettinger WJ, Kattner UR. Development of a diffusion mobility database for Ni-base superalloys. Acta Materialia 2002;50:775.

Cha PR, Kim SG, Yeon DH, Yoon JK. A phase field model for the solute drag on moving grain boundaries. Acta Materialia 2002;50:3817.

Chen LQ. A novel computer-simulation technique for modeling grain-growth. Scripta Metallurgica Et Materialia 1995;32:115.

Chen LQ. Phase field models for microstructure evolution. Annual Review of Materials Research 2002;32:113.

Chen LQ, Shen J. Applications of semi-implicit Fourier-spectral method to phase field equations. Computer Physics Communications 1998;108:147.
Chen Q, Ma N, Wu K, Wang Y. Quantitative phase field modeling of diffusion-controlled precipitate growth and dissolution in Ti-6Al-4V. Scripta Materialia 2004;50:471.
Collings EW. Materials Properties Handbook: Titanium Alloys. Materials Park, OH: ASM International, 1994.
Dobrich K, Rau C, Krill CE. Quantitative characterization of the three-dimensional microstructure of polycrystalline Al-Sn using X-ray microtomography. Metallugical and Materials Transaction A 2004;35:1953.
Elder KR, Grant M, Provatas N, Kosterlitz JM. Sharp interface limits of phase-field models. Physical Review E 2001;64:021604.
Eshelby JD. The determination of the elastic field of an ellipsoidal inclusion, and related problems. Proceedings of the Royal Society of London. Series A 1957;241.
Eshelby JD. The elastic field outside an ellipsoidal inclusion. Proceedings of the Royal Society of London. Series A 1959;252:561.
Fahrmann M, Hermann W, Fahrmann E, Boegli A, Pollock TM, Sockel HG. Determination of matrix and precipitate elastic constants in (gamma-gamma ') Ni-base model alloys, and their relevance to rafting. Materials Science and Engineering A 1999;260:212.
Fan DN, Chen LQ. Diffuse-interface description of grain boundary motion. Philosophical Magazine Letters 1997a;75:187.
Fan D, Chen LQ. Computer simulation of grain growth and ostwald ripening in alumina-zirconia two-phase composites. Journal of the American Ceramic Society 1997b;80:1773.
Fan D, Chen SP, Chen LQ. Computer simulation of grain growth kinetics with solute drag. Journal of Materials Research 1999;14:1113.
Feng WM, Yu P, Hu SY, Liu ZK, Du Q, Chen LQ. Spectral implementation of an adaptive moving mesh method for phase-field equations. Journal of Computational Physics 2006;220:498.
Foiles SM, Hoyt JJ. Computation of grain boundary stiffness and mobility from boundary fluctuations. Acta Materialia 2006;54:3351.
Gabb TP, Backman DG, Wei DY, Mourer DP, Furrer D, Garg A, Ellis DL. γ' formation in a nickel-base disk superalloy. In: Pollock TM, Kissinger RD, Bowman RR, Green KA, McLean M, Olson S, Schirra JJ, editors. Superalloys 2000. Warrendale, PA: TMS, 2000. p. 405.
Ghosh S, Bhandari Y, Groeber M. CAD based Reconstruction of three dimensional polycrystalline microstructures from FIB generated serial sections, Journal of Computer Aided Design, Vol. 40/3 pp 293–310, 2008.
Ginzburg VL, Landau LD. On the theory of superconductivity. Zhurnal Eksperimentalnoy i Teoreticheskoy Fiziki (USSR), 1950;20:10641082 (in Russian) [English translation: in Men of Physics, vol. 1. 1965. Oxford: Pergamon Press, pp. 138167].
Glavicic MG, Kobryn PA, Bieler TR and Semiatin SL. An automated method to determine the orientation of the high temperature beta phase from measured EBSD data fro the low temperature alpha phase in Ti-6Al-4V, Materials Science and Engineering A, 351, 2003: 258–264.
Grafe U, Botteger B, Tiaden J, Fries SG. Coupling of multicomponent thermodynamic database to a phase field model: application to solidification and solid state transformations of superalloys. Scripta Materialia 2000;42.
Groeber M, Ghosh S, Uchic M, Dimiduk D. Development of a robust 3D characterization-representation framework for modeling polycrystalline materials. JOM 2007;59:32.
Gronhagen K, Agren J. Grain-boundary segregation and dynamic solute drag theory – a phase-field approach. Acta Materialia 2007;55:955.
Gruber J, Ma N, Rollett AD, Rohrer GS. Sparse data structure and algorithm for the phase field method. Modelling and Simulation in Materials Science and Engineering 2006;14:1189.
Gunton JD, Miguel MS, Sahni PS. The dynamics of first-order phase transitions. In: Domb C, Lebowitz JL, editors. Phase Transitions and Critical Phenomena, vol. 8. New York: Academic Press, 1983.
Hillert M. A Theory of Nucleation of Solid Metallic Solutions. vol. Sc.D. Cambridge, MA: Massachusetts Institute of Technology, 1956.

Hirth JP, Lothe J. Theory of Dislocations. New York: Wiley, 1982.
Hoyt JJ, Asta M, Karma A. Method for computing the anisotropy of the solid-liquid interface free energy. Physical Review Letters 2001;86:5530.
Hu SY, Chen LQ. A phase-field model for evolving microstructures with strong elastic inhomogeneity. Acta Materialia 2001;49:1897.
Huang Y, Humphreys HJ, Mackenzie JK. Acta Materialia 2000;48:2017.
Johnson WC. Influence of elastic stress on phase transformations. In: Aaronson HI, editor. Lectures on the Theory of Phase Transformations. Warrendale, PA: The Minerals, Metals & Materials Society, 1999. p. 35.
Jonsson B. Ferromagnetic ordering and diffusion of carbon and nitrogen in BCC CR-FE-NI alloys. Zeitschrift fur Metallkunde 1994;85:498.
Karma A. Phase field methods. In: Buschow KHJ, Cahn RW, Flemings MC, Ilschner B, Kramer EJ, Mahajian S, editors. Encyclopedia of Materials: Science and Technology, vol. 7. Oxford: Elsevier, 2001. p. 6873.
Karma A, Rappel W-J. Phase-field method for computationally efficient modeling of solidification with arbitrary interface kinetics. Physical Review E 1996;53:3017.
Karma A, Rappel W-J. Quantitative phase-field modeling of dendritic growth in two and three dimensions. Physical Review E 1998;57:4323.
Kaufman L, Bernstein H. Computer Calculation of Phase Diagrams with Special Reference to Refractory Metals. New York: Academic Press, 1970.
Kazaryan A, Wang Y, Jin YMM, Wang YU, Khachaturyan AG, Wang LS, Laughlin DE. Development of magnetic domains in hard ferromagnetic thin films of polytwinned microstructure. Journal of Applied Physics 2002a;92:7408.
Kazaryan A, Wang Y, Dregia SA, Patton BR. Grain growth in anisotropic systems: comparison of effect of energy and mobility. Acta Materialia 2002b;50:2491.
Khachaturyan AG. Fizika Tverdogo Tela 1966;8:2710.
Khachaturyan AG. Some questions concerning the theory of phase transformations in solids. Soviet Physics – Solid State 1967;8:2163.
Khachaturyan AG. Theory of Structural Transformations in Solids. New York: Wiley, 1983.
Khachaturyan AG, Shatalov GA. Elastic interaction potential of defects in a crystal. Soviet Physics – Solid State 1969;11:118.
Khachaturyan AG, Semennovskaya S, Tsakalakos T. Elastic strain energy of inhomogeneous solids. Physical Review B 1995;52:15909.
Kim SG, Park YB. Grain boundary segregation, solute drag and abnormal grain growth. Acta Materialia 2008;56:3739.
Kim SG, Kim WT, Suzuki T. Phase-field model for binary alloys. Physical Review E 1999;60:7186.
Kim SG, Kim WT, Suzuki T, Ode M. Phase-field modeling of eutectic solidification. Journal of Crystal Growth 2004;261:135.
Kitashima T. Coupling of the phase-field and CALPHAD methods for predicting multicomponent, solid-state phase transformations. Philosophical Magazine 2008;88:1615.
Kitashima T, Harada H. A new phase-field method for simulating gamma' precipitation in multicomponent nickel-base superalloys. Acta Materialia 2009;57:2020.
Kobayashi H, Ode M, Kim SG, Kim WT, Suzuki T. Phase-field model for solidification of ternay alloys coupled with thermodynamic database. Scripta Materialia 2003;48:689.
Krill CE, Chen LQ. Computer simulation of 3-D grain growth using a phase-field model. Acta Materialia 2002;50:3057.
Krill CE, Dobrich K, Michels D, Michels A, Rau C, Weitkamp T, Snigirev A, Birringer R. In: Bonse U, editor. Developments in X-Ray Tomography III, Proc. SPIE, vol. 5335. Bellingham, WA: SPIE Press, 2004. p. 205.
Lan CW, Hsu CM, Liu CC, Chang YC. Adaptive phase field simulation of dendritic growth in a forced flow at various supercoolings. Physical Review E 2002;65:061601.
Landau L, Lifshitz E. Physikalische Zeit schrift der Sowjetunion 1935;8:153.
Langer JS. Statistical theory of the decay of metastable states. Annals of Physics 1969;54:258.
Langer JS. Theory of spinodal decomposition in alloy. Annals of Physics 1971;65:53.

Langer JS. Models of pattern formation in first-order phase transitions. In: Grinstein G, Mazenko G, editors. Direction in Condensed Matter Physics. Singapore: World Scientific, 1986. p. 165.
Langer JS. An introduction to the kinetics of first-order phase transitions. In: Godrèche C, editor. Solids Far from Equilibrium. New York: Cambridge University Press, 1992.
Lee YW, Aaronson HI. Anisotropy of coherent interphase boundary energy. Acta Metallurgica 1980;28:539.
Li DY, Chen LQ. Shape of a rhombohedral coherent Ti11Ni14 precipitate in a cubic matrix and its growth and dissolution during constrained aging. Acta Materialia 1997a;45:2435.
Li DY, Chen LQ. Computer simulation of morphological evolution and rafting of gamma' particles in Ni-based superalloys under applied stresses. Scripta Materialia 1997b;37:1271.
Ma N, Dregia SA, Wang Y. Segregation transition and drag force at grain boundaries. Acta Materialia 2003;51:3687.
Ma N, Kazaryan A, Dregia SA, Wang Y. Computer simulation of texture development during grain growth: effect of boundary properties and initial microstructure. Acta Materialia 2004;52:3869.
Ma N, Chen Q, Wang Y. Simulating microstructural evolution with high interfacial energy anisotropy using the phase field method. Scripta Materialia 2006;54:1919.
Mackenzie JK. Second paper on statistics associated with the random disorientation of cubes. Biometrika 1958;45:229.
McFadden GB, Wheeler AA, Anderson DM. Thin interface asymptotics for an energy/entropy approach to phase-field models with unequal conductivities. Physica D 2000;144:154.
Moelans N, Blanpain B, Wollants P. A phase field model for the simulation of grain growth in materials containing finely dispersed incoherent second-phase particles. Acta Materialia 2005;53:1771.
Moelans N, Blanpain B, Wollants P. Quantitative analysis of grain boundary properties in a generalized phase field model for grain growth in anisotropic systems. Physical Review B 2008;78:024113.
Onuki A. Ginzburg-Landau approach to elastic effects in the phase separation of solids. Journal of the Physical Society of Japan 1989;58:3065.
Poduri R, Chen LQ. Computer simulation of morphological evolution and coarsening kinetics of δ' (Al3Li) precipitates in Al-Li alloys. Acta Materialia 1998;46:3915.
Porter DA, Easterling KE. Phase Transformation in Metals and Alloys. New York: Van Nostrand Reinhold, 1981.
Provatas N, Greenwood M. Multiscale modeling of solidification: phase-field methods to adaptive mesh refinement. International Journal of Modern Physics B 2005;19:4525.
Provatas N, Goldenfeld N, Dantzig J. Efficient computation of dendritic microstructures using adaptive mesh refinement. Physical Review Letters 1998;80:3308.
Raabe D. Computational Materials Science: The Simulation of Materials Microstructures and Properties. Weinheim: Wiley-VCH Verlag GmbH, 1998.
Read W, Shockley W. Dislocation models of crystal grain boundaries. Physical Review 1950;78:275.
Saunders N, Miodownik AP. CALPHAD (Calculation of Phase Diagrams): A Comprehensive Guide. Oxford, New York: Pergamon, 1998.
Searles T, Tiley J, Tanner A. Rapid characterization of titanium microstructural features for specific modelling of mechanical properties. Measurement Science and Technology 2005;16:60.
Semiatin SL, Scoper JC, Sukonnik IM. Short-time beta grain growth kinetics for a conventional titanium alloy. Acta Materialia 1996;44:1979.
Semiatin SL, Fagin PN, Glavicic MG, Sukonnik IM, Ivasishin OM. Materials Science and Engineering A 2001;299:225.
Shen C, Wang Y. Coherent precipitation – phase field method. In: Yip S, editor. Handbook of Materials Modeling, Part B: Models. New York: Springer, 2005. p. 2117.
Shen C, Wang Y. "Phase field microstructure modeling," in Fundamentals of Modeling for Materials Processing, ASM Handbook, Volume 22A, Eds. D. Furrer and S.L. Semiatin, TMS (2010).

Shen C, Chen Q, Wen YH, Simmons JP, Wang Y. Increasing length scale of quantitative phase field modeling of growth-dominant or coarsening-dominant process. Scripta Materialia 2004a;50:1023.
Shen C, Chen Q, Wen YH, Simmons JP, Wang Y. Increasing length scale of quantitative phase field modeling of concurrent growth and coarsening processes. Scripta Materialia 2004b;50:1029.
Simmons JP, Shen C, Wang Y. Phase field modeling of simultaneous nucleation and growth by explicit incorporating nucleation events. Scripta Materialia 2000;43:935.
Steinbach I, Pezzolla F. A generalized field method for multiphase transformations using interface fields. Physica D 1999;134:385.
Steinbach I, Pezzolla F, Nestler B, Seesselberg M, Prieler R, Schmitz GJ, Rezende JLL. A phase field concept for multiphase systems. Physica D 1996;94:135.
Stogner RH, Carey GF, Murray BT. Approximation of Cahn-Hilliard diffuse interface models using parallel adaptive mesh refinement and coarsening with C1 elements. International Journal for Numerical Methods in Engineering 2006;64:1.
Sutton AP, Balluffi RW. Interfaces in Crystalline Material. New York: Oxford University Press, 1995.
Suwa Y, Saito Y, Onodera H. Phase field simulation of grain growth in three dimensional system containing finely dispersed second-phase particles. Scripta Materialia 2006;55:407.
Tiaden J, Nestler B, Diepers HJ, Steinbach I. The multiphase-field model with an integrated concept for modelling solute diffusion. Physica D 1998;115:73.
Uchic MD. 3-D microstructural characterization: Methods, analysis, and applications. JOM 2006;58:24.
Unocic R, Kovarik L, Shen C, Sarosi P, Wang Y, Li J, Ghosh S, Mills MJ. Deformation mechanisms in Ni-base disk superalloys at higher temperatures. In: Reed RC, Green KA, Caron P, Gabb TP, Fahrmann MG, Huron ES, Woodard SA, editors. Superalloys 2008. Warrendale, PA: TMS, 2008. p. 377.
van der Waals JD. The thermodynamik theory of capillarity under the hypothesis of a continuous variation of density. Konink. Akad. Weten. Amsterdam (Sect. 1) 1893;1:56.(in Dutch) [English translation (with commentary): J. S. Rowlinson, J. Stat. Phys. 20, 197 (1979)].
Vedantam S, Patnaik BS. Efficient numerical algorithm for multiphase field simulations. Physical Review E 2006;73:016703.
Wang YU. Computer modeling and simulation of solid-state sintering: a phase field approach. Acta Materialia 2006;54:953.
Wang Y, Chen LQ. Simulation of Microstructural Evolution Using the Field Method. Methods in Material Research. New York: Wiley, 2000. p. 2a.3.1
Wang Y, Chen LQ, Khachaturyan AG. Modeling of dynamical evolution of micro/mesoscopic morphological patterns in coherent phase transformations. In: Kirchner HO, Kubin KP, Pontikis V, editors. Computer Simulation in Materials Science – Nano/Meso/Macroscopic Space and Time Scales. Dordrecht: Kluwer Academic Publishers, 1996. p. 325.
Wang Y, Banerjee D, Su CC, Khachaturyan AG. Field kinetic model and computer simulation of precipitation of L12 ordered intermetallics from fcc solid solution. Acta Materialia 1998;46:2983.
Wang YU, Jin YM, Khachaturyan AG. Three-dimensional phase field microelasticity theory and modeling of multiple cracks and voids. Applied Physics Letters 2001;79:3071.
Wang YU, Jin YM, Khachaturyan AG. Phase field microelasticity theory and modeling of elastically and structurally inhomogeneous solid. Journal of Applied Physics 2002;92:1351.
Wang YU, Jin YM, Khachaturyan AG. Phase field microelasticity modeling of dislocation dynamics near free surface and in heteroepitaxial thin films. Acta Materialia 2003;51:4209.
Wang YU, Jin YM, Khachaturyan AG. Dislocation dynamics – phase field. In: Yip S, editor. Handbook of Materials Modeling, Part B: Models. New York: Springer, 2005a. p. 2287.
Wang Y, Ma N, Chen Q, Zhang F, Chen SL, Chang YA. Predicting of phase equilibrium, phase transformation, and microstructure evolution in advanced titanium alloys. JOM 2005b;September:32.
Warren JA, Kobayashi R, Lobkovsky AE, Carter WC, Sutton AP. Acta Materialia 2003;51:6035.

Wheeler AA, Boettinger WJ, McFadden GB. Phase-field model for isothermal phase transitions in binary alloys. Physical Review A 1992;45:7424.

Wu K, Zhou N, Pan X, Morral JE, Wang Y. Multiphase Ni-Cr-Al diffusion couple: a comparison of phase field simulations with experimental data. Acta Materialia 2008;56:3854.

Zhang F, Xie FY, Chen SL, Chang YA, Furrer D, Venkatesh V. Predictions of titanium alloy properties using thermodynamic modeling tools. Journal of Materials Engineering and Performance 2005;14:717.

Zhang F, Chen SL, Chang YA, Ma N, Wang Y. Development of thermodynamic description of a pseudo-ternary system for multicomponent Ti64 alloy. Journal of Phase Equilibria and Diffusion. 2007;28:115.

Zhang F, Yang Y, CaoWS, Chen SL,Wu K, Chang YA, Commercial Alloy Phase Diagrams and Their Industrial Applications, in ASM Handbook, Volume 22B, Modeling and Simulation: Processing of Metallic Materials, D.U. Furrer and S.L. Semiatin, editors. ASM International, 2010.

Zhou N, Shen C, Mills MJ, Wang Y. Contributions from elastic inhomogeneity and fro plasticity to γ' rafing in single-crystal Ni-Al. Acta Materialia 2008;56:6156.

Zhou N, Shen C, Mills MJ, Wang Y. to be submitted. 2010a.

Zhou N, Shen C, Mills MJ, Wang Y. Large-scale Three-Dimensional Phase Field Simulation of γ' Rafting and Creep Deformation. Philosophical Magazine, 2010b;90:405

Zhu J, Chen LQ, Shen J, Tikare V. Coarsening kinetics from a variable-mobility Cahn-Hilliard equation: Application of a semi-implicit Fourier spectral method. Physical Review E 1999;60:3564.

Zhu JZ, Liu ZK, Vaithyanathan V, Chen LQ. Linking phase-field model to CALPHAD: application to precipitate shape evolution in Ni-base alloys. Scripta Materialia 2002;46:401.

Zhu JZ, Wang T, Ardell AJ, Zhou SH, Liu ZK, Chen LQ. Three-dimensional phase-field simulations of coarsening kinetics of γ' particles in binary Ni-Al alloys. Acta Materialia 2004;52:2837.

Representation of Materials Constitutive Responses in Finite Element-Based Design Codes

Yoon Suk Choi and Robert A. Brockman

Abstract Finite element analysis codes developed originally for engineering structural analysis and design have been adopted by many investigators for materials science studies, and for development of computational material models on the continuum scale. The variety of modeling tools, solution paths, and utilities for constructing new material models make the commercial finite element codes an attractive environment for material model development. This chapter reviews several commonly used continuum mechanics codes, with emphasis on capabilities for representing important classes of material behaviors. A detailed discussion is presented of modeling anisotropic and heterogeneous material structures using representative volume elements and repeating unit cells, with particular emphasis on metallic and intermetallic engineering materials. The presentation includes numerical representations of microscopic and macroscopic material behaviors, and recent efforts to link the responses at these length scales. Numerical and phenomenological aspects of the development of material constitutive models are discussed.

1 Introduction

The finite element method (FEM) is widely used to predict mechanical or thermo-mechanical responses of solids and structures under imposed geometric, thermal, or mechanical constraints. FEM technology has advanced to the point where a number of commercial software products are available, all offering very impressive capabilities. In engineering practice, these FEM-based design codes may be used to:

- predict "nominal" mechanical properties of a selected material system
- evaluate the suitability of a selected material system for a targeted machine component

Y.S. Choi (✉)
Universal Energy Systems, Dayton, USA
e-mail: Yoon-Suk.Choi@wpafb.af.mil

S. Ghosh and D. Dimiduk (eds.), *Computational Methods for Microstructure-Property Relationships*, DOI 10.1007/978-1-4419-0643-4_6,

- optimize component geometry to maximize the mechanical performance of the selected material system, or
- predict upper and lower bounds of critical features of a component under imposed boundary conditions that simulate actual thermal and/or mechanical service conditions.

These modeling efforts are based upon the assumption that constitutive responses, which represent a selected material system, are already determined and properly implemented in design codes. Increasingly, commercial FEM codes are being used (often in conjunction with more specialized constitutive models) to investigate the mesoscopic and microscopic behaviors of nominally homogeneous materials, as well as the detailed performance characteristics of heterogeneous material systems. This materials science-oriented role of the commercial FEM tools provides the motivation for the discussion in this chapter.

The material constitutive law tells how the strain ε (and its time dependence $\dot{\varepsilon}$) is related to the stress σ under a variety of thermal, mechanical, or material conditions. The characterization of material constitutive relations requires carefully controlled experiments followed by the thorough data analysis. Even though the minimization (or elimination) of experimental uncertainties and measurement errors should guarantee the reproducibility of the constitutive response, this response also varies with the heterogeneity of intrinsic material features, and with the length scale chosen for characterizing the material response. The importance of the length scale is closely related to the commonly used concept of representative volume elements (RVE) in FEM-based models of constitutive behaviors. The details of RVE modeling are discussed in several places throughout the book (particularly in "Multiscale Characterization and Domain Partitioning for Multiscale Analysis of Heterogeneous Materials"). Strictly speaking, the RVE is a critical material volume large enough to macroscopically homogenize the material heterogeneity caused by the variability of microstructures and other fine-scale features (Nemat-Nasser and Hori 1993; Kovač and Cizelj 2005). However, in many of FEM-based material modeling practices, the RVE is often misused as a material volume arbitrarily chosen for sampling the constitutive response of interest. It would be reasonable to call such a material volume an arbitrarily chosen volume element, which strictly differs from an RVE unless it is proven to be a statistically, microstructurally, and mechanically "representative" material volume. An RVE must be chosen such that the consistency of a constitutive relation is maintained throughout the entire simulation geometry. For a given RVE, the material constitutive law can be expressed in the form

$$\dot{\varepsilon} = f\left(\sigma, V_i^E, V_i^M\right), \tag{1}$$

where V_i^E represents experimental (external) variables, such as a temperature and sample size, and V_i^M indicates internal variables, which represent key features of the material microstructure responsible for the major constitutive mechanisms.

As FEM-based mechanics codes become core tools in material design there is growing interest about the representation of material constitutive responses implemented in the codes. This chapter focuses on the review of materials constitutive models, chiefly for plastic deformation of metallic and intermetallic materials, as implemented in FEM-based design codes. First two sections begin by surveying present-day FEM-based computational tools for material design, and review the material models available in several prominent commercial codes. This survey deals exclusively with macroscopic material models, since the built-in constitutive models in production codes are aimed at engineering design use, as opposed to materials science. Next, we discuss several important aspects of constitutive modeling approaches for microscopic and macroscopic material behaviors, classified by the RVE concept, and underlying material physics for user-defined material constitutive models, particularly focused on deformation plasticity.

2 Code Survey

The finite element (FE) method provides a general and powerful building-block modeling approach, and its use in engineering design is virtually universal. After more than 50 years of vigorous development, the technique is reasonably mature, and a number of large commercial FE-based software packages have emerged as the tools of choice for design. These commercial off-the-shelf (COTS) codes provide a high level of general capability; however, industries and businesses with unusual problems to solve have limited opportunity for modifying the software or introducing specialized models. To compare the material models that exist in today's FE codes, we must realize that there are many different types of FE software packages, whose intended purposes span a tremendous range of engineering problems. We begin with a general description of the major classes of FE solutions and software.

2.1 Code Classifications

FE analysis codes often are classified as *Lagrangian* or *Eulerian*, referring to the relationship between the mesh and the material. Lagrangian codes, the most common in structural mechanics work, use a computational mesh that moves with the material. As a result, the boundaries of the mesh correspond to the material boundaries, and each element represents a particular quantity of material throughout the computation. This approach lends itself well to detailed, history-dependent material models, such as viscoelasticity, plasticity, and creep. The disadvantage of the Lagrangian approach is that a mesh that follows the material can deform so severely that the computation cannot continue. Eulerian codes use a mesh that is fixed in space, with the material of interest moving through it. The Eulerian technique is more typical in fluid dynamics computations, but is useful for solids as well when the deformations are extremely large, since excessive distortion of the computational mesh is not an issue. In the Eulerian approach, detailed material modeling is complicated by the need to advect state information through the mesh, which produces dispersion error in the numerical solution. Eulerian codes typically use relatively simple deviatoric stress models in conjunction with a nonlinear equation of state (pressure–volume–temperature relation), because a hydrodynamic response often dominates in the class of events for which these programs are used. The upshot is that the characteristics of Lagrangian meshes are better suited to fine-scale material modeling, unless large deformations threaten to distort the mesh to a degree that renders the calculation ineffective or impossible.

Another important classification of FE codes and solution procedures describes the treatment of dynamic problems as either *implicit* or *explicit*. The implicit approach, in which the dynamic solution procedure is similar to a static analysis, predominates in structural mechanics. In the implicit approach, one solves the nonlinear equilibrium equations $\mathbf{R} = \mathbf{F}^{\text{int}} - \mathbf{F}^{\text{ext}} = \mathbf{0}$ (for static analysis) or $\mathbf{R} = \mathbf{F}^{\text{int}} - \mathbf{F}^{\text{ext}} + \mathbf{M}\ddot{\mathbf{U}}$ (in dynamics) using a Newton–Raphson iteration:

$$\Delta\mathbf{U}^{(\nu+1)} = \Delta\mathbf{U}^{(\nu)} - \left[\frac{\partial \mathbf{R}}{\partial \mathbf{U}}\right]^{-1}_{(\nu)} \mathbf{R}^{(\nu)} = \Delta\mathbf{U}^{(\nu)} - \mathbf{K}^{-1}_{(\nu)}\mathbf{R}^{(\nu)}. \tag{2}$$

Here, $\Delta\mathbf{U}$ is the increment in the nodal displacements for the current increment, and ν is the iteration counter. It is important to note that the internal forces $\mathbf{F}^{\text{int}}$ and external forces $\mathbf{F}^{\text{ext}}$ are evaluated at the nodes of the finite element model, by integrating the distributed values of the body and surface forces (for the external forces) and the current stresses (for internal forces). The Newton–Raphson iteration is a search for a system of displacement increments $\Delta\mathbf{U}$ that balances the internal and external forces at the end of the current time or loading increment. The matrix $\mathbf{K}_{(\nu)}$, composed of derivatives of the internal and external forces with respect to the nodal displacements, is called the tangent stiffness matrix, and normally is recomputed and solved at each iteration. Every iteration of the implicit solution requires a solution of the complete system, and is therefore time-consuming. In a dynamic solution, the accelerations $\ddot{\mathbf{U}}$ are replaced by an implicit finite-difference approximation. Therefore, the calculation of $\mathbf{R}$ is slightly different, and $\mathbf{K}_{(\nu)}$ is augmented by a mass-dependent term that depends upon the particular finite difference approximation used in time, but the solution procedure is essentially identical. Figure 1 shows the iteration process, with typical checks for convergence and stepwise accuracy.

Because of the relatively large effort required for a single time-step, the implicit solution is most efficient when larger time-steps can be used. For static solutions, or dynamic solutions in which low-frequency motions dominate the response, the implicit solution is effective. When the response is more rapid, and smaller steps are required to follow the material behavior accurately, implicit techniques quickly become prohibitively time-consuming. The greatest advantage of the implicit approach is stability: most implicit integration schemes used in structural FE codes are unconditionally stable; that is, the integration of the time response remains stable regardless of the time-step size. This property is, of course, a double-edged sword, since one obtains a solution even if the time is too large for acceptable accuracy.

An explicit solution attacks the system $\mathbf{R} = \mathbf{F}^{\text{int}} - \mathbf{F}^{\text{ext}} + \mathbf{M}\ddot{\mathbf{U}} = \mathbf{0}$ by solving for the accelerations directly

$$\ddot{\mathbf{U}} = \mathbf{M}^{-1}(\mathbf{F}^{\text{ext}} - \mathbf{F}^{\text{int}}) \tag{3}$$

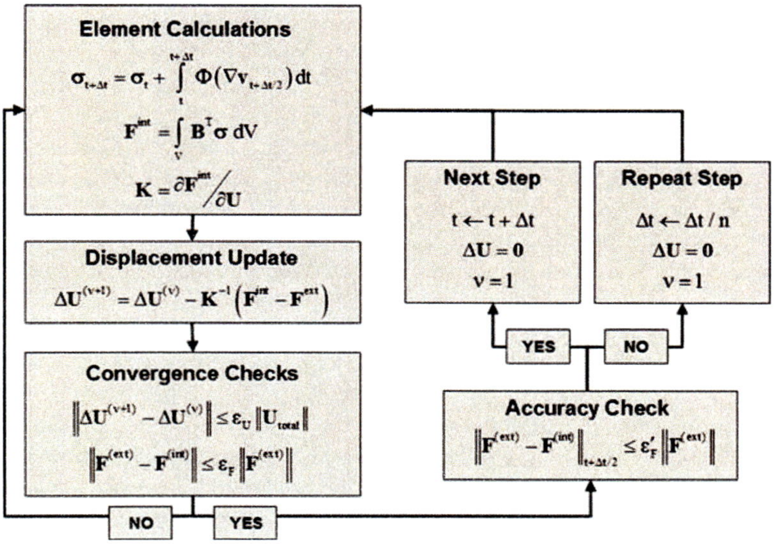

Fig. 1 Implicit finite element solution procedure with step size control

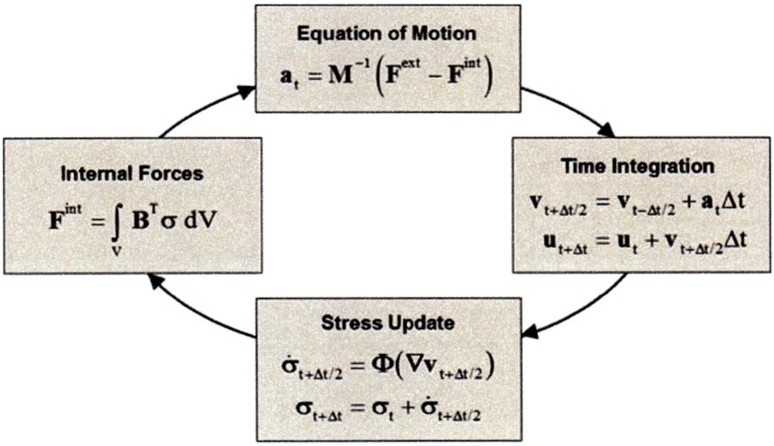

Fig. 2 Explicit finite element solution procedure

and integrating to obtain updated velocities and displacements, without iteration. Normally a lumped (diagonal) mass matrix is employed, so the equation-solving effort is negligible. The next time-step involves the computation of new external and internal forces based upon the updated velocities and displacements, followed by a solution for the new accelerations (Fig. 2). Since the integration procedure is explicit (the equations of motion at time t are used to determine the accelerations at time t, and the new deformed configuration at time $t + \Delta t$), the process is *conditionally stable*; that is, the solution remains stable only when the time-step is sufficiently small. A very close estimate of the allowable time-step can be obtained from $\Delta t_{max} = \ell_{min}/c$, where ℓ_{min} is the minimum element edge length and c is the wave velocity in the material (e.g., $c = \sqrt{E/\rho}$ for longitudinal waves). Time-step limitations for the explicit solution may be severe: for a component made of steel (c $\sim$ 5,000 m/s), and a mesh whose smallest edge length is 1 mm, the critical time-step is about 0.2 μs. Because the time-step limit is established by the mesh refinement and the material wave speed, dynamic problems in which the response of interest spans an interval of milliseconds or longer may require many millions of time-steps using the explicit approach.

The total computational work required for an explicit solution is approximately $W_E c T_{max}/\ell_{min}$, where W_E is work per time-step, c is the material wave velocity, and T_{max} is the time extent of the problem. For a given mesh, material, and total simulation time, the computation time is fixed. In an implicit solution, the work is approximately $W_I \omega_{max} T_{max}$, where ω_{max} is the highest frequency component of interest in the solution. The work W_I per implicit time step is a few orders of magnitude larger than W_E. The quantity ω_{max} is negotiable: by selecting a larger time-step in the implicit solution, a less expensive solution can be obtained at the expense of some accuracy in certain frequency components.

The explicit method essentially performs a wave propagation solution, and is best-suited to high-energy dynamic problems. In such situations, the small time-step needed for stability may also be required to adequately follow the material behavior adequately, making the explicit solution far more efficient than an implicit method. Explicit solutions also handle complicated contact constraints and material failure more gracefully than implicit methods. Explicit techniques can be used to advantage in computational materials science with finite element codes; the key concern is maintaining quasistatic conditions (if appropriate), since the explicit solution is intrinsically a dynamic analysis.

The role of the material model is very different in these two approaches. In the implicit technique, the constitutive model comes into play in computing both the internal forces $\mathbf{F}^{int}$ and the tangent stiffness matrices $\mathbf{K}_{(v)}$. Given strain increments over the current time-step, and the previous values of stresses $\boldsymbol{\sigma}^{(t)}$ and state variables $\mathbf{v}^{(t)}$, the constitutive model in an implicit solution must not only integrate the rate equations of the material model to produce $\boldsymbol{\sigma}^{(t+\Delta t)}$ and $\mathbf{v}^{(t+\Delta t)}$, but also supply the tangent material stiffness

$$\mathbf{D} = \frac{\partial(\Delta\boldsymbol{\sigma})}{\partial(\Delta\boldsymbol{\varepsilon})}. \tag{4}$$

Matrix **D**, called the *consistent tangent* or *algorithmic tangent*, refers to a derivative taken over a finite increment, in a manner consistent with the numerical algorithm used for time integration of the stresses, and is different from the instantaneous derivative of stress with respect to strain for nearly all integration methods in common use. The need to formulate and compute **D** limits the complexity of the material models used in implicit solutions; the models used in explicit codes may be arbitrarily complex, since only the calculation of the updated stresses and state variables is required. For relatively simple material models, the derivation of **D** and its computation is straightforward; when the material model involves finite rotations and strains (Sansour et al. 2008), or discontinuous behavior such as phase change, the formulation of **D** can be quite an adventure. Kirchner (2001) presents an outline of the formulation of **D** for a fairly general class of inelastic material models and time integration operators. Examples of the tangent modulus formulation, with associated user material model code for ABAQUS, may be found in the text by Dunne and Petrinik (2005).

2.2 *Specific Capabilities*

The finite element codes in common use today originated in the mechanical, aerospace, and nuclear engineering communities. The built-in material models are firmly rooted in continuum mechanics, and biased toward macroscopic behavior at a level typical of engineering laboratory measurement. Even so, many such models exist, and the relative strengths of the major codes reflect the disciplines from which the codes emerged.

Tables 1 and 2 contain a summary of some of the material modeling capabilities of several commonly used FE packages, including both Lagrangian or Eulerian codes. A checklist comparison does not expose the differences, often significant, that exist among competing codes in terms of the technical detail, quality of implementation, and numerical efficiency of a given model type; a thorough comparison can be made only in terms of suitability for specific problems. Perhaps, the most significant pattern evident in the features table is the difference between the implicit and explicit codes: the implicit packages contain a more diverse selection of material model types, with many more options than the explicit codes, because many of the finer points addressed by these model options become less important in the high-energy dynamic setting where an explicit solution is more appropriate. It is also interesting to note that *none* of the major FE codes contains built-in models for mesoscale behavior, such as crystal plasticity and strain-gradient plasticity. This dimensional scale is still the realm of materials science rather than engineering. Nonetheless, the materials science community has become quite active in developing such models in the form of user-written routines, as the capacity of computing hardware has grown to the point where highly detailed microstructural models are not only feasible, but routine.

Table 1 A selection of finite element analysis codes

Code	Primary mesh type	Solution type
ABAQUS	Lagrangian	Implicit
ABAQUS/Explicit	Lagrangian	Explicit
ANSYS	Lagrangian	Implicit
CTH	Eulerian	Explicit
LS–DYNA	Lagrangian	Explicit
MARC	Lagrangian	Implicit

Table 2 Available material model types by finite t element code

Model Class	Model feature	ABAQUS	ABAQUS/Explicit	ANSYS	CTH	LS–DYNA	MARC
Elastic	Isotropic	•	•	•	•	•	•
	Anisotropic	•	•	•	•	•	•
	Layered media	•	•	•		•	•
Hyperelastic	Neo–Hookean	•	•	•		•	•
	Mooney–Rivlin	•	•	•		•	•
	Generalized potentials	•	•	•		•	•
	Compressibility	•	•	•		•	•
	Mullins effect	•	•				•
	Foam model(s)	•	•			•	•
Viscoelastic	Linear	•	•	•		•	•
	Nonlinear	•	•	•			•
	Anisotropic	•	•	•		•	•
	Large strain	•	•	•			•
Elastic-plastic	J_2 Plasticity	•	•	•	•	•	•
	Drucker–Prager	•	•	•	•	•	•
	Rate dependence	•	•	•	•	•	•
	Isotropic/kinematic hardening	•	•	•	•	•	•
	Nonlinear hardening	•	•	•		•	•
	Anisotropy (Hill, Barlat)	•	•	•		•	•
	Finite strain kinematics	•	•	•	•	•	•
	Damage/failure models	•	•		•	•	•
	Viscoplastic (power law)	•	•	•	•	•	•
	Crystal plasticity model						
	Strain gradient model						
	Nonlinear equation of state		•		•	•	
Creep	Primary creep			•			•
	Secondary creep	•		•			•
	Swelling (volumetric creep)	•		•			•
	Solution coupled with plasticity						•
Other	Concrete	•	•	•		•	•
	Soils	•	•	•			•
	Fabrics	•				•	
	Piezoelectric materials	•		•			•
	Cohesive zone model	•	•	•			•
User	User-defined material model	•	•	•		•	•

It is worth noting that all of the major FE software products now include capabilities for modeling problems in which the deformations and rotations are large, although these capabilities may not be available for all elements in the program's library. For Lagrangian codes, the key problems that remain in the kinematic description are related to mesh entanglement in purely Lagrangian models, and the representation of failed material regions (including both cracking and rubble). In Eulerian codes, the remaining challenges related to large motions include the tracking of material points, boundaries, and interfaces, and the advection[1] of history-related material variables.

A final category of constitutive behavior that deserves mention is friction. The frictional interaction models in present-day finite element codes have become quite good in terms of searching for contact during the solution. However, the friction models invariably are based upon a form of Amonton's Law, relying on a single coefficient of friction (perhaps a function of relative velocity and temperature) to define the properties of the frictional interaction. The implicit codes typically use an artificial stiffness to introduce stick-slip contact behavior, and the results may be sensitive to iteration strategy and even details such as coordinate system orientation. For complex contact and frictional interaction problems, a common strategy is to perform an explicit solution even for static or quasistatic events, because the numerical behavior is more reliable. Despite the fact that it is now quite routine to include contact conditions and friction in finite element design models, results for problems with significant frictional effects remain decidedly qualitative in character.

2.3 User Material Models

All of the Lagrangian codes listed include "hooks" for introducing user-defined material models. These interfaces have become quite general, allowing the user to maintain a nearly unlimited collection of state variables, provide signals to the system-level solution when a time-step reduction is needed, and even maintain a local database containing multiple material points for use in gradient-based or non-local models. In finite element calculations, the constitutive models are *strain-driven*; that is, an increment in strain is calculated at each material point based on the current displacement estimate, and the constitutive model is expected to provide

---

[1] *Advection* refers to the process of moving material point state variables with respect to the computational mesh in an Eulerian solution, where the material moves relative to the mesh (and therefore with respect to the element sampling points). At each step in the solution, one must move the state variables with the material, and then redefine appropriate values of the state variables for the material points now located at the integration points of each element. Because some of the state information is tensorial, it is difficult to advect the state information in such a way that the final values still satisfy the constitutive relationships. Normally, this process is done as a separate operation at the end of a time-step or series of time-steps. A similar process is required to follow material boundaries, which often move out of one element and into another. Both of these operations are potential sources of significant accumulated error if not performed with extreme care.

a solution for the corresponding stresses and state variables at the end of the step. The reason for this organization lies in the finite element formulation, in which displacements are the primary solution variables. At the model level, the displacement solution is performed using a Newton–Raphson iteration, for which a residual (the force equilibrium error) must be computed for each new candidate displacement solution. During an iterative (implicit) solution, the material model routine is called once per iteration at each material point, always starting from the end of the last time-step where a converged solution is known. In explicit codes, a user material model must update the values of the stresses and state variables; in implicit codes, the material model must also supply a suitable tangent material stiffness for the increment. The material tangent calculation often is the most challenging component of a user-developed material model, and can destroy the rapid convergence of the iterative global solution if done incorrectly.

User material models are of two types: point stress models and unit cell models. In a point stress model, the material is treated as being homogeneous, and the material points at which the model is applied act as sampling points for the calculation of deformations, stresses, other response quantities of interest, and of element internal forces. A unit cell model represents a sample of material which may include multiple materials, phases, or microstructural components, and produces as output homogenized stress values that reflect the presence of the actual material components. Common examples of unit cell models include representative volume elements of a fiber-reinforced plastic, a woven fabric, a polymer or metal containing voids, or a metal alloy composed of multiple material phases in a regular pattern. In each case, a typical repeating volume element of the material – including reinforcing fiber, or weave pattern, or a pattern of defects or secondary material phases – is modeled explicitly at each material point, and subjected to the strain history imposed at that point. Because the stress calculation is strain-driven, the unit cell approach usually employs a Taylor assumption, with the point strain values supplied by the finite element code being applied to the boundaries of the unit cell. In effect, the stress computation for a unit cell model yields the stress and material state that would exist in an infinite grid of these cells, all subjected to a macroscopically uniform strain condition. The assumptions inherent in the unit cell approach make it difficult to include gradient effects in the material model in a realistic way. The unit cell approach is discussed in greater detail in Sect. 5, in the context of crystal plasticity models.

3 Material Modeling in Engineering Design Practice

Material modeling plays a pivotal role in modern engineering design: accurate models provide guidance for design iterations and for more selective experimentation, resulting in shorter development time, lower cost, and higher reliability. Often the ability of an engineering model to predict a correct answer is less important than its ability to predict trends and sensitivities to important design parameters, which in turn calls for material models with an accurate physical basis. This section discusses some of the key strengths and shortcomings of the current generation of finite element analysis packages with respect to engineering design needs and practices, as well as a few common pitfalls for the designer.

In most cases, the strengths of the commercial analysis codes correspond to techniques and submodels that are relatively mature, or where a reasonable consensus exists among the solid mechanics community. As computing capabilities evolve, additional features and numerical approaches become feasible and are integrated into existing models and solution algorithms. A third factor is customer interest; the major code vendors now cater to users in many industries, and competition for development resources is keen. The remainder of this section presents a snapshot of the capabilities of the current generation of commercial finite element codes.

3.1 Material Modeling: Strong Points of the Major Codes

For both historical and economic reasons, general-purpose finite element packages are best-equipped to analyze routine problems involving relatively simple material behaviors. More sophisticated capabilities in the individual packages reflect their funding history and customer-driven priorities. Accordingly, all of the major codes have very strong capabilities for analyzing metals, at least on a scale where the behavior may be considered isotropic or orthotropic. The plasticity models present in today's codes were developed chiefly with metals in mind, and are, almost without exception, limited to material descriptions whose level of detail is consistent with the data collection capability of the typical engineering test laboratory. Plasticity models that provide good modeling fidelity at the microstructural level are not yet standard fare in commercial FEA codes, which cater primarily to engineers. The modeling options available for metal components are quite good for most engineering design work, while most materials scientists will consider them quite rudimentary. For instance, no commercial finite element package includes, as of this writing, even a simple model of crystal plasticity, not for lack of a suitable model but instead for purely economic reasons: most of the customer base consists of engineers with little interest in such a capability.

The major commercial codes now include stress-deformation models that are adequate for most engineering work, when applied properly. Anisotropic materials (linear and nonlinear), linear viscoelasticity models based on Prony series (Soussou et al. 1970), hyperelasticity using numerous options for energy potentials, specialized concrete descriptions, composite laminates, and various foam materials with and without voids all may be included in finite element models using most of the leading codes. The use of nonlinear material modeling, as well as contact and other advanced features, has become quite common in routine engineering design as computing capabilities have evolved to make these analyses practical. Increasingly, the limiting factors in performing useful design analysis are the experience and training of code users, and the collection of sufficient material data to adequately define the material model(s) of interest.

The proper definition of a nonlinear material model often requires specialized technical knowledge, and often the appropriate selection of model parameters depends upon the specific application. For instance, a plasticity model might be defined quite differently if the expected strain magnitudes are less than one percent than if the strains are 50 times larger. For hyperelastic materials, the major analysis codes now include utilities that accept data from one or more types of experiments as input, and perform regression or optimization to define a suitable set of material model parameters. Similar aids for defining viscoelastic and elastic–plastic models would be a welcome addition.

Correct treatment of large deformations and the associated kinematics within nonlinear constitutive models is now commonplace. For inelastic material models, the kinematic treatment is based on the Kröner–Lee decomposition (Kröner 1960; Lee 1969) of the deformation gradient into elastic (reversible) and plastic (irreversible) components, in the form $\mathbf{F} = \mathbf{F}^{e}\,\mathbf{F}^{p}$ (Fig. 3). This kinematic model is consistent with the traditional additive rate decomposition $\dot{\boldsymbol{\varepsilon}} = \dot{\boldsymbol{\varepsilon}}^{e} + \dot{\boldsymbol{\varepsilon}}^{p}$ but provides a consistent framework for models considering large deformation and complex stress trajectories (Sansour et al. 2008). While there is ongoing discussion about the proper interpretation of this kinematic description (Gurtin and Anand 2005), the $\mathbf{F}^{e}$–$\mathbf{F}^{p}$ formulation does provide a consistent mathematical framework for large-deformation constitutive equations and their application in complex systems.

The implicit codes have almost universally adopted some version of the one-step rate integration methods (Wilkins 1964; Ponthot 2002) formerly used only in explicit codes, where the time-steps are small. In general, such an approach is efficient and accurate in implicit solutions provided suitable accuracy controls are exercised. Currently, each of the major commercial codes has its own scheme for time-step control, all of which are heuristic. Early attempts to define step control methods based on theoretical convergence properties of the Newton–Raphson iteration (c.f. Bergan et al. 1978) failed to account for the discretization error incurred in large incremental motions, which in turn precludes an accurate integration of the constitutive equations. During a time increment involving finite rotation (Fig. 4),

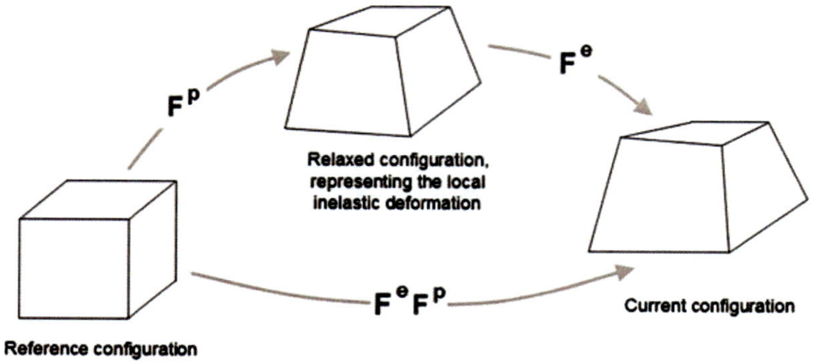

Fig. 3 Kröner–Lee decomposition of the deformation gradient into elastic and inelastic contributions

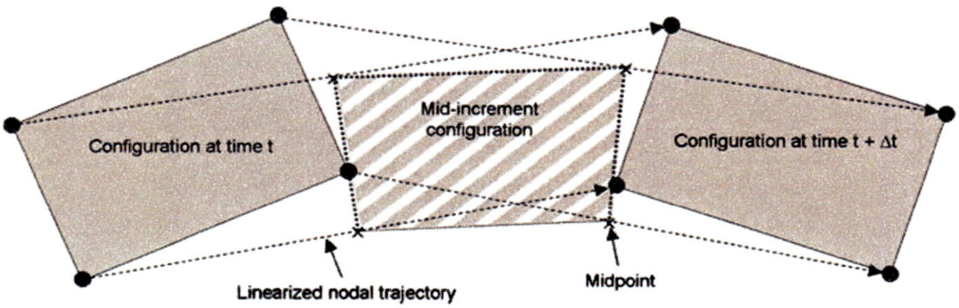

Fig. 4 Mid-increment configuration used in single-step integration of rate constitutive equations

the linearization of the motion over the current step (not to be confused with linearization of the mathematical system) may suggest that intermediate configurations within the step experience strains in excess of the actual values. The notion of an "incrementally objective" integration algorithm (Hughes and Winget 1980), together with heuristic criteria for measuring the net effect of this linearization error (Hibbitt and Karlsson 1979), has led to very effective single-step integration methods for a wide range of constitutive equations, and accompanying methods for assessing the probable accuracy of the stepwise results. Conservative step size control is a crucial factor in tracking nonlinear material behavior accurately without introducing accumulated error, and present-day finite element codes do this well in most applications.

A particularly strong feature of almost all current finite element codes is the capability to supply user-defined material (and other) submodels that can be integrated into an analysis. The availability of user subroutine "hooks" in major codes allows the specialist in a technical area to perform research or model development in that specialty, while using highly developed nonlinear solution strategies, complex element types and solution paths, and pre- and post-processing tools. The price of using this powerful toolset is that the submodel is limited to working with the data objects supported by the user subroutine interface. In some model types, notably strain gradient plasticity and nonlocal plasticity, the existing interfaces are limiting; however, some codes provide more general user routines for accessing the program's database, or implementing one's own internal database scheme. In many organizations, including universities, the availability of user-written material routines has completely transformed the performance of research in computational material modeling.

3.2 Material Modeling: Shortcomings and Challenges

Despite the significant rate at which material modeling capability in the leading finite element codes has progressed in recent years, several notable gaps remain. The most obvious shortcomings in the current generation of material models are related to issues beyond the calculation of raw stress data: conclusions about cumulative damage effects such as crack initiation and growth, delamination, and ultimate failure too often involve "throwing data over the fence" to additional computer programs that use hard-earned stress information in an overly simplistic way.

At the same time, the progressive damage modeling features built into the finite element codes often are unacceptable for serious use. For instance, simulating crack propagation by "unzipping" element boundaries provides only coarse resolution of crack growth increments, and limits attention to predefined crack paths. Most explicit codes now contain element deletion capability for eliminating material based on the constitutive relationships (including cumulative damage), which can be useful in high-energy problems. In high-energy problems, predicting failure accurately often requires only an accurate depiction of the energy deposited in the material per unit volume, and accurate failure predictions can indeed be made for elastic–plastic materials in high-velocity impact problems. One shortcoming associated with nearly all isotropic plasticity models, as implemented in production codes, is the inability to distinguish between plastic strain or work accumulated under tension from that sustained in compression. The model typically works in terms of effective plastic strain (a positive scalar), and accepts a single monotonic stress–strain (or stress–plastic strain) curve as the material characteristic for effective stress versus plastic strain magnitude regardless of the sign of the stress. The failure threshold of most metals in terms of effective plastic strain or plastic work is quite different for tensile and compressive loading, and no mechanism exists for accumulating and testing these quantities separately during the solution.

The element deletion capabilities offered in explicit codes are useful, but in some cases may yield misleading results. When failure occurs in a contact zone, complete removal of the failed material (rubble) creates voids at the contact surface, thereby reducing the contact pressure and introducing errors in the tractions along the interface. The treatment of failed elements differs between codes, and even between element types and material models in the same code. The use of such advanced features in practical applications often involves situations that lie beyond the boundaries within which the algorithms have been tested. In these circumstances, the importance of careful and critical reading of the documentation, together with the solution of test problems whose solutions are readily apparent, cannot be overestimated.

The numerical treatment of contact conditions is significantly different in implicit solutions, because the displacements are used as primary unknowns and because of the need to provide solutions for quasistatic motion. Typically, the implicit solution procedure includes a heuristic treatment of stick-slip motion based on regularization, in the form of an artificial stiffness associated with tangential motions. This numerical device permits constraints to be applied to surface-parallel displacement components until the shear tractions reach the level corresponding to the static coefficient of friction. However, the process is often sensitive to very slight perturbations in the surface geometry, and requires extremely fine step sizes for high accuracy at a load reversal. The recommended remedy for numerical problems in all but the simplest contact problems is to perform an explicit solution, where the heuristics for stick-slip contact motion are not needed.

In implicit contact solutions, a worthwhile approach is to solve the problem of interest with little to no friction, and introduce frictional effects gradually to evaluate the effects of the friction model on the solution. This procedure not only addresses the need for performing sensitivity calculations in contact solutions, where the friction coefficients are never known with high accuracy, but also makes it more likely that anomalies introduced by the stick-slip rules or other model heuristics will become apparent.

Interest in the prediction of residual stresses has increased in recent years, particularly in connection with life prediction. It is now feasible to simulate common surface treatment processes such as shot peening and laser shock processing directly; however, a stress-free initial material state is invariably assumed. For estimation of the bulk residual stress condition, modeling of the initial processing of the material is needed. The ability to perform a forming or forging calculation using a relatively simple material model on an Eulerian mesh, and later continue the simulation of shot peening or laser shock processing on the same model using a Lagrangian mesh with a more detailed material description, would be enormously useful. Numerous examples of useful material processing applications exist that are not quite possible with present-day tools, but which are quite feasible from a purely technical perspective.

Fatigue damage prediction remains a problem in all commercial finite element codes. Most codes have reasonably automated capabilities for computing path integrals (typically the J or M integral), and estimating the corresponding stress intensity factors. However, the quality of the stress intensity factors obtained from this process, which is inherently two-dimensional, is poor near free surfaces. More reliable methods for estimating energy release rates in three-dimensional problems are sorely needed.

Crack propagation under monotonic and fatigue loading presently is feasible in two dimensions using either direct mesh modifications (Jäger et al. 2008) or extended finite element (XFEM) approximations (Abdelaziz and Hamouine 2008; Giner et al. 2008), although the commercial codes have not yet introduced general capabilities for this class of problems. In three-dimensional problems, both approaches are still challenging, and no general-purpose algorithms are likely to appear in the commercial codes for several years. Even when the daunting problem of three-dimensional crack mesh generation is solved, numerical algorithms are needed for fatigue problems, to perform extrapolation of crack geometry and state variables based, say, on analysis of a typical loading cycle or a small number of cycles at each stage of the simulation.

External crack propagation codes are available that perform a series of finite element solutions with remeshing performed at each crack growth iteration (Carter et al. 2000; Chanwandi and Timbrell 2007), and currently represent the most general crack growth simulation capabilities available for use with general-purpose finite element codes. The crack growth software controls the solution, generating updated model files for the finite element solver and reading the output database to define the next increment of crack growth. The most significant limitations of this approach are that the crack growth code interacts with a finite element input deck rather than more primitive modeling data, leading to misinterpretation in some instances, and that the crack growth models typically are limited to linear elastic fracture mechanics (LEFM) concepts.

Structural plastics continue to present numerous modeling challenges. The existence of long- and short-range molecular interactions in these materials produces distinctive and complex behavior that cannot be characterized through a single standardized test type, and makes material model development a prodigious challenge. To make matters worse, polymer compounds are not standardized in the same way as metallic alloys, making users of these materials subject to the whims of material suppliers whose standard of performance, in some cases, may be limited to a simple DSC (differential scanning calorimetry) test. For decades, finite element analysts were limited to elastic–plastic or simple viscoelastic models of such materials; in the last decade, material modeling capabilities for polymeric materials are much improved. Viscoelastic material models (typically based on Prony series representation of the time-dependent modulus) may be combined with rate-sensitive elastic–plastic and nonlinear hardening models in a general way.

The process of data reduction for combined viscoelastic and viscoplastic models is challenging and time-consuming (Frank and Brockman 2001). Presently, the proper definition of combined model parameters requires both specialized technical experience and software tools that are not widely available. A further complication for models combining viscoelastic, viscoplastic, and nonlinear hardening/recovery elements is that the numerical algorithm used in applying the model is significant, in the sense that properties must be defined consistently with their role in the point stress solution. Often the algorithmic details are not published in user documentation, making it more difficult to define an appropriate set of material constants. Optimization of the material model parameters using the target analysis code is preferred, but is not always possible at present.

Polymeric material modeling also is an area in which users are particularly likely to underestimate the difficulty of assembling a reliable material model. At intermediate rates, a viscoelastic stress–strain trace often resembles elastic–plastic response, and the temptation is great to characterize the polymer by measuring modulus, yield point, and hardening slope, and to account for viscous effects by adopting a rate-dependent plasticity model. With this approach, the appropriate set of material parameters becomes a moving target, and the ability of the model to predict behavioral trends for guiding experimental or design work is lost completely. The availability of utilities for optimizing the selection of viscoelastic–viscoplastic material parameters based on multiple test types (monotonic, cyclic, creep, relaxation, stepped rates, etc.), as mentioned in the previous section, would help to encourage more responsible use of the proper constitutive models for this challenging class of materials.

4 Microscopic vs. Macroscopic Behaviors and Models for Metallic Materials

Microstructures for most of the engineering materials are intrinsically heterogeneous. Heterogeneous microstructures give rise to heterogeneous deformation responses at the microscopic and extending through mesoscopic scales. To fully understand such heterogeneities at the microscopic scale, deformation mechanisms responsible for microscopic behaviors should be clarified. Such a clear understanding also helps develop a theoretical or numerical strategy for the homogenization of microscopic responses over the length scale of an order of 2–4 for the macroscopic (homogenized) representation of constitutive responses.

The computation in most of FEM-based design codes preliminarily requires an entire machine component to be reasonably meshed by a limited number of finite-volume elements. This sets the size of a material represented point (MRP[2]) at which the material constitutive relation is actually represented and computed. It means that MRPs in many of FEM-based design codes represent relatively macroscopic volumes, such as mm^3, cm^3, or even m^3. Figure 5 schematically illustrates a general procedure for developing a constitutive model and implementing it to a machine component. Here, constitutive models described in Fig. 5 are mostly empirical and phenomenological, based upon outputs from mechanical tests of bulk samples. However, with growing concern on microstructural influences of

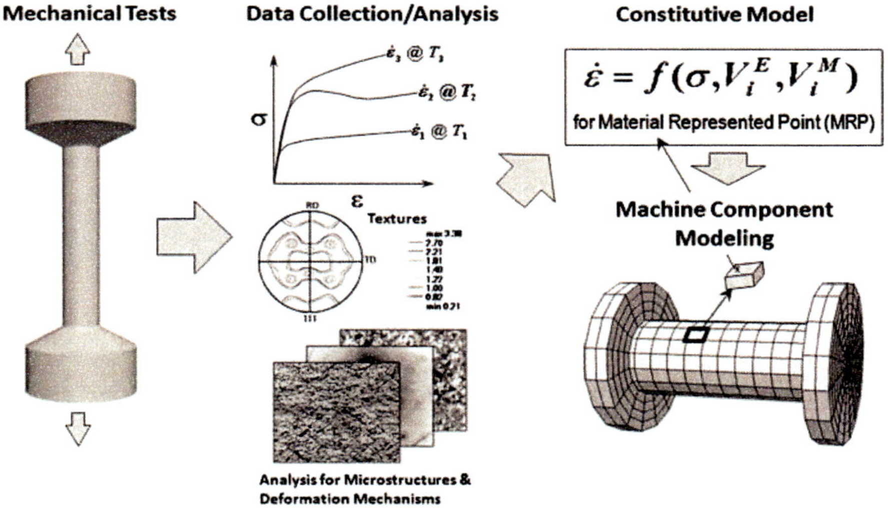

Fig. 5 Schematic illustration showing the general procedure of developing a constitutive model and implementing to a machine component through MRPs. The experimental data are collected from mechanical tests, and a new constitutive model is developed by analyzing the mechanical behavior, microstructures, and corresponding deformation mechanisms. The new constitutive model is implemented to a solid-mechanics FEM platform to model the machine component behavior

[2] In this chapter a material represented point (MRP) was used and differentiated from an RVE since we believe that an RVE can be used only for material volumes that were numerically inspected and proven for their representativeness in microstructural and mechanical responses. We utilized an MRP as a general terminology that describes a materials volume (arbitrarily) chosen for representation of the selected constitutive behavior in 3D FEM modeling without thorough inspection of its representativeness as an RVE.

macroscopic behaviors, linking the MRP responses to the microscopic features has been a major subject in constitutive modeling for design codes. There are numerous numerical techniques to reflect microstructural effects in macroscopic constitutive representations. Some of those approaches directly utilize FEM-based modeling at the microscopic scale, while others adopt different modeling techniques to capture homogenized features of microscopic deformation behaviors. This section reviews constitutive modeling efforts for the microscopic-to-macroscopic connection and discusses advantages and limitations of those modeling approaches, and their physical relevance to actual microscopic and macroscopic behaviors of single or polycrystalline materials.

4.1 FEM-based Modeling of Macroscopic Deformation Behaviors

There are numerous FEM-based modeling approaches to capture macroscopic deformation behaviors of polycrystalline materials. Even though detailed modeling strategies slightly vary by application types, those approaches can be roughly categorized into three types. Schematic illustrations of those three types are given in Fig. 6. In this section, each of those three approaches is discussed in detail.

4.1.1 Generic Modeling with a Homogeneous Bulk Continuum Medium as an MRP

Figure 6a illustrates the first type of the macroscopic modeling approach. It starts with constitutive modeling of a homogeneous bulk continuum medium (solid), based upon the analytical representation of empirical stress–strain curves. A typical example of the homogeneous continuum constitutive model is that proposed by Ramberg and Osgood (1943).

$$\varepsilon = \frac{\sigma}{E} + K\left(\frac{\sigma}{E}\right)^n \tag{5}$$

Here, E is the Young's modulus, and K and n are adjustable constants. The elastic and plastic regions of stress–strain curves for many of polycrystalline materials are well represented by the first and second terms of the RHS in (5), respectively. Please note in (5) that a Power–Law relation was used to represent stress–strain curves in the plastic regime. They further modified their model such that the stress in the Power–Law term of (5) is normalized by the yield strength σ_y (Ramberg and Osgood 1943):

$$\varepsilon = \frac{\sigma}{E} + \eta\varepsilon_o\left(\frac{\sigma}{\sigma_y}\right)^n, \tag{6}$$

where $\varepsilon_o = \sigma_y/E$ and a new parameter, $\eta = K{\varepsilon_o}^{n-1}$. The advantage of the Ramberg–Osgood model is that with two known materials properties (E and σ_y)

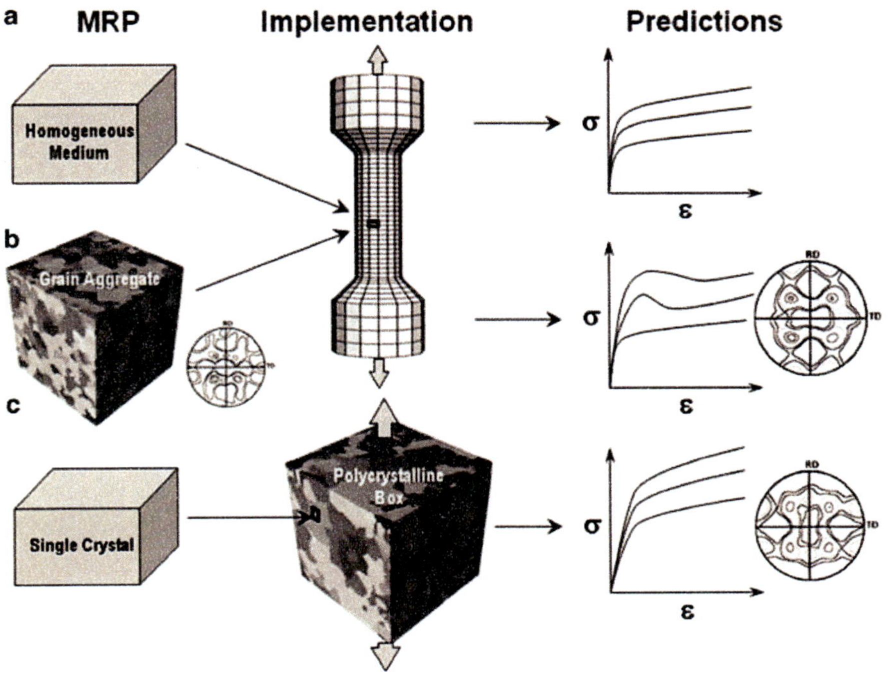

Fig. 6 Schematic illustration of three types of FEM-based macroscopic modeling approaches for constitutive behaviors of materials: (**a**) generic modeling with a homogeneous bulk continuum medium as an MRP; (**b**) polycrystal modeling with an idealized grain aggregate as an MRP; (**c**) polycrystal modeling with a single crystal as an MRP

only two adjustable constants (K and n) are adequate to represent constitutive relations of polycrystalline materials in the plastic regime. However, their model is fully phenomenological and no explicit deformation mechanisms and underlying physics were involved. Nonetheless, because of its simplicity, the Ramberg–Osgood model is one of the major constitutive models that are already built in most of commercial solid-mechanics FEM packages. A homogeneous bulk continuum medium serves as an MRP for this model since the model was built based upon a phenomenological description of macroscopic stress–strain curves. This means that no microstructural effects are involved in the FEM analysis using this model, hence neither was the effect of texturing of crystallographic orientations of polycrystalline materials (see Fig. 6a).

The Power–Law relation involves physical understandings of deformation mechanisms for the representation of stress–strain behaviors in the plastic regime and was utilized in various constitutive modeling in quasi-static (tension and compression), static (creep and relaxation) deformations. For the former, a typical kinetic equation can be written as (Kocks 1976)

$$\dot{\varepsilon}_{\rm p} = \dot{\varepsilon}_{\rm o} \left(\frac{\sigma}{\hat{\sigma}} \right)^{n_{\rm s}}. \tag{7}$$

Here, $\dot{\varepsilon}_{\mathrm{p}}$ and $\hat{\sigma}$ is the plastic strain rate and the internal variable representing the material strength against plastic flow, respectively, and $\dot{\varepsilon}_{\mathrm{o}}$ and n_{s} are temperature-dependent material parameters. In (7), n_{s} is used as a measure for the strain rate sensitivity of plastic flow (Kocks 2000). A steady-state creep rate ($\dot{\varepsilon}_{\mathrm{ss}}$) in most of Power–Law based creep models usually has a functional form of

$$\dot{\varepsilon}_{\mathrm{ss}} = AD_l\sigma^{n_c}, \tag{8}$$

where A and n_c are the microstructure-sensitive parameter and the Power–Law creep exponent, respectively. Also, D_l is the coefficient $[D_l = D_{\mathrm{o}} \exp(-Q_D/kT)]$ for the lattice self-diffusion or the grain boundary diffusion, depending upon a diffusion-assisted creep mechanism that dominates steady-state creep. It should be noted that even though frameworks for (7) and (8) were Power–Law based, each of those constitutive models was derived from completely different physical understandings of deformation mechanisms.

4.1.2 Polycrystal Modeling with a Grain Aggregate as an MRP

Figure 6b shows the second type of the macroscopic modeling approach. It is generally known that a polycrystalline material with the same chemistry, but with different processing histories tends to exhibit different macroscopic deformation behaviors. These variabilities in macroscopic responses stem from different microstructures generated from different processes. Incorporating such differences to a constitutive model has been subject to intensive study in FEM-based material modeling. In particular, the best success was made in macroscopically modeling the influence of the heterogeneity of grain-level crystallographic orientations (i.e., the texture effect).

Crystallographic texturing is due to the slip system-based deformation responses of individual grains, conforming to the geometrical compatibility with neighboring grains under the imposed macroscopic deformation constraint. Thus, in order to incorporate the effect of crystallographic texturing in macroscopic constitutive modeling, it is necessary to numerically delineate how individual slip activities contribute to the deformation (and the resulting rotation) of each grain at the grain level (i.e., single crystal level). This can be done using the kinematic relation of the slip system-based crystal deformation. For a grain k, the plastic strain rate $\dot{\varepsilon}_{\mathrm{p}(k)}$ can be expressed by the following kinematic equation:

$$\dot{\varepsilon}_{\mathrm{p}(k)} = \sum_{\alpha} \dot{\gamma}^{\alpha}_{(k)} P^{\alpha}_{(k)}, \tag{9}$$

where $\dot{\gamma}^{\alpha}_{(k)}$ is the plastic shear strain rate induced by a slip system α for a grain k,. In (9), $P^{\alpha}{}_{(k)}$ is the symmetric part of the Schmid tensor, $b^{\alpha} \otimes n^{\alpha}$, where b^{α} and n^{α} are the slip direction and the slip-plane normal for a slip system α, respectively. Most of the analytical constitutive models, such as those in (5) through (8), still provide valuable frameworks for this approach. For instance, the Power–Law relation (7) can be rewritten as a slip system-based deformation for a grain k:

$$\dot{\gamma}^{\alpha}_{(k)} = \dot{\gamma}_{\mathrm{o}} \left(\frac{\tau^{\alpha}}{\hat{\tau}^{\alpha}} \right)^{n_{\mathrm{s}}}_{(k)}, \tag{10}$$

where τ^{α} and $\hat{\tau}^{\alpha}$ are the resolved shear stress and the slip resistance for a slip system α, respectively. Here, differently oriented grains activate different slip systems and thus give different magnitudes of τ^{α} and $\dot{\gamma}^{\alpha}_{(k)}$ (also $\hat{\tau}^{\alpha}$ if anisotropic hardening is considered). The influence of crystallographic orientations is accounted for modeling in such a way.

The second type of the macroscopic modeling approach shown in Fig. 6b takes a grain aggregate as an MRP. Here, the grain aggregate is assumed to contain a large number of grains (at least on the order of 10^3 to 10^4 grains). To account for the texture effect in macroscopic constitutive modeling, one needs to perform the numerical homogenization that derives $\dot{\varepsilon}_{\mathrm{p}}$ of an MRP from $\dot{\varepsilon}_{\mathrm{p}(k)}$'s and $\dot{\gamma}^{\alpha}_{(k)}$'s of individual grains (similarly, stress σ and stiffness D of an MRP from $\sigma_{(k)}$'s and $D_{(k)}$'s of individual grains). Here, a key issue is how to numerically treat deformation compatibilities (i.e., interactions) among adjacent grains in the MRP-level deformation. This issue was indirectly addressed by adopting the Taylor hypothesis (Taylor 1938a, b), which assumes that the deformation of individual grains (hence the deformation gradient $F_{(k)}$ for a grain k) is comparable to that of an MRP (hence F for an MRP). Based upon this assumption, σ and D for an MRP can be determined by averaging all $\sigma_{(k)}$'s and $D_{(k)}$'s over the entire volume of an MRP:

$$\sigma = \frac{1}{V_{\mathrm{MRP}}} \int_k \sigma_{(k)} \mathrm{d}V_{(k)} \tag{11}$$

and

$$D = \frac{1}{V_{\mathrm{MRP}}} \int_k D_{(k)} \mathrm{d}V_{(k)}, \tag{12}$$

where V_{MRP} and $V_{(k)}$ are volumes of an MRP and a grain k, respectively. Since both $\sigma_{(k)}$ and $D_{(k)}$ are orientation dependent, the influence of the crystallographic texture and its evolution during deformation can be incorporated in an MRP of the constitutive model by accounting for all $\sigma_{(k)}$'s, $\dot{\varepsilon}_{\mathrm{p}(k)}$'s, $D_{(k)}$'s and $V_{(k)}$'s for an entire grain aggregate representing the MRP.

The advantage of this modeling approach is that the initial texture information of an MRP can be obtained from experimental measurements. Since direct measurement of $V_{(k)}$ is not feasible $V_{(k)}$ is often replaced by an orientation weight $w_{(k)}$, which is obtainable from the bulk texture analysis. In this analysis the texture of an MRP can be represented by Ω_i and w_i, where Ω_i is the ith orientation bin ($i = 1$ to n_t, where n_t is the total number of orientation bins accounted for). Here, all grains (i.e., all k's) within an MRP fall into one of Ω_i's, and their volumes (i.e., $V_{(k)}$'s) are weighted by w_i. By doing so, the calculation of $\sigma_{(k)}$ and $D_{(k)}$ in (11) and (12) can be done by calculating $\sigma_{(\Omega_i)}$ and $D_{(\Omega_i)}$.

The macroscopic modeling approach of Fig. 6b has shown success in predicting the evolution of the crystallographic texture during the deformation of polycrystals (Asaro and Needleman 1985; Mathur and Dawson 1989; Bronkhorst et al. 1992; Kalidindi et al. 1992; Kalidindi and Anand 1994; Beaudoin et al. 1993, 1994; Balasubramanian and Anand 1996; Marin and Dawson 1998). The initial texture for an RVE has been chosen in various ways. Some of those modeling studies applied the same initial texture throughout the entire simulation geometry, while others implemented initial textures, which spatially vary element by element, to capture the macroscopic heterogeneity of deformation responses (Becker 2002).

4.1.3 Polycrystal Modeling with a Single Crystal as an MRP

The Talyor hypothesis adopted in the modeling approach of Fig. 6b enforces the identical deformation of all grains. However, heterogeneous deformation incompatibilities arising from heterogeneous interactions among neighboring grains can also affect macroscopic deformation responses. The third type of the macroscopic modeling approach shown in Fig. 6c avoids the Taylor hypothesis by directly handling interactions of individual grains using a single crystal constitutive model. For this approach, an MRP is a volume of a grain (i.e., a single crystal). Thus, constitutive models that directly incorporate slip system-based kinematics [such as (9) and (10)] can be used to represent constitutive responses of the crystal-level MRP.

Here, the fidelity of polycrystalline grain structures depicted in the finite element geometry can impact predicted deformation behaviors. Polycrystalline grain structures have been discretized in finite element geometries in various ways for the representation of polycrystals. Several numerical studies utilized simple brick elements, each of which was assigned as a grain, and predicted the evolution of deformation textures (Beaudoin et al. 1995; Balasubramanian and Anand 2002). Other numerical studies used a regular shape of grains or a discretized shape of irregular grains to investigate local texture gradients due to grain interactions (Harren and Asaro 1989; Becker 1991; Becker and Panchanadeeswaran 1995; Sarma et al. 1998; Mika and Dawson 1999; Zhao et al. 2007). With the progress in acquisition of realistic 3D polycrystalline microstructures (refer to chapters title "Serial Sectioning Methods for Generating 3D Characterization Data of Grain- and Precipitate-Scale Microstructures" and "Digital Representation of Materials Grain Structure" for details), this approach was extended to investigate deformation heterogeneities caused by interactions among complex grains (Barbe et al. 2001a, b; Cailletaud et al. 2003; Lewis et al. 2006, 2008; Zeghadi et al. 2007a, b).

One may notice that the polycrystalline geometry used in the modeling approach of Fig. 6c may be too small to represent macroscopic polycrystals. However, many of numerical studies utilized such a 'small' polycrystalline box (that contains a few hundreds of grains at most) to predict macroscopic flow behaviors of polycrystals (Beaudoin et al. 2000; Brockman 2003; Cheong et al. 2005; Evers et al. 2002; Frénois et al. 2001; Hasija et al. 2003; Kok et al. 2002; Xie et al. 2004). Those attempts basically assume that crystallographic textures and the corresponding spatial distribution represented in their simulation boxes are good enough to represent bulk textures, and that grain geometries implemented in their simulations are sufficient to represent the overall complexity of grains in bulk polycrystals. This means that macroscopic flow features are comparable with those averaged over all grain-level geometrical, mechanical features and heterogeneities for the entire volume of the polycrystalline box. However, a systematic preliminary study needs to be preceded to determine the optimum size of a polycrystalline simulation box before direct macroscopic modeling of an arbitrarily chosen polycrystalline box.

The Fig. 6c approach requires the explicit description of polycrystalline grain structures in the finite element geometry. Here, it is important to clarify what level of fidelity of the grain structure is required to take a full advantage of this modeling approach, *viz.* the sensitivity of the degree of the grain structure fidelity implemented in the simulation geometry. For instance, one needs to check whether or not simulation geometries having regular hexahedral grains, regular polyhedral grains or irregular polyhedral grains produce different mechanical responses. If the answer is yes, it may be necessary to further clarify the influence of the number of accounted grains on the sensitivity of the grain structure fidelity. Many numerical studies appear to suggest that simple regular grain shapes work reasonably well as representative polycrystalline grain structures in the prediction of macroscopic quasi-static (such as simple tension or compression) stress–strain responses if a moderate number of grains are accounted for. However, such simplified grain geometries may not be a reasonable assumption for other applications, such as numerical studies on the development of heterogeneous plastic flow, its grain structure dependence, driving forces for failure models, and their influence on the macroscopic instability of plastic flow. Equally important, but essentially unaddressed, is the influence of the homogenization intrinsic to the constitutive law itself. For example, essentially all examples of using these grain-level models for elasto-viscoplastic flow employ Power Laws, such as (10), without ever addressing its validity at the slip system level.

Various issues were discussed above regarding the implementation and modeling of polycrystalline grain structures with a single crystal as an MRP. These issues are theoretically and numerically challenging. Some of these issues are treated in several chapters throughout this book, and some of them may help develop a numerical strategy to tackle such issues.

4.1.4 Final Remarks on Macroscopic Modeling Approaches

Let us compare all three modeling approaches in Fig. 6. It appears that most of the modeling approaches currently used in engineering material design still heavily rely on a type of the Fig. 6a approach. This approach is relatively simple and computationally affordable, compared to other approaches. However, there are growing concerns on the microstructural variability and its effect on the macroscopic performance of the material in many of the engineering material modeling fields. In this case, macroscopic modeling should involve numerical techniques to link microstructural effects, such as modeling approaches shown in Fig. 6b, c. Here, the modeling approach used in Fig. 6c may not be directly applicable to the simulation of a machine component for engineering design, unless the component is small enough to computationally accommodate all morphological details of grains in the finite element geometry. Otherwise, it is not computationally feasible to explicitly implement details of grains in the component geometry.

An affordable (or feasible) solution would be coupling of modeling approaches of Fig. 6b, c. The Fig. 6c approach delivers the information on how local interactions among various grains give rise to mesoscopic plastic responses. This information can be used to build a constitutive model for an MRP from grain aggregates (Fig. 6b). However, it should be noted that this coupling approach necessitates intensive and systematic numerical approaches. One first needs to identify which types of mesoscopic responses have a significant impact on what types of macroscopic behaviors. Examples of mesoscopic responses are the intergranular or intragranular flow localization and damage initiation, and their grain-boundary character dependence. Examples of macroscopic behaviors are the evolution of stress – strain hysteresis loops and damage under cyclic loading, the time-dependent plastic flow and damage behavior under static loading at elevated temperatures, and the combination of both conditions.

A few numerical studies utilized such a coupling concept to reflect microstructural effects in macroscopic modeling (Kumar et al. 2006; Nakamachi et al. 2007; Swaminathan and Ghosh 2006; Ghosh et al. 2008). The chapter titled "Multiscale Characterization and Domain Partitioning for Multiscale Analysis of Heterogeneous Materials" also deals with some aspects of such a coupling approach. This approach has a strong potential to address many issues in predicting the variability of macroscopic mechanical performances of materials, particularly when microstructural heterogeneities are sources of the variability.

4.2 Numerical Approaches for Linking Microscopic and Macroscopic Behaviors

The modeling approaches used in Fig. 6b, c already described aspects of incorporating microstructural effects (the texture effect for Fig. 6b and both texture and grain interaction effects for Fig. 6c) in macroscopic modeling. Strictly speaking, the microscopic scale involves a range of length scales, wherein dynamics of dislocations within the grain is of main interest at a fine scale, while the deformation heterogeneity due to interactions among grains is a major factor at a coarse scale. The influence of the latter was handled in the modeling approach of Fig. 6c. There are other numerical approaches that intend to reflect fine-scale effects in coarse-scale modeling or utilize fine-scale modeling to capture coarse-scale responses. This section categorizes such numerical approaches and discusses details of each approach.

4.2.1 Coarse Graining from Dislocation Behaviors

It is generally accepted that plastic responses of metallic and intermetallic materials originate from dislocation behaviors at the scale of an order of 10^{-7} to 10^{-5} m.

At such a fine scale, it is difficult to explicitly clarify the contribution of individual dislocations to the observed stress–strain response since activities of individual dislocations are spatially and statistically heterogeneous, and resulting dislocation structures also evolve with time and strain. In this section, the definition of coarse graining is limited to numerical approaches that rigorously homogenize collective behaviors of dislocations to derive kinetic and kinematic relations of crystal plasticity directly from dislocation behaviors (not from the empirical phenomenology of the bulk behavior) (Dimiduk et al. 2006).

There are roughly two types of approaches for coarse graining. The first type treats the problem under the framework of continuum mechanics, where collective dislocations are prescribed in the discretized finite element space (Acharya et al. 2006; Acharya and Roy 2006; Roy et al. 2007). Dynamics of dislocations is treated in the continuum basis such that the resulting dislocation kinetics and kinematics directly lead to plasticity of the simulation geometry. The second type of the coarse graining approach utilizes statistical mechanics and theories to numerically describe "ensemble" behaviors of collective dislocations directly from observed or simulated dislocation structures (El-Azab 2006; El-Azab et al. 2007; Rickman et al. 2005; Rickman and LeSar 2006; Zaiser and Hochrainer 2006).

Even though coarse graining is one of the several promising numerical approaches to fill the gap in length scale between the fine scale dislocation behavior and the coarse scale constitutive behavior, it is still limited to the grain level (or crystal level) homogenization of collective dislocation behaviors. It is important to develop numerical or theoretical methods to interface such outputs with macroscopic modeling to make coarse graining approaches useful for solving practical problems of engineering material design.

4.2.2 Grain Level Constitutive Modeling based upon Discrete Dislocation Dynamics Simulations

Constitutive models for single crystals usually contain internal state variables that reflect the influence of the microstructure on constitutive responses. Major internal state variables are the dislocation density ρ and the slip resistance $\hat{\tau}$. It is important to numerically describe the evolution of these two variables with strain since they are main factors controlling the constitutive response. Discrete dislocation dynamics (DD) simulations can help quantify these variables.

Numerous formulations have been proposed for the evolution of the dislocation density $(d\rho/d\gamma)$. Most of those formulations have two terms, one for the dislocation multiplication and the other for the dislocation annihilation. DD simulations were often used to quantify $d\rho/d\gamma$ and accordingly adjust two terms in the dislocation density evolution for single crystal constitutive modeling (Arsenlis and Tang 2003).

The evolution of the slip resistance $(\mathrm{d}\hat{\tau}/\mathrm{d}\gamma)$ determines strain hardening. Interactions among gliding dislocations produce various types of junctions and locks, which act as a barrier against dislocation gliding, and become a major source for strain hardening. This means that the slip resistance is determined by interactions of dislocations with those locks and obstacles. For a slip system α, $\hat{\tau}$ can be expressed by

$$\hat{\tau}^{\alpha} = \hat{\tau}_{o} + \eta \mu b \sqrt{\sum_{\beta} A^{\alpha\beta} \rho^{\beta}}, \tag{13}$$

where $\hat{\tau}_o$ and η are the intrinsic slip resistance and the constant, respectively. Also, μ and b are the shear modulus and the magnitude of burgers vector, respectively. For FCC crystals $A^{\alpha\beta}$ in (13) can be a 12×12 strengthening interaction matrix, which describes the contribution to strengthening of the slip system α due to the formation of dislocation obstacles in the slip system α by dislocation interactions with the slip system β (Franciosi and Zaoui 1982). For FCC crystals, six characteristic interaction types were indentified: self-hardening, coplanar interactions, cross-slip interactions (collinear), glissile junctions, Hirth locks, and Lomer–Cottrell locks (Franciosi and Zaoui 1982). Here, DD simulations have been often used to determine and quantify the contribution of those six types of junctions and locks (Fivel et al. 1998; Madec et al. 2002; Madec et al. 2003; Devincre et al. 2006, 2008). Values of $A^{\alpha\beta}$ used in various slip system based crystal plasticity FEM simulations of FCC crystals are summarized in Table 3. Values of $A^{\alpha\beta}$ calculated from DD simulations are also included in Table 3.

Compared to coarse graining approaches in Sect. 4.2.1, the current approach indirectly couples fine scale dislocation behaviors to crystal level constitutive modeling. However, the advantage of the current approach is that once major constitutive variables are reasonably determined based upon DD simulations the constitutive model can be utilized to directly predict polycrystalline level behaviors.

4.2.3 Unit Cell Modeling

A unit cell is the smallest morphological unit that represents a symmetric (or periodic) feature of the microstructure. The entire microstructure can be rebuilt by the rotation, translation, and reflection of a unit cell. A unit cell has been widely used as a representative microstructure in FEM modeling for the deformation of materials having relatively periodic microstructures, such as fiber or particle reinforced composites and precipitate strengthened alloys (Christman et al. 1989; Tvergaard 1990; Bhattacharya 1991; Du and Zok 1998; Niordson and Tvergaard 2002; Zong et al. 2007; Jin et al. 2007; Tirtom et al. 2008). Here, our attention for a unit cell is limited to two-phase microstructures, although the unit cell approach is also used for single-phase materials, such as materials with micropores (Murray and Dunand 2004; Potirniche et al. 2006). Unit cell FEM modeling basically assumes that, in addition to its morphological periodicity, the deformation response of the unit cell is also periodic (displacement-controlled boundary conditions are enforced to keep the unit cell parallelepiped during deformation). Based upon this assumption, the deformation response of the entire microstructure (i.e., the macroscopic response)

Table 3 Values of strength interaction coefficients ($A^{\alpha\beta}$) used for latent hardening in various slip system-based crystal plasticity FEM simulations of FCC crystals, and calculated from dislocation dynamics (DD) simulations

Author(s)	Material (modeling)	Forest density used	A^{SH}	A^{CO}	A^{CS}	A^{HL}	A^{GL}	A^{LC}
Harder 1999	Cu poly-crystal (FEM)	–	0.03	0.18	0.18	0.18	0.21	0.23
Arsenlis & Parks 2002	Al single crystal (FEM)	–	0.1	0.22	0.3 $B^{\alpha\beta}$	0.16 $B^{\alpha\beta}$	0.38 $B^{\alpha\beta}$	0.45 $B^{\alpha\beta}$
Tabourot et al. 1997	Cu single crystal (FEM)	–	0.2	0.3	0.3	0.4	0.4	1.0
Tabourot et al. 2001	Al single crystal (FEM)	–	0.96	0.96	0.96	0.96	0.96	1.0
Fivel et al. 1998	FCC crystal (DD)	$7 \times 10^{11}/m^2$	0.16	0.15	0.15	0.15	0.27	0.27
		$1.4 \times 10^{12}/m^2$	0.16	0.14	0.14	0.14	0.21	0.21
Madec et al. 2003	FCC crystal (DD)	$10^{12}/m^2$	–	–	1.265	0.051	0.075	0.084
Devincre et al. 2006	FCC crystal (DD)	$10^{12}/m^2$	–	–	0.625	0.045	0.137	0.122

A^{SH} self hardening, A^{CO} coplanar system, A^{CS} cross slip (collinear), A^{HL} Hirth lock, A^{GL} Glissile lock, A^{LC} Lomer–Cottrell lock, $B^{\alpha\beta} = |n^{\alpha} t^{\beta}|$, where n^{α} is the unit normal of a slip plane and t^{α} is the unit tangent of a dislocation line

is considered to be equivalent to the deformation response of the unit cell, which significantly saves the computational cost. In this sense, unit cell modeling is the numerical approach that directly links the microscopic response to the macroscopic response.

However, there are several limitations in unit cell modeling, and one must fully consider those aspects when utilizing the unit cell approach. First, the entire microstructure must be periodic. The unit cell prediction may significantly differ from the macroscopic response if the simulated microstructure has a low degree of periodicity. This is because of the variation of geometric constraints due to the irregularity of the microstructure.

Second, most of the unit cell simulations assume the fully elastic or incompressible rigid body behavior for the reinforcement (i.e., fibers, particles, and precipitates) and the elasto-plastic (or elasto-viscoplastic) behavior for the matrix. This indicates no plastic shearing of particles and precipitates by dislocations. The deformation compatibility between the matrix and reinforcement also holds unless matrix/reinforcement debonding (or sliding) is allowed. This induces a deformation constraint in the soft matrix. Because of these restrictions, the unit cell assumption is usually valid only for small plastic strains before precipitate shearing or particle debonding takes place (Busso et al. 2000).

Third, the information that you can get from unit cell simulations is limited. Most of the unit cell simulations are intended to investigate how the geometrical constraint induced by the hard precipitate (or reinforcement) and the soft matrix influences the stress–strain response. This includes effects of the precipitate (or reinforcement) morphology and its volume fraction, and effects of interfacial debonding or sliding on the stress–strain response. However, because of the microstructural representativeness of the unit cell, the onset of a mechanical instability within the unit cell is assumed to be the onset of the same instability at every unit cell throughout the entire microstructure. Thus, the initiation and propagation of the instability is assumed macroscopically uniform in unit cell modeling.

Last, the interfacial properties that represent the precipitate (or reinforcement) and the matrix are a key influential factor in unit cell modeling. Stress–strain responses significantly differ when different types of interfacial behaviors (such as debonding and sliding) are accounted for in unit cell modeling. There have been intensive numerical studies on the influence of the interfacial properties on the flow behavior of particle (or fiber) reinforced composites using a unit cell (McHugh and Connolly 1994; Luciano and Bisegna 1998; Li and Ellyin 1998; Ismar et al. 2001; Niordson and Tvergaard 2002; Li and Anchana 2004; Prabu et al. 2004; Ganguly and Poole 2005; Segurado and LLorca 2005; Bansal and Pindera 2006; Bonora and Ruggiero 2006; Metzger et al. 2006). Here, the reinforcement has no crystallographic relation with the matrix, *viz.* the incoherent interface between the matrix and reinforcement. Many of these studies successfully capture the phenomenology of the interfacial behavior using a unit cell.

However, there still seems to be several unsolved issues on the numerical treatment of the interfacial properties for the case of the coherent interface between the matrix and precipitate. Ni-base single crystal superalloys have been subject to numerous unit cell modeling studies because of the nearly periodic nature of their microstructures – a high volume fraction of roughly cuboidal γ' precipitates ($L1_2$ structure) regularly distributed in the γ matrix (FCC structure) with the crystallographic coherency (Nouailhas and Cailletaud 1996; Nouailhas and Lhuillier 1997; Kuttner and Wahi 1998; Busso et al. 2000; Preußner et al. 2005; Choi et al. 2005). There are two numerical issues regarding unit cell modeling of this material. The first issue is to model initial internal stresses (i.e., coherency stresses) induced by the coherency between the γ' precipitate and γ matrix. This initial internal stress state was handled in FEM modeling by applying different thermal expansion coefficients for the γ' precipitate and γ matrix and raising the temperature from RT to an elevated target temperature prior to applying the deformation (Glatzel and Feller-Kniepmeier 1989; Ganghoffer et al. 1991; Pollock and Argon 1992; Müller et al. 1993; Meissonnier et al. 2001). Strictly speaking, this FEM approach produces thermal mismatch stresses, not coherency stresses. However, these thermal mismatch stresses are often considered as coherency stresses in modeling since they produce an initial internal stress state, which looks equivalent to that induced by the γ'/γ coherency. The second issue is related with modeling the evolution of the γ'/γ coherency with the strain and time. As the plastic deformation proceeds, the gradual loss of the γ'/γ coherency is accompanied by gradual building of interfacial dislocations, which releases the γ'/γ coherency, hence the coherency stress state. The complete loss of the coherency may accelerate plastic flow of the γ matrix around the γ' precipitate. However, there is no systematic numerical study to validate thermal mismatch stresses as coherency stresses and their relaxation behavior in conjunction with the gradual loss of the γ'/γ coherency with the strain and time. The proper numerical implementation of the γ'/γ coherency effect within the unit cell will provide crucial information to understand some static and quasistatic behaviors of Ni-based single crystal superalloys at elevated temperatures.

Unit cell modeling is a simplified approach to save computational costs. Even though it may provide the meaningful information on how microstructural effects influence the macroscopic response, this approach cannot be directly used for modeling of a machine component. The best use of the unit cell approach can be found in utilizing its results, particularly microstructural effects, to build an upper scale constitutive model for the prediction of macroscopic behaviors.

5 User-Defined Material Constitutive Models for Crystal Plasticity

Most of the commercial solid-mechanics FEM programs have an interface to link user-defined constitutive behaviors through user subroutines. Modeling approaches described in Fig. 6b, c were implemented within commercial FEM codes through those user subroutines (Bronkhorst et al. 1992; Kalidindi et al. 1992; Kalidindi and Anand 1994; Balasubramanian and Anand 1996, 2002; Becker 1991; Becker and Panchanadeeswaran 1995; Lewis et al. 2006, 2008; Brockman 2003; Cheong et al. 2005; Evers et al. 2002; Hasija et al. 2003; Xie et al. 2004). Here, the user can implement a new material constitutive model through a user subroutine. There are a large number of constitutive models proposed for elasto-plastic or elasto-viscoplastic behaviors of different types of materials and crystals. Some of them were purely based upon dislocation micro-mechanisms and micro-mechanics, while others heavily relied on the theory of solid mechanics at the continuum scale.

Development of a new constitutive model is numerically and mechanistically challenging. In particular when it is intended for use in 3D FEM modeling, extra caution must be taken since extending nondimensional constitutive modeling may give rise to numerical issues that the modeler should account for. This section is intended to discuss various numerical and phenomenological aspects that modelers may overlook or ignore when developing and verifying constitutive models for 3D FEM. Particular attention is paid to discussion of constitutive modeling for elasto-plastic or elasto-viscoplastic behaviors of metallic and intermetallic materials. Some of these aspects which will be discussed in the following subsections are critical and may lead to wrong results unless they are fully accounted for in constitutive modeling.

5.1 Determination of an MRP

It is important to determine the scope of the MRP that a new constitutive model will be based upon. Here, the modeler needs to clearly identify the length scale of an MRP as well as its corresponding microstructural or structural features. For instance, suppose that a modeler plans to develop a new constitutive model for creep of an Ni-base single crystal superalloy, which is a key material for turbine blades. Figure 7 schematically illustrates several points that the modeler needs to consider. Let us assume that a series of creep data was obtained from creep tests of bulk specimen (Fig. 7a), and the modeler decided to use the same specimen geometry for FEM modeling (Fig. 7e) using a new creep constitutive model. Now, the size of the MRP for the new constitutive model appears to be macroscopic as the modeler decided to use the entire geometry of the test specimen (Fig. 7a, e). After the extensive literature survey, the modeler identified that several microstructural features at different length scales influence creep behaviors of the bulk specimen (Fig. 7a) through different creep mechanisms. Those microstructural features are illustrated at three length scales in Fig. 7b–7d along with detailed descriptions.

It is solely the modeler's responsibility to decide the microstructural features that will be included in the MRP via the new constitutive model. For instance, let us say that one only chooses the microstructural feature of Fig. 7d for the representation of the MRP and develops a constitutive model based upon creep mechanisms for Fig. 7d. In this case, it is questionable if the direct comparison of FEM predictions (Fig. 7e) with the creep data obtained from bulk tests (Fig. 7a) reasonably establishes the validity of the new constitutive model since other microstructural features (Fig. 7b and 7c) and their contributions are still missing from the MRP and from the constitutive model. The modeler needs to clearly identify the scope of the MRP and

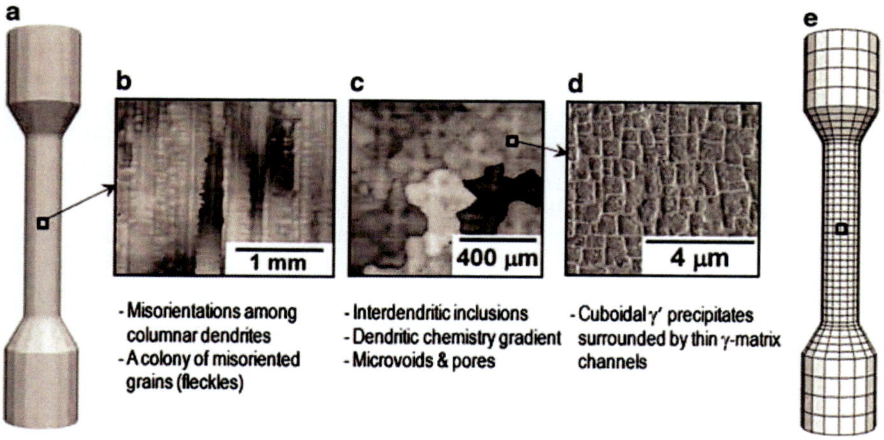

Fig. 7 3D FEM-based constitutive modeling for creep of an Ni-base single crystal superalloy: **(a)** creep test specimen; defects and microstructural features; **(b)** at the macroscopic scale; **(c)** at the intermediate scale; **(d)** at the microscopic scale; **(e)** creep test specimen meshed for FEM modeling

the corresponding constitutive model, and clarify the limitation of modeling results, compared to the experimental data from bulk tests. This may be one of the reasons that current interest in mechanical tests are aiming toward the subscale and even the microscale, where the influence of microstructural features at coarser scales can be minimized. Hence, a better understanding of deformation responses purely at the microscopic scale of interest may be provided from such tests. Some of those experimental approaches are discussed in chapter titled "Emerging Methods for Matching Simulation and Experimental Scales".

It is always a good practice to thoroughly review microstructural details and corresponding deformation mechanisms of the material at different length scales before jumping into the features and mechanisms of interest, and to reasonably account for those contributions in the MRP and constitutive modeling. In engineering practice, the development of a new constitutive model is usually intended for its direct application for the prediction of the machine component behavior. Here, the scope of the original MRP is also transferred to that of machine component modeling. This means that the description of the MRP should be consistent between the two modeling cases. Otherwise, the use (or application) of the new constitutive model (and the MRP) should be limited to thermal, mechanical conditions that keep the consistency with the scope of the original MRP.

5.2 *Strain Rate Sensitivity and Hardening Laws: Intrinsic Flow Responses*

In a material constitutive model, elasto-plastic or elasto-viscoplastic behaviors can be described by the contribution of elasticity and plasticity to the total strain:

$$\varepsilon = \varepsilon_{\mathrm{e}} + \varepsilon_{\mathrm{p}}, \tag{14}$$

where ε_{e} and ε_{p} are the elastic and plastic strains, respectively. Here, the kinetic relation of ε_{p}, viz. $\dot{\varepsilon}_{\mathrm{p}}$ is the one of the important factors controlling the plastic flow response. As described in (1), $\dot{\varepsilon}_{\mathrm{p}}$ is a function of the stress, the temperature, and the other external or internal material variables. Various types of descriptions for $\dot{\varepsilon}_{\mathrm{p}}$ have been used for constitutive modeling of different types of materials. The representative examples are the Power–Law equation shown in (7), an equation based upon thermally activated plastic flow

$$\dot{\varepsilon}_{\mathrm{p}} = \dot{\varepsilon}_{\mathrm{o}} \exp\left(-\frac{Q_p}{kT}\right), \tag{15}$$

and the linear stress dependence of $\dot{\varepsilon}_{\mathrm{p}}$

$$\dot{\varepsilon}_{\mathrm{p}} = b\frac{(\sigma - \hat{\sigma})}{B}, \tag{16}$$

where Q_p and B are the activation energy for plastic flow and the material constant, respectively: Q_p is often expressed as a function of the effective stress and the activation volume, depending on the nature of thermally surmounted obstacles (Kocks et al. 1975). All these descriptions are derived from phenomenological or physical interpretations of experimental results and mechanisms. In particular, the validity of these equations is limited to different specific ranges of temperature, stress, and strain [one can refer to Kocks et al. (1975) and Nadgorny (1988) for detailed discussion]. Deformation mechanisms hypothesized for each equation are also different. Thus, each of these equations yields different strain rate sensitivities and stress dependences. The modeler should fully consider the underlying physics and phenomenologies of all those descriptions when selecting the right kinetic relation for ε_p to start with for a new constitutive model. Failing to do so often results in bringing many material parameters having physically unacceptable values to fit the experimental data.

Strain hardening ($d\hat{\sigma}/d\varepsilon_p$) is also an important factor in determining the plastic flow behavior. Based upon the type of the yield surface and its evolution with strain various hardening laws were proposed, such as isotropic, latent, and kinematic hardening laws. It should be noted that strain hardening is an intrinsic material property, which is purely from dislocations micromechanisms and their interactions with microstructures, such as obstacles, grains, etc. Here, strain hardening can be differentiated into two types, the grain-level (i.e., crystal-level) hardening and the polycrystal-level (i.e., aggregate-level) hardening. The former is based upon dislocation glide and its interaction with local obstacles and dislocations on different slip systems, while the latter results from interactions of grains having different dislocation activities. Most hardening laws are phenomenologically derived from results of mechanical tests of bulk samples. This means that experimental strain hardening curves, even from carefully controlled experiments, always bear a certain degree of extrinsic constraint effects induced by mechanical tests. This somewhat contradicts a strict interpretation of strain hardening in FEM constitutive modeling being an intrinsic material property. One of the efforts to address this issue is to utilize DD simulations to numerically quantify intrinsic strain hardening behaviors associated purely with dislocation behaviors (Arsenlis and Tang 2003). Although the method still limited to the early stage of the deformation (i.e., the early stage I behavior), this approach has improved understanding of the strain hardening behavior, particularly the latent hardening behavior, at the crystal level (Fivel et al. 1998; Madec et al. 2002; Madec et al. 2003; Devincre et al. 2006, 2008).

5.3 NonDimensional Analytical Modeling, 3D FEM Modeling, and Designing the Simulation Geometry and Boundary Conditions

Many of the crystal plasticity constitutive equations for FEM modeling were adopted from nondimensional analytical models of crystal plasticity. Even for constitutive equations intended for FEM modeling, a preliminary parametric study can be done analytically without using FEM. However, unlike FEM modeling, nondimensional analytical modeling has no way of systematically tracking the change of the simulation geometry and the deformation incompatibility caused by the imposed mechanical constraint. In the case that a nonuniform geometrical change is expected due to imposed deformation constraints, the nondimensional analytical result may significantly differ from that of FEM modeling. Softening and the rotation are usually accompanied by the local (or nonuniform) geometry change, and this influences the resulting stress–strain response. Thus, one needs to clearly understand the limitation of nondimensional crystal plasticity models when trying to extend those models to the FEM application.

Softening and the rotation of the local geometry are also significantly influenced by the simulation geometry and boundary conditions applied in FEM. This means that in crystal plasticity FEM the stress–strain output is also sensitive to the geometry and boundary conditions. This sensitivity is particularly significant when modeling the asymmetric deformation behavior, which is frequently observed in axial loading of single crystals oriented in a low symmetry direction. It is suggested that the modeler should carefully introduce boundary conditions, such as symmetric and periodic boundary conditions that do not directly reflect constraints imposed in actual mechanical tests, but are frequently used in FEM modeling to save on the computational cost. Figure 8 shows examples.

Tensile creep tests of a single crystal rod having a length-to-diameter ratio of 6 were simulated using two different representations of the simulation geometry and boundary conditions (Choi 2009, unpublished work). In the first case (case I), the simulation geometry was represented by an eighth of the entire rod sample geometry with a symmetric boundary condition, which assumes that the deformation of the entire sample geometry can be constructed by the rotation and reflection of its octant. The second case (case II) uses the entire rod sample geometry without applying the symmetric boundary conditions. Tensile creep simulations were performed for a single crystal oriented in a low symmetry crystallographic direction $[\overline{50}\ 77\ 996]$. The creep constitutive model was implemented such that $\{111\}\langle 11\bar{2}\rangle$ slip systems are activated at 750°C and 750 MPa. Figure 8a and 8b shows initial and deformed geometries of cases I and II, and the resulting creep curves, respectively. Tensile creep loading in the $[\overline{50}\ 77\ 996]$ activates single $(1\bar{1}\bar{1})$ $[\bar{1}1\bar{2}]$ slip, which causes the rotation and the asymmetric deformation of the simulation geometry. Resulting creep curves are quite different between cases I and II. One can see that the rotation and resulting geometric softening caused by single $(1\bar{1}\bar{1})$ $[\bar{1}1\bar{2}]$ slip is better represented when the entire sample geometry is fully described without invoking the symmetric boundary condition. This is a particularly important practice when one intends to model slip system-based, orientation-dependent softening behaviors of single crystals (MacLachlan and Knowles 2002; Ma et al. 2008).

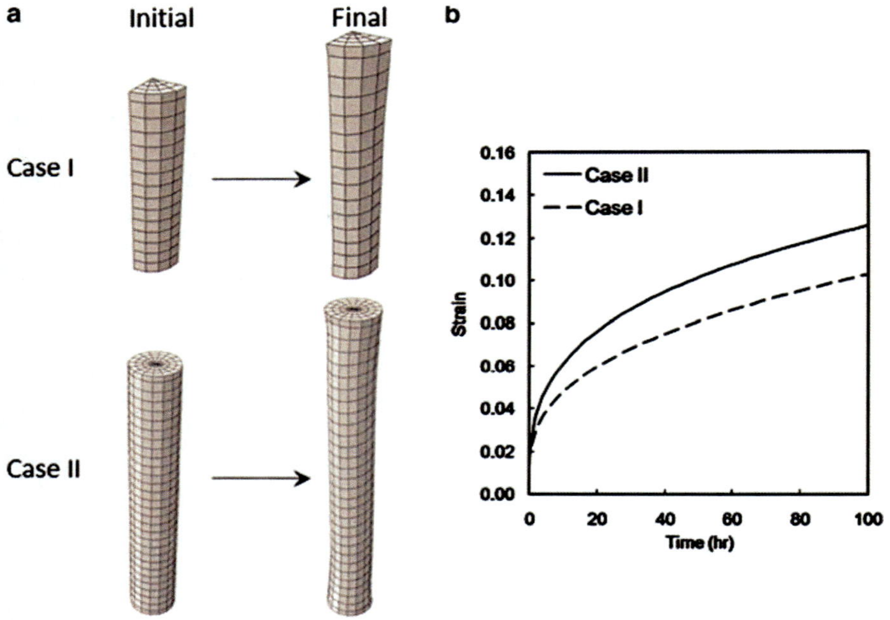

Fig. 8 (**a**) two representations of a creep simulation geometry for a [$\overline{50}$ 77 996] oriented single crystal: an partial octant representation of the entire rod sample with the symmetric boundary condition (case I) and a full representation of the entire rod sample (case II), final geometries after [$\overline{50}$ 77 996] creep are also shown; (**b**) creep curves for cases I and II calculated from crystal plasticity FEM. The creep constitutive model was implemented such that {111}<11$\bar{2}$> slip systems are activated at 750°C and 750 MPa (Choi 2009, unpublished work)

6 Concluding Remarks

The current generation of FEM-based continuum mechanics codes focuses on providing the macroscopic material modeling tools needed for routine engineering design practice, and does so very effectively. However, the growing use of finite element technology for materials science investigations, as well as advanced engineering applications, calls for a more detailed level of material description. This need extends much further than materials science: detailed analysis of the material interface behavior and constitutive modeling are becoming the province of engineering designers who work with polymeric and ceramic composites, micro- and nano-reinforced materials, and functionally graded materials. Despite the availability of very reliable and effective models for the analysis of heterogeneous materials and microstructures, this class of models (with the exception of composite laminates) currently is sorely underrepresented in FEM-based mechanics codes.

Detailed material response analysis will become increasingly common in normal engineering design practice to better understand the influence of fine-scale heterogeneity, inherent material defects, and localized interface behaviors on part performance, reliability, and useful life. Not so long ago, elastic–plastic analysis and contact modeling were used relatively rarely in engineering design, and now are quite routine.

The use of large commercial codes as a platform for implementing advanced material models is a significant advantage in many cases, since the solution capabilities and options offered in the major codes would be difficult or impossible to reproduce in a simple research code. However, acceptance of new types of material models as standard features in the commercial codes has been quite slow. In part, this reluctance to adopt more detailed models is tied to the need for property information that is correspondingly more detailed. That fact runs counter to the widespread notion that "material characterization" involves the determination of only a handful of properties that are easily obtained from macroscale laboratory tests. However, much of the needed information exists already in the materials science community. Improved communication between engineers, code developers, and materials scientists is needed to reach a turning point for the adoption of improved materials modeling technology. The selective use of more highly detailed constitutive modeling tools, especially in conjunction with now-commonplace capabilities for submodeling or "zooming" has the potential to provide improved function, lower cost, and higher reliability in many areas of engineering design.

References

Abdelaziz Y, Hamouine A (2008) A survey of the extended finite element. Comput Struct 86: 1141–1151

Acharya A, Roy A, Sawant A (2006) Continuum theory and methods for coarse-grained, mesoscopic plasticity. Scripta Mater 54:705–710

Acharya A, Roy A (2006) Size effects and idealized dislocation microstructure at small scales: predictions of a phenomenological model of mesoscopic field dislocation mechanics: part I. J Mech Phys Solids 54:1687–1710

Arsenlis A, Parks DM (2002) Modeling the evolution of crystallographic dislocation density in crystal plasticity. J Mech Phys Solids 50:1979–2009

Arsenlis A, Tang M (2003) Simulations on the growth of dislocation density during state 0 deformation in BCC metals. Modell Simul Mater Sci Eng 11:251–264

Asaro RJ, Needleman A (1985) Texture development and strain hardening in rate dependent polycrystals. Acta Metall 33:923–953

Balasubramanian S, Anand L (1996) Single crystal and polycrystal elasto-viscoplasticity: application to earing in cup drawing of FCC materials. Comput Mech 17:209–225

Balasubramanian S, Anand L (2002) Plasticity of initially textured hexagonal polycrystals at high homologous temperatures: application to titanium. Acta Mater 50:133–148

Bansal Y, Pindera M-J (2006) Finite-volume direct averaging micromechanics of heterogeneous materials with elastic-plastic phases. Int J Plast 22:775–825

Barbe F, Decker L, Jeulin D, Cailletaud G (2001a) Intergranular and intragranular behavior of polycrystalline aggregates. Part 1: FE model. Int J Plast 17:513–536

Barbe F, Forest S, Cailletaud G (2001b) Intergranular and intragranular behavior of polycrystalline aggregates. Part 2: results. Int J Plast 17:537–563

Beaudoin AJ, Mathur KK, Dawson PR, Johnson GC (1993) Three-dimensional deformation process simulation with explicit use of polycrystal plasticity models. Int J Plast 9:833–860

Beaudoin AJ, Dawson PR, Mathur KK, Kocks UF, Korzekwa DA (1994) Application of polycrystal plasticity to sheet forming. Comput Methods Appl Mech Eng 117:49–70
Beaudoin AJ, Dawson PR, Mathur KK, Kocks UF (1995) A hybrid finite element formulation for polycrystal plasticity with consideration of macrostructural and microstructural linking. Int J Plast 11:501–521
Beaudoin AJ, Acharya A, Chen SR, Korzekwa DA, Stout MG (2000) Consideration of grain-size effect and kinetics in the plastic deformation of metal polycrystals. Acta Mater 48:3409–3423
Becker R (1991) Analysis of texture evolution in channel die compression-I. Effects of grain interaction. Acta Metall Mater 39:1211–1230
Becker R, Panchanadeeswaran S (1995) Effects of grain interactions on deformation and local texture in polycrystals. Acta Metall Mater 43:2701–2719
Becker R (2002) Developments and trends in continuum plasticity. J Comput Aided Mater Des 9:145–163
Bergan PG, Horrigmoe G, Bråkeland B, Søreide TH (1978) Solution techniques for non-linear finite element problems. Int J Numer Methods Eng 12:1677–1696
Bhattacharya AK (1991) A composite model to predict plastic flow of a superalloy based on its constituent properties. Scripta Metall Mater 25:1663–1667
Bonora N, Ruggiero A (2006) Micromechanincal modeling of composites with mechanical interface – Part II: damage mechanics assessment. Compos Sci Technol 66:323–332
Brockman RA (2003) Analysis of elastic-plastic deformation in TiAl polycrystals. Int J Plast 19:1749–1772
Bronkhorst CA, Kalidindi SR, Anand L (1992) Polycrystalline plasticity and the evolution of crystallographic texture in FCC metals. Philos Trans R Soc Lond A 341:443–477
Busso EP, Meissonnier FT, O'Dowd NP (2000) Gradient-dependent deformation of two-phase single crystals. J Mech Phys Solids 48:2333–2361
Cailletaud G, Diard O, Feyel F, Forest S (2003) Computational crystal plasticity: from single crystal to homogenized polycrystals. Technische Mechanik 23:130–145
Carter BJ, Wawrzynek PA, Ingraffea AR (2000) Automated 3D crack growth simulation. Int J Numer Methods Eng 47:229–253
Chanwandi R, Timbrell C (2007) Simulation of 3-D non-planar crack propagation. Proc NAFEMS World Congress, Vancouver, British Columbia
Cheong KS, Busso EP, Arsenlis A (2005) A study of microstructural length scale effects on the behaviour of FCC polycrystals using strain gradient concepts. Int J Plast 21:1797–1814
Choi YS, Parthasarathy TA, Dimiduk DM, Uchic MD (2005) Numerical study of the flow responses and the geometric constraint effects in Ni-base two-phase single crystals using strain gradient plasticity. Mater Sci Eng A397:69–83
Christman T, Needleman A, Suresh S (1989) An experimental and numerical study of deformation in metal-ceramic composites. Acta Metall 37:3029–3050
Devincre B, Kubin L, Hoc T (2006) Physical analysis of crystal plasticity by DD simulations. Scripta Mater 54:741–746
Devincre B, Hoc T, Kubin L (2008) Dislocation mean free paths and straining hardening of crystals. Science 320:1745–1748
Dimiduk DM, Koslowski M, LeSar R (2006) Preface to the viewpoint set on: statistical mechanics and coarse graining of dislocation behavior for continuum plasticity. Scripta Mater 54:701–704
Du Z-Z, Zok FW (1998) Limit stress conditions for weakly bonded fiber composites subject to transverse biaxial tensile loading. Int J Solids Struct 35:2821–2842
Dunne F, Petrinik N (2005) Introduction to computational plasticity. Oxford University Press, New York
El-Azab (2006) Statistical mechanics of dislocation systems. Scripta Mater 54:723–727
El-Azab, Deng J, Tang M (2007) Statistical characterization of dislocation ensembles. Philos Mag 87:1201–1223
Evers LP, Parks DM, Brekelmans WAM, Geers MGD (2002) Crystal plasticity model with enhanced hardening by geometrically necessary dislocation accumulation. J Mech Phys Solids 50:2403–2424

Fivel M, Tabourot E, Rauch E, Canova G (1998) Identification through mesoscopic simulations of macroscopic parameters of physically based constitutive equations for the plastic behaviour of fcc single crystals. J Phys IV France 8:Pr8.151–Pr8.158
Franciosi P, Zaoui A (1982) Multislip in FCC crystals: a theoretical approach compared with experimental data. Acta Metall 30:1627–1637
Frank G, Brockman RA (2001) A viscoelastic-viscoplastic constitutive model for glassy polymers. Int J Solids Struct 38:5149–5164
Frénois S, Munier E, Feaugas X, Pilvin P (2001) A polycrystalline model for stress-strain behaviour of tantalum at 300K. J Phys IV France 11:Pr5.302–Pr5.308
Ganghoffer JF, Hazotte A, Denis S, Simon A (1991) Finite element calculation of internal mismatch stresses in a single crystal nickel base superalloy. Scripta Metall Mater 25:2491–2496
Ganguly P, Poole WJ (2005) Rearrangement of local stress and strain fields due to damage initiation in a model composite system. Comput Mater Sci 34:107–122
Ghosh S, Dakshinamurthy V, Hu C, Bai J (2008) Multi-scale characterization and modeling of ductile failure in cast aluminum alloys. Int J Comput Meth Eng Sci Mech 9:1–18
Giner E, Sukumar N, Denia FD, Fuenmayor FJ (2008) Extended finite element method for fretting fatigue crack propagation. Int J Solids Struct 45:5675–5687
Glatzel U, Feller-Kniepmeier M (1989) Calculations of internal stresses in the γ/γ' microstructure of a nickel-base superalloy with high volume fraction of γ'-phase. Scripta Metall 23: 1839–1844
Gurtin ME, Anand L (2005), The decomposition F^eF^p, material symmetry, and plastic irrotationality for solids that are isotropic-viscoplastic or amorphous, Int J Plast 21:1686–1719
Harder J (1999) A crystallographic model for the study of local deformation processes in polycrystals. Int J Plast 15:605–624
Harren SV, Asaro SV (1989) Nonuniform deformations in polycrystals and aspects of the validity of the Taylor model. J Mech Phys Solids 37:191–232
Hasija V, Ghosh S, Mills MJ, Joseph DS (2003) Deformation and creep modeling in polycrystalline Ti-6Al alloys. Acta Mater 51:4533–4549
Hibbitt HD Karlsson BI (1979) Analysis of pipe whip, Paper 79-PVP-122, ASME Pressure Vessels and Piping Conference, San Francisco, California
Hughes TJR, Winget J (1980) Finite rotation effects in numerical integration of rate constitutive equations arising in large deformation analysis, Int J Numer Methods Eng 15:1862–1867
Ismar H, Schroter F, Streicher F (2001) Effects of interfacial debonding on the transverse loading behaviour of continuous fibre-reinforced metal matrix composites. Comput Struct 79: 1713–1722
Jäger P, Steinmann P, Kuhl E (2008), Modeling three-dimensional crack propagation – a comparison of crack path tracking strategies, Int J Numer Methods Eng 76:1328–1352
Jin K-K, Oh J-H, Ha S-K (2007) Effect of fiber arrangement on residual thermal stress distributions in a unidirectional composite. J Compos Mater 41:591–611
Kalidindi SR, Bronkhorst CA, Anand L (1992) Crystallographic texture evolution in bulk deformation processing of FCC metals. J Mech Phys Solids 40:537–569
Kalidindi SR, Anand L (1994) Macroscopic shape change and evolution of crystallographic texture in pre-textured FCC metals. J Mech Phys Solids 42:459–490
Kirchner E (2001) Modeling single crystals: time integration, tangent operators, sensitivity analysis and shape optimization. Int J Plast 17:907–942
Kocks UF (1976) Laws for work-hardening and low-temperature creep. J Eng Mater Technol 98:76–85
Kocks UF (2000) Kinematics and kinetics of plasticity. In: Kocks UF, Tomé CN, Wenk H-R (eds) Texture and anisotropy. Cambridge University Press, Cambridge
Kocks UF, Argon AS, Ashby MF (1975) Thermodynamics and kinetics of slip. Prog Mater Sci 19:1–288
Kok S, Beaudoin AJ, Tortorelli DA (2002) On the development of stage IV hardening using a model based on the mechanical threshold. Acta Mater 50:1653–1667

Kovač M, Cizelj L (2005) Modeling elasto-plastic behavior of polycrystalline grain structure of steels at mesoscopic level. Nucl Eng Des 235:1939–1950
Kröner E (1960) Allgemeine kontinuumstheorie der versetzungen und eigenspannungen. Arch Rational Mech Anal 4:273–334
Kumar RS, Wang A-J, McDowell DL (2006) Effects of microstructure variability on intrinsic fatigue resistance of nickel-base superalloys – a computational micromechanics approach. Int J Fract 137:173–210
Kuttner T, Wahi RP (1998) Modelling of internal stress distribution and deformation behaviour in the precipitation hardened superalloy SC16. Mater Sci Eng A242:259–267
Lee EH (1969) Elastic plastic deformation at finite strain. Trans ASME J Appl Mech 36:1–6
Lewis AC, Suh C, Stukowski M, Geltmacher AB, Spanos G, Rajan K (2006) Quantitative analysis and feature recongnition in 3-D microstructural data sets. JOM 58 (12):52–56
Lewis AC, Jordan KA, Geltmacher AB (2008) Determination of critical microstructural features in an austenitic stainless steel using image-based finite element modeling. Metall Mater Trans A 39:1109–1117
Li C, Ellyin F (1998) A micro-macro correlation analysis for metal matrix composites undergoing multiaxial damage. Int J Solids Struct 35:637–649
Li S, Anchana W (2004) Unit cells for micromechanical analyses of particle-reinforced composites. Mech Mater 36:543–572
Luciano R, Bisegna P (1998) Bounds on the overall properties of composites with debonded frictionless interfaces. Mech Mater 28:23–32
Ma A, Dye D, Reed RC (2008) A model for the creep deformation behavior of single-crystal superalloy CMSX-4. Acta Mater 56:1657–1670
MacLachlan DW, Knowles DM (2002) The effect of material behavior on the analysis of single crystal turbine blades: Part II – component analysis. Fatigue Fract Eng Mater Struct 25: 399–409
Madec R, Devincre B, Kubin LP (2002) From dislocation junctions to forest hardening. Phys Rev Lett 89:255508-1–25508-4
Madec R, Devincre B, Kubin LP, Hoc T, Rodney D (2003) The role of collinear interaction in dislocation-induced hardening. Science 301:1879–1882
Marin EB, Dawson PR (1998) On modeling the elasto-viscoplastic response of metals using polycrystal plasticity. Comput Methods Appl Mech Eng 165:1–21
Mathur KK, Dawson PR (1989) On modeling the development of crystallographic texture in bulk forming processes. Int J Plast 5:67–94
McHugh PE, Connolly P (1994) Modelling the thermo-mechanical behaviour of an Al alloy-SiC_p composite. Effects of particle shape and microscale fracture. Comput Mater Sci 3:199–206
Meissonnier FT, Busso EP, O'Dowd NP (2001) Finite element implementation of a generalized non-local rate-dependent crystallographic formulation for finite strains. Int J Plast 17:601–640
Metzger DR, Duan X, Jain M, Wilkinson DS, Mishra R, Kim S, Sachdev AK (2006) The influence of particle distribution and volume fraction on the post-necking behaviour of aluminum alloys. Mech Mater 38:1026–1038
Mika DP, Dawson PR (1999) Polycrystal plasticity modeling of intracrystalline boundary textures. Acta Mater 47:1355–1369
Müller L, Glatzel U, Feller-Kniepmeier M (1993) Calculation of the internal stresses and strains in the microstructure of a single crystal nickel-base superalloy during creep. Acta Metall Mater 41:3401–3411
Murray NGD, Dunand DC (2004) Effect of thermal history on the superplastic expansion of argon-filled pores in titanium: part II modeling of kinetics. Acta Mater 52:2279–2291
Nadgorny E (1988) Dislocation dynamics and mechanical properties of crystals. Prog Mater Sci 31:1–530
Nakamachi E, Tam NN, Morimoto H (2007) Multi-scale finite element analysis of sheet metals by using SEM-EBSD measured crystallographic RVE models. Int J Plast 23:450–489
Nemat-Nasser S, Hori M (1993) Micromechanics: overall properties of heterogeneous materials. North-Holland, Amsterdam

Niordson CF, Tvergaard V (2002) Nonlocal plasticity effects on fibre debonding in a whisker-reinforced metal. Eur J Mech A-Solid 21:239–248
Nouailhas D, Cailletaud G (1996) Finite element analysis of the mechanical behavior of two-phase single-crystal superalloys. Scripta Mater 34:565–571
Nouailhas D, Lhuillier S (1997) On the micro-macro modeling of γ/γ' single crystal behavior. Comput Mater Sci 9:177–187
Pollock TM, Argon AS (1992) Creep resistance of CMSX-3 nickel base superalloy single crystals. Acta Metall Mater 40:1–30
Ponthot JP (2002), Unified stress update algorithms for the numerical simulation of large deformation elastic-plastic and elasto-viscoplastic processes, Int J Plast 18:91–126
Potirniche GP, Hearndon JL, Horstemeyer MF, Ling XW (2006) Lattice orientation effects on void growth and coalescence in fcc single crystals. Int J Plast 22:921–942
Prabu SB, Karunamoorthy L, Kandasami GS (2004) A finite element analysis study of micromechanical interfacial characteristics of metal matrix composites. J Mater Proc Tech 153–154:992–997
Preußner J, Rudnik Y, Völkl R, Glatzel U (2005) Finite-element modeling of anisotropic single-crystal superalloy creep deformation based on dislocation densities of individual slip systems. Z Metallkd 96:595–601
Ramberg W, Osgood WR (1943) Description of stress-strain curves by three parameters (Technical Note No. 902). National Advisory Committee for Aeronautics, Washington DC
Rickman JM, Vinals J, LeSar R (2005) Unified framework for dislocation-based defect energetics. Philos Mag 85:917–929
Rickman JM, LeSar R (2006) Issues in the coarse-graining of dislocation energetic and dynamics. Scripta Mater 54:735–739
Roy A, Puri S, Acharya A (2007) Phenomenological mesoscopic field dislocation mechanics, lower-order gradient plasticity, and transport of mean excess dislocation density. Modell Simul Mater Sci Eng 15:S167–S180
Sansour C, Karšaj I, Sorić J (2008) On a numerical implementation of a formulation of anisotropic continuum elastoplasticity at finite strains. J Comput Phys 227:7643–7663
Sarma GB, Radhakrishnan B, Zacharia T (1998) Finite element simulations of cold deformation at the mesoscale. Comput Mater Sci 12:105–123
Segurado J, LLorca J (2005) A computational micromechanics study of the effect of interface decohesion on the mechanical behavior of composites. Acta Mater 53:4931–4942
Soussou JE, Moavenzadeh F, Gradowczyk MH (1970) Application of Prony series to linear viscoelasticity. J Rheol 14:573–584
Swaminathan S, Ghosh S (2006) Statistically equivalent representative volume elements for unidirectional composite microstructures: part I – without damage. J Compos Mater 40:583–604
Tabourot L, Dumoulin S, Balland P (2001) An attempt for a unified description from dislocation dynamics to metallic plastic behaviour. J Phys IV France 11:Pr5.111–Pr5.118
Tabourot L, Fivel M, Rauch E (1997) Generalized constitutive laws for fcc single crystals. Mater Sci Eng A 234–236:639–642
Taylor GI (1938a) Plastic strain in metals. J Inst Metals 62:307–324
Taylor GI (1938b) Analysis of plastic strain in a cubic crystal, S. Timoshenko 60th Anniversary Volume. Macmillan, New York
Tirtom I, Güden M, Yildiz H (2008) Simulation of the strain rate sensitive flow behavior of SiC-particulate reinforce aluminum metal matrix composites. Comp Mater Sci 42:570–578
Tvergaard V (1990) Analysis of tensile properties for a whisker-reinforced metal-matrix composite. Acta Metall Mater 38:185–194
Wilkins M (1964) Calculation of elastoplastic flows. In: Alder B (ed) Methods in computational physics, vol 3. Academic Press, New York, pp 211–263
Xie CL, Ghosh S, Groeber M (2004) Modeling cyclic deformation of HSLA steels using crystal plasticity. J Eng Mater Tech 126:339–352
Zaiser M, Hochrainer T (2006) Some steps towards a continuum representation of 3D dislocation systems. Scripta Mater 54:717–721

Zeghadi A, N'guyen F, Forest S, Gourgues A-F, Bouaziz O (2007a) Ensemble averaging stress-strain fields in polycrystalline aggregates with a constrained surface microstructure – part 1: anisotropic elastic behaviour. Philos Mag 87:1401–1424

Zeghadi A, Forest S, Gourgues A-F, Bouaziz O (2007b) Ensemble averaging stress-strain fields in polycrystalline aggregates with a constrained surface microstructure – Part 2: crystal plasticity. Philos Mag 87:1425–1446

Zhao Z, Kuchnicki S, Radovitzky R, Cuitiño (2007) Influence of in-grain mesh resolution on the prediction of deformation textures in fcc polycrystals by crystal plasticity FEM. Acta Mater 55: 2361–2373

Zong BY, Zhang F, Wang G, Zuo L (2007) Strengthening mechanism of load sharing of particulate reinforcements in a metal matrix composite. J Mater Sci 42:4215–4226

Accounting for Microstructure in Large Deformation Models of Polycrystalline Metallic Materials

C.A. Bronkhorst, P.J. Maudlin, G.T. Gray III, E.K. Cerreta,
E.N. Harstad, and F.L. Addessio

Abstract Microstructures of metallic polycrystalline materials are varied and evolve with mechanical deformation. The influence of microstructure on mechanical behavior is discussed in the context of material model development. Several modeling approaches have been developed over the past 80 years which have acknowledged the importance of accounting for microstructural details and these are discussed. Examples of two approaches to the large deformation coupled thermo-mechanical modeling of metallic materials are presented and their differences are compared. First, a macroscale continuum internal state variable-based model is presented for tantalum, which also allows for damage evolution. Next, a multi-scale polycrystal plasticity approach is presented, which explicitly represents the polycrystal aggregate. Experiments necessary for both material parameter evaluation (simple compression tests at different strain rates and temperatures) and model validation (dynamic forced shear) are given and discussed. Results from both modeling approaches are compared against results from the forced shear experiments. Both models predict a temperature increase in the shear zone of the sample of 400 K due to plastic work and assuming adiabatic conditions. The continuum model performs better than the mesoscale crystal plasticity approach at predicting the load-displacement responses. Although the single crystal model is 3D, the numerical model is 2D and is believed to be restrictive to the deformation response of the polycrystal. This point-of-view is also supported by comparisons between experimental and predicted crystallographic texture in the shear region. Distributions of vonMises stress, temperature, equivalent plastic strain, and equivalent plastic strain rate in the shear region of the sample as predicted by the polycrystal plasticity model are presented. Simulations like this can assist in our understanding of how materials behave and allow us to develop more physically realistic internal state variable theories for use in engineering applications.

C.A. Bronkhorst (✉)
Theoretical Division, Los Alamos National Laboratory, Los Alamos, NM 87545, USA
e-mail: cabronk@lanl.gov

S. Ghosh and D. Dimiduk (eds.), *Computational Methods for Microstructure-Property Relationships*, DOI 10.1007/978-1-4419-0643-4_7,

1 Introduction

Life as we know it today would not be possible without the existence of metallic materials. In either single or polycrystalline form, pure metallic materials or metallic alloys give us a tremendously broad range of performance properties for use in products and equipment which we rely upon each moment of the day. Those essential services include transportation (roadways, vehicles, and aircraft), electrical power (generation, storage, and transmission), telecommunications, utility infrastructure, medicine, and energy. Not only are service properties of the material or materials within an application component important but the materials must also have the characteristics which allow manufacture of a final component, in a cost-effective manner. The manufacturing process can also be used to modify the characteristics of the material itself to provide properties which are more suitable for end-user performance. This in turn can facilitate improved efficiency in the design of material components, and thereby, reduce the level of overdesign used in a system. This is a process of continual closed-loop iterative improvement, which is as much a function of advancement of our understanding as it is advancement of material and manufacturing technology.

Customization of microstructure is one of the ways in which metallic polycrystal material properties can be altered and optimized for a given application. This is most often accomplished by a combination of mechanical working and heat treatment. For manufacturing operations and those cases where a component is expected to inelastically deform during service, the materials microstructure evolves during this deformation process. Evolution occurs in the forms of grain distortion (morphological texture), preferred grain rotation (crystallographic texture), formation of twins, and dislocation cell structures; slip banding, and evolution of dislocation density. At very small grain sizes and higher deformation rates, the role of grain boundaries becomes more prevalent and so the deformation response of grain boundaries can dominate its mechanical response (Gurtin and Anand 2008). At high rates of deformation and shock or impact loadings, damage and failure processes can also be a means by which a material accommodates imposed loading. The role of microstructure in how these processes evolve is also very important. It is expected that when damage and failure processes become prevalent, they are increasingly dependent upon weak link statistics involving microstructure, defect distribution, loading, and time.

Central to our understanding of the complex physical processes discussed above is our ability to model such events since understanding and modeling success are directly coupled. There is a dual need for the role of modeling in representation of large deformation material behavior; 1) constitutive model development to allow for intelligent engineering design; 2) computational physics work in tandem with good experiments to facilitate our learning about complex physical processes exposed to complex loading. The former is an established process of model formulation based upon supposed dominant physics, small-scale experiments to evaluate material parameters resident in the proposed model, followed by an additional suite of experiments of more complex nature to test the ability of the model to predict the material's more complex loading history response, i.e., validation. The latter approach is more in-line with a computational materials science view where material theory and computational physics are used to assist in our interpretation and understanding of experimental results. This approach can be used to link details of microstructural characteristics to a materials performance. This is especially true for events associated with damage and failure. The nucleation and growth of damage leading to ultimate failure within materials is a stochastic process and is extremely difficult to quantify. In this regime of behavior, mean response is no longer an adequate goal, but rather being able to tangibly link statistical variations within space and time is critical to our being able to understand and interpret these complex relationships. The discussion embarked upon in this chapter has in mind the final goal of modeling the complex processes involved within an efficient theoretical framework. Because of this, we restrict our discussion to finite deformation theory of a local nature.

The concepts of material length scale and representative volume elements are relevant to our discussion here and their relationship is an important one in the context of materials modeling. A representative volume element in general is an amount of material over which a computational solution is meant to represent. For example, a computational cell as part of a finite element model represents a certain volume of material. Within that computational cell are contained a certain number of Gauss points and so the equations being solved over the amount of material associated with a Gauss point is meant to be an average material response over that amount of material. Therefore, a representative volume of material is associated with a characteristic length – say zone size. In general, microstructural features and physical processes which occur during the deformation of metallic polycrystalline materials do so at characteristic lengths. For example, grain size distribution within a polycrystalline aggregate will define a certain length scale of variability in stress and strain associated with that microscopic feature. In addition, phenomenon such as twinning and dislocation substructural development do so at characteristic length scales which are significantly smaller than a single crystal. Because of the assumed averaging over a representative volume element, the size of the computational zone must be chosen consistent with the material model used to represent the material. In addition, as this discussion points out, material theory development in general cannot be divorced from numerical aspects.

Ultimately, the goal of our theoretical and computational work is to provide predictive ability to describe a materials deformation response (or other behaviors but our focus here is on mechanical deformation). Regardless of the length scale of interest, material models should be physically motivated to best represent behaviors and increase the probability for predictability. Many material models are based upon the concept of internal state variable as a way to represent important properties of a material and at any given time provide a unique description of the materials current state at the microstructural level. Proper local internal state variables include porosity, crystallographic orientation, dislocation density, etc. (Haghi 1995). Other state variables establishing the instantaneous state of the material (not directly related to microstructure) are stress, strain rate, and temperature. A constitutive model is then built upon evolution equations for each state variable employed and a thermodynamically sound theory is developed which characterizes relationships between the variables which adheres to both first and second laws of thermodynamics (Rice 1975; Anand 1985; Lemaitre and Chaboche 1990). In the context of continuum models for materials, we will focus here on two types of constitutive models, one which averages over the entire polycrystalline aggregate and one which represents mean single crystal response.

Although not dealt with directly in this chapter, homogenization models fall between the full continuum treatment of polycrystalline metallic materials and the direct polycrystal modeling of these aggregates (Kocks et al. 1998). Each of these approaches attempts in different ways to account for the materials microstructure and microstructural evolution. Perhaps the simplest of these theories is commonly termed the Taylor model (Taylor 1938). In this aggregate theory, the polycrystal is assumed to be comprised of a certain number population of grains, a model of the single crystal is put forward and a macroscopic deformation (gradient) is imposed upon each of the grains in the population. Each grain responds in stress space and the response of the aggregate is simply the number average (weighted or not) of the resultant stress in each grain. In this case, compatibility is preserved but stress continuity is not. This will under most situations of monotonic loading result in an upper bound stress–strain response from the aggregate polycrystal. In the other extreme is what has been called the Sachs model (Sachs 1928) in which rather than imposing a consistent deformation field across all grains, a consistent stress is imposed and the aggregate deformation response is returned. This yields for most monotonic loading situations a lower bound to the polycrystalline aggregate stress–strain response. Of the two approaches, the Taylor type approach has found more success in representing the deformation response of polycrystal materials. Of course both approaches neglect intergranular interactions. Self-consistent models account for these interactions in an average sense by solving the Eshelby inclusion problem for each grain by imposing the iteratively derived aggregate average homogeneous effective medium response (Lebensohn and Tome 1993; Lebensohn et al. 2007). In this approach, there still is defined a statistically significant number of grains representing the material but the effective medium field imposed on each grain is derived from the granular response of the grains themselves so the interaction effect between grains is that of each individual grain with the average aggregate response (effective medium). Although more accurate, this approach is more computationally demanding than that of the Taylor or Sachs models since the homogeneous effective medium must also be derived and imposed upon each grain, requiring solving for the response of the entire population of grains a number of times until a converged solution is achieved. Stochastic Taylor models have also been proposed as have relaxed constraint models as simple ways in which to account for intergranular interactions inside a Taylor model framework (Tonks et al. 2008; Kocks et al. 1998).

As will be discussed in more detail later within this chapter, classical crystal plasticity theory has been used to directly simulate the mechanical response of metallic polycrystals. A recent review of this category of work has been presented by Bronkhorst et al. (2007). In general, a numerical code is used to explicitly represent a virtual microstructure with the goal of linking structure and single crystal plastic behavior in simulations for not only low rate responses (Diard et al. 2005; Venkataramani et al. 2006; Ghosh and Moarthy 2004; Li and Ghosh 2006; Sinha and Ghosh 2006) but also high deformation rate responses (Becker 2004; Radovitzky and Cuitino 2003; Case and Horie 2007; Bronkhorst et al. 2007; Vogler and Clayton 2008). Individual grains are resolved explicitly with computational zones in a size range which is as small as possible but still several times larger than the characteristic length of a dislocation subcell (Hansen et al. 2009).

Recently, there has been progress in development of adaptive modeling techniques in which the type of model used and the length scale at which a continuum assumption is applied at any given time is determined by the nature of the deformation and its history (Arsenlis et al. 2006; Knap et al. 2008; Barton et al. 2008). In general, more severe forms of deformation will demand that more physics be represented and the continuum length scale to be smaller. Since accounting for more physics always leads to more numerical demand, a technique to evolve represented physics as a function of severity of deformation is essential. These techniques have used adaptive physics algorithms to embed regions of polycrystal microstructures represented by direct polycrystal plasticity. If there is a need for direct representation of polycrystal deformation response, then it is equally important to be able to represent the microstructure of the material beyond simply imposing single point statistics (grain size distribution) onto the constructed virtual microstructures (Groeber et al. 2008a, b; Rollett et al. 2007; Adams 1994; Adams and Olson 1998). Consideration also needs to be given to the role of grain boundaries, the relationship between neighboring grains and the shape of the individual grains as well as the degree of variability of the microstructure as a function of position within the macroscopic body – perhaps due to physical or thermal processing (Gao et al. 2006; Fullwood et al. 2008).

This chapter continues by introduction of a continuum-based internal state-variable model which has been successfully developed to represent the large deformation behavior of polycrystalline tantalum. This model is anisotropic in its elastic and plastic response which has been informed by Taylor model calculations relating initial nonrandom crystallographic texture to the topology of the initial yield surface. This model is then applied to represent the deformation of forced shear experiments and the results are compared favorably against the results of these experiments. Due to the small size of the forced shear experiments, we are afforded the opportunity to represent their response by direct polycrystal modeling as well. Section 3 presents a coupled thermo-viscoplastic single crystal model. This model is then applied within a virtual microstructural representation of the shear zone of the forced shear sample. Simulations are performed and results are compared directly to experimental results. Statistical evolution of the microstructure is also examined within the shear zone of the numerical simulations. This chapter is concluded with discussion and conclusion sections.

2 Experimental

2.1 Tantalum Material

The commercial purity tantalum used in this study was upset forged, rolled, and annealed at 1,773 K for 1 h to yield a 7.62-mm thick plate, with an equiaxed grain size of approximately 42 μm. Tantalum is known to maintain a residual crystallographic texture upon annealing, so the initial texture of this plate material has been characterized by Maudlin et al. (1999a, b). The initial crystallographic texture of this particular tantalum material will be used throughout this work to represent the initial plastic anisotropy for both continuum and crystal plasticity-based modeling of large deformation histories. Equal area pole figures representing the initial crystallographic texture state as measured by electron backscatter diffraction (EBSD) techniques is given in Fig. 1. Based upon these measurements, a population of 512 crystallographic orientations was derived to best match the initial crystallographic state. Equal-area pole figures derived from this population are also given in Fig. 1. This population of grains will be used later in polycrystal plasticity modeling of experiments performed on this material.

2.2 Experiments

Both constitutive model parameter evaluation and validation experiments are used in this study. The models presented in this chapter are designed to represent the dynamic large deformation response of metallic polycrystal materials so the experiments were performed under dynamic loading conditions. The mechanical tests used here are simple compression experiments performed dynamically in a Split Hopkinson Pressure Bar test system, Taylor anvil experiments where a right circular cylinder of high aspect ratio is projected against a rigid smooth surface at high velocity, and an axisymmetric forced shear (tophat) experiment (Fig. 2; Tables 1 and 2) also performed dynamically in a Split Hopkinson Pressure Bar test system. The simple compression experiments support determination of plasticity model parameters, whereas the Taylor anvil and forced shear tophat experiments are used for validation purposes. Other experimental results found in the literature are also used for parameter evaluation purposes and will be discussed in our description of the evaluation process for the material models.

3 Material Modeling

One of the fundamental principles behind the approach to materials modeling espoused in this work is the development of mathematics that represent the relevant physical processes involved. The foundation for this principle is based on the idea

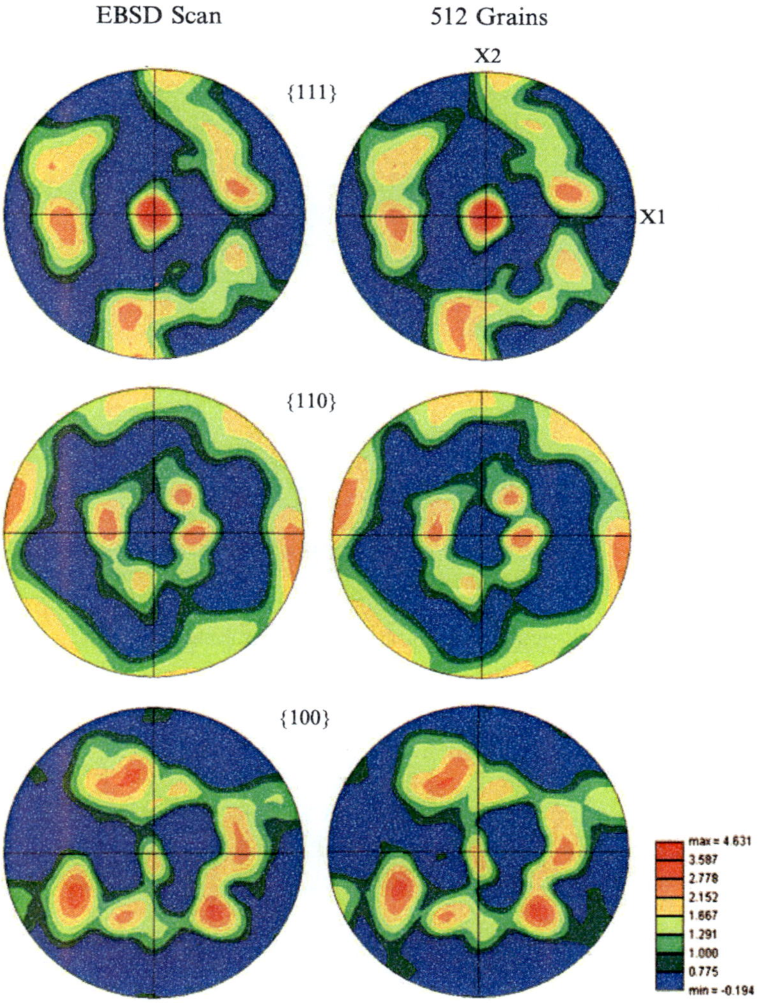

Fig. 1 EBSD measured initial crystallographic texture state of the tantalum used for this study and the 512-grain approximation representing this initial state

that physically based models will prove more predictive in representing the events involved during the deformation, damage, and failure process in these materials than purely phenomenological approaches. Physically based constitutive models are also more capable for extension beyond the regimes for which experiments are available. For this reason, internal state variable theories are developed which endeavor to select variables which best characterize the physical state of the material at any given time. These theories must also be thermodynamically consistent in the way that preserves the first and second laws in the evolution of strain, stress, temperature,

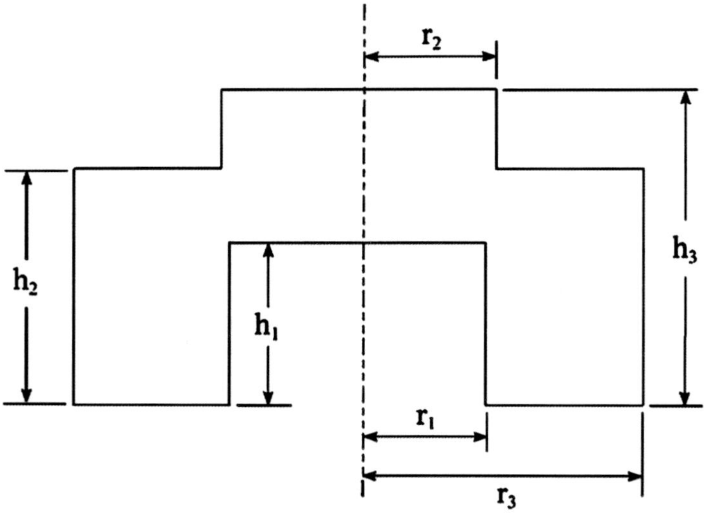

Fig. 2 Schematic drawing of the axisymmetric hat-shaped sample. Drawing is only approximately to scale

Table 1 Dimensions (mm) for the tantalum forced shear samples

Material	r_1	r_2	r_3	h_1	h_2	h_3
Ta	2.090	2.280	4.300	2.600	3.470	5.110

The dimension variables correspond to those shown in Fig. 2

Table 2 Test conditions for each of the tantalum forced shear experiments performed on the Split Hopkinson Pressure Bar system

Test	Initial temperature (K)	Pressure (kPa)	Striker length (cm)
Ta 1356	298	83	8.89
Ta 1357	298	138	8.89
Ta 1358	298	172	6.35
Ta 1359	298	83	15.24

and internal state variables, such as porosity. Since the focus of our work is the dynamic response of metallic materials, the theories must also be coupled thermo-mechanically, which also accounts for the rate and temperature sensitivity inherent in material behavior.

In this section, we present two approaches to the continuum modeling of polycrystalline metallic materials. First, we discuss a macroscale continuum model, where each numerical element is meant to represent a statistically significant number of grains over which the state of the material can be adequately averaged. Second, we discuss a multiscale polycrystal approach, which employs both continuum and polycrystal models to describe macroscale behavior.

3.1 Nomenclature

Standard direct notation is used throughout this paper. Second rank tensors are denoted by boldface uppercase letters. Fourth rank tensors are denoted by underscored boldface uppercase letters. The following variables are used: $\mathbf{I}$ identity, $\mathbf{F}$ deformation gradient, $\mathbf{D}$ stretching, $\mathbf{T}$ Cauchy stress, ρ density, and θ temperature. The prime symbol $\mathbf{A}'$ indicates a deviatoric quantity. The inner product of two second rank tensors $\mathbf{A}$ and $\mathbf{B}$ is defined by $\mathbf{A} \cdot \mathbf{B} = \text{trace}(\mathbf{A}^{\text{T}}\mathbf{B})$. The dyadic product of two vectors, $\otimes$, leads to a second rank tensor. The over-tilde $\tilde{\mathbf{A}}$ represents the quantity $\mathbf{A}$ in the undamaged material.

3.2 Continuum-Based Material Modeling

The continuum model presented here is an internal state variable formulation, which accounts for ductile damage evolution in the form of porosity. This model also accounts for the time and temperature sensitivity of plastic flow through a model of flow stress which is based upon physics dominated by thermally activated dislocation processes (Kocks et al. 1975). The materials initial crystallographic texture state is represented by an initially anisotropic yield surface. The topology of the initial yield surface was evaluated through the use of Taylor model calculations (Maudlin et al. 1996, 1999a, b, 2003a). Based on this same initial crystallographic texture state, the elasticity tensor was also evaluated to be anisotropic (Mason and Maudlin 1999).

3.2.1 Continuum Constitutive Model

Since large material stretches are generally observed in the deformed samples of interest to us here, we include the possibility for material damage to occur. We do not attempt to capture any possible damage nucleation but rather assume an initial defect population distribution, which may then grow in severity. The constitutive model used is from the work of Addessio and Johnson (1993), Maudlin et al. (1999a, b), and Maudlin et al. (2003b). The model is summarized here.

The Cauchy stress in the damaged state is given by

$$\mathbf{T} = \underline{\mathbf{M}}\tilde{\mathbf{T}}, \tag{1}$$

where the stress in the undamaged material is $\tilde{\mathbf{T}}$ and the fourth rank isotropic damage tensor defined in relationship to porosity ϕ is given by

$$\underline{\mathbf{M}} = (1 - \phi)\underline{\mathbf{I}}. \tag{2}$$

Time integration is performed in the material frame relative to the laboratory frame defined by the rotation **R** given by the polar decomposition

$$\mathbf{F} = \mathbf{RU} = \mathbf{VR}. \tag{3}$$

The material time rate for the Cauchy stress is given by

$$\dot{\mathbf{T}} = \underline{\underline{\mathbf{M}}}\,\tilde{\underline{\underline{\mathsf{L}}}}'\left(\mathbf{D}' - \mathbf{D}^{p'}\right) + \underline{\underline{\mathbf{M}}}\left(\frac{\rho_0}{\tilde{\rho}}\tilde{K}_s \mathrm{tr}\mathbf{D}^e - \frac{\tilde{\rho}}{\rho}\Gamma\mathbf{T}\cdot\mathbf{D}^p\right)\underline{\mathbf{I}} + \underline{\underline{\dot{\mathbf{M}}}}\,\underline{\underline{\mathbf{M}}}^{-1}\mathbf{T}, \tag{4}$$

where $\tilde{\underline{\mathsf{L}}}$ is the fourth order elasticity tensor, $\tilde{K}_s$ is the isentropic bulk modulus, Γ is the Gruneisen coefficient and

$$\underline{\underline{\mathsf{L}}} = \underline{\underline{\mathbf{M}}}\,\tilde{\underline{\underline{\mathsf{L}}}}\,\underline{\underline{\mathbf{M}}}. \tag{5}$$

The deformation is additively decomposed as

$$\mathbf{D} = \mathbf{D}^e + \mathbf{D}^p = \mathbf{D}^e + (\mathbf{D}^d + \mathbf{D}^{p'}), \tag{6}$$

where the plastic contribution to deformation is separated into spherical and deviatoric components. The contribution due to damage is given by

$$\mathbf{D}^d = \frac{\dot{\phi}}{3(1-\phi)}\mathbf{I}, \tag{7}$$

where the scalar quantity ϕ represents damage as porosity (Addessio and Johnson 1993).

The flow rule is given by

$$\mathbf{D}^p = \frac{1}{\tau_r}(\mathbf{T} - \mathbf{T}^{\mathrm{proj}}). \tag{8}$$

In general, we allow the stress state to reside off the yield surface during plastic deformation. The overstress model of (8) uses a relaxation constant τ_r with the tensorial quantity $\mathbf{T}^{\mathrm{proj}}$ being the stress state on the yield surface corresponding to the current Cauchy stress projected back onto the plastic flow surface by radial return. A value of $\tau_r = 70\,\mathrm{Pa}\cdot\mathrm{s}$ was found to work adequately (Mason 2004, private communication). The plastic flow surface utilized here is that developed by Gurson (1977) and later modified by Tvergaard (1981, 1982), Tvergaard and Needleman (1984), and Maudlin et al. (2003b) is given by

$$\tau - \sigma_f(\dot{\varepsilon}, \theta)^2[1 + q_3\phi^2 - 2q_1\phi\cosh\delta] = 0, \tag{9}$$

where

$$\tau = \frac{1}{2}\mathbf{T}'\cdot\underline{\boldsymbol{\alpha}}T', \tag{10}$$

is a quadratic relationship allowing for plastic anisotropy, $\sigma_f(\dot{\varepsilon}, \theta)$ is the rate and temperature sensitive flow stress and

$$\delta = -\frac{3q_2\tilde{P}}{2\sigma_s}, \tag{11}$$

where

$$\tilde{P} = -\frac{1}{3}\mathrm{tr}\tilde{\mathbf{T}}. \tag{12}$$

The quantities q_1, q_2, and q_3 are material parameters and the saturation flow stress σ_s is defined below [below (23)].

A polynomial Mie–Gruneisen equation of state is used

$$\tilde{P} = (K_1\tilde{\beta} + K_2\tilde{\beta}^2 + K_3\tilde{\beta}^3)\left(\frac{1-\Gamma\tilde{\beta}}{2}\right) + \Gamma\tilde{E}_s(1+\tilde{\beta}), \tag{13}$$

where

$$\tilde{\beta} = \frac{\tilde{\rho}}{\tilde{\rho}_0} - 1 \tag{14}$$

$$\rho = (1-\phi)\tilde{\rho} \tag{15}$$

$$\dot{\tilde{E}}_s = (\bar{\sigma}\,\dot{\bar{\varepsilon}}^p - P\,\mathrm{tr}\mathbf{D})\frac{\rho_0}{\tilde{\rho}} \tag{16}$$

$$\bar{\sigma} = \sqrt{\tfrac{3}{2}\mathbf{T}'\cdot\mathbf{T}'}, \quad \dot{\bar{\varepsilon}}^p = \sqrt{\tfrac{2}{3}\mathbf{D}^{p'}\cdot\mathbf{D}^{p'}}. \tag{17}$$

K_1, K_2, and K_3 are polynomial coefficients; Γ is the Gruneisen coefficient and $\tilde{E}_s$ is the internal energy. The quantity $\tilde{\rho}$ is the current density of the undamaged material.

The strain rate and temperature sensitivity of the plastic deformation response is represented through the flow stress. The deformation of tantalum at rates observed here has been shown to be well represented by several models (Chen and Gray 1996; Nemat-Nasser and Isaacs 1997; Kothari and Anand 1998), all of which are based upon thermal activation kinetics developed by Kocks et al. (1975). We employ here the isotropic mechanical threshold stress (MTS) model, which has been well established for Tantalum and is due to the work of Follansbee and Kocks (1988), Chen and Gray (1996), and Maudlin et al. (1999a, b). The MTS model is based on the concept of a superposition of resistances to the glide of dislocations. Generally, they are grouped as athermal barriers (e.g., grain boundaries) and thermally influenced barriers (e.g., Peierls stress–intrinsic lattice resistance, forest dislocations, dislocation structure, solute atoms). The MTS is the deformation resistance at 0 K. The flow stress used here is that stress adjusted to current temperature and strain rate. The reader is referred to Follansbee and Kocks (1988) and Chen and Gray (1996) for more details.

The relationship for flow stress is given by

$$\sigma_f(\dot{\varepsilon},\theta) = \sigma_a + \frac{\mu}{\mu_0}[S_i(\dot{\varepsilon},\theta)\hat{\sigma}_i + S_\varepsilon(\dot{\varepsilon},\theta)\hat{\sigma}_\varepsilon], \tag{18}$$

where σ_a is the constant athermal resistance, $\hat{\sigma}_i$ is the constant intrinsic lattice resistance at 0 K, and $\hat{\sigma}_\varepsilon$ is the resistance due to dislocation structure at 0 K, which evolves with deformation. The relationship for shear modulus as a function of temperature is given as

$$\mu = \mu_0 - \frac{D_0}{\exp(\theta_0/\theta) - 1}, \tag{19}$$

which was first proposed by Varshni (1970). The rate and temperature kinetics are represented by the two premultiplying terms

$$S_i\,(\dot{\varepsilon},\theta) = \left\{1 - \left[\frac{k\theta}{\mu b^3 g_{0i}} \ln\left(\frac{\dot{\varepsilon}_{0i}}{\dot{\varepsilon}}\right)\right]^{1/q_i}\right\}^{1/p_i} \tag{20}$$

and

$$S_\varepsilon\,(\dot{\varepsilon},\theta) = \left\{1 - \left[\frac{k\theta}{\mu b^3 g_{0\varepsilon}} \ln\left(\frac{\dot{\varepsilon}_{0\varepsilon}}{\dot{\varepsilon}}\right)\right]^{1/q_\varepsilon}\right\}^{1/p_\varepsilon}, \tag{21}$$

where $\dot{\varepsilon}$ is the equivalent strain rate, k is Boltzmann's constant, b is the magnitude of the Burgers vector, g_0 are normalized activation energies, $\dot{\varepsilon}_0$ are reference strain rates and p and q are exponents, which determine the shape of the energy barrier profile. Kocks et al. (1975) suggest that $p \in [0, 1]$ and $q \in [1, 2]$.

The resistance due to the evolution of the dislocation structure changes with strain as

$$\frac{\mathrm{d}\hat{\sigma}_\varepsilon}{\mathrm{d}\varepsilon} = h_0\left(1 - \frac{\hat{\sigma}_\varepsilon}{\hat{\sigma}_{\varepsilon s}}\right)^\kappa, \tag{22}$$

where the saturation stress as a function of rate and temperature is given by (Kocks 1976)

$$\hat{\sigma}_{\varepsilon s} = \hat{\sigma}_{\varepsilon s 0}\left(\frac{\dot{\varepsilon}}{\dot{\varepsilon}_{0_{\varepsilon s}}}\right)^{(k\theta)/(\mu b^3 g_{0_{\varepsilon s}})}. \tag{23}$$

The saturation stress σ_s used in (11) is taken as the current value of the flow stress given by (18) with the quantity $\hat{\sigma}_\varepsilon$ replaced by its saturation value $\hat{\sigma}_{\varepsilon s}$ given by (23).

The local mechanical work done to the material changes the local temperature by the following relationship

$$\dot{\theta} = \frac{1}{\rho C_p}\dot{\tilde{E}}_s, \tag{24}$$

where $\tilde{E}_s$ is the internal energy in the undamaged material and is given by (16).

Equation (10) allows for the opportunity to represent anisotropy in the plastic flow response of the material due to either crystallographic or morphological texture. The tensorial quantity $\boldsymbol{\alpha}$ can in principal be evolutionary and quantified by experimental data for particular loading histories or evaluated by linking directly to polycrystal plasticity calculations. Maudlin et al. (1996, 1999a, b, 2003a) have used a Taylor model to evaluate the initial yield surface shape based upon the initial crystallographic texture state of a tantalum plate material. The tensor $\boldsymbol{\alpha}$ was then evaluated based upon the initial state yield surface and was not evolved during simulations of Taylor cylinder experiments performed on this material. A fixed $\boldsymbol{\alpha}$ is used here in this work.

3.2.2 Continuum Model Material Parameter Evaluation

Mason and Maudlin (1999) have characterized the elastic behavior of this plate material. The values of the symmetric elasticity tensor $\tilde{L}$ used here in Voigt notation are $\tilde{L}_{11} = 283.3$ GPa, $\tilde{L}_{22} = 282.6$ GPa, $\tilde{L}_{33} = 283.0$ GPa, $\tilde{L}_{12} = 144.9$ GPa, $\tilde{L}_{13} = 144.5$ GPa, $\tilde{L}_{23} = 145.2$ GPa, $\tilde{L}_{44} = 139.0$ GPa, $\tilde{L}_{55} = 137.5$ GPa, and $\tilde{L}_{66} = 138.2$ GPa. These values are assumed to remain constant for all temperatures considered here.

The material parameters for the equation of state given in (13) and (14) are $K_1 = 196.8$ GPa, $K_2 = 259.8$ GPa (=0 for hydrostatic tensile loading), $K_3 = 256.6$ GPa (=0 for hydrostatic tensile loading), $\Gamma = 1.60$, and $\tilde{\rho}_0 = 16,640\,\mathrm{kg/m^3}$.

The three parameters for the Gurson flow surface given in (9) were evaluated recently by Mason and Maudlin (2004, private communication) for the same material under explosively loaded conditions. These values are used here and are $q_1 = 1.5$, $q_2 = 0.625$, and $q_3 = 2.25$. Even upon annealing, tantalum retains significant residual crystallographic texture and therefore the resulting inelastic behavior is anisotropic. Maudlin et al. (1999a, b) have characterized this anisotropic response through a crystal plasticity derived flow surface. This is represented by the quadratic equation given in (10). The tensor $\boldsymbol{\alpha}$, evaluated by Maudlin et al. (1999a, b) and used here is

$$\boldsymbol{\alpha} = \begin{bmatrix} 2.23 & -1.23 & -1 & 0 & 0 & 0 \\ -1.23 & 2.23 & -1 & 0 & 0 & 0 \\ -1 & -1 & 2 & 0 & 0 & 0 \\ 0 & 0 & 0 & 4.12 & 0 & 0 \\ 0 & 0 & 0 & 0 & 4.12 & 0 \\ 0 & 0 & 0 & 0 & 0 & 3.35 \end{bmatrix}. \quad (25)$$

Note that since we have assumed axisymmetry in all simulations, the actual values for $\alpha_{44} = \alpha_{55} = 4.12$ have been averaged and the result used in both positions.

As noted above, to first-order (7) represents the evolution of damage through a relationship originally developed for porosity growth. We make no attempt to model any damage nucleation but rather assume an initial damaged state, i.e., defect population and distribution. This is represented through an initial value for porosity. For pure materials such as the tantalum used in this study, this value is expected to be rather small. In the work of Addessio and Johnson (1993), they found that a value of $\phi_0 = 0.0003$ worked sufficiently well for OFHC copper. We have adopted the same initial value here.

The tantalum parameters for (18)–(24), the rate and temperature sensitive flow stress are given in Table 3. Note that there are two values listed for the parameters $\hat{\sigma}_i$ and g_{0i}. As discovered by empirical evidence presented by Chen and Gray (1996) and Maudlin et al. (1999a, b), the first number is valid up to a value of 0.161 for the normalized activation energy

$$\left[\frac{k\theta}{\mu b^3}\ln\left(\frac{\dot{\varepsilon}_{0i}}{\dot{\varepsilon}}\right)\right]^{1/q_i}, \tag{26}$$

and the second is valid for values of (26) in excess of 0.161. Melting temperature for tantalum was taken as 3,270 K. The quality of representation by the MTS model for the tantalum material can be found in Fig. 3.

Table 3 Tantalum material parameter values for the MTS model given in equations (18)–(24)

Parameter	Tantalum
μ_0	65.25 GPa
D_0	0.38 GPa
θ_0	40 K
σ_a	40 MPa
$\hat{\sigma}_i$	1,203, 167 MPa
Initial $\hat{\sigma}_\varepsilon$	0 MPa
b	2.863 Å
g_{0i}	0.1236, 5.1463
$g_{0\varepsilon}$	1.6
$\dot{\varepsilon}_{0i} = \dot{\varepsilon}_{0\varepsilon} = \dot{\varepsilon}_{0_{\varepsilon s}}$	$10^7\ \mathrm{s}^{-1}$
p_i	1/2
q_i	3/2
p_ε	2/3
q_ε	1
h_0	2.0 GPa
h_1	0
κ	3
$\hat{\sigma}_{\varepsilon s_0}$	350 MPa
$g_{0_{\varepsilon s}}$	1.6
C_p	150 J/kg-K
ρ_0	16,640 kg/m^3
k	54.4 W/m-K

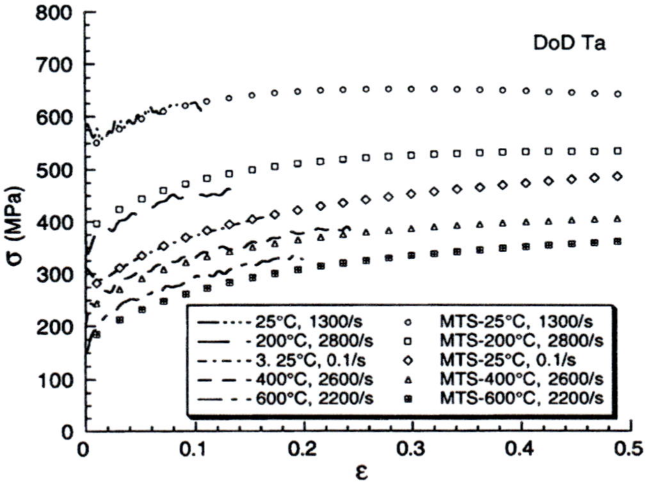

Fig. 3 Representation of the stress–strain response of tantalum at various deformation rates and initial temperatures (Chen and Gray 1996; Maudlin et al. 1999a, b)

3.2.3 Continuum Model Results

As a validation step for the MTS flow stress model discussed above, Taylor cylinder experiments were performed on the tantalum material. These experiments are ideal as a model validation tool since they involve a large gradation of deformation – very large strains and strain rates at the foot and tapering to zero as one approaches the free end of the sample. With this gradation also comes finite material rotations. Comparison between the experimental results and the predicted response can be found in Fig. 4. The simulation produces a footprint which has an eccentricity ratio (major to minor diameters) of approximately 1.20, which compares favorably to experimental values ranging from 1.18 to 1.23. The major and minor side profiles compared in Fig. 4 indicate that the final length agrees well with experiment, and that the axial dimension variation compares well to the experimental results. Detailed examination by Maudlin et al. (1996, 1999a, b, 2003a) showed very little subsequent crystallographic texture evolution in the recovered and sectioned Taylor cylinder sample. Overall, the MTS model performs well in representing the large deformation response of the tantalum material examined here.

The constitutive model was also used to model forced shear experiments performed on the same tantalum material. These axisymmetric experiments were represented by the numerical mesh shown in Fig. 5 along with the applied boundary conditions. The experimentally measured top-to-bottom velocity (velocity difference between the top and bottom surfaces of the sample as it is being compressed) versus time profile for the four experiments examined here are given in Fig. 6. Piecewise linear representations of each of these curves were developed and applied to the models for each of the four tests (Bronkhorst et al. 2006).

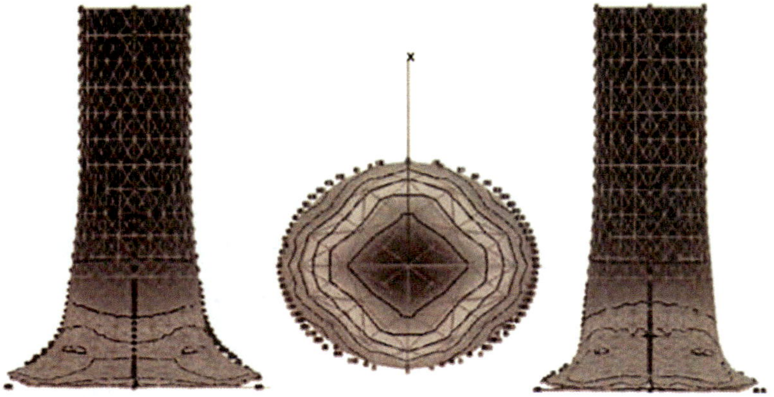

Fig. 4 Comparison between experimental and predicted Taylor cylinder results showing major and minor side profiles and the impact-interface footprint with digitized experimental posttest shapes from three shots (Maudlin et al. 1999a)

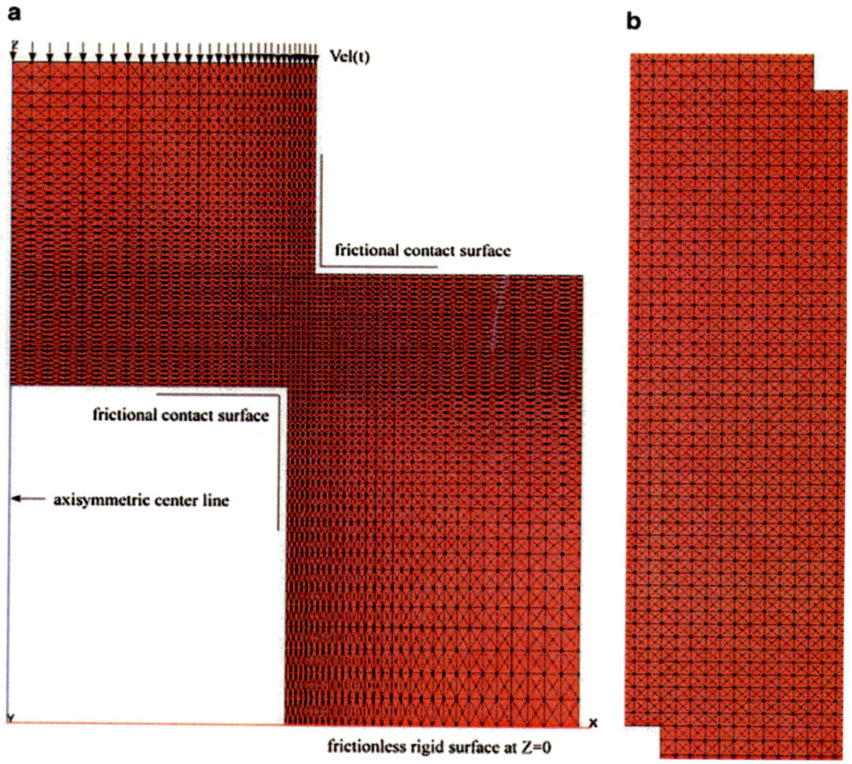

Fig. 5 Mesh geometry used for the continuum simulations: (**a**) low-density mesh, (**b**) shear zone in the high-density mesh. Note mesh used in the simulations does not illustrate properly due to the high mesh density so the low-density mesh is shown in (**a**)

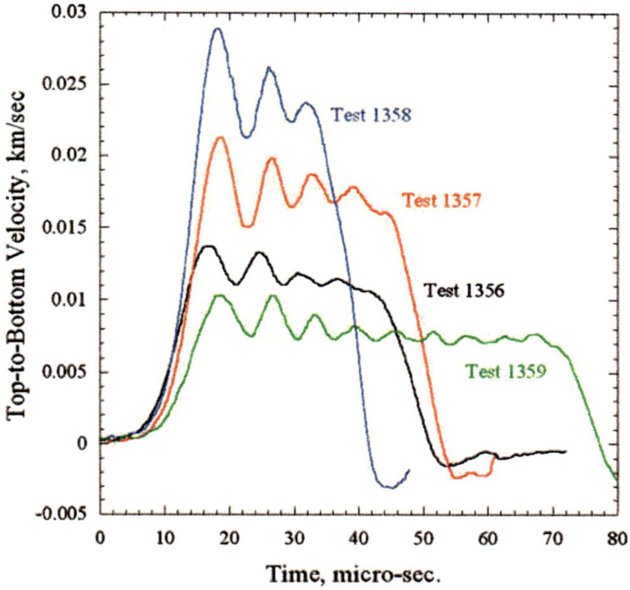

Fig. 6 Top-to-bottom surface velocity results of tophat experiments performed on tantalum at an initial temperature of 298 K

Comparison between the experimental and simulated forced shear results can be found in Fig. 7 for those experiments performed at an initial temperature of 298 K. Note that each of the curves gives top surface engineering stress versus top-to-bottom surface displacement. The models are seen to accurately predict the mean stress response of the samples. The experimental ringing is not matched by the simulations since it would be prohibitive to simulate the entire SHPB test system. Adiabatic conditions have been assumed for these simulations. The accuracy of this assumption has been verified to be accurate by Bronkhorst et al. (2006). The explicit finite element code EPIC (Johnson et al. 2003) was used for all the continuum simulations. A qualitative comparison between an experimentally deformed sample 1356 and the simulated sample 1357 at a time of 40 μs can be found in Fig. 8. The simulated deformed geometry compares reasonably well to the recovered samples. Evidence of mesh channeling is evident in the neighborhood of the lower corner of the numerical model and is strictly a numerical artifact but does not play a significant role in the results. For all the specimens tested, the deformation was always confined to the shear zone, with the bulk of the specimen undeformed after testing. An extensive optical characterization of the region of shear was performed to determine the extent of the damage in that region. An examination of the shear zone is given in Fig. 8. No shear banding was evident and instead there was a diffuse area of shear localization. Extensive grain rotation and elongation, as observed in sample 1356, accommodated the shear and no voids or cracks were observed optically

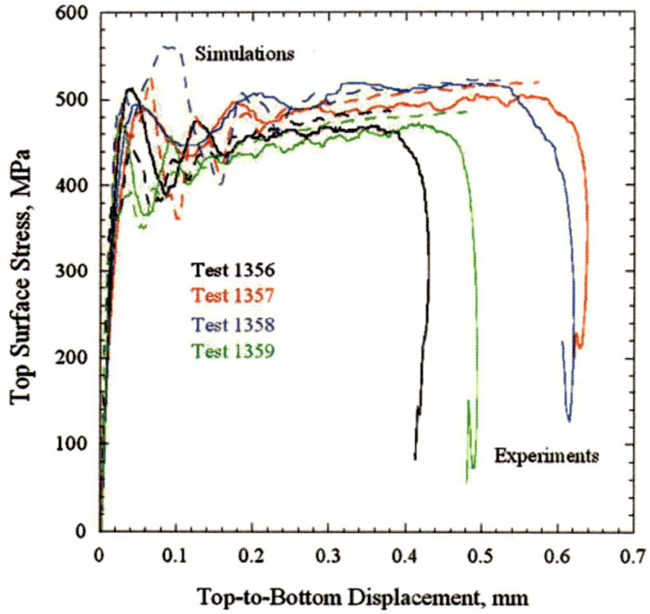

Fig. 7 Experimental and simulation results for tantalum samples 1356–1359 at an initial temperature of 298 K. The solid curves are experimental results

or through electron microscopy within the localization region in each of the specimens. The simulation results also suggest that damage is not an appreciable factor in the deformation history, particularly in the corner regions of the sample where it is somewhat difficult to ascertain experimentally since there is so much surface contact. A constant coefficient of friction of 0.2 was used for the contact surface in both corners of the numerical models.

3.3 Polycrystal-Based Material Modeling

This section differs from Sect. 3.2 in that the length scale at which the continuum assumption is applied is substantially smaller than the size of the single crystal. More specifically, computational zones are chosen in size so that they are much larger than individual dislocations or dislocation structures but also much smaller than the size of an individual crystal in the polycrystalline aggregate. In the previous section, we presented an example, where a Taylor-type homogenized polycrystalline model was used to evaluate the initial shape of the three-dimensional yield surface of the initially textured material. The characteristic length scale for the continuum assumption in that model was the single crystal.

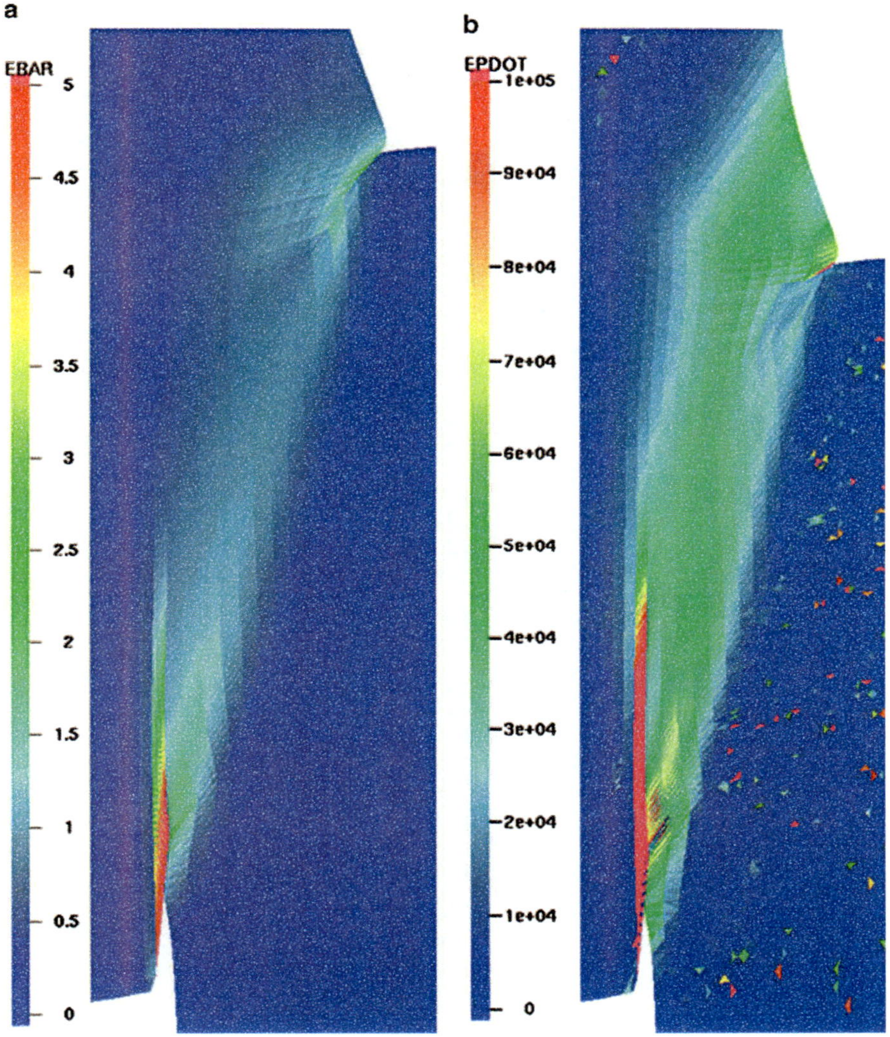

Fig. 8 Simulation results for tantalum adiabatic forced shear simulation 1357 at a time of 40 μs: (**a**) equivalent strain, (**b**) equivalent plastic strain rate, (**c**) temperature K, and (**d**) experimental deformed shear zone

For reasons that will be discussed later, two constitutive models were used in this study. The first is a simple isotropic elasto-viscoplastic constitutive model based upon the mechanical threshold stress (MTS) flow stress theory. The second is a thermomechanically coupled elasto-viscoplastic single crystal constitutive model to examine more closely the behavior in the shear zone of the tophat experiments. We will use these two material models in a simple multiscale approach to the modeling of the tophat experiments presented and discussed in the previous section.

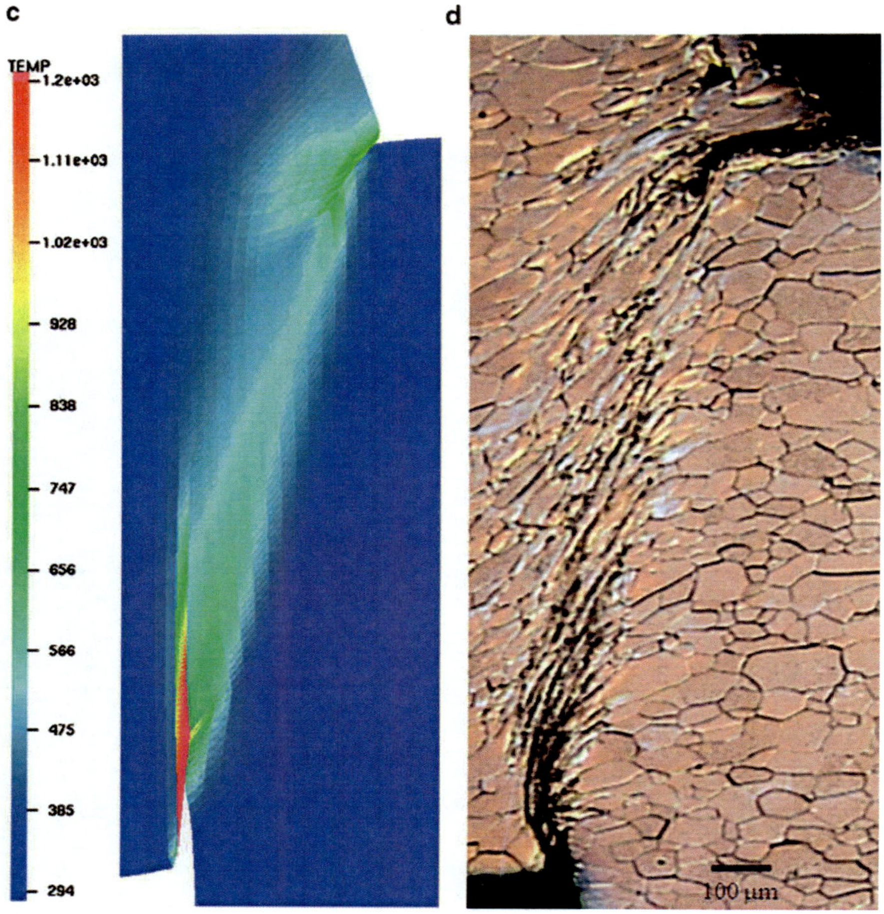

Fig. 8 (continued)

3.3.1 Isotropic Constitutive Model

An isotropic elasto-viscoplastic constitutive model is employed in our study of localization behavior for those regions of the analysis which are far away from the shear zone within the tophat sample geometry. This is accomplished through the use of the ABAQUS (2005) user subroutine UHARD, which uses standard J_2 flow theory and is briefly summarized here.

The constitutive relationship for the Cauchy stress is given by the hypoelastic relationship

$$\dot{\sigma} = \underline{\mathbb{C}}\dot{\varepsilon}^e, \tag{27}$$

where $\underline{\mathbb{C}}$ is the fourth order isotropic elasticity tensor. The total logarithmic strain rate is additively decomposed into elastic and plastic parts

$$\dot{\boldsymbol{\varepsilon}} = \dot{\boldsymbol{\varepsilon}}^e + \dot{\boldsymbol{\varepsilon}}^p. \tag{28}$$

The normality flow rule is given as

$$\dot{\boldsymbol{\varepsilon}}^p = \frac{3}{2}\frac{\boldsymbol{\sigma}'}{\bar{\sigma}}\dot{\bar{\varepsilon}}^p. \tag{29}$$

The yield surface is given by

$$\bar{\sigma} - \sigma_f(\dot{\bar{\varepsilon}}^p, \theta) = 0, \tag{30}$$

where $\bar{\sigma}$ is equivalent stress, σ_f is the flow stress, $\dot{\bar{\varepsilon}}^p$ is the equivalent plastic strain rate, and θ is the temperature. The mechanical threshold stress (MTS) model as presented in the previous section was used as the model for flow stress $\sigma_f(\dot{\bar{\varepsilon}}^p, \theta)$ in (30).

3.3.2 Single Crystal Constitutive Model

The historical basis for the single crystal constitutive model presented here can be found in the works by Rice (1971), Hill and Rice (1972), Asaro and Rice (1977), Asaro (1983a, b), Kocks et al. (1975). Within the deformation rates examined here, and in the same vane as the continuum plasticity model considered earlier, the coupled thermomechanical, elasto-viscoplastic formulation assumes that thermally activated slip is the dominant mechanism for plastic deformation.

The constitutive equation for the intragranular stress is taken as

$$\mathbf{T}^* = \mathbf{C}[\mathbf{E}^e - \mathbf{A}(\theta - \theta_0)], \tag{31}$$

where $\mathbf{C}$ is the fourth order crystal elasticity tensor, $\mathbf{A}$ is the second order crystal thermal expansion tensor, and θ_0 is the initial material temperature. This is the same relationship used by Kothari and Anand (1998). The elastic strain measure $\mathbf{E}^e$ is defined as

$$\mathbf{E}^e \equiv \frac{1}{2}(\mathbf{F}^{e^T}\mathbf{F}^e - \mathbf{1}), \tag{32}$$

where the elastic deformation gradient is given by

$$\mathbf{F}^e = \mathbf{F}\mathbf{F}^{p^{-1}}, \quad \det \mathbf{F}^e > 0 \tag{33}$$

The measure of stress which is elastic work conjugate to the elastic strain measure $\mathbf{E}^e$ is defined by

$$\mathbf{T}^* \equiv (\det \mathbf{F}^e)\mathbf{F}^{e^{-1}}\mathbf{T}\mathbf{F}^{e^{-T}}, \quad (34)$$

where $\mathbf{T}$ is the symmetric Cauchy stress. Based upon the work of Simmons and Wang (1971), the terms in the crystal elasticity tensor are made linearly temperature dependent by the following relationship

$$C_{ijkl} = C_{ijkl_0} + m_{C_{ijkl}}\theta, \quad (35)$$

where C_{ijkl_0} is the appropriate value at 0 K. The diagonal thermal expansion coefficient tensor for the crystal is given by

$$A_{ij} = a_{ij}\delta_{ij} \text{ (no sum)}. \quad (36)$$

The plastic velocity gradient is related to the rate of slip on each of the crystallographic slip systems in the crystal $\dot{\gamma}^{\alpha}$, by the relationship

$$\mathbf{L}^p = \dot{\mathbf{F}}^p\mathbf{F}^{p^{-1}} = \sum_{\alpha} \dot{\gamma}^{\alpha}\mathbf{S}_0^{\alpha}, \quad \mathbf{S}_0^{\alpha} \equiv \mathbf{m}_0^{\alpha} \otimes \mathbf{n}_0^{\alpha} \quad (37)$$

where $\mathbf{m}_0^{\alpha}$ and $\mathbf{n}_0^{\alpha}$ are the vectors representing the slip direction and slip plane normal, respectively, for slip system α in the reference configuration.

The lattice rotation as a result of deformation evolves and the resulting crystallographic texture is determined by

$$\mathbf{m}^{\alpha} = \mathbf{F}^e\mathbf{m}_0^{\alpha} \quad (38)$$

$$\mathbf{n}^{\alpha} = \mathbf{F}^{e-T}\mathbf{n}_0^{\alpha} \quad (39)$$

The flow rule for the slip rate on the α slip system is given by a relationship introduced by Busso (1990) and based upon thermally activated dislocation motion represented in the MTS model discussed above and developed by Kocks et al. (1975),

$$\dot{\gamma}^{\alpha} = \dot{\gamma}_0 \exp\left[-\frac{F_0}{k\theta}\left\langle 1 - \left\langle \frac{|\tau^{\alpha}| - s_{\rho}^{\alpha}\mu/\mu_0}{s_l^{\alpha}\mu/\mu_0} \right\rangle^p \right\rangle^q\right] sgn\,(\tau^{\alpha}), \quad (40)$$

where s_{ρ}^{α} is the deformation resistance due to evolving dislocation density and s_l^{α} is the constant intrinsic lattice resistance on slip system α. The magnitude of each of these two quantities is defined at 0 K in the same way as is found in (18). The brackets $\langle x \rangle$ indicate simply that $\langle x \rangle = x$ for $x > 0$ and $\langle x \rangle = 0$ for $x \leq 0$. Temperature sensitivity of the deformation resistances in (40) is represented to first order by scaling each quantity by the shear modulus ratio

$$\frac{\mu}{\mu_0} \cong \frac{C_{12}}{C_{12_0}} = 1 + \frac{m_{12}}{C_{12_0}}\theta \quad (41)$$

The resolved shear stress drives the shear rate and is defined by the relationship

$$\tau^\alpha \equiv \left(\mathbf{C}^e\mathbf{T}^*\right)\cdot\mathbf{S}_0^*, \quad \mathbf{C}^e = \mathbf{F}^{e^T}\mathbf{F}^e \tag{42}$$

The evolution equation for the slip system deformation resistance is given by

$$\dot{s}_\rho^\alpha = \sum_\beta h^{\alpha\beta}\left|\dot{\gamma}^\beta\right|. \tag{43}$$

The total hardening rate which includes the effects of forest hardening is taken as

$$h^{\alpha\beta} = \left[r + (1-r)\,\delta^{\alpha\beta}\right]h^\beta. \tag{44}$$

The self-hardening rate h^β is given by the Voce-type relationship used by Acharya and Beaudoin (2000), which includes effects of both dislocation generation and annihilation and is given by

$$h^\beta = h_o\left(\frac{s_s^\beta - s_\rho^\beta}{s_s^\beta - s_0^\beta}\right), \tag{45}$$

where the saturation stress parameter s_s^β as a function of temperature and shear rate, as proposed by Kocks (1976), is given as

$$s_s^\beta = \hat{s}_s^\beta(\dot{\gamma},\theta) = s_{s0}\left(\frac{\dot{\gamma}^\beta}{\dot{\gamma}_0}\right)^{k\theta/A}. \tag{46}$$

For the high rate applications examined here, adiabatic conditions were assumed where the relationship between plastic work and temperature is given by

$$\rho c_p\dot{\theta} = \eta\sum_\alpha \tau^\alpha\dot{\gamma}^\alpha, \tag{47}$$

where $\eta \in [0, 1]$ is the thermal conversion factor, taken here as 0.0 for quasistatic conditions and 0.95 for dynamic conditions.

This single crystal model was implicitly implemented into a user subroutine UMAT in ABAQUS (2005) following the work of Kalidindi et al. (1992) and Bronkhorst et al. (1992).

3.3.3 Crystal Material Parameters for Tantalum

There are a total of 20 parameters required for evaluation of the single crystal elasto-viscoplastic constitutive model. This list of parameters and their values is given in Table 4. The material density, ρ; specific heat, c_p; and thermal conversion factor, η values used in (47) were taken from the values used by Bronkhorst et al. (2006).

Table 4 Tantalum material parameters for the single crystal model given in equations (31)–(47)

Parameter	Tantalum
ρ	16,640 kg/m^3
c_p	150 J/kg-K
a	6.5 μm/m-K
η	0.0, 0.95
m_{11}	−24.5 MPa/K
C_{11_0}	268.5 GPa
m_{12}	−11.8 MPa/K
C_{12_0}	159.9 GPa
m_{44}	−14.9 MPa/K
C_{44_0}	87.1 GPa
r	1.4
$\dot{\gamma}_0$	10^7 s^{-1}
s_0	50 MPa
s_l	550 MPa
F_0	2.1×10^{-19} J
p	0.34
q	1.66
s_{s0}	125 MPa
h_{s0}	300 MPa
A	10^{-18} J

The thermal expansion coefficients $a_{ij} = a$, contained in (36), were taken as a constant for all temperatures and evaluated from Boyer and Gall (1985). The six elastic constants m_{11}, C_{11_0}, m_{12}, C_{12_0}, m_{44}, and C_{44_0} contained in (35) were evaluated from the single crystal data of Simmons and Wang (1971). The value for r contained in (44) was taken as the value used by Bronkhorst et al. (1992). The reference shear strain rate $\dot{\gamma}_0$, contained in the flow rule of (40), was taken as the same value as used in the isotropic model (20), (21), and (23).

The remaining plastic parameters s_0, s_l, F_0, p, q, s_{s0}, h_{s0}, and A contained in (40) and (43)–(46) were evaluated by performing polycrystal simulations and matching the simple compression results of Chen and Gray (1996). Maudlin et al. (1999a, b) found that the tantalum material examined here and by Bronkhorst et al. (2006) was in an initial crystallographically textured state. This initial state was characterized in a set of 512 initial crystallographic orientations (Fig. 1). In the same way as Bronkhorst et al. (1992) and Kalidindi et al. (1992), a cubic polycrystal of $8 \times 8 \times 8 = 512$ elements, where each element was a distinct crystal, was deformed in simple compression using the initially textured state. Values for η in (47) were taken as 0.0 for the strain rate cases of 0.001 s^{-1} and 0.1 s^{-1} and 0.95 for all others. The final results are compared to the experimental data in Fig. 9. Since the results of Chen and Gray (1996) were only performed to strains of approximately −0.25, the results of the final material parameter fit were also compared to higher strain experiments published by Kothari and Anand (1998) and were found to agree well. The finite element code ABAQUS (2005) was used for these simulations.

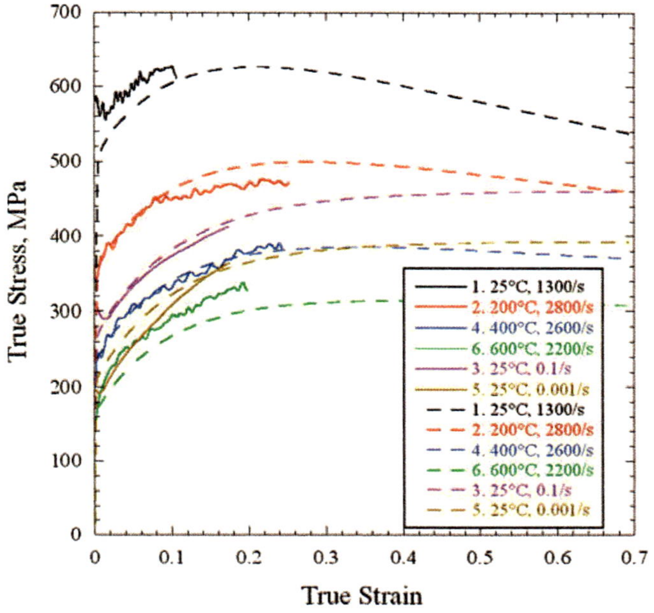

Fig. 9 Representation of the simple compression data of Chen and Gray (1996) by the single crystal model within a cubic polycrystal aggregate. The simulations fitted to the data are the broken curves

3.3.4 Numerical

The intent of this study was to examine the localized shear behavior of tantalum using a more realistic representation of material microstructure. For the case of the tophat sample and the localized deformation, most of the sample mass is not exposed to large deformation. As shown by Bronkhorst et al. (2006), the regions away from the shear zone deform very little and it is not necessary to represent those regions of material with polycrystal detail. We therefore use a two-dimensional model of the axisymmetric tophat sample geometry, which represents the material of the shear zone as a discrete polycrystal region using the single crystal model outlined above in Sect. 3.3.2. Outside of the shear zone however, the material is represented by the simple isotropic constitutive model introduced in Sect. 3.3.1. A schematic representation of the entire numerical model is shown in Fig. 10. Note that two-dimensional axisymmetric elements are used and axisymmetric conditions are assumed at the center line of the model. Prescribed displacement conditions are applied equally to all nodes of the top loading surface while allowing frictionless movement horizontally. The base surface is assumed to be rigid and frictionless. The two inside corners are frictional contact surfaces. Following Bronkhorst et al. (2006), a constant friction coefficient of 0.2 was used for all simulations. The finite element code ABAQUS (2005) was used for these simulations.

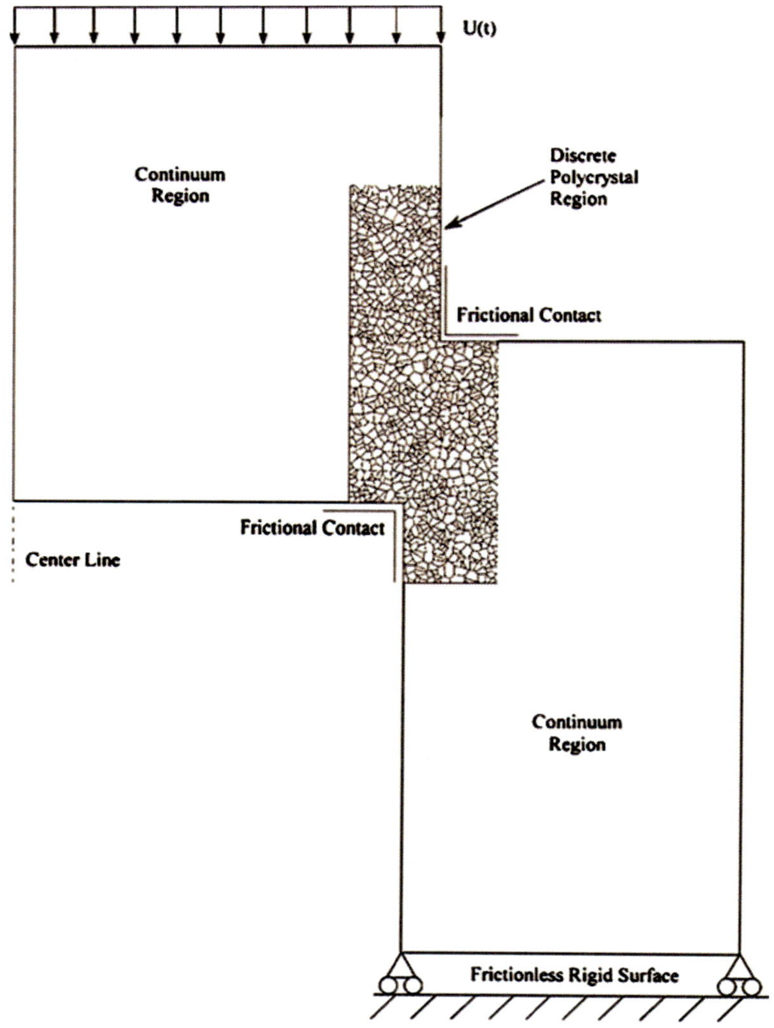

Fig. 10 Two-length scale numerical model used to simulate the tophat sample. This figure is drawn to scale

The final result for the shear zone used in this study can be found in Fig. 11. It contains 1,091 grains. The resulting mean grain size was 37 μm, which is slightly smaller than the experimental value of 42 μm. The meshing process resulted in the 1,091 grains being represented by a total of 52,912 linear triangular elements – or approximately 50 elements per averaged size grain. The initial crystallographic orientation of each grain was assigned randomly from the set of 512 orientations (used three times for a total set size of 1,536) used for material parameter evaluation and shown in Fig. 1. More detail can be found in Bronkhorst et al. (2007).

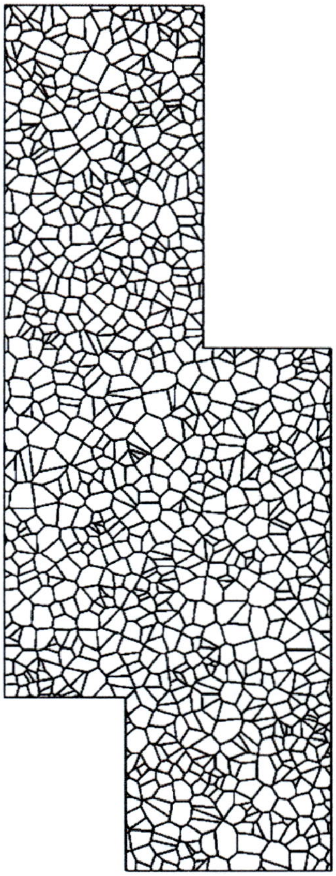

Fig. 11 Polycrystal microstructure used to represent the shear zone. This was generated by a Voronoï tessellation algorithm (Bronkhorst et al. 2007)

3.3.5 Polycrystal Model Results

Within this section, we present results of the polycrystal simulations of the forced shear experiments and where possible compare against experimental results. The first of such is given in Fig. 12, where top surface stress versus displacement results are compared against initial room temperature experiments 1356–1359. The oscillation in the stress signal is not described since we are not attempting to model the entire SHPB loading system. The simulations pick up the proper rate sensitivity, but the stress magnitude is too high. Prominent among the possible explanations for this discrepancy is the fact that these two-dimensional simulations are perhaps over-constraining the single crystal deformation. Three-dimensional simulations could possibly bring the stress responses more in line with those observed experimentally. In addition, these calculations were performed using linear triangular elements, which could also contribute to the discrepancy.

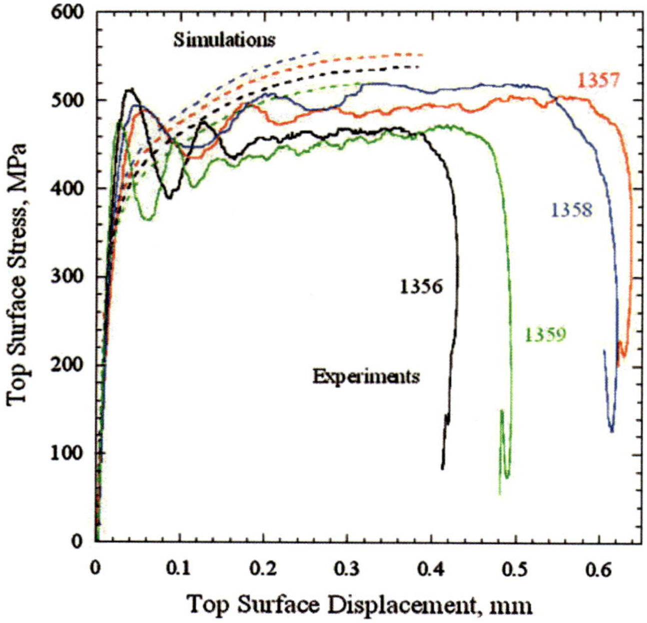

Fig. 12 Comparison between experimental results for tests 1356–1359 and the corresponding polycrystal simulation results

Figures 13 and 14 contain contour plots of vonMises stress, and temperature at the maximum point in time for the simulation of test 1356. Figure 13 nicely illustrates the significant stress heterogeneity and the nature of the aggregate response of the polycrystal material. The final geometry of the sample in the shear zone region compares well to the micrograph image of the sectioned sample 1356 after deformation (Fig. 8d). Significant contact occurs in the corner regions and the contour plots demonstrate the significant material pile-up, which occurs in the top corner region.

A small region of elements within the shear zone was isolated for closer examination. These elements are identified in Fig. 15 and encompass 3,897 material points. Histograms showing vonMises stress, equivalent plastic strain, equivalent plastic strain rate, and temperature for this isolated region of material points is shown in Figs. 16–19. These results illustrate quantitatively the heterogeneity in material state within the polycrystal aggregate for this small region of space. The data in Fig. 16 indicate a near normal distribution of vonMises stress with a factor of greater than two difference between minimum and maximum stress states. The equivalent plastic strain results of Fig. 17 suggest a nonsymmetric distribution with a major peak at approximately 0.7 and a minor peak at approximately 0.9 for this realization. The difference between minimum and maximum points of the distribution are greater than a factor of 3 with a minimum at 0.4 and maximum at 1.5 equivalent strain magnitudes. The distribution of equivalent plastic strain rate

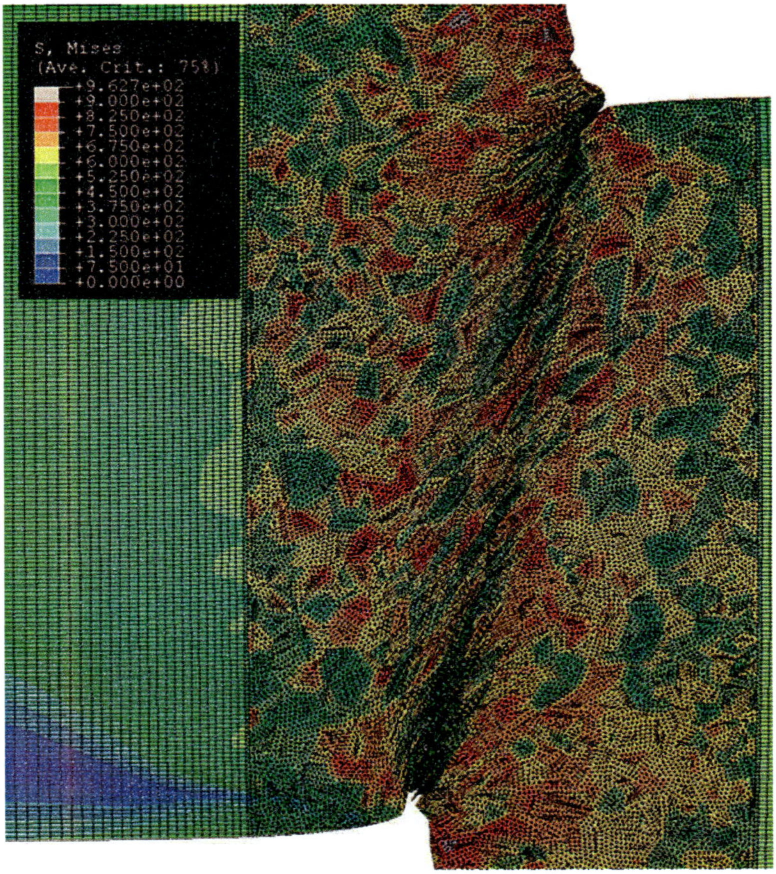

Fig. 13 vonMises stress (MPa) in the deformed mesh of the polycrystal model at a top surface displacement of 0.383 mm

magnitudes given in Fig. 18 is even more asymmetric with a factor of approximately 8 difference between minimum and maximum equivalent strain rate magnitudes – with a maximum at 8×10^4 l/s. The temperature distribution is given in Fig. 19, where the minimum temperature is at 400 K and the maximum is at approximately 650 K. It is important to note that it is very likely that these distributions would change with different microstructural realizations.

The crystallographic texture was measured in a region similar to that described in Fig. 15 of the deformed and sectioned sample 1356. The results are compared against simulation predicted crystallographic texture in Fig. 20 showing equal area pole figures. The simulation results clearly show two-dimensional grain rotations, it describes with some degree of success the crystallographic texture evolution of the microstructure given that the simulations are two-dimensional. Recall that the initial texture for the numerical simulations was given in Fig. 1.

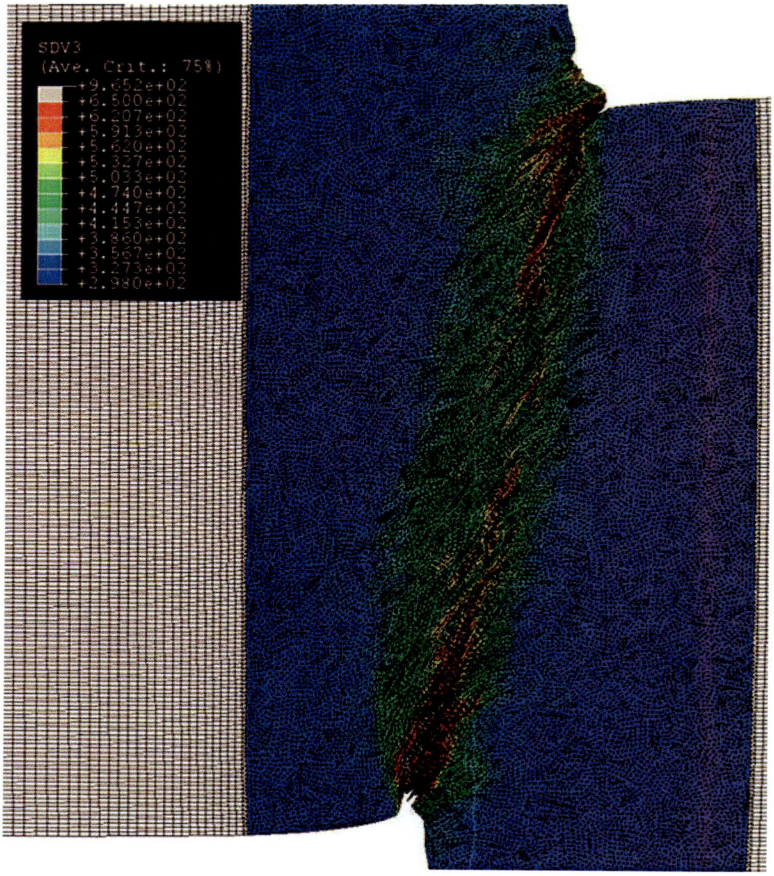

Fig. 14 Temperature (K) in the deformed mesh of the polycrystal model at a top surface displacement of 0.383 mm

4 Discussion

Two approaches to material modeling were presented in this chapter. The first was one where the material was represented within a purely continuum framework and the representative volume element represented the macroscale polycrystal. In this approach, Taylor-type polycrystal models were used to account for initial crystallographic texture of the material through building of an anisotropic yield surface and elastic modulus tensor. Accounting for anisotropy in this way was validated through the successful representation of Taylor cylinder impact experiments. This constitutive model employed the internal state variable porosity φ to represent the microstructural evolution of ductile damage. This constitutive model was successfully applied to the problem of a forced shear experiment and was able to predict the macroscopic load displacement response of the small-scale experiment very well.

Fig. 15 Shear zone elements from the test 1356 polycrystal model selected for statistical analysis

The model correctly predicted that no significant damage evolution took place in the forced shear experiments. Although the mean response of the experiment was properly captured, this approach was not able to represent the local field fluctuations of the material due to the polycrystal aggregate nature of the material.

The second constitutive modeling approach presented was one where the continuum assumption was applied on a length scale which was on the order of 5 μm. Individual grains were resolved numerically and the constitutive model employed was once again continuum in nature but the representative volume element was substantially smaller than a single crystal but many times larger than dislocation tangles and subcells which develop in the material as it deforms and contributes to small-scale fluctuations in the stress and strain field on the order of 1 μm. The single crystal model used in this work is a traditional local theory and therefore cannot represent grain size effects on the plastic flow response of the material (Hansen et al. 2009; Lele and Anand 2009). Nonlocal theories for single crystal plasticity

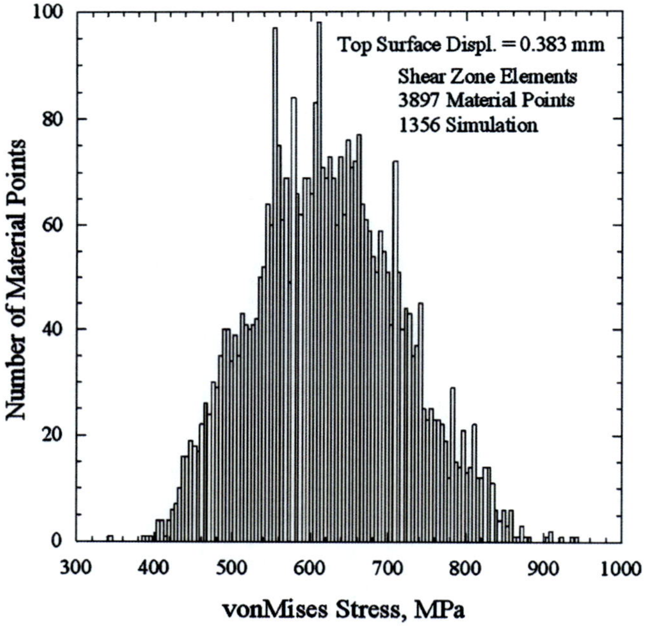

Fig. 16 Histogram of vonMises stress for the isolated shear zone elements (Fig. 15) within the polycrystal model at a top surface displacement of 0.383 mm for the test 1356 simulation

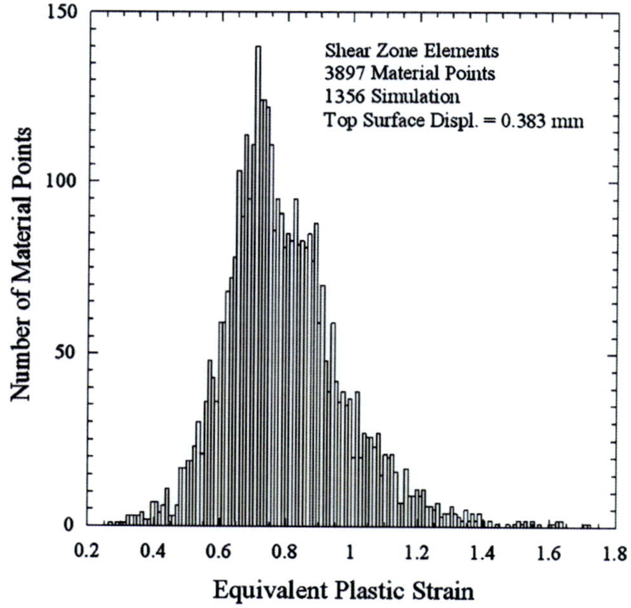

Fig. 17 Histogram of equivalent plastic strain for the isolated shear zone elements (Fig. 15) within the polycrystal model at a top surface displacement of 0.383 mm for the test 1356 simulation

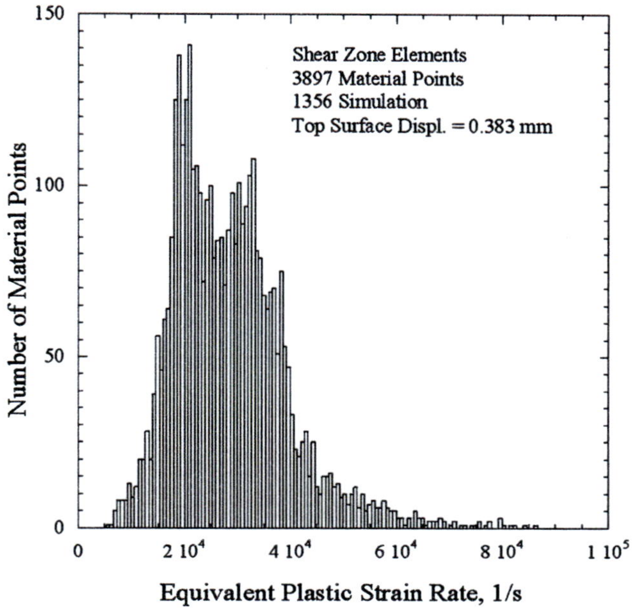

Fig. 18 Histogram of equivalent plastic strain rate for the isolated shear zone elements (Fig. 15) within the polycrystal model at a top surface displacement of 0.383 mm for the test 1356 simulation

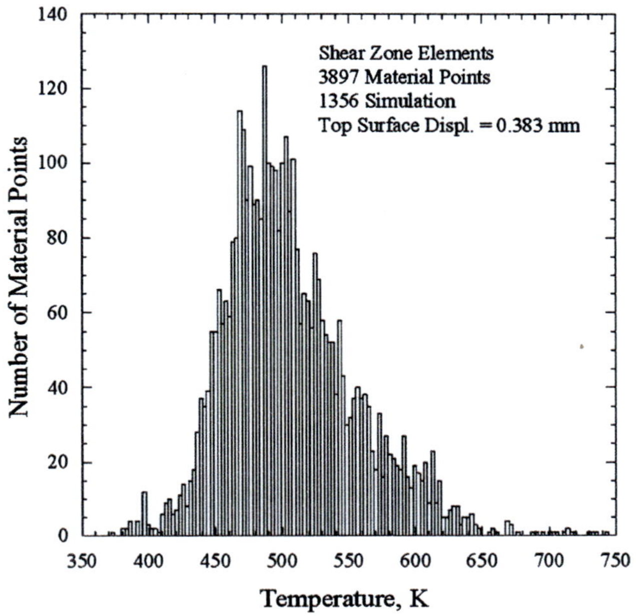

Fig. 19 Histogram of temperature for the isolated shear zone elements (Fig. 15) within the polycrystal model at a top surface displacement of 0.383 mm for the test 1356 simulation

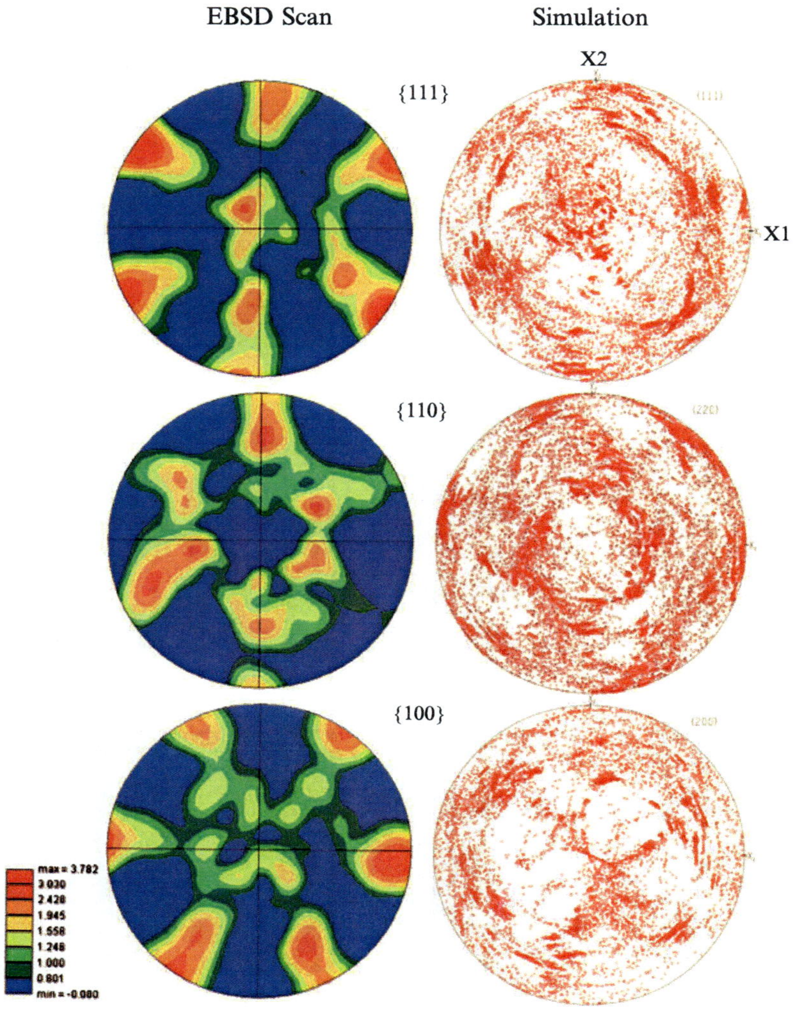

Fig. 20 Test 1356 experimental and predicted shear zone equal area pole figures

are covered elsewhere in this publication. They are also at present too demanding computationally to be applied to the problem size examined here. Nevertheless, the multiscale polycrystal approach discussed does a reasonably good job in representing the microstructural evolution of the material and affords the opportunity to examine the statistical response of polycrystal aggregates. Although outside the scope of the problems examined here, this is believed to be of particular importance when one is required to understand and model the damage and failure response of materials when exposed to cases of extreme mechanical loading. In these situations, the nucleation and growth of these damage and failure processes are believed to be stochastic and heavily influenced by details about the materials microstructure.

Comparison of the results between these two approaches shows substantial differences. Polycrystal approaches in general are more computationally intensive and therefore are not as yet suitable for many engineering applications. In addition, substantial expertise is required of the user to adequately employ most polycrystal (direct as shown here or homogenized theories) models. Therefore, there will always be a need to develop and foster constitutive models that are simple to use and are computationally efficient yet account for the operative and controlling physics of the problem. For many problems of engineering interest, this includes those which involve large deformation, damage, and failure of materials. As such, lower length scale models and in our example here, polycrystal plasticity models accounting for a materials microstructure can be used effectively to statistically link structural characteristics of a materials microstructure (grain size distribution, grain shape distribution, grain boundary shape distribution, inclusion content distribution, initial defect distribution, etc.) to the response of materials exposed to arbitrary states of deformation. In essence, these lower length scale physics models are tools of learning and a detailed link back to experiments. The insight gained through this process can then be used to advance physically based constitutive model development.

5 Conclusion

In this chapter, we have demonstrated two ways in which a polycrystalline metallic material can be represented within a local theoretical framework. The two approaches represent, to some extent, an extreme in the way in which microstructural detail of a material is represented. The material's microstructure and its evolution becomes a more prevalent factor in mechanical behavior for problems involving large deformations and severe loading conditions. This is particularly true when damage and failure events occur. Many times these events are spatially controlled by material microstructure. Our understanding and therefore our ability to model such stochastic processes are still timid in comparison with the complexity of the materials we are attempting to describe. Progress will be made through efforts combining innovative (derivative and validation) experiments, advances in continuum materials theory, and work in computational materials science at multiple length scales.

Acknowledgements This work was conducted under both the DoE Advanced Simulation and Computing program and the joint DoD/DoE Munitions Technology Development Program.

References

ABAQUS, 2005, Version 6.5–4 User's manual, ABAQUS Inc., Providence RI

Acharya, A., Beaudoin, A. J., 2000, Grain size effect in viscoplastic polycrystals at moderate strains, J. Mech. Phys. Solids 48, 2213–2230.

Adams, B. L., 1994, Orientation imaging of the microstructure of polycrystalline materials, Proc. Mats. Res. Soc. 343, 23–32.

Adams, B. L., Olson, T., 1998, The microstructure properties linkage in polycrystals, J. Prog. Mats. Sci. 42, 1–87.
Addessio, F. L., Johnson, J. N., 1993, Rate-dependent ductile failure model, J. Appl. Phys. 74, 1640–1648.
Anand, L., 1985, Constitutive equations for the hot-working of metals, Int. J. Plasticity 1, 213–234.
Arsenlis, A., Barton, N. R., Becker, R., Rudd, R. E., 2006, Generalized in-situ adaptive tabulation for constitutive model evaluation in plasticity, Comput. Meth. Appl. Mech. Eng. 196, 1–13.
Asaro, R. J., Rice, J. R., 1977, Strain localization in ductile single crystals, J. Mech. Phys. Solids 25, 309–338.
Asaro, R. J., 1983a, Micromechanics of crystals and polycrystals, Adv. Appl. Mech. 23, 1–115.
Asaro, R. J., 1983b, Crystal plasticity, ASME J. Appl. Mech. 50, 921–934.
Barton, N. R., Knap, J., Arsenlis, A., Becker, R., Hornung, R. D., Jefferson, D. R., 2008, Embedded polycrystal plasticity and adaptive sampling, Int. J. Plasticity 24, 242–266.
Becker, R., 2004, Effects of crystal plasticity on materials loaded at high pressures and strain rates, Int. J. Plasticity 20, 1983–2006.
Boyer, H. E., Gall, T. L., (Eds.), 1985, Metals handbook – desk edition, ASM, Metals Park, OH.
Bronkhorst, C. A., Kalidindi, S. R., Anand, L., 1992, Polycrystalline plasticity and the evolution of crystallographic texture in FCC metals, Phil. Trans. R. Soc. Lond. A, 341, 443–477.
Bronkhorst, C. A., Cerreta, E. K., Xue, Q., Maudlin, P. J., Mason, T. A., Gray III, G. T., 2006, An experimental and numerical study of the localization behavior of tantalum and stainless steel, Int. J. Plasticity 22, 1304–1335.
Bronkhorst, C. A., Hansen, B. L., Cerreta, E. K., Bingert, J. F., 2007, Modeling the microstructural evolution of metallic polycrystalline materials under localization conditions, J. Mech. Phys. Solids 55, 2351–2383.
Busso, E. P., 1990, Cyclic deformation of monocrystalline nickel aluminide and high temperature coatings, Ph.D. Thesis, MIT.
Case, S., Horie, Y., 2007, Discrete element simulation of shock wave propagation in polycrystalline copper, J. Mech. Phys. Solids 55, 589–614.
Chen, S. R., Gray III, G. T., 1996, Constitutive behavior of tantalum and tantalum-tungsten alloys. Metall. Mater. Trans. A 27A, 2994–3006.
Diard, O., Leclercq, S., Rousselier, G., Cailletaud, G., 2005, Evaluation of finite element based analysis of 3D multicrystalline aggregates plasticity – Application to crystal plasticity model identification and the study of stress and strain fields near grain boundaries. Int. J. Plasticity 21, 691–722.
Follansbee, P. S., Kocks, U. F., 1988, A constitutive description of the deformation of copper based on the use of the mechanical threshold stress as an internal state variable. Acta Metall. 36, 81–93.
Fullwood, D. T., Niezgoda, S. R., Kalidindi, S. R., 2008, Microstructure reconstructions from 2-point statistics using phase-recovery algorithms, Acta Mater. 56, 942–948.
Gao, X., Przybyla, C. P., Adams, B. L., 2006, Methodology for recovering and analyzing two-point pair correlation functions in polycrystalline materials, Metall. Mater. Trans. 37A, 2379–2387.
Ghosh, S., Moarthy, S., 2004, Three dimensional voronoi cell finite element model for microstructures with ellipsoidal heterogeneities, Comput. Mech. 34, 510–531.
Groeber, M., Ghosh, S., Uchic, M. D., Dimiduk, D. M., 2008a, A framework for automated analysis and simulation of 3D polycrystalline microstructures. Part 1: Statistical characterization, Acta Mater. 56, 1257–1273.
Groeber, M., Ghosh, S., Uchic, M. D., Dimiduk, D. M., 2008b, A framework for automated analysis and simulation of 3D polycrystalline microstructures. Part 2: Synthetic structure generation, Acta Mater. 56, 1274–1287.
Gurson, A. L., 1977, Continuum theory of ductile rupture by void nucleation and growth: Part 1 – Yield criteria and flow rules for porous ductile media. J. Eng. Mater. Technol. 99, 2–15.
Gurtin, M. E., Anand, L., 2008, Nanocrystalline grain boundaries that slip and separate: a gradient theory that accounts for grain-boundary stress and conditions at a triple-junction, J. Mech. Phys. Solids 56, 184–199.

Haghi, M., 1995, A framework for constitutive relations and failure criteria for materials with distributed properties, with application to porous viscoplasticity, J. Mech. Phys. Solids 43, 573–597.
Hansen, B. L., Bronkhorst, C. A., Ortiz, M., 2009, Dislocation subgrain structures and modeling the plastic hardening of metallic single crystals, J. Mech. Phys. Solids (submitted).
Hill, R., Rice, J. R., 1972, Constitutive analysis of elastic-plastic crystals at arbitrary strain, J. Mech. Phys. Solids 20, 401–413.
Johnson, G. R., Beissel, S. R., Stryk, R. A., Gerlach, C. A., Holmquist, T. J., 2003, User instructions for the 2003 version of the EPIC code, Network Computing Services Inc., Minneapolis, MN.
Kalidindi, S. R., Bronkhorst, C. A., Anand, L., 1992, Crystallographic texture evolution in bulk deformation processing of FCC metals, J. Mech. Phys. Solids 40, 537–569.
Knap, J., Barton, N. R., Hornung, R. D., Arsenlis, A., Becker, R., Jefferson, D. R., 2008, Adaptive sampling in hierarchical simulation, Int. J. Numer. Meth. Eng. 76, 572–600.
Kocks, U. F., 1976, Laws for work-hardening and low-temperature creep. J. Eng. Mater. Technol. 98, 76–85.
Kocks, U. F., Argon, A. S., Ashby, M. F., 1975, Thermodynamics and kinetics of slip, Progress in materials science, Pergamon, Oxford.
Kocks, U. F., Tome, C. N., Wenk, H.-R., 1998, Texture and anisotropy, Cambridge University Press, Cambridge, UK.
Kothari, M, Anand, L., 1998, Elasto-viscoplastic constitutive equations for polycrystalline metals: application to tantalum, J. Mech. Phys. Solids 46, 51–83.
Lebensohn, R. A., Tome, C. N., 1993, A self-consistent anisotropic approach for the simulation of plastic deformation and texture development of polycrystals: Applications to zirconium alloys, Acta Metall. Mater. 41, 2611–2624.
Lebensohn, R. A., Tome, C. N., Ponte Casteneda, P., 2007, Self-consistent modelling of the mechanical behavior of viscoplastic polycrystals incorporating intragranular field fluctuations, Philos. Mag. 87, 4287–4322.
Lele, S. P., Anand, L., 2009, A large-deformation strain-gradient theory for isotropic viscoplastic materials, Int. J. Plasticity 25, 420–453.
Lemaitre, J., Chaboche, J. L., 1990, Mechanics of solid materials, Cambridge University Press, Cambridge, UK.
Li, S., Ghosh, S., 2006, Multiple cohesive crack growth in brittle materials by the extended voronoi cell finite element model, Int. J. Fract. 141, 373–393.
Mason, T. A., Maudlin, P. J., 1999, Effects of higher-order anisotropy elasticity using textured polycrystals in three-dimensional wave propagation problems, Mech. Mater. 31, 861–882.
Maudlin, P. J., Wright, S. I., Kocks, U. F., Sahota, M. S., 1996, An application of multi-surface plasticity theory: yield surfaces of textured materials, Acta Metall. Mater. 44, 4027–4032.
Maudlin, P. J., Bingert, J. F., House, J. W., Chen, S. R., 1999a, On the modeling of the Taylor cylinder impact test for orthotropic textured materials: experiments and simulations, Int. J. Plasticity 15, 139–166.
Maudlin, P. J., Gray III, G. T., Cady, C. M., Kaschner, G. C., 1999b, High rate material modeling and validation using the Taylor cylinder impact test, Phil. Trans. R. Soc. Lond. A, 357, 1707–1729.
Maudlin, P. J., Bingert, J. F., Gray III, G. T., 2003a, Low-symmetry plastic deformation in BCC tantalum: experimental observations, modeling and simulations, Int. J. Plasticity 19, 483–515.
Maudlin, P. J., Mason, T. A., Zuo, Q. H., Addessio, F. L., 2003b, TEPLA-a: Coupled anisotropic elastoplasticity and damage. The Joint DoD/DOE Munitions Technology Program progress report. Vol. 1. LA-UR-14015-PR.
Nemat-Nasser, S., Isaacs, J. B., 1997, Direct measurement of isothermal flow stress of metals at elevated temperatures and high strain rates with application to Ta and Ta-W alloys, Acta. Mater. 45, 907–919.
Radovitzky, R., Cuitino, A., 2003, Direct numerical simulation of polycrystals. In: Collection of technical papers – structures, structural dynamics and materials conference, Vol. 3, April 7–10, Norfolk, VA, 1920–1928.

Rice, J. R., 1971, Inelastic constitutive relations for solids: an internal variable theory and its applications to metal plasticity, J. Mech. Phys. Solids 19, 433–455.
Rice, J. R., 1975, Continuum mechanics and thermodynamics of plasticity in relation to microscale deformation mechanisms. In: Constitutive equations in plasticity (A. Argon, ed.), MIT Press, Cambridge, MA.
Rollett, A. D., Lee, S.-B., Campman, R., Rohrer, G. S., 2007, Three-dimensional characterization of microstructure by electron backscatter diffraction, Annu. Rev. Mater. Res. 37, 627–658.
Sachs, G., 1928, Zur Ableitung einer Fliessbedingung, Z. Ver. Dtsch. Ing. 72, 734–736.
Simmons, G., Wang, H., 1971, Single crystal elastic constants and calculated aggregate properties: A handbook, The MIT Press, Cambridge, MA.
Sinha, S., Ghosh, S., 2006, Modeling cyclic ratcheting based fatigue life of HSLA steels using crystal plasticity FEM simulations and experiments, Int. J. Fatigue 28, 1690–1704.
Taylor, G. I., 1938, Plastic strain in metals, J. Inst. Metals 62, 307–324.
Tonks, M. R., Bingert, J. F., Bronkhorst, C. A., Harstad, E. N., Tortorelli, D. A., 2008, Two stochastic mean-field polycrystal plasticity methods, J. Mech. Phys. Solids (submitted)
Tvergaard, V., 1981, Influence of voids on shear band instabilities under plane strain conditions, Int. J. Fract. 17, 389–407.
Tvergaard, V., 1982, On localization in ductile materials containing spherical voids, Int. J. Fract. 18, 237–252.
Tvergaard, V., Needleman, A., 1984, Analysis of the cup-cone fracture in a round tensile bar, Acta Metall. 32, 157–169.
Varshni, Y. P., 1970, Temperature dependence of the elastic constants. Phys. Rev. B 2, 3952–3958.
Venkataramani, G., Deka, D., Ghosh, S., 2006, Crystal plasticity based Fe model for understanding microstructural effects on creep and dwell fatique in Ti-6242, J. Eng. Mater. Technol., Transactions of the ASME, July, 356–365.
Vogler, T. J., Clayton, J. D., 2008, Heterogeneous deformation and spall of an extruded tungsten alloy: plate impact experiments and crystal plasticity modeling, J. Mech. Phys. Solids 56, 297–335.

Dislocation Mediated Continuum Plasticity: Case Studies on Modeling Scale Dependence, Scale-Invariance, and Directionality of Sharp Yield-Point

Claude Fressengeas, A. Acharya, and A.J. Beaudoin

1 Introduction

Plasticity of crystalline solids is a dynamic phenomenon resulting from the motion under stress of linear crystal defects known as dislocations. Such a statement is grounded on numerous convincing observations, and it is widely accepted by the scientific community. Nevertheless, the conventional plasticity theories use macroscopic variables whose definition does not involve the notion of dislocation. This paradoxical situation arises from the enormous range covered by the length scales involved in the description of plasticity, from materials science to engineering. It may have seemed impossible to account for the astounding complexity of the (microscopic) dynamics of dislocation ensembles at the (macroscopic) scale of the mechanical properties of materials. Justifications offered for such a simplification usually reside in perfect disorder assumptions. Namely, plastic strain is regarded as resulting from a large number of randomly distributed elementary dislocation glide events, showing no order whatsoever at intermediate length scales. Hence, deriving the mechanical properties from the interactions of dislocations with defects simply requires averaging on any space and time domain. The existence of grain boundaries in polycrystals is of course affecting this averaging procedure, but it does not change it fundamentally.

This straightforward jump from microscopic to macroscopic scale has long been the prevailing point of view in the mechanical science as well as in the materials science community. It may be justified, for example, in bcc materials at low temperature, where the motion of dislocations is subject to large lattice friction. It reaches its limits when elastic interactions between dislocations become the order of the interactions with other obstacles to their motion (lattice friction, solute atoms, precipitates....). Since dislocation densities commonly increase during material loading, such a situation is met sooner or later when strain increases. The field of elastic interactions between dislocations then becomes able to generate

C. Fressengeas (✉)
LPMM, Universite Paul Verlaine-Metz/CNRS Ile du Saulcy, 57045 Metz Cedex 01, France
e-mail: claude.fressengeas@univ-metz.fr

S. Ghosh and D. Dimiduk (eds.), *Computational Methods for Microstructure-Property Relationships*, DOI 10.1007/978-1-4419-0643-4_8,

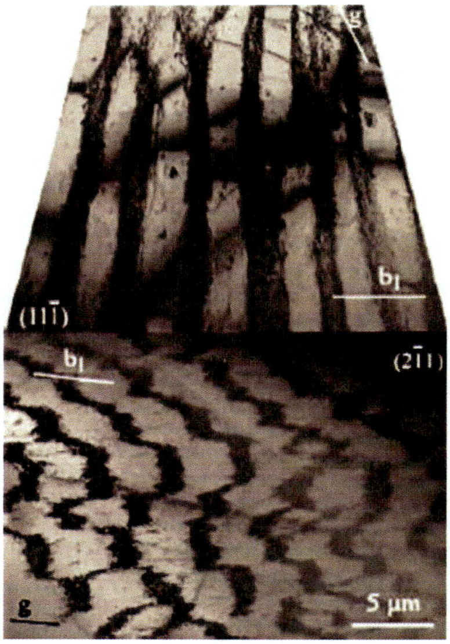

Fig. 1 Dislocation walls in Si single crystal cyclically loaded in tension – compression at high temperature; after (Legros et al. 2004)

collective behavior and self-organized phenomena at some intermediate length scale. Collective phenomena include dislocation patterning and the emergence of complex dynamic regimes (Kubin et al. 2002). Numerous examples of dislocation patterns, involving dislocation–rich and dislocation–poor areas, are observed in optical or electronic microscopy. Such is the case of the dislocation walls formed in cyclic loading (see Fig. 1), of dislocation cells (Fig. 2) and localized slip bands on the surface of single crystals (Fig. 3). Similar spatial structures can also be inferred from the complex temporal behavior inherent to deformation curves in certain metallic alloys (Portevin–Le Chatelier effect, Lüders bands...) (Kubin et al. 2002). In such conditions, the simple averaging procedures alluded to above are no longer justified. Thus, the conventional theories of plasticity are no longer valid and they are unable to account for the emerging patterns, because they lack the relevant internal length scales.

In attempting to account for self-organization phenomena, a first approach consisted in using strain gradients and rotation gradients in the description of the kinematics of the transformation and introducing the necessary length scales in a phenomenological way into the constitutive equations of plasticity (Aifantis 1986; Fleck et al. 1994; Forest et al. 1997; Nix and Gao 1998). Such approaches can be referred to as "nonlocal" theories, as opposed to the "local" conventional plasticity theories and they are known as "strain gradient" plasticity theories. They may be useful in the characterization of the emerging patterns, but the identification of the involved length scales, as well as their physical justification may raise difficulties. Further, additional boundary conditions of higher order may be required. The notion that necessary ingredients for the dynamic description of the emerging patterns are the areal dislocation densities defined in a continuous manner over surfaces of

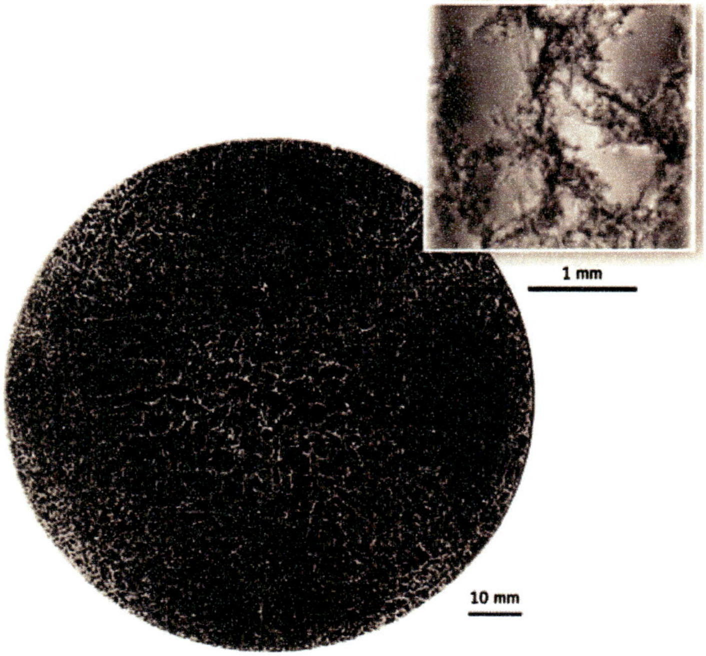

Fig. 2 Optical micrography of giant dislocation cells after GaAs crystal growth. Note that the average cell size varies in inverse proportion to stress. Inset: dislocation cells through X-ray imaging: dark areas are images of lattice distortion around dislocations; after (Neubert and Rudolph 2001)

Fig. 3 Slip lines on the surface of Cu30at%Zn single crystal strained in tension at 19.4% and 77 K; after (Zaiser 2006)

appropriate dimensions is recent (Acharya 2001), although these dislocation densities were known for a much longer time (Nye 1953; Kröner 1958; Mura 1963; Kosevich 1979; Kröner 1980). In a way to be documented hereafter, a change of scale must be performed to proceed from individual dislocations to dislocation densities (Acharya and Roy 2006). Clearly, the scale of resolution must be smaller than the characteristic length scale of the dislocation patterns to be described. However, there is no mandatory rule, and the choice of the resolution length scale depends on the accuracy demanded on the description. Hence, a phenomenon deemed "non-local" in a fine scale resolution scheme may well be classified as "local" when the scale of resolution is sufficiently enlarged. Further, when envisioned on such intermediate resolution length scales, dislocation motion can be construed as transport of the areal dislocation densities, which confers propagative character to these variables, in connection with the equation for dislocation transport (Kröner 1958; Mura 1963). Fundamental changes in the mathematical nature of the governing equations derive from the properties of transport and impact on the algorithms devoted to the solution of boundary value problems (Roy and Acharya 2005, 2006; Varadhan et al. 2006).

The stress field responsible for the nucleation and motion of dislocations derives in the first place from the tractions and displacements imposed on the sample boundaries. When the distribution of dislocations becomes inhomogeneous at some chosen scale, their long-range internal stress field brings about a redistribution of stresses and dislocations, in a manner accounting for the elastic properties of the material, stress equilibrium, and boundary conditions. Along this process, the evolution of the dislocation density and internal stress fields depends on the strain path and anisotropy is induced. The objective assigned to field dislocation dynamics theories is to account for the emergence of inhomogeneous dislocation distributions at some mesoscopic (intermediate) length scale, as well as their consequences on mechanical behavior. Three such problems will be reviewed in this chapter. First, field dislocation dynamics theories are obviously well suited for small size systems: nanostructures, microsystems... Here, the overall dimensions of the sample are not much larger than the characteristic length scale of the dislocation patterns and, consequently, effects of sample size on mechanical response are to be observed. Interpretation of these size effects through field dislocation dynamics (Taupin et al. 2007) will be discussed in the following. Second, since dislocations are often conveniently viewed as distinct objects, dislocation activity appears to be inhomogeneous at a finer scale of resolution. Because dislocation glide is controlled by local obstacles in a large class of materials, it is also intermittent in time. In these materials, dislocation motion consists in successive fast runs of dislocation segments from one obstacle to the next one, with the flight time of dislocations being much smaller than their arrest time on obstacles. Although intermittency of plasticity was described as early as 1932 in Zn single crystals (Becker and Orowan 1932), the prevailing interpretation has been perfect disorder. In average over sufficiently large length and time scales, intermittent fluctuations have been regarded as adding at random to a smooth in time and homogeneous in space net response. A fundamentally different understanding emerged during the last few years when statistical analysis of these fluctuations became available, that of a scale-invariant phenomenon characterized by power law distributions of fluctuation size, and correlations in space and time (Weiss and Grasso 1997; Miguel et al. 2001; Dimiduk et al. 2006; Brinckmann et al. 2008; Weiss et al. 2007). Because they feature correlations in space due to both the long-range internal stresses and the short-range interactions involved in dislocation transport, field dislocation dynamics theories are candidates for the interpretation of scale-invariant intermittency (Fressengeas et al. 2009), and the results will be reported herein. Finally, the anisotropy of strain hardening induced by the emergence of internal stress fields will be reviewed. The directionality of the sharp yield point in strain-aged steels and the occurrence of a Bauschinger effect after a sequence of forward-reverse straining will receive interpretation within the framework of a field dislocation theory coupling the evolution of statistical and polar dislocation densities with that of point defects due to strain-aging (Taupin et al. 2008).

By considering internal stresses due to dislocation–dislocation interactions, alternative modeling approaches such as statistical mechanics (Zaiser 2006), phase field (Koslowski et al. 2004) and discrete dislocation dynamics methods (Miguel et al. 2001; Csikor et al. 2007; Devincre et al. 2008) reproduce the scale-invariance of plastic activity. They have a potential for retrieving length-scale dependence of material properties, but usually consider periodic boundary conditions over small domains. Further, both phase field and statistical mechanics methods have not been shown to retrieve the propagative character of dislocation dynamics related with dislocation transport. In discrete dislocation dynamics simulations, transport of dislocation densities is present, but fully resolved into the motion of individual dislocations. As a rule of thumb, using present day computing facilities, most dislocation dynamics codes are able to handle a tenfold increase of the initial number of dislocations (Devincre, private communication). Hence, dislocations dynamics simulations are still limited to small size/small strain systems, with a simulation box size of the order of $10\,\mu\text{m}^3$ and a plastic strain achieved amounting to a few percents. Thus, if not for the treatment of boundary conditions, geometric and elastic nonlinearity, and inertia, the chances to tackle large-scale engineering problems in the future with discrete dislocation dynamics simulations are slim, and field dislocation theories seem to be more fitted for real scale boundary value problems.

The chapter is organized as follows. In Sect. 2, we provide an overview of the current field dislocation dynamics theories, augmented with recent developments in macroscopic polycrystal response. Section 3 deals with the effects of sample size on mechanical response, and is illustrated with the example of ice single crystals submitted to torsion creep where robust size effects are observed. Section 4 is devoted to scale-invariance and transport effects in the intermittency of crystal plasticity. Examples include the behavior of copper single crystals in tension. Section 5 deals with the anisotropy in mechanical properties induced by complex strain paths, with the example of the directionality of the sharp yield point and the occurrence of a Bauschinger effect in strain-aged polycrystalline steels. The concluding section provides insights into the flexibility of the theory regarding the scale of resolution and its ability to deal with fine-scale vs. engineering-scale simulations.

2 Field Dislocation Dynamics Theory

The theory uses the continuum description of dislocations based upon Nye's dislocation density tensor $\boldsymbol{\alpha}$ (Nye 1953). Operating on the normal $\mathbf{n}$ to a unit surface S, $\boldsymbol{\alpha}$ provides the net Burgers vector $\mathbf{b} = \boldsymbol{\alpha}.\mathbf{n}$ of all dislocations lines threading S, i.e., the incompatibility in plastic displacement found along the Burgers circuit C surrounding S. When surface S is so small that it is threaded by a single dislocation with Burgers vector $\mathbf{b}$ and line vector $\mathbf{t}$, $\boldsymbol{\alpha} = \mathbf{b} \otimes \mathbf{t}$ and the involved dislocation is labeled as a "polar dislocation". When the size of S, i.e., the resolution length scale, is increased to the point where S is threaded by a large number of dislocations, $\mathbf{b}$ may be zero if all individual Burgers vectors statistically offset. Then $\boldsymbol{\alpha}$ is zero, the dislocations are unresolved and they are deemed "statistical". In intermediate cases, the net Burgers vector $\mathbf{b}$ is nonzero, but part of the dislocations threading S may remain unresolved. The subscripts in the density components α_{ij} then indicate, respectively, the net Burgers vector and line vector directions of polar dislocations, whereas the remaining statistical dislocations are not accounted for in tensor $\boldsymbol{\alpha}$. Due to lattice incompatibility, the plastic distortion tensor $\mathbf{U_p}$ is not a gradient; it is written as a sum of a gradient and an incompatible part that cannot be expressed as a gradient

$$\mathbf{U_p} = \mathbf{grad\ z} - \boldsymbol{\chi} \tag{1}$$

The incompatible part results from the distribution $\boldsymbol{\alpha}$ through the fundamental geometrical equation of incompatibility

$$\mathbf{curl\ U_p} = -\mathbf{curl}\ \boldsymbol{\chi} = -\boldsymbol{\alpha} \tag{2}$$

augmented with the side conditions

$$\mathbf{div}\ \boldsymbol{\chi} = 0 \tag{3}$$

and $\boldsymbol{\chi}.\mathbf{n} = 0$ on the boundary with unit normal $\mathbf{n}$ to ensure that when $\boldsymbol{\alpha} = 0$ the incompatible part $\boldsymbol{\chi}$ vanishes identically on the body. The compatible part depends upon the history of plastic straining and records the compatible increments of the plastic strain rate produced by the motion of dislocations through the equation

$$\mathbf{div\ grad\ \dot{z}} = \mathbf{div}\ (\boldsymbol{\alpha} \times \mathbf{V}), \tag{4}$$

where the field $\mathbf{V}$ represents the velocity of an infinitesimal dislocation segment at any spatiotemporal location. In this model of dislocation mechanics, the total displacement field, $\mathbf{u}$, does not represent the actual physical motion of atoms involving topological changes but only a consistent shape change and hence is not required to be discontinuous. However, the stress produced by these topological changes in the lattice is adequately reflected in the theory through the utilization of incompatible elastic/plastic distortions. As usual in continuum plasticity, the elastic distortion (nonsymmetric) is assumed to be the difference of the total displacement gradient and the plastic distortion

$$\mathbf{U_e} := \mathbf{grad\ u} - \mathbf{U_p} \tag{5}$$

and the stress is a function of the elastic distortion (in the linear elastic case given by $\mathbf{T} = \mathbf{C} : \mathbf{U_e}$) satisfying the equation of equilibrium

$$\mathbf{div}\ \mathbf{T} = \mathbf{0}. \tag{6}$$

Finally, $\boldsymbol{\alpha}$ evolves according to the fundamental transport law, which derives from the conservation of Burgers vector content

$$\dot{\boldsymbol{\alpha}} = -\mathbf{curl}\ (\boldsymbol{\alpha} \times \mathbf{V}). \tag{7}$$

Gathering all equations, the complete theory reads as

$$\mathbf{curl}\ \boldsymbol{\chi} = \boldsymbol{\alpha} \tag{8}$$

$$\mathbf{div}\ \boldsymbol{\chi} = 0 \tag{9}$$

$$\mathbf{div}\ \mathbf{grad}\ \dot{\mathbf{z}} = \mathbf{div}\ (\boldsymbol{\alpha} \times \mathbf{V}) \tag{10}$$

$$\mathbf{div}\ [\mathbf{C} : \{\mathbf{grad}\ (\mathbf{u} - \mathbf{z}) + \boldsymbol{\chi}\}] = 0 \tag{11}$$

$$\dot{\boldsymbol{\alpha}} = -\mathbf{curl}\ (\boldsymbol{\alpha} \times \mathbf{V}). \tag{12}$$

To derive the structure of an averaged mesoscopic theory, we adapt an averaging procedure commonly used in the study of multiphase flows (see, for example, Babic 1997). For a microscopic field f given as a function of space and time, we define the mesoscopic space-time averaged field $\bar{f}$ as

$$\bar{f}(\mathbf{x}, t) = \frac{1}{\int_{I(t)} \int_{\Omega(\mathbf{x})} w(\mathbf{x} - \mathbf{x}', t - t')\mathrm{d}\mathbf{x}'\mathrm{d}t'} \int_{\Im} \int_{B} w(\mathbf{x} - \mathbf{x}', t - t') f(\mathbf{x}', t')\mathrm{d}\mathbf{x}'\mathrm{d}t', \tag{13}$$

where B is the body and $\Im$ a sufficiently large interval of time. In the above, $\Omega(\mathbf{x})$ is a bounded region within the body around point $\mathbf{x}$ with linear dimension of the order of the spatial resolution of the macroscopic model we seek, and $I(t)$ is a bounded interval in $\Im$ containing t. The weighting function w is nondimensional, assumed to be smooth in the variables $(\mathbf{x}, \mathbf{x}', t, t')$ and, for fixed $\mathbf{x}$ and t, has support (i.e., is non-zero) only in the domain $\Re = \Omega(\mathbf{x}) \times I(t)$ when viewed as a function of $(\mathbf{x}', t')$. The averaged field $\bar{f}$ is simply a weighted, space-time running average of the microscopic field f over $\Re$, whose scale is determined by the scale of spatial and temporal resolution of the averaged model one seeks. Applying this operator to (8–12), we obtain (Acharya and Roy 2006) an exact set of equations for the averages given as

$$\mathbf{curl}\ \bar{\boldsymbol{\chi}} = \bar{\boldsymbol{\alpha}} \tag{14}$$

$$\mathbf{div}\ \bar{\boldsymbol{\chi}} = 0 \tag{15}$$

$$\mathbf{div}\ \mathbf{grad}\ \dot{\bar{\mathbf{z}}} = \mathbf{div}\ (\bar{\boldsymbol{\alpha}} \times \bar{\mathbf{V}} + \mathbf{L_p}) \tag{16}$$

$$\mathbf{div}\ [\mathbf{C} : \{\mathbf{grad}\ (\bar{\mathbf{u}} - \bar{\mathbf{z}}) + \bar{\boldsymbol{\chi}}\}] = 0 \tag{17}$$

$$\dot{\bar{\boldsymbol{\alpha}}} = -\mathbf{curl}\ (\bar{\boldsymbol{\alpha}} \times \bar{\mathbf{V}} + \mathbf{L_p}), \tag{18}$$

where $\mathbf{L_p}$, defined as

$$\mathbf{L_p} := \overline{(\boldsymbol{\alpha} - \bar{\boldsymbol{\alpha}}) \times \mathbf{V}(\mathbf{x}, t)} = \overline{\boldsymbol{\alpha} \times \mathbf{V}(\mathbf{x}, t)} - \bar{\boldsymbol{\alpha}} \times \bar{\mathbf{V}}(\mathbf{x}, t), \tag{19}$$

and $\bar{\mathbf{V}}$ are the terms that require closure. Physically, $\mathbf{L_p}$ is representative of a portion of the average slip strain rate produced by the "microscopic" dislocation density; in particular, it can be non-vanishing even when $\bar{\boldsymbol{\alpha}} = 0$ and, as such, it is to be physically interpreted as the strain-rate produced by the so-called "statistical dislocations", as is also indicated by the extreme right-hand side of (19). The variable $\bar{\mathbf{V}}$ has the obvious physical meaning of being a space-time average of the pointwise, microscopic dislocation velocity. Initial and boundary conditions for (8–12) are important from the physical modeling point of view (Acharya and Roy 2006), particularly in the context of triggering inhomogeneity under boundary conditions corresponding to homogeneous deformation in conventional plasticity theory (Roy and Acharya 2006).

It is possible to eliminate the fields $(\boldsymbol{\chi}, \mathbf{z})$ in the set of (14–18), *provided that one retains the continuity conditions implied by the set of equations across any arbitrary surface moving through the body*. Such a surface is called a material surface when its normal velocity coincides with the material velocity of the particle surface it is coincident with. Interesting properties derive from these continuity conditions when discontinuity of the material properties and/or field variables occurs across the surface (as in grain boundaries). Considering for simplicity material surfaces, and dropping overhead bars for convenience when $\mathbf{L_p} \neq \mathbf{0}$, a reduced set of equations can be written as

$$\mathbf{div\,T} = 0 \tag{20}$$

$$\mathbf{T} = \mathbf{C} : \mathbf{U_e} \tag{21}$$

$$\mathbf{U_e} = \mathbf{grad\,u} - \mathbf{U_p} \tag{22}$$

$$\dot{\mathbf{U}}_\mathbf{p} = \boldsymbol{\alpha} \times \mathbf{V} + \mathbf{L_p} \tag{23}$$

$$\dot{\boldsymbol{\alpha}} = -\mathbf{curl}\,\dot{\mathbf{U}}_\mathbf{p} \tag{24}$$

$$\left[\dot{\mathbf{U}}_\mathbf{p}\right] \times \mathbf{n} = 0, \tag{25}$$

where $[\mathbf{A}]$ represents a jump of $\mathbf{A}$ at the surface of discontinuity and $\mathbf{n}$ is the unit normal to the surface, with arbitrarily chosen orientation. Equation (25) has the practical implication (say for finite element calculations) that, unlike conventional plasticity, the plastic distortion field has to satisfy a "hard" partial continuity constraint, i.e., the tangential action of the plastic distortion rate has to be continuous at a material surface of discontinuity. Equation (24) provides for the evolution and transport of polar dislocation densities. Through the curl of the total plastic distortion rate tensor $\dot{\mathbf{U}}_\mathbf{p}$, it couples the polar and statistical dislocation densities for the nucleation of polar dislocations. Complementing the above equations with a constitutive relation for the average dislocation velocity $\mathbf{V}$ as a function of stress and dislocation orientation, and with phenomenological evolution equations for the statistical densities involved in the conventional velocity gradient $\mathbf{L_p}$, one obtains a closed theory in the sense that it contains enough statements to derive uniquely the dynamics of stress and dislocation densities in a bounded domain from boundary and initial conditions. In particular, the direction $\mathbf{g}$ of velocity $\mathbf{V}$ is prescribed as (Acharya and Roy 2006)

$$\mathbf{V} = v\frac{\mathbf{g}}{|\mathbf{g}|}$$

$$\mathbf{g} := \mathbf{f} - \left(\mathbf{f}.\frac{\mathbf{d}}{|\mathbf{d}|}\right)\frac{\mathbf{d}}{|\mathbf{d}|}$$

$$\mathbf{f} := \mathbf{X}(\mathbf{T}'\boldsymbol{\alpha});\ f_i = e_{ijk}T'_{jr}\alpha_{rk};\ \mathbf{d} = \mathbf{X}(tr(\mathbf{T})\boldsymbol{\alpha});\ d_i = \left(\frac{1}{3}T_{mn}\right)e_{ijk}\alpha_{jk}, \tag{26}$$

where $\mathbf{X}$ is the alternating Levi–Civita tensor and $\mathbf{T}'$ the deviatoric stress tensor. The definition of $\mathbf{g}$ can be approached from two points of view. In the situation when the dislocation density may not be expressed as an elementary dyad formed from a Burgers vector direction and a line direction, the definition (26) arises as a sufficient condition for pressure independence of the polar dislocation plastic strain rate and ensuring positive dissipation. The dissipation D in the model can be written as

$$D = \int_B (\mathbf{X}(\mathbf{T}\boldsymbol{\alpha}).\mathbf{V} + \mathbf{T} : \mathbf{L_p})dv \tag{27}$$

Focusing on the dissipation due to polar dislocation motion, $\mathbf{X}(\mathbf{T}\boldsymbol{\alpha}).\mathbf{V}$, and writing

$$\mathbf{X}(\mathbf{T}\boldsymbol{\alpha}) = \mathbf{f} + \mathbf{d} \tag{28}$$

where $\mathbf{f}$ is a pressure-independent term, it makes physical sense to require $\mathbf{V}$ to be in the direction of $\mathbf{f}$. However, this does not guarantee that the dissipation due to polar dislocation motion is independent of pressure and neither that $\mathbf{X}(\mathbf{T}\boldsymbol{\alpha}).\mathbf{f} \geq 0$; subtracting the component of $\mathbf{f}$ in the direction of $\mathbf{d}$ ensures the latter fact:

$$(\mathbf{f}+\mathbf{d}).\left(\mathbf{f} - \left(\mathbf{f}.\frac{\mathbf{d}}{|\mathbf{d}|}\right)\frac{\mathbf{d}}{|\mathbf{d}|}\right) = \mathbf{f}.\mathbf{f} - \frac{(\mathbf{f}.\mathbf{d})^2}{|\mathbf{d}|^2} + \mathbf{d}.\mathbf{f} - \mathbf{f}.\mathbf{d} = \mathbf{f}.\mathbf{f} - \left(\mathbf{f}.\frac{\mathbf{d}}{|\mathbf{d}|}\right)^2 \geq 0 \tag{29}$$

by the Cauchy–Schwarz inequality (Pythagoras' theorem). Alternatively, a compelling mechanistic interpretation arises when $\boldsymbol{\alpha}$ may be interpreted as an elementary dyad formed from a Burgers vector direction and a line direction. Then the direction of $\mathbf{d}$ represents the direction of climb whenever $\boldsymbol{\alpha}$ represents a dislocation segment of pure edge or mixed character, and it is degenerate when $\boldsymbol{\alpha}$ is of pure screw character. Thus, $\mathbf{g}$ represents the fact that mixed or edge dislocations cannot climb or cross-slip whereas screw dislocations are unrestricted in their motion.

It is perhaps insightful to evoke analogies between dislocation dynamics and eddy dynamics in turbulent flow (Marcinkowsky 1989). This analogy extends to transport of polar/statistical dislocation densities, as expressed through (18, 23, 24), and Large Eddy Simulations (LES) in the analysis of turbulence (Varadhan et al. 2006). Turbulent flow is characterized by eddies at all scales. Averaging in space and time the Navier–Stokes equations provides equations for large resolved eddies, while unresolved ones are dealt with using additional sub-grid-scale variables. Closure of the theory is obtained through subgrid phenomenological models featuring scaling character (Meneveau and O'Neil 1994). In dislocation dynamics, averaging in space secures equations for polar dislocations while providing the link with conventional plasticity: closure for the unresolved variables $\mathbf{L_p}$ derives from well-established models for the viscoplasticity of crystalline materials, i.e., relations for forest hardening and lattice rotation having received decades of attention and experimental validation (see below (30, 31, 40)). Also in contrast to turbulence, scaling behavior is associated with grid scale level, not sub-grid scale, as we show in Sect. 4.

Two types of solutions are offered in what follows, to provide various insights. First we conduct full 3-D numerical solutions of (20–25) in single crystals, by using a Galerkin – Least Squares finite element method appropriate for transport problems (see Acharya and Roy 2006; Varadhan et al. 2006 for details). In these simulations, the plastic velocity gradient $\mathbf{L_p}$ follows from the activity of the statistical mobile dislocations on all slip systems

$$\mathbf{L_p} = \sum_s \rho_m b V_s \mathbf{b_s} \otimes \mathbf{n_s}, \tag{30}$$

where ρ_m is the mobile statistical dislocation density, $\mathbf{b_s}$ and $\mathbf{n_s}$ are the slip system Burgers vector and glide plane normal, respectively. V_s is the ensemble dislocation velocity, which follows the power law relationship

$$V_s = V_0 \mathrm{sgn}(\sigma_s) \left(\frac{|\sigma_s|}{\sigma_0 + \sigma_h} \right)^n, \quad \sigma_s = \mathbf{b_s} \otimes \mathbf{n_s} : \mathbf{T}. \tag{31}$$

Here, σ_s is the resolved shear stress on a glide plane, with reference velocity V_0, athermal stress σ_0, and stress exponent n as material parameters. The threshold stress σ_h reflects short range obstacle overcoming. It relates to the statistical forest density ρ_f through the usual Taylor relation $\tau_h = \bar{\alpha}\mu b\sqrt{\rho_f}$, where $\bar{\alpha}$ is a non dimensional parameter. Large n values reflect abruptness of dislocation unpinning from obstacles. The velocity V of polar dislocations is taken as the average of the statistical slip velocity absolute values $|V_s|$ over all slip systems. Hence, the same physics applies to both dislocation species. We shall also consider simplified situations with dislocations pertaining to, and gliding in a single slip plane with no out-of-plane motion. The latter simulations provide for representative behavior of some portion of a slip plane in a single crystal experiment. In such situations, the constitutive behavior (31) may be phenomenologically altered (as in (34) below for example).

3 Effects of Sample Size on Mechanical Response

In a torsion test, the shear stress increases from the axis to the exterior of the sample. When investigating the plastic response of materials, this gradient is commonly viewed as a drawback of torsion testing. It becomes beneficial when the material behavior involves internal length scales associated with emerging dislocation microstructures. The inhomogeneity of the boundary conditions then generates polar dislocations, which give rise to long-range elastic stress fields. Hence, torsion is a challenging case for theories of plasticity with internal length scales. Thin copper wires with diameters in the range 12–170 μm were twisted to probe into such theories (Fleck et al. 1994). Size effects were reported, and the trend is that the greater is the imposed gradient, the greater is the degree of hardening. However, the large strains achieved, the polycrystalline character of the material, the texture evolution and varying grain size of the samples may have complicated the interpretation. In this chapter, the creep response of ice single crystals in torsion, a much simpler material and experimental configuration, is described with focus on the effects of the sample dimensions on this response. As an hcp material with a strong anisotropy of plasticity, ice is a choice material in this respect. It deforms plastically by the activity of basal slip systems almost exclusively, (Duval et al. 1983) and it is characterized by a low Peierls stress (Shearwood and Whitworth 1991). Plastic anisotropy and a low lattice friction favor long-range elastic interactions and dislocation transport, as well as their interactions and, indeed, the creep response of ice single crystals oriented for basal slip in torsion exhibits spectacular size effects in the centimeter range (Taupin et al. 2007).

In (Taupin et al. 2007), the torsion creep tests were carried out on cylindrical samples machined from laboratory grown single crystals. The applied torque M was such that the average shear stress across a sample cross-section $\bar{\tau} = 3M/2\pi R^3$ (R is the sample radius) remained constant throughout the experiments. Figure 4 shows the forward creep curves, i.e., the evolution in time of $\gamma = \kappa R$, the strain on the outer surface (κ is the constant twist per unit length of the sample). A forward and reverse creep curve, with the torque sign changed at reversal, is also shown below in Fig. 7. Dimensional analysis shows that, for a material devoid of internal length scales, creep curves gathered from samples with varying radius superpose if the average shear stress $\bar{\tau}$ and height/radius ratio are kept constant. Conversely, distinct curves in this plot are evidence for an effect of size on the plastic response. Figure 4 suggests that the time needed to achieve a given strain decreases with the diameter, which indicates a softening effect of diameter reduction on the sample response, a trend opposite to that reported in (Fleck et al. 1994) for polycrystalline Cu samples. Dispersion in the curves is existing, but limited. It results mainly from uncontrollable fluctuations in the initial dislocation microstructure, which lead to uneven initial creep strain rates.

Hard X-ray diffraction analyses performed on slices extracted from the strained samples show that plasticity is almost exclusively due to polar dislocations of screw character gliding in basal planes, with very few mobile statistical dislocations (Montagnat et al. 2003; Chevy et al. 2010). The initial density of dislocations

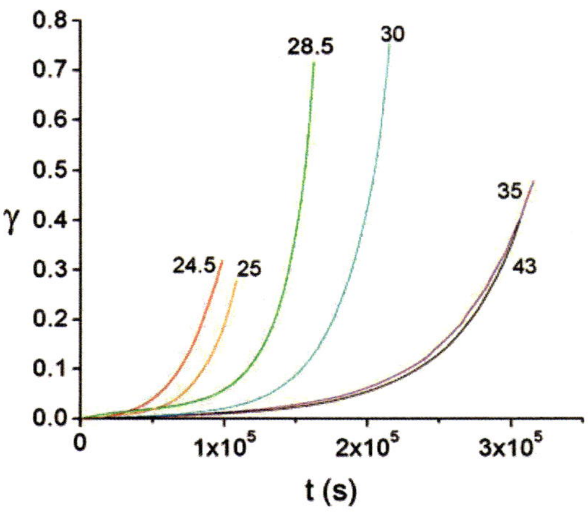

Fig. 4 Creep strain on outer surface vs. time, for various diameter values. The average shear stress is $\bar{\tau} = 0.12\,\mathrm{MPa}$. In each sample, height and diameter are equal (their common value is given in millimeter), in order to avoid any bias due to end effects. Courtesy of Juliette Chevy (2008)

present in the samples, mostly sessile dislocations, was shown to be small (less than $10^8\,\mathrm{m}^{-2}$). In addition, the analyses reveal a scale invariant arrangement of polar dislocations along the torsion axis suggesting propagation of slip in this direction. The latter can be explained by the occurrence of double cross-slip of screw dislocations driven by the internal stress field through prismatic planes (Chevy et al. 2010; Montagnat et al. 2006).

Interpretation in terms of dislocation dynamics of these observations is now provided on the basis of the model described in Sect. 2. We begin with a simplified 1-D model designed for twofold purpose: to illustrate the critical aspects of the theory; to allow for effective parametric study of size effects. In this idealization for deformation under a gradient of simple shear, we consider screw dislocation density of infinite extent in the (x_1, x_3) tangential and axial directions, line and Burgers vector along the tangential direction x_1 and transport in the radial direction x_2. The distributions of shear stress σ_{13}, polar screw density α_{11}, and mobile statistical density ρ_m along a sample radius are the unknowns. The resulting equations derived from (14–18) in creep reduce to

$$\sigma_{13,1} = \sigma_{13,3} = 0 \tag{32}$$

$$\dot{\alpha}_{11} + (\alpha_{11} v_2)_{,2} = -(\rho_m b v)_{,2} \tag{33}$$

Here, b is the length of the Burgers vector, and a comma indicates a partial derivative. Since the concern is on transient primary creep and the sample remains mostly elastic in its central part, as will be discussed below, an elastic approximation is used for the shear stress, leading to : $\sigma_{13} = (x_2/R)\tau$. It was checked that the latter differs from the stress distribution expected for a fully viscoplastic response by less than 15%. Equation (33) is a transport equation. It represents the transport of screw dislocations along the radius with a source term due to gradients in statistical dislocation mobility. In an attempt to offset the model simplifications, account of the physics of dislocation velocity and of the history of straining is now made through phenomenological statements. Following (Duval et al. 1983), we assume a power law relationship for the average polar and statistical dislocation velocities (v_2, v) in the form

$$v_2 = v = v_0 \operatorname{sgn}(\sigma_{13} - \sigma_\mu) \left(\frac{|\sigma_{13} - \sigma_\mu|}{\sigma_0 + \sigma_h} \right)^n \tag{34}$$

with $n = 2$. Parameters (v_0, σ_0) are reference velocity and stress values, respectively. They are identified from the experimental data (Duval et al. 1983; Shearwood and Whitworth 1991). An isotropic statistical hardening is derived from the sessile density ρ_s in the Taylor form: $\sigma_h = \bar{\alpha}\mu b\sqrt{\rho_s}$, where μ denotes the elastic shear modulus and $\bar{\alpha}$ is a constant. Only a fraction $(1 - \beta)$ of the nucleated screws glides in the basal planes. They induce a back-stress, with rate of formation

$$\dot{\sigma}_\mu = \tilde{\alpha}\mu\alpha_{11}v_2 - \frac{|v_2|}{\hat{\alpha}b}\sigma_\mu, \tag{35}$$

where $(\tilde{\alpha}, \hat{\alpha})$ are constants. Relation (35) is similar to the Armstrong-Frederick law for kinematic hardening (Armstrong and Frederick 1966), but here the back-stress builds up from polar dislocation mobility only. Note that the involved relaxation time $\tau_r = \hat{\alpha}b/|v_2|$ is inversely proportional to the polar dislocation velocity. It is such that, at equilibrium $(\dot{\sigma}_\mu = 0)$, the back-stress value does not depend on the dislocation velocity, but only on the polar dislocation density. The complementary fraction β of nucleated screws experiences out-of-plane motion induced by the internal stress field. Therefore, the statistical sessile density increases – due to the formation of edge segments in prismatic planes, assumed to be proportional to the rate of screw nucleation

$$\dot{\rho}_s = \beta|(\alpha_{11}v_2)_{,2}|. \tag{36}$$

In our calculations, σ_h remains smaller than the reference stress σ_0, implying that statistical hardening is relatively insignificant, whereas the back-stress σ_μ can be of the order of the applied stress τ. The statistical mobile dislocation density ρ_m has a very low initial value. It increases due to dislocation sources associated with edge jogs in prismatic planes (Louchet 2004). Its nucleation rate is supposed to be proportional to the shear strain rate, with coefficient C_1. Saturation of mobile dislocations results from their mutual annihilation, with coefficient C_2.

$$\dot{\rho}_m = \left(\frac{C_1}{b^2} - C_2\rho_m \right)\dot{\Gamma}, \quad \dot{\Gamma} = |\alpha_{11}v_2 + \rho_m b v|. \tag{37}$$

Note that the evolution law (37) is also used in the upcoming 3-D computations, with $\dot{\Gamma} = |\dot{\mathbf{U}}_\mathbf{p}|$, but that the latter automatically include out-of-plane dislocation motion and back-stress build-up. Their presence in the phenomenology of the 1-D formulation through (35, 36) is an offset for the assumed invariance in the c-axis direction. Numerical constants and initial conditions are given in Table I. Figure 5 shows the (locallyresolved) polar screw density and Fig. 6 the shear component of

Table 1 Numerical constants and initial conditions used in the model

b	v_0	σ_0	n	μ	β	C_1	C_2
4.5×10^{-10} m	3.6×10^{-7} m/s	0.1 MPa	2	3 GPa	0.1	10^{-8}	17
$\bar{\alpha}$	$\tilde{\alpha}$	$\hat{\alpha}$		ρ_m	ρ_s	$\alpha_{11}(R)$	
0.133	0.666×10^{-2}	10^5		10^6 m^{-2}	10^8 m^{-2}	0.32 m^{-1}	

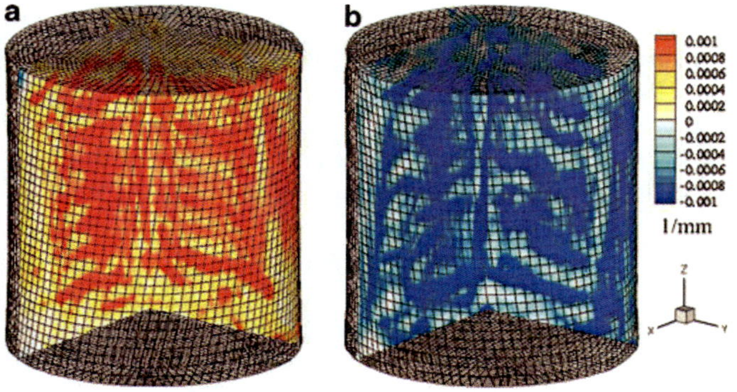

Fig. 5 Polar screws α_{11} piercing the plane with normal in the x_1 direction and α_{22} piercing the plane with x_2 normal: (**a**) just before reversal in Fig. 7; (**b**) the structure developed after reversing the direction of creep, at the end of the curve

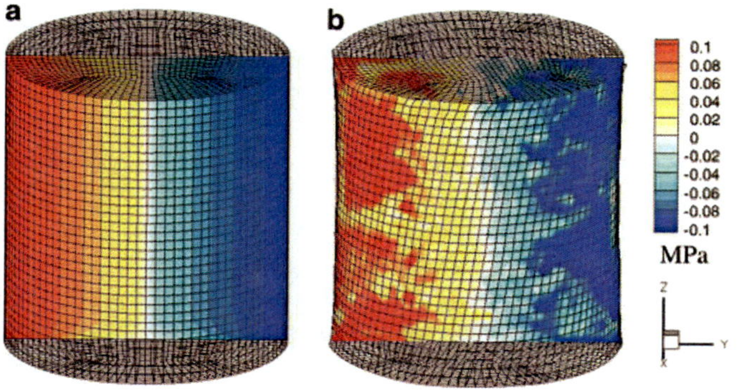

Fig. 6 Shear stress component σ_{13} shown at the beginning (**a**) and end (**b**) of the dashed curve in Fig. 7. The figure highlights the development of stress due to the multiplication of polar dislocations, from the (effectively) elastic solution with low mobile density. Plot A also shows end effects in the distribution of stress

the stress obtained from the 3-D model. Note that the 3-D stress distribution supports the assumption made in the 1-D idealization. Under a positive torque, an outstanding feature of both models is the nucleation of positive screw dislocations close to the edge of the sample, their transport toward its axis and, as stress and velocity decrease in this area, the formation of pile-ups. As seen in Fig. 7, the continuous

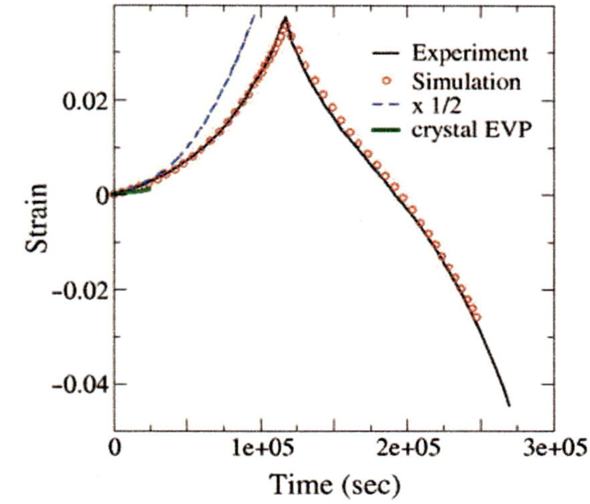

Fig. 7 Creep curves in forward/reverse torsion from experiments, 1-D and 3-D models. The dashed line shows the forward creep curve for a sample with halved radius and height. It is seen that the acceleration of creep increases when the sample size is reduced. The thick continuous line interrupted after 2.5 10^4 s is obtained for conventional elasto-viscoplastic (EVP) treatment. It shows that the latter is unable to retrieve the acceleration of creep

increase in the forward creep rate is retrieved from both models. The reverse torsion behavior is also shown in the figure. At torque reversal, an increase in the creep rate absolute value is observed in the experiments. It is fully retrieved by both the 1-D and 3-D models. This asymmetry of slopes at the reversal point can be attributed to the positive screw dislocation pile-ups built in forward torsion. Whereas they were opposing dislocation motion in forward loading, the resulting internal stresses are helping reverse dislocation motion after torque reversal, hence the "instantaneous" creep acceleration at reversal. Screw dislocations of negative sign are nucleated in reverse loading, which progressively annihilate with the positive screw pile-ups created in forward loading. Hence, the total polar screw density decreases, with the consequence that the positive screw pile-ups are dismantled and that the creep rate absolute value decreases. The minimum creep rate value is reached at the inflexion of the creep curve. At this point, creep is mostly accommodated by statistical dislocations. In the rest of the reverse creep curve, creep keeps accelerating while negative screw pile-ups are created, in a fashion similar to forward loading, though obviously with reversed sign. Hence, the anisotropy of creep behavior derives from the nucleation, transport, and annihilation of inhomogeneous polar screw density distributions. The excellent agreement between experimental and simulated creep curves suggests that these ingredients are indeed key aspects of the physical response.

Sample size effects on the creep response are obtained. By reducing the sample diameter, screw nucleation is promoted and acceleration of the creep rate increases, in close agreement with experimental data (see Figs. 7 and 8). Thus, the greater is the imposed gradient, the greater is the degree of softening, not hardening as would have been expected if polar screw dislocations had been contributing to isotropic statistical hardening in a way similar to statistical dislocations. Rather, the polar screw dislocations induce softening due to larger rates of plastic distortion.

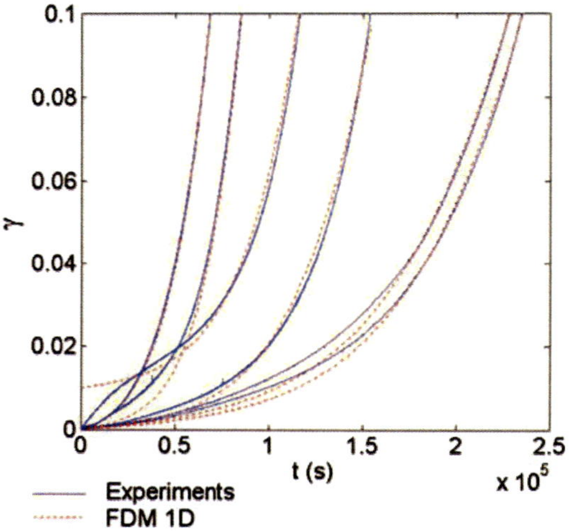

Fig. 8 Experimental data from the creep tests shown in Fig. 4 and simulated creep curves from the 1-D model. In the model, initial conditions on screw dislocation density are consistent with the observed initial strain rate $\dot{\gamma}$. Courtesy Juliette Chevy (2008)

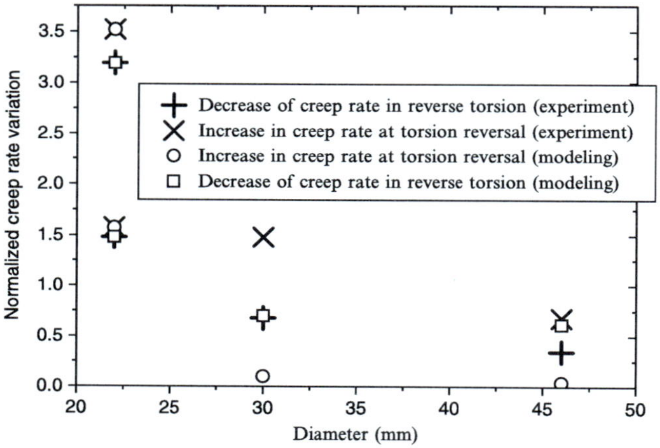

Fig. 9 Sample diameter effects on acceleration of creep rate at torsion reversal and on creep rate deceleration during reverse torsion, from reversal to inflexion of the creep curve

Several other effects of sample size on mechanical response were predicted by the model and observed in the experiments. We report here on two such effects, observed in reverse torsion and shown in Fig. 9. First, the larger is the imposed gradient (the smaller is the sample diameter), the larger is the increase in the creep rate absolute value at torque reversal. Second, the larger is the imposed gradient, the larger is the decrease in the creep rate, from torque reversal to inflexion of the creep curve. Note that the latter is a hardening effect on the creep response. As mentioned above, the asymmetry of the creep curve at torsion reversal reflects screw dislocation pile-ups and internal stresses built up during forward torsion. Hence, the larger is the imposed gradient, the larger are the internal stress level and creep rate acceleration at reversal. The interpretation suggested by the model for the second effect is as follows. Since the inflexion point corresponds to the instant when the polar screw density is closer to zero, a larger deceleration in the creep rate is representative of a larger nucleation of polar screw density, which in turn is due to a larger imposed gradient.

In pure Cu, there is experimental evidence of sample size effects on mechanical response. In polycrystals, the work by Fleck et al. (1994) seems to be pointing at response hardening when the sample diameter is reduced, but the interpretation might be complex, as already mentioned. In single crystals in tension, the extent of easy glide distinctly increases with decreasing crystal radius (response softening), while the rate of hardening in stages I and II is only weakly dependent on crystal size (Basinski and Basinski 1979). The extent of easy glide is also orientation dependent (Diehl 1956), but it was shown to decrease with size for selected orientations: near [110] (Suzuki et al. 1956), for single glide orientation (Garstone et al. 1956) and for various other orientations (Fourie 1967). We note that observations similar in spirit to ours (Weertman 2002) were made to explain the "anomalous hardening effect" observed in (Fleck et al. 1994). In this reference, the reduction in the density of polar screw dislocations in the center region of the sample is seen as the origin of hardening. As its plastic distortion is reduced near the axis, the metal behaves more like an elastic solid and, as such, it becomes harder. The difference in behavior with ice single crystals might then be attributed to the difference in the elastic constant values (the elastic shear modulus in ice is of the order of 3 GPa, much smaller than the 40 GPa Cu value). On the basis of the above simulations, dislocation transport and long-range internal stress build-up appear as the controlling mechanisms for the rarefaction of polar screw dislocations in the center of the sample, through polar screw pile-up formation.

4 Intermittency of Crystal Plasticity: Scale Invariance and Transport Effects

Transport is a convective process, pervasive in many branches of physics, by which certain species, or variations in certain quantities, propagate in a medium. For example, it serves as a cornerstone in the theory of fluid dynamics. When envisioned on length scales over which areal densities of dislocations may be envisaged, dislocation motion is amenable to the transport of these densities. The fundamental equation for dislocation transport (7) has been known for half a century (Mura 1963; Kosevich 1979), mostly as a curiosity, and only recently has it been effectively used for dislocation dynamics predictions (Acharya 2001). Similarly, the relevant length scale for the observation in dislocation dynamics of the propagative features associated with transport has remained elusive, and only recently has experimental evidence been provided (Fressengeas et al. 2009), although observation of strain waves (Zuev 2007) could perhaps have given a clue earlier.

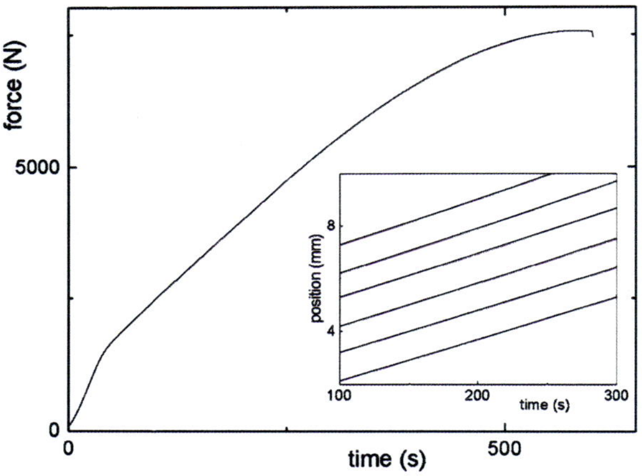

Fig. 10 Cu single crystal oriented for multislip under uniaxial tension (Gauge length: 30 mm, width: 5.5 mm, thickness: 5.5 mm, Schmid's factor: 0.3, temperature: 20°C, driving strain rate: $\dot{\epsilon}_a = 5 \times 10^{-4}\,\mathrm{s}^{-1}$), macroscopic force vs. time (main graph) and displacement vs. time in six locations distant by 1 mm (inset)

An inherent connection between dislocation transport and the intermittency of plastic activity is revealed in (Fressengeas et al. 2009) by applying high resolution extensometry to Cu single crystals in tension. When oriented for multislip, Cu single crystals represent the truly emblematic situation where material instability can be ruled out and homogeneous straining in a traditional (mechanical) sense expected at small strains (see loading curve in Fig. 10). Yet, the inhomogeneous dislocation microstructure and the intermittent dislocation activity at a microscopic scale may well induce intermittency and inhomogeneity in dislocation transport at some intermediate scale. Hence, Cu crystals represent the perfect case for evidencing intermittency and dislocation transport properties. The extensometry method is based on a digital image correlation technique in one-dimensional setting. The sample surface is painted with alternated black and white strips, which perfectly reflect the material displacement underneath. A high-resolution CCD camera with recording frequency 10^3 Hz and pixel size 1.3 μm captures the longitudinal displacements of the black–white switches, from which the axial strain rate field is rendered. Subtracting the constant driving strain rate $\dot{\epsilon}_a$ from the local strain rate at a material point leaves the variations shown in Fig. 11. Despite smoothness of the loading curve in Fig. 10, Fig. 11 displays jerks well above experimental noise level. The figure suggests that the intermittency of dislocation motion at the microscopic scale shows up at a somewhat larger scale. The probability density for the size of jerks in Fig. 11 shows power law scaling (see Fig. 12), with scaling exponent $\tau \approx 2$. This exponent is consistent with the scaling law reported for the associated acoustic emission (Weiss et al. 2007). Such scaling is evidence for self-organization of the observed fluctuations, which is also demonstrated by Fig. 13. The figure features a space-time diagram for local fluctuations about the driving strain-rate during the elasto-plastic transition in

Fig. 11 Variations of axial strain rate about the driving strain rate $\dot{\epsilon}_a == 5 \times 10^{-4}\,s^{-1}$, as obtained from the lowest displacement curves in the stack in Fig. 10. Note that the maximum size of the fluctuations $2.5 \times 10^{-3}\,s^{-1}$ is much larger than $\dot{\epsilon}_a$

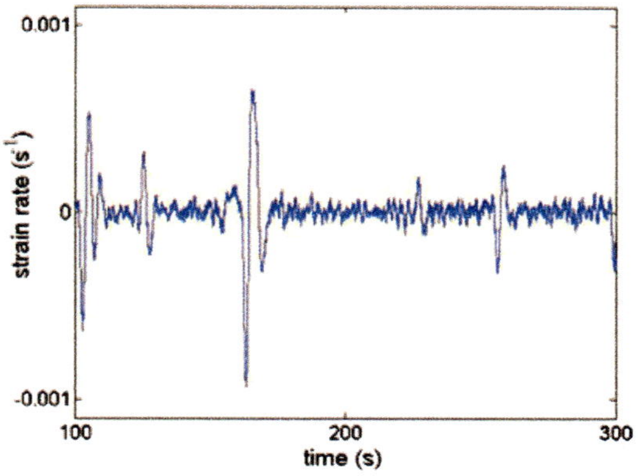

Fig. 12 Probability density (normalized to bin size) for event size (amplitude of strain rate jerks) in the time series shown in Fig. 11. The dashed line indicates the power-law trend with slope $\tau = 2$

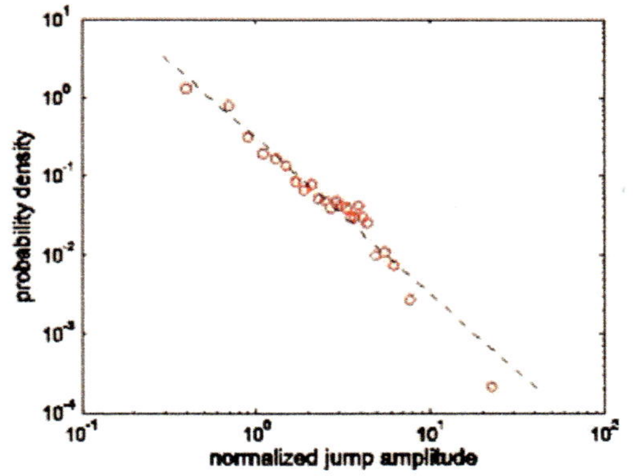

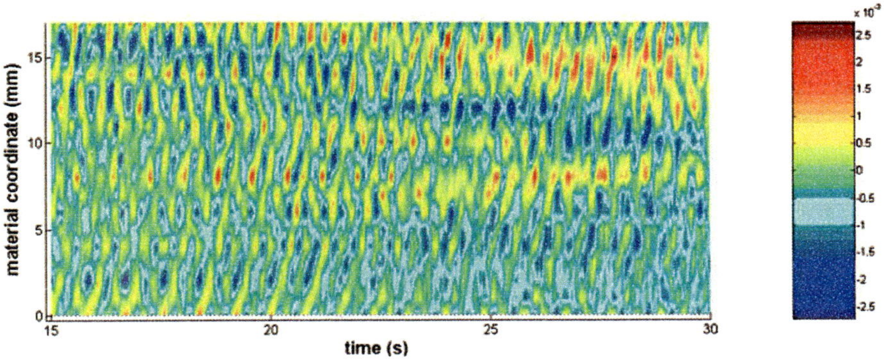

Fig. 13 Longitudinal fluctuations about the imposed strain rate in a space – time diagram during the elasto-plastic transition. Dotted characteristic lines run from the left and right of the gauge length, reflecting intermittency and transport. The imposed strain rate is $\dot{\epsilon}_a = 5 \times 10^{-4}\,s^{-1}$. Fluctuations can be as high as $2.5 \times 10^{-3}\,s^{-1}$

the test shown in Fig. 10. It displays spots of intense activity dotted along straight lines, suggesting wave propagation at constant average speed. The average wave velocity measured from the slope of the characteristic lines (about 10^{-2} m.s^{-1}) is five orders of magnitude smaller than the velocity of elastic waves, but much larger than the material particles velocity. It is of the order of the average velocity of dislocation ensembles, which suggests that the observed waves reflect the underlying collective motion of dislocations. In this interpretation, the dotted pattern of spots along the characteristic lines is manifestation of the intermittency of collective dislocation motion. It is also proof to the wavy structure of plastic activity, when envisioned at appropriate length scale. At larger strains, this wavy pattern is seen on shorter time and length scales, because the dislocation mean free path decreases in relation with the multiplication of forest obstacles.

A 3-D generic simulation of the above experiments using field dislocation dynamics is now briefly outlined. The reader is referred to (Fressengeas et al. 2009) and (Acharya et al. 2008) for further details. A flat Cu whisker is clamped to the left end, while the right end has constant velocity. The applied strain rate is $\dot{\epsilon}_a = 10^{-3}\,\mathrm{s}^{-1}$. The elastic response is taken to be anisotropic with constants C_{11}, C_{12}, and C_{44}, and the evolution equations for ρ_m and ρ_f are

$$\dot{\rho}_m = (C_1/b^2 - C_2\rho_m)\dot{\Gamma} \tag{38}$$

$$\dot{\rho}_f = (C_0/b)|b| + C_2\rho_m)\dot{\Gamma} \tag{39}$$

as outlined in (Varadhan et al. 2006), with simplifications deemed appropriate for the low strain level achieved in the experiments. C_0, C_1, and C_2 are material parameters accounting for the interaction between polar and forest dislocations, the mobile dislocation generation and loss, respectively. The material parameters are listed in Table 2. Note that there is no inhomogeneity introduced in either the material parameters or the initial conditions.

Elastic loading of the sample is followed by a yield drop associated with plastic activity localized near the clamped end, then by a stress plateau shown in Fig. 14. Thus, inhomogeneity of plastic straining clearly stems from the inhomogeneity of the boundary conditions. This prediction of a yield drop is in full agreement with experimental data on Cu whiskers (Nittono 1971; Brenner 1957; Gotoh 1974). During the plateau, the plastic activity propagates along the sample through the motion of a plastic front, before linear homogeneous strain hardening takes place. Propagation does not occur in this flat sample if transport is turned off in the equations. For further details on the propagation of slip along the plateau, the reader is referred to

Table 2 Material parameters used in the 3D Cu whisker simulation

α	b	n	V_0	σ_0	
0.35	2.5×10^{-10} m	20	3.5×10^{-8} m/s	3.7 MPa	
C_0	C_1	C_2	C_{11}	C_{12}	C_{44}
25	2.43 10^{-5}	3.03	170 GPa	123 GPa	75.2 GPa

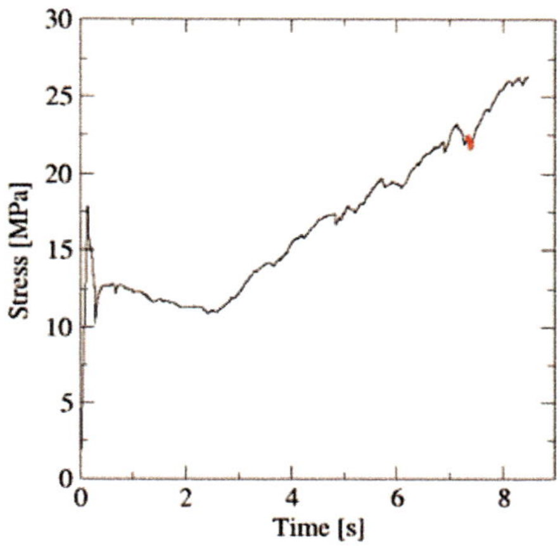

Fig. 14 Simulation of the tensile test of a flat Cu whisker of dimensions $200 \times 30 \times 2,400\,\mu$m; stress vs. time response during the elasto-plastic transition and stage II linear hardening. The highlighted portion corresponds to the strain rate plots shown in Fig. 15

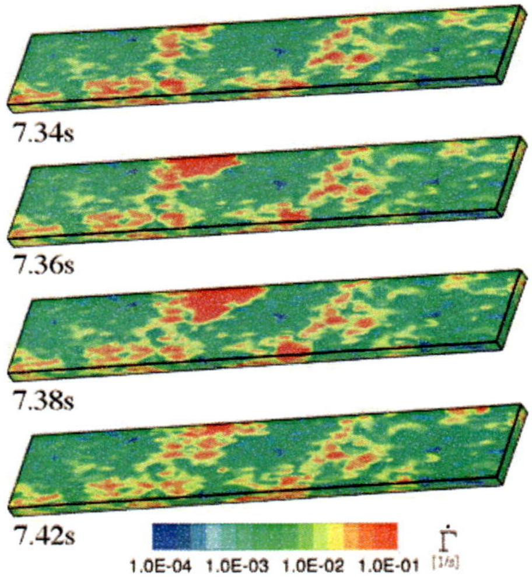

Fig. 15 Successive frames of the strain-rate spatiotemporal field along the sample showing intermittent plasticity through dislocation transport, and general progression of plastic slip from left to right. The sequence highlights a single plastic burst shown in Fig. 14

(Acharya et al. 2008). Here, we focus on the intermittency of plastic activity during the eventual linear hardening period. Bursts in stress-rate (an average measure of plastic activity over the sample) are seen all along the curve during this period. One particular sequence, highlighted in Fig. 14, corresponds to the plots of plastic strain rate shown in Fig. 15. In this figure, intermittent events and transport are clearly seen, with a general progression of plastic activity from left to right of the sample.

In view of these results, 2D simulations (of course more tractable than 3D) were carried out to check for scaling behavior at a smaller scale and for possible variation of the scaling exponent under diverse material and experimental conditions. In these simulations, a rectangle subjected to constant shear rate $v_{1,3}$ at boundaries is considered in a glide plane of a Cu single crystal. Only dislocations pertaining to, and gliding in this plane are considered. Out-of-plane motion by cross-slip and climb is not considered, and single slip activity is assumed. As all gradients normal to the slip plane are ignored, out-of-plane features of lattice incompatibility and internal stresses are lost in this simplified description. Elasticity is taken to be isotropic with shear modulus μ. The average velocity V of dislocations in the plane is described with the thermally activated constitutive law

$$V = V_0 \exp \frac{-\Delta G_0}{kT} \exp \left(\frac{V^* \sigma_s}{kT(1 + \sigma_h/\sigma_0)} \right), \tag{40}$$

where V_0 is a reference velocity, $(\Delta G_0, V^*, k, T)$ a reference enthalpy, the activation volume, the Boltzmann constant, and the temperature. σ_s is again the resolved shear stress and σ_h the threshold stress for obstacle overcoming. Equation (40) is an alternative to (31), used to describe weak rate-sensitivity of the shear stress. The imposed strain rate is $\dot{\epsilon}_a = 5 \times 10^{-4}\,\mathrm{s}^{-1}$. In the initial configuration, polar dislocations are absent and the statistical mobile density is chosen at random about an average value. Since the boundary conditions are homogeneous (in contrast to the above 3D simulation), the incompatibility arising from the distribution of statistical dislocations is initially the only source for polar dislocations. The information on material parameters, initial and boundary conditions is summed up in Table 3.

Figure 16 shows a space-time diagram for the strain-rate fluctuations. In qualitative agreement with Fig. 13, spots of intense plastic activity dotted along straight lines are seen. This pattern follows naturally from the development of polar dislocation density, by virtue of dislocation transport and internal stress. The velocity

Table 3 Initial and boundary conditions, complementary material parameters in 2D simulations

$\alpha_{ij}(0)$	$\rho_m(0)$	μ	σ_h	$\nu_{1,3}$	kT/V^*
0	$10^8\,\mathrm{m}^{-2}$	40 GPa	50 MPa	$5 \times 10^{-4}\,\mathrm{s}^{-1}$	2.27 MPa

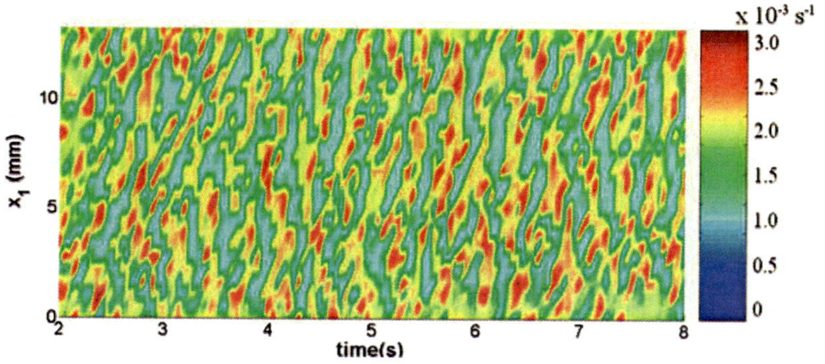

Fig. 16 Model predictions for axial (x_1) strain-rate fluctuations in a space–time diagram. The sample is a 13 mm × 13 mm square in a glide plane subjected to equal shear rates $5 \times 10^{-4}\,\mathrm{s}^{-1}$ on both sides. The figure shows the evolution in time of the strain rate profile seen along the x_2 direction

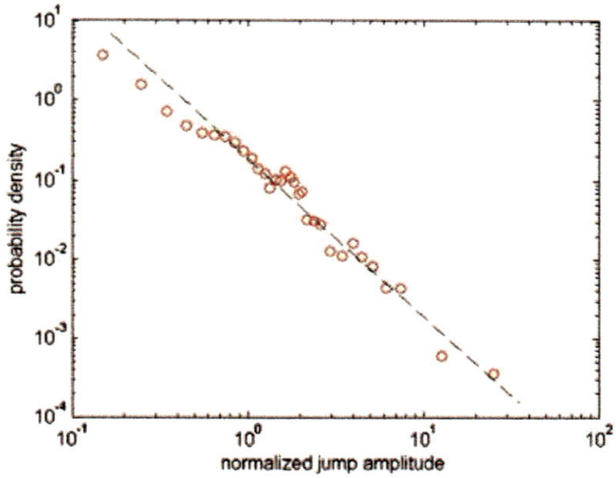

Fig. 17 Probability density of event size in a 2-D simulation. The event size is defined as the maximum strain-rate value during the event. The dotted line shows a $\tau = 2$ slope

Table 4 Numerical constants used in the model

b	ν_0	S_0	μ	$\bar{\alpha}$	$\tilde{\alpha}$	$\hat{\alpha}$
2.7×10^{-10} m	8.5×10^8 s^{-1}	2.27 MPa	80 GPa	0.3	150	10^3
f_0	τ	C_0	C_1	C_3	C_4	
40 MPa	10^6 s	25	1.45×10^{-4}	5.4×10^{-2}	20	

obtained from the slopes in Fig. 16 has the order of magnitude observed in experiments. Figure 17 shows the probability density for event size computed from the time series obtained for the net shear strain rate at several material points in the plane. A scaling distribution is seen, with exponent $\tau \approx 2$ in agreement with both the exponent found from the stress vs. time response and the experimental value. This result confirms that the fluctuations in Fig. 16 are not numerical noise, but reflect instead correlations due to polar dislocation development, long-range stress and dislocation transport.

As reported in (Fressengeas et al. 2009), the statistics of intermittency, and in particular the exponent value $\tau \approx 2$, seem to be insensitive to sample size and shape, or to the driving strain rate, to the extent that velocity gradients remain large enough to induce polar dislocation development, however. With unchanged geometry and loading conditions, possible influence of material behavior was also investigated by switching from a thermally activated law in Cu ((40) with material parameters in Table 3) to viscous drag in ice ((31) with material parameters from Table 4) in the simulations. Despite these differences, a scaling regime with exponent $\tau \approx 2$ in the event size distribution is still found. Thus, the scaling behavior of intermittency obtained from the model features a rather universal scaling exponent. In the interpretation suggested by the present model, intermittency of plastic activity requires abruptness of the unpinning transition on short-range obstacles as described through the weakly rate-sensitive stress-velocity relationships (31, 40) at the present scale of resolution. The model implies that both dislocation transport and long-range interactions play a role in the emergence of the scale-invariant behavior of intermittency. As dislocation transport involves such mechanisms as double cross-slip of screw dislocations by-passing short-range obstacles, it follows from this remark that short-range interactions play a significant role in the intermittency of plastic activity. Such a conclusion is fully consistent with the observations of dislocation avalanches arrested on obstacles made in dislocation dynamics simulations (Devincre et al. 2008).

5 Internal Stresses and Anisotropy of Mechanical Behavior

The sharp yield point phenomenon (Piobert 1842; Lüders 1860) primarily occurs in b.c.c. polycrystals at room temperature. In a tensile sample loaded at constant cross-head velocity, it is associated with a band of localized dislocation activity, the so-called Lüders band, traveling along the sample. The band nucleation, usually at one grip, corresponds to a drop in stress, from the Upper Yield Point (UYP) to the Lower Yield Point (LYP). The plastically strained area then spreads along the sample. A clear-cut front separates this area from the unstrained one, into which it propagates, until the sample is uniformly stretched. From this point (referred to as the Lüders strain) onward, the deformation proceeds uniformly in the sample. It is commonly accepted that strain aging is responsible for this behavior: solute atoms tend to diffuse to arrested dislocations, which increases the unpinning stress up to the UYP level. Dislocations are collectively unpinned at the UYP, but since the stress needed to accommodate the imposed strain rate is substantially lower, an abrupt multiplication of dislocations takes place, along with elastic relaxation of the rest of the sample. This unpinning mechanism has intricate connections with the spatial correlations responsible for band propagation. According to the Cottrell assumption (Cottrell 1953), propagation occurs once stress concentration due to dislocation pile-ups at grain boundaries is able to activate new dislocation sources in neighboring grains. The long-range internal stresses due to incompatibilities in plastic strain in the vicinity of the band provide the mechanism for band propagation.

In a low-carbon steel, evidence of the role of internal stresses in the unpinning mechanism is also provided by the directionality of the yield point. If such a material is deformed beyond the Lüders strain, then aged and further strained, a sharp yield point phenomenon reappears provided straining is pursued in the same direction. Such a behavior can be explained within the framework of a local model coupling aging properties with isotropic strain hardening (Kubin et al. 1992). However, if the sample is strained in the direction opposite to that before aging, a Bauschinger effect is observed and the sharp yield point phenomenon is usually absent (Tipper 1952; Wilson and Ogram 1968; Elliott et al. 2004). This phenomenon is shown in the tension–compression of a mild steel in Fig. 18. Such directionality of the yield point is of considerable practical importance. It may be useful, as it curbs the return of the sharp yield point in temper-rolled or bake-hardened steels, but it may also limit the benefits of strain aging as a strengthening mechanism. Further, it demonstrates that the strain aging and unpinning mechanisms are dependent on the gradients of the distribution of dislocations, which challenges local interpretations. In this chapter, an interpretation of the yield point directionality is presented by

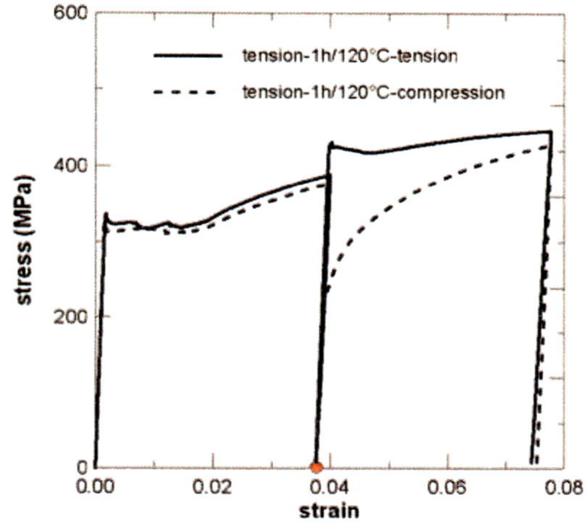

Fig. 18 Stress–strain curves during tension–aging–tension and tension–aging–compression experiments on a 1020 mild steel. The aging period is marked with a solid circle. For convenience the sign of the compression stress is reversed. Both tension and compression stresses are plotted against the same cumulative strain

using the framework of a field dislocation dynamics model. We strive to understand this phenomenon by coupling the evolution of polar and statistical dislocations with the kinetics of strain aging (Taupin et al. 2008).

The model for strain aging is developed and applied to pure torsion to set up the simplest possible configuration. In parallel, the model is also used in a three-dimensional approach, with some minor variations to be indicated below. In using the pure torsion model, the goal is to provide heuristic modeling, fit for the illustration of the critical aspects of the approach, i.e., the connections between strain aging and dislocation microstructures. The unique active slip plane, $(\mathbf{e_1}, \mathbf{e_2})$, is assumed to be normal to the torsion axis. It is also assumed that plastic strain is accommodated by "circumferential" screw dislocations of infinite extent in the $\mathbf{e_1}$ direction only. For a positive torque, positive screw dislocations move from the edge of the sample toward its axis, with velocity v_2 along the radial direction $\mathbf{e_2}$. A radial screw density along the $\mathbf{e_2}$ direction does exist, to verify equilibrium (Weertman 2002), but it does not contribute to torsion accommodation. Hence, it is ignored all together. The basic equations then reduce to

$$\dot{\sigma}_{13} = \mu(v_{1,3} - \rho_m b v - \alpha_{11} v_2) \tag{41}$$

along with (32, 33). Equation (41) expresses the shear stress rate in the glide plane as a function of polar screw and statistical dislocation mobility, with μ denoting the elastic shear modulus. Account of the physics of dislocation velocity, strain aging and straining history is now made through phenomenological statements. An Arrhenius dependence is assumed for the polar and statistical dislocation velocities (v_2, v) in the form

$$v_2 = v = V_0 \exp((|\sigma_{13}| - \sigma_\mu \mathrm{sgn}(\sigma_{13}) - \sigma_h - \sigma_s)/S_0). \tag{42}$$

Here, V_0 is a reference velocity, S_0 denotes the strain rate sensitivity of the flow stress in the absence of solute effects, and the numerator in the exponential represents an effective stress for dislocation glide. The reference velocity is taken in the form $V_0 = \nu_0 \rho_f^{-1/2}$, where ν_0 is a constant reference frequency for dislocation unpinning. Hence, the current waiting time $t_w = \rho_f^{-1/2}/\nu$ of dislocations on their obstacles is

$$t_w = \nu_0^{-1} \exp(-(|\sigma_{13}| - \sigma_\mu \mathrm{sgn}(\sigma_{13}) - \sigma_h - \sigma_s)/S_0) \tag{43}$$

In this relation, ν_0^{-1} appears as the waiting time under zero effective stress. Further, during elasto-plastic loading, t_w evolves as a function of stress, from its initial (large) value to its current, much smaller value. σ_μ is the back-stress (or internal stress), σ_h represents statistical hardening and σ_s is the additional stress due to solute hardening. The back-stress is induced by the polar screw density, with rate of formation provided by (35). Although it is an offset for the assumed invariance in the other two directions, the phenomenological treatment of the internal stresses in the 1-D model provides a result of general utility, i.e., specific insight into the constitutive specification of back-stress evolution, to be contrasted with the Armstrong–Frederick kinematic hardening specification (Armstrong and Frederick 1966). Isotropic statistical hardening is assumed in the Taylor form: $\sigma_h = \bar{\alpha}\mu b\sqrt{\rho_f}$, where $\bar{\alpha}$ is a constant. The evolution of statistical mobile and forest densities follows the Kubin–Estrin model (Kubin and Estrin 1990)

$$\dot{\rho}_m = ((C_1/b^2) - (C_3/b)\sqrt{\rho_f}\,)\dot{\Gamma} \tag{44}$$

$$\dot{\rho}_f = ((C_0/b) \mid \underline{\alpha} \mid +(C_3/b)\sqrt{\rho_f} - C_4\rho_f)\dot{\Gamma}, \tag{45}$$

where C_1 stands for the multiplication of dislocation line, C_3 represents mobile dislocation immobilization, and C_4 dynamic recovery. Already introduced in Eq. (39), the term C_0 added to the model stands for the contribution to statistical hardening of polar dislocations, through pile-ups at grain boundaries (Acharya and Beaudoin 2000). Following (Louat 1981), the additional stress due to aging is expressed as

$$\sigma_s = f_0(1 - \exp(-(t_a/\tau)^{2/3})), \tag{46}$$

where t_a denotes the aging time, τ is a characteristic time for solute diffusivity, and f_0 represents the maximum pinning stress. The exponent 2/3 stands for bulk diffusion, but other types of pinning kinetics could be considered as well (Kubin et al. 1992). The evolution of the aging time follows that of the waiting time with some delay, because solute concentration cannot change instantly. Following (McCormick 1988), these circumstances are described by allowing the aging time to relax to the current waiting time according to the first-order kinetics

$$\dot{t}_a = 1 - t_a/t_w \tag{47}$$

Computation of the solutions to (41–47) uses numerical constants provided in Table 4 (see justifications in Taupin et al. 2008). The 3-D simulations also use the statistical and solute hardening modeling shown above in (44–47), though written in terms of slip system strength and resolved shear stress. Slip system geometry is taken for bcc crystal symmetry. As development of the back-stress is inherent in the 3-D calculations, the form of the kinetic relation, (43), is also different; a constant athermal strength replaces the signed back-stress σ_μ and a constant reference velocity v_0 is used. The boundary conditions are written in terms of dislocation fluxes. Inward flux of dislocations is not permitted, although dislocation sources on boundaries are allowed. Outgoing of dislocations is permitted without constraint. The initial conditions are summed up in Table 5. Note that the initial distribution of dislocations is chosen to be uniform, with no polar dislocations. The initial aging time reflects saturation of dislocations with solute atoms.

Tension–compression tests were simulated using the full 3-D model. A polycrystal of dimensions $2 \times 2 \times 10$ mm was clamped to the left end, while the right end was submitted to constant velocity. The sample was first deformed in tension until the strain reaches 0.04, then unloaded. Each element was assigned a single crystallographic orientation taken from the sampling of a uniform distribution. The results are shown in Figs. 19, 20.

A UYP associated with dislocation unpinning and multiplication is seen in Fig. 19, followed by a Lüders plateau corresponding to the propagation of a band in equivalent plastic strain rate from the left end to the right end of the sample. A realistic rendering of the band is achieved, as shown in Fig. 20. In this figure, the band is at angle with both the axial direction and the transverse direction, consistent with experimental observations in CuAl and CuMn single crystals (Ziegenbein et al. 1995). The figure also shows a trailing amount of residual strain rate left behind the band. When the band reaches the right end, the sample is uniformly stretched at

Table 5 Initial conditions

ρ_m	ρ_f	α_{11}	σ_{13}	t_a
$10^{12}\,\mathrm{m}^{-2}$	$10^{11}\,\mathrm{m}^{-2}$	0	0	2×10^6 s

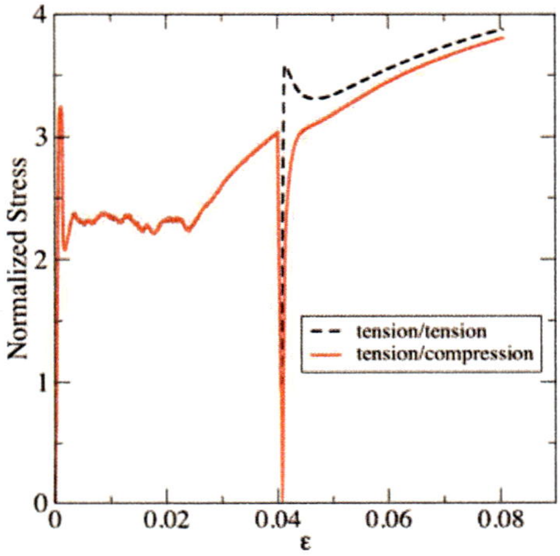

Fig. 19 Stress (normalized by the initial slip system strength) vs. strain curves in tension–aging–forward tension and tension–aging–compression 3-D simulations. The applied strain rate is $6 \times 10^{-5}\,\mathrm{s}^{-1}$. For convenience the stress sign is reversed in compression

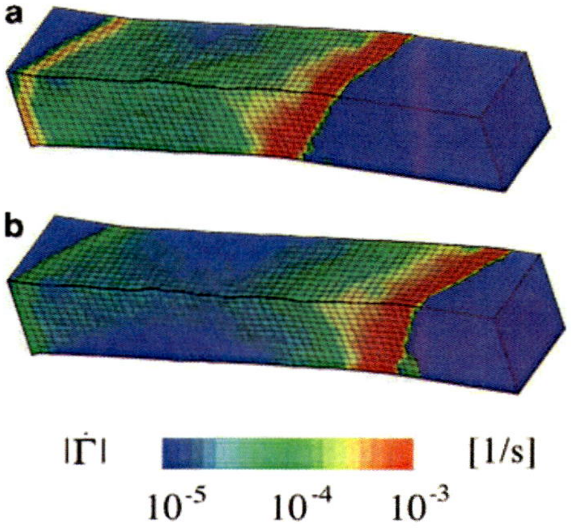

Fig. 20 Strain-rate contours for the Lüders band corresponding to the stress plateau in Fig. 19 at two instants A, B. After nucleation to the left, the band propagates to the right end of the sample. Note the inclination of the band, at angle with both the axial and the transverse directions

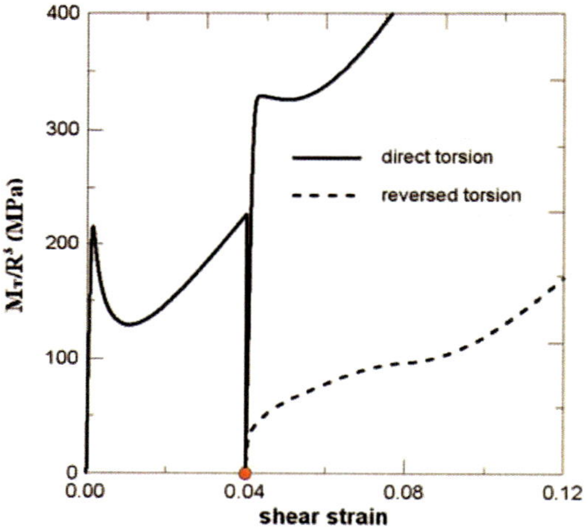

Fig. 21 Torque evolution during forward torsion–aging–forward torsion and forward torsion–aging–reverse torsion simulations. The aging period is marked with a solid circle. For convenience the torque sign is reversed in reverse torsion. Both forward and reverse torque are plotted against the same cumulative strain

the Lüders strain, and deformation proceeds uniformly. At unloading, aging of the material is carried out by augmenting the aging time. When from this point the sample is loaded forward in tension, restoration of a UYP is predicted, though strain localization is hardly seen, whereas the UYP is missing if compression is applied. In the latter case, a Bauschinger effect is also predicted, as well as a transient inflexion in strain hardening. Figure 19 shows qualitative agreement with the experimental trend in mild steel seen in Fig. 18.

We now proceed with simulations using the 1-D model to reveal the interplay between evolution of polar density, back-stress, and aging. A thin walled tube is first strained in forward torsion with a positive torque until the shear strain reaches 0.04, then unloaded. A UYP is obtained, as can be seen in Fig. 21. Since axial invariance is assumed and only dislocation transport in the plane normal to the torsion axis is considered, Fig. 21 does not feature a Lüders plateau. Indeed, the latter goes with axial band propagation. Dislocation unpinning and multiplication shift from the

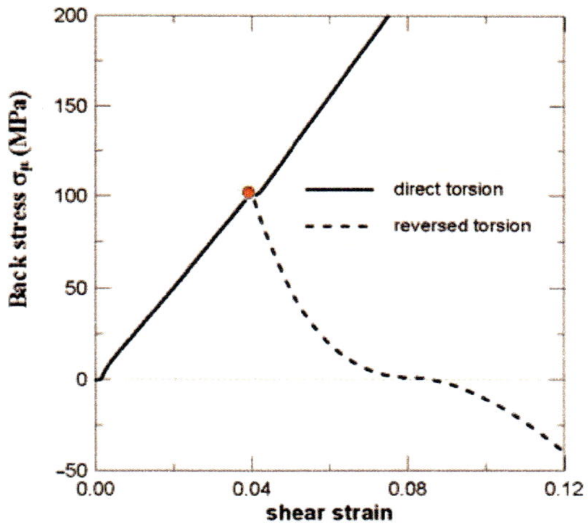

Fig. 22 Back-stress evolution during forward torsion–aging–forward torsion and forward torsion–aging–reverse torsion simulations. The aging period is marked with a solid circle

outer edge, where the stress is high, to the interior where it lessens, which generates gradients in plastic strain rate. These gradients act as sources for polar dislocations. As the torque is positive, positive screws are nucleated, and a back-stress associated with these dislocations builds up. At unloading, aging of the material is performed as detailed above in 3-D computations. When, from this point, the sample is loaded in forward torsion, restoration of a UYP is again predicted, whereas it is absent if the sense of torsion is reversed. A strong Bauschinger effect is also seen in that case. The interpretation derives from the evolution of the polar screw density and back-stress shown in Fig. 22. Indeed, the back-stress opposes unpinning of aged dislocations in forward torsion because it lowers the effective stress and the dislocation velocity. Thus, a UYP is needed for unpinning, but the drop in stress is reduced with respect to its first occurrence due to the concomitant reduction in the multiplication rate. In contrast, the back-stress favors unpinning and dislocation multiplication in reverse torsion, because it enhances the effective stress and the dislocation velocity. Hence, a UYP now becomes unnecessary for unpinning.

Dislocation rearrangement and back-stress relaxation eventually play a pivotal role. During reverse torsion, gradients in the plastic distortion generate negative polar screw dislocations, which annihilate with positive screws formed in forward torsion. Hence, the total screw density and the associated back-stress drop down to zero. This drop in back-stress gradually hardens the material because it limits the dislocation velocity as well as the rate of multiplication of mobile dislocations. The dependence of the relaxation time τ_r on the dislocation velocity is such that back-stress relaxation is initially fast, and that it slows down as it comes to an end. If reverse torsion is further pursued to larger strains, a new structure of negative polar screw dislocations develops and an inflexion of the (now negative) back-stress is obtained (Fig. 22). This inflexion transfers to the torque history, which therefore shows transient curbing of strain hardening after path reversal (Fig. 21). As mentioned earlier, this feature is also apparent in Fig. 19 from tension–aging–compression 3-D simulations. It is consistent with observations of a similar transient behavior in polycrystalline aluminum samples at various temperatures (Hasegawa et al. 1975) and in IF-steels when the strain path is reversed (Peeters et al. 2000). The microstructural analyses and qualitative arguments provided in these references, i.e., the annihilation of the structure of polar dislocations formed in pre-deformation and the development of a new structure along the eventual strain path, are in full agreement with our own interpretations.

6 Conclusions

Conventional plasticity is rather ineffective in the understanding of the directionality of the yield point, because it does not deal with long-range internal stresses. Arguably, conventional kinematic hardening as expressed by the Armstrong–Frederick law for back-stress evolution has an ability to phenomenologically describe anisotropic hardening. However, spatial correlations due to the lattice distortion and internal stress fields are missed, which must have implications on back-stress build up. In contrast to standard plasticity treatment, the 3-D field dislocation model provides for polar dislocation microstructure and internal stress field development, and it naturally features induced anisotropic hardening. The formulation of the heuristic 1-D model provides insight into a more adequate specification of back-stress evolution in terms of polar dislocations through the constitutive equation (35). Phenomena related to strain path dependence of strain hardening (such as the inflexion in strain hardening after path reversal in strain-aged steels or the inflexion of creep of ice single crystals in reverse torsion) can only be retrieved from the nucleation, transport, and annihilation of polarized dislocation distributions. Conventional local treatment using statistical dislocations, whose multiplication mechanisms are unaffected by strain path orientation reproduce such phenomena only by having recourse to multiple dislocation species and algorithmic rules (Peeters et al. 2002). This model also deals separately with dislocation density and dislocation velocity, which, in contrast, are merged into plastic strain-rate in standard plasticity treatment. This feature proves to be useful in the understanding of the directionality of hardening, because back-stress relaxation effects on dislocation velocity and dislocation density appear to be very much involved. It is consistent with common material science practice, which provides material data in terms of dislocation velocity and density.

The prediction of the propagation of plastic fronts has been a long-time goal of the theory of plasticity. However, its realization has been hampered by the mathematical structure of the traditional approach. Within the latter, propagating fronts of plastic deformation can be obtained only in the presence of spatial inhomogeneity in material properties (see for example, Kok et al. 2003, for the propagation of Portevin–Le Chatelier (PLC) bands in polycrystals), or when spatial coupling due to 3-D stress field equilibrium is strong enough (see Hutchinson and Neale 1983) for the study of neck propagation in polymers). In contrast, in the presence of spatially homogeneous material characteristics or when the 3-D character of the stress field is weak (in 1-D situations or in flat samples), the capability of the field dislocation dynamics framework in representing propagating plastic fronts derives from the description of transport through partial differential equations. Then, inhomogeneity in boundary conditions may or may not trigger the propagation of plastic fronts depending upon material behavior, as illustrated by a study of PLC band propagation in single crystals (Varadhan et al. 2009). Because they feature correlations in space due to both the long-range internal stresses and the short-range interactions involved in dislocation transport, field dislocation dynamics theories also provide interpretation for the scale-invariant intermittency of dislocation transport. Size effects derive as well from the presence of material length scales linked with long-range stresses and dislocation transport. Other features such as dislocation wall and grain boundary formation were also predicted using this theoretical structure, although with a different numerical implementation (Limkumnerd and Sethna 2008).

In the formulation of a mesoscopic theory, the linear dimension of the support $\Omega(\mathbf{x})$ for spatial averaging (see relation (13)) is arbitrarily chosen. Hence, flexibility is left in the order of magnitude of the spatial resolution sought for the model. Depending on the objectives of modeling, the formulation can span from a physical theory with a short resolution length scale, where only a small number of dislocations is involved in the averaging procedure, to engineering codes applying to large scale systems, where blurring of dislocation ensembles needs to be more extensive due to computational costs. Hence, as mentioned earlier, phenomena deemed "nonlocal" in the former approach may be labeled "local" in the latter. However, if the resolution length scale is kept small enough and can be compared with the linear characteristic length of the relevant dislocation microstructures, length-scale dependence of the results can be retained.

Acknowledgments AA gratefully acknowledges the support from the National Science Foundation through the CMU MRSEC, grant no. DMR-0520425, the LMA-CNRS, Marseille and the Dept. of Civil and Env. Engineering at CMU. AB received support under US Dept. of Energy grant DEFG03-02-NA00072 and the Center for Simulation of Advanced Rockets at the University of Illinois at Urbana-Champaign (UIUC), US DOE subcontract B341494. AB and CF benefited from exchanges under a joint agreement between Centre National de la Recherche Scientifique and UIUC. We thank Juliette Chevy for providing Figs. 4, 8 and Russell J. McDonald for his help in carrying out the experiments in Fig. 18.

References

L.P. Kubin, C. Fressengeas and G. Ananthakrishna, *Collective Behaviour of Dislocations in Plasticity*, in Dislocations in Solids, vol 11, Eds. F.R.N. Nabarro and M.S. Duesbery, Elsevier Science B.V., 100–192 (2002).

M. Legros, A. Jacques and A. George, Mat. Sci. Eng. A **387–389**, 495 (2004).

M. Neubert and P. Rudolph, Prog. Cryst. Growth Charact. Matr. **43**, 119 (2001).

M. Zaiser, Adv. Phys. **55**, 185 (2006).

E.C. Aifantis, Mat. Sci. Eng. **81**, 563 (1986).

N.A. Fleck, G.M. Muller, M.F. Ashby and J.W. Hutchinson, Acta Metall. Mater. **42**, 475 (1994).

W.D. Nix and H. Gao, J. Mech. Phys. Solids, **46**, 411 (1998).
F. Forest, G. Cailletaud and R. Sievert, Arch. Mech. **49**, 705 (1997).
A. Acharya, J. Mech. Phys. Solids **49**, 761 (2001).
J.F. Nye, Acta Metall. **1**, 153 (1953).
E. Kröner, Erg. Angew. Math. **5**, 1–179 (1958).
T. Mura, Phil. Mag. **89**, 843 (1963).
A.M. Kosevich, *Crystal dislocations and the theory of elasticity*, in Dislocations in Solids, Ed. F.R.N. Nabarro, North-Holland, Amsterdam, 33–141 (1979).
E. Kröner, *Continuum theory of defects*, in Physics of Defects, Ed. R. Balian et al., North-Holland, Amsterdam, 218–314 (1980).
A. Acharya and A. Roy, J. Mech. Phys. Sol. **54**, 1687 (2006).
A. Roy and A. Acharya, J. Mech. Phys. Sol. **53**, 43–170 (2005).
A. Roy and A. Acharya, J. Mech. Phys. Sol. **54**, 1711–1743 (2006).
S. Varadhan, A.J. Beaudoin, A. Acharya and C. Fressengeas, Modelling Simul. Mater. Sci. Eng. **14**, 1 (2006).
V. Taupin, S. Varadhan, J. Chevy, C. Fressengeas, A.J. Beaudoin, M. Montagnat and P. Duval, Phys. Rev. Lett. **99**, 155507 (2007).
R. Becker and E. Orowan, Z. Phys. **79**, 566 (1932).
J. Weiss and J.R. Grasso, J. Phys. Chem. **101**, 6113 (1997).
M.C. Miguel, A. Vespignani, S. Zapperi, J. Weiss and J.R. Grasso, Nature (London) **410**, 667 (2001).
D.M. Dimiduk, C. Woodward, R. LeSar and M.D. Uchic, Science **312**, 1188 (2006).
S. Brinckmann, J.Y. Kim and J.R. Greer, Phys. Rev. Lett. **100**, 155502 (2008).
J. Weiss, T. Richeton, F. Louchet, F. Chmelik, P. Dobron, D. Entemeyer, M. Lebyodkin, T. Lebedkina, C. Fressengeas and R.J. McDonald, Phys. Rev. B. **76**, 224110 (2007).
C. Fressengeas, A.J. Beaudoin, D. Entemeyer, T.Lebedkina, M. Lebyodkin, and V. Taupin, Phys. Rev. B, **79**, 014108 (2009).
V. Taupin, S. Varadhan, C. Fressengeas and A.J. Beaudoin, Acta Mater. **56**, 3002 (2008).
M. Koslowski, R. LeSar and R. Thomson, Phys. Rev. Lett. **93**, 125502 (2004).
F.F. Csikor, C. Motz, D. Weygand, M. Zaiser and S. Zapperi, Science **318**, 251 (2007).
B. Devincre, T. Hoc and L.P. Kubin, Science **320**, 1745 (2008).
M. Babic, Int. J. Eng. Sci. 35, 523–548 (1997).
A. Acharya, Proc. Roy. Soc. **A 459**, 1343 (2003).
M.J. Marcinkowsky, Phys. Stat. Sol. B **152**, 9 (1989).
C. Meneveau and J. O'Neil, Phys. Rev. E **49**, 2866 (1994).
P. Duval, M.F. Ashby and I. Anderman, J. Phys. Chem. **87**, 4066 (1983).
C. Shearwood and R.W. Whitworth, Phil. Mag. **64**, 289 (1991).
J. Chevy, Ph. D. Thesis, Institut Polytechnique de Grenoble (2008).
M. Montagnat, P. Duval, P. Bastie and B. Hamelin, Scripta Mater. **49**, 411 (2003).
J. Chevy, C. Fressengeas, M. Lebyodkin, V. Taupin, P. Bastie, P. Duval, Acta Mater. **58**, 1837 (2010).
M. Montagnat, J. Weiss, J. Chevy, P. Duval, H. Brunjail, P. Bastie and J. Gil Sevillano, Phil. Mag. **86**, 4259 (2006).
P.J. Armstrong and C.O. Frederick, *A mathematical representation of the multiaxial Bauschinger effect*, Technical Report RD/B/N/731, Central Electricity Generating Board (1966).
F. Louchet, C.R. Physique, **5**, 687 (2004).
S.J. Basinski, Z.S. Basinski, *Plastic deformation and work hardening*, in Dislocations in Solids, Ed. F.R.N. Nabarro, Vol 4, 261–362, North-Holland, Amsterdam (1979).
J. Diehl, **47**, 331–343 (1956).
H. Suzuki, S. Ikeda and S. Takeuchi, J. Phys. Soc. Jpn. **11**, 382 (1956).
J. Garstone, R.W.K. Honeycombe and G. Greetham, Acta Met. **4**, 485 (1956).
J.T. Fourie, Phil. Mag. **15**, 187 (1967).
J. Weertman, Acta Mater. **50**, 673 (2002).
L.B. Zuev, Ann. Phys. **16**, 286–310 (2007).

A. Acharya, A.J. Beaudoin and R. Miller, Math. Mech. Solids **13**, 292 (2008).
S. Varadhan, A.J. Beaudoin and C. Fressengeas, Proc. of Science, SMPRI2005, 004 (2006).
O. Nittono, Jap. J. Appl. Phys. **10**, 188 (1971).
S.S. Brenner, J. Appl. Phys. **28**, 1023 (1957).
Y. Gotoh, Phys. Stat. Sol. A **24**, 305 (1974).
A. Piobert, Mémoires de l'artillerie **5**, 502 (1842).
W. Lüders, Dinglers Polytech. J. **155**, 18 (1860).
A.H. Cottrell, *Dislocations and Plastic Flow in Crystals*, University Press, Oxford, (1953).
L.P. Kubin, Y. Estrin and C. Perrier, Acta Metall. Mater. **40**, 1037 (1992).
C.F. Tipper, J. Iron Steel Inst. **2**, 143 (1952).
D.V. Wilson and G.R. Ogram, J. Iron Steel Inst. 911–920 (1968).
R.A. Elliott, E. Orowan, T. Udoguchi, A.S. Argon, Mech. Mat. **36**, 1143 (2004).
L.P. Kubin and Y. Estrin, Acta Metall. Mater. **38**, 697 (1990).
A. Acharya and A.J. Beaudoin, J. Mech. Phys. Sol. **48**, 2213 (2000).
N. Louat, Scripta Metall. **15**, 1167 (1981).
P.G. McCormick, Acta Metall. **36**, 3061 (1988).
A. Ziegenbein, Ch. Achmus, J. Plessing and H. Neuhäuser, in *Plastic and Fracture Instabilities in Materials*, **200**, 101, ASME-AMD (1995).
T. Hasegawa, T. Yakou and S. Karashima, Mat. Sci. Eng. **20**, 267 (1975).
B. Peeters, S.R. Kalidindi, P. Van Houtte, E. Aernoudt, Acta Mater. **48**, 2123 (2000).
B. Peeters, S.R. Kalidindi, C. Teodosiu, P. Van Houtte, E. Aernoudt, J. Mech. Phys. Sol. **50**, 783 (2002).
S. Kok, M.S. Bharathi, A.J. Beaudoin, C. Fressengeas, G. Ananthakrishna, L.P. Kubin, M. Lebyodkin, Acta Mater. **51** 3651–3662 (2003).
J.W. Hutchinson and K. Neale, J. Mech. Phys. Sol. **31**, 405–426 (1983).
S. Varadhan, A.J. Beaudoin and C. Fressengeas, J. Mech. Phys. Sol. **57**, 1733–1748 (2009).
S. Limkumnerd and J.P. Sethna, J. Mech. Phys. Sol. **56**, 1450 (2008).

Dislocation-Mediated Time-Dependent Deformation in Crystalline Solids

Michael Mills and Glenn Daehn

Abstract The time-dependent plastic deformation of crystalline solids has been the subject of much academic and practical interest for at least 100 years. Many studies have emphasized the phenomenological quantitative macroscopic relationships between stress, time, and strain rate, while other studies have focused on the dislocation structures and microstructures that develop while deforming over time under stress at elevated temperature. This review attempts to unify these two, largely separate schools of thought by using microstructural information to develop simple but broad quantitative mechanistic relationships that match the observed phenomenology. Dislocation processes that plastically deform crystals are modeled. Grain boundary sliding and related processes are not considered for simplicity and clarity. Two common classes of deformation are recognized. In mobility-controlled systems, dislocations move through the crystal under stress as controlled by their mobility. This may be the result of a frictional interaction with the lattice; interactions with mobile solute species or the diffusion-controlled motion of jog segments on screw dislocations. The other broad class is based on the interaction of dislocations with discrete obstacles. This is argued to control the deformation of a range of alloys spanning pure metals, many engineering alloys, to oxide dispersion strengthened metals. It is the stability of the obstacles against recovery and coarsening that is the key difference between these materials. The unifying theme in both materials classes is that dislocation-level mechanics is directly used to derive equations for creep.

1 Introduction

The time-dependent plastic deformation of crystalline solids has been studied for many years by a large number of researchers. It has importance as a topic for pure scientific study and practical significance in the manufacture of metal products as well as in life prediction and failure studies at low and high temperatures.

M. Mills (✉)
Department of Materials Science and Engineering, The Ohio State University, Columbus, OH 43210, USA
e-mail: mills.108@osu.edu

S. Ghosh and D. Dimiduk (eds.), *Computational Methods for Microstructure-Property Relationships*, DOI 10.1007/978-1-4419-0643-4_9,

The approach in this work is to describe the most general models for the physics of deformation within individual crystals. Crystals most commonly deform by the motion of dislocations. In short, materials are made stronger by hindering the motion of dislocations. In fact, there are a very limited number of ways to hinder dislocation motion. First, discrete obstacles may exist in the dislocation slip plane. Continued motion of the dislocation lines requires that the obstacles are either overcome or the line loops around them. Alternatively, the dislocations may have an intrinsic resistance. The Peierls stress represents an intrinsic lattice resistance at low temperatures. At elevated temperatures, diffusional processes may be required for continued dislocation motion. By limiting the scope of this chapter to dislocation-dominated mechanisms, we are intentionally ignoring several localized deformation processes such as twinning and grain boundary sliding, and the related phenomenon of diffusional creep.

Here, we hope to provide relatively general models that are based on simple representations of physically defendable and measurable, rate-limiting processes. Rather than presenting models that are "tuned" for specific materials, the intent is to elucidate the general approaches available and provide a scientific basis for modeling of both mobility-controlled and obstacle-controlled metals.

2 Broad Phenomenology and Commonality in Creep and Plasticity

The goal of this work is to present consider plastic deformation based as closely as possible directly on the motion of dislocations through a crystal. The strength and time dependence are then related to the resistance to dislocation motion, which can come in one of twoforms: (1) discrete and relatively immobile obstacles, such as precipitates or the stress fields or junctions with other dislocations, or (2) the intrinsic resistance provided by the lattice, either through a Peierls stress or a chemical interaction such as provided by local ordering or the development of a solute atmosphere around the dislocation. In the first idealized case, dislocations are thought to spend most of their time immobilized at obstacles, but they will sometimes break free quickly moving to the next obstacle. In the latter case, dislocations are often moving through the lattice at a finite speed that is a function of stress and temperature. The resistance to motion is provided by the lattice itself, by interactions with mobile solutes or by the diffusion-controlled motion of the dislocation itself.

In the case pure metals, the preponderance of data shows clear trends with respect to strain and strain-rate hardening over a range of temperatures. These are often characterized by the strain-hardening exponent, N, and strain rate-sensitivity exponent, m. There are orderly trends in these parameters that are roughly obeyed for pure polycrystalline metals over a wide range of temperatures. Figure 1 schematically depicts the strain-hardening exponent and strain rate-sensitivity exponent plotted as a function of temperature for pure metals. The changes in these characteristic

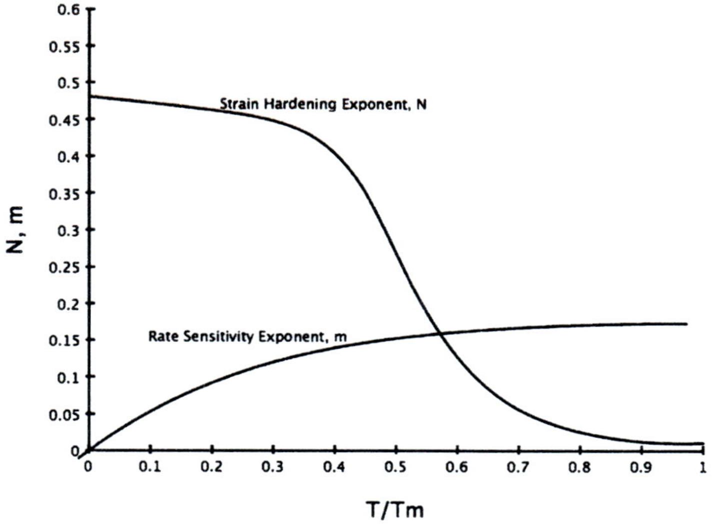

Fig. 1 Schematic representation of the trend approximately seen for the strain rate-sensitivity and strain-hardening exponents for pure metals over the full range of homologous temperatures. At the highest temperatures, strain hardening vanishes and a limiting value of rate hardening of about 0.2 is reached. At very low temperatures, the rate-sensitivity exponent rises nearly linearly with temperature. For source data, see for example, Luhban and Felgar (1961)

exponents are continuous, not showing the kinds of abrupt changes we might expect if mechanisms vary distinctly with temperature. The rate sensitivity generally increases nearly linearly with temperature at low temperature and comes to a plateau at a value of about 0.2 beyond about 0.7 of the absolute melting temperature, T_m. The strain-hardening exponent monotonically decreases with temperature, dropping rapidly over the range, where dynamic recovery becomes important, and if a material is undergoing true "steady-state" creep (Sherby et al. 1977; Sherby and Burke 1967), there is no strain hardening, by definition. The modeling presented in Sect. 4 can capture the salient features of deformation over the entire range of temperature from low to high without resort to multiple mechanisms. Wilshire (Wilshire and Scharning 2008; Wilshire et al. 2009) has recently argued the point that the creep stress exponent naturally is a weak function of stress and temperature, and that throughout for most structural metals the "mechanism" remains the same; the motion of dislocations through the structure and over or through obstacles. This theme is continued in the present development of obstacle-controlled creep.

To keep the scope of this chapter manageable, creep deformation at temperatures exceeding 0.5 T_m will be emphasized, but the arguments and models presented for obstacle-controlled creep are also sufficiently broad and flexible that they can be adapted to the full range of temperature from zero to melting. The salient features of creep of pure metals have been known for some time and are clearly described in the classic paper by Sherby and Burke (1967). They are:

1. The activation energy for creep is similar to that for pure metals.
2. The stress exponent is commonly near 5 over an intermediate range of stress. At low stresses, this exponent can decrease to a value near one (and this is often referred to as the Harper–Dorn creep mechanism) and at high stresses, the stress exponent often increases in what is often termed power–law breakdown.
3. There is a remarkable scaling property where if strain rate normalized by diffusion is plotted versus stress normalized by elastic modulus, the experimental data for a wide range of metals collapses to nearly a single trend, as shown in Fig. 2 (Sherby 1962). The variation in terms of what makes some metals "stronger" in this normalized way may be partly explained by staking fault energy. It has been

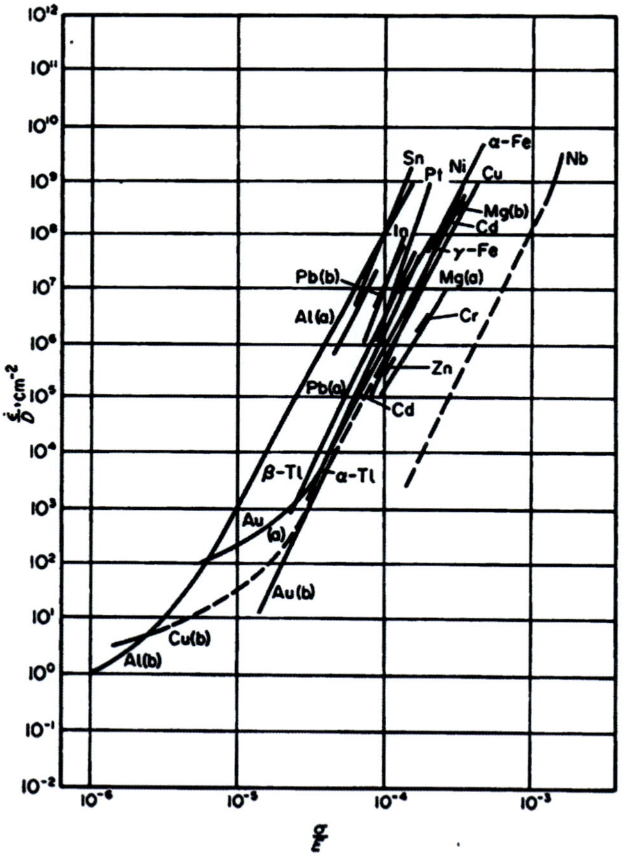

Fig. 2 Normalized plot of creep rate normalized by self-diffusivity as a function of applied stress normalized by elastic modulus from Sherby (Sherby and Burke 1967). The broad sweep of similar data suggests very similar mechanisms control the deformation of a wide range of pure metals

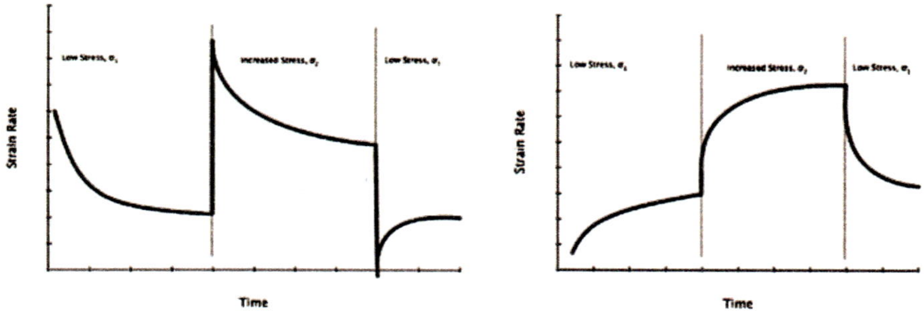

Fig. 3 Schematic representation of transients in strain rate in stress change experiments in metals at elevated temperature. Class II or pure metal type behavior is shown on the *left*, and alloy-type behavior is shown on the *right*. The changes in strain-rate after a stress change give insight into the "constant structure" behavior of the material. Continued strain and time modify the dislocation structure, and this must be accounted for in a complete theory of creep

shown that single-phase metals with low-stacking fault energies generally have better creep resistance (Barrett and Sherby 1965).

4. Upon stress changes in pure metals there is a characteristic transient wherein upon a stress increase the strain rate immediately after the change is much greater than the steady-state increase in strain rate. Upon a stress decrease, this "constant structure stress-exponent" is also much greater than that in "steady-state." The opposite trend is observed in metals that are solid-solution strengthened. This behavior is well known. The alloy behavior has been referred to Class I behavior by Sherby and Burke (1967), while pure metal behavior is referred to as Class II. This is illustrated in Fig. 3.

While the forgoing discussion is based on pure metals, engineering metals where the microstructure has been manipulated by adding strength-giving second phases, follow similar trends in creep with some important differences. These differences include:

1. The activation energy for creep is often significantly larger than that for self-diffusion.
2. The stress exponents are often much greater than 5, except for strong solute strengthening, which can result in stress exponents closer to 3.
3. The material strengths are significantly greater than those for pure metals.
4. The form of the transients is often unavailable, or has not been reported for many engineering metals.

Both pure metals and engineering materials can be scaled similarly using modulus and activation energy as normalizing parameters and relating strain rate to a power–law of stress. This is the basis behind what is often referred to as the "Dorn equation" (Dorn 1975), which can be written in its most simple form as:

$$\dot{\gamma} = k\left(\frac{\tau}{\mu}\right)^{n} \exp\left(\frac{Q_{\mathrm{c}}}{RT}\right), \tag{1}$$

where $\dot{\gamma}$ represents shear plastic strain rate. The materials parameters are the free constants: k, the shear modulus, μ, the stress exponent, n, and the creep activation energy Q_c. Temperature and the gas constant are represented by T and R, respectively, and τ represents the shear stress applied to the sample. This equation is ultimately based on phenomenological fitting of data to a form rather than being the result of any theory. It has been remarkably successful and guides the way creep is thought of. Also, it is routinely modified to add elements, such as threshold stresses and backstresses, to allow fitting to a wider array of data by adding adjustable parameters. There is not a strong theoretical basis for these modifications, but they are often justified with some theoretical ideas. Creep mechanisms are often thought of as being ranges of behavior where a fixed value of n and Q_c can be identified – and implicitly a mechanism is often thought of as field in temperature and stress space where n and Q_c are nearly fixed. The book "Deformation Mechanism Alloys: The Plasticity and Creep of Metals and Ceramics" (Frost and Ashby 1982) nicely summarizes this school of thought. We wish to concur with the argument put forth by Wilshire that this focus on n and Q_c fixing a creep mechanism has sometimes impaired our ability to develop a proper and comprehensive understanding of creep.

In this chapter, we will present another very simple framework for obstacle–controlled creep, where the governing assumption is that dislocations will move when they can overcome the resistance provided by the obstacle field. As deformation takes place, further obstacles, in the form of dislocation density, are added and this provides further resistance.

In mobility-controlled deformation, the dislocation velocity is inhibited by lattice friction, the drag of solutes with the dislocations, the drag of jogs along screw segments, the restoration of order or climb. As a consequence, dislocation multiplication may occur over extended time/strain, leading to gradually increasing creep rates (for constant stress conditions) or decreasing flow stress (for constant strain rate conditions). In the limit of true mobility control, mobile segments can all move in a similar way and their collective motion provides the strain. The Orowan equation provides the key governing relation:

$$\dot{\gamma} = \rho_m b v, \tag{2}$$

where v is the average dislocation velocity and ρ_m is the mobile dislocation density, which has often been related to the applied stress through the Taylor relation:

$$\tau = \alpha \mu b \sqrt{\rho} \tag{3}$$

even for cases in which the predominant strengthening process is not from dislocation interactions. A detailed discussion of this mode of deformation is the subject of the next section.

3 Mobility-Controlled Dislocation Creep

Modeling of mobility-controlled deformation conventionally has been based upon the Orowan equation. Importantly, implicit assumptions in utilizing this approach are that (a) an "average" dislocation velocity can be used to characterize the mobile dislocation content and (b) grain-to-grain load-shedding effects in polycrystals are negligible. The latter assumption may be a reasonable assumption for FCC metals, but is indeed questionable for BCC and especially HCP metals which may be more anisotropic in their flow properties as a function of crystal (i.e., grain) orientation. The power of this approach is that it is possible in principle to directly measure the modelparameters to calibrate their values and their dependence upon applied stress (or strain rate) and strain.

Several examples for which this approach has been used with some success will now be described. This will include discussion of thermally activated deformation controlled by lattice friction, climb and solute interactions. It is noted that there are several excellent recent reviews treating several of these topics (Argon 2008; Caillard and Martin 2005), and an emphasis is placed here on the important dislocation mobility modes that have not been as thoroughly treated in these reviews.

3.1 *Peierls Friction-Controlled Deformation*

For covalently bonded solids, dislocations move from one Peierls valley to the next by propagation of kink pairs, as illustrated in Fig. 4. The deep Peierls valleys exist along <110> directions. Two regimes of behavior have been identified: one in which kink pairs are nucleated while existing kink pairs are still present on the dislocation segment [case (a) in Fig. 4]. This regime is considered to exist at high temperature and low stress levels, and is also denoted as the "length-independent" regime since the spacing between kinks (X) is the controlling parameter, rather than the length of the dislocation (L). Dislocations can take on curvature due to the large population of kinks, and the velocity of the dislocation segment under these conditions is given by:

$$v = 2h\sqrt{v_{\mathrm{k}} P_{\mathrm{kp}}}, \tag{4}$$

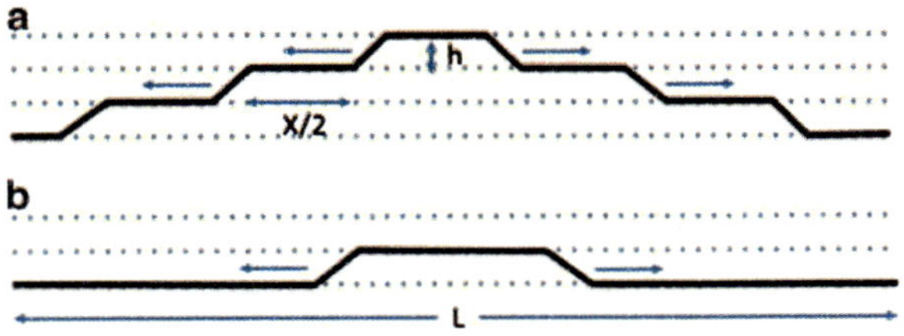

Fig. 4 Dislocation motion limited by Peierls potential for (**a**) the length-independent ($X < L$) regime and (**b**) the length-dependent ($X > L$) regime

where h is the kink height, v_k is the kink velocity, and P_{kp} is the probability of kink-pair nucleation. In the second regime, the kink-pair traverses the length of the dislocation segment prior to formation of a new kink pair [case (b) in Fig. 4]. This regime is expected at low temperature and high stress, where the force acting on the kinks is large but the probability of forming kink pairs via thermal activation is smaller, so that X is less than L. For this segment-length-independent regime, the velocity is given by:

$$v = P_{kp} L h. \tag{5}$$

In the Hirth and Lothe (1968) model, in which the kinks are assumed to move viscously with an activation energy of U_m and an activation energy for kink-pair nucleation of U_{kp}, the velocities are:

$$v = 2 f_D \frac{\tau h^2 b^2}{kT} \exp\left(-\frac{1/2 U_{kp}^{(c)} + U_m}{kT}\right) \tag{6a}$$

in the length-independent regime, and

$$v = f_D \frac{\tau h^2 b L}{kT} \exp\left(-\frac{U_{kp}^{(c)} + U_m}{kT}\right) \tag{6b}$$

in the length-dependent regime, where f_D is the Debeye frequency. There have been a number of proposed modifications and refinements to this model, but as discussed in a recent review article by Caillard and Martin (2005), the model of Hirth and Lothe (1968) appears to provide an adequate description of the stress and temperature dependence of dislocations velocities, as measured using in situ X-ray topography experiments. The data for a linear stress dependence of velocity is very convincing for high purity Si (Imai and Sumino 1983), while the stress dependence increases at low stress for Si-containing impurities, indicating an effect from solute interaction. The Peierls' mechanism has also been proposed to explain the strong temperature dependence of yield strength at lower temperatures in BCC (Dorn and Rajnak 1964) and HCP (Caillard and Martin 2005) metals. However, in comparison with covalent solids containing very low dislocation density, the ability to quantitatively test dislocation velocity models for metallic materials is very challenging.

As in all the mobility-controlled deformation models, predicting the macroscopic mechanical behavior is dependent upon an understanding of dislocation multiplication and annihilation. Unfortunately, these processes remain poorly understood in metals. While the Taylor expression is widely employed to treat mobile dislocation density, it is at best an approximation applicable to "steady-state" conditions. The Taylor form will be used for other models to be discussed below. However, the ability to predict yield point phenomena and transient response requires a description of multiplication and annihilation processes. Early studies in covalent solids, for which the initial dislocation density is small and more readily measured (Alexander and Haasen 1968), have resulted in the following phenomenological description:

$$\frac{\mathrm{d}\rho_{\mathrm{m}}}{\mathrm{d}t} = K_1 \dot{\gamma}_p \tau^*, \tag{7}$$

where K_1 is a constant, while more recent 3D dislocation modeling of Frank–Read source operation has indicated a modified version of this multiplication law (Moulin et al. 1997):

$$\frac{\mathrm{d}\rho_{\mathrm{m}}}{\mathrm{d}t} = K_2 \dot{\gamma}_p \sqrt{\frac{\tau^*}{\rho_{\mathrm{m}}}}. \tag{8}$$

A combination of 3D dislocation modeling and in situ TEM experiments would appear to be a very powerful approach to understanding these physical processes, and developing more quantitative rules for dislocation evolution. While beyond the scope of this chapter, the difficult problem of treating the exhaustion of mobile dislocations by substructure formation, solute interaction, or annihilation events with other dislocations or grain boundaries has also been reviewed recently by Caillard and Martin (2005), indicating the existence of a considerable gap in our present understanding of these processes.

3.2 Climb-Controlled Creep

Dislocation movement at high temperature can be controlled by pure climb of dislocations or climb of jogs that limit the movement of gliding dislocation segments. Several examples of these mechanisms are described in this section, with emphasis on the jogged-screw dislocation mechanism that has been recently revisited and appears to have merit for several important engineering alloys.

3.2.1 Pure Dislocation Climb

Nabarro (1967) proposed a steady-state model by which single crystals can deform by the climb of individual edge dislocations (with the applied stress providing the driving force). In tension, prismatic dislocation loops would climb by the production of vacancies, as illustrated in Fig. 5. These vacancies create an osmotic force for climb on other dislocation families which otherwise would have no forces acting upon them. It is further argued that the network spacing between climbing dislocations will scale inversely with applied stress in the steady-state limit, which yields a strain rate that is proportional to the third power of stress, with an activation energy equal to that for bulk diffusion (Nabarro 1967):

$$\dot{\varepsilon} = \frac{Db\sigma^3 / \pi k T \mu^2}{\ln(4\mu/\pi\sigma)}. \tag{9}$$

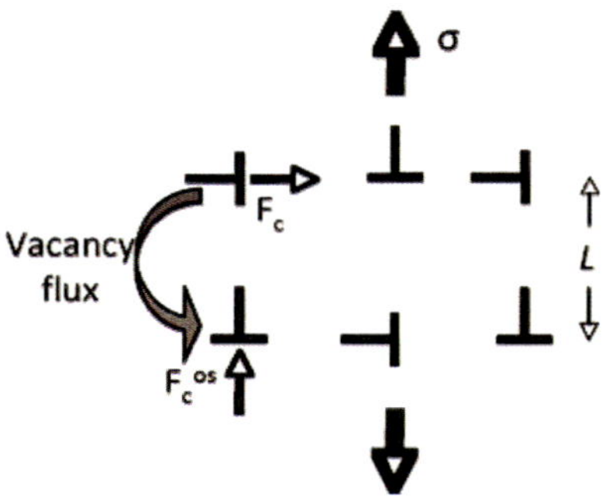

Fig. 5 Nabarro climb creep model of dislocation network with pure climb force due to the applied tensile stress (F_c) and osmotic force (F_{os}) with an example of vacancy flux between different dislocations. The dislocation spacing L is inversely dependent on stress

There are only a few examples in which deformation occurs by pure climb since for high symmetry crystals and most orientations, dislocations will experience both glide and climb forces. Edelin and Poirier (1973) claim to have achieved a climb-controlled creep condition in Mg single crystals oriented along [0001], for which deformation proceeds by climb of c-type dislocations. Their dislocation velocity measurements indicate an activation energy that is larger than self-diffusion and a stress dependence that is significantly larger ($m = \delta \ln v / \delta \ln \tau = 2.8$) than the value of unity predicted by the Nabarro model. Caillard and Martin (Caillard and Martin 2005) have recently suggested that this discrepancy may be understood through refinement of the Nabarro model through consideration of jog-pair nucleation. Groves and Kelly (1969) and Firestone and Heuer (1976) have studied the creep of sapphire single crystal oriented along [0001]. Due to a restructuring of the dislocation core 1/3< −1101> prismatic dislocations are sessile in glide (Chang et al. 2003; Bodur et al. 2005), but TEM evidence indicates that they move by pure climb. Quantitative agreement with the Nabarro climb creep model is obtained in these studies. This model has also been used to model creep of the [0001]-oriented alumina matrix of a directionally solidified ceramic eutectic of Al_2O_3/c-ZrO_2(Y_2O_3) (Yi et al. 2006). A distinctly different variant on the concept of climb-controlled dislocation creep has been proposed by Mills (Mills and Miracle 1993) to explain the high temperature deformation of [001]-oriented single crystals of the intermetallic compound NiAl having the B2 (CsCl) crystal structure. High-resolution TEM has revealed that the active <110> dislocations indeed comprise two coupled <100> dislocations that they propose to move by vacancy exchange process. Coordinated dislocation climb has more recently been invoked by Epishin and Link (2004) to explain creep and creep cavity formation in [001] oriented superalloy single crystals. In this case, the climb-controlled movement of ½<110> dislocations at the "horizontal" γ/γ' interfaces is assumed to produce vacancies that drive dislocation climb in the "vertical" channels, promoting "rafting" of the cuboidal particles into plates. Alternatively, the vacancies produced by the climbing dislocations aggregate into voids that grow in number and size with creep strain.

3.2.2 Harper–Dorn Creep

There have been a number of reports of creep in a variety of pure metals and some ceramics which exhibit a stress exponent close to 1, activation energy near that for self-diffusion and normal primary transients, but for which the creep rates are too large to be explained by pure diffusional creep [i.e., Nabarro (1948) or Coble (1963)]. The Harper–Dorn model (Harper and Dorn 1957) is based on the assumption that dislocation climb is rate limiting, but that the dislocation density is not significantly influenced by the applied stress–a condition that might only occur at very low stress levels. This model and its existence is extremely controversial, with several reviews and rebuttals having been written in the last several years which have been recently reviewed by Kassner and Perez-Prado (2004). Indeed, it has been suggested that the stress exponent values may be inaccurate due to the difficulty in attaining a "steady-state" or minimum strain rate condition at very low stress values.

3.2.3 Jog-Dragging-Controlled Creep

There are several important alloy systems that appear to have the macroscopic characteristics of a "dislocation creep" process (i.e., stress exponents of about 5), but in fact do not exhibit a strong tendency for subgrain formation. As an alternative, a new dislocation-level approach has been developed that is fundamentally based on a mobility-controlled picture of creep. The core of this physically based model is the concept that jogged-screw dislocations are limiting the creep rate. Jog motion can be rate limiting since they can move conservatively only along the dislocation line, but must climb to keep up with the rest of the dislocation. While this is not a new concept, as it goes back to the early work of Mott (1956) and later by Barrett and Sherby (1965), it has traditionally been believed that the jogs on the screw dislocations are a result of the intersection between dislocations on different slip systems.

As a result of direct TEM observations of <110> dislocations in γ-TiAl following creep deformation (Viswanathan et al. 1999), it was recognized that the jogs in this case were quite tall (up to hundreds of Burgers vectors in height). It was also noticed that jogged screw dislocations could be observed even in grains that were deforming by one predominant slip system. As a consequence, it was proposed that these tall "super-jogs" form as a natural result of dislocation motion due to multiple cross-slip events as illustrated in Fig. 6. This was also consistent with other studies of cross-slip and jog-formation related to the anomalous increase of yield strength with temperature exhibited in TiAl (Sriram et al. 1997). Subsequent studies of several HCP alloy systems, including α-Ti(Al) solid solution (Moon et al. 2009) the α-phase of α/β Ti alloys such as Ti-6242 (Karthikeyan et al. 2004) and the α-phase of the Zr-based alloy, Zircaloy-4 (Moon et al. 2006), have also demonstrated similarities in terms of jogged-screw configurations. Examples of TEM observations of jogged-screw dislocations in these alloys are shown in Fig. 7.

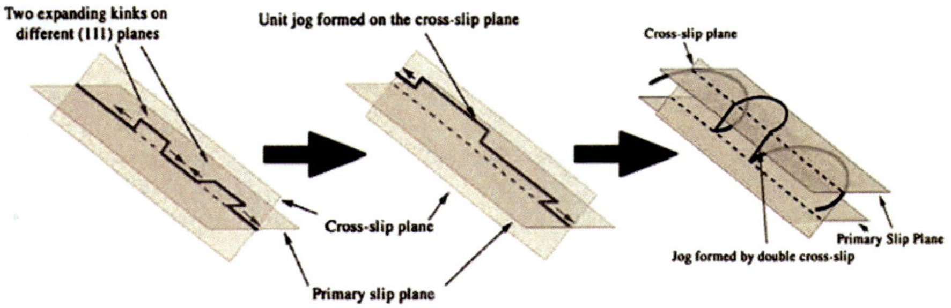

Fig. 6 Formation of short kinks and jogs, and eventually tall jogs, by multiple cross-slip events

It may be asked why similar dislocation behavior may be observed in the crystallographically distinct alloy systems, such as the $L1_0$ and HCP structures? The important commonality for these systems and the basic criteria for the "natural" development of super-jogs on screws are hypothesized to be the following:

1. Screw dislocation cores are spread slightly on multiple glide planes, such that cross-slip is competitive to glide movement
2. Screw orientation is prominent due to strong lattice friction, as a consequence of nonplanar cores
3. Jog-pair and kink-pair expansion is sluggish due to lattice or solute friction

Criteria (1) and (2) insure that cross-slip events are frequent and that there are long lengths of dislocations in screw orientation, thereby increasing the probability of forming super-jogs. If kink-pair and jog-pair expansion is very rapid (relative to their rate of creation), then super-jog creation is less likely, which necessitates Criterion (3). These conditions appear to prevail in γ-TiAl due to nonplanar core spreading for both screw and 60° dislocation orientations based on atomistic calculations by Rao et al. (Simmons et al. 1998). Relative to the Ti alloys, it has been suggested by several groups that a-type (1/3<1120>) dislocations have nonplanar cores that are spread simultaneously on several planes (Legrand 1985; Naka et al. 1988) with some evidence for this via HRTEM observations in a Ti-6Al alloy (Neeraj et al. 2005). Thus, the conditions for the formation of superjogs are satisfied. Similar atomistic studies have not been reported for Zr, but the similarity in structure and lattice parameter between Ti and Zr suggests that similar, nonplanar cores exist also for Zr alloys. With respect to Criteria (3), the Ti and Zr alloys that have been studied in detail are solid solution strengthened, which may make lateral motion of super-jogs more viscous.

The principal modification of the original jogged screw model is the allowance for a stress-dependent jog height. It is assumed that in the course of motion that jogs will rapidly accumulate and grow in height. However, if the jog height exceeds a critical value at a given stress associated with the bypass of the edge dislocations attached to either end of the jog, then the jog would no longer be dragged, but could become stationary and act as a double-ended dislocation source (see Fig. 8).

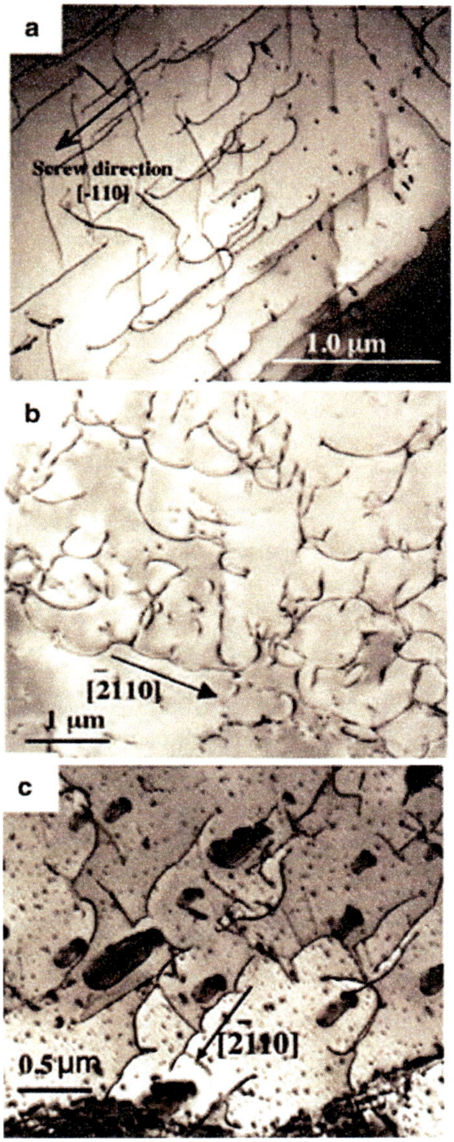

Fig. 7 Examples of deformation substructures in (**a**) equiaxed, ϒ-phase ($L1_2$ structure) Ti48Al, (**b**) Ti-6Al with the α-phase (HCP) crystal structure, and (**c**) Zircaloy, an α-phase (HCP) crystal structure Zr alloy

Assuming this critical bypass condition to be an upper bound on the possible jog height, Viswanathan et al. (1999) employed a "characteristic" jog height to model the average jog height at a given stress. It is also possible that the jogs become static relative to the rest of the dislocation, and that a dipole is then extended from either end of the jog (see Fig. 8). The conditions for which each of these processes may become dominant are discussed in detail by Karthikeyan et al. (2004).

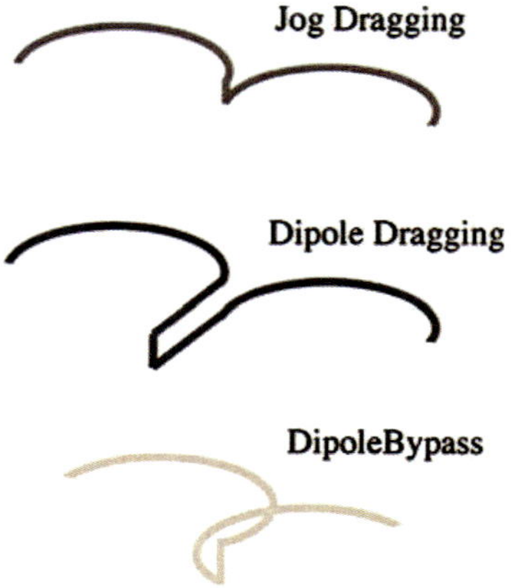

Fig. 8 Three possible processes involving jogs of varying heights

Table 1 Key microstructural parameters for the jogged-screw model for several alloys

Material	α	H_o (m-Pa^2)	L$_1$ (m)	L$_o$ (m-Pa)
Ti-48Al	2	3×10^8	3.5×10^{-8}	20
Ti-6Al	3.7	1.5×10^6	0	38
Zircaloy	1.15	3×10^7	0	27

Karthikeyan et al. (2004) have also argued that the tall jog cannot be considered simply as a point source/sink of vacancies, magnified by its length, but rather must be treated as a finite line source/sink. This correction results in the following dislocation velocity expression:

$$v_s = \left(\frac{4\sqrt{2}\pi D_s}{b\delta(h)} \right) \left[\sinh \left(\frac{\tau \Omega l}{hkT} \right) \right], \tag{10}$$

where v_s is dislocation velocity and $\delta(h)$ is a logarithmic function of h and b:

$$\delta(h) = 0.17 + 4.528 \ln \left(\frac{h}{b} \right). \tag{11}$$

The steady-state creep rate is then related to the dislocation velocity via the Orowan equation (2). As found for creep of other metals (Karthikeyan et al. 2004), the density of mobile dislocations is assumed to follow the Taylor relation (3), which states the material flow stress scales with the square root of dislocation density. This has been found to provide good approximation based on actual TEM measurement of densities for TiAl (Karthikeyan et al. 2004) Ti-6Al (Moon et al. 2009) and Zircaloy (Moon et al. 2006). The rather large value for α (see Table 1) compared with that found for creep of face-centered cubic metals (Karthikeyan et al. 2004) indicates that the applied stress is indeed supported by considerable friction on the operative dislocations, presumably due to the presence of the jogs on the dislocations. Because of the relatively homogeneous distribution of dislocations and the general absence of subgrain structures that could limit dislocation motion, it seems reasonable that the measured dislocation density can represent the mobile dislocation density.

These modifications then provide the final strain-rate equation for the modified jogged-screw model as:

$$\dot{\gamma} = \left(\frac{\pi D_s}{\beta \mathbf{b} h_d}\right)\left(\frac{\tau}{\alpha\mu}\right)^2 \left[\sinh\left(\frac{\tau\Omega l}{4\beta h_d kT}\right)\right], \tag{12}$$

where D_s is the self-diffusion coefficient, β is a parameter that relates characteristic jog height to the critical jog height (usually assumed to be a value between 0.5 and 1), $\mathbf{b}$ is the Burgers vector, h_d is the critical jog height, τ is the applied shear stress, α is the Taylor factor, μ is the shear modulus, Ω is the atomic volume, l is the spacing between jogs, k is the Boltzman's constant, and T is the temperature.

The most important parameter in the model, and unfortunately the one most difficult to measure directly, is the critical jog height, h_d, and its dependence on stress. A distribution of jog heights is typically observed and it is crucial to determine how this distribution is affected by stress and other deformation conditions. Karthikeyan et al. (2004) have derived a form based on theoretical considerations of the processes competing with jog dragging provided a reasonable match to limited experimental data. This leads to the following approximate stress dependence for the critical jog height:

$$h_d = H_0 \tau^{-2}, \tag{13}$$

where H_0 is a constant. Table 1 presents a summary table of this parameter for several alloys system. At sufficiently small jog velocities, the creep mechanism directly changes from a jog dragging to a dipole bypass mechanism as the jog height increases. While a distribution of jog heights is expected, the present form of the model assumes a "characteristic" jog height, which has been assumed to be equal to the critical jog height ($\beta = 1$), since these would be the most difficult to drag and therefore would be rate limiting.

Based on TEM measurements and theoretical considerations, Karthikeyan et al. (2004) suggested that jog spacing is inversely proportional to the applied stress according to:

$$L = L_1 + \frac{L_o}{\tau}. \tag{14}$$

This expression appears to provide a reasonable description of experimental data for TiAl (Karthikeyan et al. 2004) Ti-6Al (Moon et al. 2009) and Zircaloy (Moon et al. 2006), with the constant L_1 unnecessary for the latter two cases as indicated in Table 1. Simulations of the movement of jogged-screw dislocations (Karthikeyan et al. 2004) suggest that conservative glide of jogs on the cross-slip plane, driven by unbalanced line tension forces, leads to jog coalescence (or annihilation, depending on the sign of jogs) and in effect, jog spacing coarsening. Such coarsening is offset

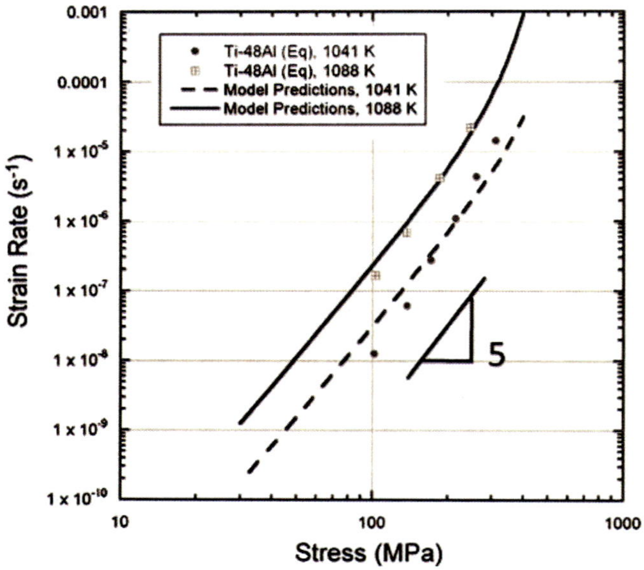

Fig. 9 Double-logarithmic plot of creep rate versus stress for Ti-48Al showing data and model predictions at two different temperatures. Note that the jogged screw predicts a stress exponents near 5 at lower stresses, with increasing stress dependence at larger stresses

by the fact that coarse jog spacings have a higher probability for jog nucleation and this in turn refines the jog spacing. The result is that the jog spacing is inversely proportional to the applied stress.

Equation (12) can predict the steady-state strain rate at a given applied shear stress if all the microstructural parameters such as jog height, jog spacing, and dislocation density are known. Good agreement between the model predictions and the observed minimum strain rate is obtained for γ-TiAl intermetallic, as shown in Fig. 9, and α-Ti-6Al solid solution. Ongoing research is being performed to adapt the model to a-Zr alloys, such as Zircaloy (Moon et al. 2009). The model predicts a stress exponent close to 5 at smaller stresses, with increasing stress dependence at larger stresses, and an activation energy equal to that for self-diffusion. The present version of the model does not account for the possibility of short-circuit or pipe-diffusion between jogs, and therefore is limited by bulk vacancy diffusion.

The modified jogged screw model has been further refined to treat cases for which there exists a finite length-scale associated with the deforming phase. For instance, it has been applied to creep of the fine-scale, two-phase (γ/α_2) "fully lamellar" (FL) microstructures that are possible in Ti-Al alloys (Karthikeyan et al. 2001) as well as to the α/β Widmanstatten type structures in the commercial titanium alloy, Ti-6242 (Hayes et al. 2002). In future work, crystal plasticity approaches to modeling the polycrystalline single-phase and lamellar microstructure response are clearly needed. Even the single phase systems for which the model has been applied thus far are anisotropic in their plastic response, with "hard" orientations corresponding to the c-axis for HCP and $L1_0$ structures. A proper accounting of the flow response for the "hard-mode" systems, as well as the load-shedding that should be pronounced in these systems is necessary to make further progress. In addition, some model parameters are difficult to measure, particularly their dependence on deformation conditions (i.e., stress and temperature). Their rigorous determination could be an extremely tedious task. Dislocation dynamics simulations could be extremely useful in exploring these essential dependencies. Finally, this model is a "steady-state" representation. This need for more complete understanding is exemplified by the fact that the materials described above all exhibit a normal primary transient–an unexpected characteristic of mobility-controlled deformation. A possible explanation for this response lies in the evolution of jogs as a function of strain. Freshly generated dislocations will lack a developed jog population. As the jogs develop in frequency and height, the dislocation velocity will dramatically decrease, causing a drop in the creep rate. The evolution of the dislocation density and other model parameters as a function of strain needs to be better understood to treat the transient creep response.

It is interesting to consider whether the jogged-screw model may have application in other materials systems. Analogous dislocation core structures, involving the simultaneous spreading on more than one crystallographic plane, have been predicted in several other systems on the basis of atomistic and first-principles calculations for screw dislocations. The operation of the jogged-screw mechanism of creep may therefore be expected for these systems. BCC solid solutions are among the likely candidates for super-jog formation. The preferential alignment of $^1/_2$<111> dislocations along screw orientation at lower temperatures in BCC metals has been attributed to a large Peierls stress for this dislocation core (Vitek 1974). Thus, the requirements for super-jog formation are present. In situ TEM observations in pure Nb (Garret-Reed and Taylor 1979; Ikeno and Furubayashi 1972, 1975) and post-mortem observations in Fe-3%Si (Furubayashi 1969; Low and Turkalo 1962) indicate that super-jogs are frequently formed. It is therefore postulated that the modified jogged-screw model may also be applicable for this broad class of alloys. It is also important to emphasize situations for which this model is not expected to be applicable. For example, it is not expected to be valid for FCC metals and alloys since dislocations tend to be dissociated in a planar manner, and the screw orientation is not strongly favored. In covalently bonded solids such as Si, strong alignment of a/2<110> dislocations along screw orientation (as well as 60° line directions) is observed. However, the modest stacking fault energy in Si may also prevent sufficiently frequent cross slip for superjogs to develop in abundance.

3.3 Solute Drag Creep

Solute strengthening is one of the most ubiquitous of strengthening mechanisms for metallic alloys. Advantages of this engineering approach include improvement in creep strength while decreasing the stress exponent relative to the pure-metal counterpart. Since phenomenology indicates that the uniform ductility should scale with the stress exponent, solute strengthening is unique in that improved ductility can also be achieved.

Several models for creep of class I (A) solid solution alloys have been proposed in the literature. These models acknowledge that at higher temperature, solute atoms are no longer static barriers to dislocation motion, but rather provide dynamic resistance to dislocation motion by forming mobile solute atmospheres (Low and Turkalo 1962; Hirth and Lothe 1968; Barnett et al. 1974). The solute interaction due to elastic and size misfit is strongest for a pure edge dislocation, and is appreciable for dislocations with any edge character in an elastically isotropic crystal. The interaction can be significant even on a screw dislocation for an elastically anisotropic medium. Nevertheless, the creep models are based on the premise that edge dislocations move viscously and more slowly than screw dislocations. The viscous nature of dislocation motion has been demonstrated by in situ observation of dislocation movements in Al-Mg alloys while the premise that edge dislocations move more slowly than screws is supported by ex situ TEM observation of dislocation line shapes (Mills 1985). An example of dislocation configurations after relatively small creep strain in Al-5.5%Mg is shown in Fig. 10. Note the relatively homogeneous distribution of dislocations, as well as the elongation of dislocations along their edge orientation (note that the majority of dislocation in this field of view have the same Burgers vector). Frank–Read source configuration is labeled "FR" in the image. An example of a creep curve for the Al-Mg solid solution alloy is shown in Fig. 11. Note the modest, inverted primary creep transient indicative of a dislocation multiplication stage of deformation. A plot of the stress dependence on creep rate for the "steady-state" regime is shown in Fig. 12, which exhibits the classic 3-power law behavior.

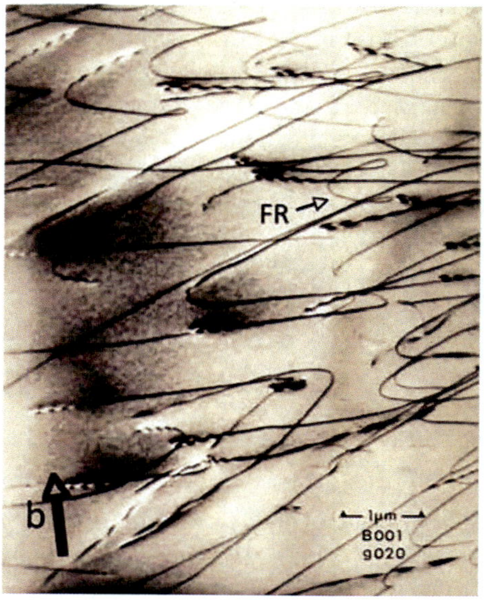

Fig. 10 TEM micrograph showing the dislocation substructure present in Al-5Mg after small creep strain (0.002) at 573 K and 47.2 MPa. The Burgers vector of the majority of dislocations present is indicated. Note the pronounced elongation of the dislocation loops perpendicular to the Burgers vector. A Frank–Read-type source is indicated as FR

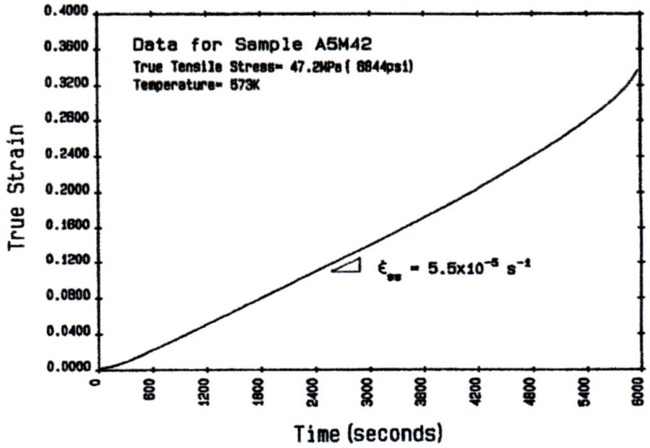

Fig. 11 Example of the inverse primary transient, followed by secondary and tertiary creep in Al-5at%Mg

Fig. 12 Double-logarithmic plot of creep rate versus stress at 673 K for Al-5.5at%Mg. "Steady-state" data as well as constant structure creep rates for constant initial applied stress and constant reduced applied stress conditions. Note the difference in stress exponents for these two constant structure experiments (Mills 1985)

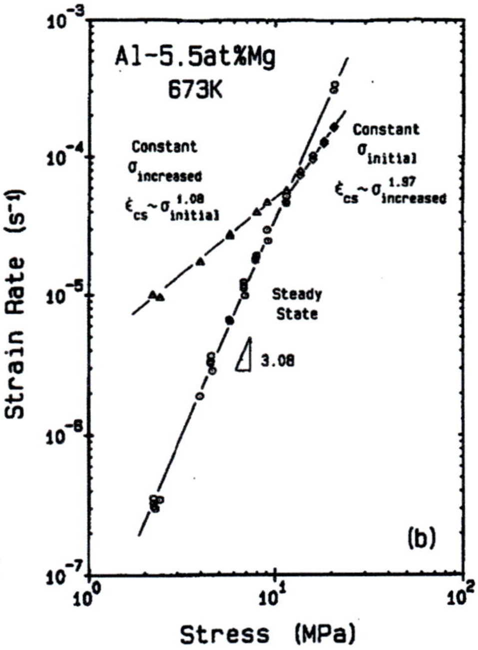

The early model of Weertman (1957) is based on the concept of edge dislocation pile-ups. This model reproduces the experimentally observed stress exponent of about 3, which is characteristic of the Class I (A) alloys, when a Taylor-type equation for the dislocation density dependence on stress is assumed. However, that dislocation pile-ups are the essential features of the substructure is a basic

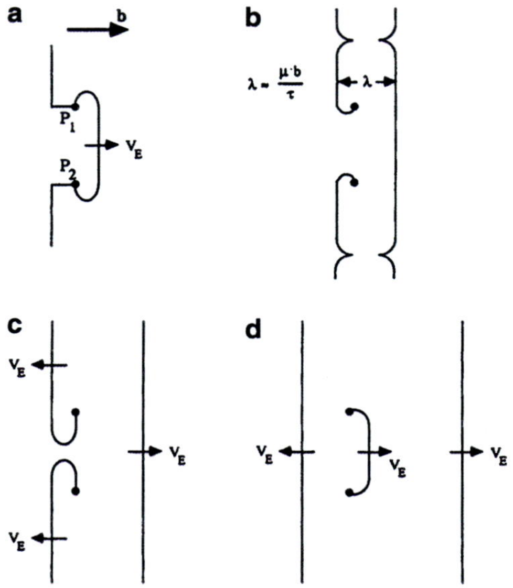

Fig. 13 Schematic representation of Takeuchi and Argon model for dislocation multiplication from a Frank–Read-type source for a solid solution alloy, with more rapid screw dislocation motion relative to edges. From the initial source (**a**), the screws are envisaged to run out rapidly (**b** and **c**) and annihilate (**d**) with screws from adjacent sources (Mills 1985)

assumption which is at odds with many TEM observations that indicate dislocations to be distributed in a relatively homogeneous manner (Mills et al. 1985; Weckert 1985), as exemplified in Fig. 10.

A model that is more faithful to the observed dislocation substructure was offered by Takeuchi and Argon (1976). In this model, the relative rates of dislocation creation and annihilation are calculated based on the idea that the rate of dislocation generation from Frank–Read sources is dependent on the glide velocity of edge dislocations, while the screw dislocations are assumed to annihilate with screws from neighboring sources after the edges have traveled a characteristic distance equal to $\mu b/\tau$, as shown in Fig. 13. A "steady-state" condition is established by enabling edge dislocations to be "captured" and annihilated by dislocations of opposing sign from nearby sources. Since the velocity for viscous glide should be similar to that for climb in these alloys, the capture width will be large (perhaps larger by a factor of 10 or more relative to the capture distance in a pure metal where $v_{\mathrm{g}} >> v_{\mathrm{c}}$). The resulting steady-state creep rate law is:

$$\dot{\varepsilon}_s = \frac{1}{8c_o e_{\mathrm{a}}^2}\left(\frac{kT}{\mu b^3}\right)\left(\frac{D_{\mathrm{sol}}}{b^2}\right)\left(\frac{\sigma^3}{\mu}\right), \tag{15}$$

where c_o is the solute concentration, e_{a} is the atomic size misfit between solute and matrix atoms, and D_{sol} is the bulk solute diffusivity. Note that the predicted stress dependence $n = 3$ and the calculated strain rate is within a factor of about 3 with the empirically derived strain rate law (Laks et al. 1957; Horiuchi et al. 1965; Sherby and Burke 1967).

While the models of Weertman and Takeuchi and Argon are fundamentally quite different, they both succeed in predicting the magnitude and stress dependence of the steady-state strain rate. This suggests that the description of steady-state behavior alone is insufficient to decide on the validity of these models. An additional criterion for judging these models would be the ability to predict the transient response following an abrupt change from steady-state conditions. The forward "constant structure" creep rate, which is observed soon after a stress change, is known to vary as the square of the reduced stress (Jones and Sellars 1970; Northwood and Smith 1984) and increased (Jones and Sellars 1970; Mills et al. 1985) stress, as shown in Fig. 12. This is not an obvious result since for a "constant structure" condition following a stress drop one might expect a linear dependence of creep rate with reduced stress since the velocity under solute drag conditions is expected to depend linearly with stress.

Mills et al. (1985, 1986) developed a dislocation-based model that can explain transient as well as steady-state behavior in the alloy-type creep regime. In this model, the dislocation substructure developed during creep in the Class I or alloy-type regime is represented as a set of isolated dislocation loops. The different glide mobilities of the edge and screw components of the loop are also incorporated by approximating the loops as ellipses with the minor axis, a_1, dictated by the edge mobility and the major axis, a_2, related to the screw mobility. The expansion of the loop is controlled by the viscous motion of the straighter edge segments, while the pure screw segments of the loop are highly bowed since line tension alone inhibits the glide motion. The relationship between a_1 and a_2 is given by:

$$a_2 = \frac{2\tau}{\mu b} a_1^2. \tag{16}$$

As in the Takeuchi and Argon model, the characteristic length a_1 is determined by the applied stress through a Taylor-like expression, $\mu b/\tau$. The additional stress dependence for "constant structure" conditions following a sudden stress change is explained as follows. While the distance a_1 remains constant immediately after the stress change, the distance a_2 decreases rapidly upon stress reduction due to an elastic contraction of the loops along screw orientation. Thus, even though the density of dislocation loops, determined by a_1, remains "constant" after the stress reduction, there is nevertheless a reduction in the actual dislocation density. The model successfully predicts a steady-state stress exponent of 3, and creep rates in close agreement with experiment. The major difference between the creep model and those developed previously (Takeuchi and Argon 1976; Weertman 1957) is that both steady-state and transient creep properties are described quantitatively.

An intrinsic anelastic response is also a natural part of the loop model. Upon a stress decrease, for example, the screw segments of the loop will run back rapidly to assume the curvature corresponding to the new stress level, while the straight edge segments, which are subjected to negligible bowing forces, remain stationary. The backstrain should be linear with stress change and, for a unit stress change, should be about *1.8/E*, where E is the elastic modulus. These predictions are far different from pure metal behavior in which the measured backstrains are much larger and more highly dependent on $\Delta\sigma$. A novel set of experiments, called "dual-drop" creep tests, were performed which validated this prediction (Mills et al. 1986).

A final important implication of this work is related to the use of the concept of "internal stresses" in modeling the constitutive material response. Creep theories for pure metals and alloys have often incorporated the idea that forward flow is driven by an "effective stress," which is the difference between the applied stress and the internal stress. The basic concept is that large internal stresses may arise from structural inhomogeneities, such as subgrains or cell boundaries (Gibeling and Nix 1981; Hasegawa et al. 1972), or even from the presence of forest dislocations (Yoshinaga et al. 1976). Upon a stress drop, the reduced stress at which zero creep is subsequently observed has been a standard criterion for determining the internal stress (Solomon 1969; Ahlquist and Nix 1969). The viscous glide mechanism, which has been associated with alloy-type creep, implies that most of the applied stress is supported by solute strengthening. The internal stress level should be small in these alloys since solute strengthening is not a directional hardening mechanism. However, attempts to directly measure σ_{internal} using single drop experiments yield values ranging from 0.5 to 0.75 of the applied stress. In contrast, the loop model accurately predicts the total anelastic response without resorting to a large internal stress. Their analysis led Mills et al. (1986) to conclude that the plastic backflow measured using single drop experiments had been erroneously interpreted as due to a long range internal stress, but rather were due to anelastic backflow of elongated loops, as described above.

3.4 Reordering Controlled Creep at Intermediate Temperatures in Superalloys

Recent investigation of creep in several polycrystalline superalloys has revealed that an important deformation process at elevated temperatures may also fall into the category of mobility-controlled deformation. Most remarkably, Kolbe (Kolbe 2001;Viswanathan et al. 2005) reported on a novel microtwinning mechanism that operates during exposure to moderate stresses and temperatures in the range of 650–750°C, where these superalloys become more strongly temperature and rate sensitive. The microtwins shear both the γ matrix phase (FCC) and strengthening γ' phase ($L1_2$ structure) and are commonly seen to traverse entire grains. The viscous nature of this process has been observed directly via in situ TEM deformation studies (Legros et al. 2002). Microtwinning has been observed at relatively low stress levels, and appears to depend strongly on temperature. These are particularly unusual characteristics for deformation twinning in general, and the thermally activated nature of this mechanism in the superalloys is linked to the diffusion-mediated reordering of the ordered structure within the precipitates after being sheared by the a/6<112> twinning partials. The reordering process removes wrong nearest-neighbors at the complex stacking faults created by pairs of twinning partials acting on adjacent {111} planes. Reordering enables pairs of a/6<112> partials to shear the secondary γ' particles at modest stress levels–a process not possible in the absence of reordering. Detailed analysis of the atomic steps of reordering by first principles calculations by Kovarik et al. (2009) indicates that the diffusion coefficient for reordering should be similar to that for Ni self-diffusion in the bulk, under the simplifying assumption of ordered Ni_3Al being representative of the γ' precipitate composition. A dislocation level model incorporating these ideas has been proposed by Karthikeyan et al. (2006), which appears to provide a prediction of creep rates that are in reasonable agreement with experiment. Kovarik et al. (2009) have also hypothesized that several other γ' shearing processes that have been reported for this intermediate temperature regime, involving the formation of both intrinsic and extrinsic stacking faults, may also be fundamentally limited by the same reordering process.

Mechanisms that depend on reordering appear to initiate at temperatures that are lower than those required for general climb of dislocations. Reordering is a conservative process involving diffusion distances of atomic dimensions, while climb bypass requires long-range diffusion over distances comparable to the precipitate size. The high temperature creep of single crystal superalloys, during which the formation of interfacial dislocation networks is the predominant process (Pollock and Field 2002), is postulated later in this chapter to be an obstacle-controlled creep process, distinct from the reordering-controlled shearing mechanisms discussed here.

4 Obstacle-Controlled Dislocation Creep

In creep when the stress exponent is near 5 and the creep activation energy is similar to that for self-diffusion, the creep literature often broadly refers to this as "climb-controlled dislocation creep" with a mechanistic explanation implicitly assumed. The most common mechanistic description usually involves assumption that dislocations must surmount inert or repulsive obstacles and they bypass them by climb. A variant of the schematic of Fig. 14 often illustrates this mechanism. It is routinely argued that because the activation energy for creep deformation is the same as (or close to) that for self-diffusion, diffusion should be an important part of the deformation process, and dislocation climb is the simplest and most compelling way to connect deformation with diffusion. The idea of climb as the rate-controlling process in creep was formalized and given a theoretical basis with the climb-creep model(s) of J. Weertman (Weertman 1968; Sherby and Weertman 1979). In summary, his models postulate that a series of dislocation loops that emanate from a fixed population of dislocation sources. The dislocations become trapped in the strain fields of dislocations produced by other sources and recovery

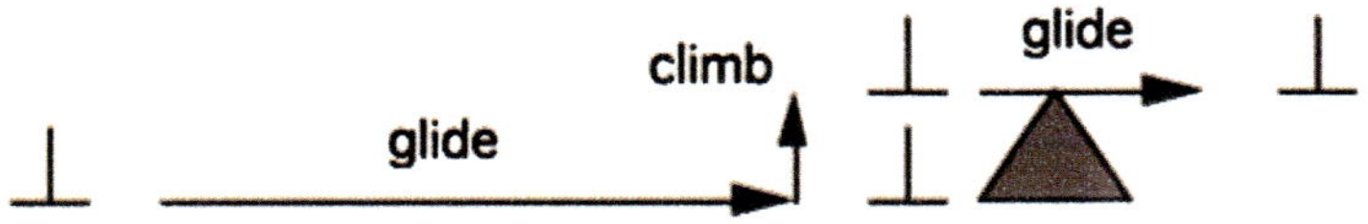

Fig. 14 Illustration of the most common representation of "climb-controlled" dislocation creep

happens by dislocation climb and annihilation. This enables further deformation. With this model, Weertman is able to develop a stress-exponent of 4.5, which is close to experimental observations of the steady-state stress exponent for pure metals. While the Weertman climb-creep model is widely cited and probably represents the de-facto theory for climb-controlled glide creep, it has some serious shortcomings that have been discussed in depth by a variety of authors, including in Nabarro's recent reviews of creep in nominally pure metals (Nabarro 2004, 2006).

Among the most important problems with Weertman's models is that it does not agree well with the full range of experimental observation. For example, the kinds of transients shown in Fig. 3 are an essential feature of pure-metal creep and not represented by the model. Also, microstructural observations are not consistent with the model. It is clear that dislocation density increases and subgrain size decreases as creep stress is increased, while the model assumes a fixed source density. Further, the microstructure that Weertman postulates – piled-up loops of dislocations climbing toward each other and annihilating – is not seen in observations of crept pure metals. Most commonly pure metals show dislocation tangles and subgrain formation.

4.1 Postulates for Modeling Obstacle-Controlled Creep

In the following sections, a simple model based on a single evolving parameter, in this case a length scale, will be developed. In many ways, this approach is similar to the approaches developed by Kocks, Mecking, Follansbee, and their collaborators (Kocks, Argon and Ashby 1975; Mecking and Kocks 1981; Kocks 1976). The MTS model of Follansbee and Kocks (1988) is based largely on the evolution of a single state variable that they refer to as $\hat{\sigma}$. This term is thought to scale largely with dislocation density in the usual Taylor formulation, where $\hat{\sigma}$ is proportional to the square root of dislocation density. They implicitly assume that there is one dominant thermally activated dislocation obstacle that can be activated over which dislocations must pass. This is represented by a stress-free energy barrier of $g_0 \mu b^3$. Where g_0 is a free parameter, μ is the material shear modulus and b is the magnitude of Burgers' vector. The mechanical threshold stress, $\hat{\sigma}_t$, is the main variable that evolves. It represents the stress to overcome short-range obstacles in the absence of thermal activation. The remaining energetic barrier that must be surmounted, ΔG, is related to the applied stress, σ_t, as:

$$\Delta G = g_o \mu b^3 \left[1 - \left(\frac{\sigma_t}{\hat{\sigma}_t} \right)^p \right]^q . \tag{17}$$

In this Mechanical Threshold Stress (MTS) model, plastic deformation increases the density of obstacles (dislocations) increases and therefore $\hat{\sigma}_t$ is a single-parameter measure of the strength of the material. This approach follows a similar tack where, an inter obstacle spacing, λ, is the single-state variable and it is presently assumed that a single barrier is operative and this barrier can be overcome by thermal activation in a manner similar to that described by Follansbee and Kocks. The length scale λ is chosen as a single characterizing parameter for a few practical reasons. The present goal is to characterize the high-temperature time-dependent behavior of the material. The microstructures of high-strength materials are always somewhat unstable, and at high-temperature they can recover, recrystallize, or coarsen. In essence, the following sections will modify the single-parameter approach used in the MTS model. A coarsening equation will be added to permit thermally activated coarsening to reduce the obstacle density as a function of temperature and time.

The foundation for this model can be written as a series of postulates for obstacle-controlled creep:

- *Obstacles are generally attractive* – dislocations will generally find attractive pinning points rather than repulsive ones. Rosler and Arzt (1990) based a very physically realistic and successful model for the creep of dispersion strengthened metals on the key idea that dislocations more commonly reside in attractive traps because the dislocation line reduces its energy by lying at the interface between the matrix and an incoherent or semi-coherent dispersoid. This is in contrast to the repulsive or inert obstacles assumed in Fig. 15.

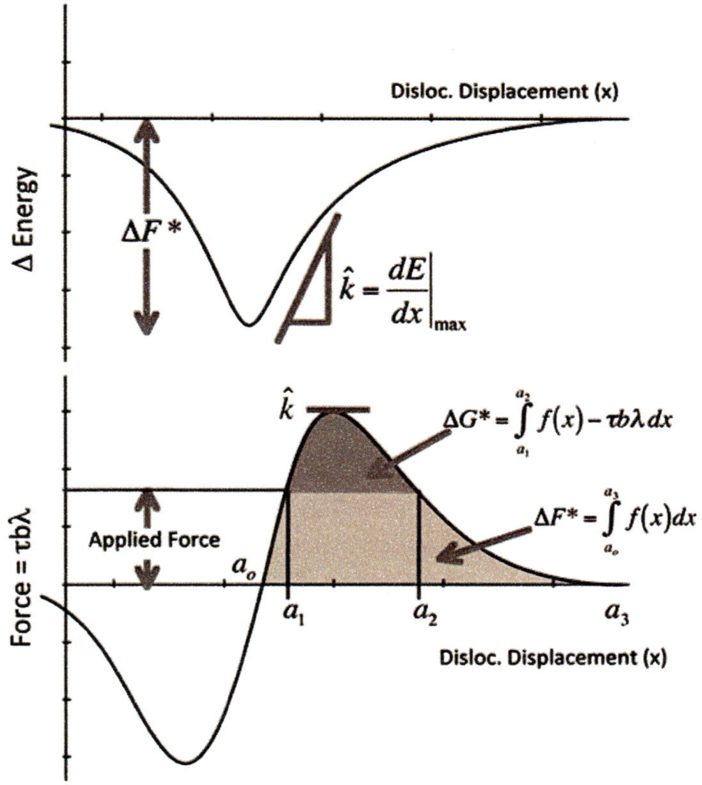

Fig. 15 *Top* shows the energetic profile for a dislocation interacting with an attractive obstacle as a function of distance. In the stress-free condition, the energy of ΔF^* must be thermally attained for thermally activated release from the obstacle. As the dislocation line is stressed, the force $\tau b\lambda$ acts on the obstacle and reduces the energy barrier to the value ΔG^*

- *Barrier traps are often relatively strong* – The activation energy for releasing dislocations from their attractive obstacles can be significantly greater than that for self-diffusion. The energetic profiles for dislocation traps can be estimated by simple closed-form estimates, as demonstrated later in this paper, or by atomistic modeling. Trap energies for dislocation-tangles are usually above 0.5 μb^3 and those for precipitate interactions can be several μb^3 (Baker 2009). On this scale, for most structural metals diffusion activation energies are on the order of 0.25 μb^3 (Shewmon 1969) experimental estimates of energies for dislocation release for copper are about 1.6 μb^3 (Follansbee and Kocks 1988).
- *Interobstacle spacing controls strength* – The current elevated temperature yield stress at an ordinary strain rate can be used to estimate obstacle density. For example, for materials that are hardened by their own dislocation field, Taylor's relation (3) has been shown effective even at high homologous temperatures (Kassner 1990), and the Orowan's relation

$$\tau_o = \frac{\mu b}{\lambda} \tag{18}$$

is very effective at modeling the strength of materials with obstacle fields due to precipitates or dispersoids. In either case, the strength is modeled as the inverse of the mean free spacing between obstacles. If obstacles are relatively weak and/or temperature is high, rate and temperature effects should be taken into account.
- *Scale will increase with time due to coarsening, and can be decreased due to dislocation generation* – The dominant time dependence for deformation is assumed to come from the recovery or coarsening of the obstacle field. In the case of pure metals, the obstacles are dislocations themselves (in the forms of tangled dislocations, cell walls or subgrain walls). In many engineered materials, second-phase particles may be the obstacles, but often they are too coarse to provide the primary strengthening and can have a strong role in stabilizing a dislocation structure. Recovery of the dislocation structure is assumed to be a standard coarsening process. Coarsening has become a fairly well-developed subdiscipline in physics (Siegert 1998; Sung et al. 1996; Sholl and Skodje 1995; Smilauer and Vvedensky 1995; Bray 1994; Chakrabarti et al. 1993; Yurke et al. 1993; Durian et al. 1991). The general form of the coarsening equation is given as:

$$\mathrm{d}\lambda^m = K\,R(T)\,\mathrm{d}t, \tag{19}$$

where λ represents a feature size, m is a coarsening exponent, $R(T)$ is the relevant fundamental rate controlling process as a function of temperature and t is time. Often diffusion is the rate controlling atomistic process. For example, this form covers Ostwald ripening and classical Burke and Turnbull grain growth. These processes have coarsening exponents of 3 and 2 with rate-controlling processes of diffusion and boundary mobility, respectively.

4.2 Limiting Solution Methods

The key idea that emerges from these postulates is that the time and temperature dependence from depinning from obstacles is often rather weak and dislocation-obstacle depinning can be described simply as a critical force. In such cases, the major time and temperature dependencies can instead come from the thermally activated, diffusion-controlled coarsening of the dislocation or obstacle field. Equations describing plasticity can be developed in one of two primary ways using these postulates. First, if the kinetics of depinning from obstacles is simplified into a single parameter that is the force required to free a dislocation from an obstacle, $\hat{k}$, which may be temperature or rate dependent, the flow stress of the material can be approximated as:

$$\tau_o = \frac{\hat{k}}{b\lambda}. \tag{20}$$

Here, λ varies with time, decreasing with increasing strain, by dislocation generation, and coarsening increases it with time. This has been developed into an approximate model elsewhere (Daehn et al. 2004).

Alternately, fewer assumptions can be made and dislocation generation by plastic flow, recovery and the release from obstacles can be considered simultaneously (Brehm and Daehn 2002). So long as the assumptions in the postulates listed are adhered to, this produces results similar to the very simple model.

The effects of varied obstacle sizes have also been considered and this, in itself, can lead to interesting transient behavior (Daehn 2001). This approach recognizes that the local stress at any location in a microstructure deviates from the remotely applied value because of load shedding from one grain to another and one dislocation to another. This can be modeled by cellular automaton, dislocation dynamics, crystal plasticity, finite elements, or other methods.

4.3 Scaling Assumptions

In this work, we wish to broadly compare materials with different crystal structures, melting points, and also those with varied detailed dislocation structures with different possible strengthening interactions. A general framework can be developing a description for a general material and obstacle. These are considered separately and the minimum parameters required for description are developed.

4.3.1 Basic Material Parameters

At a very simplistic level, a material can be described by four terms: its melting point, T_m; the magnitude of Burgers' vector, b, the shear modulus, μ, and the

Debye frequency, ν. The homologous temperature, T_h, is defined as the absolute test temperature, T divided by melting point, T_m. T_h ranges from zero to one. Burger's vector magnitude, b, represents the closest interatomic spacing in metals and it also varies only very modestly for all structural metals. It ranges from about 0.2 to 0.4 nm with many metals having a value of about 0.25 nm.

Shear modulus varies slowly with temperature relative to other materials properties. Sherby et al. (1977) have shown that normalized stress, defined as applied stress normalized by elastic modulus, $\tau/\mu(T)$, where τ is the applied shear stress and $\mu(T)$ is the elastic shear modulus, is an effective scaling parameter. This term is similar in quantity to the axially applied stress normalized by Young's modulus. These normalizing parameters have been commonly used by Sherby and coworkers and have been shown to provide good scaling in Frost and Ashby's analysis (Frost and Ashby 1982).

The attempt frequency, ν, is taken as the Debye Frequency. This value is in the range of 10^{11} to 10^{13} s^{-1} for transition metals and will be assumed to be a constant of 10^{11} in this work.

To facilitate broad comparisons of materials, it is important to link the energetic barrier $s_1 \mu b^3$ with the material's melting point. Crystals with higher melting points generally have nearly proportionately higher values of μ, while values of b are quite similar for most transition metals, so that a scaling can be created comparing the two energy terms kT_m and μb^3, by:

$$\mu b^3 = AkT_m. \tag{21}$$

In this treatment, generic values that are approximately correct for a wide range of metals are used presently. The term A Equation (20) varies only modestly over most structural metals and a value of 45 will be used in this chapter.

4.3.2 Obstacles

The prototypical interaction between a dislocation and an attractive obstacle in terms of force and energy is illustrated in Fig. 15. The energetic barrier that must be surmounted to remove a dislocation from an obstacle trap in the absence of stress is known as ΔF^* and is given as

$$\Delta F^* = s_1\, \mu b^3 s \tag{22}$$

where s_1 is one of the two adjustable parameters that describes the obstacle. The force required in the absence of any thermal activation required to remove a dislocation from an obstacle trap is defined as $\hat{k}$, and is scaled to base materials properties by s_2 as:

$$\hat{k} = F_{\max} = s_2\, \mu b^2. \tag{23}$$

The elastic interaction between a solute and dislocation is the prototype for a "small" barrier. The intersection of two dislocations is taken as the prototype of a "medium" barrier and a dislocation on a dispersoid is considered as a "large" barrier example. General scaling will be introduced after these examples are introduced. The key idea in each of these examples is that the athermal strength of a material can often be simply approximated as the product of obstacle strength and the density of the obstacles along the dislocation line.

A goal of this approach is to be able to develop a common description that can function over a very wide range of temperatures and materials.

Example – Dislocation Solute Interaction

As a baseline, one can consider the interaction of a misfitting solute atom and a dislocation through the energetic interaction $P\Delta V$ for a volumetric interaction and $\tau\Delta\gamma\Omega$ for a shear interaction, where P or τ represent the pressure or shear stress due to a dislocation and ΔV and $\Delta\gamma$ represent the volumetric change or shear strain change that is introduced by the presence of a solute atom. The atomic volume is represented by Ω. In either case, the dislocation will interact with the solute when the energy of the system decreases and the energy reduction can be expressed in the form of (18), where s_1 is an interaction parameter that is on the order of 0.1, or less (Argon 2008). Assuming the interaction can be considered a local energy well, the local force required to detach the solute from the dislocation in the absence of any thermal activation is the derivative of energy with respect of position. This yields values of s_2 between about 0.05 and 0.1. By simple force balance dislocation lines will form soft cusps at these points of attractive interaction and the average angle of inclination at detachment without thermal activation will be approximately $\sin^{-1}(s_2)$.

Example – Intersecting Dislocations

Two intersecting dislocations will form a node and a new dislocation segment in accord with Franks rule. These well-known reactions will reduce the energy of the system (Argon et al. 1960; Baird and Gale 1965). Such structures will be stable over a given range of interaction angles, which can also be expressed as an athermal breaking angle or force. In a very simple example of a common reaction in BCC metals, two $1/2\,\langle 111\rangle$ type dislocations may react to form a <100> type dislocation. In that reaction, by simple elastic considerations between parallel initial dislocations, there is a loss in energy per length of dislocation of over $0.25\,\mu b^2$, per length of dislocation reacted. As the reacted segments are often many Burgers' vectors in length, these defects are very energetically stable ($>100\,\mu b^3$). Hence, once these intersections form, they are not subject to release by thermal activation, except at stresses close to what is required for decohesion of the junction. Hence, s_1 values can greatly exceed unity and s_2 values can range from small to large.

Example – Dispersoid Interaction

Precipitation and dispersoid hardening has been treated in detail by a number of authors (Ashby 1966). The most common assumption is that the dislocation must either cut through or loop around the particle. Experimental observations at elevated temperature show that as stress approaches a lower limit "threshold stress" the strain-rate tends quickly to zero.

If dislocation line tension is assumed to be equal to 0.5 μb^2, then the maximum effective value for s_2 is unity or else bypass by looping will take place. Thus, the upper bound of s_2 is 1.0. There can be a significant energy reduction when a dislocation situates at the dispersoid–matrix interface. This is well described by the theory of high temperature creep of dispersion hardened metals developed by Rosler and Arzt (1990). They model the reduction in dislocation line tension, T, by an amount k when it resides on an interphase boundary. Numerically, k takes on values from 0 to 1, where 0 represents full relaxation and 1 indicates no energy loss. Typical values for k are between 0.7 and 0.95. Thus, the total energy reduction per unit length of dislocation line by the interaction is $(1-k)T$ (where the line tension is often approximated as 0.5 μb^2). Hence,

$$s_1 = \frac{n}{2}(1-k). \tag{24}$$

Here, n represents the length of the dislocation line along the particle–matrix interface, as expressed as a number of Burger's vectors. Because n is typically on the order of thousands, even at large k values, s_1 can be very significant, meaning that these interactions are not easily subject to thermally activated detachment.

Relative Sizes of Obstacle and Diffusion Barriers

For most structural metals, diffusion activation energy is about 0.25 μb^3, or the diffusion activation energy can be represented by an s_1 value of 0.25. This is consistent with a self-diffusivity at a metal's melting point of about 10^8 cm^2/s, as seen in many metals (Shewmon 1963). If a dislocation release process has a much higher activation energy than that for diffusion, diffusion may remove or modify the obstacle rather than the dislocation bypassing it as-is.

It can be useful to bear in mind the absolute size of the term inside the Arrhenius exponential. If $\Delta G/kT$ (or $s_1 A/T_h$, in its normalized form) is less than 30, the rate-limiting events are relatively rapid, taking place many times per second. On the other hand if the term $\Delta G/kT$ is greater than 50, thermal activation is nearly impossible, periods of weeks to years may take place between discrete enabling events. Thus, events that are frequent, possible, and very unlikely by thermal activation can be easily separated. The general magnitudes of these barriers is summarized in Table 2.

Table 2 Normalized parameters for obstacles and diffusion

	Small: solute strengthening	Medium: dislocation Intersections	Large: dispersoid Hardening	Diffusion
s_1	0.1	>1	>1	0.25
s_2	< 0.1	~0.5	~1	N/A

4.4 Examples of Use of the Obstacle Controlled Creep Approach

Using the scaling arguments above, very simple models will be built that can describe the broad behavior of materials for three common classes of strengthening mechanisms. The models will build on each other in terms of complexity, but they will illustrate the broad behavior seen in actual materials. The three classes of models are:

1. A fixed field of obstacles to dislocation motion where there is neither strain hardening nor recovery. Oxide dispersion strengthened alloys are the prototype of this class. If obstacles are strong, high stress exponents and deformation activation energies can be expected.
2. Pure metal behavior is argued to be controlled by the balance between dislocation motion, breeding new dislocations that give the material strength, while at elevated temperature recovery removes dislocation density, concurrently reducing material strength. The balance between these two processes provides the essence of the creep process for pure metals. If recovery is diffusion controlled, the activation energy for diffusion may be expected to be similar to that for creep.
3. Engineered metals, including metal matrix composites and superalloys, behave in manner that is intermediate between these two limits, and as such may have activation energies and stress exponents intermediate between these pure metals and engineered systems with fixed obstacles.

4.4.1 A Model for Temperature-Dependent Strength at Fixed Structure Applied to Oxide Dispersion Strengthened (ODS) Alloys

In this section, a material with a fixed array of obstacles is considered. New obstacles are not created by strain hardening, nor does the obstacle field diminish by coarsening. This set of assumptions may be appropriate for a dispersion-hardened material. In later sections, the model will be extended to include structural evolution of the obstacle field.

When subjected to an applied shear stress, τ, dislocations are subjected to a forward force per unit length assumed to be equal to τb. Further, it is assumed that their motion is resisted by interaction with a uniformly spaced array of obstacles, in a manner similar to the classic Orowan strengthening model. If the obstacles are spatially discrete, but infinitely strong, bypass can take place at the Orowan stress(with a line tension assumed to be as $\mu b^2/2$), which represents the maximum possible flow stress for the material with this interobstacle spacing:

$$\tau_o = \frac{\mu b}{\lambda}. \tag{25}$$

Gross deformation can take place if the dislocations can overcome the obstacles, at a stress just beyond the Orowon stress. Weaker obstacles, such as solutes, short range ordered regions, strain fields or small shearable precipitates, or dislocations attached to dispersoids by attractive interaction, can often be overcome or cut with forces applied to the obstacle that are much less than the force required for Orowan looping. Equating the forward force on the dislocation line with the restraint provided by the obstacles, the flow stress in the absence of thermal activation is then

$$\tau_f^o = \frac{s_2 \mu b}{\lambda} = \frac{\hat{k}}{\lambda} = \hat{\tau}, \tag{26}$$

where s_2 is in the range of zero to unity, as described earlier. This equation provides the flow strength of the material in the absence of thermal activation (or the ideal strength at a temperature of zero Kelvin). This provides a term very similar to the strength at zero Kelvin known as $\hat{\sigma}$ in the Mechanical Threshold Stress (MTS) model (Follansbee and Kocks 1988).

Of course, thermal activation can assist overcoming obstacles. In this case, a description of how the energetic barrier changes with the force applied to the obstacle is required. The common Arrhenius form is based on the probabilistic activation over a barrier and sets the rate at which a given obstacle will be overcome. The rate of release rate from a particular obstacle, R, can be expressed as:

$$R = \nu \exp\left(\frac{-\Delta G^*}{kT}\right), \tag{27}$$

where ΔG^* is the barrier energy. The energy barrier depends on the structure of the obstacle and is a function of the applied stress. This approach is schematically illustrated in Fig. 15 and similar to the formalisms developed by Kocks, Argon, and Ashby (KAA) (Kocks et al. 1975), which is also summarized in Argon's recent book (Argon 2008).

The process of dislocation release can be simply developed into a strain rate. First, it is assumed that the dislocation release rate is rate controlling, and negligible time is required for the dislocation to get to the next obstacle. This is reasonable for most metals at high homologous temperature as the Peierls stress will be ineffective and if solute drag may be weak. If the dislocation density is small enough that there is less than one dislocation poised on each obstacle (i.e., $\rho < \lambda^{-2}$) a dislocation moves a distance about equal to λ when released from an obstacle, the strain rate can be derived from Orowan's equation as:

$$\gamma = \rho_{\mathrm{m}} b \lambda \nu \exp\left(\frac{-\Delta G^*}{kT}\right). \tag{28}$$

The most important term in this equation is ΔG^*, which is of course a strong function of the applied stress and temperature modifies the exponential term strongly. This can be fully expanded into the normalized form as:

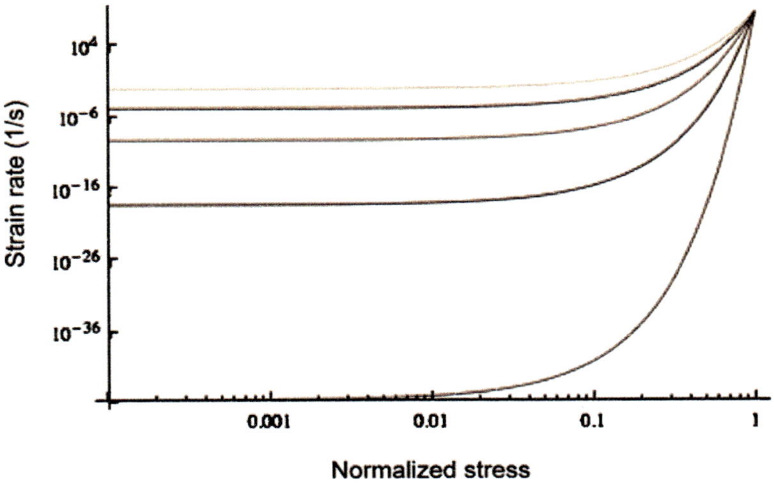

Fig. 16 Strain rate as a function of applied stresses calculated from (28) for homologous temperatures of 0.2, 0.4, 0.6, 0.8 and 1.0 T_m. Note that typically strain rates below about $10^{-12}\,s^{-1}$ are experimentally immeasurable. Stresses are normalized to the flow stress at 0 K

$$\gamma = \nu \rho b \lambda \exp\left\{ -\frac{s_1 A\left[1 - (\tau\lambda/s_2\mu b)^p\right]^q}{T_h} \right\}. \tag{29}$$

Again, the utility of this form is that standard values exist that are roughly describe all metals ($\nu = 10^{11}$, $b = 0.25$ nm, $A = 45$), and situation specific parameters that describe the material and the obstacles (λ, ρ, μ, s_1, s_2). This allows broad trends to be easily analyzed over many materials. The behavior of this equation is shown in Fig. 16 using barrier energy and strength parameters, $s_1 = 0.5$ and $s_2 = 1.0$. This corresponds to a barrier with a high mechanical threshold force with a modest, but significant, energetic barrier. All the curves show a similar form, if stress is very low, barriers will be surmounted thermally and the stress assist plays a very small role. The rate is represented by (26) with the full value of ΔF^* with the barrier, $s_1 \mu b^3$, used for ΔG^*. At the high-stress, low-temperature limit, the material will flow at the Orowan stress, even in the absence of thermal activation. The form of the curve between these limits depends on many factors, such as the values of p and q. The stress exponent, $n = \partial \log \dot{\gamma} / \partial \log \tau$, increases rapidly as temperature decreases and the Orowan stress is approached. It is also worthy of note that strain rates beneath about $10^{-11}\,s^{-1}$ are basically immeasurably small due to experimental limitations, so the low stress shelf seen in the graph may be quite common, but experimentally inaccessible.

If one is to engineer a material for high strength at high temperature, the key clearly is to have relatively deep energetic wells for dislocation interaction. This will produce measured activation energies that are much larger than those for self-diffusion. Further, following this model, for significant obstacles (on the order of

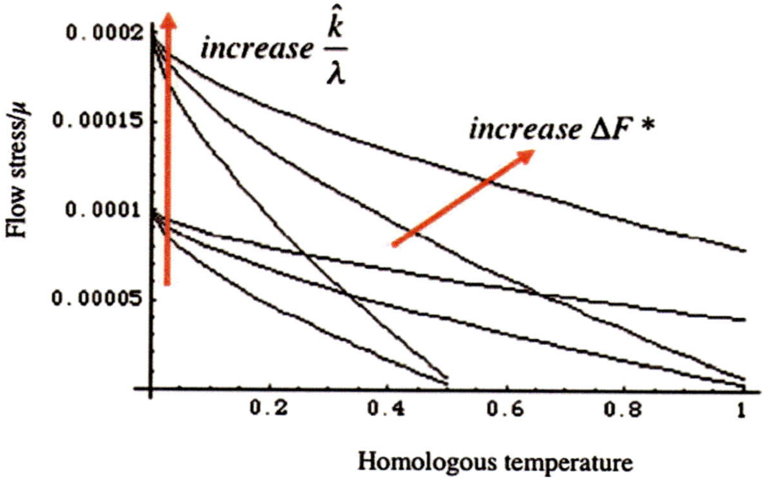

Fig. 17 The numerical solution to (28) is seen to give flow stress as a function of homologous temperature. The two intercepts at the left correspond to s_2 values of 1.0 (higher) and 0.5, in each case three s_1 values are considered: 25, 0.5, and 1.0. The smallest s_1 value gives the highest slope. In all cases, the strain rate is $10^{-3}\ \mathrm{s}^{-1}$ and the l/b value is set at 5,000

μb^3, or above), activation energies will generally be quite large, stress exponents can be quite high and both activation energies and stress exponents will be strong functions of stress as seen experimentally.

Equation (26) can also be solved to provide flow stress for a given material as a function of temperature for a given strain rate. This is shown in Fig. 17. The intercept at absolute zero is $\hat{\tau}$ from (22). As temperature increases, strength decreases rapidly for small energetic barriers and slowly for large energetic barriers. Mechanisms such as dispersoid interactions can have much deeper energetic wells and they maintain strength as temperature increases.

The prediction in this section has no arbitrary parameters; it is broadly consistent with the important and successful models of strength – Orowan's relation and the Taylor relation – but extends these in such a way to develop a temperature dependence of strength. The reduction in strength with temperature very much resembles experimental measurements. The predictions are particularly compelling when considering that the strength of an alloy is usually the result of more than one strengthening mechanism. Then the full strength-temperature curve can be thought as being made up of a large population of obstacles with a small barrier (such as a Peierls barrier or solutes) and a smaller density of large obstacles, such as dispersoids or precipitates. Then a "knee" in an experimental strength versus temperature curve can be justified.

This approach to modeling the deformation of dispersion hardened materials is similar in many ways to that developed by Rösler and Arzt. However, their approach has one important fundamental difference in that an additional term, D_v, representing the diffusivity of vacancies, is included as a multiplier in the final strain rate equation. In their work, this is justified by the assumption that vacancy absorption is required for dislocation release. It is presently argued that stress-assisted thermal activation is all that is required. With this, the major observations seen in the creep of oxide dispersion strengthened materials are captured.

A recent example from the work of Schneibel et al. (2009) demonstrates the fundamental behavior of this model to a system, where the obstacle structure is fixed by very small but very stable dispersoids of modest density. They have created a steel alloy that has a ferritic matrix and a high density of extremely small yttrium oxide nanoclusters about 2–4 nm in diameter. The nanoclusters are remarkably resistant to coarsening to temperatures up to 1,200°C. Examination via TEM shows clear attractive interactions between dislocations and the nanocluster dispersoids, as shown in Fig. 18. With respect to creep behavior for 800°C, at sufficiently high stress (beyond about 350 MPa) the material exhibits threshold-like behavior which monotonically decreases with decreasing stress to a stress exponent of about 1.5. This is very much in accord with the behavior schematically shown in Fig. 16. While the low stress-exponent at low stress is often attributed to diffusional creep, the strain rates seen are far too low to be consistent with these mechanisms. Here, it is proposed that the strainrate at low stress is due to the dislocations overcoming obstacles by thermal activation. While this model predicts a similar plateau where a strain rate persists as stress approaches zero, this has not been seen in many other systems. It is presently postulated that this is because in most cases the dispersoids are an order of magnitude (or more) larger in size and this produces a much larger energy well. This puts the stress-free strain-rate at immeasurably low values. Clearly, this hypothesis requires additional exploration.

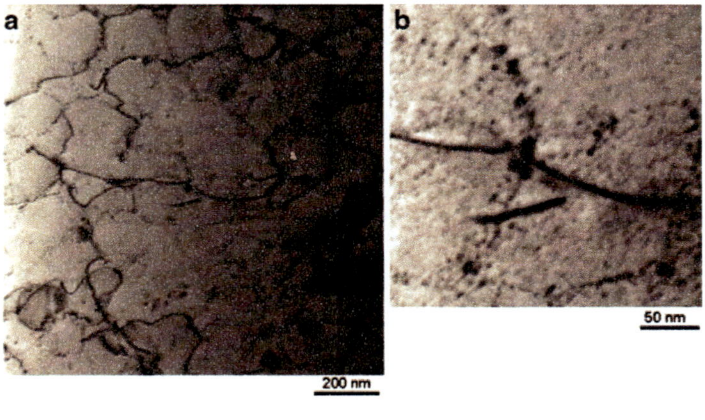

Fig. 18 (**a**) Bright-field STEM micrograph for a nanocluster-strengthened ferritic steel sample following creep deformation, showing strong dislocation pinning by the nanoclusters [magnified view in (**b**)], which appear dark in this image Schneibel et al. (2009)

4.4.2 Pure Metal Behavior and the Role of Structural Change

There are two first-order effects that cause significant structural change during metal deformation at elevated temperature: strain hardening by dislocation storage, which increases the obstacle density, and recovery or coarsening processes, which reduce obstacle density. The following section gives an account of a model that can reproduce the salient features of pure metal creep that are described in the introductory section of this paper with a very simple model. This is a somewhat abbreviated and simplified version of a model that has appeared elsewhere (Daehn et al. 2004; Brehm and Daehn 2002). For simplicity's sake, it is assumed that one type of obstacle (represented by s_1 and s_2) is present in the structure. In doing this, the structure evolution can be described by the evolution of a single structure parameter that represents the average distance between obstacles, λ. However, this approach can be extended to consider multiple populations of strengthening obstacles in the microstructure, which produces complex and interesting deformation transients (Daehn 2001; Suri et al. 1997).

Strain hardening decreases λ, while coarsening processes increase λ. At any moment, assuming obstacles are large, simply simply measuring the yield strength of the material at low temperature and solving (26), with an appropriate estimate of s_2, provide an estimate λ. Strain hardening, has been studied extensively at low temperature and there is a rich literature in this area that can greatly add to the first-order ideas presented here. Dynamic recovery is affected by stress, the current state of the material and possibly strain rate. However, these important details will be dispensed to provide a very simple model that can capture the physical essence of the process by assuming that time-independent, strain-based dislocation storage and time-dependent recovery (coarsening) are separate, distinguished, and independent processes and these processes can be described by separate simple equations.

The strain-hardening or dislocation storage part is modeled by an equation of the form:

$$\frac{d\rho}{d\lambda} = M\rho^c, \qquad (30)$$

where M and c are empirical constants, and ρ is the dislocation density; where c is typically in the range of 0–0.5, where a value of 0 represents a linear rate of dislocation accumulation with strain that may be expected if the density of features of responsible for dislocation generation do not change with strain. A C value of 0.5 is expected if the rate of dislocation storage is inversely related to the inter-dislocation spacing. Both of these exponents have been associated with Stage II hardening. And in Stage III and IV hardening the rate of hardening generally reduces presumably due to dynamic recovery. This has been discussed extensively (Honeycombe 1968; Argon 2008). However, the most extensive dataset, by Gilman (1969), tends to favor a value for c of 0 for most pure metals, which will result in parabolic hardening over a range of strains. It is assumed that this form holds true, and a time-dependent recovery mechanism causes the actual rate of hardening to appear to decrease with further strain.

The dislocations generated generally find their way into lower energy configurations including networks, cell systems, subgrains, and so forth. At elevated temperature, these structures will coarsen to reduce the total energy of the structure (seen as line or surface tension). While most models of creep have emphasized the dislocation climb process as enabling deformation, here that coarsening of the dislocation structure is envisioned to be the enabler of further flow. Atkinson (1988) compared the theories of grain growth in pure metal, single-phase systems to experimental observation. It was found that over a time increment Δt the initial average grain size $\overline{R}_0$ will grow to $\overline{R}_t$ as:

$$\overline{R}_t^m - \overline{R}_0^m = K(T)\Delta t, \tag{31}$$

where $\overline{R}$ is the average grain size, $K(T)$ is a temperature-dependent fitting constant that contains the temperature dependence of the rate-limiting fundamental process. The grain growth or coarsening exponent, m, is particularly interesting. The classical value of m is 2 and is expected from the seminal work of Burke and Turnbull (1952). However, both experiments and models that consider the interplay of the topological requirements for space filling with growth kinetics show a wider range of exponents are available. Experimental data shows the same range, with several metals showing grain growth exponents near 2.5 to 3 and aluminum has shown a value as high as 4 for subbrain growth. We presently assume that grains, subgrains, and even dislocation networks behave similarly in that they follow the same generalized coarsening equation, which has been shown to be quite general (Krill and Chen 2002; Siegert 1998; Sung et al. 1996; Sholl and Skodje 1995; Smilauer and Vvedensky 1995; Bray 1994; Fradkov and Udler 1994; Siegert and Plischke 1994; Chakrabarti et al. 1993; Yurke et al. 1993; Durian et al. 1991). One theme that persists in this work is that a wide range of coarsening exponents are available depending on detailed conditions (Ratke and Voorhees 2002).

In our implementation of the model, we adapt (29) to describe the coarsening of the substructure parameters, λ, as:

$$\lambda_t^m - \lambda_o^m = D(T)\Delta t, \tag{32}$$

where $D(T)$ represents temperature-dependent self-diffusivity and may include dislocation pipe diffusion and boundary diffusion in addition to lattice diffusion. There is ample evidence in the literature that diffusion may be the rate-limiting step in recovery and/or coarsening behavior (Oden et al. 1972; Masing and Raffelsieper 1949).

There are two important issues related to the coarsening equation, one is its form and the other is the calibration to a given material and mechanism. The coarsening literature shows that depending on whether the process is conservative or not; depending on long-range diffusion or not and depending on the topology of the defects, a range of exponents (that are clearly linked to the coarsening mechanism) are available ranging from $m = 2$ to about $m = 6$. Calibrating to a given material and defect type is more problematic, but we note here that data for subgrain growth by Huang and Humphreys (2000) show a coarsening exponent of about 3 and a temperature dependence similar to that for self-diffusion. Based on this evidence, (30) is adopted to describe recovery with diffusion as the preferred temperature-dependent process.

There are two relatively straightforward ways to develop a creep model based on the equations for refinement and coarsening presented above. First, one may assume that in a given time-step, a strain is imposed and decreases λ due to dislocation generation and strain hardening. Also, at the same time, increment λ increases due to coarsening. If temperature is low or λ is large, the strain hardening term tends to dominate. For small values of λ and higher temperatures, coarsening can be significant. In general if one starts with an annealed pure metal, as dislocation density increases, λ will decrease, increasing the coarsening rate until they balance, providing a pseudo-steady-state value of (Daehn et al. 2004):

$$\lambda_{ss} = \left(\frac{2K\, D(T)\, g^{2-2c}}{M\gamma \;\; Y} \right)^{1/(2+m_c-2c)} . \tag{33}$$

Then the strength of the material can be calculated in one or two ways. The first is by simply using the Taylor equation with a value for α which may reflect temperature (Daehn et al. 2004), or second, by using the thermally activated dislocation release equation (Brehm and Daehn 2002). Both approaches give very similar results and so long as the activation energy for dislocation release is significantly greater than that for self-diffusion, the coarsening process, limited by self-diffusion will set the temperature dependence for the creep process. The stress–strain rate relation using the closed-form approach, as:

$$\dot{\gamma} = BD(T) \left(\frac{\tau}{\mu} \right)^n = \left(\frac{2\, K\, g^{(2-2c)}\, (b\, s)^{-(2+m_c-2c)}}{M} \right) D(T) \left(\frac{\tau}{\mu} \right)^{(2+m_c-2c)} . \tag{34}$$

The equation provides a steady-state creep stress exponent equal to $n = (2 + m_c - 2c)$, where m_c is often observed to be in the range of 3–4 for most coarsening data. Strain-hardening experiments show c in the range of 0–0.5, with zero showing the better fit to experimental data. This immediately explains the commonly observed steady-state creep stress exponent between 4 and 6. Values of the stress exponent that are considerably higher than 5 can also be rationalized in this framework.

This model can provide quantitative predictions of steady-state strain rates. Parameters measured from noncreep experiments with the model provide a trendline that is near the center of the data set shown in Fig. 2 (Daehn et al. 2004). Deviations above or below can be explained by the relative facility of coarsening processes changing with factors, such as stacking fault energy and nascent dislocation and boundary structures. Also applied stress may increase the coarsening rate, which may also elevate the predicted curve.

This model suggests that in the creep of relatively pure metals, the dislocations and their intersections are the obstacles to further flow. These interactions may be too large to be broken by thermal activation. Instead of these breaking, they simply dissolve, or are consumed in the coarsening process and the newly released dislocation segments move through the structure, causing strain hardening. The key reason that the activation energy for creep is typically similar to that for diffusion is that diffusion is typically the rate-controlling step in this process is recovery or coarsening.

Thus, pure metals and ODS alloys represent two limiting cases. Both provide strong obstacles to dislocation motion, but in the pure metal, these obstacles are quite unstable over the time scale of the creep experiment and the structures are increasingly unstable as they become finer. With ODS alloys, the obstacles are very persistent over time and strain. Next, the case is made that many engineered alloys are intermediate between these limits.

4.4.3 Other High Temperature Engineered Alloys

Alloys engineered for high temperature strength usually fall between the two limiting cases above. Pure metals are very poor with respect to high temperature properties, and single-phase solid-solution strengthened alloys are only modestly better. Alloys for high temperature service can almost be universally described as multiphase materials with a significant volume fraction of second-phase particles. However, these alloys often have interparticle spacings that are much larger than those seen in ODS alloys and as a result, their high strengths cannot be described using the Orowan equation.

The primary role of the second-phase particles is argued to be to help stabilize a dislocation structure that has been developed by creep deformation or thermal-mechanical processing. This can be done by pinning some of the dislocation content to matrix–particle interfaces. The resulting network structure is therefore much more stable than it would be in the case of a pure metal, but is not nearly so stable as that in an ODS alloy. Because of the wide range of particle morphologies, interface conditions, volume fractions, dislocation structures that can develop, detailed or general approaches are very difficult to provide and a much more qualitative description is provided. In general, the primary obstacles to dislocation motion are assumed to be relatively strong interdislocation interactions as may be created by junctions. Strain hardening will increase the density of these obstacles, and diffusion-controlled recovery or coarsening processes will reduce their density and increase interobstacle spacing. The hardening and coarsening processes are much more complex in the presence of a significant population of interacting dispersoids. In particular, there is likely a lower-bound dislocation density that is approached even after very long-term annealing, which is somewhat analogous to a minimum grain size obtained in Zener pinning. Thus, at low stresses and long creep times the structure becomes somewhat fixed again. At this point, thermal activation may be required to move dislocations from the relatively deep energetic wells formed by interactions. At high stress, the substructure may be much finer than the interparticle spacing and the process may again be rate-limited by diffusion-controlled coarsening of the substructure. These essential points are illustrated in two examples from metal matrix composites and nickel-based superalloys.

Metal Matrix Composites

The 1990s saw intense development of aluminum composites reinforced with silicon carbide particles. These were typically created by powder-metallurgy methods with either equiaxed or whisker form fibers and processing usually involved extrusion to a final shape, which aligned with whiskers in the axial direction of the composite. Typical reinforcement dimensions were on the order of 1 μm in diameter and lengths from equiaxed to 20 μm in length. Volume fractions are typically on the order of 5–20%. For the finest-typical case of a 20% dispersion of 1 μm spheroidal particles, the interparticle distance is approximately 0.85 μm and this leads to an ideal Orowan stress of about 5 MPa in shear or about 10 MPa in tension, but the experimentally measured strengths are much greater. Further simple computation includes the assumption that the particles are uniformly distributed, which is seldom seen in these alloys, making these strength estimates upper bounds based on the Orowan mechanism.

The creep behavior of this class of materials has been studied by a number of researchers, and a number of clear trends are apparent. First, the composites are much stronger than a comparable un-reinforced alloy. Continuum modeling (Bao et al. 1991) shows that some strength increase can be expected due to load-shedding onto the stronger SiC particles and because additional local strain is needed to impart a given external deformation. But the continuum mechanics assumptions dictate that this should not change the stress exponent, or activation energy and the magnitude of these continuum strengthening is too small to explain the experimental observations. The experimentally measured stress exponents are much greater than those seen in the un-reinforced material. The activation energies are much greater than those for self-diffusion. These aspects all suggest a fundamental change in mechanism and are illustrated in the strain rate vs. stress plot shown in Fig. 19 from the PhD thesis of Y-C Chen (Chen 1991). The data and compilations in this document form a significant basis for the discussion in this section.

It is important to look at the transient behavior of these materials in order to gain insight into their deformation mechanisms. Fig. 20 shows the initial strain-time portion of a sample subjected to initial loading and stress-changes. This shows that after in an initial loading transient, the material responds to a stress-change by changing its strain rate almost instantly with no transient of the type that dominate in pure metals and alloys (seen in Fig. 20). This indicates that over the time-scale of the experiment the microstructure is effectively fixed and dislocations are moving over what is effectively a fixed set of barriers. The stress exponent is much greater than that for pure-metal-like creep and in many ways this is approaching what is expected for the fixed obstacles that are characteristic of ODS alloys.

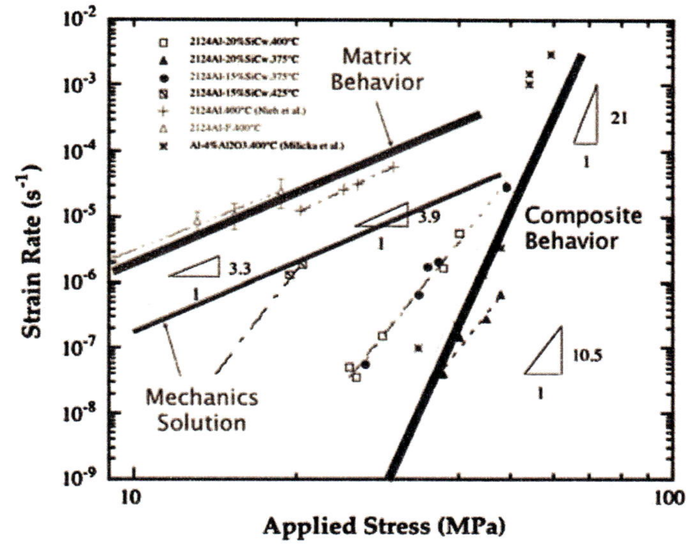

Fig. 19 Strain rate versus applied stress for a number of Al-SiC composites and the base alloys. The compilation is from Chen (1991) and contains work from (Nieh et al. 1989) and (Milicka et al. 1970)

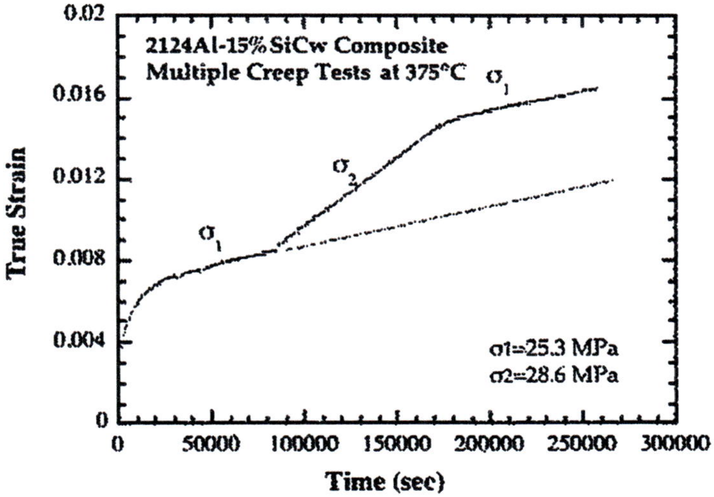

Fig. 20 Primary creep transient and stress change transients in a typical Al-SiC Composite (Chen 1991). Strain rates change nearly instantaneously after a stress change, suggesting that there is no significant structural change to the material due to the accumulation of plastic strain or time

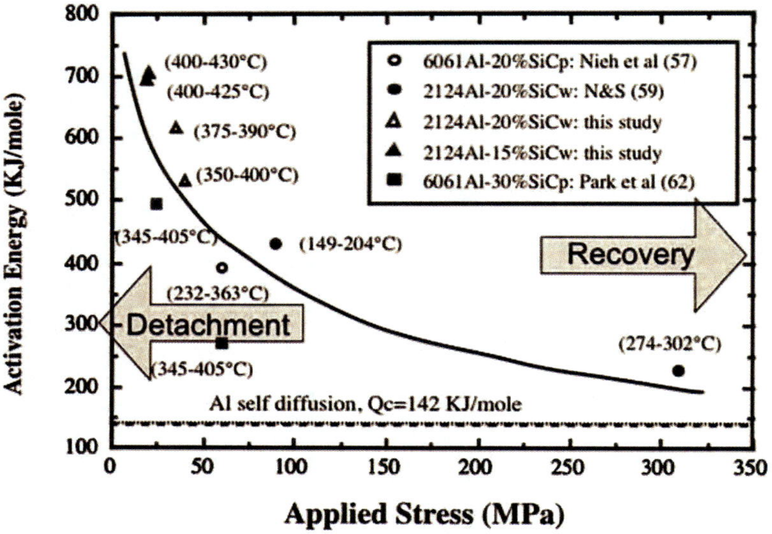

Fig. 21 Compilation of creep activation energies for a number of Al-SiC Composites as a function of applied (Chen 1991). It is argued that at low stresses detachment of dislocations from a relatively stable substructure is rate controlling (Nieh et al. 1989; Nardone and Strife 1987; Park et al. 1990)

Figure 21 shows a compilation of data from a number of studies, including the data shown in Figs. 19 and 20. These shed further light on the potential deformation mechanisms. This shows that while the creep activation energy remains above that for self-diffusion for a wide range of stresses, there is a systematic decrease in activation energy with increasing stress and at the highest stresses, the creep activation energy approaches that for self-diffusion. The present interpretation of this is that the particles strongly stabilize a dislocation structure with an interdislocation spacing that is quite stable and considerably smaller than the interparticle spacing. At low stresses, the material behaves as if the obstacle distribution is fixed, similarly to the ODS alloy case. As stress increases, the network becomes finer and is subject to diffusionally controlled coarsening and the activation energy decreases tending to that for self-diffusion. A detailed model for this is necessarily complicated and system-specific. Other explanations for this kind of behavior often include the invocation of a temperature-dependent threshold stress, beneath which deformation will not proceed (Li and Mohamed 1997; Park and Mohamed 1995; Mohamed et al. 1992; Cadek et al. 1995). This class of model require experimental determination of a threshold stress. These threshold stresses are typically higher than can be justified by the Orowan equation and measured interparticle spacings.

Creep of Superalloys at High Temperature

While superalloys have seen much more commercial development than metal matrix composites, there are many aspects of their deformation behavior that are still not well understood. Nickel-based superalloys are based on two primary phases, an FCC phase known as γ and an ordered $L1_2$ structure of Ni_3Al known as γ'. Key variables include the volume fraction of the γ' phase and the lattice mismatch between the two phases. These materials have a microstructure that is considerably less stable than either the ODS alloy materials or MMCs in that the γ' phase can coarsen and change morphology over time at high temperatures. Morphological changes can include rafting, where the γ' precipitates align themselves perpendicular to the stress direction and full morphological inversion, where the γ' phase becomes continuous.

From a macroscopic point of view, creep is anomalous with respect to pure metal creep. The stress exponents (usually based on minimum creep rates) are in the range of 7 or greater and activation energies are generally significantly greater than that for self-diffusion (Reed 2006; McKamey et al. 1998; Song et al. 2000). Even without delving into the fine mechanistic details of how deformation takes place, the general approach theory developed above may be useful to explain the high temperature creep behavior of superalloys. Here, we propose that a primary role of the γ' particles is stabilizing the dislocation network and dramatically inhibiting dislocation network coarsening. The γ' particles dramatically reduce the degrees of freedom available for the dislocations to engage in a coarsening process. Dislocation line energy is reduced by locating at the $\gamma - \gamma'$ interface and can again produce an effect much like particles do in promoting Zener pinning.

There is evidence for this basic postulate. Increasing the misfit between γ and γ' is known to stabilize a reduced dislocation network spacing. A series of papers by Koizumi and co-workers (Zhang et al. 2003, 2004, 2005a, b) has shown the close correlation between the inter-dislocation spacing and minimum creep rate. In particular with a series of related superalloys, they were able to reduce the steady-state strain rate by a factor of over 30 by systematically increasing the misfit to reduce the inter-dislocation spacing in the network from about 65 to less than 25. This supports the idea that attractive dislocation interactions to the interfaces are a primary strengthening mechanism and by stabilizing a fine network, the material is strengthened.

It is proposed that an important role of the γ' particles and $\gamma - \gamma'$ interfaces in particular is in stabilizing a fine dislocation network that would otherwise be quickly coarsened. Strain-transient, stress-drop creep data presently gives support for this claim. Consider that in pure metals, the creep rate following a stress drop typically recovers rapidly to the expected "steady-state" value (Sherby and Burke 1967; Wilshire and Battenbough 2007). Conversely, similar experiments were reported by Han and Chaturvedi on IN-718, a superalloy strengthened with a relatively low volume fraction of γ'' precipitates, at an intermediate homologous temperature of 0.67 T_m (Han and Chaturvedi 1987a, b; Chen and Chaturvedi 1994). They found that upon stress drop, a near-zero creep rate persists for many hours without return to the expected steady-state value at the reduced stress. They invoked the idea of a threshold or backstress to explain this remarkable behavior. An alternative explanation is that the most important strengthening element in the superalloys is the second-phase-stabilized dislocation substructure that coarsens more slowly than in a single-phase metal. Further systematic transient experiments over a range of temperatures may enable a more quantitative extraction of these critical coarsening rates for the high temperature deformation of the superalloys.

4.5 A General Approach to Obstacle-Controlled Strengthening

The three examples listed above are all variants of obstacle-controlled time-dependent plasticity. In all cases, it is assumed that the pinning points for the obstacles are very strong, based on attractive dislocation–defect interactions. The difference between the three cases is the stability of the obstacles. A simplified approach to creep in pure metals, particle-strengthened alloys, and oxide dispersion strengthened alloys is given in Fig. 22. The first of these represents pure metal creep. To aid an intuitive understanding of this process, assume that a strain rate can be

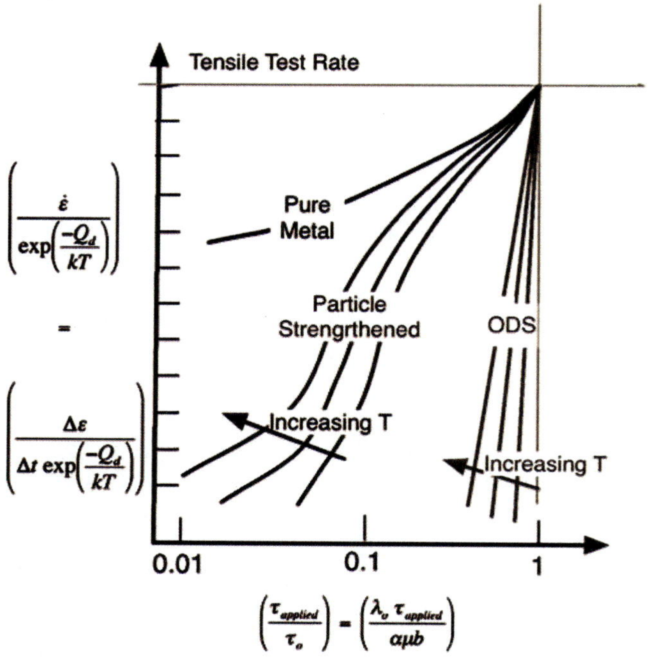

Fig. 22 Schematic diagram of scaled creep pure metals, oxide dispersion strengthened alloys, and pure metals show on log–log axes. The intent of this is to emphasize that it is the evolution of the obstacle spacing l with time that is the dominant aspect in creep. This can be qualitatively understood that by traveling downward on the y-axis time increases and unstable obstacles such as dislocation networks in pure metals will coarsen quickly (and this may be diffusionally controlled). Stable oxide dispersoids will not significantly coarsen, so these materials will largely maintain strength with time but strain rates will not scale with diffusivity

thought of as a separate strain increment, $\Delta\varepsilon$ and divided by a time increment Δt. Initially, the imposed strain increment refines the average interdislocation spacing to a value λ_o. The obstacle density can give the strength by an equation can be approximated as Taylor's equation (16) and the yield strength of the material is therefore simply inversely related to the distance between dislocations. The initial strength at short time can be plotted in normalized form on the graph as:

$$\left(\frac{\tau_{\text{applied}}}{\tau_o}\right) = \left(\frac{\lambda_o \tau_{\text{applied}}}{\alpha\mu b}\right), \tag{35}$$

where α is a term similar to s_2 that represents the effective pin strength at the temperature and strain rate of study. In the plot, decreasing strain rate is equivalent to increasing the time available for coarsening, and this can be shown on the abscissa label in the log–log plot. Over time, the interobstacle spacing will increase in accord with the coarsening law (29), which is presumably diffusion controlled, as much data suggests. The slope of the curve can be developed by considering the yield strength of the structure after coarsening of the dislocation network has taken place, and this can be seen as the material becoming much weaker with time due to a recovery annealing process. That is to say λ is increasing relatively rapidly with increasing with time. In this case, even if there is no thermal activation of release from obstacles, the approximate 5-power-law behavior can be predicted, and strain rate will scale with the rate of self-diffusion, so the usual scaling with the activation energy for self-diffusion will cause creep data taken at different temperatures to collapse to one curve.

Of the three cases shown, the simplest is the ODS alloy behavior. In the case of the ODS alloy, this is effectively "constant-structure" creep as was defined and studied by Gibeling and co-workers (Nix et al. 1985; Nakayama and Gibeling 1990). Dislocations are pinned at very stable oxide particles. The pin spacing, λ, is effectively fixed with time because of the very high stability of dispersed oxide particles. In this case, flow is activated by force on obstacles and stress, and hence, the effective activation energy decreases with increasing stress, because the remnant activation barrier decreases. Therefore, the scaling with diffusion activation energy will not hold and at lower stresses, the stress exponent may increase somewhat because thermal detachment dominants. Broadly, the mechanics of detachment of a dislocation from an obstacle may be very similar to that in pure metals, but coarsening of the obstacle field is absent. The idea that the obstacle density in creep is fixed in ODS alloys but a function of the applied stress in pure metals is supported by stress-change data (Biberger and Gibeling 1995) that demonstrates that the true activation areas for lithium fluoride, aluminum, and copper all scale inversely with the creep stress prior to the stress change. This again indicates that the Taylor–Orowan relation is valid in creep – at least over time periods where the structure is stable.

The multiphase alloys that are most common in elevated temperature service fall between these two bounds. The particles partially stabilize a dislocation structure by encouraging the stable tangling of dislocation structures that may not be able to unwind with the particles present. The particles themselves may be subject to coarsening themselves, and this may happen by a different type of diffusion control and over a different time-scale than that of the creep experiment. At low stresses, for stable particles, the creep activation energy may be expected to be much greater than self-diffusion as high stress exponents as thermally activated detachment controls. Here, the alloys resemble the ODS case. At sufficient high stresses, a fine structure may form between the particles and diffusionally controlled coarsening of a substructure that is much finer than the interparticle spacing, may again be the dominant factor in the creep process, and pure metal behavior is approximated. Also at very long times, diffusional coarsening of the particles may also be rate controlling. For these reasons, particular materials and situations will always be very microstructure specific.

In summary, thermally activated depinning of obstacles is a pervasive mechanism of deformation of crystals that is important from very low to high temperatures. This reduces our need to label a large number of specific "mechanisms." The formalisms developed by Argon, Kocks, and colleagues can be well adapted to modeling obstacle–dislocation interactions. The challenge is in modeling the structure evolution of the pertinent obstacles, both in strain hardening and in coarsening or recovery. While models are presented for the limiting cases of pure-metal and constant structure creep, there is much room for further research in understanding microstructure evolution in the creep of engineering materials.

5 Conclusions

In this chapter, we have reviewed the variety of dislocation-mediated, high-temperature creep modes that are broadly classified into mobility- and obstacle-controlled deformation categories. The understanding of creep deformation has been dominated by the phenomenological approach exemplified by Sherby and Burke. In this chapter, we have presented several examples for which the macroscopic deformation phenomenology can be rationalized based on relatively simple, physically based models using characteristic dislocation processes. For the mobility-controlled category, these processes include deformation controlled by Peierls barriers, pure climb, jogged-screw dislocations, and solute atmospheres. For the obstacle-controlled category, we describe a framework whereby thermal activation from various attractive barriers, including solute atoms, dislocations and dispersoids, can yield a wide range of creep phenomenology. While these models offer significant promise, they rely upon underlying assumptions for dislocation characteristic lengths/densities, without specific treatment of dislocation multiplication, exhaustion, and annihilation processes. It is suggested that experimental measurment and modeling of these processes are now required to inform similar models capable of treating even more general (e.g., transient strain, stress, strain rate or temperature) deformation conditions.

References

Ahlquist, C. N., & Nix, W. D. (1969). *Scripta Metallurgica, 3*, 679.
Alexander, H., & Haasen, P. (1968). *Solid State Physics, 22*, 28.
Argon, A., Hirth, J., & Saada, G. (1960). *Acta Metallurgica, 8*, 841.
Argon, A. S. (2008). *Strengthening Mechanism in Crystal Plasticity*. New York: Oxford University Press.
Ashby, M. F. (1966). Work hardening of dispersion-hardened crystals. *Philosophical Magazine, 14*(132), 1157.
Atkinson, H. V. (1988). *Acta Metallurgica, 36*, 469–491.
Baird, J., & Gale, B. (1965). *Philosophical Transactions of the Royal Society, 257*, 68.
Baker, I. (2009). Deformation mechanism maps. http://engineering.dartmouth.edu/defmech/
Bao, G., Hutchinson, J. W., & McMeeking, R. M. (1991). *Acta Metallurgica, 39*, 1871–1882.
Barrett, C., & Nix, W. (1965). *Acta Materialia, 13*, 1247.
Barrett, C. R., & Sherby, O. D. (1965). Influence of stacking-fault energy on high-temperature creep of pure metals. *Transactions of the Metallurgical Society of Aime, 233*(6), 1116.
Barnett, D. M., Wong, G., & Nix, W. D. (1974). *Acta Materialia, 22*, 2035.
Biberger, M., & Gibeling, J. C. (1995). Analysis of creep transients in pure metals following stress changes. *Acta Metallurgica Et Materialia, 43*(9), 3247–3260.
Bodur, C. T., Chiang J., Argon, A.S. (2005). *Journal of the European Ceramic Society, 25*, 1431.
Bray, A. J. (1994). Theory of phase-ordering kinetics. *Advances in Physics, 43*(3), 357–459.
Brehm, H., & Daehn, G. S. (2002). A framework for Modeling creep in pure metals. *Metallurgical and Materials Transactions A – Physical Metallurgy and Materials Science, 33*(2), 363–371.
Burke, J. E., & Turnbull, D. (1952). *Progress in Metal Physics* (Vol. 3). London: Pergamon Press.
Cadek, J., Oikawa, H., & Sustek, V. (1995). Threshold creep-behavior of discontinusous aluminum and aluminum-alloy matrix composites - an overview. *Materials Science and Engineering A – Structural Materials Properties Microstructure and Processing, 190*(1–2), 9–23.
Caillard, D., & Martin, J. L. (2005). Dislocation motion controlled by interactions with crystal lattice: modelling and experiments. *International Materials Reviews, 50*(6), 366–384.
Chakrabarti, A., Toral, R., & Gunton, J. D. (1993). Late-stage coarsening for off-critical quenches: scaling functions and the growth law. *Physical Review E, 47*(5), 3025–3038.
Chang, J., Bodur, C. T., & Argon, A. S. (2003). Pyramidal edge dislocation cores in sapphire. *Philosophical Magazine Letters, 83*(11), 659–666.
Chen, Y.-C. (1991). Elevated temperature deformation and forming of aluminum-matrix composites, *Ph.D. Dissertation*, The Ohio State University.
Chen, W., & Chaturvedi, C. (1994). The effect of grain boundary precipitates on the creep behavior of Inconel 718. *Materials Science and Engineering, A183*, 81–89.
Coble, R. L. (1963). *Journal of Applied Physics, 34*, 1679.
Daehn, G. S. (2001). Modeling thermally activated deformation with a variety of obstacles, and its application to creep transients. *Acta Materialia, 49*(11), 2017–2026.
Daehn, G. S., Brehm, H., Lee, H., & Lim, B. S. (2004). A model for creep based on microstructural length scale evolution. *Materials Science and Engineering A – Structural Materials Properties Microstructure and Processing, 387–89*, 576–584.
Dorn, J. E. (1975). Paper presented at the Rate Processes in Plastic Deformation, Cleveland, OH.
Dorn, J. E., & Rajnak, S. (1964). Nucleation of kink pairs and the peierls mechanism of plastic deformation. *Transactions of the Metallurgical Society of AIME, 230*, 1052.
Durian, D. J., Weitz, D. A., & Pine, D. J. (1991). Scaling behavior in shaving cream. *Physical Review A, 44*(12), R7902–R7905.
Edelin, G., & Poirier, J. P. (1973). Study of dislocation climb by means of diffusional creep experiments in magnesium. 1. Deformation mechanism. *Philosophical Magazine, 28*, 1203.
Epishin, A., & Link, T. (2004). Mechanisms of high-temperature creep of nickel-based superalloys under low applied stresses. *Philosophical Magazine, 84*(19), 1979–2000.

Firestone, R. F., & Heuer, A. H. (1976). Creep deformation of 0∞ degree sapphire. *Journal of the American Ceramic Society, 59*, 24–29.
Follansbee, P. S., & Kocks, U. F. (1988). A constitutive description of the deformation of copper based on the use of the mechanical threshold stress as an internal state variable. *Acta Metallurgica, 36*(1), 81–93.
Fradkov, V. E., & Udler, D. (1994). Two-dimensional normal grain growth: topological aspects. *Advances in Physics, 43*(6), 739–789.
Frost, H. J., & Ashby, A. F. (1982). *Deformation-Mechanism Maps: The Plasticity and Creep of Metals and Ceramics*: London: Pergamon Press.
Furubayashi, E. (1969). *Journal of the Physical Society of Japan, 27*, 130.
Garret-Reed, A., & Taylor, G. (1979). *Philosophical Magazine, 39*, 597.
Gibeling, J. C., & Nix, W. D. (1981). Observations of anelastic backflow following stress reductions during creep of pure metals. *Acta Metallurgica, 29*(10), 1769–1784.
Gilman, J. J. (1969). *Micromechanics of Flow in Solids*. New York: McGraw Hill.
Groves, G. W., & Kelly, A. (1969). Change of shape due to dislocation climb. *Philosophical Magazine, 19*, 977.
Han, Y., & Chaturvedi, C. (1987a). A study of back stress during creep deformation of a superalloy inconel 718. *Materials Science and Engineering, 85*, 59–65.
Han, Y., & Chaturvedi, C. (1987b). Steady state creep deformation of superalloy inconel 718. *Materials Science and Engineering, 89*, 25–33.
Harper, J. D., Dorn J. E. (1957). *Acta Materialia, 5*, 654.
Hasegawa, T., Ikeuchi, Y., & Karashima, S. (1972). *Metal Science Journal, 6*, 72.
Hayes, R. W., Viswanathan, G. B., & Mills, M. J. (2002). Creep behavior of Ti-6Al-2Sn-4Zr-2Mo: I. The effect of nickel on creep deformation and microstructure. *Acta Materialia, 50*(20), 4953–4963.
Hirth, J. P., & Lothe, J. (1968). *Theory of Dislocations*. New York: Mc Graw Hill.
Honeycombe, R. W. K. (1968). *The Plastic Deformation of Solids*. London: Arnold Press.
Horiuchi, R., Yoshinaga, H., & Hama, S. (1965). *Transactions of the Japan Institute of Metals, 6*, 123.
Huang, Y., & Humphreys, F. J. (2000). Subgrain growth and low angle boundary mobility in aluminium crystals of orientation {110}{001}. *Acta Materialia, 48*(8), 2017–2030.
Ikeno, S., & Furubayashi, E. (1972). *Physica Status Solidi A, 12*, 611.
Ikeno, S., & Furubayashi, E. (1975). *Physica Status Solidi A, 27*, 581.
Imai, M., & Sumino, K. (1983). *Philosophical Magazine, A47*, 599.
Jones, B. L., & Sellars, C. M. (1970). *Metal Sciience Journal, 4*, 96.
Karthikeyan, S., Viswanathan, G. B., Vasudevan, V. K., Kim, Y. W., & Mills, M. J. (2001). *Mechanisms and Effect of Microstructure on Creep of TiAl-Based Alloys*. Paper presented at the Strucural Intermetallics 2001.
Karthikeyan, S., Viswanathan, G. B., & Mills, M. J. (2004). Evaluation of the jogged-screw model of creep in equiaxed gamma-TiAl: identification of the key substructural parameters. *Acta Materialia, 52*(9), 2577–2589.
Karthikeyan, S., Unocic, R. R., Sarosi, P. M., Viswanathan, G. B., & Mills, M. J. (2006). Modeling microtwinning during creep in Ni-based superalloys. *Scripta Materialia, 54*(6), 1157–1162.
Kassner, M. E. (1990). A case for Taylor Hardening during primary and steady-state creep in aluminum and type-304 stainless-steel. *Journal of Materials Science, 25*(4), 1997–2003.
Kassner, M. E., & Perez-Prado, M. T. (2004). *Fundamentals of Creep in Metals and Alloys*. Oxford: Elsevier Publications.
Kocks, U. F. (1976). Laws for work-hardening and low-temperature creep. *Journal of Engineering Materials and Technology-Transactions of the ASME, 98*(1), 76–85.
Kocks, U. F., Argon, A. S., & Ashby, M. F. (1975). Thermodynamics and kinetics of slip. *Progress in Materials Science, 19*, 1–281.
Kolbe M. (2001). The high temperature decrease of the critical resolved shear stress in nickel-based superalloys. *Material Science and Engineering*, 383, 319–321.

Kovarik, L., Unocic, R. R., Li, J., Sarosi, P., Shen, C., Wang Y., & Mills, M. J. (2009). Microtwinning and other shearing mechanisms in Ni base superalloys at intermediate temperatures. *Progress in Materials Science*, 54, 839–873.
Krill, C. E., & Chen, L. Q. (2002). Computer simulation of 3-D grain growth using a phase-field model. *Acta Materialia, 50*(12), 3057–3073.
Laks, I. H., Wiseman, C., Sherby, O. D., & Dom, J. E. (1957). *Journal of Applied Mechanics, 24*, 207.
Legrand, B. (1985). Core structure of the screw dislocations 1/3[1120] in titanium. *Philosophical Magazine A – Physics of Condensed Matter Structure Defects and Mechanical Properties, 52*(1), 83–97.
Legros M., Clement N., Caron P., & Coujou A. (2002). In-site observation of deformation micromechanisms in a rafted γ/γ' superalloy at 850 °C. *Materials Science and Engineering A 337*, 160–169.
Li, Y., & Mohamed, F. A. (1997). An investigation of creep behavior in an SiC-2124 Al composite. *Acta Materialia, 45*(11), 4775–4785.
Low, J., & Turkalo, A. (1962). *Acta Materialia, 10*, 215.
Luhban, J. D., & Felgar, R. P. (1961). *Plasticity and Creep of Metals*. New York: Wiley.
Masing, G., & Raffelsieper, J. (1949). Mechanische Erholung von Aluminium-Einkristallen, Zeitschrift für Metallkunde, Band 41, Heft 3.
McKamey, C. G., Carmichael, C. A., Cao, W. D., & Kennedy, R. L. (1998). Creep properties of phosphorous+boron modified alloy 718. *Scripta Materialia, 38*, 485–491.
Mecking, H., & Kocks, U. F. (1981). Kinetics of flow and strain-hardening. *Acta Metallurgica, 29*(11), 1865–1875.
Mills, M. J. (1985), A new theoretical interpretation of the high temperature deformation of solid solution alloys based on the steady state and transient creep properties of Al-5.5 at% Mg. *Ph.D. Dissertation*, Stanford University.
Mills, M. J., Gibeling, J. C., & Nix, W. D. (1985). A dislocation loop model for creep of solid-solutions based on the steady-state and transient creep-properties of A1–5.5 at percent-Mg. *Acta Metallurgica, 33*(8), 1503–1514.
Mills, M. J., Gibeling, J. C., & Nix, W. D. (1986). Measurement of anelastic creep strains in A1–5.5 at percent Mg using a new technique – implications for the mechanism of class-I creep. *Acta Metallurgica, 34*(5), 915–925.
Mills, M. J., & Miracle, D. B. (1993). The structure of A(100) and A(110) dislocation cores in NiAl. *Acta Metallurgica Et Materialia, 41*(1), 85–95.
Milicka, K., Cadek, J., & Rys, P. (1970). *Acta Metallurgica, 18*, 733–746.
Mohamed, F. A., Park, K. T., & Lavernia, E. J. (1992). Creep behavior of discontinuous SiC-Al composites. *Materials Science and Engineering A – Structural Materials Properties Microstructure and Processing, 150*(1), 21–35.
Moon, J. H., Cantonwine, P. E., Anderson, K. R., Karthikeyan, S., & Mills, M. J. (2006). Characterization and modeling of creep mechanisms in Zircaloy-4. *Journal of Nuclear Materials, 353*(3), 177–189.
Moon, J. H., Karthikeyan, S., Morrow, B. M., Fox, S. P., & Mills, M. J. (2009). High-temperature creep behavior and microstructure analysis of binary Ti-6Al alloys with trace amounts of Ni. *Materials Science and Engineering A* 510–511, 35–41.
Mott, N. F. (1956). *Creep and fracture of metals at high temperatures*. Paper presented at the Proc. NPL Symp, London.
Moulin, A., Condat, M., & Kubin, L. P. (1997). Simulation of Frank-Read sources in silicon. *Acta Materialia, 45*(6), 2339–2348.
Nabarro, F. R. N. (1948). *Rept. Conf. Strength of Solids*, Univ. Bristol.
Nabarro, F. R. N. (1967). Steady state diffusional creep. *Philosophical Magazine, 16*, 231.
Nabarro, F. R. N. (2004). Do we have an acceptable model of power-law creep? *Materials Science and Engineering A 387–389*, 659–664.
Nabarro, F. R. N. (2006). Creep in commercially pure metals. *Acta Materialia, 54*(2), 263–295.

Naka, S., Lasalmonie, A., Costa, P., & Kubin, L. P. (1988). The low-temperature plastic-deformation of alpha-titanium and the core structure of A-type screw dislocations. *Philosophical Magazine A – Physics of Condensed Matter Structure Defects and Mechanical Properties, 57*(5), 717–740.

Nakayama, G. S., & Gibeling, J. C. (1990). Creep of copper under constant structure conditions. *Scripta Metallurgica Et Materialia, 24*(11), 2031–2035.

Nardone, V.C., & Strife, J.R. (1987). Analysis of the creep behavior of silicon carbide whisker reinforced 2124 Al (T4). *Metallurgical Transactions A, 18A*, 109–114.

Neeraj, T., Savage, M. F., Tatalovich, J., Kovarik, L., Hayes, R. W., & Mills, M. J. (2005). Observation of tension-compression asymmetry in alpha and alpha/beta titanium alloys. *Philosophical Magazine, 85*, 279–295.

Nieh, T.G., Xia, K., & Longdon, T.G. (1989). *Scripta Metallurgica, 23*, 851–854.

Nix, W. D., Gibeling, J. C., & Hughes, D. A. (1985). Time-dependent deformation of metals. *Metallurgical Transactions A – Physical Metallurgy and Materials Science, 16*(12), 2215–2226.

Northwood, D. O., & Smith, I. O. (1984). Instantaneous strain and creep transients in an Al-7.72 at percent-Mg alloy. *Materials Science and Engineering, 66*(2), 205–212.

Oden, A., Lind, E., & Lagneborg, R. (1972). Creep Strength in Steel and High-temperature Alloys. Sheffield. American Society for Metals, Cleveland OH.

Park, K. T., & Mohamed, F. A. (1995). Creep strengthening in a discontinuous SiC-Al composite. *Metallurgical Transactions A, 26(12)*, 3119–3129.

Park, K. T., Lavernia, E. J., & Mohamed, F. A. (1990). High temperature creep of silicon carbide particulate reinforced aluminum. *Acta Metallurgica, 38*(11), 2149–2159.

Pollock, T. M, & Field, R. D. (2002). Dislocations and high temperature plastic deformation of superalloys single crystal. In: Nabarro FRN, Duesbery MS, Hirth J, editors. Dislocations in Solids. Amsterdam: Elsevier.

Ratke, L., & Voorhees, P. W. (2002). *Growth and Coarsening: Ostwald Ripening in Material Processing*. Berlin: Springer.

Reed, R. C. (2006). *The Superalloys: Fundamentals and Applications*. New York: Cambridge University Press.

Rosler, J., & Arzt, E. (1990). A new model-based creep equation for dispersion strengthened materials. *Acta Metallurgica Et Materialia, 38*(4), 671–683.

Schneibel, J. H., Liu, C. T., Miller, M. K., Mills, M. J. (2009). "Ultrafine-grained nanocluster-strengthened alloys with unusually high creep strength," *Scripta Materialia, 61*, 793–796.

Sherby, O. D. (1962). *Acta Metallurgica, 10*, 135–141.

Sherby, O., & Burke, P. (1967). Mechanical behavior of crystalline solids at elevated temperature. *Progress in Materials Science, 13*(7), 325.

Sherby, O. D., Klundt, R. H., & Miller, A. K. (1977). Flow stress, subgrain size and subgrain stability at elevated temperature. *Metallurgical Transactions A – Physical Metallurgy and Materials Science, 8*(6), 843–850.

Sherby, O. D., & Weertman, J. (1979). Diffusion-controlled dislocation creep: A defense. *Acta Metallurgica, 27*(3), 387–400.

Shewmon, P. G. (1963). *Diffusion in Solids*. New York: McGraw Hill.

Shewmon, P. G. (1969). *Transformation in Metal*. New York: McGraw Hill.

Sholl, D. S., & Skodje, R. T. (1995). Diffusion of clusters of atoms and vacancies on surfaces and the dynamics of diffusion-driven coarsening. *Physical Review Letters, 75*(17), 3158–3161.

Siegert, M. (1998). Coarsening dynamics of crystalline thin films. *Physical Review Letters, 81*(25), 5481–5484.

Siegert, M., & Plischke, M. (1994). Slope selection and coarsening in molecular beam epitaxy. *Physical Review Letters, 73*(11), 1517–1520.

Simmons, J. P., Rao, S. I., & Dimiduk, D. M. (1998). Simulation of dislocation single kinks in gamma-TiAl using embedded-atom method potentials. *Philosophical Magazine Letters, 77*(6), 327–336.

Smilauer, P., & Vvedensky, D. D. (1995). Coarsening and slope evolution during unstable epitaxial growth. *Physical Review B, 52*(19), 14263–14272.

Solomon, A. A. (1969). *Review of Scientific. Instruments, 40*, 1025.

Song, H. W., Guo, S. R., Lu, D. Z., Xu, Y., Wang, Y. L., Lin, D. L., & Hu, Z. Q. (2000). Compensation effect in creep of conventional polycrystalline alloy 718. *Scripta Materialia, 42*, 917–922.

Sriram, S., Dimiduk, D. M., Hazzledine, P. M., & Vasudevan, V. K. (1997). The geometry and nature of pinning points of 1/2 (110) unit dislocations in binary TiAl alloys. *Philosophical Magazine A – Physics of Condensed Matter Structure Defects and Mechanical Properties, 76*(5), 965–993.

Sung, L., Karim, A., Douglas, J. F., & Han, C. C. (1996). Dimensional crossover in the phase separation kinetics of thin polymer blend films. *Physical Review Letters, 76*(23), 4368–4371.

Suri, S., Neeraj, T., Daehn, G. S., Hou, D. H., Scott, J. M., Hayes, R. W., et al. (1997). Mechanisms of primary creep in alpha/beta titanium alloys at lower temperatures. *Materials Science and Engineering A – Structural Materials Properties Microstructure and Processing, 234*, 996–999.

Takeuchi, S., & Argon, A. S. (1976). Steady state creep of alloys due to viscous motion of dislocations. *Acta Metallurgica, 24*, 883.

Viswanathan, G. B., Vasudevan, V. K., & Mills, M. J. (1999). Modification of the jogged-screw model for creep of gamma-TiAl. *Acta Materialia, 47*(5), 1399–1411.

Viswanathan G. B., Sarosi P. M., Henry M. F., Whitis D. D., Milligan W. W., & Mills M. J. (2005). Investigation of creep deformation mechanisms at intermediate temperatures in Rene 88DT Superalloys. *Acta Materialia*, 53, 3041–3057.

Vitek, V. (1974). Theory of the Core Structure of Dislocations in BCC metals *Crystal Lattice Defects, 5*, 1–34.

Weckert, E. (1985). *Strength of Metals and Alloys*. Paper presented at the ICSMA7.

Weertman, J. (1957). Steady state creep through dislocation climb. *Journal of Applied Physics, 28*, 1185.

Weertman, J. (1968). Dislocation climb theory of steady-state creep. *ASM Transactions Quarterly, 61*(4), 681.

Wilshire, B., & Battenbough, A. J. (2007). Creep and creep fracture of polycrystalline copper. *Materials Science and Engineering A – Structural Materials Properties Microstructure and Processing, 443*(1–2), 156–166.

Wilshire, B., & Scharning, P. J. (2008). Creep and creep fracture of commercial aluminium alloys. *Journal of Materials Science, 43*(12), 3992–4000.

Wilshire, B., Scharning, P. J., & Hurst, R. (2009). A new approach to creep data assessment. *Materials Science and Engineering A 510–511*, 3–6.

Yi, J., Argon, A. S., & Sayir, A. (2006). Internal stresses and the creep resistance of the directionally solidified ceramic eutectics. *Materials Science and Engineering A, 421*, 86–102.

Yoshinaga, H., Toma, K., & Morozumi, S. (1976). *Japan Institute of Metals, 17*, 559.

Yurke, B., Pargellis, A. N., Kovacs, T., & Huse, D. A. (1993). Coarsening dynamics of the XY model. *Physical Review E, 47*(3), 1525–1530.

Zhang, J. X., Murakumo, T., Koizumi, Y., & Harada, H. (2003). The influence of interfacial dislocation arrangements in a fourth generation single crystal TMS-138 superalloy on creep properties. *Journal of Materials Science, 38*(24), 4883–4888.

Zhang, J. X., Koizumi, Y., Kobayashi, T., Murakumo, T., & Harada, H. (2004). Strengthening by gamma/gamma′ interfacial dislocation networks in TMS-162 – toward a fifth-generation single-crystal superalloy. *Metallurgical and Materials Transactions A – Physical Metallurgy and Materials Science, 35A*(6), 1911–1914.

Zhang, J. X., Koizumi, Y., & Harada, H. (2005a). Strengthening mechanisms in some single-crystal superalloys. 5th Pacific Rim International Conference on Advanced Materials and Processing, *475–479*, 623–626.

Zhang, J. X., Wang, J. C., Harada, H., & Koizumi, Y. (2005b). The effect of lattice misfit on the dislocation motion in superalloys during high-temperature low-stress creep. *Acta Materialia, 53*(17), 4623–4633.

Modeling Heterogeneous Intragrain Deformations Using Finite Element Formulations

Paul Dawson, Jobie Gerken, and Tito Marin

Abstract Polycrystalline materials exhibit deformation patterns that are heterogeneous both between and within crystals. The deformation heterogeneity within crystals can arise from variations of the crystallographic slip due to spatial variations in the stress driven by interactions among neighboring crystals. Typically, misorientations develop across crystals if the slip is not homogeneous. Furthermore, dislocations may accumulate within crystals, causing lattice distortion (elastic straining) and contributing to the stress. In this chapter, we summarize basic and extended crystal elastoplasticity formulations to address these effects. Finite element methodologies for both formulations are presented and examples of their use are discussed.

1 Introduction

Metallic alloys are the mainstay of structural materials. These alloys are complex systems that typically are polycrystalline and often are polyphase. Individual crystals comprising the materials exhibit well-defined lattice structures that possess certain structural symmetries dependent on the packing arrangements of atoms. Accompanying the symmetries are anisotropies of the mechanical properties, both elastic and plastic. Consequently, a crystal's response to applied loads depends on the spatial orientation of its lattice; so in an aggregate of crystals displaying a range of orientations, there exists a range of directional properties in relation to the load. This orientation-dependent behavior at the crystal level implies that stress and deformation are not uniform over the volume of a loaded polycrystalline aggregate.

P. Dawson (✉)
Sibley School of Mechanical and Aerospace Engineering, Cornell University,
Ithaca, New York 14853, USA
e-mail: prd5@cornell.edu

S. Ghosh and D. Dimiduk (eds.), *Computational Methods for Microstructure-Property Relationships*, DOI 10.1007/978-1-4419-0643-4_10,

The heterogeneity of stress and deformation has broad implications for the performance of a material, affecting average (macroscopic) properties ranging from the apparent stiffness of the material to the modes of failure it exhibits.

The finite element method offers a powerful tool for investigating the complexity of polycrystalline alloys. The method provides approximate solutions to a designated set of model equations and boundary conditions. For materials research, it allows the modeler to address many aspects associated with the structure of polycrystalline systems as well as with the physical behavior of the crystals, or grains, themselves. In principle, one can construct samples that reflect a material's grain morphology and phase topology in great detail. However, there are practical limits to the size of a simulation that constrain the volume of material that may be modeled if fine detail is sought. Within the same formulation, one can use complex constitutive models of the elastic and plastic behaviors to mathematically describe the behavior of the individual grains. Here, we consider plastic flow by crystallographic slip in which deformation is accomplished by the movement of dislocations causing a shift of close-packed crystallographic planes in close-packed directions. The number of slip systems defined by the crystallographic structure is relatively small, invoking the characterization of plastic flow as restricted slip, and endowing the individual crystals with a level of plastic anisotropy that can be quite high in some alloys. Furthermore, the model for slip that is implemented accounts for the strong nonlinearity in the kinetics of slip that itself promotes a greater degree of heterogeneity of the deformation within crystals.

The spatial heterogeneity of the deformation implies that the underlying slip and lattice rotation processes vary spatially as well. Gradients in slip and lattice rotation can produce several consequences. Because the lattice rotations accompany slip in most deformation modes, misorientations of the lattice are intimately tied to slip gradients. A grain discretized with many finite elements each having a common initial lattice orientation typically will display the distribution of orientations following some amount of strain. The misorientation across the boundary of regions of like orientation reflects the presence of geometrically necessary dislocations. Geometrically necessary dislocations leave a material volume in a distorted and stressed configuration locally even when external loads are absent. This stress acts together with the stress generated by externally applied loads in maintaining equilibrium and thus biases the activation of slip processes themselves. Thus, slip gradients passively affect the material through the formation of misorientation fields and actively alter the process of plastic flow through the stress field accompanying residual static dislocations.

In this chapter, we present methodologies for addressing these two effects of slip gradients associated with intragrain deformation heterogeneity. First, we summarize the governing equations and numerical formulation for a crystal-scale elastoplastic model based on a simple kinematic decomposition of slip, rotation, and stretch. We then quantify the orientational dependence of the misorientation field within grains of a polycrystal subjected to a macroscopically homogeneous deformation. Next, we introduce a modification of the kinematic decomposition that accommodates distortions arising from the presence of a static dislocation distribution.

The complete system of equations for this extended model is discussed along with the modifications needed in the numerical methodology for performing a simulation. The influence of static dislocation population on asymmetry of the yield stress is demonstrated through the example of the flexing of a thin foil. Finally, we discuss implications of the two examples in regards to issues in modeling the heterogeneity of strain with the grains of a deforming polycrystal.

2 Crystal Elastoplasticity Model Equations

Metals are capable of deforming elastically and plastically by a number of different physical mechanisms. Plastic flow occurs by different combinations of slip, twinning, and diffusion depending on the regime of temperature and stress (or strain rate) (Honeycombe 1984). A processing window can be defined based on the ranges of strain rate and temperature, which in turn determines the dominant deformation modes. Here we limit our attention to a processing window in which slip is the dominant mode of plastic deformation. This window is characterized by moderate strain rate and moderate homologous temperature (Ashby and Jones 1980).

2.1 Lattice Orientations and Orientation Distributions

The crystallographic lattices of crystals are oriented in space with respect to a reference. We specify orientations using the Rodrigues angle-axis parameterization, as described by Frank (1988)

$$\mathbf{r} = \mathbf{n} \tan \frac{\phi}{2}, \tag{1}$$

where $\mathbf{n}$ is the rotation axis and ϕ is the angle of rotation about this axis.[1] Here, a Rodrigues vector specifies the rotation needed to take base vectors aligned with the reference axes to coincidence with axes embedded with the lattice of a crystal in its spatial orientation. The rotation tensor equivalent to the Rodrigues vector is given by

$$\mathbf{R} = \frac{1}{1 + \mathbf{r} \cdot \mathbf{r}} \left(\mathbf{I} \left(1 - \mathbf{r} \cdot \mathbf{r}\right) + 2 \left(\mathbf{r} \otimes \mathbf{r} + \mathbf{I} \times \mathbf{r} \right) \right). \tag{2}$$

When the probability distribution of the lattice orientations of a population of crystals is not uniform (meaning that all orientations do not have equal likelihood of occurring in a sample), that population is said to have preferred orientation or texture. Mathematically, this is represented by an orientation distribution function

[1] The following notation convention is used: plain fonts are used for scalars and math bold fonts are used for vectors, higher-order tensors, and matrices. A superscript $'$ refers to the deviatoric part of a tensor and a superposed dot indicates material time differentiation.

(ODF), $A(\mathbf{r})$ (Bunge 1982). The ODF is defined over an appropriately reduced fundamental region of the orientation space. Fundamental regions are subregions of orientation space, constructed to contain only one of every set of orientations that are equivalent under crystal symmetries. Consequently, these are free of the ambiguity associated with assigning crystals to orientations taken from all of orientation space. Precisely, the ODF describes the local values of the probability distribution over the fundamental region, so that the crystal volume fraction enclosed within region Ω_{fr} of the fundamental region is given by

$$v_f = \int_{\Omega_{fr}} A(\mathbf{r})\,\mathrm{d}\Omega, \tag{3}$$

where $\mathrm{d}\Omega = \sqrt{\det \mathbf{g}}\,\mathrm{d}r_1 \mathrm{d}r_2 \mathrm{d}r_3$ is a volume element of orientation space and $\mathbf{g}$ is the metric tensor of the space. $A(\mathbf{r})$ is normalized to satisfy the condition that its integral over the fundamental region is unity.

2.2 *Crystal-Scale Elastic and Plastic Behaviors*

When loaded to a sufficiently high stress level, the deformation of crystals consists of elastic and plastic strains, as well as rotation (Kocks et al. 1998). Within the specified processing window, plastic deformation occurs by crystallographic slip between atomic planes of the crystal lattice, while elastic straining occurs by lengthening or shortening of the interatomic distances. In reality, slip is accomplished by a complex process of dislocation motion. In crystal plasticity, slip is idealized as a much simpler process in which only the net effect of dislocation motion on changes in shape and orientation of a body is retained, as depicted in Fig. 1. This elementary description gives rise to the representation of the kinematics of crystal deformations with a multiplicative decomposition of the deformation gradient, $\mathbf{f}$,

$$\mathbf{f} = \mathbf{f}^\flat\,\mathbf{f}^\star\,\mathbf{f}^\sharp = \mathbf{v}^\flat\,\mathbf{r}^\star\,\mathbf{f}^\sharp, \tag{4}$$

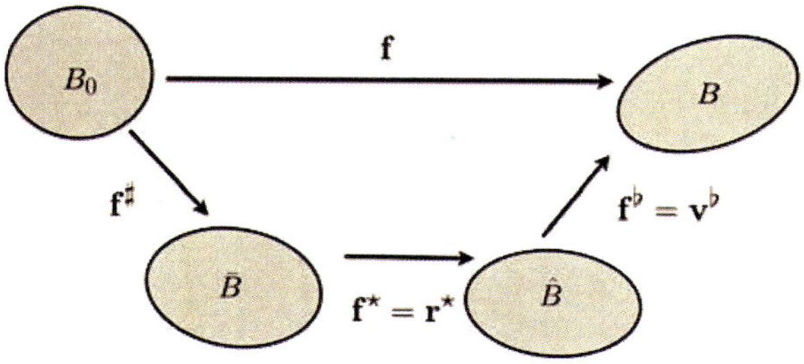

Fig. 1 Elementary kinematic decomposition for motion at the crystal level with plastic flow occurring via slip

where $\mathbf{f}^{\sharp}$ is that portion of $\mathbf{f}$ arising from slip, $\mathbf{f}^{\star}$ is a rotation coincident with the lattice reorientation, and $\mathbf{f}^{\flat}$ is elastic, arising from stretching of the lattice (Dawson and Marin 1998).[2] To emphasize that $\mathbf{f}^{\star}$ is purely rotational, it is written as $\mathbf{r}^{\star}$. The deformation gradient $\mathbf{f}^{\sharp}$ can be used to define an intermediate configuration, $\hat{B}$, which is a configuration obtained by unloading without rotation from the current configuration, B. Using this interpretation of $\hat{B}$, the symmetric left elastic stretch tensor, $\mathbf{v}^{\flat}$, is introduced. For the case of small elastic strains, $\mathbf{v}^{\flat} = \mathbf{I} + \mathbf{e}^{\flat}$, where the infinitesimal elastic (lattice) strain, $\mathbf{v}^{\flat}$, satisfies the constraint that $||\mathbf{e}^{\flat}|| << 1$ and $\mathbf{I}$ is the second order identity tensor.

From this decomposition, the kinematics then are expressed in rate form as

$$\mathbf{l} = \dot{\mathbf{f}}\,\mathbf{f}^{-1} = \mathbf{d} + \mathbf{w}, \tag{5}$$

where $\mathbf{d}$ is the deformation rate tensor and $\mathbf{w}$ the spin tensor, expressed in the current configuration B. These terms may be split into spherical and deviatoric parts, respectively, as

$$\mathrm{tr}(\mathbf{d}) = \mathrm{tr}(\dot{\mathbf{e}}^{\flat}) \tag{6}$$

and

$$\mathbf{d}' = \dot{\mathbf{e}}^{\flat\prime} + \hat{\mathbf{d}}^{\sharp\prime} + \mathbf{e}^{\flat\prime}\hat{\mathbf{w}}^{\sharp} - \hat{\mathbf{w}}^{\sharp}\mathbf{e}^{\flat\prime} \tag{7}$$

$$\mathbf{w} = \hat{\mathbf{w}}^{\sharp} + \mathbf{e}^{\flat\prime}\hat{\mathbf{d}}^{\sharp\prime} - \hat{\mathbf{d}}^{\sharp\prime}\mathbf{e}^{\flat\prime}, \tag{8}$$

where $\mathbf{d}'$ and $\mathbf{e}^{\flat\prime}$ are the deviatoric components of $\mathbf{d}$ and $\mathbf{e}^{\flat}$, respectively. The $\hat{\cdot}$ superscript indicates mapping forward by $\mathbf{r}^{\star}$ according to

$$\hat{\mathbf{w}}^{\sharp} = \mathbf{r}^{\star}\mathbf{w}^{\sharp}\mathbf{r}^{\star T} \tag{9}$$

$$\hat{\mathbf{d}}^{\sharp\prime} = \mathbf{r}^{\star}\mathbf{d}^{\sharp\prime}\mathbf{r}^{\star T} \tag{10}$$

to define the plastic deformation rate tensor, $\hat{\mathbf{d}}^{\sharp\prime}$, and the plastic spin tensor, $\hat{\mathbf{w}}^{\sharp}$, in the unloaded configuration $\hat{B}$.

The elastic response follows a linear relation

$$\boldsymbol{\tau} = \mathbf{c}\mathbf{e}^{\flat} \tag{11}$$

where

$$\mathbf{c} = \mathbf{c}(\mathbf{r}). \tag{12}$$

$\mathbf{c}$ is the tensor containing elastic moduli for cubic crystal symmetry which depends on the orientation of the crystallographic lattice, $\mathbf{r}$. The Kirchhoff stress, $\boldsymbol{\tau}$, is related to the Cauchy stress, $\boldsymbol{\sigma}$, through $\boldsymbol{\tau} = \beta\boldsymbol{\sigma}$, where $\beta = \det(\mathbf{v}^{\flat})$.

[2] Superscript symbols, $\flat, \sharp, \natural$, and $\star$ designate configurations associated with the mapping of coordinates as the crystal deforms.

The viscoplastic flow rule is written to mimic the crystallographic slip and is defined as

$$\hat{\mathbf{l}}^{\#} = \hat{\mathbf{d}}^{\#'} + \hat{\mathbf{w}}^{\#} = \dot{\mathbf{r}}^{\star}\mathbf{r}^{\star\mathrm{T}} + \sum_{\alpha} \dot{\gamma}^{\alpha}(\hat{\mathbf{b}}^{\alpha} \otimes \hat{\mathbf{m}}^{\alpha}), \tag{13}$$

where $\hat{\mathbf{b}}^{\alpha}$ is the slip direction and $\hat{\mathbf{m}}^{\alpha}$ the normal to the slip plane along the α-slip system in configuration $\hat{B}$. The assumed slip systems for the FCC crystals are the 12 systems with (110) directions and $\langle 111 \rangle$ normals (Hosford 1993). The symmetric and skew symmetric parts of the plastic velocity gradient, $\hat{\mathbf{d}}^{\#'}$ and $\hat{\mathbf{w}}^{\#}$, respectively, are defined as

$$\hat{\mathbf{d}}^{\#'} = \sum_{\alpha} \dot{\gamma}^{\alpha}\hat{\mathbf{p}}^{\alpha} \tag{14}$$

$$\hat{\mathbf{w}}^{\#} = \dot{\mathbf{r}}^{\star}\mathbf{r}^{\star\mathrm{T}} + \sum_{\alpha} \dot{\gamma}^{\alpha}\hat{\mathbf{q}}^{\alpha} \tag{15}$$

where

$$\hat{\mathbf{p}}^{\alpha} = \hat{\mathbf{p}}^{\alpha}(\mathbf{r}) = \mathrm{sym}\,(\hat{\mathbf{b}}^{\alpha} \otimes \hat{\mathbf{m}}^{\alpha}) \tag{16}$$

$$\hat{\mathbf{q}}^{\alpha} = \hat{\mathbf{q}}^{\alpha}(\mathbf{r}) = \mathrm{skw}\,(\hat{\mathbf{b}}^{\alpha} \otimes \hat{\mathbf{m}}^{\alpha}). \tag{17}$$

The plastic shearing rate on the α-slip system, $\dot{\gamma}^{\alpha}$, is related to the crystal stress by the phenomenological relation

$$\dot{\gamma}^{\alpha} = \dot{\gamma}_0 \left(\frac{|\tau^{\alpha}|}{g}\right)^{\frac{1}{m}} \mathrm{sgn}(\tau^{\alpha}) = f(\tau^{\alpha}, g), \tag{18}$$

where g is the slip system hardness, $\dot{\gamma}_0$ is a reference shear rate, and m is the rate sensitivity of slip. The resolved shear stress, τ^{α}, is the plastic work rate conjugate to $\dot{\gamma}^{\alpha}$, and is the projection of the deviatoric part of the Kirchhoff stress, $\boldsymbol{\tau}'$ on the α-slip system as

$$\tau^{\alpha} = \mathrm{tr}(\hat{\mathbf{p}}^{\alpha}\boldsymbol{\tau}'). \tag{19}$$

The slip system hardness, g, is assumed to be the same for all slip systems and evolves according to a phenomenological hardening rule

$$\dot{g} = h_0 \left(\frac{g_s(\dot{\gamma}) - g}{g_s(\dot{\gamma}) - g_0}\right)^{n} \dot{\gamma}, \tag{20}$$

where

$$g_s(\dot{\gamma}) = g_1 \left(\frac{\dot{\gamma}}{\dot{\gamma}_1}\right)^{m'} \quad \text{and} \quad \dot{\gamma} = \sum_{\alpha} |\dot{\gamma}^{\alpha}|. \tag{21}$$

Here, $\dot{\gamma}^{\alpha}$ is the net shear strain rate in the crystal, $g_s(\dot{\gamma})$ is the saturation hardness, and h_0, g_0, n, g_1, $\dot{\gamma}_1$, and m' are slip system hardening parameters, which are the same for all slip systems. The lattice orientation evolves as a consequence of the spin and is given by

$$\mathbf{v} = \frac{1}{2}\boldsymbol{\omega} + (\boldsymbol{\omega} \cdot \mathbf{r})\mathbf{r} + \boldsymbol{\omega} \times \mathbf{r}, \tag{22}$$

where $\mathbf{v}$ is the reorientation velocity ($\dot{\mathbf{r}}$) and

$$\boldsymbol{\omega} = \text{vect}\left(\hat{\mathbf{w}}^{\#} - \sum_{\alpha} \dot{\gamma}^{\alpha} \hat{\mathbf{q}}^{\alpha}\right). \tag{23}$$

3 Crystal Elastoplasticity Simulation Methodology

The mechanical response at either the crystal or continuum scale is determined using a finite element formulation (Marin and Dawson 1998a,b). For the crystal scale, a weighted residual is formed on the equilibrium equation as

$$R_u = -\int_B \text{tr}\left(\boldsymbol{\sigma}'^{\mathrm{T}} \, \text{grad}\boldsymbol{\psi}\right) \mathrm{d}B + \int_B \pi \, \text{div}\boldsymbol{\psi} \, \mathrm{d}B + \int_{\partial B} \mathbf{t} \cdot \boldsymbol{\psi} \, \mathrm{d}\Gamma + \int_B \boldsymbol{\iota} \cdot \boldsymbol{\psi} \, \mathrm{d}B, \tag{24}$$

where $\boldsymbol{\psi}$ are vector weighting functions, $\mathbf{t}$ is the traction vector, $\boldsymbol{\iota}$ is a body force per unit volume, B is the volume of the body, and ∂B is its surface. The deviatoric Cauchy stress, $\boldsymbol{\sigma}'$, and mean stress (negative of the pressure, π) sum to the total Cauchy stress

$$\boldsymbol{\sigma} = \boldsymbol{\sigma}' - \pi \mathbf{I}. \tag{25}$$

Traction or velocity is specified over the boundary.

The stress is replaced ultimately with the velocity field through introduction of the constitutive equations and the kinematic relation defining the velocity gradient. The first step in this process is to introduce a difference expression for the elastic strain rate as

$$\left\{\dot{\mathsf{e}}^{\mathsf{b}}\right\} = \frac{1}{\Delta t}\left(\left\{\mathsf{e}^{\mathsf{b}}\right\} - \left\{\mathsf{e}^{\mathsf{b}}_{0}\right\}\right). \tag{26}$$

This is separated into the volumetric part

$$-\pi = \frac{\kappa \Delta t}{\beta} \text{tr}\left\{\mathsf{d}\right\} + \frac{\kappa}{\beta} \text{tr}\left\{\mathsf{e}^{\mathsf{b}}_{0}\right\} \tag{27}$$

and the deviatoric part

$$\left\{\mathsf{d}'\right\} = \frac{1}{\Delta t}\left\{\mathsf{e}^{\mathsf{b}\prime}\right\} + \left\{\hat{\mathsf{d}}^{\#}\right\} + \left[\hat{\mathsf{w}}^{\#}\right]\left\{\mathsf{e}^{\mathsf{b}\prime}\right\} - \frac{1}{\Delta t}\left\{\mathsf{e}^{\mathsf{b}\prime}_{0}\right\}. \tag{28}$$

Inverting the equation for the elastic behavior in (11),

$$\left\{e^{b\prime}\right\} = \left[c\right]^{-1}\left\{\tau'\right\} \tag{29}$$

and combining it with a relation obtained from merger of (14), (18) and (19)

$$\left\{\hat{d}^{\#}\right\} = \left[m\right]\left\{\tau'\right\} \tag{30}$$

$$\left[m\right] = \sum_{\alpha}\left(\frac{f(\tau^{\alpha}, g)}{\tau^{\alpha}}\right)\left\{p^{\alpha}\right\}\left\{p^{\alpha}\right\}^{T} \tag{31}$$

results in a matrix equation for the stress in terms of the total deformation rate

$$\left\{\sigma'\right\} = \left[s\right]\left(\left\{d'\right\} - \left\{h\right\}\right), \tag{32}$$

where

$$\left[s\right]^{-1} = \frac{\beta}{\Delta t}\left[c\right]^{-1} + \beta\left[m\right] \tag{33}$$

and

$$\left\{h\right\} = \left[\hat{w}^{\#}\right]\left\{e^{b\prime}\right\} - \frac{1}{\Delta t}\left\{e_0^{b\prime}\right\}. \tag{34}$$

Equations (26)–(33) are substituted into (24) to eliminate the explicit appearance of the stress.

The trial functions are introduced for the velocity, leading to a matrix equations for the nodal point velocities at the end of a time step. This is a nonlinear system, involving the elastic strain at the end of the time step as well as the velocity, which is solved iteratively. The lattice orientations and slip system strengths are updated over the time step by numerically integrating (22) and (20), respectively.

4 Lattice Misorientations from Geometrically Necessary Dislocations

4.1 *Intragrain Lattice Misorientations*

The formulation outlined in Sects. 2 and 3 can be used to examine the heterogeneity of deformation that develops within grains over the course of large plastic deformations in polycrystals. The illustration we choose for this purpose is the evolution of intragrain misorientation distributions. Development of substructure within grains can take the form of cells defined by a misorientation of the lattice above some threshold. Of interest are the average misorientation angle, the frequency distribution of the misorientation angles, and whether or not some spatial correlation exists between the misorientation and the external loading axes.

Recalling that an orientation is represented as a rotation that takes reference axes to an alignment with axes fixed to the crystal lattice, we write a misorientation as taking axes aligned with the lattice at one point within a crystal to axes aligned with the lattice at another point. Adapting the rotation tensor defined in (2)

$$\mathbf{R}^{A/B} = \mathbf{R}^{A^T} \cdot \mathbf{R}^{B}. \tag{35}$$

One of these orientations is an average taken over all the points within a grain. The other is the orientation of some point within the grain. Working via an exponential map defined by

$$\exp \boldsymbol{\Omega} = \mathbf{R}^{A/B}, \tag{36}$$

we can write a misorientation between the average and any other point in the grain in terms of an axial vector as

$$\mathbf{w} = \mathrm{vect}(\boldsymbol{\Omega}). \tag{37}$$

Using the method developed in (Barton and Dawson 2001a), a tensorial quantity is defined from the aggregate of axial vector dyads

$$\mathbf{A}^{-1} = \sum_{i=1}^{n} \phi^i (\mathbf{w}^i \otimes \mathbf{w}^i), \tag{38}$$

from which an average misorientation angle, θ^s, may be extracted as

$$\theta^s = \sqrt{\mathrm{tr}(\mathbf{A}^{-1})}. \tag{39}$$

$\mathbf{A}$ can be interpreted as an ellipsoid that fits the swarm of misorientation vectors. In a similar manner a tensor, $\mathbf{M}$, can be defined based on the relative position vectors, $\mathbf{x}$, defined as the vector connecting a grain's centroid to a position within the grain

$$\mathbf{M}^{-1} = \sum_{i=1}^{n} \phi^i (\mathbf{x}^i \otimes \mathbf{x}^i). \tag{40}$$

This tensor is an ellipsoid that approximates the grain's volume. A spatial correlation between the position within a grain relative to its centroid and the lattice misorientation relative to an average orientation of the grain is constructed using the quantities defined by (38) and (40) after normalizing the position and misorientation vectors by

$$\bar{\mathbf{w}}^i = \mathbf{A}^{\frac{1}{2}} \cdot \mathbf{w}^i \tag{41}$$

and

$$\bar{\mathbf{x}}^i = \mathbf{M}^{\frac{1}{2}} \cdot \mathbf{x}^i. \tag{42}$$

Using $\bar{\mathbf{w}}^i$ and $\bar{\mathbf{x}}^i$ to form a correlation, $\mathbf{X}$, we obtain

$$\mathbf{X}^{-1} = \sum_{i=1}^{n} \phi^i (\bar{\mathbf{w}}^i \otimes \bar{\mathbf{x}}^i). \tag{43}$$

As we are interested in the directions associated with the correlation, $\mathbf{X}$ is decomposed into eigenvectors and eigendirections as

$$\mathbf{X} = \mathbf{U} \cdot \mathbf{S} \cdot \mathbf{V}^T. \tag{44}$$

The eigenvalues, s_1, s_2, and s_3, are the normal components of $\mathbf{S}$. There are two sets of eigenvectors for $\mathbf{X}$. One set, $\mathbf{U}$, lie in the space of $\mathbf{w}$ while the other, $\mathbf{V}$, lies in the space of $\mathbf{x}$. The values of s_1, s_2, and s_3 range between zero and unity. The larger the value of s_1, s_2, or s_3, the stronger the spatial correlation in the associated direction.

4.2 *Misorientations Developed Under Tension of an FCC Polycrystal*

We consider the evolution of intragrain misorientations in FCC polycrystals subjected to uniaxial tension. The focus is on determining whether the misorientations within crystals exhibit spatial correlations and whether the directions associated with the correlations collectively are related to the loading axes. Three virtual polycrystals, identical in all respects except for the aspect ratio of the grains (Marin 2006), are built and subjected to the same loading history. From the results, we compute the misorientation quantities outlined in Sect. 4.1 and explore possible spatial trends.

The three polycrystals, referred to as Virtual Polycrystals A, B, and C (VP-A, VP-B and VP-C, respectively), are shown in Fig. 2. The relative lengths of the grains for the three samples are: 1:1:1, 1:1:1.5, and 1:1:2. There are a total of 559 grains in each sample. Of these, 341 are full grains of truncated octahedral shape. The remainder are partial grains with at least one facet on the sample surface. All of the full grains are discretized by 184 ten-noded tetrahedral elements. The total number of elements for all grains is 79,488. We have chosen to use more highly resolved crystals over larger numbers of crystals to give better resolved intragrain

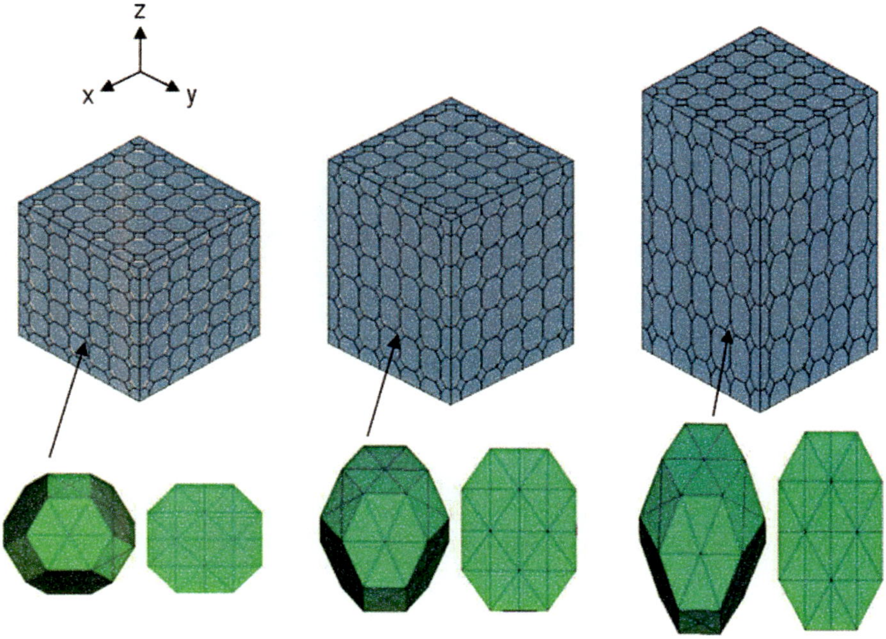

Fig. 2 Virtual polycrystal used to examine the evolution of intragranular lattice misorientations of FCC metal of differing grain shape

deformations at the expense of better average trends for the polycrystal. The single crystal properties correspond to an aluminum alloy, and the set of orientations assigned to the grains are chosen randomly from a uniform distribution. The samples are pulled in tension to a nominal strain of $\approx 20\%$ and exhibit similar stress–strain curves.

The textures that evolve over the course of the deformation for all of the virtual samples are similar to each other. The texture at 20% nominal strain is shown in Fig. 3 for VP-A. Although the texture remains relatively weak after the modest amount of strain imposed, it does show strengthening primarily along the $< 111 > \|$ z-axis fiber and to a lesser extent along the $< 100 > \|$ z-axis fiber. There are no stable equilibrium points along these fibers; so the tendency at larger strains would be for the texture to strengthen uniformly along the fibers (Kumar and Dawson 2009). The simulated textures are qualitatively consistent with textures observed experimentally under uniaxial tension (Wenk 1985; Kocks et al. 1998).

Individual grains do not deform homogeneously and, as a consequence, the initially uniform lattice orientation tends to develop into a distribution of values that is apparent as a spread, or swarm, of points within orientation space. We can depict this trend for an individual grain by plotting the lattice orientations in the fundamental region. Shown in Fig. 4 is a swarm of orientations from one grain after 20% nominal strain. To the right of the fundamental region are plotted the swarm at increasing amounts of strain. The red dot indicates the average of all the orientations in the

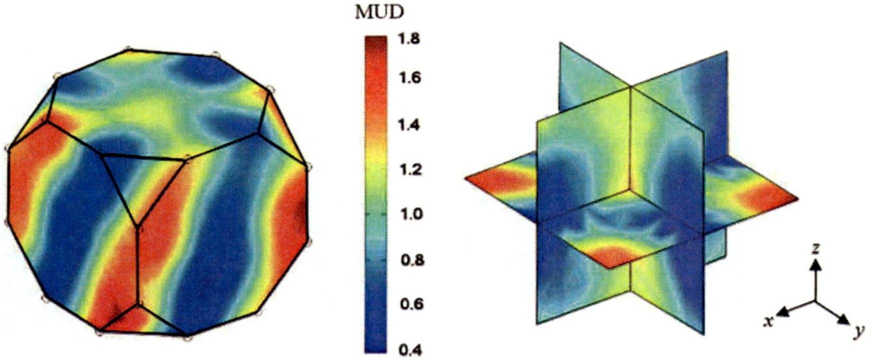

Fig. 3 The ODF at 20% nominal strain for VP-A sample shown over the fundamental region for FCC crystal symmetry

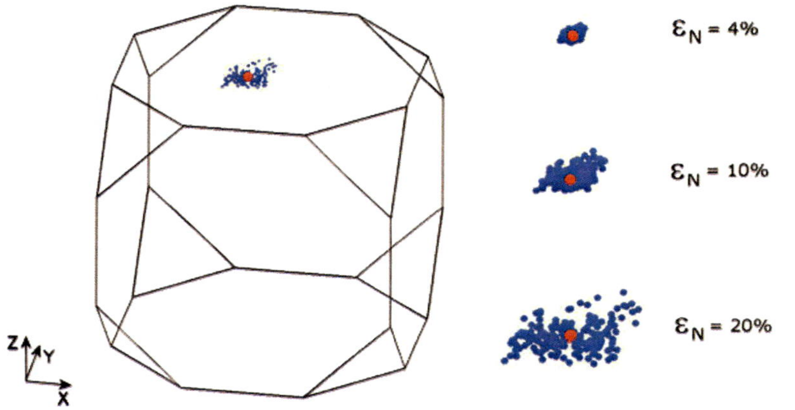

Fig. 4 Orientations of lattice orientations for elements originally within one crystal from VP-A following several amounts of strain

swarm. Different measures of the average may be computed. The approach adopted by Mika et al. (1999) computes the misorientation across all element boundaries internal to the grain; Barton et al. (2001a) defined an average lattice orientation as the orientation that minimizes the sum of the misorientations between the average and the lattice orientation of every element within a grain. These measures were determined to be essentially equivalent. Figure 5 shows the evolution of the average of the misorientation angles (as computed from (39)) across all the grains as a function of the nominal strain. The standard deviation is given as well, showing that the spread in orientations is increasing with increasing strain. All of the samples display similar behavior in this regard. This trend compares to that observed experimentally from TEM measurements (Dawson et al. 2002).

To consider the possible spatial correlations of the misorientations of the lattice, we examine the misorientations at the centroids of the finite elements within the

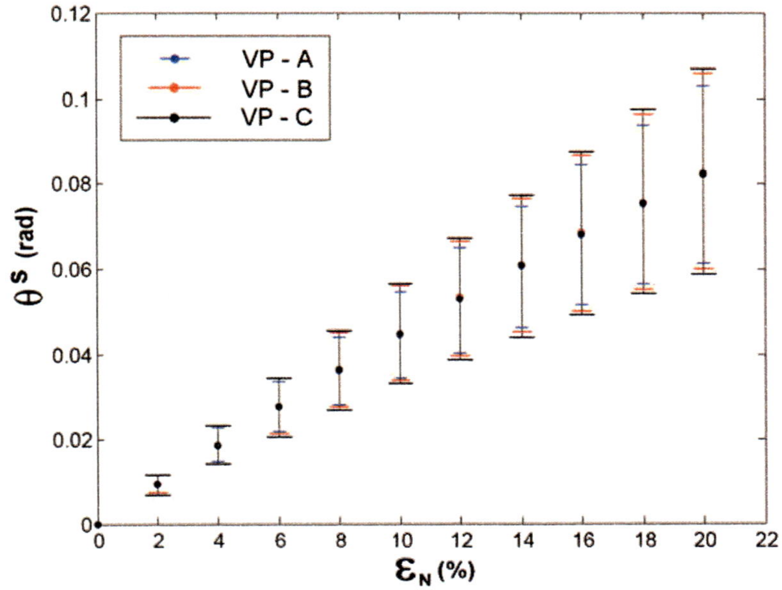

Fig. 5 Evolution of the average misorientation angle and its standard deviation for VP-A, VP-B and VP-C

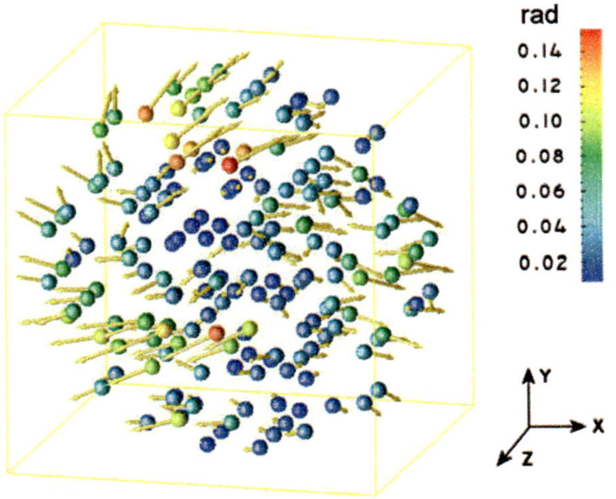

Fig. 6 Misorientations vectors for lattice orientations within one crystal from VP-A at 20% nominal strain. Color scale displays the magnitude of the misorientations in radians

same grain. Figure 6 shows misorientation vectors within one grain (the same grain discussed in Fig. 4) again after 20% nominal strain. Every misorientation is plotted using a combination of a color-coded point and an arrow. The arrows are aligned with the misorientation axis and the points are colored according to the magnitude of the misorientation from the average orientation, θ^s. Qualitatively, one can observe that in the central core of the grain the misorientation is small, while to either side of the center, the lattice is rotated in opposite directions from the core. There appears to be much less variation in orientation from top to bottom than from side to side.

Using the equations developed in Sect. 4.1, we can quantitatively determine the nature of the spatial correlation between the misorientation and position. Specifically, we compute the correlation tensor, $\mathbf{X}$, and decompose it into its eigenvalues and eigenvectors, as indicated in (44), for all the grains in the polycrystal. The three values of the eigenvalues are indicated in Fig. 7 for VP-A, which indicates that there tends to be directionality within grains that remains fairly constant over the deformation. The spatial trends for the dominate eigendirection are shown in Fig. 8 where the projection of the misorientation vector onto the first eigenvector in orientation space, $\mathbf{w} \cdot \mathbf{e}_1^{\mathrm{U}}$, is shown for each point within a crystal after 20% nominal strain. One can observe that there is strong correlation along the $\mathbf{e}_1^{\mathrm{V}}$ axis (colored blue) for all of the polycrystal models, VP-A, VP-B, and VP-C. Correlations in other directions exist, but are much weaker. To determine whether the eigendirections for all the grains in a polycrystal are similarly oriented, the directions are plotted as pole distributions, as shown in Fig. 9. For this FCC system, we see that by 20% nominal strain a weak preference develops along the loading (z) direction for polycrystals with elongated grains, but that very little, if any, preference develops in the sample with equiaxed grains. Thus, the grains tend to develop a pattern having a misorientation gradient across the grains in one direction, but that direction differs from grain to grain without strong correlation to the loading axes.

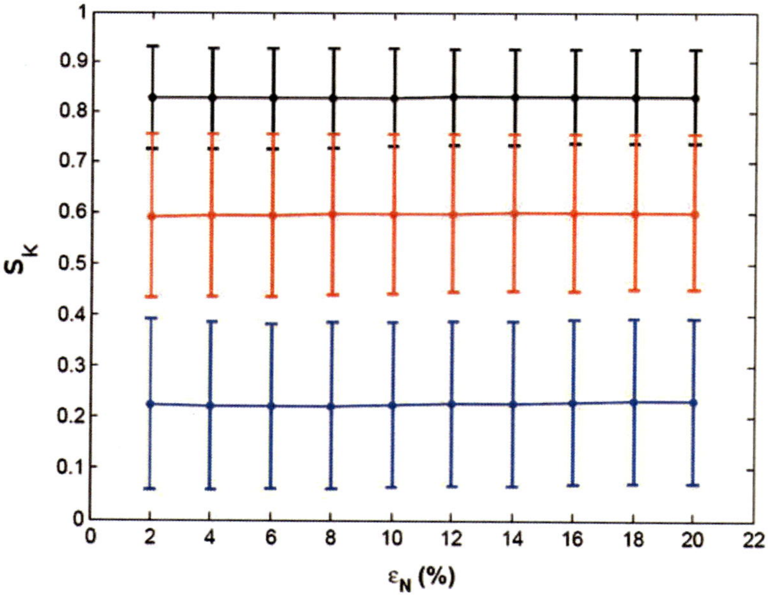

Fig. 7 Evolution of three principal values of **S** for VP-A, showing the average value and associated standard deviation for each eigenvalue. Similar trends were found for VP-B and VP-C

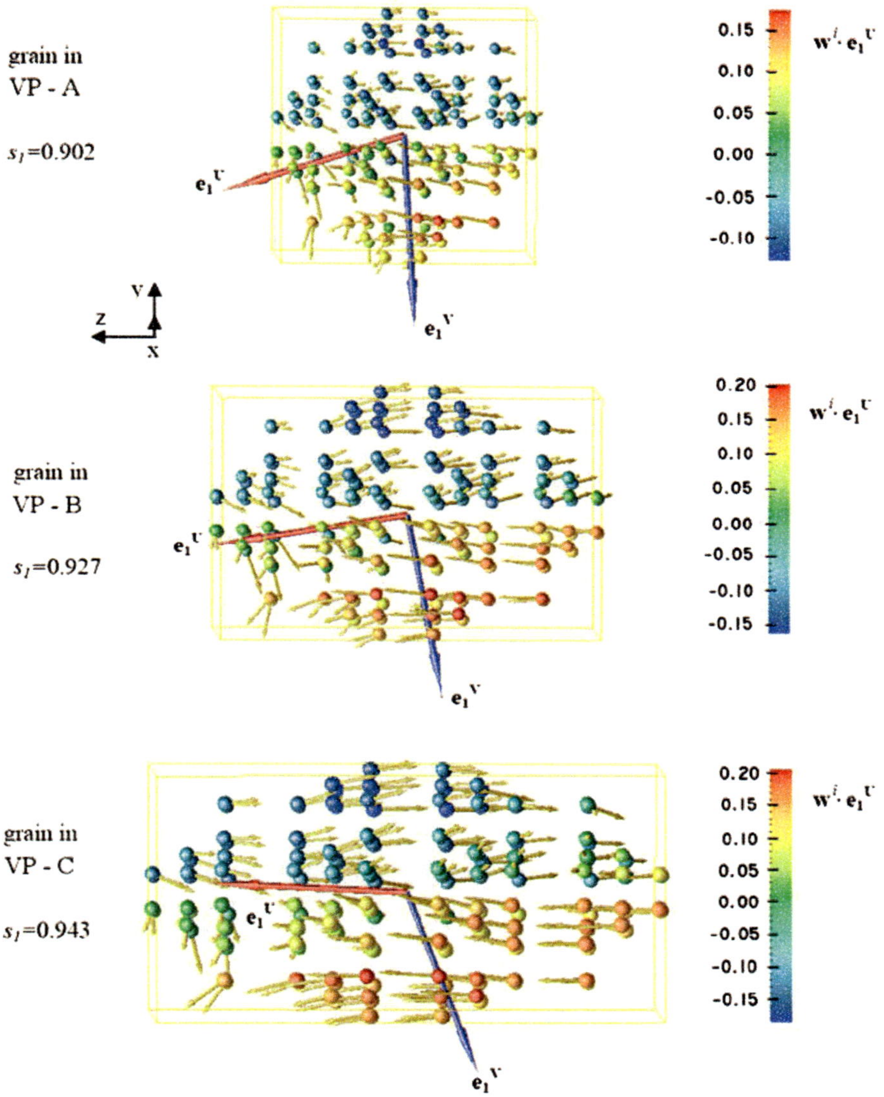

Fig. 8 Misorientation vectors with strongest spatial correlation at 20% nominal strain for the same grain from each of the three virtual polycrystals. Color scale displays the magnitude of the misorientations in radians

This result is specific only to FCC undergoing unaxial extension as we have not explored other loading modes for FCC polycrystals. Loading modes with well-defined texture components might be more inclined to exhibit spatial correlations. Barton et al. (2001b) did observe a correlation for HCP polycrystals subjected to plane strain compression (idealized rolling). The HCP system considered in that

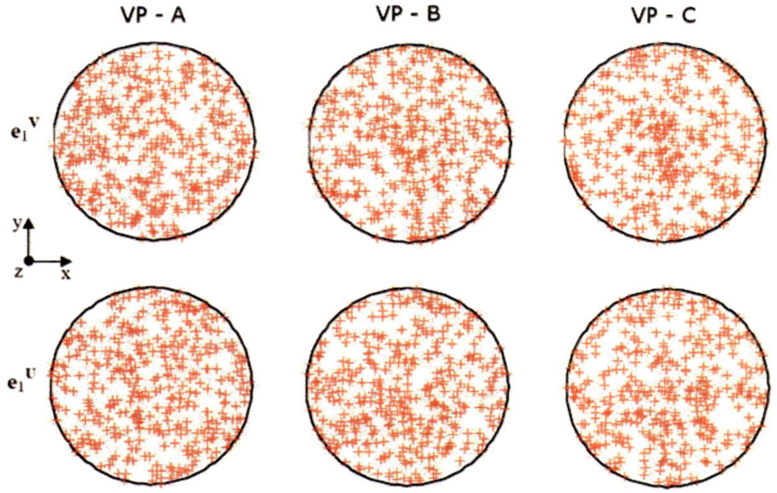

Fig. 9 Pole distributions of the singular vector axes for three virtual polycrystals at 20% nominal strain

investigation possessed a high degree of anisotropy in the slip system strengths, with c-axis deformation being more difficult to achieve than other deformation modes. A correlation was found between the eigendirections of the spatial correlation tensor and the transverse direction in rolling.

5 Extending the Kinematic Model for Incomplete Slip

The elementary kinematic decomposition that forms the basis of the basic crystal elastoplasticity formulation is idealized in many respects. Simply stated, it accommodates small elastic distortions of the lattice, large plastic deformation from the combined action of simple shear on multiple slip systems, and lattice rotations. Actual deformations deviate from this idealized model in all three of the ways motion is accommodated by the elementary decomposition. Here we discuss an extension to the basic formulation that addresses one of the limitations of the elementary decomposition. The simple model for slip does not account for the incomplete movement of dislocations through the representative volume associated with the kinematic decomposition. To accommodate incomplete slip, a process in which there is a net change in the dislocation content of the representative volume, we can insert a new configuration in the kinematic decomposition that represents the state of the representative volume possessing a dislocation population. This population distorts the volume and is quantified by a deformation gradient induced by long range strains. Long range strains are the elastic strains present in the lattice due to dislocations. Under deformations in which there exists a gradient in the slip activity, the dislocation content can evolve locally thereby changing the long range strains and the configuration of the representative volume associated with those strains.

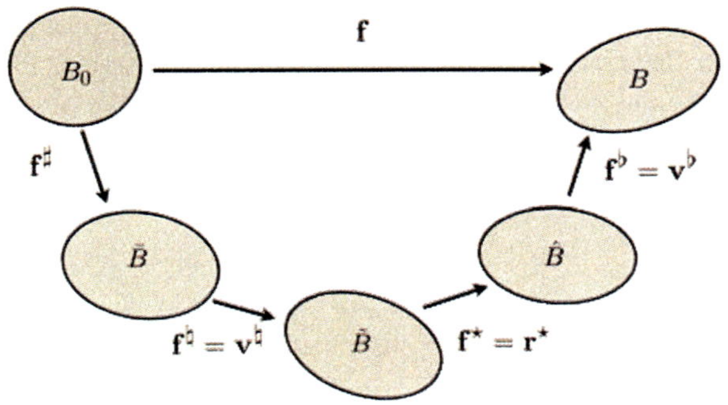

Fig. 10 Extension of the elementary kinematic decomposition to include a configuration associated with the long range strain induced by a distribution of static dislocations

Following the work of (Hartley 2003), Fig. 10 shows the revised multiplicative decomposition of the deformation gradient. In this revised decomposition, we assume that the plastic deformation is composed of both a permanent deformation due to dislocation motion along crystallographic slip planes, $\mathbf{f}^{\sharp}$, and a long range deformation, $\mathbf{f}^{\natural}$, due to dislocations in the lattice distributed throughout the body. As in the elementary decomposition, a polar decomposition of the elastic deformation is used to yield the left stretch tensor, $\mathbf{f}^{\flat} = \mathbf{v}^{\flat}$, and a rotation, $\mathbf{f}^{*} = \mathbf{r}^{*}$. The multiplicative decomposition of the deformation gradient is then given by

$$\mathbf{f} = \mathbf{f}^{\flat}\,\mathbf{f}^{\star}\,\mathbf{f}^{\natural}\,\mathbf{f}^{\sharp}. \tag{45}$$

Assuming that the elastic deformation has the form $\mathbf{v}^{\flat} = \mathbf{I} + \mathbf{e}^{\flat}$ and the long range deformation $\mathbf{f}^{\natural} = \mathbf{I} + \mathbf{e}^{\natural}$, where $\mathbf{e}^{\flat}$ and $\mathbf{e}^{\natural}$ are small, symmetric, elastic lattice strains (Gerken and Dawson 2008a), the approximate rate of deformation and spin tensors become

$$\mathbf{d} = \dot{\mathbf{e}}^{\flat} + \dot{\mathbf{e}}^{\natural} + \hat{\mathbf{d}}^{\sharp} + \mathbf{e}^{\natural}\hat{\mathbf{w}}^{\sharp} - \hat{\mathbf{w}}^{\sharp}\mathbf{e}^{\natural} + \mathbf{e}^{\flat}\mathbf{w} - \mathbf{w}\mathbf{e}^{\flat} \tag{46}$$

and

$$\mathbf{w} = \hat{\mathbf{w}} + \mathbf{e}^{\flat}\hat{\mathbf{d}} - \hat{\mathbf{d}}\mathbf{e}^{\flat}, \tag{47}$$

where $\hat{\mathbf{w}}$ and $\hat{\mathbf{d}}$ are the rate of deformation and spin from the reference configuration to $\hat{X}$ and $\hat{\mathbf{d}}^{\sharp}$ and $\hat{\mathbf{w}}^{\sharp}$ are the plastic deformation rate and spin tensors.

Assuming linear elasticity, the Kirchhoff stress is given by

$$\boldsymbol{\tau} = \mathbf{c}\mathbf{e} = \mathbf{c}(\mathbf{e}^{\flat} + \hat{\mathbf{e}}^{\natural}) = \boldsymbol{\tau}^{e} + \boldsymbol{\tau}^{b}, \tag{48}$$

where $\mathbf{c}$ is the fourth-order isotropic elasticity tensor and $\mathbf{e} = \mathbf{e}^{\flat} + \hat{\mathbf{e}}^{\natural}$ is the total lattice strain tensor.

Detailed in (Gerken and Dawson 2008a), using Volterra's solution for the stress field of an edge dislocation and a bi-linear distribution of edge dislocations in a square region around the representative volume, the long range strain on a slip system is given by

$$(\mathbf{e}^{\natural})^{\alpha}_{xx} = \frac{\left(\nu - \frac{\pi}{4}\right)}{\pi\,(1-\nu)} a^2 \frac{\partial \gamma^{\alpha}}{\partial b^{\alpha} \partial m^{\alpha}} \tag{49}$$

$$(\mathbf{e}^{\natural})^{\alpha}_{yy} = -\frac{\left((1-\nu) - \frac{\pi}{4}\right)}{\pi\,(1-\nu)} a^2 \frac{\partial \gamma^{\alpha}}{\partial b^{\alpha} \partial m^{\alpha}} \tag{50}$$

$$(\mathbf{e}^{\natural})^{\alpha}_{xy} = \frac{\left(1 - \frac{\pi}{4}\right)}{\pi\,(1-\nu)} a^2 \frac{\partial \gamma^{\alpha}}{\partial b^{\alpha} \partial b^{\alpha}}, \tag{51}$$

where a is the dimension of the square region, ν is Poisson's ratio, and γ^{α} is the slip on slip system α defined by slip plane normal m^{α} and slip direction b^{α}. The assumptions of the Volterra solution, notably that of isotropy, are not strictly true for a crystal scale continuum. However, to first-order approximation, the Volterra model captures the spatial stress and strain variation and provides a first attempt at incorporating these effects.

Using superposition to construct the total long range strain tensor

$$\mathbf{e}^{\natural} = \sum_{\alpha=1}^{n} \mathbf{R}^{\alpha} (\mathbf{e}^{\natural})^{\alpha} (\mathbf{R}^{\alpha})^{T}, \tag{52}$$

where n is the number of slip systems and $\mathbf{R}^{\alpha}$ is a change of basis tensor from the basis of $(\mathbf{e}^{\natural})^{\alpha}$ to the basis of $\mathbf{e}^{\natural}$.

The slip system hardness, g^{α}, is given by

$$g^{\alpha} = g = \mathcal{G}_1(\gamma) + \mathcal{G}_2\left(\frac{\partial \gamma}{\partial x_i}\right), \tag{53}$$

where γ is the total slip on all slip systems. The function $\mathcal{G}_1(\gamma)$ is the hardness due to statistically stored dislocations and evolves using hardening model given above. The function $\mathcal{G}_2\left(\frac{\partial \gamma}{\partial x_i}\right)$ is the gradient hardness due to the population of dislocations due to slip gradients. Assuming a Taylor-like hardening similar to that due to statistically stored dislocations, the gradient hardness function is

$$\mathcal{G}_2\left(\frac{\partial \gamma}{\partial x_i}\right) = \beta \mu \sqrt{\rho_b} \quad \text{where} \quad \rho_b = \sum_{\alpha=1}^{n} \left| \frac{\partial \gamma^{\alpha}}{\partial b^{\alpha}} \right|. \tag{54}$$

Here β is a material parameter and μ is the shear modulus (Gerken and Dawson 2008a). Other functional forms of $\mathcal{G}_2\left(\frac{\partial \gamma}{\partial x_i}\right)$ are readily incorporated into the hardening slip system hardness. For example, see the recent work of (Guruprasad et al. 2008) in which they model the effects of discrete dislocations and show that Taylor-like hardening behavior is inconsistent with their observations.

6 Simulation Methodology for Nonlocal Crystal Constitutive Equations

Simulating the deformation of a polycrystal using the constitutive theory outlined in Sect. 5 is more complex than that for the elementary decomposition. The inclusion of the long-range elastic strains introduces a nonlocal character to the constitutive model owing to the terms associated with the gradients of slip. This precludes reduction of the system of equations to a local equation involving the velocity field by direct substitution for the stress in the equilibrium residual. Instead, we must also treat the constitutive model as a spatially-dependent partial differential equation as detailed in (Gerken and Dawson 2008b). Again, we enforce equilibrium with a weighted residual given by (24). In addition, we construct the weak form of the deformation rate given by (46) in a smooth continuous region B

$$\int_B \left[\dot{\mathbf{e}}^\flat + \dot{\mathbf{e}}^\natural + \hat{\mathbf{d}}^\sharp + \mathbf{e}^\natural\hat{\mathbf{w}}^\sharp - \hat{\mathbf{w}}^\sharp\mathbf{e}^\natural + \mathbf{e}^\flat\mathbf{w} - \mathbf{w}\mathbf{e}^\flat - \mathbf{d}\right]\boldsymbol{\Gamma}\, \mathrm{d}B = 0, \tag{55}$$

where $\boldsymbol{\Gamma}$ are weight functions. Forward Euler integration for the elastic and long range strain rates are

$$\dot{\mathbf{e}}^\flat = \frac{1}{\Delta t}\left(\mathbf{e}^\flat - \mathbf{e}^\flat{}_0\right) \tag{56}$$

and

$$\dot{\mathbf{e}}^\natural = \frac{1}{\Delta t}\left(\mathbf{e}^\natural - \mathbf{e}^\natural{}_0\right), \tag{57}$$

where $\mathbf{e}^\flat{}_0$ and $\mathbf{e}^\natural{}_0$ are the elastic and long range strains from the previous time step. Rewriting the long range strain rate gives

$$\dot{\mathbf{e}}^\natural = \sum_{\alpha=1}^{n} \nabla\hat{\mathbf{H}}^\alpha\hat{\mathbf{b}}^\alpha, \tag{58}$$

where $\hat{\mathbf{H}}^\alpha$ is a tensor containing the slip rate gradients. Then, using integration by parts, (55) becomes

$$\int_B \left[\frac{1}{\Delta t}\mathbf{c}\mathbf{t}^e - \frac{1}{\Delta t}\mathbf{e}^\flat{}_0 + \mathbf{t}^e\mathbf{c}\boldsymbol{\varphi} + \hat{\mathbf{d}}^\sharp + \mathbf{e}^\natural\boldsymbol{\vartheta}\right]\boldsymbol{\Gamma}\,\mathrm{d}B - \int_B \sum_{\alpha=1}^{n} \hat{\mathbf{H}}^\alpha\nabla\left(\hat{\mathbf{b}}^\alpha\boldsymbol{\Gamma}\right)\mathrm{d}B$$
$$+ \int_{\partial B} \sum_{\alpha=1}^{n} \hat{\mathbf{H}}^\alpha\boldsymbol{\Gamma}\left(\hat{\mathbf{b}}^\alpha\cdot\mathbf{n}\right)\mathrm{d}\Gamma = 0, \tag{59}$$

where $\boldsymbol{\varphi}$ and $\boldsymbol{\vartheta}$ are symmetric product matrices constructed from the components of $\mathbf{w}$ and $\hat{\mathbf{w}}^\sharp$, respectively, and $\mathbf{n}$ is the unit normal to surface ∂B. This is a weak form of (55) with the boundary conditions

$$\mathbf{t}^e = \mathbf{t}^* \quad \text{on } \partial B_\tau \tag{60}$$

and

$$\sum_{\alpha=1}^{n} \hat{\mathbf{H}}^{\alpha} \left(\hat{\mathbf{b}}^{\alpha} \cdot \mathbf{n} \right) = \mathbf{H}^{*} \quad \text{on } \partial B_{v}, \tag{61}$$

where $\mathbf{t}^*$ is the traction on the surface and $\mathbf{H}^*$ is the tensor containing the slip rate gradients on the surface.

Using a finite element method, solution of this system of equations gives the elastic stress at the nodal points which is used to determine the material stiffness. Then, similar to the elementary model above, equilibrium is determined using the finite element method. A full description is available in (Gerken and Dawson 2008b).

7 Yield Asymmetry From Long-Range Strains Associated with Excess Dislocations

Bending deformation naturally includes a variation in the deformation ranging from a high tension or compression deformation on the outer surface to zero deformation at the neutral axis. The behavior of the constitutive model in a bending deformation is investigated by simulating the bending of a thin foil of single crystal aluminum with a thickness of $t = 25\,\mu\text{m}$. A comparison of the yielding behavior with simulations of foils of other thickness shows that both isotropic and kinematic type hardening behavior results from the slip gradient effects. These observed behaviors are connected to the increase in dislocations that appear due to slip gradients and the stress field that results from the build up of these dislocations.

7.1 Bending of a Thin Foil

The specimen, shown in Fig. 11, is a single FCC crystal and the crystal axes are aligned with the specimen axes. A velocity is applied to the lower right edge of the specimen, inducing a bending moment and a small amount of compression in the gage section. The velocity, $V_0 = 10\,\mu\text{m/s}$, is applied to the loaded edge of the specimen for 29.5 s and then reversed over a second and applied in the opposite direction for 29.5 s to straighten the gage section. This velocity pattern is repeated three times. The gage section is 0.5 mm long, 0.25 mm wide, and $t = 25\,\mu\text{m}$ thick (simulations using 50, and 100 μm thick samples are reported in (Gerken and Dawson 2007)). The slip rate gradient boundary condition $\mathbf{H}^* = 0$ has been specified on the entire surface. This is equivalent to specifying zero flux of geometrically necessary dislocations across the surface.

A mesh of the sample also is shown in Fig. 11. It consists of 8,217 nodes defining 1,488 quadratic 20 node hexahedral elements. The gage section is partitioned into three elements through the thickness, 30 elements along its length, and 8 elements along its width. This mesh was chosen such that the element scale is significantly

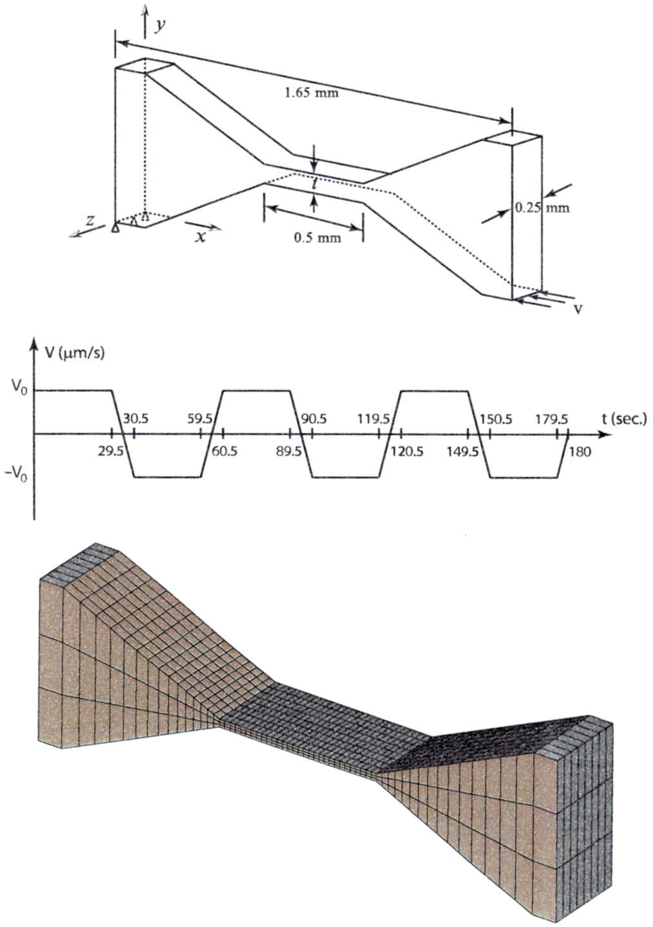

Fig. 11 Single crystal foil bending sample. A constant velocity, V_0, is applied to the loaded edge for approximately 30 s, bending the specimen so that the top surface of the gage section strains to 0.04. The velocity is then reversed to straighten the specimen back to (nearly) the original position. This loading cycle is performed three times. Mesh used to simulate the bending sample. The mesh contains 1,488 quadratic 20 node hexahedral elements and a total of 8,217 nodes

larger than the scale chosen for the length parameter in the long range stress. If this was not the case, the physical interpretation of the long-range formulation would conflict with the numerical model represented by the mesh.

The plasticity and hardening material parameters shown in Table 1 were chosen to approximately match the experimental data reported by Dumoulin and Tabourot (2005) for tensile tests on 99.99% pure aluminum single crystal samples.

Table 1 Material parameters chosen to match experimental data on a tensile test of single crystal 99.99% pure aluminum

Elasticity		
Young's Modulus	E	70 GPa
Poisson's Ratio	ν	0.33
Plasticity & Hardening		
Initial hardness	g_0^α	11 MPa
Slip rate sensitivity	n	0.1
Hardening coefficient	h_0	4 MPa
Nominal saturation	g_0	133 MPa
Saturation coefficient	g_1	160 MPa
Saturation slip rate	$\dot{\gamma}_1$	$5 \times 10^9\,\mathrm{s}^{-1}$
Saturation exponent	m'	0.005
Long Range & Gradient		
Length parameter	a	1.0 μm
Hardness coefficient	β	$1.0 \times 10^{-6}\sqrt{\mathrm{m}}$

7.2 *Development of Asymmetries from Long Range Strain Gradients*

For results presented here, the length parameter and gradient hardness coefficient are $a = 1\,\mu\mathrm{m}$ and $\beta = 2 \times 10^{-6}\sqrt{\mathrm{m}}$, respectively. The length parameter was selected to be below a typical crystal size but larger than the scale relevant to individual dislocations. This range of length scales then would capture the *average* effects of subcrystal behavior such as a distribution of dislocation pile-ups. The gradient hardness coefficient was chosen to provide demonstrative simulation results.

Shown in Fig. 12 are contours of the evolving Voce hardness at the end of each bending and straightening cycle. The value contoured is the element average Voce hardness ($\mathscr{G}_1(\gamma)$) calculated by volume averaging the values at each element's integration points. During each subsequent cycle of deformation, the relative pattern does not change appreciably; however, the magnitude continues to evolve. The relative hardness magnitude indicates more plastic deformation near the front and back edges of the foil. Simulations using other sets of parameters that include long range and/or gradient hardness effects show very similar relative patterns with the hardness generally varying less than 5% from the results shown in Fig. 12.

The gradient hardness $\left(\mathscr{G}_2\left(\frac{\partial \gamma}{\partial x_i}\right)\right)$ at the end of each bending and straightening cycle is shown in Fig. 13. These results show that the gradient hardness is on the order of 10% of the Voce hardness. Increasing the gradient hardness coefficient to $\beta = 2 \times 10^{-6}\sqrt{\mathrm{m}}$ results in a gradient hardness about twice that shown in Fig. 13. The gradient hardness pattern shows a large gradient effect at the ends of the foil strip near the tapered sections. The patterns in the foil away from the ends have little correlation among the models or to the Voce hardness patterns. There are notable changes between the fully bent and fully straightened deformations that indicate

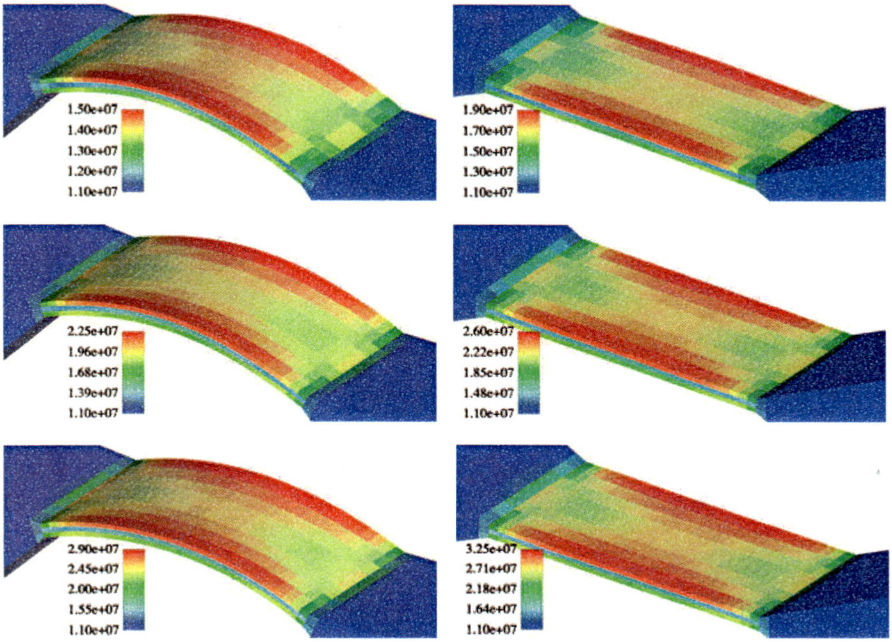

Fig. 12 Contours of the Voce hardness at points of maximum and minimum flexures over the three bending cycles. Distributions are shown for values of the length parameter and gradient hardness coefficient of $a = 1\,\mu\text{m}$ and $\beta = 2 \times 10^{-6}\sqrt{\text{m}}$, respectively. Units are Pa

a somewhat uniform gradient hardness along the width direction at fully bent and a relative reduction in gradient hardness near the edges at fully straightened. The evolution of the pattern at fully straightened for each of the simulations indicates less variation in the gradient hardness with each subsequent straightening cycle.

Figure 14 shows the von Mises effective long-range strain, $\eta_{\text{eff}} = \sqrt{\frac{3}{2}\mathbf{e}^{\natural} \cdot \mathbf{e}^{\natural}}$. In general, these figures show that the bent specimen has elevated long-range strain near the center of the gage section, and for the straightened specimen, the pattern shifts to elevated magnitude near the edges of the gage section. The patterns continue to evolve throughout the three cycles but generally maintain these regions of relative magnitude. For the same long range strain length parameter, $a = 1.0\,\mu\text{m}$, but different values of gradient hardness parameter, the results indicate that the gradient hardness has a noticeable effect on the long range strain. Additionally, changing the long-range strain length parameter to $a = 0.9\,\mu\text{m}$ and $a = 1.1\,\mu\text{m}$ results in a change in the magnitude of the effective long range strain in proportion to the change in length parameter and a slight change in the pattern of effective long range strain (Gerken and Dawson 2007).

Shown in Fig. 15 is the scaled moment (M/bh^2) versus the strain on the top surface of the foil for six different simulations using different long range strain length parameters and gradient hardness coefficients. The scaled moment is

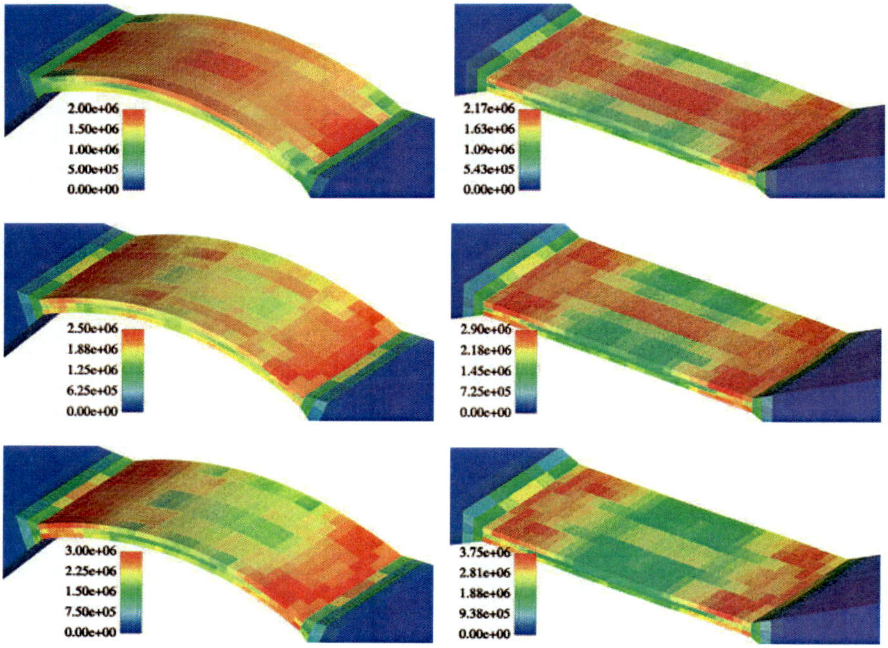

Fig. 13 Contours of the gradient hardness at points of maximum and minimum flexures over the three bending cycles. Distributions are shown for values of the length parameter and gradient hardness coefficient of $a = 1\,\mu\text{m}$ and $\beta = 2 \times 10^{-6}\sqrt{\text{m}}$, respectively. Units are Pa

thickness-independent measure of bending stress and is the applied moment M scaled by the width times height squared. The plot shows overlaid results for each individual bending/straightening cycle.

In the first cycle, an initial yield is the scaled moment when the strain on the top surface is 0.001, and the yield at the start of straightening is the scaled moment when the strain on the top surface is 0.038. The unloading scaled moment is the maximum shown in the figure. These results are given in Table 2. The initial, primarily elastic, behavior of each of the six simulations is very similar. However, during the initial stages of yielding, differences in the results become apparent. The initial yield point for the model without long range and gradient hardness effects is the lowest of the six simulations at 2.73 MPa, while the highest initial yield is for the model with a size parameter of $a = 1\,\mu\text{m}$ and a gradient hardness coefficient of $\beta = 2\times10^{-6}\sqrt{\text{m}}$. Of the models that include long range and/or gradient hardness effects, the lowest initial yield is for the model with zero gradient hardness coefficient and the highest corresponds to the highest gradient hardness coefficient. For the three simulations with a gradient hardness coefficient of $\beta = 1 \times 10^{-6}\sqrt{\text{m}}$, there is little difference in the behavior near the initial yield point. However, the models quickly distinguish themselves.

The significant influence of the long range and gradient hardness on the material behavior is readily apparent. The model without these effects requires a scaled moment of 4.29 MPa at maximum curvature, while the models including these

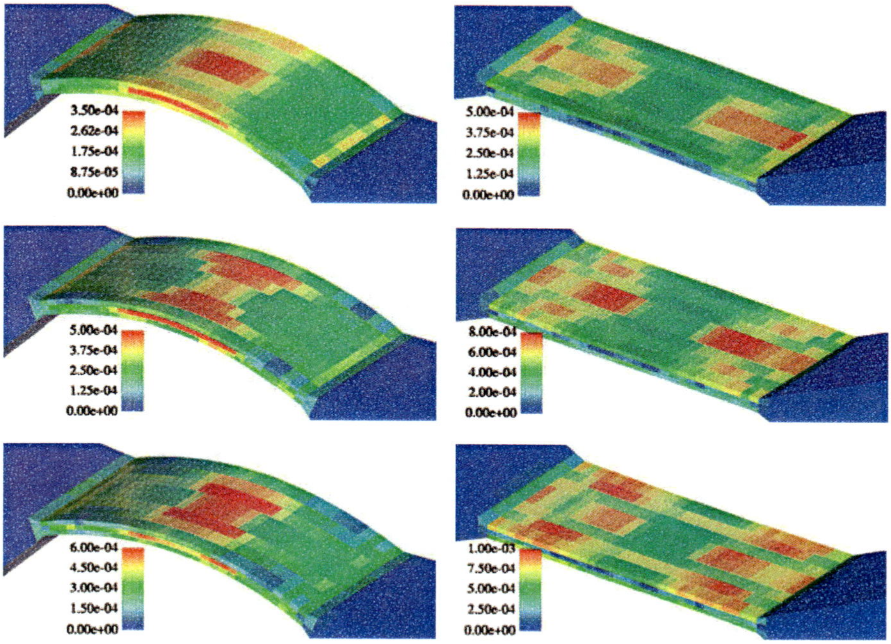

Fig. 14 Contours of the effective values of the long range strain at points of maximum and minimum flexures over the three bending cycles. Distributions are shown for values of the length parameter and gradient hardness coefficient of $a = 1\,\mu\text{m}$ and $\beta = 2 \times 10^{-6}\sqrt{\text{m}}$, respectively. Dimensionless scales

effects require, on average, 8.5 MPa. The relative differences between the models with long range and gradient hardness effects developed during initial yield continue through the beginning of unloading. In other words, the model with the highest gradient hardness coefficient shows the highest unloading scaled moment and the model with zero gradient hardness shows the lowest unloading scaled moment. After unloading, the yield behavior at the start of the straightening (i.e., reverse yield) shows a shift in the yield compared to the model without long range and gradient hardness effects. The reverse yield for the model without long range or gradient hardness effects is nearly the same magnitude as the scaled moment at unloading, while the models with these effects show a shift in yield of approximately 4 MPa. Upon reverse yielding, each of the three models with $\beta = 1 \times 10^{-6}\sqrt{\text{m}}$ shows similar behavior, with the distance between the unloading scaled moment and the reverse yield value being nearly identical. The model with $a = 1.1\,\mu\text{m}$ has a reverse yield that is notably higher than these three while the reverse yield for the model with $a = 0.9\,\mu\text{m}$ is notably lower.

These results indicate that the gradient hardness coefficient causes an isotropic-type hardening effect, while the long-range strain causes a kinematic type hardening effect. This effect can be seen in the difference between the unloading scaled moment and the reverse yielding scaled moment. The data in Table 2 show that the models with a gradient hardness coefficient $\beta = 1 \times 10^{-6}\sqrt{\text{m}}$ have nearly the

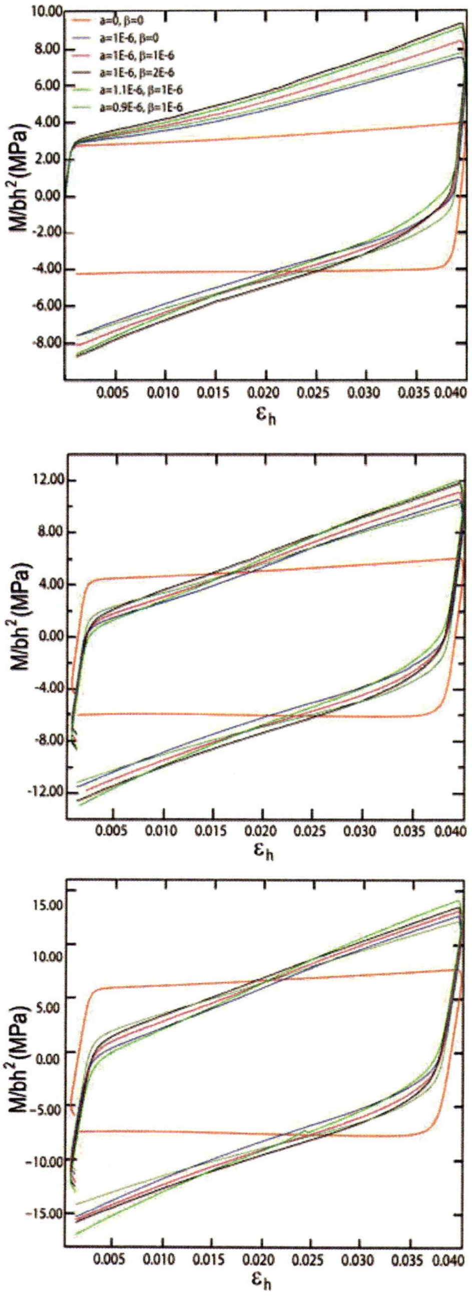

Fig. 15 Moment versus maximum strain for a 25 μm thick gage section: overlaid results for the six simulations comparing scaled moment versus top surface strain. The plots show, from top to bottom, cycles 1, 2, and 3, respectively

Table 2 25 μm thick gage section: scaled moment, in MPa, at yield, the start of unloading, yield in reverse loading and difference between the start of unloading and yield in reverse loading

a (μm)	β ($10^{-6}\sqrt{\text{m}}$)	Yield	Unload	Reverse	Δ
0.0	0.0	2.73	4.29	−3.78	8.07
1.0	0.0	2.82	7.55	−0.30	7.84
1.0	1.0	2.85	8.45	−0.19	8.64
1.0	2.0	2.93	9.36	−0.10	9.46
1.1	1.0	2.84	9.17	+0.52	8.64
0.9	1.0	2.82	7.80	−0.87	8.67

same distance between unloading and reverse yield, and the difference is directly correlated to the gradient hardness coefficient for the three models with $\beta = 0 \times 10^{-6}\sqrt{\text{m}}$, $\beta = 1 \times 10^{-6}\sqrt{\text{m}}$, and $\beta = 2 \times 10^{-6}\sqrt{\text{m}}$.

The relative differences between the five long range strain and gradient hardness models observed in the first bend are not maintained in each of the subsequent straightening and bending cycles. However, comparing the three models with a length parameter $a = 1$ μm shows that during plastic deformation, a larger gradient hardness coefficient always requires a higher magnitude of scaled moment. The differences between the three models becomes less significant during each cycle. Comparing the three models with a gradient hardness coefficient $\beta = 1 \times 10^{-6}\sqrt{\text{m}}$ shows that, upon change in loading direction, there is an inverse correlation of the length parameter with the re-yielding magnitude. However, at the end of the bend or straightening deformation, there is a direct correlation between the unloading scaled moment and the length parameter.

8 Discussion

The two examples presented illustrate different means by which heterogeneous deformations develop during loading of crystalline materials. In the first, the anisotropic properties at the single crystal level couple with the spatial heterogeneity of a polycrystal to produce variations in the loading on individual crystals. A simple constitutive model for the crystal behavior is assumed. In the second, a single crystal, although spatially homogeneous, is subjected to boundary conditions (bending) that deliberately introduce variations in the stress and deformation. A more complex model is used that accounts for the effects of the spatially varying slip on the subsequent mechanical behavior.

The simple model of single crystal plasticity applies in the context of average crystal behavior, incorporating plastic deformation via complete slip of dislocations along slip planes and elastic deformation via stretching and shearing of the atomic lattice. However, during loading of a polycrystal, the deformation within each crystal develops inhomogeneously due to a combination of the loading conditions, interaction with neighboring crystals, and orientational dependence of the material properties. An inhomogeneous deformation leaves dislocations within the material. Some dislocations contribute to lattice misorientations (referred to as geometrically necessary or excess dislocations), but other dislocations do not (referred to as statistically-stored dislocations). In either case, the net effect is that, when viewed over a sufficient volume, slip is not complete in that all dislocations do not pass through the volume to leave the shape changed but the lattice elastically undistorted. Any effect of incomplete slip is not directly incorporated into the simple model of crystal plasticity (only indirectly through enforcement of equilibrium and continuity). With currently available computing resources, our ability to resolve crystals within a polycrystal remains a limiting factor for simulation, which implies that at some level the details of incomplete slip cannot be sufficiently captured to incorporate the underlying phenomenology into the constitutive model. If these effects are significant, the only way to capture the behavior in the simple model is to adjust the empirical relationships that underlie the elastic and slip models. A comprehensive theory has been advanced by Archarya and Roy (2006) formulated in terms of continuously distributed dislocations and using the dislocation density as the primary variable. Motivated by this theory, Mach et al. (2009) have demonstrated the importance of continuity of the rotational (lattice orientation) field on the evolving microstructure under deformation. Another possibility is to investigate the details of slip gradients to determine whether they might have a significant effect on the constitutive behavior and to include the phenomenology of the slip gradients into the material response, as has been presented here.

Slip gradients manifest as orientation gradients and result in excess dislocations. To quantify the slip gradients in a single crystal, we use a misorientation tensor (or misorientation ellipsoid) that measures the deviation of the pointwise orientation from the average orientation in a single crystal. The size of the ellipsoid is a measure of the misorientation present within a single crystal. The introduction of the spatial correlation tensor and its singular value decomposition gives a measure of the spatial variation of the misorientation. As the misorientation ellipsoid increases in size, the ability of the simple single crystal model to capture the phenomenology of the material behavior should be examined, as the dislocation content is increasing in a manner that suggests the presence of greater content of excess dislocations. Furthermore, as the misorientation correlation strengthens, the ellipsoid may indicate if regions of the single crystal have segregated into cells which tend to behave as a collection of smaller homogeneous single crystals. More work is needed to better understand the implications of the measures of lattice defects relative to the assumptions underlying the simple model.

The magnitude and spatial correlation of the misorientation only provide an indication how to interpret behaviors computed with the simple single crystal model. We have also presented an extension to the simple model that incorporates the effects of slip gradients. The foundation of this extension is a "first-order" treatment of the excess dislocations that result from misorientations. These excess dislocations are directly related to the lattice misorientation via slip gradients. Their presence at some point in a crystal induces a lattice distortion and corresponding stress. We have shown that the stress field results in a kinematic type hardening during cyclic loading. For a monotonic loading, it is not possible to distinguish between isotropic and kinematic hardening behaviors, and the simple model of crystal plasticity could be adjusted to match monotonic behavior. However, the simple model cannot capture the kinematic hardening effects that are caused by excess dislocations. This behavior requires a theory in which the excess dislocations play a more direct role, as accomplished with the extended theory.

9 Summary and Conclusions

In this chapter, we discuss the modeling of heterogeneous strains that develop within the grains of a polycrystal during deformation. In the case of a relatively simple model of the crystal elastoplasticity, the heterogeneity of the straining is accompanied by the development of intragrain lattice misorientations. We examine the lattice misorientations in deformed FCC polycrystals using two tensors, one that describes the misorientation cloud and the other that quantifies the spatial correlation. The example indicates that as misorientations develop, there is one direction in most grains that has a higher degree of spatial correlation, but that direction is only weakly associated with a sample direction in the case of elongated grains. To include the influence of the excess dislocations, which are implied by the presence of intragrain lattice misorientations in the simpler model, directly on the mechanical response, a more complex model is presented. This model incorporates a kinematic degree of freedom related to the distortion of the lattice by static dislocations, in particular ones introduced as excess dislocations from gradients in slip. Associated with the elastic strains derived from these lattice distortions are stresses, acting as back stresses and biasing the response of the material when subjected to externally applied stresses. The example of the bending of a single crystal foil is used to illustrate the yield asymmetry that is captured with the more complex model but not observed using the simpler model. Together, the two models and examples present the theory and implementation for quantifying the heterogeneity of the strains over crystals within a polycrystal via its impact on the uniformity of the lattice, and for incorporating one mechanism by which the heterogeneity directly alters the mechanical behavior of the polycrystal through the lattice distortion it causes.

Acknowledgments Support for this work has been provided by the *Office of Naval Research* under contract N00014-06-1-0241. Large scale simulations were conducted at the *Cornell Theory Center.*

References

Archarya A, Roy A (2006) Size effects and idealized dislocation microstructure at small scales: Predictions of phenomenological model of mesoscopic field dislocation mechanics: Part i, Journal of the Mechanics and Physics of Solids 54:1687–1710

Ashby MF, Jones DRH (1980) Engineering Materials 1: An Introduction to their properties and applications. Pergamon

Barton NR, Dawson PR (2001a) A methodology for determining average lattice rotations and its application to the characterization of grain substructure. Metallurgical and Materials Transactions 32A:1967–1975

Barton NR, Dawson PR (2001b) On the spatial arrangement of lattice orientations in hot rolled multiphase titanium. Modeling and Simulation in Materials Science and Engineering 9:433–463

Bunge H (1982) Texture Analysis In Materials Science. Butterworth, London

Dawson PR, Marin EB (1998) Computational mechanics for metal deformation processes using polycrystal plasticity. In: van der Giessen E, Wu TY (eds) Advances in Applied Mechanics, Academic, vol 34, pp 78–169

Dawson PR, Mika DP, Barton NR (2002) Finite element modeling of lattice misorientations in aluminum alloys. Scripta Materialia 47:713–717

Dumoulin S, Tabourot L (2005) Experimental data on aluminium single crystals behaviour. Proceedings of the Institution of Mechanical Engineers, Part C: Journal of Mechanical Engineering Science 219:1159–1167

Frank F (1988) Orientation mapping. In: S KJ, Gottstein G (eds) Eighth International Conference on Textures of Materials, The Metallurgical Society, Warrendale, PA, pp 3–13

Gerken JM, Dawson PR (2007) Bending of a single crystal thin foil material with slip gradient effects. Modeling and Simulation in Materials Science and Engineering 15:799–822

Gerken JM, Dawson PR (2008a) A crystal plasticity model that incorporates stresses and strains due to slip gradients. Journal of the Mechanics and Physics of Solids 56:1651–1672

Gerken JM, Dawson PR (2008b) A finite element formulation to solve a non-local constitutive model with stresses and strains due to slip gradients. Computer Methods in Applied Mechanics and Engineering 197:1343–1361

Guruprasad PJ, Carter WJ, Berzerga AA (2008) A discrete dislocation analysis of the bauschinger effecit in microcrystals. Acta Materialia 56:5477–5491

Hartley CS (2003) A method for linking thermally activated dislocation mechanisms of yielding with continuum plasticity theory. Philosophical Magazine 83:3783–3808

Honeycombe R (1984) The Plastic Deformation of Metals, 2nd edn. Edward Arnold

Hosford WF (1993) The Mechanics of Crystals and Textured Polycrystals. Oxford Science Publications

Kocks UF, Tome CN, Wenk HR (1998) Texture and Anisotropy. Cambridge University Press

Kumar A, Dawson PR (2009) Dynamics of texture evolution in face-centered cubic polycrystals. Journal of the Mechanics and Physics of Solids 57:422–445

Mach JC, Beaudoin AJ, Archarya A (2009) Continuity in the plastic strain rate and its influence on texture evolution submitted for publication

Marin EB, Dawson PR (1998a) Elastoplastic finite element analysis of metal deformations using polycrystal constititive models. Computer Methods in Applied Mechanics and Engineering 165:23–41

Marin EB, Dawson PR (1998b) On modeling the elasto-viscoplastic response of metals using polycrystal plasticity. Computer Methods in Applied Mechanics and Engineering 165:1–21

Marin T (2006) Elastoplasticity in polycrystalline metals: Experiments and computational modeling. PhD thesis, University of Parma (Italy)

Mika DP, Dawson PR (1999) Polycrystal plasticity modeling of intracrystalline boundary textures. Acta Materialia 47(4):1355–1369

Wenk HR (1985) Preferred Oreintations of Deformed Metals and Rocks: An Introduction to Modern Texture Analysis. Academic

Full-Field vs. Homogenization Methods to Predict Microstructure–Property Relations for Polycrystalline Materials

R.A. Lebensohn, P. Ponte Castañeda, R. Brenner, and O. Castelnau

Abstract In this chapter, we review two recently proposed methodologies, based on crystal plasticity, for the prediction of microstructure–property relations in polycrystalline aggregates. The first, known as the second-order viscoplastic self-consistent (SC) method, is a mean-field theory, while the second, known as the fast Fourier transform (FFT)-based formulation, is a full-field method. The main equations and assumptions underlying both formulations are presented, using a unified notation and pointing out their similarities and differences. Concerning mean-field SC homogenization theories for the prediction of mechanical behavior of nonlinear viscoplastic polycrystals, we carry out detailed comparisons of the different linearization assumptions that can be found in the literature. Then, after validating the FFT-based full-field formulation by comparison with available analytical results, the effective behavior of model material systems predicted by means of different SC approaches are compared with ensemble averages of full-field solutions. These comparisons show that the predictions obtained by means of the second-order SC approach – which incorporates statistical information at grain level beyond first-order, through the second moments of the local field fluctuations inside the constituent grains – are in better agreement with the FFT-based full-field solutions. This is especially true in the cases of highly heterogeneous materials due to strong nonlinearity or single-crystal anisotropy. The second-order SC approach is next applied to the prediction of texture evolution of polycrystalline ice deformed in compression, a case that illustrates the flexibility of this formulation to handle problems involving materials with highly anisotropic local properties. Finally, a full three-dimensional implementation, the FFT-based formulation, is applied to study subgrain texture evolution in copper deformed in tension, with direct input and validation from orientation images. Measurements and simulations agree in that grains with initial orientation near <110> tend to develop higher misorientations. This behavior

---

R.A. Lebensohn (✉)
Materials Science and Technology Division, Los Alamos National Laboratory, MS G755, Los Alamos, NM 87545, USA
e-mail: lebenso@lanl.gov

S. Ghosh and D. Dimiduk (eds.), *Computational Methods for Microstructure-Property Relationships*, DOI 10.1007/978-1-4419-0643-4_11,

can be explained in terms of attraction toward the two stable orientations and grain interaction. Only models like the FFT-based formulation that account explicitly for interaction between individual grains are able to capture these effects.

Keywords Antiplane deformation · Crystal plasticity · Fast Fourier transform · Field fluctuations · Green function method · Mean-field vs. full-field models · Micromechanics · Misorientation · Orientation imaging microscopy formulation · Polycrystal · Second-order homogenization · Texture · Viscoplastic self-consistent

1 Introduction

An accurate prediction of the mechanical behavior of polycrystalline aggregates undergoing plastic deformation based on the directional properties and evolving substructure of their constituent single-crystal grains is an indispensable tool to establish the relationship between microstructure and properties of this large and ubiquitous class of materials. On one hand, advances in the theories that link microstructures and properties of nonlinear heterogeneous materials have enabled the development of new concepts and algorithms for the prediction of the effective plastic response of statistically defined classes of polycrystalline aggregates using crystal plasticity-based *mean-field approaches*. On the other hand, novel and very efficient *full-field approaches* also based on crystal plasticity have been proposed and applied to the prediction of the actual micromechanical fields that develop inside the grains of polycrystals with particular microstructures. In this chapter, we will review two of the most recent crystal plasticity-based mean-field [i.e., the second-order (SO) viscoplastic self-consistent (VPSC) theory] and full-field [i.e., the fast Fourier transform (FFT)-based formulation] models, establishing the connections existing between the two formulations, and showing applications of both approaches to the prediction of microstructure–property relations of polycrystalline aggregates.

Concerning mean-field approximations, the computation of effective mechanical response and texture evolution of polycrystalline materials using homogenization approaches has a long tradition (Sachs 1928; Taylor 1938). At present, self-consistent approximations are extensively used to deal with this problem. The 1-site viscoplastic (VP) self-consistent (SC) theory of polycrystal deformation can be traced back to the seminal work of Molinari et al. (1987). Later, Lebensohn and Tomé (1993) implemented this formulation numerically to fully account for polycrystal anisotropy, developing the first version of the *VPSC code*. In the last decade, this code has experienced several improvements and extensions (Tomé and Lebensohn 2008), and it is nowadays extensively used to simulate plastic deformation of polycrystalline aggregates and to interpret experimental evidence on metallic, geological, and polymeric materials (see Lebensohn et al. 2007 for a comprehensive list of material systems studied with the VPSC theory and code).

The self-consistent approximation, one of the most commonly used homogenization methods to estimate the mechanical response behavior of polycrystals, was originally proposed by Hershey (1954) for linear elastic materials. For nonlinear aggregates (as those formed by grains deforming in the viscoplastic regime), the several self-consistent approximations that were proposed subsequently differ in the procedure used to linearize the nonlinear local mechanical behavior, but eventually all of them end up making use of the original linear self-consistent theory. Among the nonlinear SC formulations, we can mention: the *secant* (SEC) (Hill 1965; Hutchinson 1976), the *tangent* (TG) (Molinari et al. 1987; Lebensohn and Tomé 1993), and the *affine* (AFF) (Ponte Castañeda 1996; Masson et al. 2000) approximations. All these are *first-order* SC approximations since they are based on linearization schemes that, at grain level, make use of information on field averages only, disregarding higher-order statistical information inside the grains. However, the above assumption may be questionable specially when strong directionality and/or large variations in local properties are to be expected. Such is the case for low rate-sensitivity materials, aggregates made of highly anisotropic grains, and multiphase polycrystals. In all those cases, strong deformation gradients are likely to develop inside grains because of the contrast in properties between neighboring grains.

To overcome the above limitations, Ponte Castañeda and coauthors have developed over the last two decades more accurate nonlinear homogenization methods, using linearization schemes at grain level that also incorporate information on the second moments of the field fluctuations in the grains. These more elaborate SC formulations are based on the use of so-called linear comparison methods, which express the effective potential of the nonlinear VP polycrystal in terms of that of a linearly viscous aggregate with properties that are determined from suitably designed variational principles. Ponte Castañeda's first *variational* method was originally proposed for nonlinear composites (Ponte Castañeda 1991) and then extended to VP polycrystals (deBotton and Ponte Castañeda 1995). It makes use of the SC approximation for linearly viscous polycrystals to obtain bounds and estimates for nonlinear VP polycrystals. The most recent *second-order* method, proposed for nonlinear composites (Ponte Castañeda 2002), and later extended to VP polycrystals (Liu and Ponte Castañeda 2004), uses the SC approximation for a more general class of linearly viscous polycrystals, having a non-vanishing strain-rate at zero stress, to generate even more accurate SC estimates for VP polycrystals. The implementation of a fully anisotropic second-order approach inside the VPSC code has been a necessary step toward improving its predictive capability for polycrystalline materials that exhibit high contrast in local properties. Unavoidably, this improved capability comes at the expense of more complex and numerically demanding algorithms.

In what concerns full-field approaches, in the last 20 years, crystal plasticity-based finite element (FE) implementations have been extensively applied to obtain solutions for the plastic deformation of polycrystalline materials with intracrystalline resolution (Becker 1991; Mika and Dawson 1998; Delaire et al. 2001; Barbe et al. 2001; Raabe et al. 2001; Bhattacharyya et al. 2001; Delannay et al. 2003, 2006; Cheong and Busso 2004; Diard et al. 2005; Musienko et al. 2007). However, the large number of degrees of freedom required by such FE calculations limits the size of the microstructures that can be investigated by these methods. Conceived as a very efficient alternative to FE methods, a formulation inspired by image-processing techniques and based on the FFT algorithm, originally proposed by Moulinec and Suquet (1994), for the prediction of the micromechanical behavior of plastically deforming heterogeneous materials. The latter includes both composites (Moulinec and Suquet 1998; Michel et al. 2000; Idiart et al. 2006), in which the source of heterogeneity is related to the spatial distribution of phases with different mechanical properties, and polycrystals (Lebensohn 2001; Lebensohn et al. 2004a, b, 2005, 2008), in which the heterogeneity is related to the spatial distribution of crystals with directional mechanical properties.

The plan of this chapter is as follows. In Sect. 2, we describe the implementation of the second-order formulation inside the VPSC code (Lebensohn et al. 2007), and present the FFT-based formulation, specialized to the case of viscoplastic polycrystals (Lebensohn et al. 2008). In Sect. 3, we first show a validation of the FFT-based approach by comparison with an exact analytical result and then discuss the differences between the first- and second-order VPSC formulations by comparing their predictions with corresponding FFT-based full-field solutions. We do so for crystals with different symmetries, as a function of anisotropy, number of independent slip systems, and degree of nonlinearity. In this comparison, the second-order estimates show the best overall agreement with the full-field solutions. The different SC approaches are then applied to the prediction of texture evolution in a strongly heterogeneous system (i.e., polycrystalline ice deforming in uniaxial compression) (Lebensohn et al. 2007). This comparison shows that the second-order formulation yields results in better agreement with experimental evidence than the first-order approximations, predicting a substantial and persistent accommodation of deformation by basal slip, even when the basal poles become strongly aligned with the compression direction. Section 3 also shows an application of the FFT-based formulation to the prediction of subgrain texture and microstructure evolution in polycrystalline copper deformed under tension, with direct input from orientation imaging microscopy (OIM) images (Lebensohn et al. 2008). Average orientations and misorientations predicted after 11% tensile strain are directly compared with OIM measurements. Experiments and simulations agree in that grains with initial orientation near <110> tend to develop higher misorientations. This behavior can be explained in terms of attraction toward the two stable orientations and grain interaction. Only models that account explicitly for interaction between individual grains, like the FFT-based formulation, are able to capture these effects.

2 Models

2.1 *Viscoplastic Self-Consistent Formalism*

In this section, the incompressible viscoplastic self-consistent formulation (Lebensohn et al. 1998) is first presented using the affine linearization scheme (Ponte Castañeda 1996; Masson et al. 2000), and the second-order linearization procedure (Ponte Castañeda 2002; Liu and Ponte Castañeda 2004) is described next.

The self-consistent formulation consists in representing a polycrystal by means of weighted, ellipsoidal, statistically representative (SR) grains. Each of these SR grains represents the average behavior of all the grains with a particular crystallographic orientation and morphology, but different environments. These SR grains should be regarded as representing the behavior of *mechanical phases*, i.e., all the single crystals with a given orientation (r) belong to mechanical phase (r) and are represented by SR grain (r). (Note the difference between "mechanical phases," which differ from each other only in terms of crystallographic orientation and/or morphology, and actual "phases" differing from each other in crystallographic structure and/or composition). In what follows, "SR grain (r)" and "mechanical phase (r)" will be used interchangeably. The weights represent volume fractions. The latter are chosen to reproduce the initial texture of the material. In turn, each representative grain will be treated as an ellipsoidal viscoplastic inclusion embedded in an effective viscoplastic medium. Both inclusion and medium have fully anisotropic properties. Deformation is carried by crystal plasticity mechanisms: slip and twinning systems activated by a resolved shear stress.

2.1.1 Local Constitutive Behavior and Homogenization

Let us consider a macroscopic velocity-gradient $V_{i,j}$ applied to an polycrystalline aggregate, which can be decomposed into an average symmetric strain-rate $\dot{E}_{ij} = \frac{1}{2}\left(V_{i,j} + V_{j,i}\right)$ and an average antisymmetric rotation-rate $\dot{\Omega}_{ij} = \frac{1}{2}\left(V_{i,j} - V_{j,i}\right)$. Let us assume that the plastic component of the deformation is much larger than the elastic part and therefore the flow is incompressible. The viscoplastic constitutive behavior at each material point $\mathbf{x}$ (in what follows, Cartesian and Fourier vectors are indicated in boldface, second and fourth-rank tensors, either in components or not, are not) is described by means of the following nonlinear, *rate-sensitive equation*:

$$\dot{\varepsilon}\left(\mathbf{x}\right) = \sum_{k=1}^{N_k} m^k\left(\mathbf{x}\right)\dot{\gamma}^k\left(\mathbf{x}\right) = \dot{\gamma}_o \sum_{k=1}^{N_k} m^k\left(\mathbf{x}\right)\left(\frac{\left|m^k\left(\mathbf{x}\right) : \sigma'\left(\mathbf{x}\right)\right|}{\tau_o^k\left(\mathbf{x}\right)}\right)^n \times \mathrm{sgn}\left(m^k\left(\mathbf{x}\right) : \sigma'\left(\mathbf{x}\right)\right), \tag{1}$$

where the symbol ":" indicates double contraction of indices, and the sum runs over all N_k slip and twinning systems. τ_o^k and $m^k(\mathbf{x}) = \frac{1}{2}\left(n^k(\mathbf{x}) \otimes b^k(\mathbf{x}) + b^k(\mathbf{x}) \otimes n^k(\mathbf{x})\right)$ are the threshold resolved shear stress and the symmetric Schmid tensor associated with slip or twinning system (k) (with $\mathbf{n}^k$ and $\mathbf{b}^k$ being the normal and Burgers vector direction of such slip or twinning system), $\dot{\varepsilon}$ and σ' are the deviatoric strain-rate and stress tensors, $\dot{\gamma}^k$ is the local shear-rate on slip or twinning system (k), $\dot{\gamma}_o$ is a normalization factor, and n is the rate-sensitivity exponent. Note that although (1) can be used to deal with crystal deforming by slip and twinning, in the examples that follows (in the context of both homogenization and full-field approaches), we will only consider crystal deformation by slip. In this way, we avoid the additional complication of having to deal with twinning reorientation.

Also note that the constitutive behavior described in (1) does not consider other high temperature crystal deformation mechanisms, such as climb, grain-boundary sliding, or recrystallization, and that elastic effects are neglected.

For later use, the plastic rotation-rate associated with a material point $\mathbf{x}$ contributing to the crystallographic lattice rotation is given by:

$$\dot{\omega}_{ij}^{p}(\mathbf{x}) = \sum_{k} \alpha_{ij}^{k}(\mathbf{x})\,\dot{\gamma}^{k}(\mathbf{x}), \tag{2}$$

where $\alpha^s(\mathbf{x}) = \frac{1}{2}\left(n^s(\mathbf{x}) \otimes b^s(\mathbf{x}) - b^s(\mathbf{x}) \otimes n^s(\mathbf{x})\right)$ is the antisymmetric Schmid tensor.

Let us assume that the following linear relation [i.e., an approximation of the actual nonlinear relation (1)] holds between the strain-rate and stress in the SR grain (r):

$$\dot{\varepsilon}(\mathbf{x}) = M^{(r)} : \sigma'(\mathbf{x}) + \dot{\varepsilon}^{o\,(r)}, \tag{3}$$

where $M^{(r)}$ and $\dot{\varepsilon}^{o(r)}$ are, respectively, the viscoplastic compliance and the back-extrapolated term of SR grain (r). Depending on the linearization assumption, $M^{(r)}$ and $\dot{\varepsilon}^{o(r)}$ can be chosen differently (some possible choices are discussed below). Taking a volumetric average, we obtain:

$$\dot{\varepsilon}^{(r)} = M^{(r)} : \sigma'^{(r)} + \dot{\varepsilon}^{o\,(r)}, \tag{4}$$

where $\dot{\varepsilon}^{(r)}$ and $\sigma'^{(r)}$ are average magnitudes in the volume of SR grain (r). Let us homogenize the behavior of a linear heterogeneous medium whose local behavior is described in (3) assuming an analogous linear relation at the effective medium (macroscopic) level:

$$\dot{E} = \bar{M} : \Sigma' + \dot{E}^{o}, \tag{5}$$

where $\dot{E}$ and Σ' are the overall (macroscopic) deviatoric strain-rate and stress tensors and $\bar{M}$ and $\dot{E}^o$ are, respectively, the viscoplastic compliance and back-extrapolated term of an a priori unknown homogeneous equivalent medium (HEM). The usual procedure to obtain the homogenized response of a linear polycrystal is the linear self-consistent method. The problem underlying the self-consistent method is that of an inhomogeneous domain (r) of moduli $M^{(r)}$ and $\dot{\varepsilon}^{o(r)}$, embedded in an infinite medium of moduli $\bar{M}$ and $\dot{E}^o$. Invoking the concept of the equivalent inclusion (Mura 1987), the local constitutive behavior in domain (r) can be rewritten as:

$$\dot{\varepsilon}(\mathbf{x}) = \bar{M} : \sigma'(\mathbf{x}) + \dot{E}^{o} + \dot{\varepsilon}^{*}(\mathbf{x}), \tag{6}$$

where $\dot{\varepsilon}^*(\mathbf{x})$ is an eigen-strain-rate field, which follows from replacing the inhomogeneity by an equivalent inclusion. Rearranging and subtracting (5) from (6) gives:

$$\tilde{\sigma}'(\mathbf{x}) = \bar{L} : \left(\tilde{\dot{\varepsilon}}(\mathbf{x}) - \dot{\varepsilon}^{*}(\mathbf{x})\right). \tag{7}$$

The symbol "∼" denotes local deviations from macroscopic values of the corresponding magnitudes, and $\bar{L} = \bar{M}^{-1}$. Combining (7) with the equilibrium condition gives:

$$\sigma_{ij,j}(\mathbf{x}) = \tilde{\sigma}_{ij,j}(\mathbf{x}) = \tilde{\sigma}'_{ij,j}(\mathbf{x}) + \tilde{\sigma}^{m}_{,i}(\mathbf{x}), \tag{8}$$

where σ_{ij} and σ^m are the Cauchy stress tensor and the mean stress, respectively. Using the relation $\tilde{\dot{\varepsilon}}_{ij}(\mathbf{x}) = \frac{1}{2}\left(\tilde{v}_{i,j}(\mathbf{x}) + \tilde{v}_{j,i}(\mathbf{x})\right)$ between the strain-rate and velocity-gradient deviations, and adding the incompressibility condition associated with plastic deformation, we obtain:

$$\left|\begin{array}{l} \bar{L}_{ijkl}\,\tilde{v}_{k,lj}(\mathbf{x}) + \tilde{\sigma}^{m}_{,i}(\mathbf{x}) + \varphi_{ij,j}(\mathbf{x}) = 0 \\ \tilde{v}_{k,k}(\mathbf{x}) = 0 \end{array}\right., \tag{9}$$

where

$$\varphi_{ij}(\mathbf{x}) = -\bar{L}_{ijkl}\dot{\varepsilon}^{*}_{kl}(\mathbf{x}) \tag{10}$$

is a heterogeneity or *polarization field*, and its divergence: $f_i(\mathbf{x}) = \varphi_{ij,j}(\mathbf{x})$ is a fictitious volumetric force field. System (9) consists of four differential equations with four unknowns: three are the components of velocity deviation vector $\tilde{v}_i(\mathbf{x})$, and one is the mean stress deviation $\tilde{\sigma}^m(\mathbf{x})$. A system of N linear differential equations with N unknown functions and a polarization term can be solved using the Green function method. Let us call $G_{km}(\mathbf{x})$ and $H_m(\mathbf{x})$ the Green functions associated with $\tilde{v}_i(\mathbf{x})$ and $\tilde{\sigma}^m(\mathbf{x})$, respectively, which solve the auxiliary problem of a unitary volumetric force, with a single non-vanishing m-component:

$$\left|\begin{array}{l} \bar{L}_{ijkl}\,G_{km,lj}(\mathbf{x}-\mathbf{x}') + H_{m,i}(\mathbf{x}-\mathbf{x}') + \delta_{im}\,\delta(\mathbf{x}-\mathbf{x}') = 0, \\ G_{km,k}(\mathbf{x}-\mathbf{x}') = 0. \end{array}\right. \tag{11}$$

Once the solution of (11) is obtained, the solution for the velocity field is given by the convolution integral:

$$\tilde{v}_k(\mathbf{x}) = \int_{R^3} G_{ki}\left(\mathbf{x}-\mathbf{x}'\right) f_i\left(\mathbf{x}'\right) \mathrm{d}x'. \tag{12}$$

System (11) can be solved using the Fourier transform method. Expressing the Green functions in terms of their inverse Fourier transforms, the differential system (11) can be transformed into an algebraic system:

$$\left|\begin{array}{l} \alpha_j\,\alpha_l\,\bar{L}_{ijkl}\,k^2\hat{G}_{km}(\boldsymbol{\xi}) + \alpha_i\,ik\hat{H}_m(\boldsymbol{\xi}) = \delta_{im}, \\ \alpha_k\,k^2\hat{G}_{km}(\boldsymbol{\xi}) = 0, \end{array}\right. \tag{13}$$

where k and $\boldsymbol{\alpha}$ are the modulus and the unit vector associated with a point of Fourier space $\boldsymbol{\xi} = k\boldsymbol{\alpha}$, respectively. Calling $A'_{ik}(\boldsymbol{\alpha}) = \alpha_j\,\alpha_l\,\bar{L}_{ijkl}$, system (13) can be expressed as a matrix product $\mathbf{A} \times \mathbf{B} = \mathbf{C}$, where $\mathbf{A}$, $\mathbf{B}$, and $\mathbf{C}$ are the matrices given by:

$$\begin{array}{c|ccc|c} & k^2\hat{G}_{11} & k^2\hat{G}_{12} & k^2\hat{G}_{13} & \\ & k^2\hat{G}_{21} & k^2\hat{G}_{22} & k^2\hat{G}_{23} & = \mathbf{B} \\ & k^2\hat{G}_{31} & k^2\hat{G}_{32} & k^2\hat{G}_{33} & \\ & ik\hat{H}_1 & ik\hat{H}_2 & ik\hat{H}_3 & \\ \hline A(\boldsymbol{\alpha}) = \begin{array}{cccc} A'_{11} & A'_{12} & A'_{13} & \alpha_1 \\ A'_{21} & A'_{22} & A'_{23} & \alpha_2 \\ A'_{31} & A'_{32} & A'_{33} & \alpha_3 \\ \alpha_1 & \alpha_2 & \alpha_3 & 0 \end{array} & \begin{array}{c} 1 \\ 0 \\ 0 \\ 0 \end{array} & \begin{array}{c} 0 \\ 1 \\ 0 \\ 0 \end{array} & \begin{array}{c} 0 \\ 0 \\ 1 \\ 0 \end{array} & = \mathbf{C} \end{array}. \tag{14}$$

Using the explicit form of matrix $\mathbf{C}$, we can write:

$$\mathbf{B} = \mathbf{A}^{-1} \times \mathbf{C} = \begin{bmatrix} A_{11}^{-1} & A_{12}^{-1} & A_{13}^{-1} \\ A_{21}^{-1} & A_{22}^{-1} & A_{23}^{-1} \\ A_{31}^{-1} & A_{32}^{-1} & A_{33}^{-1} \\ A_{41}^{-1} & A_{42}^{-1} & A_{43}^{-1} \end{bmatrix}. \tag{15}$$

Finally, comparing (14) and (15):

$$k^2\hat{G}_{ij}(\boldsymbol{\xi}) = A_{ij}^{-1}(\boldsymbol{\alpha}) \quad (i, j = 1, 3). \tag{16}$$

Since the components of $\mathbf{A}$ are real functions of $\boldsymbol{\alpha}$, so are those of $k^2\hat{G}_{ij}(\boldsymbol{\xi})$. This property leads to real integrals in the derivation that follows.

Knowing the Green tensor expression in Fourier space, we can write the solution of our eigen-strain-rate problem using the convolution integral. Taking partial derivatives to (12), we obtain:

$$\tilde{v}_{k,l}(\mathbf{x}) = \int_{R^3} G_{ki,l}(\mathbf{x} - \mathbf{x}') f_i(\mathbf{x}') \, d\mathbf{x}'. \tag{17}$$

Replacing the expression of the fictitious volumetric force field in (17), recalling that $\partial G_{ij}(\mathbf{x} - \mathbf{x}')/\partial \mathbf{x} = -\partial G_{ij}(\mathbf{x} - \mathbf{x}')/\partial \mathbf{x}'$, integrating by parts, and using the divergence theorem (Mura 1987), we obtain:

$$\tilde{v}_{k,l}(\mathbf{x}) = \int_{R^3} G_{ki,jl}(\mathbf{x} - \mathbf{x}') \varphi_{ij}(\mathbf{x}') \, d\mathbf{x}'. \tag{18}$$

The integral equation (18) provides an exact implicit solution to the problem. Furthermore, it is known from Eshelby's elastic inclusion formalism that if the eigen-strain is uniform over an ellipsoidal domain where the stiffness tensor is uniform, then the stress and the strain are constant over the domain of the inclusion (r). The latter suggests to use an a priori unknown constant polarization within the volume Ω of the ellipsoidal inclusion. This allows us to average the local field (18) over the domain Ω and obtain an average strain-rate inside the inclusion of the form:

$$\tilde{\nu}_{k,l}^{(r)} = \left(-\frac{1}{\Omega}\int_{\Omega}\int_{\Omega} G_{ki,jl}\left(\mathbf{x}-\mathbf{x}'\right)\mathrm{d}\mathbf{x}\,\mathrm{d}\mathbf{x}'\right)\bar{L}_{ijmn}\,\dot{\varepsilon}_{mn}^{*(r)}, \tag{19}$$

where $\tilde{\nu}_{k,l}^{(r)}$ and $\dot{\varepsilon}_{mn}^{*(r)}$ have to be interpreted as average quantities inside the inclusion. Expressing the Green tensor in terms of the inverse Fourier transform and taking derivatives, we obtain:

$$\tilde{\nu}_{k,l}^{(r)} = \left(\frac{1}{8\pi^3\Omega}\int_{\Omega}\int_{\Omega}\int_{R^3}\alpha_j\alpha_l\left(k^2\hat{G}_{ki}(\boldsymbol{\xi})\right)\exp\left[-i\boldsymbol{\xi}\left(\mathbf{x}-\mathbf{x}'\right)\right]\mathrm{d}\boldsymbol{\xi}\mathrm{d}\mathbf{x}\mathrm{d}\mathbf{x}'\right)\bar{L}_{ijmn}\dot{\varepsilon}_{mn}^{*(r)}$$

$$= T_{klij}\,\bar{L}_{ijmn}\,\dot{\varepsilon}_{mn}^{*(r)}. \tag{20}$$

Writing $d\boldsymbol{\xi}$ in spherical coordinates: $d\boldsymbol{\xi} = k^2\sin\theta\,dk\,\mathrm{d}\theta\,\mathrm{d}\varphi$ and using relation (16), the Green interaction tensor T_{klij} can be expressed as:

$$T_{klij} = \frac{1}{8\pi^3\Omega}\int_0^{2\pi}\int_0^{\pi}\alpha_j\,\alpha_l\,A_{ki}^{-1}(\boldsymbol{\alpha})\,\Lambda(\boldsymbol{\alpha})\sin\theta\,\mathrm{d}\theta\,\mathrm{d}\varphi, \tag{21}$$

where

$$\Lambda(\boldsymbol{\alpha}) = \int_0^{\infty}\left(\int_{\Omega}\int_{\Omega}\exp\left[-i\boldsymbol{\xi}\left(\mathbf{x}-\mathbf{x}'\right)\right]d\mathbf{x}\,d\mathbf{x}'\right)k^2\mathrm{d}k. \tag{22}$$

Integrating (22) inside an ellipsoidal grain of radii (a, b, c) (Berveiller et al. 1987) and replacing in (21) gives:

$$T_{klij} = \frac{abc}{4\pi}\int_0^{2\pi}\int_0^{\pi}\frac{\alpha_j\,\alpha_l\,A_{ki}^{-1}(\boldsymbol{\alpha})}{[\rho(\boldsymbol{\alpha})]^3}\sin\,\theta\,\mathrm{d}\theta\,\mathrm{d}\varphi, \tag{23}$$

where $\rho(\boldsymbol{\alpha}) = \left[(a\alpha_1)^2 + (b\alpha_2)^2 + (c\alpha_3)^2\right]^{1/2}$. The symmetric and antisymmetric Eshelby tensors (functions of $\bar{L}$ and the shape of the ellipsoidal inclusion, representing the morphology of the SR grains) are defined as:

$$S_{ijkl} = \frac{1}{4}\left(T_{ijmn} + T_{jimn} + T_{ijnm} + T_{jinm}\right)\bar{L}_{mnkl}, \tag{24}$$

$$\Pi_{ijkl} = \frac{1}{4}\left(T_{ijmn} - T_{jimn} + T_{ijnm} - T_{jinm}\right)\bar{L}_{mnkl}. \tag{25}$$

Taking symmetric and antisymmetric components to (20) and using (24) and (25), we obtain the average strain-rate and rotation-rate deviations of the ellipsoidal domain:

$$\tilde{\dot{\varepsilon}}^{(r)} = S : \dot{\varepsilon}^{*(r)}, \tag{26}$$

$$\tilde{\dot{\omega}}^{(r)} = \Pi : \dot{\varepsilon}^{*(r)} = \Pi : S^{-1} : \tilde{\dot{\varepsilon}}^{(r)}, \tag{27}$$

where $\tilde{\dot{\varepsilon}}^{(r)} = \dot{E} - \dot{\varepsilon}^{(r)}$ and $\tilde{\dot{\omega}}^{(r)} = \dot{\Omega} - \dot{\omega}^{(r)}$ are deviations of the average strain-rate and rotation-rate inside the inclusion, with respect to the corresponding overall magnitudes, and $\dot{\varepsilon}^{*(r)}$ is the average eigen-strain-rate in the inclusion.

2.1.2 Interaction and Localization Equations

Taking volume averages over the domain of the inclusion on both sides of (7) gives:

$$\tilde{\sigma}'^{(r)} = \bar{L} : \left(\tilde{\dot{\varepsilon}}^{(r)} - \dot{\varepsilon}^{*(r)}\right). \tag{28}$$

Replacing the eigen-strain-rate given by (26) into (28), we obtain the *interaction equation*:

$$\tilde{\dot{\varepsilon}}^{(r)} = -\tilde{M} : \tilde{\sigma}'^{(r)}, \tag{29}$$

where the interaction tensor is given by:

$$\tilde{M} = (I - S)^{-1} : S : \bar{M}. \tag{30}$$

Replacing the constitutive relations of the inclusion and the effective medium in the interaction equation and after some manipulation, one can write the following *localization equation*:

$$\sigma'^{(r)} = B^{(r)} : \Sigma' + b^{(r)}, \tag{31}$$

where the localization tensors are defined as:

$$B^{(r)} = \left(M^{(r)} + \tilde{M}\right)^{-1} : \left(\bar{M} + \tilde{M}\right), \tag{32}$$

$$b^{(r)} = \left(M^{(r)} + \tilde{M}\right)^{-1} : \left(\dot{E}^{o} - \dot{\varepsilon}^{o(r)}\right). \tag{33}$$

2.1.3 Self-Consistent Equations

The derivation presented in the previous sections solves the problem of an equivalent inclusion embedded in an effective medium. In this section, we use the previous result to construct a polycrystal model, consisting in regarding each SR grain (r) as an inclusion embedded in an effective medium that represents the polycrystal. The properties of such medium are not known a priori but have to be found through an iterative procedure. Replacing the stress localization equation (31) in the average local constitutive equation (4), we obtain:

$$\varepsilon^{(r)} = M^{(r)} : B^{(r)} : \Sigma + M^{(r)} : b^{(r)} + \varepsilon^{o(r)}. \tag{34}$$

Taking volumetric average to (34), enforcing the condition that the average of the strain-rates over the aggregate has to coincide with the macroscopic quantities, i.e.:

$$\dot{E} = \left\langle \dot{\varepsilon}^{(r)} \right\rangle, \tag{35}$$

where the brackets "⟨ ⟩" denote average over the SR grains, weighted by the associated volume fraction, and using the macroscopic constitutive relation (5), we obtain the following *self-consistent equations* for the HEM's compliance and back-extrapolated term:

$$\bar{M} = \left\langle M^{(r)} : B^{(r)} \right\rangle, \tag{36a}$$

$$\dot{E}^{o} = \left\langle M^{(r)} : b^{(r)} + \dot{\varepsilon}^{o(r)} \right\rangle. \tag{36b}$$

These self-consistent equations are derived imposing the average of the local strain-rates to coincide with the applied macroscopic strain-rate (35). If all the SR grains are represented by ellipsoids that have the same shape and orientation, it can be shown that the same equations are obtained from the condition that the average of the local stresses coincides with the macroscopic stress. If the SR grains have different morphologies, they have associated different Eshelby tensors, and the interaction tensors cannot be factored from the averages. In such case, the following generalized self-consistent expressions should be used (Walpole 1969):

$$\bar{M} = \left\langle M^{(r)} : B^{(r)} \right\rangle : \left\langle B^{(r)} \right\rangle^{-1}, \tag{37a}$$

$$\dot{E}^{o} = \left\langle M^{(r)} : b^{(r)} + \dot{\varepsilon}^{o(r)} \right\rangle - \left\langle M^{(r)} : B^{(r)} \right\rangle : \left\langle B^{(r)} \right\rangle^{-1} : \left\langle b^{(r)} \right\rangle. \tag{37b}$$

2.1.4 Linearization Assumptions

As stated earlier, different choices are possible for the linearized behavior at grain level, and the results of the homogenization scheme depend on this choice. In what follows, we present several first-order linearization schemes, defined in terms of the stress first-order moment (average) inside SR grain (r).

The *secant approximation* (Hill 1965; Hutchinson 1976) consists in assuming the following linearized moduli:

$$M_{\text{sec}}^{(r)} = \dot{\gamma}_{o} \sum_{k} \frac{m^{k\,(r)} \otimes m^{k\,(r)}}{\tau_{o}^{k\,(r)}} \left(\frac{m^{k\,(r)} : \sigma'^{(r)}}{\tau_{o}^{k\,(r)}} \right)^{n-1}, \tag{38}$$

$$\dot{\varepsilon}_{\text{sec}}^{o(r)} = 0, \tag{39}$$

where the index (r) in $m^{k(r)}$ and $\tau_o^{k(r)}$ indicates uniform (average) values of these magnitudes, corresponding to a given orientation and hardening state associated with SR grain (r).

Under the *affine approximation* (Ponte Castañeda 1996; Masson et al. 2000), the moduli are given by:

$$M_{\text{aff}}^{(r)} = n\dot{\gamma}_o \sum_k \frac{m^{k(r)} \otimes m^{k(r)}}{\tau_o^{k(r)}} \left(\frac{m^{k(r)} : \sigma'^{(r)}}{\tau_o^{k(r)}} \right)^{n-1}, \tag{40}$$

$$\dot{\varepsilon}_{\text{aff}}^{o(r)} = (1-n)\,\dot{\gamma}_o \sum_k \left(\frac{m^{k(r)} : \sigma'^{(r)}}{\tau_o^{k(r)}} \right)^{n} \times \operatorname{sgn}\left(m^{k(r)} : \sigma'^{(r)} \right). \tag{41}$$

In the case of the *tangent approximation* (Molinari et al. 1987; Lebensohn and Tomé 1993), the moduli are, formally, the same as in the affine case: $M_{\text{tg}}^{(r)} = M_{\text{aff}}^{(r)}$ and $\dot{\varepsilon}_{\text{tg}}^{o(r)} = \dot{\varepsilon}_{\text{aff}}^{o(r)}$. However, instead of using these moduli, and to avoid the iterative adjustment of the macroscopic back-extrapolated term, Molinari et al. (1987) used the secant SC compliance (38) to adjust $\bar{M}$ (to be denoted $\bar{M}_{\text{sec}}$), in combination with the tangent–secant relation: $\bar{M}_{\text{tg}} = n\bar{M}_{\text{sec}}$ (Hutchinson 1976). Then, the expression of the interaction tensor is given by:

$$\tilde{M} = (I-S)^{-1} : S : \bar{M}_{\text{tg}} = n\,(I-S)^{-1} : S : \bar{M}_{\text{sec}}. \tag{42}$$

Qualitatively, the interaction equation (29) indicates that the larger the interaction tensor, the smaller the deviation of grain stresses with respect to the average stress should be. As a consequence, for $n \to \infty$, the tangent approximation tends to a uniform stress state [Sachs (1928) or *lower-bound approximation*]. This rate-insensitive limit of the tangent formulation is an artifact created using the above tangent–secant relation of the nonlinear polycrystal in the self-consistent solution of the linear comparison polycrystal. On the other hand, the secant interaction has been proven to tend to a uniform strain-rate state [Taylor (1938) or *upper-bound approximation*] in the rate-insensitive limit.

2.1.5 Second-Order Formulation

The more sophisticated *second-order approximation* to linearize the behavior of the mechanical phases is based on the calculation of average fluctuations of the stress distribution inside the linearized SR grains. The methodology to obtain these fluctuations were derived by Bobeth and Diener (1987), Kreher (1990), and Parton and Buryachenko (1990), and reads as follows. The effective stress potential $\bar{U}_T$ of a linearly viscous polycrystal described by (5) may be written in the form (Laws 1973; Willis 1981):

$$\bar{U}_T = \frac{1}{2}\bar{M} :: \left(\Sigma' \otimes \Sigma' \right) + \dot{E}^o : \Sigma' + \frac{1}{2}\bar{G}, \tag{43}$$

where $\bar{G}$ is the power under zero applied stress. Let us rewrite the self-consistent expression for $\bar{M}$ and $\dot{E}^o$ (36) as:

$$\bar{M} = \left\langle M^{(r)} : B^{(r)} \right\rangle = \sum_r c^{(r)} M^{(r)} : B^{(r)}, \tag{44}$$

$$\dot{E}^o = \left\langle M^{(r)} : b^{(r)} + \dot{\varepsilon}^{o(r)} \right\rangle = \sum_r c^{(r)} \left(M^{(r)} : b^{(r)} + \dot{\varepsilon}^{o(r)} \right) = \sum_r c^{(r)}\, \dot{\varepsilon}^{o(r)} : B^{(r)}, \tag{45}$$

where $c^{(r)}$ is the volume fraction associated with SR grain (r). The corresponding expression for $\bar{G}$ is:

$$\bar{G} = \sum_r c^{(r)}\, \dot{\varepsilon}^{o\,(r)} : b^{(r)}. \tag{46}$$

The average second-order moment of the stress field over a SR grain (r) of this polycrystal is a fourth-rank tensor given by:

$$\left\langle \sigma' \otimes \sigma' \right\rangle^{(r)} = \frac{2}{c^{(r)}} \frac{\partial \tilde{U}_T}{\partial M^{(r)}}. \tag{47}$$

Replacing (44–46) in (47), we obtain:

$$\left\langle \sigma' \otimes \sigma' \right\rangle^{(r)} = \frac{1}{c^{(r)}} \frac{\partial \bar{M}}{\partial M^{(r)}} :: \left(\Sigma' \otimes \Sigma' \right) + \frac{1}{c^{(r)}} \frac{\partial \dot{E}^o}{\partial M^{(r)}} : \Sigma' + \frac{1}{c^{(r)}} \frac{\partial \bar{G}}{\partial M^{(r)}}. \tag{48}$$

Using matrix notation for symmetric deviatoric tensors (Lequeu et al. 1987), the first derivative in the right term can be obtained solving the following equation:

$$\Omega_{ijkl} \frac{\partial \bar{M}_{kl}}{\partial M_{uv}^{(r)}} = \pi_{ij}^{(r,uv)}, \tag{49}$$

where i,j,k,l and $u, v = 1, 5$. The expressions for Ω_{ijkl} and $\pi_{ij}^{(r,uv)}$ are given in the Appendix. Expression (49) is a linear system of 25 equations with 25 unknowns (i.e., the components of $\partial \bar{M}_{kl} / \partial M_{uv}^{(r)}$). In turn, the other two derivatives appearing in (48) can be calculated as:

$$\frac{\partial E_i^o}{\partial M_{uv}^{(r)}} = \zeta_{ikl} \frac{\partial \bar{M}_{kl}}{\partial M_{uv}^{(r)}} + \kappa_i^{(r,uv)}, \tag{50}$$

$$\frac{\partial \bar{G}}{\partial M_{uv}^{(r)}} = \varphi_{ij} \frac{\partial \bar{M}_{ij}}{\partial M_{uv}^{(r)}} + \vartheta_i \frac{\partial E_i^o}{\partial M_{uv}^{(r)}} + \eta^{(r,uv)}, \tag{51}$$

where ζ_{ikl}, φ_{ij}, ϑ_i, $\kappa_i^{(r,uv)}$ and $\eta^{(r,uv)}$ are given in Appendix.

Once the average second moments of the stress are obtained, the corresponding second moments of the strain-rate can be calculated as:

$$\langle \dot{\varepsilon} \otimes \dot{\varepsilon} \rangle^{(r)} = \left(M^{(r)} \otimes M^{(r)} \right) :: \left\langle \sigma' \otimes \sigma' \right\rangle^{(r)} + \dot{\varepsilon}^{(r)} \otimes \dot{\varepsilon}^{o(r)} + \dot{\varepsilon}^{o(r)} \otimes \dot{\varepsilon}^{(r)} - \dot{\varepsilon}^{o(r)} \otimes \dot{\varepsilon}^{o(r)}. \tag{52}$$

The average second moments can be used, for instance, to generate the average second moment of the equivalent stress and strain-rate in mechanical phase (r) as:

$$\bar{\bar{\sigma}}_{\mathrm{eq}}^{(r)} = \left(\frac{3}{2} I :: \left\langle \sigma' \otimes \sigma' \right\rangle^{(r)} \right)^{1/2}, \tag{53}$$

$$\bar{\bar{\dot{\varepsilon}}}_{\mathrm{eq}}^{(r)} = \left(\frac{2}{3} I :: \langle \dot{\varepsilon} \otimes \dot{\varepsilon} \rangle^{(r)} \right)^{1/2}, \tag{54}$$

where I is the fourth-order identity tensor. The standard deviations of the equivalent magnitudes over the whole polycrystal are defined as:

$$\mathrm{SD}\left(\sigma_{\mathrm{eq}}\right) = \sqrt{\bar{\bar{\Sigma}}_{\mathrm{eq}}^{2} - \Sigma_{\mathrm{eq}}^{2}}, \tag{55}$$

$$\mathrm{SD}\left(\dot{\varepsilon}_{\mathrm{eq}}\right) = \sqrt{\bar{\bar{\dot{E}}}_{\mathrm{eq}}^{2} - \dot{E}_{\mathrm{eq}}^{2}}, \tag{56}$$

where

$$\bar{\bar{\Sigma}}_{\mathrm{eq}}^{2} = \left\langle \left(\bar{\bar{\sigma}}_{\mathrm{eq}}^{(r)} \right)^{2} \right\rangle = \sum_{r} c^{(r)} \left(\bar{\bar{\sigma}}_{\mathrm{eq}}^{(r)} \right)^{2}, \tag{57}$$

$$\bar{\bar{\dot{E}}}_{\mathrm{eq}}^{2} = \left\langle \left(\bar{\bar{\dot{\varepsilon}}}_{\mathrm{eq}}^{(r)} \right)^{2} \right\rangle = \sum_{r} c^{(r)} \left(\bar{\bar{\dot{\varepsilon}}}_{\mathrm{eq}}^{(r)} \right)^{2}. \tag{58}$$

Once the average second-order moments of the stress field over each SR grain (r) are obtained, the implementation of the second-order procedure follows the work of Liu and Ponte Castañeda (2004). The covariance tensor of stress fluctuations in the SR grains of the linear comparison polycrystal is given by:

$$C_{\sigma'}^{(r)} = \left\langle \sigma' \otimes \sigma' \right\rangle^{(r)} - \sigma'^{(r)} \otimes \sigma'^{(r)}. \tag{59}$$

The average and the average fluctuation of resolved shear stress on slip system (k) of SR grain (r) are given by:

$$\bar{\tau}^{k(r)} = m^{k(r)} : \sigma'^{(r)}, \tag{60}$$

$$\hat{\tau}^{k(r)} = \bar{\tau}^{k(r)} \pm \left(m^{k(r)} : C_{\sigma'}^{(r)} : m^{k(r)} \right)^{1/2}, \tag{61}$$

where the positive (negative) branch should be selected if $\bar{\tau}^{k(r)}$ is positive (negative). The slip potential associated with slip system (k) of the nonlinear polycrystal is defined as:

$$\phi^k(\tau) = \frac{\tau_o^k}{n+1}\left(\frac{|\tau|}{\tau_o^k}\right)^{n+1}. \tag{62}$$

Two scalar magnitudes associated with each slip system (k) of each SR grain (r) are defined by:

$$\alpha^{k(r)} = \frac{\phi'^{k(r)}\left(\hat{\tau}^{k(r)}\right) - \phi'^{k(r)}\left(\bar{\tau}^{k(r)}\right)}{\hat{\tau}^{k(r)} - \bar{\tau}^{k(r)}}, \tag{63}$$

$$e^{k(r)} = \phi'^{k(r)}\left(\bar{\tau}^{k(r)}\right) - \alpha^{k(r)}\bar{\tau}^{k(r)}, \tag{64}$$

where $\phi'^k(\tau) = \mathrm{d}\phi^k/\mathrm{d}\tau(\tau)$. The linearized local behavior associated with SR grain (r) is then given by:

$$\dot{\varepsilon}^{(r)} = M_{\mathrm{SO}}^{(r)} : \sigma'^{(r)} + \dot{\varepsilon}_{\mathrm{SO}}^{o(r)} \tag{65}$$

with

$$M_{\mathrm{SO}}^{(r)} = \sum_k \alpha^{k(r)}\left(m^{k(r)} \otimes m^{k(r)}\right), \tag{66}$$

$$\dot{\varepsilon}_{\mathrm{SO}}^{o(r)} = \sum_k e^{k(r)} m^{k(r)}. \tag{67}$$

Once the linear comparison polycrystal is defined by (66–67), different second-order estimates of the effective behavior of the nonlinear aggregate can be obtained. Approximating the potential of the nonlinear polycrystal in terms of the potential of the linear comparison polycrystal and a suitable measure of the error, Liu and Ponte Castañeda (2004) generated the following expression (corresponding to the so-called energy version of the second-order theory) for the effective potential of the nonlinear polycrystal:

$$\bar{U}(\Sigma') = \sum_r c^{(r)} \sum_k \left\{\phi^{k(r)}\left(\hat{\tau}^{k(r)}\right) + \phi'^{k(r)}\left(\bar{\tau}^{k(r)}\right)\left(\bar{\tau}^{k(r)} - \hat{\tau}^{k(r)}\right)\right\} \tag{68}$$

from where the effective response of the homogenized polycrystal can be obtained as $\dot{E} = \partial\bar{U}(\Sigma')/\partial\Sigma'$. The alternate *constitutive equation* version of the second-order theory simply consists in making use of the effective stress–strain-rate relations for the linear comparison polycrystal, in which case, e.g., the effective strain is obtained as:

$$\dot{E} = \sum_r c^{(r)} \sum_k m^{k(r)} \phi'^{k(r)}\left(\bar{\tau}^{k(r)}\right). \tag{69}$$

Both versions of the SO theory give slightly different results, depending on nonlinearity and local anisotropic contrast. Such gap is relatively small compared with the larger differences obtained with the different SC approaches. The "constitutive equation" version is in principle less rigorous since it does not derive from a potential function, but has the advantage that can be obtained by simply following the affine algorithm described in the previous sections, using the linearized moduli defined in (66–67). Therefore, it is the adequate choice to be implemented in the VPSC code.

2.1.6 Numerical Implementation

To illustrate the use of the self-consistent formulation, we describe here the steps required to predict the local and overall viscoplastic response of a polycrystal. Starting for convenience with an initial Taylor guess, i.e., $\dot{\varepsilon}^{(r)} = \dot{E}$ for all grains, we solve the following nonlinear equation to get $\sigma'^{(r)}$:

$$\dot{E} = \dot{\gamma}_{o} \sum_{k} m^{k(r)} \left(\frac{m^{k(r)} : \sigma'^{(r)}}{\tau_{o}^{k(r)}} \right)^{n} \times \operatorname{sgn}\left(m^{k(r)} : \sigma'^{(r)} \right), \tag{70}$$

and we use an appropriate first-order linearization scheme to obtain initial values of $M^{(r)}$ and $\dot{\varepsilon}^{o(r)}$, for each SR grain (r). Next, initial guesses for the macroscopic moduli $\bar{M}$ and $\dot{E}^{o}$ are obtained (usually as simple averages of the local moduli). With them and the applied strain-rate, the initial guess for the macroscopic stress Σ' can be obtained (5), while the Eshelby tensors S and Π can be calculated using the macroscopic moduli and the ellipsoidal shape of the SR grains, by means of the procedure described in Sect. 2.1.1. Subsequently, the interaction tensor $\tilde{M}$ (30), and the localization tensors $B^{(r)}$ and $b^{(r)}$ (32 and 33), can be calculated as well. With these tensors, new estimates of $\bar{M}$ and $\dot{E}^{o}$ are obtained by solving iteratively the self-consistent equations (36) (for a unique grain shape) or (37) (for a distribution of grain shapes). After achieving convergence on the macroscopic moduli (and, consequently, also on the macroscopic stress and the interaction and localization tensors), a new estimation of the average grain stresses can be obtained, using the localization relation (31). If the recalculated average grain stresses are different (within certain tolerance) from the input values, a new iteration should be started, until convergence is reached. If the chosen linearization scheme is the second-order formulation, an additional loop on the linearized moduli is needed, using the improved estimates of the second-order moments of the stress in the grains, obtained by means of the methodology described in Sect. 2.1.5 and the Appendix. This additional loop roughly increases the calculation time by one order of magnitude with respect to first-order linearizations. When the iterative procedure is completed, the average shear-rates on the slip system (k) in each grain (r) are calculated as:

$$\dot{\gamma}^{k(r)} = \dot{\gamma}_{o} \left(\frac{m^{k(r)} : \sigma'^{(r)}}{\tau_{o}^{k(r)}} \right)^{n} \times \operatorname{sgn}\left(m^{k(r)} : \sigma'^{(r)} \right). \tag{71}$$

These average shear-rates are in turn used to calculate the lattice rotation-rates associated with each SR grain:

$$\dot{\omega}_{ij}^{(r)} = \dot{\Omega}_{ij} + \dot{\tilde{\omega}}_{ij}^{(r)} - \dot{\omega}_{ij}^{p(r)}, \tag{72}$$

with [c.f. (2)]:

$$\dot{\omega}_{ij}^{p(r)} = \sum_k \alpha_{ij}^{k(r)} \dot{\gamma}^{k(r)}, \tag{73}$$

where $\alpha_{ij}^{k(r)}$ is the uniform antisymmetric Schmid tensor of system (k) in SR grain (r).

It is worth noting that in the case of first-order approximations, although the second-order moments are not needed to readjust iteratively the linearized behavior of the SR grains, the average field fluctuations associated with the converged values of the effective moduli can be obtained as well, after convergence is reached.

The above numerical scheme can be used to predict texture development, by applying viscoplastic deformation to the polycrystal in incremental steps. The latter is done by assuming constant rates during a time interval Δt (such that $\dot{E}\ \Delta t$ corresponds to a macroscopic strain increment of the order of a few percents) and using: (1) the strain-rates and rotation-rates (times Δt) to update the shape and orientation of the SR grains, and (2) the shear-rates (times Δt) to update the critical stress of the deformation systems due to strain hardening, after each deformation increment. Using extended Voce law (Tomé et al. 1984), the evolution of the threshold stress with accumulated shear strain in each grain is given by:

$$\tau^{*k(r)} = \tau_{oo}^k + \left(\tau_1^k + \theta_1^k \Gamma^{(r)}\right)\left(1 - \exp\left(-\Gamma^{(r)} \left|\theta_o^k / \tau_1^k\right|\right)\right), \tag{74}$$

where $\Gamma^{(r)}$ is the total accumulated shear in the grain; τ_{oo}^k, τ_1^k, θ_o^k, and θ_1^k are the initial threshold stress, initial hardening rate, asymptotic hardening rate, and back-extrapolated threshold stress, respectively. In addition, we allow for the possibility of “self” and “latent” hardening by defining coupling coefficients $h^{kk'}$, which empirically account for the obstacles that new dislocations (or twins) associated with system k' represent for the propagation of dislocations (or twins) on system k. The increase in the threshold stress is calculated as:

$$\Delta\tau_o^{k(r)} = \frac{d\tau^{*k(r)}}{d\Gamma^{(r)}} \sum_{k'} h^{kk'} \dot{\gamma}^{k'(r)} \Delta t. \tag{75}$$

Note that the above explicit update schemes rely on the fact that the orientation and hardening variables evolve slowly within the adopted time interval. Otherwise, Δt should be chosen smaller.

2.2 FFT-Based Formalism

The FFT-based full-field formulation for viscoplastic polycrystals is conceived for periodic unit cells, provides an "exact" solution (within the limitations imposed by the unavoidable discretization of the problem and the iterative character of the numerical algorithm, see below) of the governing equations, and has better numerical performance than a finite element calculation for the same purpose and resolution (at least when comparing sequential implementations of both methods). It was originally developed (Moulinec and Suquet 1994, 1998; Michel et al. 2000) as a fast algorithm to compute the elastic and elastoplastic effective and local response of composites, and later adapted (Lebensohn 2001; Lebensohn et al. 2004b, 2008) to deal with the viscoplastic deformation of three-dimensional (3D) power–law polycrystals. It shares some common characteristics with the phase field method, although it is limited to what in phase field jargon is known as long-range interactions (Chen 2004), since no heterogeneous chemical energy term is involved in the mechanical response and/or microstructure evolution of a single-phase polycrystal. Recently, a similar kind of phase field analysis was proposed (Wang et al. 2002) to obtain the local fields in elastically heterogeneous polycrystals. The FFT-based approach, however, is not restricted to linear behaviors. Problems involving nonlinear materials (e.g., viscoplastic polycrystals) are treated similarly to a linear problem, using the concept of *linear reference material.*

Briefly, the viscoplastic FFT-based formulation consists in iteratively adjusting a compatible strain-rate field, related to an equilibrated stress field through a constitutive potential, such that the average of local work-rates is minimized. The method is based on the fact that the local mechanical response of a heterogeneous medium can be calculated as a convolution integral between Green functions associated with appropriate fields of a linear reference homogeneous medium and the actual heterogeneity field. For periodic media, use can be made of the Fourier transform to reduce convolution integrals in real space to simple products in Fourier space. Thus, the FFT algorithm can be used to transform the heterogeneity field into Fourier space and, in turn, to get the mechanical fields by transforming that product back to real space. However, the actual heterogeneity field depends precisely on the a priori unknown mechanical fields. Therefore, an iterative scheme has to be implemented to obtain, upon convergence, a compatible strain-rate field and a stress field in equilibrium.

2.2.1 Periodic Unit Cell: Green Function Method

A periodic unit cell representing the polycrystal is discretized into $N_1 \times N_2 \times N_3$ Fourier points. This discretization determines a regular grid in the Cartesian space $\{\mathbf{x}^{\mathbf{d}}\}$ and a corresponding grid in Fourier space $\{\boldsymbol{\xi}^{\mathbf{d}}\}$. Velocities and tractions along the boundary of the unit cell are left undetermined. A velocity-gradient $V_{i,j}$ (which can be decomposed into a symmetric strain-rate and a antisymmetric rotation-rate: $V_{i,j} = \dot{E}_{ij} + \dot{\Omega}_{ij}$) is imposed to the unit cell. The local strain-rate field is a function of the local velocity field, i.e., $\dot{\varepsilon}_{ij}(\nu_k(\mathbf{x}))$, and can be split into its average and a fluctuation term: $\dot{\varepsilon}_{ij}(\nu_k(\mathbf{x})) = \dot{E}_{ij} + \tilde{\dot{\varepsilon}}_{ij}(\tilde{\nu}_k(\mathbf{x}))$, where $\nu_i(\mathbf{x}) = \dot{E}_{ij}x_j + \tilde{\nu}_i(\mathbf{x})$. By imposing periodic boundary conditions, the velocity fluctuation field $\tilde{\nu}_k(\mathbf{x})$ is assumed to be periodic across the boundary of the unit cell, while the traction field is antiperiodic, to meet equilibrium on the boundary between contiguous unit cells.

The local constitutive relation between the strain-rate $\dot{\varepsilon}_{ij}(\mathbf{x})$ and the deviatoric stress $\sigma'_{ij}(\mathbf{x})$ is given by the same rate–sensitivity relation used within the VPSC framework (1). Let us choose a fourth-order tensor L^o to be the stiffness of a linear reference medium (the choice of L^o can be quite arbitrary, but the speed of convergence of the method will depend on this choice) and define the polarization field $\varphi_{ij}(\mathbf{x})$ [c.f. (10)] as:

$$\varphi_{ij}(\mathbf{x}) = \tilde{\sigma}'_{ij}(\mathbf{x}) - L^o_{ijkl}\,\tilde{\dot{\varepsilon}}_{kl}(\mathbf{x}). \tag{76}$$

Then, the Cauchy stress deviation can be written as:

$$\tilde{\sigma}_{ij}(\mathbf{x}) = L^o_{ijkl}\tilde{\dot{\varepsilon}}_{kl}(\mathbf{x}) + \varphi_{ij}(\mathbf{x}) + \tilde{\sigma}^m(\mathbf{x})\,\delta_{ij}. \tag{77}$$

Combining (77) with the equilibrium ($\sigma_{ij,j}(\mathbf{x}) = \tilde{\sigma}_{ij,j}(\mathbf{x}) = 0$), the incompressibility condition, and the relation $\tilde{\dot{\varepsilon}}_{ij}(\mathbf{x}) = \frac{1}{2}\left(\tilde{\nu}_{i,j}(\mathbf{x}) + \tilde{\nu}_{j,i}(\mathbf{x})\right)$:

$$\left|\begin{array}{l} L^o_{ijkl}\tilde{\nu}_{k,lj}(\mathbf{x}) + \tilde{\sigma}^m_{,i}(\mathbf{x}) + \varphi_{ij,j}(\mathbf{x}) = 0, \\ \tilde{\nu}_{k,k}(\mathbf{x}) = 0. \end{array}\right. \tag{78}$$

This system of differential equations is formally equivalent to system (9). However, both systems actually differ in that: (1) the HEM's stiffness modulus $\bar{L}$ of (9) is replaced in (78) by the stiffness of a linear reference medium L^o, and (2) the polarization field in (78) has in general nonvanishing values throughout the unit cell and is periodic (owing to the unit cell's periodicity), while the polarization field in (9) vanishes outside the domain of the inclusion. The auxiliary system involving Green functions is then given by [c.f. (13)]:

$$\left|\begin{array}{l} L^o_{ijkl}\,G_{km,lj}(\mathbf{x}-\mathbf{x}') + H_{m,i}(\mathbf{x}-\mathbf{x}') + \delta_{im}\,\delta(\mathbf{x}-\mathbf{x}') = 0, \\ G_{km,k}(\mathbf{x}-\mathbf{x}') = 0. \end{array}\right. \tag{79}$$

After some manipulation, the convolution integrals that give the velocity and velocity-gradient deviation fields are:

$$\tilde{\nu}_k(x) = \int_{R^3} G_{ki,j}\left(x - x'\right)\varphi_{ij}\left(x'\right)\mathrm{d}x', \tag{80}$$

$$\tilde{\nu}_{i,j}(x) = \int_{R^3} G_{ik,jl}\left(x - x'\right)\varphi_{kl}\left(x'\right)\mathrm{d}x'. \tag{81}$$

Convolution integrals in direct space are simply products in Fourier space:

$$\hat{\tilde{v}}_k(\boldsymbol{\xi}) = \left(-i\xi_j\right)\hat{G}_{ki}(\boldsymbol{\xi})\,\hat{\varphi}_{ij}(\boldsymbol{\xi}), \tag{82}$$

$$\hat{\tilde{v}}_{i,j}(\boldsymbol{\xi}) = \hat{\Gamma}_{ijkl}(\boldsymbol{\xi})\,\hat{\varphi}_{kl}(\boldsymbol{\xi}), \tag{83}$$

where the symbol "^" indicates a Fourier transform. The Green operator in (83) is defined as $\Gamma_{ijkl} = G_{ik,jl}$. The tensors $\hat{G}_{ij}(\boldsymbol{\xi})$ and $\hat{\Gamma}_{ijkl}(\boldsymbol{\xi})$ can be calculated by taking Fourier transform to system (79):

$$\left|\begin{array}{l} \xi_l\xi_j\; L^o_{ijkl}\hat{G}_{km}(\boldsymbol{\xi}) + i\xi_i\hat{H}_m(\boldsymbol{\xi}) = \delta_{im} \\ \xi_k\,\hat{G}_{km}(\boldsymbol{\xi}) = 0 \end{array}\right. . \tag{84}$$

Defining the 3×3 matrix $\mathbf{A}'_{ik}(\boldsymbol{\xi}) = \xi_l\xi_j L^o_{ijkl}$, and the 4×4 matrix $\mathbf{A}(\boldsymbol{\xi})$:

$$\mathbf{A}(\boldsymbol{\xi}) = \begin{vmatrix} A'_{11} & A'_{12} & A'_{13} & \xi_1 \\ A'_{21} & A'_{22} & A'_{23} & \xi_2 \\ A'_{31} & A'_{32} & A'_{33} & \xi_3 \\ \xi_1 & \xi_2 & \xi_3 & 0 \end{vmatrix}, \tag{85}$$

we obtain from (84) [c.f. (14 and 15)]:

$$\hat{G}_{ij}(\boldsymbol{\xi}) = \mathbf{A}^{-1}_{ij}\ (i, j = 1, 3) \tag{86}$$

and

$$\hat{\Gamma}_{ijkl}(\boldsymbol{\xi}) = -\xi_j\xi_l\hat{G}_{ik}(\boldsymbol{\xi}) \tag{87}$$

2.2.2 FFT-Based Algorithm

Assigning initial guess values to the strain-rate field in the regular grid $\{\mathbf{x^d}\}$ (e.g., $\tilde{\dot{\varepsilon}}^o_{ij}(\mathbf{x^d}) = 0 \Rightarrow \dot{\varepsilon}^o_{ij}(\mathbf{x^d}) = \dot{E}_{ij}$), and computing the corresponding stress field $\sigma'^o_{ij}(\mathbf{x^d})$ from the local constitutive relation (1) (which requires to know the initial values of the critical stresses $\tau^s_o(\mathbf{x^d})$, and the Schmid tensors $m^s_{ij}(\mathbf{x^d})$, e.g., from an orientation image, in which the image's pixels coincide with the Fourier grid), allow us to obtain an initial guess for the polarization field in direct space $\varphi^o_{ij}(\mathbf{x^d})$ (76), which in turn can be Fourier-transformed to obtain $\hat{\varphi}^o_{ij}(\boldsymbol{\xi}^\mathbf{d})$. Furthermore, assuming that $\lambda^o_{ij}(\mathbf{x^d}) = \sigma^o_{ij}(\mathbf{x^d})$ is the initial guess for a field of Lagrange multipliers associated with the compatibility constraints, the iterative procedure based on Augmented Lagrangians proposed by Michel et al. (2000) reads as follows. With the polarization field after iteration n being known, the $n+1$-th iteration starts by computing the new guess for the kinematically admissible strain-rate deviation field:

$$\hat{\tilde{d}}^{n+1}_{ij}(\boldsymbol{\xi}^\mathbf{d}) = -\hat{\Gamma}^{\mathrm{sym}}_{ijkl}(\boldsymbol{\xi}^\mathbf{d})\,\hat{\varphi}^n_{kl}(\boldsymbol{\xi}^\mathbf{d}), \quad \forall\boldsymbol{\xi}^\mathbf{d} \neq 0; \quad \text{and} \quad \hat{\tilde{d}}^{n+1}_{ij}(\mathbf{0}) = 0, \tag{88}$$

where $\hat{\Gamma}^{\mathrm{sym}}_{ijkl}$ is the Green operator, appropriately symmetrized. The corresponding field in real space is thus obtained by application of the inverse FFT, i.e.,

$$\tilde{d}^{n+1}_{ij}\left(\mathbf{x^d}\right) = \mathit{fft}^{-1}\left\{\hat{\tilde{d}}^{n+1}_{ij}\left(\boldsymbol{\xi}^{\mathbf{d}}\right)\right\}, \tag{89}$$

and the new guess for the deviatoric stress field is calculated from (omitting subindices):

$$\begin{aligned}\sigma'^{n+1}\left(\mathbf{x^d}\right) + L^o : \dot{\gamma}_o \sum_k m^k\left(\mathbf{x^d}\right)\left(\frac{m^k\left(\mathbf{x^d}\right) : \sigma'^{n+1}\left(\mathbf{x^d}\right)}{\tau^k\left(\mathbf{x^d}\right)}\right)^n \times \operatorname{sgn}\left(m^k\left(\mathbf{x^d}\right) : \sigma'^{n+1}\left(\mathbf{x^d}\right)\right) \\ = \lambda^n\left(\mathbf{x^d}\right) + L^o : \left(\dot{E} + \tilde{d}^{n+1}\left(\mathbf{x^d}\right)\right).\end{aligned} \tag{90}$$

The iteration is completed with the calculation of the new guess of the Lagrange multiplier field:

$$\lambda^{n+1}\left(\mathbf{x^d}\right) = \lambda^n\left(\mathbf{x^d}\right) + L^o : \left(\tilde{\dot{\varepsilon}}^{n+1}\left(\mathbf{x^d}\right) - \tilde{d}^{n+1}\left(\mathbf{x^d}\right)\right). \tag{91}$$

Equations (90 and 91) guarantee the convergence of: $\dot{\varepsilon}\left(\mathbf{x^d}\right)$ (i.e., the strain-rate field related with the stress through the constitutive equation) toward $d\left(\mathbf{x^d}\right)$ (i.e., the kinematically admissible strain-rate field) to fulfill compatibility, and the Lagrange multiplier field $\lambda\left(\mathbf{x^d}\right)$ toward the stress field $\sigma'\left(\mathbf{x^d}\right)$ to fulfill equilibrium.

Upon convergence, the microstructure can be updated using an explicit scheme, as follows. The resulting strain-rate field, and the shear-rate field, i.e.

$$\dot{\gamma}^k\left(\mathbf{x^d}\right) = \dot{\gamma}_o\left(\frac{m^k\left(\mathbf{x^d}\right) : \sigma'\left(\mathbf{x^d}\right)}{\tau^k\left(\mathbf{x^d}\right)}\right)^n \times \operatorname{sgn}\left(m^k\left(\mathbf{x^d}\right) : \sigma'\left(\mathbf{x^d}\right)\right) \tag{92}$$

can be assumed to be constant during a time interval $[t, t + \Delta t]$. The macroscopic and local strain increments are then calculated as: $\Delta E_{ij} = \dot{E}_{ij} \times \Delta t$ and $\Delta\varepsilon_{ij}\left(\mathbf{x^d}\right) = \dot{\varepsilon}_{ij}\left(\mathbf{x^d}\right) \times \Delta t$, and the local crystallographic orientations are updated according to the following local lattice rotation:

$$\omega_{ij}\left(\mathbf{x^d}\right) = \left(\dot{\Omega}_{ij}\left(\mathbf{x^d}\right) + \tilde{\dot{\omega}}_{ij}\left(\mathbf{x^d}\right) - \dot{\omega}^p_{ij}\left(\mathbf{x^d}\right)\right) \times \Delta t, \tag{93}$$

where $\dot{\omega}^p_{ij}\left(\mathbf{x^d}\right)$ can be obtained from (2) and (92), and $\tilde{\dot{\omega}}\left(\mathbf{x^d}\right)$ is obtained back-transforming the converged antisymmetric field:

$$\hat{\tilde{\dot{\omega}}}_{ij}\left(\boldsymbol{\xi}^{\mathbf{d}}\right) = -\hat{\Gamma}^{\mathrm{antisym}}_{ijkl}\left(\boldsymbol{\xi}^{\mathbf{d}}\right)\hat{\varphi}_{kl}\left(\boldsymbol{\xi}^{\mathbf{d}}\right), \quad \forall\boldsymbol{\xi}^{\mathbf{d}} \neq 0; \quad \text{and} \quad \hat{\tilde{\dot{\omega}}}_{ij}\left(\mathbf{0}\right) = 0. \tag{94}$$

The critical resolved shear stresses of the deformation systems associated with each material point should also be updated after each deformation increment due to strain hardening, e.g., in an analogous way as explained in Sect. 2.1.6 for the VPSC case (in terms a phenomenological Voce law) or with more sophisticated hardening laws based directly on dislocation densities. Note that, in the latter case, the possibility of calculating the intragranular misorientations would allow us to track the evolution of geometrically necessary dislocations (GND) densities explicitly, and, at the same time, introduce a length scale in the formulation (see, e.g., Acharya et al. 2003). This more elaborated treatment of hardening, however, is not going to be discussed further in this work.

After each time increment, the new position of the Fourier points can be determined calculating the velocity fluctuation term $\tilde{v}_k\left(\mathbf{x^d}\right)$ back-transforming (82), and:

$$X_i\left(\mathbf{x^d}\right) = x_i^d + \left(\dot{E}_{ij}x_j^d + \tilde{v}_i\left(\mathbf{x^d}\right)\right) \times \Delta t. \tag{95}$$

Evidently, due to the heterogeneity of the medium, the set of convected Fourier points no longer forms a regular grid, after the very first deformation increment. A rigorous way of dealing with this situation was proposed by Lahellec et al. (2001) based on the particle-in-cell (PIC) method (Sulsky et al. 1995). In the example presented in Sects. 3 and 4 below, however, the following simplification was adopted. Neglecting the velocity fluctuation term in (95), the updated coordinates of the Fourier points can be approximated by:

$$X_i\left(\mathbf{x^d}\right) \tilde{=} x_i^d + \dot{E}_{ij}x_j^d \times \Delta t. \tag{96}$$

In this way, the Fourier grid remains regular after each deformation increment. The distances between adjacent Fourier points, however, do change, following the variations of the unit cell dimensions, thus determining an "average stretching" of the grains, following the macroscopic deformation.

3 Results

3.1 Validation of the Full-Field Formulation Using an Analytical Result

Let us consider a model polycrystal consisting of columnar orthorhombic grains with symmetry axes aligned with the x_3 axis, such that, when loaded in antiplane mode with shearing direction along x_3, the only two slip systems that can be activated in the grains are those defined by the following Schmid tensors:

$$m^{\mathrm{s}} = \frac{(\mathbf{e}_1 \otimes \mathbf{e}_3 + \mathbf{e}_3 \otimes \mathbf{e}_1)}{2}, \quad m^{\mathrm{h}} = \frac{(\mathbf{e}_2 \otimes \mathbf{e}_3 + \mathbf{e}_3 \otimes \mathbf{e}_2)}{2}, \tag{97}$$

where $\{\mathbf{e}_1, \mathbf{e}_2, \mathbf{e}_3\}$ is an orthonormal basis of crystallographic axes, and "s" and "h" stand for soft and hard slip systems, respectively. If we further consider that

$\mathbf{e}_3$ lies parallel to x_3, and the material is incompressible, the problem becomes two-dimensional (2D). The local stress and strain-rate are characterized by the 2D vectors with components σ_{13} and σ_{23}, and $\dot{\varepsilon}_{13}$ and $\dot{\varepsilon}_{23}$ (denoted hereafter σ_1 and σ_2, and $\dot{\varepsilon}_1$ and $\dot{\varepsilon}_2$, respectively), and the viscous stiffness tensor $L = 2\mu$, by a 2D symmetric second-order tensor with diagonal components $2\mu_{1313}$ and $2\mu_{2323}$, and off-diagonal components $2\mu_{1323}$ (denoted $2\mu_{11}$, $2\mu_{22}$, and $2\mu_{12}$, respectively). In addition, let us assume that the constituent grains exhibit a linear response:

$$\dot{\varepsilon}(\mathbf{x}) = L^{-1} : \sigma(\mathbf{x}) = \left(\frac{1}{\tau_o^s} m^s \otimes m^s + \frac{1}{\tau_o^h} m^h \otimes m^h \right) : \sigma(\mathbf{x}), \tag{98}$$

with τ_o^s and τ_o^h being the viscosities of the soft and hard slip systems $\left(\tau_o^s < \tau_o^h\right)$. It can be shown that the behavior of such polycrystal is characterized by an effective 2D viscous stiffness tensor $\bar{L} = 2\bar{\mu}$ such that $\dot{E} = \bar{L}^{-1} : \Sigma$ (where Σ and $\dot{E}$ are the 2D effective stress and strain-rate, respectively), such that (Dykhne 1970; Lurie and Cherkaev 1984):

$$\det(\bar{\mu}) = \bar{\mu}_{11} \times \bar{\mu}_{22} - \bar{\mu}_{12}^2 = \tau_o^s \times \tau_o^h. \tag{99}$$

In the particular case of an isotropic 2D polycrystal, $\bar{\mu}_{11} = \bar{\mu}_{22}\ (= \bar{\mu})$ and $\bar{\mu}_{12} = 0$, so that the effective shear modulus becomes:

$$\bar{\mu} = \sqrt{\tau_o^s \times \tau_o^h}. \tag{100}$$

Note that the above result is independent of the 2D microstructure as far as it remains isotropic. This analytical result can be used for validating the FFT-based formulation. Let us consider the periodic 2D two-phase composite shown in Fig. 1a (Lebensohn et al. 2005), whose unit cell consists of four square grains, with the crystallographic orientations of the two pairs of opposite grains (i.e., each pair shearing only the central vertex) being characterized by angles +45° and −45°, respectively (note that the orientation of each 2D crystal is fully characterized by the angle between the crystal direction $\mathbf{e}_1$ and the sample direction x_1). The antiplane deformation of this unit cell for an applied strain-rate of the form $\dot{E} = \left(\dot{E}_{13}, 0\right)$ was solved numerically using different discretizations: 64, 128, 256, and 512 Fourier points along each direction (i.e., 1,024, 4,096, 16,384, 65,536 Fourier points per grain), for a contrast of $\tau_o^h / \tau_o^s = 25$, which gives theoretical polycrystal viscosity of $\bar{\mu} / \tau_o^s = 5$. Figure 2 shows the relative deviations of the polycrystal viscosities calculated with the FFT-based model from the theoretical value, as the number of iterations of the FFT-based method increases. It is seen that: (1) the convergence of $\bar{\mu}^{FFT}$ toward its theoretical value is rather good, although it saturates at different levels, depending on the number of discretization points used; (2) the precision of the FFT solution can be increased by refining appropriately the Fourier grid. This is due to the fact that a more refined grid provides a higher spatial resolution to represent the strong gradients and jumps of the local fields, localized at grain boundaries (see discussion of Fig. 1c–f below).

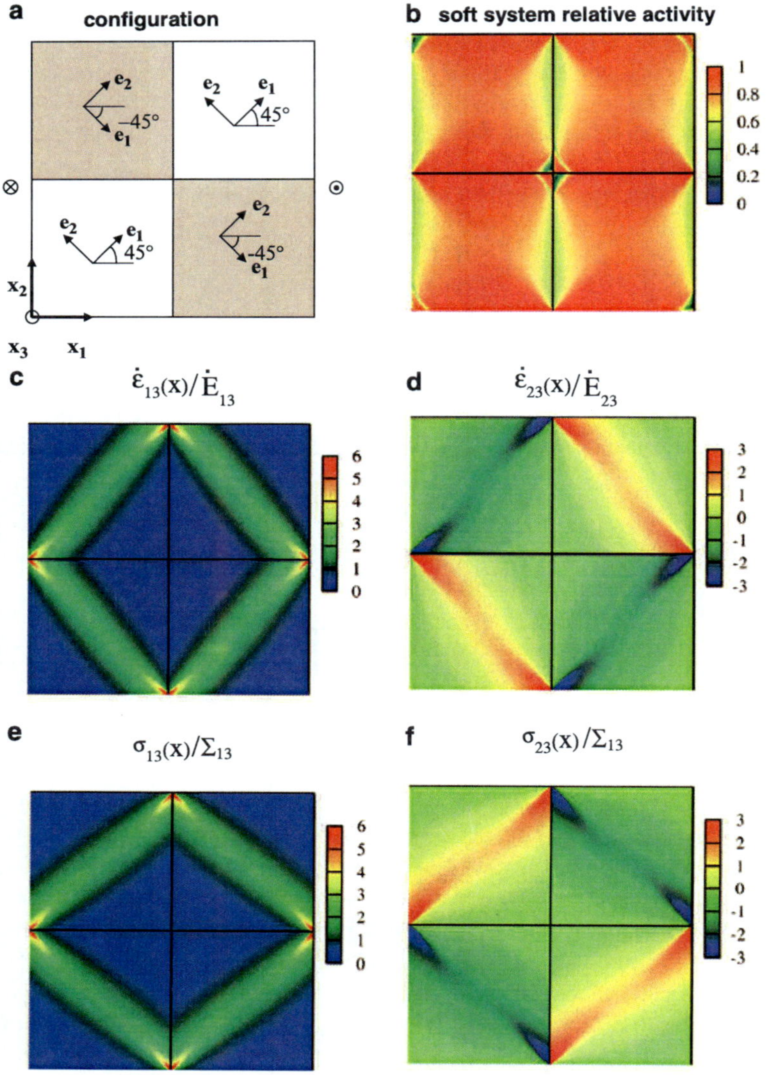

Fig. 1 **(a)** Two-dimensional two-phase isotropic unit cell undergoing antiplane deformation. FFT-based results of: **(b)** relative activity field of the soft slip system. Shear components *13* and *23* of the **(c–d)** strain-rate and **(e–f)** stress components, normalized with the value of the corresponding macroscopic component

On the other hand, since one of our goals is to understand the influence of the microstructure on the distribution of the stress and strain-rate fields, it is important to assess the precision of the FFT-based results also at the local scale. In this context, a great advantage of microstructures with only two phases is that the phase averages of

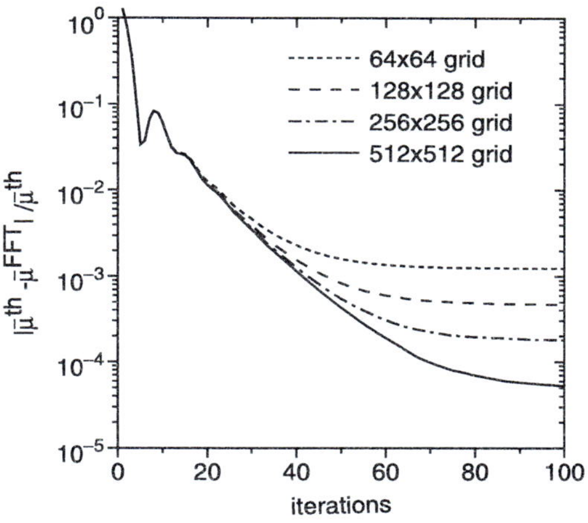

Fig. 2 Relative deviations from the theoretical (th) value $\left(\bar{\mu}/\tau_o^s = 5\right)$ of the polycrystal's viscosity predicted with the present formulation (FFT) for different grid refinements, in the case 2D antiplane deformation of the isotropic unit cell of Fig. 1a, consisting of grains having linear behavior (98) with $\tau_o^h/\tau_o^s = 25$

the localization tensors $A(\mathbf{x})$ and $B(\mathbf{x})$, defined by the expressions $\dot{\varepsilon}(\mathbf{x}) = A(\mathbf{x}) : \dot{E}$ and $\sigma(\mathbf{x}) = B(\mathbf{x}) : \Sigma$ can be easily calculated analytically as (Lebensohn et al. 2005):

$$\langle A \rangle_1 = \frac{1}{c_1}(L_1 - L_2)^{-1} : \left(\bar{L} - L_2\right), \tag{101a}$$

$$\langle B \rangle_1 = \frac{1}{c_1}(M_1 - M_2)^{-1} : \left(\bar{M} - M_2\right), \tag{101b}$$

where c_i and $\langle\cdot\rangle_i$ denote volume fraction and average over phase $i = 1, 2$, respectively, and the local and effective compliance tensors are given by $M_i = L_i^{-1}$ and $\bar{M} = \bar{L}^{-1}$, respectively (similar relations can be obtained for $\langle A \rangle_2$ and $\langle B \rangle_2$ by interchanging indices 1 and 2. For isotropic microstructures with linear behavior as the one considered here, since $\bar{L}$ is microstructure-independent, the above expressions are also microstructure-independent. Using (100), the phase-average localization tensors for the considered microstructure are given by:

$$\langle A \rangle_1 = \langle B \rangle_2 = \begin{pmatrix} 1 & +\alpha \\ +\alpha & 1 \end{pmatrix}, \langle A \rangle_2 = \langle B \rangle_1 = \begin{pmatrix} 1 & -\alpha \\ -\alpha & 1 \end{pmatrix}, \tag{102}$$

where the indices 1 and 2 were used for phases at angles +45° and −45°, respectively, and:

$$\alpha = 1 - \frac{2}{\left(1 + \sqrt{\tau_o^h/\tau_o^s}\right)}. \tag{103}$$

The above analytical expressions can be used to evaluate the accuracy reached with the FFT-based simulations at phase-average level. Table 1 shows the values of α obtained for different grid refinements (Lebensohn et al. 2005). The agreement

Table 1 Value of parameter α (103) predicted for different grid refinements and relative error with respect to the theoretical value ($\alpha = 2/3$)

Grid	α	Relative error
64×64	0.666021	9.69×10^{-4}
128×128	0.666340	4.90×10^{-4}
256×256	0.666452	1.87×10^{-4}
512×512	0.666734	1.01×10^{-4}

between the FFT-based predictions and the theoretical values is as good as for the corresponding effective viscosities shown in Fig. 2 and, like before, is better for more refined Fourier grids.

Finally, we show the predicted local strain-rate and stress fields. Figure 1c, d shows, respectively, the *13* and *23* components of the strain-rate field (normalized with the value of the applied macroscopic shear-rate), while Fig. 1e, f shows the analogous stress components (normalized with the resulting macroscopic shear stress). The first observation concerns the formation of localization bands (both of stress and strain-rate), which are normal to $\mathbf{e}_1$ (i.e., the normal to the shear plane of the soft slip system) in every grain. These bands go through quadruple points, where the *13* components (*13* is the only nonvanishing component at polycrystal level) of the stress and strain-rate fields reach their maximum values. Meanwhile, the *23* components of the local fields also have non-negligible values along the localization bands, with alternating signs in the phases, such that the strain-rate is negative where the stress is positive and vice versa. This alternation is consistent with the plus and minus signs preceding α in (102) and also give vanishing average *23* stress and strain-rate components at macoscopic level. It is also worth noting that the corresponding stress and strain-rate field components are related by a 90° rotation. Such symmetry is evidently related to the fact that a 2D divergence-free field has the property of transforming into a curl-free field when rotated by 90° (Dykhne 1970). Note for instance that while stress equilibrium requires continuity of the *23* stress component through the "horizontal" grain boundaries (as in Fig. 1f), strain compatibility requires continuity of the *23* strain-rate component through the "vertical" grain boundaries. To complete the analysis, Fig. 1b shows the complicated pattern of the field of relative activity of the soft slip system, associated with the local and macroscopic response discussed above.

3.2 Validation of Mean-Field Formulations Using Full-Field Computations

The advantage of using field fluctuation information in nonlinear homogenization schemes to get improved predictions of the mechanical behavior and texture development of viscoplastic polycrystals becomes evident as the heterogeneity (contrast in local properties) increases. The two possible sources of heterogeneity in single-phase viscoplastic aggregates are the nonlinearity of the material's response and the local anisotropy of the constituent single crystals. To study the influence of both sources of heterogeneity, we show here examples of self-consistent calculations on different material systems: (1) fcc aggregates (compatible with, e.g., polycrystalline copper) with fixed local anisotropy (given by the – rather mild – range of variation of the Taylor factor of individual grains) and variable rate-sensitivity, and (2) hexagonal polycrystals with four and two soft independent slip systems, and orthorhombic aggregates (compatible with Ti deforming at high temperature, ice, and olivine, respectively), with mild nonlinear behavior and variable local anisotropy, given by the ratio between the threshold resolved shear stresses associated with hard and soft slip modes (Lebensohn et al. 2007).

The prediction of the effective properties of a random fcc polycrystal as the rate-sensitivity of the material decreases is a classical benchmark for the different nonlinear SC approaches. Figure 3a shows a comparison between average Taylor

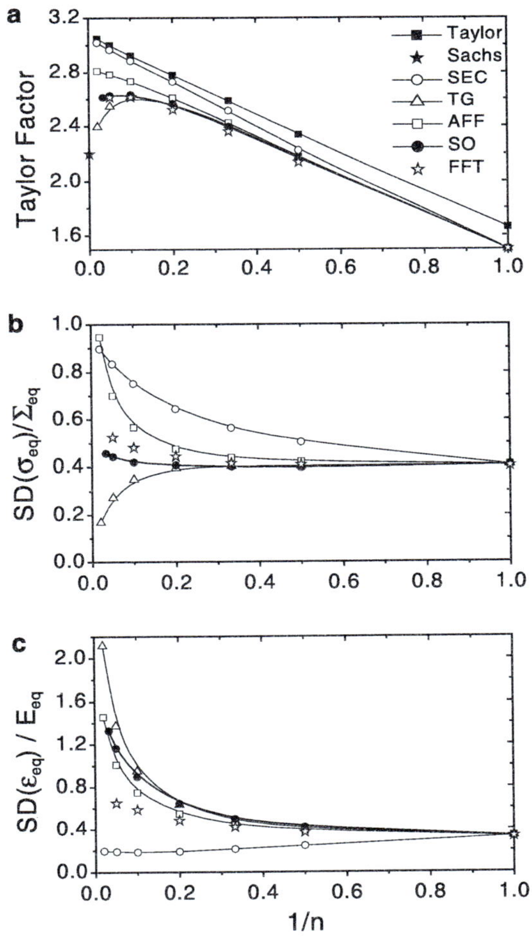

Fig. 3 (**a**) Average Taylor factor and normalized overall (**b**) stress and (**c**) strain standard deviations vs. rate-sensitivity, for a random fcc polycrystal under uniaxial tension, calculated with the different SC approaches (lines + symbols), and "exact" values (stars) from ensemble averages of FFT-based solutions (Lebensohn et al. 2007)

factor vs. rate-sensitivity ($1/n$) curves, for a random fcc polycrystal under uniaxial tension. The Taylor factor was calculated as Σ_{eq}^{ref}/τ_o, where τ_o is the threshold stress of the (111)<110> slip systems, and Σ_{eq}^{ref} is the macroscopic equivalent stress corresponding to an applied uniaxial strain-rate with a Von Mises equivalent value $\dot{E}_{eq}^{ref} = 1$. The curves in Fig. 3 correspond to the Taylor model, the different first-order SC approximations, and the second-order procedure. The solid star indicates the rate-insensitive Sachs estimate. The open stars correspond to the "exact" solution, obtained from *ensemble averages* of FFT-based full-field solutions performed on random polycrystals. These ensemble averages were calculated over the outcomes of "numerical experiments" performed on 100 specimens generated alike, i.e., by random assignation of orientations to a given array of grains, but which differ at microlevel due to the inherent stochastic character of such generation procedure. To obtain the results that follow, we have considered periodic 3D polycrystals consisting of $8 \times 8 \times 8 = 512$ cubic grains with randomly chosen orientations. These unit cells were in turn discretized using a $64 \times 64 \times 64$ Fourier grid, resulting in $8 \times 8 \times 8 = 512$ Fourier points per grain. The averages over a sufficiently large number of configurations should give the effective properties of a polycrystal with random microstructure. It should be noted that the microstructures of these polycrystals generated for ensemble averaging are random only in a restricted sense, since the grain orientations were chosen randomly but the morphology was set a priori to be equiaxed. The generation of fully random microstructures would require grains with both random orientation and morphology (Kanit et al. 2003). However, for our purposes, the above restricted random procedure allows us to reduce the number of configurations needed to obtain an isotropic ensemble response.

From the comparison between the different mean-field and the full-field estimates, it can be observed that: (1) the Taylor approach gives the stiffest response, consistent with the upper-bound character of this model; (2) all the SC estimates coincide for $n = 1$, i.e., the linear SC case; (3) in the rate-insensitive limit, the secant and tangent models tend to the upper and lower bounds, respectively, while the affine and second-order approximations remain intermediate with respect to the bounds; (4) except for the tangent model for $n>10$, the second-order procedure gives the lowest Taylor factor among the SC approaches. This softer macroscopic response (i.e., a lower stress is needed to induce a given strain-rate) is a consequence of the softer behavior at grain level in the linear comparison polycrystal that results when the average field fluctuations are considered for the determination of the linearized behavior of the SR grains; (5) the best match with the exact solutions (at least for rate-sensitivity exponents up to 20, i.e., the highest value we were able to use in the full-field computations, without losing accuracy) corresponds to the second-order estimates.

Concerning the overall heterogeneity of the mechanical fields, reflected in the standard deviations of the equivalent magnitudes over the whole polycrystal (55–56), the SC predictions (including the second-order approximation) are less accurate. Figure 3b, c shows these overall SDs (normalized, for an unbiased comparison, by the corresponding effective magnitudes) as a function of the rate-sensitivity. It can be observed that: (1) at high nonlinearities, only the SC models that do not tend to the bounds in the rate-insensitive limit (i.e., AFF and SO) show the expected increases in both stress and strain-rate heterogeneity. In the TG case, the stress heterogeneity decreases as the rate heterogeneity increases, while the SEC approach predicts the opposite trend; (2) both the AFF and SO approximations overestimate the strain heterogeneity; (3) the SO gives the best match with the full-field predictions for the stress heterogeneity, although it remains below the exact solution. In connection with the SO estimates, the use of the field fluctuations in the linear comparison material to estimate the corresponding fluctuations in the VP polycrystal has recently been shown (Idiart and Ponte Castañeda 2007) to be inconsistent. In fact, improved estimates can be generated by taking into account certain correction terms that are associated with the lack of full stationarity of these estimates with respect to the reference stresses. Still, the SC methods would not be expected to yield accurate estimates for the higher-order statistics of the fields, which become increasingly more sensitive to the details of the microstructure as the order increases. For example, the third-order moments, which contain information on the asymmetry of the distributions, are likely to become relatively important in low rate-sensitivity materials (Moulinec and Suquet 2003), since the strain tends to localize in deformation bands inside or across grains.

The next example concerns predictions of the effective behavior of random aggregates of grains with less than five linearly independent soft slip systems (Lebensohn et al. 2007). In this case, we analyze the dependence with the local contrast C, given by the ratio between the critical stresses associated with the hard and the soft slip modes. Figure 4 shows the predicted effective stress, relative to the threshold stress of the soft slip systems $\Sigma_{eq}^{ref}/\tau_o^{soft}$ (where Σ_{eq}^{ref} corresponds to an applied uniaxial strain-rate, with a Von Mises equivalent $\dot{E}_{eq}^{ref} = 1$), as a function the local contrast C, predicted by different homogenization approaches, and by averaging 100 FFT-based solutions, for the following cases:

1. A random hcp aggregate with four linearly independent soft slip systems, given by a suitable combination of {1010}<1120> prismatic (pr) slip, and (0001)<1120> basal (bas) slip (such that $\tau_o^{soft} = \tau_o^{pr} = \tau_o^{bas}$). The hard slip mode is {1011}<1123> pyramidal-<*c* + *a*> of the first-type (pyr1), and the contrast parameter is therefore given by $C = \tau_o^{pyr1}/\tau_o^{pr} = \tau_o^{pyr1}/\tau_o^{bas}$. Assuming a rate-sensitivity exponent $n = 4$ and a c/a ratio of 1.587 makes the above material model appropriate for a Ti aggregate deforming at elevated temperatures (Semiatin and Bieler 2001)
2. A random orthorhombic aggregate, with three linearly independent soft slip systems, given by a suitable combination of (010)[100], (001)[100], (010)[001], (100)[001]. The hard mode, which closes the single crystal yield surface, is assumed to be {111}<110>. All the soft systems were assumed to have the same threshold stress τ_o^{soft}, resulting in a contrast parameter $C = \tau_o^{\{111\}}/\tau_o^{soft}$. With a rate-sensitivity exponent $n = 4$ and b/a and c/a ratios of 2.122 and 1.245, respectively, this material model is consistent with the behavior of an olivine polycrystal, deforming under conditions found in the Earth's upper mantle (Wenk and Tomé 1999; Castelnau et al. 2008).

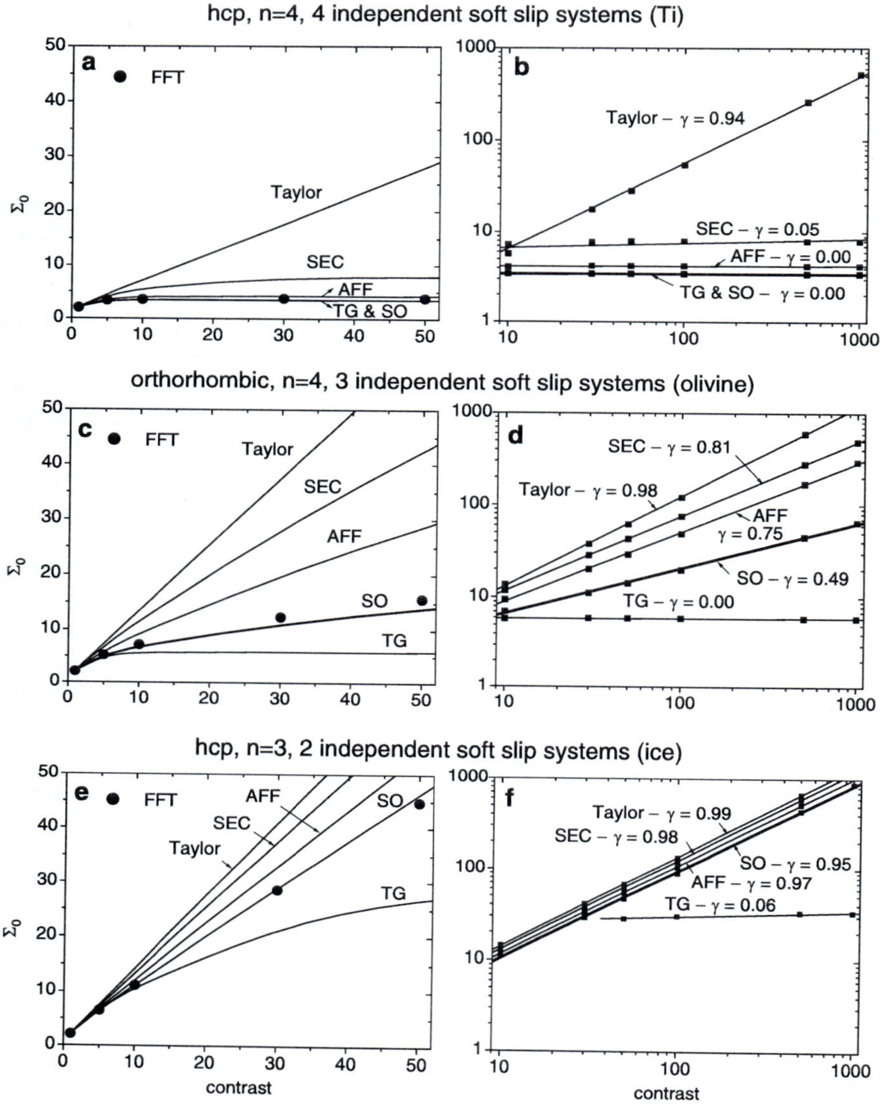

Fig. 4 Plots of reference stress vs. contrast, for random polycrystals with different number of independent soft slip systems, obtained with different SC approaches (lines) and from ensemble averages of FFT-based solution (symbols) for (**a**, **b**) four, (**c**, **d**) three (**e**, **f**) two independent soft slip systems, obtained with different SC approaches (lines) and from ensemble averages of FFT-based solution (symbols). *Left column*: linear scale plots, up to a contrast of 50. *Right column*: log–log plots, up to a contrast of 1,000. The value of γ corresponds to the slope of the logarithmic line (Lebensohn et al. 2007)

3. A random hcp aggregate with two linearly independent soft systems, corresponding to {0001}<1120> basal slip (i.e. $\tau_o^{soft} = \tau_o^{bas}$). The hard slip modes are the {1122}<1123> pyramidal-<c + a> of the second-type (pyr2), and the contrast parameter is given by $C = \tau_o^{pr}/\tau_o^{bas} = \tau_o^{pyr2}/\tau_o^{bas}$. Assuming a rate-sensitivity exponent $n = 3$ and a c/a ratio of 1.629, this material model is relevant for ice polycrystals deforming under conditions found in glaciers (Castelnau et al. 1996).

Figure 4a, c, e shows the curves (plotted in linear scale) of reference stress (i.e., $\Sigma_o = \Sigma_{eq}^{ref}/\tau_o^{soft}$, for $\dot{E}_{eq}^{ref} = 1$) vs. contrast C, predicted with the different SC approximations, the Taylor model, and the full-field FFT-based solution, for C up to 50. The agreement between the SO estimates and the exact solutions is apparent. Figure 4b, d, f shows log–log plots of the effective stress obtained with the different homogenization models, for contrasts up to 1,000, with the corresponding regression lines superimposed. It is evident that the results for all models can be described by scaling laws of the form $\Sigma_o \sim C^{\gamma}$ (Nebozhyn et al. 2000). In every case analyzed ($i = 2$, 3, and 4, where i is the number of linearly independent soft systems) $\gamma \cong 1$ for the Taylor model and $\gamma \cong 0$ for the tangent SC approach (note that the latter exponent also corresponds to the lower-bound Sachs model), while the secant, affine, and second-order SC models give different exponents, depending on the value of i. Interestingly, the exponents corresponding to the second-order approach follow the relation proposed by Nebozhyn et al. (2000): $\gamma \cong (4 - i)/2$, in the context of Ponte Castañeda's (1991) variational approach. The asymptotic trend to the lower-bound that the tangent SC approach exhibits when the contrast increases due to the increase of the exponent n is also obtained when the heterogeneity increases due to local anisotropy, even for relatively low values of n. This observation sheds light on why the tangent SC approach has been favored to predict mechanical behavior of low-symmetry materials, which have "open" single crystal yield surfaces with three or less independent deformation systems. In such cases, the tangent SC approach allows accommodation of the local deformation with the available slip systems, without need of "artificial" systems to close the single crystal yield surface. While these artificial hard systems make a very small contribution to strain, they have a strong influence on the predicted macroscopic behavior (effective viscosity) in these low-symmetry systems, unless a saturated behavior, like the one displayed by the tangent predictions in Fig. 4, is obtained.

3.3 Overall Texture Development Predictions Using Mean-Field Approaches

Almost 100% of plastic deformation in the ice single crystals is carried by basal dislocations. Since basal slip provides only two independent slip systems, the prediction of texture development of polycrystalline ice is a challenging problem that allows us to discriminate among the different SC approaches. Moreover, an understanding of the deformation mechanisms and the microstructural evolution of ice deforming in compression is relevant in glaciology, since compression (together with shear) is one of the main deformation modes of glaciers. In what follows, we will use the *basal texture factor* along the axial direction to characterize the evolving texture of ice in compression. The basal texture factor is defined as the weighted average of the projections of the c-axis along the axial direction, i.e., $\langle \cos^2 \alpha^{(r)} \rangle$, where $\alpha^{(r)}$ is the angle between the basal pole of SR grain (r) and the axial compression direction.

In fact, the stiff Taylor and SC secant models are not suitable to simulate plastic deformation of polycrystalline ice because the strong constraints that these models impose upon strain are incompatible with the shortage of independent slip systems in ice. On the other hand, the compression textures of ice typically exhibit a strong basal pole component aligned with the axial direction (Castelnau et al. 1996). The formation of this component is related to the crystallographic plastic rotations associated with basal slip. However, as the basal poles become more aligned with the axial direction, the basal systems become unfavorably oriented to accommodate deformation. Therefore, at large strains, even a "soft" first-order approximation like the tangent SC fails in reproducing the observed texture with only basal slip activity (Castelnau et al. 1996). Up to now, the Sachs model (which completely disregards strain compatibility) has been the only approach able to give a reasonably effective behavior with predominant basal slip at large strains, when the basal texture along the compressive direction becomes very strong.

Figure 5 shows the compression texture evolution (in terms of the basal texture factor), effective stress, relative basal activity, and average number of active slip systems (AVACS) per grain, for the case of an initially random ice polycrystal (Lebensohn et al. 2007). Results were obtained using the TG, AFF, and SO approaches, under the assumption of $n = 3$ and $\tau_o^{pr} = 20 \times \tau_o^{bas}$ and $\tau_o^{pyr2} = 200 \times \tau_o^{bas}$ (Castelnau et al. 1996), with no strain-hardening, up to a compressive strain of 1.5.

As expected, all models predict a prevalence of basal slip, with a consequent increase of the basal texture factor along the axial direction, and a progressive geometric hardening. While the alignment of basal poles along the compression direction predicted by all three models is similar, they differ in other indicators. At around 0.8 strain, the tangent predictions show a sudden drop in the basal activity, together with a rapid increase in the effective stress and in the number of active deformation systems, which indicates that the strain accommodation starts requiring the activation of the 200 times harder pyramidal systems. In other words, under the tangent SC approach, the basal slip by itself is not enough to accommodate the compressive deformation when the basal poles become strongly aligned with the compression direction.

The SO and AFF models, on the other hand, do a better job at accommodating large strain mostly with basal slip. The SO results, however, are superior to the AFF results in this respect. This superior performance of the second-order SC approximation can be explained in terms of its intrinsic adaptability to microstructural changes. Figure 6 shows the evolution (as predicted with the SO formulation) of the normalized standard deviations of the equivalent stress and strain rate over the whole

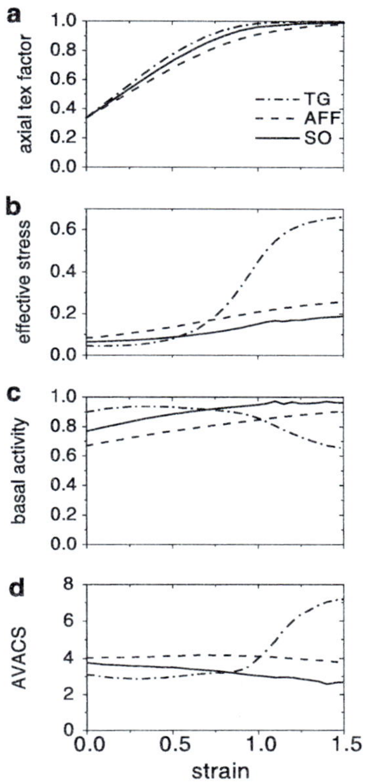

Fig. 5 Simulation of compression of an ice polycrystal. (**a**) Basal texture factor along the compression direction, (**b**) effective stress, (**c**) relative basal activity, and (**d**) the average number of active slip systems per grain, as predicted with the tangent (TG), affine (AFF), and second-order (SO) SC approaches (Lebensohn et al. 2007)

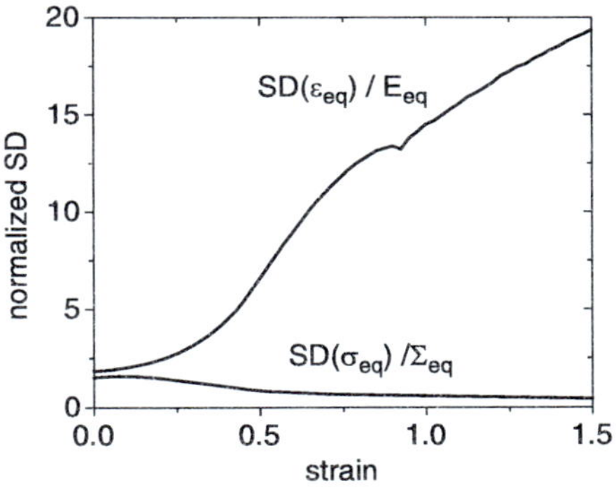

Fig. 6 Evolution of the normalized overall standard deviations of the equivalent stress and strain-rate, as predicted with the second-order formulation, for the case of ice in compression (Lebensohn et al. 2007)

polycrystal, defined by (55–56). Note that the above magnitudes are indicators not only of intergranular but also of intragranular heterogeneity (as a matter of fact, these average scalar magnitudes reflect the collective contribution of every component of the fluctuation tensors in each SR grain). Evidently, as the basal texture concentrates along the axial direction, the stress becomes more uniform and the strain-rate becomes more heterogeneous. This trend toward a uniform stress state obviously indicates a trend toward the Sachs condition. Therefore, given that the aforementioned local fluctuation information is contained in the second-order linearization, the SO results approach the lower-bound as deformation proceeds, allowing a substantial accommodation of deformation by basal slip at those large strains.

3.4 Local Texture Development Predictions Using the FFT-Based Full-Field Approach

Owing to its image-processing lineage, the FFT-based formulation is particularly suitable for use with direct input from actual images of the material, e.g., optical or scanning electron microscopy (SEM) images that show the phase distribution in the case of composites (Moulinec and Suquet 1998), or orientation images in the case of polycrystals (Lebensohn et al. 2008). The latter will be used here for a quantitative study of the average orientations and intragranular misorientations developed in a Cu polycrystal deformed in tension. Electron back-scattering diffraction (EBSD)-based OIM was used to characterize the local orientations measured in an area of about 500 × 500 μm, located on one of the flat surfaces of a recrystallized copper sample. The spatial resolution (given by the distance between two consecutive pixels) was 2 μm in each direction. Two OIM images were taken, one from the undeformed sample, and another after 11% tensile strain along the y-direction (deformation was carried out at room temperature). The scanned area of the deformed sample (332 × 445 pixels) was larger than that of the initial microstructure (274 × 339 pixels), and contained more orientations (2,429 vs. 1,585 grains). This allowed us to register the images, and thus to identify the ID numbers given by the OIM software to each individual grain. Once the two images were appropriately registered, a correlation table allowed us to identify the ID numbers of the grains with largest areas, in the pre- and postdeformation images, for further comparisons.

Next, a 2D 256 × 256 image, containing information on the local (pixel by pixel) crystallographic orientation and a total of 1,124 grains, was cropped from the original OIM image, obtained from the free surface of the undeformed Cu sample, consisting originally of 274 × 339 pixels and 1,585 grains. The average grain size (in units of length) of the cropped image, which can be roughly estimated, is: $\sqrt{256 \times 256/1124} \times 2\,\mu\text{m} = 7.64 \times 2\,\mu\text{m} = 15.28\,\mu\text{m}$. Note, however, that, since the grains in the 2D image are not necessarily sliced across their largest extent as projected onto the observation plane, the above grain size is a low estimate of the true grain size in 3D.

Since the actual 3D microstructure of the bulk of the sample was not known, a 3D unit cell was built assuming a randomly generated distribution of bulk grains underneath the measured surface grains (i.e., a "3D substrate"), having same average grain size and overall crystallographic orientation distribution as the surface grains. For this, a 3D Voronoi was generated (note that since the FFT-based calculation requires a discrete description of the microstructure on a regularly spaced grid, the procedure is simpler than in the case of having to determine the exact position of the boundaries between Voronoi cells in a continuum), as follows: (1) the number of Fourier points in the third dimension (z-direction) was chosen to be 32, resulting in a unit cell of $256 \times 256 \times 32 = 2,097,152$ Fourier points. Note that this choice gives, in average, about four grains along third dimension; (2) the number of grains of the Voronoi structure was calculated as $2{,}097{,}152/(7{,}64)^3 = 4{,}703$; (3) then, 4,703 points were randomly distributed in a 3D unit cell. This Poisson distribution of points constitutes the nuclei of the random grains; (4) the sides of the unit cell were divided into equispaced $256 \times 256 \times 32$ Fourier points, or voxels. Each Fourier point was assigned to its nearest nucleus (accounting for periodic boundary conditions across the unit cell limits), determining 4,703 different domains (grains). Next, the measured 2D and the numerically generated 3D microstructures were merged as follows. First, every 3D grain having a voxel on the first z-layer was removed, and every voxel corresponding to these removed grains was assigned with the crystallographic orientation of the pixel of the 2D OIM image having the same x- and y-coordinates. These replacements determined a structure of "extruded" (columnar) grains of variable depth in the z-direction (from one to several layers), with its first ("surface") layer having the same topology as the OIM image, lying on a 3D substrate. The number of grains of this intermediate configuration decreased to 3,965 grains. Subsequently, to obtain more realistic grain shapes, especially in the transition zone between the columnar grains and the 3D substrate, the microstructure was "annealed" using a standard 3D Monte Carlo (MC) grain growth model with isotropic boundary properties (Rollett and Manohar 2004). The voxels in the surface layer that corresponded to the measured OIM scan (reproduced on the bottom layer, with periodic boundary conditions) were fixed and not allowed to evolve. All other parts of the microstructure were allowed to evolve with the result that grain boundaries moved to minimize their areas. The annealing was run for 1,000,000 MC steps, at which time evolution had essentially ceased because of the pinning effect of the surface layers. The number of grains was further decreased to 3,697 in the final annealed microstructure.

As already pointed out, after carrying out this numerical treatment of the unit cell's microstructure, the first layer of the resulting representative volume element turned out to have the exact same topology as the OIM image. However, without any further manipulation of this configuration, the measured "surface" grains would become bulk grains, upon the imposition of periodic boundary conditions across the unit cell. Therefore, to reproduce the actual free surface condition on the measured grains, the bottom five z-layers (z-layers 28–32) of Fourier points were replaced by a "buffer zone," or "gas phase," with infinite compliance (i.e., identically zero local stress). Such gas phase allowed us to consider the presence of surface grains (corresponding precisely to the grains whose local orientations were actually measured by OIM) while keeping, at the same time, the periodicity across the unit cell (this buffer "disconnects" the surface from the bottom of the periodic repetition of the unit cell, located immediately above). A similar technique was used in phase field simulations of microstructure evolution in thin films (Hu and Chen 2004). The resulting configuration of the 3D unit cell, including the zero-stress buffer zone, is shown schematically in Fig. 7 (Lebensohn et al. 2008). In the next section, we show

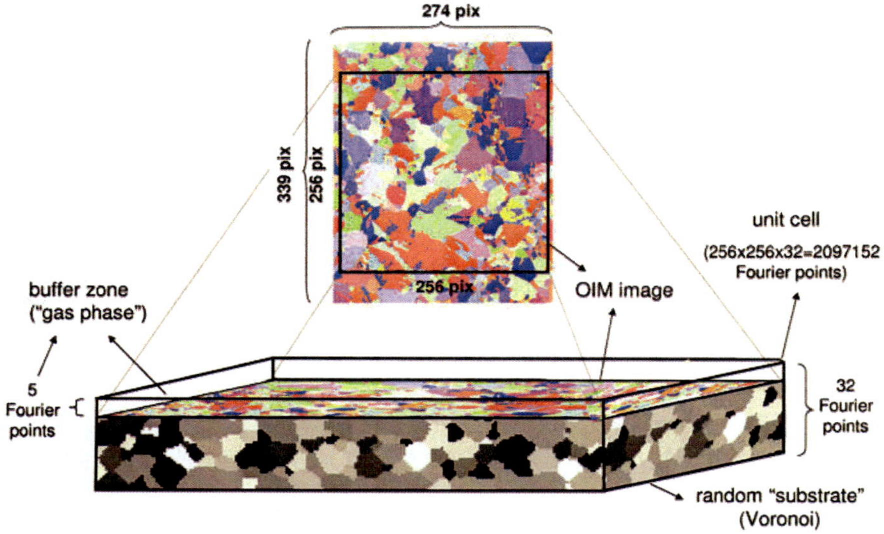

Fig. 7 Schematic representation of the 3D unit cell used in the FFT-based simulations of local orientation and misorientation evolution, with direct input from OIM images (Lebensohn et al. 2008)

and compare results of both unit cell configurations, i.e., the original one resulting from the merging of the OIM and Voronoi structures plus the MC annealing, with no buffer zone (amounting to neglect the surface character of the grains whose orientations were measured by OIM), and the one including the gas phase, for a direct comparison with the OIM measurements.

FFT-based simulations of the tensile deformation of the polycrystalline copper sample were performed using the two above-described unit cells (with and without the buffer zone). The rate-sensitive crystal plasticity relation (1) was used as the local constitutive relation, assuming glide on the twelve (111)<110> systems as the active slip mode, and a viscoplastic exponent $n = 20$. The initial distribution of critically resolved shear stresses was assumed to be uniform. The extended Voce law hardening parameters (74), adjusted to match the experimental macroscopic stress–strain curve measured during the tensile deformation of the copper sample, were: $\tau^s_{oo} = 11$ MPa, $\tau^s_1 = 15.5$ MPa, $\theta^s_0 = 430$ MPa, $\theta^s_1 = 110$ MPa, $(s = 1,12)$ and $h^{ss'} = 1$, for all ss'.

Figure 8a shows the registered initial and 11% strain OIM images (the latter is shown already cropped), measured on the surface of the copper sample (Lebensohn et al. 2008). The postdeformation image clearly indicates the development of intragranular misorientations, in terms of noticeable color grades inside several grains. The location and number of pixels of the 13 largest ("marked") grains are shown in Fig. 8b. The initial orientations of the marked grains can be seen in Fig. 9, in an inverse pole figure representation. Figure 9 also shows the trajectories of the mean orientations of these marked grains (except for grain #5, which is very close

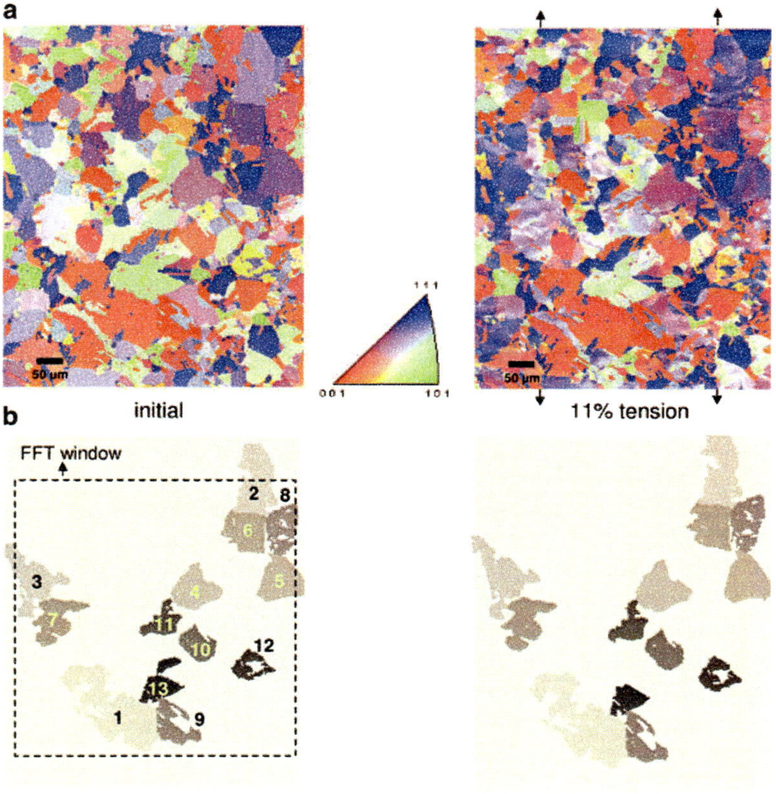

Fig. 8 (**a**) Registered OIM images of the copper polycrystal before deformation and after 11% tensile strain. (**b**) Location and morphology of the 13 largest grains, before and after deformation. The "FFT window" shows the 256 × 256 pixel region that was actually used to construct the unit cell (Lebensohn et al. 2008)

Fig. 9 Inverse pole figure of the measured initial orientation and the final average orientation of the largest grains, and trajectories predicted with the FFT-based approach (Lebensohn et al. 2008)

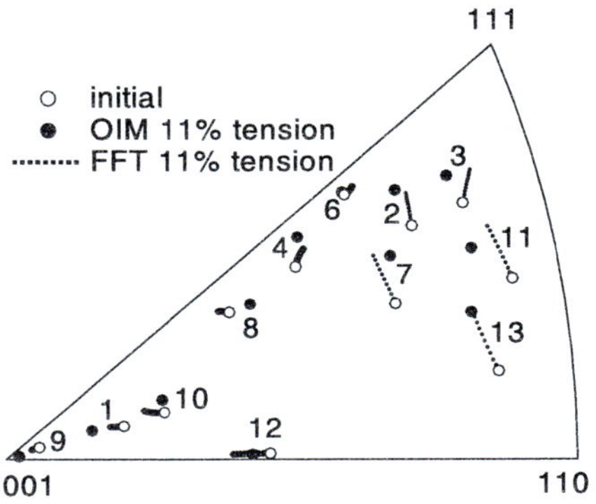

and behaves very similarly to grain #6, and was not plotted for sake of clarity), as predicted by the FFT-based model (with buffer zone). The small crosses defining these trajectories were obtained in increments of 1% overall plastic strain. The actual final average orientations, measured with OIM, are shown as well. In the region close to the <001>-corner, grains #1, #9, and #12 rotate toward the stable orientation <001>, a trend that is predicted, at least qualitatively, by the model. Grains with initial orientations close to the upper half of the <001>–<111>-line (#4, #5, and #6) exhibit rotations along this line toward the other stable orientation, i.e., <111> (also well reproduced by the model). The grains starting near the <110>-corner (#11 and #13), or in intermediate orientations between <110> and the midsection of the <001>–<111>-line (#2, #3 and #7) rotate toward this line, presumably in their way toward the stable orientation <111>. The total rotations of these grains are the largest. All these features are acceptably reproduced by our simulations, except for the reorientation of grain #3, which is predicted to be directly toward <111>. Finally, the trends of grain #10, and especially grain #8, which starts close a well-known transition point on the <001>–<111>-line, close to <113> (Chin et al. 1967), are not adequately reproduced by the model. It is also interesting to compare (while keeping in mind the obvious differences, i.e., Cu vs. Al, 2D vs. 3D) the above trends and the 3D X-ray diffraction characterization done by Winther et al. (2004) of the rotations of bulk grains of an Al polycrystal deformed in tension. The rotations measured on Cu surface grains (and well predicted by our model) with initial orientations belonging to three of the four "regions" characterized by Winther et al. (2004) (i.e., region 1: close to <111>, region 2: near the upper half of the <110>–<111>-line, and region 4: near the <100>-corner) are also in acceptable agreement with the observations of Winther et al. (2004). No grains with orientations in Winther et al.'s region 3 (close to <221>) were large enough in the Cu sample to be used for this analysis.

Although interesting, the above-described reasonable agreement between the measured and predicted average orientation evolution is not surprising, since almost every (either full-field or mean-field) model based on crystal plasticity qualitatively predicts the development of two stable texture components in <001> and <111> in fcc materials deformed under tension, e.g., see comparisons of the above-mentioned measurements with corresponding Taylor and self-consistent predictions in Winther et al. (2004). A much less investigated aspect of the local texture evolution of these materials is reported in Table 2, which shows the comparison between the measured and the predicted values (in degrees) of the average misorientations (defined as the average over every pixel belonging to a given grain with respect to the average orientation of that grain, a quantity that can be readily calculated using quaternion algebra) inside the marked grains (Lebensohn et al. 2008). Together with the size (in pixels) of the grains, two predicted values are reported for each grain: the fourth column shows the predictions obtained using the unit cell shown in Fig. 7, i.e., including the buffer zone, therefore considering the effect of measuring misorientations in surface grains. The fifth column displays the predictions obtained for a unit cell without the buffer zone, giving an idea of the misorientation values that would be measured if the grains were bulk. Evidently, the proper consideration in

Table 2 Area (in pixels) and average misorientation of the 13 largest grains after 11% tensile strain, measured by OIM and predicted with the FFT-based approach, with and without buffer zone (Lebensohn et al. 2008)

Grain no.	Number of pixels	Average misorientation OIM 11% tension (deg)	Average misorientation FFT 11% tension (deg)	Average misorientation FFT 11% tension (no buffer) (deg)
1	3,625	2.89	2.18	1.81
2	2,019	2.52	2.05	1.64
3	1,842	2.92	2.97	2.25
4	1,479	2.86	2.33	2.27
5	1,466	2.26	2.35	1.76
6	1,380	3.14	2.70	1.91
7	1,331	3.37	3.06	2.89
8	1,113	3.21	2.62	2.12
9	955	2.65	2.80	2.22
10	830	2.92	2.37	1.81
11	692	2.22	2.79	1.99
12	639	4.33	3.36	2.39
13	596	3.09	3.26	2.83

the model of the surface character of the grains, whose average misorientations were measured by OIM, leads to a good agreement with the corresponding experimental values. On the other hand, the artificial assumption of the bulk character of these grains tends to underestimate the actual average misorientations of surface grains. The reason why the predictions obtained under the bulk assumption fall short is related to this being a different configuration, compared with the actual traction-free boundary conditions imposed on the surface grains.

Another interesting observation is that except for grain #12, which exhibits the largest average misorientation, the initial orientations of other grains with measured misorientations larger than 3.0° (#6, #7, #8, and #13) lie in a region of the stereographic triangle spanning from <110> toward the midsection (from 1/3 to 2/3) of the <001>–<111>-line. While in the case of grain #12, the high average misorientation seems to be related to its particular morphology (note in Fig. 8b that this grain has a large "hole" in its center, filled by another grain with a completely different orientation, a configuration that may determine a relative "disconnection" between different parts of grain #12); for the rest of the grains, their initial orientations belonging to the above region may be related to their relatively large average misorientation.

To elucidate whether this orientation dependence does exist (and if our model is capable of reproducing it), we investigated the behavior of a larger number of representative grains. Figure 10 shows the average orientations (given by each pole projected in the inverse pole figure) and the average misorientations (given by the different symbols used) of the largest 306 grains, as measured by OIM and predicted with the FFT-based approach, after 11% tensile strain (Lebensohn et al. 2008). The misorientation values of the grains were grouped into six bins of equal size, and different symbols were assigned to each bin. The first observation is that, as expected,

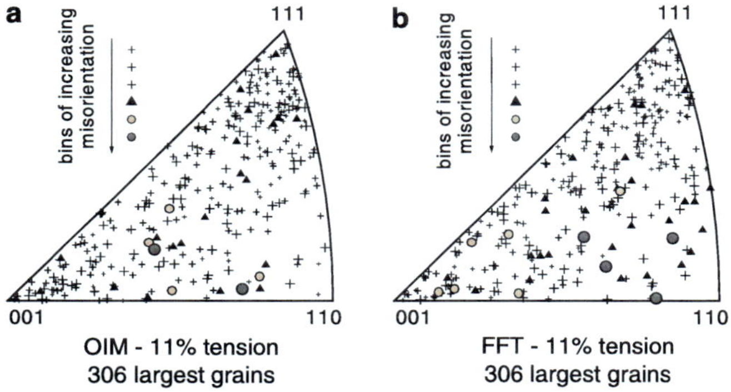

Fig. 10 Inverse pole figures of the average orientations and misorientations of the 306 largest grains after 11% tensile strain, (**a**) measured by OIM, and (**b**) predicted with the FFT-based approach. The misorientations were grouped in bins of equal size, and different symbols were assigned to each bin (Lebensohn et al. 2008)

after 11% tension, there is already a mild but noticeable trend of large grains to rotate toward one of the stable <001> and <111> orientations (the region near <110> is mildly depleted of orientations). Moreover, it is evident (from both the experiments and the simulations) that most of the grains with the highest average misorientation are grains transitioning from their initial orientation near <110> toward the stable orientations. This observation can be explained in the following terms: depending on their initial orientation, the grains of an fcc polycrystal in tension are attracted toward one of the two stable orientations, i.e., <001> or <111> (the latter, directly or via the <001>–<111>-line). Grains with orientations in a region of the orientation space, spanning from near <110> toward the midsection of the <001>–<111>-line, can be pulled simultaneously toward both stable orientations. In this case, the instability of the initial grain orientation and the contribution of interactions with neighbor grains may define the preference of different portions of these "indecisive" grains to rotate toward different stable orientations. This conflicting attraction toward two completely different orientations may be accommodated by the development of relatively higher misorientations between different grain's subdomains. This corresponds to the transition band concept documented by Dillamore et al. (1972) and Dillamore and Katoh (1974) in polycrystals. Note also that this orientation split is also observed in deformed single crystal with initial unstable orientations [thus implying no grain interaction effect (Becker et al. 1991)].

4 Conclusions

In this chapter, we have thoroughly reviewed recently proposed some novel crystal plasticity-based methods for the prediction of microstructure–property relations in polycrystalline aggregates. The first method is of the mean-field type and is known as the second-order VPSC theory, while the second is a full-field method and is known as the FFT-based formulation. We have comprehensively presented the equations and assumptions underlying both formulations, using a unified notation and pointing out their similarities and differences. Mean-field approaches are in general much more efficient than full-field computations. However, models like the second-order VPSC formulation, which incorporates more statistical information, require more complex and numerically demanding algorithms, but still are much faster than full-field approaches.

Concerning the mean-field theories, we have carried out detailed comparisons of the different self-consistent approximations for viscoplastic polycrystals. We have also discussed the numerical implementation of the different SC approaches in the VPSC code, together with results obtained using different linearization strategies. The comparison of the effective behavior of model material systems predicted by different SC approaches has shown that the second-order SC predictions are in better agreement with the ensemble averages of FFT-based full-field solutions. The latter is especially true in the cases of highly heterogeneous materials (due to a strong nonlinearity or local anisotropy), a case in which the gap between the Taylor and the Sachs bounds is large. With regards to applications of the second-order SC approximation, we have studied the texture evolution of polycrystalline ice (a material characterized by a strong local anisotropy, due a strong contrast of plastic properties at single crystal level) deformed in compression, to illustrate the flexibility of the second-order formulation to handle these highly anisotropic problems.

The FFT-based full-field formulation for plastically deforming polycrytals has been conceived as a very efficient alternative to crystal-plasticity FE methods. In this work, FFT-based computations were first applied to the antiplane deformation of isotropic, linearly viscous 2D polycrystals. In this case, our numerical implementation was validated by comparison with analytical results for the effective and per-phase properties of such special configuration. Next, the full 3D implementation was applied to the study of the subgrain texture evolution in a copper aggregate deformed under tension. Direct input was obtained from OIM images for the construction of the representative volume element. A methodology to build such 3D unit cell, including the 2D OIM data, a 3D substrate, and the presence of a free surface, was given. The average orientations and misorientations of large grains, predicted with the FFT-based approach after 11% tensile strain, were directly compared with OIM measurements. The experimental data and the predictions showed good agreement. The orientation dependence of the average misorientations was also studied. Again, measurements and predictions showed reasonable agreement. Grains with initial orientation near <110> tend to develop higher misorientations, as deformation proceeds. Attraction toward the two different stable orientations (i.e., corresponding to the alignment of the <001> and the <111> crystal orientations with the tensile axis) of different subdomains inside these grains, influenced by interactions with different neighbors, may be responsible for this behavior. Only full-field models such as the FFT-based formulation, which account for topological information and grain interaction in the determination of the local micromechanical fields, can capture these effects.

Acknowledgments We wish to thank our colleagues Carlos Tomé (LANL, Los Alamos, USA), Tony Rollett (CMU, Pittsburgh, USA), Pierre Gilormini (ENSAM, Paris, France), and Pierre Suquet (LMA, Marseille, France) for fruitful discussions.

References

Acharya A, Bassani JL, Beaudoin A (2003) Geometrically necessary dislocations, hardening, and a simple gradient theory of crystal plasticity. Scr Mater 48: 167–172.

Barbe F, Decker L, Jeulin D, Cailletaud G (2001) Intergranular and intragranular behavior of polycrystalline aggregates. Part 1: F.E. model. Int J Plast 17: 513–536.

Becker R (1991) Analysis of texture evolution in channel die compression. 1. Effects of grain interaction. Acta Metall Mater 39: 1211–1230.

Becker R, Butler JF, Hu H, Lalli LA (1991). Analysis of an aluminum single-crystal with unstable initial orientation (001)[110] in channel die compression. Metall Trans A 22:45–58.

Berveiller M, Fassi-Fehri O, Hihi A (1987) The problem of 2 plastic and heterogeneous inclusions in an anisotropic medium. Int J Eng Sci 25: 691–709.

Bhattacharyya A, El-Danaf E, Kalidindi SR, Doherty RD (2001) Evolution of grain-scale microstructure during large strain simple compression of polycrystalline aluminum with quasi-columnar grains: OIM measurements and numerical simulations. Int J Plast 17: 861–883.

Bobeth M, Diener G (1987) Static elastic and thermoelastic field fluctuations in multiphase composites. J Mech Phys Solids 35:137–149.

Castelnau O, Duval P, Lebensohn RA, Canova GR (1996) Viscoplastic modelling of texture development in polycrystalline ice with a self-consistent approach: comparison with bound estimates. J Geophys Res B 101: 13851–13868.

Castelnau O, Blackman DK, Lebensohn RA, Ponte Castañeda P (2008) Micromechanical modelling of the viscoplastic behavior of olivine. J Geophys Res B 113: B09202.

Chen LQ (2004) Introduction to the phase-field method of microstructure evolution. In: Raabe D, Roters F, Barlat F, Chen LQ (eds). Continuum scale simulations of engineering materials. Wiley, Wenheim, pp. 37–51.

Cheong KS, Busso EP (2004) Discrete dislocation density modelling of single phase FCC polycrystal aggregates. Acta Mater 52: 5665–5675.

Chin GY, Mammel WL, Dolan MT (1967) Taylor's theory of texture for axisymmetric flow in body-centered cubic metals. Trans Met Soc AIME 239: 1854–1855.

deBotton G, Ponte Castañeda P (1995) Variational estimates for the creep-behavior of polycrystals. Proc R Soc Lond A 448: 121–142.

Delaire F, Raphanel JL, Rey C (2001) Plastic heterogeneities of a copper multicrystal deformed in uniaxial tension: experimental study and finite element simulations. Acta Mater 48: 1075.

Delannay L, Logé RE, Chastel Y, Signorelli JW, Van Houtte P (2003) Measurement of in-grain orientation gradients by EBSD and comparison with finite element results. Adv Eng Mater 5: 597–600.

Delannay L, Jacques PJ, Kalidindi SR (2006) Finite element modeling of crystal plasticity with grains shaped as truncated octahedrons. Int J Plast 22: 1879–1898.

Diard O, Leclercq S, Rousselier G, Cailletaud G (2005) Evaluation of finite element based analysis of 3D multicrystalline aggregates plasticity Application to crystal plasticity model identification and the study of stress and strain fields near grain boundaries. Int J Plast 21: 691.

Dillamore IL, Morris PL, Smith CJE, Hutchinson WB (1972) Transition bands and recrystallization in metals. Proc R Soc Lond A 329: 405–420.

Dillamore IL, Katoh H (1974) Mechanisms of recrystallization in cubic metals with particular reference to their orientation-dependence. Met Sci J 8: 73–83.

Dykhne AM (1970) Conductivity of a two-dimensional two-phase system. Dokl Akad Nauk SSSR 59: 110–115.

Hershey AV (1954) The elasticity of an isotropic aggregate of anisotropic cubic crystals. J Appl Mech 21: 236–240.
Hill R (1965) Continuum micro-mechanics of elastoplastic polycrystals. J Mech Phys Solids 13: 89.
Hutchinson JW (1976) Bounds and self-consistent estimates for creep of polycrystalline materials. Proc R Soc Lond A 348: 101–127.
Hu SY, Chen LQ (2004) Spinodal decomposition in a film with periodically distributed interfacial dislocations. Acta Mater 52: 3069–3074.
Idiart MI, Ponte Castañeda P (2007) Field statistics in nonlinear composites. I. Theory. Proc R Soc Lond A 463: 183–202.
Idiart MI, Moulinec H, Ponte Castaneda P, Suquet P (2006) Macroscopic behavior and field fluctuations in viscoplastic composites: second-order estimates versus full-field simulations. J Mech Phys Solids 54: 1029–1063.
Kanit T, Forest S, Galliet I, Mounoury V, Jeulin D (2003) Determination of the size of the representative volume element for random composites: statistical and numerical approach. Int J Solids Struct 40: 3647–3679.
Kreher W (1990) Residual-stresses and stored elastic energy of composites and polycrystals. J Mech Phys Solids 38: 115–128.
Lahellec N, Michel JC, Moulinec H, Suquet P (2001) Analysis of inhomogeneous materials at large strains using fast Fourier transforms. In: Miehe C (ed). IUTAM symposium on computational mechanics of solids materials. Kluwer Academic, Dordretcht, pp. 247–268.
Laws N (1973) On the thermostatics of composite materials. J Mech Phys Solids 21: 9–17.
Lebensohn RA, Tomé CN (1993) A selfconsistent approach for the simulation of plastic deformation and texture development of polycrystals: application to Zirconium alloys. Acta Metall Mater 41: 2611–2624.
Lebensohn RA, Turner PA, Signorelli JW, Canova GR, Tomé CN (1998) Calculation of intergranular stresses based on a large strain viscoplastic selfconsistent polycyrstal model. Model Simul Mater Sci Eng 6: 447–465.
Lebensohn RA (2001) N-site modelling of a 3D viscoplastic polycrystal using fast Fourier transform. Acta Mater 49: 2723–2737.
Lebensohn RA, Liu Y, Ponte Castañeda P (2004a) On the accuracy of the self-consistent approximation for polycrystals: comparison with full-field numerical simulations. Acta Mater 52: 5347–5361.
Lebensohn RA, Liu Y, Ponte Castañeda P (2004b) Macroscopic properties and field fluctuations in model power-law polycrystals: full-field solutions versus self-consistent estimates. Proc R Soc Lond A 460: 1381–1405.
Lebensohn RA, Castelnau O, Brenner R, Gilormini P (2005) Study of the antiplane deformation of linear 2-D polycrystals with different microstructures. Int J Solids Struct 42: 5441–5449
Lebensohn RA, Tomé CN, Ponte Castañeda P (2007) Self-consistent modeling of the mechanical behavior of viscoplastic polycrystals incorporating intragranular field fluctuations. Phil Mag 87: 4287–4322.
Lebensohn RA, Brenner R, Castelnau O, Rollett AD (2008) Orientation image-based micromechanical modelling of subgrain texture evolution in polycrystalline copper. Acta Mater 56: 3914–3926.
Lequeu P, Gilormini P, Montheillet F, Bacroix B, Jonas JJ (1987) Yield surfaces for textured polycrystals. 1. Crystallographic approach. Acta Metall 35: 439–451.
Liu Y, Ponte Castañeda P (2004) Second-order theory for the effective behavior and field fluctuations in viscoplastic polycrystals. J Mech Phys Solids 52: 467–495.
Lurie KA, Cherkaev AV (1984). G-closure of some particular sets of admissible material characteristics for the problem of bending of thin elastic plates. J Optim Theor Appl 42: 305–316.
Masson R, Bornert M, Suquet P, Zaoui A (2000) Affine formulation for the prediction of the effective properties of nonlinear composites and polycrystals. J Mech Phys Solids 48: 1203–1227.

Michel JC, Moulinec H, Suquet P (2000) A computational method based on augmented Lagrangians and fast Fourier transforms for composites with high contrast. Comput Model Eng Sci 1: 79–88.

Mika DP, Dawson PR (1998) Effects of grain interaction on deformation in polycrystals. Mater Sci Eng A 257: 62–76.

Molinari A, Canova GR, Ahzi S (1987) Self consistent approach of the large deformation polycrystal viscoplasticity. Acta Metall 35: 2983–2994.

Moulinec H, Suquet P (1994) A fast numerical method for computing the linear and nonlinear mechanical properties of composites. C R Acad Sci Paris II 318: 1417–1423.

Moulinec H, Suquet P (1998) Numerical method for computing the overall response of nonlinear composites with complex microstructure. Comput Methods Appl Mech Eng 157: 69–94.

Moulinec H, Suquet P (2003) Intraphase strain heterogeneity in nonlinear composites: a computational approach. Eur J Mech Solids 22: 751–770.

Mura T (1987) Micromechanics of defects in solids. Martinus-Nijhoff Publishers, Dordrecht.

Musienko A, Tatschl A, Schmidegg K, Kolednik O, Pippan R, Cailletaud G (2007) Three-dimensional finite element simulation of a polycrystalline copper specimen. Acta Mater 55: 4121–4136.

Nebozhyn MV, Gilormini P, Ponte Castañeda P (2000) Variational self-consistent estimates for viscoplastic polycrystals with highly anisotropic grains. C R Acad Sci Paris IIb 328: 11–17.

Parton VZ, Buryachenko VA (1990) Stress fluctuations in elastic composites. Sov Phys Dokl 35(2):191–193.

Ponte Castañeda P (1991) The effective mechanical properties of nonlinear isotropic composites. J Mech Phys Solids 39: 45–71.

Ponte Castañeda P (1996) Exact second-order estimates for the effective mechanical properties of nonlinear composites. J Mech Phys Solids 44: 827–862.

Ponte Castañeda P (2002) Second-order homogenization estimates for nonlinear composites incorporating field fluctuations: I- theory. J Mech Phys Solids 50: 737–757.

Raabe D, Sachtleber M, Zhao Z, Roters F, Zaefferer S (2001) Micromechanical and macromechanical effects in grain scale polycrystal plasticity experimentation and simulation. Acta Mater 49: 3433–3441.

Rollett AD, Manohar P (2004) The Monte Carlo method. In: Raabe D, Roters F, Barlat F, Chen LQ (eds) Continuum scale simulations of engineering materials. Wiley, Wenheim, pp.77–111

Sachs G (1928). On the derivation of a condition of flowing. Z Verein Deut Ing 72: 734–736.

Semiatin SL, Bieler TR (2001) The effect of alpha platelet thickness on plastic flow during hot working of Ti-6Al-4V with a transformed microstructure. Acta Mater 49: 3565–3573.

Sulsky D, Zhou SJ, Schreyer HL (1995) Application of a particle-in-cell method to solid mechanics. Comput Phys Commun 87: 236–252.

Taylor GI (1938) Plastic strain in metals. J Inst Met 62: 307–324.

Tomé CN, Canova GR, Kocks UF, Christodoulou N, Jonas JJ (1984) The relation between macroscopic and microscopic strain-hardening in fcc polycrystals. Acta Metall 32: 1637–1653.

Tomé CN, Lebensohn RA (2008). Manual for Code Viscoplastic Self-Consistent (version 7).

Walpole LJ (1969) On the overall elastic moduli of composite materials. J Mech Phys Solids 17: 235–251.

Wang YU, Jin YMM, Khachaturyan AG (2002) Phase field microelasticity theory and modeling of elastically and structurally inhomogeneous solid. J Appl Phys 92: 1351–1360.

Wenk HR, Tomé CN (1999) Modeling dynamic recrystallization of olivine aggregates deformed in simple shear. J Geophys Res B 104: 25513–25527.

Willis JR (1981) Variational and related methods for the overall properties of composites. Adv Appl Mech 21: 1–78.

Winther G, Margulies L, Schmidt S, Poulsen HF (2004) Lattice rotations of individual bulk grains Part II: correlation with initial orientation and model comparison. Acta Mater 52: 2863–2872.

Appendix: Calculation of Effective Moduli Derivatives

We give here the expressions and algorithm for the calculation of the derivatives of the effective moduli within the context of the VPSC formulation (Lebensohn et al. 2007).

A.1 Calculation of $\partial B_{kj}^{(s)}/\partial M_{uv}^{(r)}$

From (32), we have [in matrix notation, all indices running from 1 to 5, except the grain indices (r) and (s)]:

$$\frac{\partial B_{kj}^{(s)}}{\partial M_{uv}^{(r)}} = -\frac{1}{2}\left[\left(M^{(s)}+\tilde{M}\right)_{ku}^{-1}\delta_{rs}B_{vj}^{(s)} + \left(M^{(s)}+\tilde{M}\right)_{kv}^{-1}\delta_{rs}B_{uj}^{(s)}\right] + $$
$$+\left(M^{(s)}+\tilde{M}\right)^{-1}\left[\frac{\partial\tilde{M}}{\partial M_{uv}^{(r)}}\left(I-B^{(s)}\right)+\frac{\partial\tilde{M}}{\partial M_{uv}^{(r)}}\right]. \tag{A1}$$

In order not to clutter the notation, the first and second term on the right are written in explicit and implicit index notation, respectively. In the second term, the indices (uv) (i.e., the component of the local compliance with respect to which the derivatives are calculated) appear only to indicate such derivative, while in the first term they appear mixed with the indices that contract. In what follows, we will use this mix of explicit indices and implicit notation, when necessary for the sake of clarity.

Deriving (30), we obtain:

$$\frac{\partial\tilde{M}_{ij}}{\partial M_{uv}^{(r)}} = (I-S)_{ik}^{-1}\frac{\partial S_{kl}}{\partial M_{uv}^{(r)}}\psi_{lj} + F_{ip}^{S}\frac{\partial\bar{M}_{pj}}{\partial M_{uv}^{(r)}}, \tag{A2}$$

where $F^S = (I-S)^{-1}S$ and $\psi = F^S\bar{M} + \bar{M}$. Using the chain rule to express the first derivative on the right, we can write:

$$\frac{\partial\tilde{M}_{ij}}{\partial M_{uv}^{(r)}} = (I-S)_{ik}^{-1}\frac{\partial S_{kl}}{\partial\bar{M}_{pq}}\frac{\partial\bar{M}_{pq}}{\partial M_{uv}^{(r)}}\psi_{lj} + F_{ip}^{S}\frac{\partial\bar{M}_{pj}}{\partial M_{uv}^{(r)}} = \theta_{ijpq}\frac{\partial\bar{M}_{pq}}{\partial M_{uv}^{(r)}}, \tag{A3}$$

where

$$\theta_{ijpq} = (I-S)_{ik}^{-1}\frac{\partial S_{kl}}{\partial\bar{M}_{pq}}\psi_{lj} + F_{ip}^{S}\,\delta_{jq}. \tag{A4}$$

The algorithm for the calculation of $\partial S/\partial\bar{M}$ is given below. Replacing (A3) in (A1) and after some manipulation, we obtain:

$$\frac{\partial B_{kj}^{(s)}}{\partial M_{uv}^{(r)}} = -\frac{1}{2}\left(\chi_{ku}^{(s)}\delta_{rs}B_{vj}^{(s)} + \chi_{kv}^{(s)}\delta_{rs}B_{uj}^{(s)}\right) + \chi^{(s)}\alpha^{(s)} : \frac{\partial\bar{M}}{\partial M_{uv}^{(r)}}, \quad \text{(A5)}$$

where

$$\chi^{(s)} = \left(M^{(s)} + \tilde{M}\right)^{-1}, \quad \text{(A6)}$$

$$\alpha_{ijkl}^{(s)} = \theta_{imkl}\left(I - B^{(s)}\right)_{mj} + \delta_{ik}\delta_{jl}. \quad \text{(A7)}$$

A.2 Calculation of $\partial\bar{M}_{ij}/\partial M_{uv}^{(r)}$

Deriving (44):

$$\frac{\partial\bar{M}_{ij}}{\partial M_{uv}^{(r)}} = \sum_s c^{(s)}\delta_{iu}\delta_{kv}\delta_{rs}B_{kj}^{(s)} + \sum_s c^{(s)}M_{ik}^{(s)}\frac{\partial B_{kj}^{(s)}}{\partial M_{uv}^{(r)}}. \quad \text{(A8)}$$

Using (A5) and calling $\beta^{(s)} = M^{(s)}\chi^{(s)}$, we get:

$$\frac{\partial\bar{M}_{ij}}{\partial M_{uv}^{(r)}} = \frac{c^{(r)}}{2}\left(\delta_{iu}B_{vj}^{(r)} + \delta_{iv}B_{uj}^{(r)}\right) - \frac{c^{(r)}}{2}\left(\beta_{iu}^{(r)}B_{vj}^{(r)} + \beta_{iv}^{(r)}B_{uj}^{(r)}\right) +$$
$$+\left(\sum_s c^{(s)}\beta^{(s)}\alpha^{(s)}\right)\frac{\partial\bar{M}}{\partial M_{uv}^{(r)}}. \quad \text{(A9)}$$

From where:

$$\Omega_{ijkl}\frac{\partial\bar{M}_{kl}}{\partial M_{uv}^{(r)}} = \pi_{ij}^{(r,uv)}, \quad \text{(A10)}$$

with

$$\pi_{ij}^{(r,uv)} = \frac{c^{(r)}}{2}\left[\left(\delta_{iu} - \beta_{iu}^{(r)}\right)B_{vj}^{(r)} + \left(\delta_{iv} - \beta_{iv}^{(r)}\right)B_{uj}^{(r)}\right]. \quad \text{(A11)}$$

$$\Omega_{ijkl} = \delta_{ik}\delta_{jl} - \sum_s c^{(s)}\beta^{(s)}\alpha^{(s)} \quad \text{(A12)}$$

A.3 Calculation of $\partial E_j^o / \partial M_{uv}^{(r)}$

Deriving (45):

$$\frac{\partial E_i^o}{\partial M_{uv}^{(r)}} = \sum_s c^{(s)} \varepsilon_k^{o(s)} \frac{\partial B_{ki}^{(s)}}{\partial M_{uv}^{(r)}}. \tag{A13}$$

Using (A5), we obtain:

$$\frac{\partial E_i^o}{\partial M_{uv}^{(r)}} = \zeta_{ikl} \frac{\partial \bar{M}_{kl}}{\partial M_{uv}^{(r)}} + \kappa_i^{(r,uv)}, \tag{A14}$$

where

$$\zeta_{ijk} = \sum_s c^{(s)} \varepsilon_m^{o(s)} \chi_{ml}^{(s)} \alpha_{lijk}^{(s)} \tag{A15}$$

$$\kappa_i^{(r,uv)} = -\frac{c^{(r)}}{2} \varepsilon_k^{o(r)} \left[\chi_{ku}^{(r)} B_{vi}^{(r)} + \chi_{kv}^{(r)} B_{ui}^{(r)} \right] \tag{A16}$$

A.4 Calculation of $\partial \bar{G} / \partial M_{uv}^{(r)}$

Deriving (46):

$$\frac{\partial \bar{G}}{\partial M_{uv}^{(r)}} = \sum_s c^{(s)} \varepsilon_i^{o(s)} \frac{\partial b_i^{(s)}}{\partial M_{uv}^{(r)}}. \tag{A17}$$

Deriving (33):

$$\frac{\partial b_i^{(s)}}{\partial M_{uv}^{(r)}} = -\frac{1}{2} \delta_{rs} \left[\chi_{iu}^{(s)} \chi_{vl}^{(s)} \left(E_l^o - \varepsilon_l^{o(s)} \right) + \chi_{iv}^{(s)} \chi_{ul}^{(s)} \left(E_l^o - \varepsilon_l^{o(s)} \right) \right]$$

$$- \chi^{(s)} \frac{\partial \tilde{M}}{\partial M_{uv}^{(r)}} \chi^{(s)} \left(E^o - \varepsilon^{o(s)} \right) + \chi^{(s)} \frac{\partial E^o}{\partial M_{uv}^{(r)}}. \tag{A18}$$

Replacing (A17) in (A16) and using (A3):

$$\frac{\partial \bar{G}}{\partial M_{uv}^{(r)}} = \varphi_{ij} \frac{\partial M_{ij}^o}{\partial M_{uv}^{(r)}} + \vartheta_i \frac{\partial E_i^o}{\partial M_{uv}^{(r)}} + \eta^{(r,uv)}, \tag{A19}$$

where

$$\varphi_{ij} = -\left[\sum_s c^{(s)} \varepsilon_k^{o(s)} \chi_{kl}^{(s)} \chi_{pq}^{(s)} \left(E_q^o - \varepsilon_q^{o(s)}\right)\right] \theta_{lpij}, \tag{A20}$$

$$\vartheta_i = \sum_s c^{(s)} \varepsilon_k^{o(s)} \chi_{ki}^{(s)}, \tag{A21}$$

$$\eta^{(r,uv)} = -\frac{c^{(r)}}{2} \varepsilon_i^{o(r)} \left[\chi_{iu}^{(r)} \chi_{vl}^{(r)} \left(E_l^o - \varepsilon_l^{o(r)}\right) + \chi_{iv}^{(r)} \chi_{ul}^{(r)} \left(E_l^o - \varepsilon_l^{o(r)}\right)\right]. \tag{A22}$$

A.5 Calculation of $\partial S/\partial \bar{M}$

The derivative of Eshelby tensor with respect to the effective compliance appearing in (A4) can be obtained as follows. From (23) and (24), the (symmetric) Eshelby tensor of an ellipsoidal inclusion of radii (a,b,c) embedded in an incompressible homogenous medium of stiffness $\bar{L} = \bar{M}^{-1}$ is given by (in tensor notation, all indices running from 1 to 3):

$$S_{ijmn} = T_{ijkl}^{sym} \bar{L}_{klmn}, \tag{A23}$$

where

$$T_{ijkl}^{sym} = \frac{abc}{16\pi} \int_0^{2\pi} \int_0^{\pi} \frac{\lambda_{ijkl}(\alpha)}{[\rho(\alpha)]^3} \sin\theta \, d\theta \, d\varphi, \tag{A24}$$

with

$$\lambda_{ijkl} = \alpha_j \alpha_l A_{ik}^{-1} + \alpha_i \alpha_l A_{jk}^{-1} + \alpha_j \alpha_k A_{il}^{-1} + \alpha_i \alpha_k A_{jl}^{-1}, \tag{A25}$$

where the matrix 4 × 4 A is given by (A5). In particular:

$$A_{ik} = \bar{L}_{ijkl} \alpha_j \alpha_l \quad (i, j = 1, 3). \tag{A26}$$

Deriving (A23):

$$\frac{\partial S_{ijmn}}{\partial \bar{M}} = \frac{\partial T_{ijkl}^{sym}}{\partial \bar{M}} \bar{L}_{klmn} + T_{ijkl}^{sym} \frac{\partial \bar{L}_{klmn}}{\partial \bar{M}}. \tag{A27}$$

The first derivative on the right is obtained as:

$$\frac{\partial T_{ijkl}}{\partial \bar{M}} = \frac{abc}{16\pi} \int_0^{2\pi} \int_0^{\pi} \frac{\partial \lambda_{ijkl}}{\partial \bar{M}} \frac{\sin\theta \, d\theta \, d\varphi}{[\rho(\alpha)]^3}. \tag{A28}$$

Using (A25) and (A26), $\partial\lambda_{ijkl}/\partial\bar{M}$ is calculated as:

$$\frac{\partial\lambda_{ijkl}}{\partial\bar{M}} = \alpha_j\,\alpha_l\,\frac{\partial A_{ik}^{-1}}{\partial\bar{M}} + \alpha_i\,\alpha_l\,\frac{\partial A_{jk}^{-1}}{\partial\bar{M}} + \alpha_j\,\alpha_k\,\frac{\partial A_{il}^{-1}}{\partial\bar{M}} + \alpha_i\,\alpha_k\,\frac{\partial A_{jl}^{-1}}{\partial\bar{M}}, \tag{A29}$$

where

$$\frac{\partial A_{ij}^{-1}}{\partial\bar{M}} = -A_{ik}^{-1}\left(\frac{\partial\bar{L}_{klmn}}{\partial\bar{M}}\alpha_l\alpha_n\right)A_{mj}^{-1}. \tag{A30}$$

The expression $\partial\bar{L}_{klmn}/\partial\bar{M}$ appearing in (A27) and (A30) is simply the derivative of a tensor with respect to its inverse. In matrix notation (indices running from 1 to 5):

$$\frac{\partial\bar{L}_{ij}}{\partial\bar{M}_{pq}} = -\frac{1}{2}\left(\bar{L}_{ip}\,\bar{L}_{qj} + \bar{L}_{iq}\,\bar{L}_{pj}\right). \tag{A31}$$

Stochastic Upscaling for Inelastic Material Behavior from Limited Experimental Data

Sonjoy Das and Roger Ghanem

Abstract A stochastic upscaling approach based on random matrix theory has recently been proposed to characterize a coarse scale (continuum) constitutive elasticity matrix of heterogeneous materials (Das, "Model, identification & analysis of complex stochastic systems: Applications in stochastic partial differential equations and multiscale mechanics", PhD thesis, University of Southern California, Los Angeles, USA, 2008; Das and Ghanem, "SIAM Multiscale Modeling and Simulation", 8(1):296–325, 2009). The approach, originally developed for linear elastic behavior, is adapted in the present work to characterize nonlinear elastic and strain-hardening plastic behavior. This is achieved by assuming that the heterogeneous material is locally elastic at any given point on the nonlinear constitutive stress–strain curve. The associated constitutive tangential elasticity matrix or tangential elastoplastic matrix is treated as a random matrix that evolves with the effective strain which depends on the current strain state of the material. The uncertainty characterized by such constitutive tangential matrix can be construed as a reflection, on the coarse scales, of fluctuations of the fine scale features from which the constitutive matrices are constructed. Under certain conditions, such constitutive matrix can be shown to be symmetric, positive-definite and bounded from below and above, in the positive-definite sense, by two symmetric and positive-definite matrices. The condition of boundedness follows from the applications of the principles of minimum complementary energy and minimum potential energy. The lower and upper matrix-valued bounds can be obtained fairly accurately from a limited amount of fine scale experimental data. They are typically computed by carrying out micromechanical analyses on a small volume element of the heterogeneous material, the size of which depends on the particular application of interest. A probability measure of the random matrix that reflects the constraints consistent with these energy-based bounds is constructed, and a sampling scheme is developed to synthesize realizations of the random matrix.

R. Ghanem (✉)
University of Southern California, Los Angeles, CA 90089
e-mail: ghanem@usc.edu

S. Ghosh and D. Dimiduk (eds.), *Computational Methods for Microstructure-Property Relationships*, DOI 10.1007/978-1-4419-0643-4_12,

1 Introduction

Proper characterization of coarse scale material properties from fine scale features within a multiscale framework and the assessment of its impact on the end performance have recently gained a tremendous research interest (To et al. 2008; Liu et al. 2009). In the present work, we particularly consider the problem of characterizing the *local* constitutive continuum (coarse scale) material properties of the heterogeneous material, namely, the fourth-order constitutive tangential elasticity tensor for the nonlinear elastic process or the fourth-order constitutive tangential elastoplastic tensor for the inelastic process, from a limited amount of fine scale information. Consider a body of the heterogeneous material over domain $\mathscr{B} \subseteq \mathbb{R}^3$, where $\mathbb{R}^3$ represents the three-dimensional (3-D) Euclidean space. In the following development, we will consider the tangential elastoplastic tensor $\mathbf{C}_{\text{ep}}(\mathbf{x})$, $\mathbf{x} \in \mathscr{B}$. The constitutive tangential elasticity tensor for the nonlinear elastic deformation process can be similarly characterized. The matrix representation of $\mathbf{C}_{\text{ep}}$ will be referred to as the constitutive tangential elastoplastic matrix which is typically symmetric and positive-definite. For the sake of simplicity, we will denote both the constitutive tangential elastoplastic tensor and its matrix representation by $\mathbf{C}_{\text{ep}}$. The constitutive matrix $\mathbf{C}_{\text{ep}}$ is constructed from the fine scale features of the heterogeneous material. By fine scale features, we allude to the morphological (or textural) and structural properties of the microconstituents of the heterogeneous material at the scale of a material grain. A limited amount of such fine scale information can be experimentally identified using microstructural-crystallographic techniques (e.g., electron backscattered diffraction) and nanoindentation tests.

The constructed $\mathbf{C}_{\text{ep}}$ depends on the spatial extent and characteristics of the microstructural fields used in the process. The present work thus characterizes $\mathbf{C}_{\text{ep}}$ by treating it as a random matrix. The uncertainty characterized by the random matrix can be construed as the effects of fluctuations of the fine scale features from which $\mathbf{C}_{\text{ep}}$ is constructed. In practice, $\mathbf{C}_{\text{ep}}$ is typically deterministically estimated via inverse analysis and shows dependence on the boundary and loading conditions used in the inverse analysis. It can be shown that $\mathbf{C}_{\text{ep}}$ is bounded from below and above, in the positive-definite sense, by two symmetric and positive-definite matrices regardless of the boundary and loading conditions provided the interfaces between the microconstituents of the heterogeneous material neither allow any slips nor contain any cracks (perfectly bonded interfaces). This follows from the applications of the principles of minimum complementary energy and minimum potential energy on a small material volume element, V, of the heterogeneous material (see Sect. 4). These lower and upper matrix-valued bounds are generally determined by carrying out micromechanical analyses on the volume element V. Based on a previous work (Das 2008; Das and Ghanem 2009), a probability measure, which reflects constraints consistent with these energy-based bounds, is proposed in the present work along with a sampling scheme to synthesize realizations of $\mathbf{C}_{\text{ep}}$. The characteristic features of the resulting probability measure of $\mathbf{C}_{\text{ep}}$ depends on the size of V which is often dictated by the scope of a particular application.

The fundamental line-of-inquiry in developing the probability model of $\mathbf{C}_{ep}$ as proposed in the present work is similar in spirit to the concept of the nonparametric model developed earlier for stochastic mechanics applications (Soize 2001). Within the framework of this nonparametric approach, based on random matrix theory (RMT), the probability models are *directly* developed for the global-level or higher-level system matrices, e.g., mass matrix, stiffness matrix (Soize 2001) or frequency response function (FRF) matrix (Ghanem and Das 2009) or the constitutive elasticity tensor for linear elastic material behavior (Soize 2006). The notion of the conventional local system parameters (e.g., Young's modulus and Poisson's ratio) does not exist within this formalism. The effort in developing the probabilistic descriptions of the conventional system parameters within the usual parametric framework is now alternatively spent on developing directly the probabilistic descriptions of the associated operators (stiffness matrix, FRF matrix, constitutive elasticity tensor, etc.). This probabilistic modeling approach has a distinct advantage over the conventional parametric approach. The nonparametric model turns out to be more efficient, particularly, in modeling complex stochastic systems involving a large number of random conventional parameters. It greatly simplifies the modeling phases of uncertainties, especially when the underlying uncertainties are homogeneously spread over the spatial domain of the mechanical structure of interest. The probability models of the global-level random system matrices are deduced by employing the principle of maximum entropy (MaxEnt) (Jaynes 1957; Kapur and Kesavan 1992; Jaynes 2003) while imposing very general knowledge-based constraints. The MaxEnt-based probability model, when constrained to the set of symmetric and positive definite matrices, turns out to be the Wishart or matrix-variate Gamma distribution supported over the entire interior of the positive-semidefinite cone (Murihead 1982; Gupta and Nagar 2000).

The present work extends the above MaxEnt- and RMT-based nonparametric model and explicitly enforces the lower and upper bounds that reflect the fine scale information of the heterogeneous material in some sense. The current work is an adaptation of recent work on the constitutive random elasticity matrix (Das 2008; Das and Ghanem 2009) to the nonlinear elasticity and strain-hardening plastic behavior. Another recent related effort in constructing the probability model of the constitutive random elasticity matrix from fine scale features is worth mentioning at this point, where the size of a representative volume element (RVE) of the heterogeneous material is determined based on the spatial correlation lengths of the underlying microstructural random field (Soize 2008). The microstructural random field is modeled as a spatially homogeneous (stationary) random field based on the Wishart or matrix-variate Gamma probability model and results in macroscopic elasticity tensors only constrained by the positive-definiteness property.

We remark here that the present work and the previous work (Das 2008; Das and Ghanem 2009) are strikingly different from many other works in the literature on stochastic upscaling or homogenization in one particular aspect. Instead of computing the optimal size of an RVE, our objective is to deduce a stochastic model that is consistent with a *specified* size of a small volume element V of the heterogeneous material. The size of V is often application-specific and reflects the associated experimental and financial constraints (Liu et al. 2009). From this perspective, the present approach bears certain similarities to another approach proposed earlier by Drugan and Willis, where the ensemble average of a response of interest is captured accurately at the cost of increased variability in the statistical features of the actual random response (Drugan and Willis 1996; Gusev 1997). While their approach is originally developed by assuming that the material volume element V is cut from an infinite body so that an explicit analytical solution can be obtained, the idea of capturing the mean response at the cost of increased variability is one of the motivation factors for proposing the current work. As highlighted earlier, the energy-based lower and upper bounds on $\mathbf{C}_{\text{ep}}$ in the context of our work will account for the issues related to the boundary value problem as encountered in practice.

To properly exploit the features of our approach, the size of V should, therefore, be smaller than the size of an RVE in the classical sense (Hill 1963). However, even if the size of V is bigger than the size of such an RVE, the mathematical formulation developed for constructing the probability model of $\mathbf{C}_{\text{ep}}$ in the current work or that of the constitutive elasticity matrix in the previous work (Das 2008; Das and Ghanem 2009) is still valid since the two bounds will converge to each other and the resulting probability distribution function will tend to a delta function (and therefore no stochastic analysis of the associated mechanical system would be necessary).

It should be noted that within the conventional framework, the typical objective of determining these energy-based bounds is to derive the appropriate bounds for the conventional parameters, e.g., Young's modulus and Poisson's ratio (Hill 1963; Hashin and Shtrikman 1963; Hazanov and Huet 1994; Torquato 2002). The effects of increasing the size of the material volume element V on the proximity between the two bounds for the linear and nonlinear material behaviors have been extensively studied in the context of several different applications (Ogden 1978; Nemat-Nasser and Hori 1999; Kanit et al. 2003; Ostoja-Starzewski 2008). Attempt has also been made to use these energy-based constitutive bounds in determining the bounds of the global response of a mechanical system (at the coarse scale or macroscale level) by having recourse to two different finite element (FE) formulations based on the minimum potential energy and minimum complementary energy principles (Ostoja-Starzewski and Wang 1999; Ostoja-Starzewski 1999). The difference between the global responses based on these two different FE formulations can be theoretically shown to be negligible, provided the FE mesh size of the mechanical system is sufficiently fine (de Veubeke 1964; Zienkiewicz 2001) (of course, the FE mesh size still must be larger than the size of the small material volume element V). Thus, any discrepancy arising in the global responses because of using two different FE formulations can be practically attributed entirely to the energy-based bounds on the constitutive material properties that are determined based on V. No attempt has been made until very recently in using these constitutive bounds to *quantify* the associated uncertainty stemming from the fine scale fluctuations over V. The first attempt was made in the context of the constitutive random elasticity matrix (Das 2008; Das and Ghanem 2009). Building on this recent work, the current work further investigates how such bounds, conditioned on the given size of V, can be incorporated in constructing the probability model of $\mathbf{C}_{\rm ep}$ to appropriately reflect the underlying fine scale material features (in the sense of Drugan and Willis), the boundary condition, and the operational loading condition.

We explain the proposed approach in Sect. 4 after introducing the basic notation and assumptions in Sect. 2 and reviewing the conventional parametric formulation of the inelastic material behavior in Sect. 3. The proposed approach is numerically illustrated in Sect. 5 and the conclusion is finally presented in Sect. 6.

2 Basic Notation and Assumptions

While the basic formulation in the current work is presented under the premise of infinitesimal strain theory, it can be adapted to large strain theory by considering the *work-conjugate pair* of stress and strain tensors in the total Lagrangian framework or an appropriate *objective stress rate* and the *rate of deformation* tensor in the updated Lagrangian framework.

Under the conjecture of the infinitesimal strain, all the strain tensors, which are typically used in the local theory of continuum mechanics, reduce to the *Cauchy strain tensor* or the (usual) *infinitesimal strain tensor*, $\boldsymbol{\varepsilon}$ (Lubarda 2002, Sect. 2.3.3). All the stress measures also reduce to the (usual) *Cauchy (true) stress tensor*, $\boldsymbol{\sigma}$. Both $\boldsymbol{\varepsilon}$ and $\boldsymbol{\sigma}$ are symmetric and second-order tensors. It may be recalled here that symmetry of the stress tensor ($\sigma_{ij} = \sigma_{ji}$) follows from the angular momentum equilibrium condition of an infinitesimal material volume element, provided there are no body moments and couple stress components. The symmetry of the strain tensor ($\varepsilon_{ij} = \varepsilon_{ji}$) follows from the conventional definition of the Green-Lagrange finite (or large) strain tensor. In component form, $\boldsymbol{\varepsilon}$ is given by the following strain-displacement relationship,

$$\varepsilon_{ij} = \frac{1}{2}\left(\frac{\partial u_i(\mathbf{x})}{\partial x_j} + \frac{\partial u_j(\mathbf{x})}{\partial x_i}\right), \tag{1}$$

where u_i's are the components of the displacement vector field $\mathbf{u}(\mathbf{x}) \equiv (u_1(\mathbf{x}), u_2(\mathbf{x}), u_3(\mathbf{x}))$, $\mathbf{x} = (x_1, x_2, x_3) \in \mathscr{B}$.

Within the local theory of continuum mechanics, consider the constitutive law that relates $\boldsymbol{\sigma}$ and $\boldsymbol{\varepsilon}$. Denote the fourth-order constitutive elasticity tensor in 3-D Euclidean space by $\mathbf{C}$ with its element being denoted by C_{ijkl}, $i, j, k, l = 1, \cdots, 3$. We can now state the relation between $\boldsymbol{\sigma}$ and $\boldsymbol{\varepsilon}$, but we first need to define the tensor product which would be handy to state the constitutive relation. We particularly need the contracted tensor product between a fourth-order tensor, say $\mathbf{B}$, and a second-order tensor, say $\mathbf{a}$. In the theoretical exposition discussed here, we will also need the contracted tensor product between two second-order tensors, say, $\mathbf{a}$ and $\mathbf{b}$. If the contracted tensor product is denoted by $:$, then $\mathbf{B} : \mathbf{a}$ is a second-order tensor (i.e., a matrix), and $\mathbf{a} : \mathbf{b}$ is a scalar. The (i, j)th elements of $\mathbf{B} : \mathbf{a}$ and $\mathbf{a} : \mathbf{b}$ in the d-dimensional Euclidean space are given by,

$$(\mathbf{B}:\mathbf{a})_{ij} = \sum_{l=1}^{d}\sum_{k=1}^{d} B_{ijkl}\, a_{lk}, \quad \text{and} \quad \mathbf{a}:\mathbf{b} = \sum_{i=1}^{d}\sum_{j=1}^{d} a_{ij}\, b_{ij}, \tag{2}$$

where B_{ijkl}, a_{ij}, and b_{ij} are, respectively, the elements of $\mathbf{B}$, $\mathbf{a}$, and $\mathbf{b}$.

Now we are in a position to state the generalized Hooke's law that relates the *local* stress and strain tensors,

$$\boldsymbol{\sigma}(\mathbf{x}) = \mathbf{C}(\mathbf{x}) : \boldsymbol{\varepsilon}(\mathbf{x}). \tag{3}$$

Here, the dependence on $\mathbf{x}$ of the local continuum elasticity tensor $\mathbf{C}$, the stress tensor $\boldsymbol{\sigma}$, and the strain tensor $\boldsymbol{\varepsilon}$ is emphasized to imply that these variables depend on the underlying microstructural fields of the heterogeneous material at $\mathbf{x} \in \mathscr{B}$. This dependence would be implied throughout the present work, but we will often suppress this dependence in the ensuing discussion for the sake of notational convenience.

Before proceeding further, we note that the tensorial notation is elegant for theoretical discussions, while the matrix and vector representations of the tensor-valued variables are more useful for implementation into the numerical algorithms. The matrix–vector representation is more practical (memory-saving) too when the symmetry of the tensor-valued variables are appropriately considered. For the sake of simplicity, we will denote both the tensor-valued variables and their matrix or vector representations by the same symbolic notation. Following this convention, the symmetric stress and strain tensors and their vector representations are noted below.

$$\boldsymbol{\sigma} = \begin{bmatrix} \sigma_{11} & \sigma_{12} & \sigma_{13} \\ \sigma_{12} & \sigma_{22} & \sigma_{23} \\ \sigma_{13} & \sigma_{23} & \sigma_{33} \end{bmatrix} \rightarrow \boldsymbol{\sigma} = \begin{Bmatrix} \sigma_{11} \\ \sigma_{22} \\ \sigma_{33} \\ \sigma_{12} \\ \sigma_{13} \\ \sigma_{23} \end{Bmatrix}, \quad \text{and} \quad \boldsymbol{\varepsilon} = \begin{bmatrix} \varepsilon_{11} & \varepsilon_{12} & \varepsilon_{13} \\ \varepsilon_{12} & \varepsilon_{22} & \varepsilon_{23} \\ \varepsilon_{13} & \varepsilon_{23} & \varepsilon_{33} \end{bmatrix} \rightarrow \boldsymbol{\varepsilon} = \begin{Bmatrix} \varepsilon_{11} \\ \varepsilon_{22} \\ \varepsilon_{33} \\ 2\varepsilon_{12} \\ 2\varepsilon_{13} \\ 2\varepsilon_{23} \end{Bmatrix}. \tag{4}$$

Here, in vector notation (also called Voigt notation), the shear strain components refer to the *engineering shears* which are twice the respective tensorial shear components. Emphasizing this distinction is necessary to obtain the consistent results. For example, consider the state of pure elastic shear described by,

$$\boldsymbol{\sigma} = \begin{bmatrix} 0 & \sigma_{12} & \sigma_{13} \\ \sigma_{12} & 0 & \sigma_{23} \\ \sigma_{13} & \sigma_{23} & 0 \end{bmatrix}, \quad \text{and} \quad \boldsymbol{\varepsilon} = \begin{bmatrix} 0 & \varepsilon_{12} & \varepsilon_{13} \\ \varepsilon_{12} & 0 & \varepsilon_{23} \\ \varepsilon_{13} & \varepsilon_{23} & 0 \end{bmatrix}. \tag{5}$$

The strain energy density is given by the contracted tensor product between $\boldsymbol{\varepsilon}$ and $\boldsymbol{\sigma}$,

$$\frac{1}{2}\boldsymbol{\varepsilon}:\boldsymbol{\sigma} = \frac{1}{2}\begin{bmatrix} 0 & \varepsilon_{12} & \varepsilon_{13} \\ \varepsilon_{12} & 0 & \varepsilon_{23} \\ \varepsilon_{13} & \varepsilon_{23} & 0 \end{bmatrix} : \begin{bmatrix} 0 & \sigma_{12} & \sigma_{13} \\ \sigma_{12} & 0 & \sigma_{23} \\ \sigma_{13} & \sigma_{23} & 0 \end{bmatrix} = (\varepsilon_{12}\sigma_{12}+\varepsilon_{13}\sigma_{13}+\varepsilon_{23}\sigma_{23}). \quad (6)$$

On the other hand, the strain energy density in terms of the corresponding vector representations of $\boldsymbol{\varepsilon}$ and $\boldsymbol{\sigma}$ is defined by the dot product or the scalar product as shown below,

$$\frac{1}{2}\boldsymbol{\varepsilon}^T\boldsymbol{\sigma} = [0 \quad 0 \quad 0 \quad 2\varepsilon_{12} \quad 2\varepsilon_{13} \quad 2\varepsilon_{23}]\begin{Bmatrix} 0 \\ 0 \\ 0 \\ \sigma_{12} \\ \sigma_{13} \\ \sigma_{23} \end{Bmatrix} = (\varepsilon_{12}\sigma_{12} + \varepsilon_{13}\sigma_{13} + \varepsilon_{23}\sigma_{23}), \quad (7)$$

where $(\cdot)^T$ denotes the transpose operator. It should be clear that for any stress and strain state (not just the pure elastic shear state), we have,

$$\boldsymbol{\varepsilon}:\boldsymbol{\sigma} = \boldsymbol{\varepsilon}^T\boldsymbol{\sigma}, \quad (8)$$

where the left-hand side (lhs) contains the tensor-valued variables and the right-hand side (rhs) is expressed in Voigt notation. In the following discussion, we will use both the tensorial notation and the Voigt notation for $\boldsymbol{\varepsilon}$ and $\boldsymbol{\sigma}$. It will be clear from the context whether we are referring to the tensor-valued variables or their vector representations.

Let us next consider the elasticity tensor $\mathbf{C}$ consisting of 81 components. The symmetry of the stress tensor implies that $C_{ijkl} = C_{jikl}$ while symmetry of the strain tensor implies that $C_{ijkl} = C_{ijlk}$. This reduces 81 components to 36 independent components. Assume now that a strain energy density function exists. The strain energy density is given by,

$$\frac{1}{2}\boldsymbol{\varepsilon}:\boldsymbol{\sigma} = \frac{1}{2}\boldsymbol{\varepsilon}:\mathbf{C}:\boldsymbol{\varepsilon} = \frac{1}{2}\sum_{ijkl}\varepsilon_{kl}C_{ijkl}\varepsilon_{ij}, \quad (9)$$

which is a scalar-valued function *symmetric* in the strain components ε_{ij}'s. Hence, $\mathbf{C}$ must additionally have the symmetry $C_{ijkl} = C_{klij}$. This means that the number of independent components of $\mathbf{C}$ further reduces to 21. For the sake of notational simplicity, C_{ijkl} components can be re-indexed using the following rules to introduce a one-to-one mapping between the C_{ijkl} components of the fourth-order elasticity tensor $\mathbf{C}$ and the elements C_{pq} of the matrix representation of $\mathbf{C}$,

$$ij \text{ (or } kl) \leftrightarrow p \text{ (or } q) : 11 \leftrightarrow 1, \quad 22 \leftrightarrow 2, \quad 33 \leftrightarrow 3, \quad 12 \text{ or } 21 \leftrightarrow 4, \quad 13 \text{ or } 31 \leftrightarrow 5, \quad 23 \text{ or } 32 \leftrightarrow 6. \quad (10)$$

The matrix representation of $\mathbf{C}$ will be referred to as the constitutive elasticity matrix, and both of them will be denoted by $\mathbf{C}$ for notational simplicity. Based on the symmetry properties as discussed above and the re-indexing rules as just introduced, the elasticity matrix $\mathbf{C}$ can be written as follows by explicitly showing its elements C_{pq}s,

$$\mathbf{C} = \begin{bmatrix} C_{11} & C_{12} & C_{13} & C_{14} & C_{15} & C_{16} \\ & C_{22} & C_{23} & C_{24} & C_{25} & C_{26} \\ & & C_{33} & C_{34} & C_{35} & C_{36} \\ & & & C_{44} & C_{45} & C_{46} \\ & \text{Sym} & & & C_{55} & C_{56} \\ & & & & & C_{66} \end{bmatrix}, \tag{11}$$

where the symmetry of $\mathbf{C}$ is emphasized. Akin to the stress and strain variables, we will use both the tensorial notation and the matrix representation of $\mathbf{C}$ in the following discussion. The context will make it clear whether the tensorial notation or the matrix representation is in use. From the consideration of the uniqueness and stability of the solution of the boundary value problem, it is required that $\mathbf{C}$ be a positive-definite matrix (Ogden 1984; Lubarda 2002). The positive-definiteness of $\mathbf{C}$ ensures that the strain energy in (9) is positive for all non-zero strain $\boldsymbol{\varepsilon}$.

It should be clear by now that the contracted tensor product between the symmetric second-order stress and strain tensors is equivalent to the scalar product between their corresponding vector representations. The contraction between the fourth-order constitutive elasticity tensor and the symmetric second-order strain tensor can be similarly inferred to be equivalent to the matrix-vector multiplication of their respective matrix and vector representations as defined by (11) and (4). The generalized Hooke's law in (3) then assumes the following equivalent form,

$$\boldsymbol{\sigma} = \mathbf{C} : \boldsymbol{\varepsilon} \quad \Longleftrightarrow \quad \boldsymbol{\sigma} = \mathbf{C}\boldsymbol{\varepsilon} \tag{12}$$

where the variables in the lhs of $\Longleftrightarrow$ need to be understood in the tensorial notation and the variables in the rhs of $\Longleftrightarrow$ are their matrix and vector counterparts as defined by (4) and (11).

Finally, let us introduce the space of the symmetric second-order tensors in the d-dimensional Euclidean space (i.e., the space of the symmetric $d \times d$ matrices) by,

$$\mathbb{S}_d = \left\{ \mathbf{a} \text{ is } d \times d \text{ matrix} : \mathbf{a} = \mathbf{a}^T \right\} \tag{13}$$

which we will use frequently in the following exposition. When expressed in the Voigt notation, $\mathbf{a}$ belongs to $\mathbb{R}^{d(d+1)/2}$. Clearly, in the tensorial notation, the stress and strain are $\mathbb{S}_3$-valued variables, and in the Voigt notation they are $\mathbb{R}^{3(3+1)/2} \equiv \mathbb{R}^6$-valued variables.

3 Parametric Formulation of Material Plasticity

In the current literature, two kinds of plasticity formulations are typically found: a stress-space formulation and a strain-space formulation. In the stress-space formulation, stress is taken as the independent variable and strain as the dependent variable. The role is reversed in the strain-space formulation. Under certain general conditions, it can be shown that the strain-space and stress-space formulations are equivalent to each other provided the corresponding model parameters are appropriately interrelated (Yoder 1980). Both the formulations have been proven useful in modeling plastic behavior across a wide range of material characteristics. For the discussion purpose, we would consider the stress-space formulation in the following development. The end product of our proposed work, however, does not differentiate between the strain-space and stress-space formulations. In fact, it has more resemblance to the underlying concept of the strain-space formulation than that of the stress-space formulation. The stress increment is treated as a function of the strain and its increment.

A review of plasticity theory is next presented to provide context and to motivate the proposed random matrix formulation by accentuating phenomenological features of the conventional plasticity theory (Hill 1950; Simo and Hughes 1998; Dunne and Petrinic 2005).

When a material is plastically deformed, then total strain $\boldsymbol{\varepsilon}$ is axiomatically decomposed into the elastic strain $\boldsymbol{\varepsilon}^{\mathrm{e}}$ and the plastic strain $\boldsymbol{\varepsilon}^{\mathrm{p}}$,

$$\boldsymbol{\varepsilon} = \boldsymbol{\varepsilon}^{\mathrm{e}} + \boldsymbol{\varepsilon}^{\mathrm{p}}. \tag{14}$$

The stress variable $\boldsymbol{\sigma}$ is explicitly related only to the elastic strain part $\boldsymbol{\varepsilon}^{\mathrm{e}}$ via the generalized Hooke's law,

$$\boldsymbol{\sigma} = \mathbf{C} : \boldsymbol{\varepsilon}^{\mathrm{e}}. \tag{15}$$

In practice, one typically controls the total strain $\boldsymbol{\varepsilon}$, i.e., from known $\boldsymbol{\varepsilon}$, one seeks to determine $\boldsymbol{\sigma}$. The current state of $\boldsymbol{\sigma}$ can be determined using (14) and (15), $\boldsymbol{\sigma} = \mathbf{C} : \boldsymbol{\varepsilon}^{\mathrm{e}} = \mathbf{C} : (\boldsymbol{\varepsilon} - \boldsymbol{\varepsilon}^{\mathrm{p}})$, provided one can compute the plastic strain component $\boldsymbol{\varepsilon}^{\mathrm{p}}$. If a mechanical system is stress-controlled (i.e., from known $\boldsymbol{\sigma}$, one aims to find out $\boldsymbol{\varepsilon}$), then $\boldsymbol{\varepsilon}^{\mathrm{e}}$ can be computed using (15), and the current total strain $\boldsymbol{\varepsilon}$ can only be computed (via (14)), provided one can also determine $\boldsymbol{\varepsilon}^{\mathrm{p}}$. A way to compute $\boldsymbol{\varepsilon}^{\mathrm{p}}$ must be devised in either case.

In the current plasticity theory, the determination of $\boldsymbol{\varepsilon}^{\mathrm{p}}$ is fundamentally based on three *phenomenological* constitutive models that typically relate the stress increment $\mathrm{d}\boldsymbol{\sigma}$ to the plastic strain increment $\mathrm{d}\boldsymbol{\varepsilon}^{\mathrm{p}}$ via three essential notions: a *yield criterion*, a *flow rule*, and a *hardening law*. The yield criterion determines an onset of the plastic state (i.e., yielding impends) from the current state of the material (i.e., stress $\boldsymbol{\sigma}$ in the stress-space formulation). The flow rule controls the "direction" of the plastic flow (i.e., the direction of the plastic strain increment $\mathrm{d}\boldsymbol{\varepsilon}^{p}$). The hardening law determines how the yield criterion is altered by increasing the stress beyond the material's initial yield strength.

Under plastic deformation, yielding continues until unloading. Yielding is accompanied by the plastic flow and hardening of the material both of which follow only after yielding emanates. However, it would be easier, for the sake of clarity, to discuss the flow rule and the hardening law before we define the yield criterion. We will describe these notions by concise and general mathematical forms which will be followed by providing more specific model-form examples for better understanding of the discussion.

The flow rule is stated by introducing a $\mathbb{S}_3$-valued function $(\boldsymbol{\sigma}, \mathbf{q}) \longmapsto \mathbf{r}(\boldsymbol{\sigma}, \mathbf{q})$ which may be a non-smooth function. Here, $\mathbf{q}$ represents a set of internal scalar-valued or/and tensor-valued (stress-like and strain-like) variables called hardening variables to be explained more clearly while stating the hardening law. The direction of the plastic flow is then postulated by,

$$\mathrm{d}\boldsymbol{\varepsilon}^p = \mathrm{d}\lambda \, \mathbf{r}(\boldsymbol{\sigma}, \mathbf{q}), \tag{16}$$

where $\mathrm{d}\lambda$ is scalar-valued variable that determines the magnitude of the plastic strain increment $\mathrm{d}\boldsymbol{\varepsilon}^p$ and is called the *plastic multiplier* or *consistency parameter*. The direction of $\mathrm{d}\boldsymbol{\varepsilon}^p$ is dictated by $\mathbf{r}(\boldsymbol{\sigma}, \mathbf{q})$ which is conjectured by introducing the notion of another scalar-valued function called *plastic potential* $(\boldsymbol{\sigma}, \mathbf{q}) \longmapsto F(\boldsymbol{\sigma}, \mathbf{q})$. The function $\mathbf{r}$ is thus assumed to be given by,

$$\mathbf{r}(\boldsymbol{\sigma}, \mathbf{q}) = \frac{\partial F(\boldsymbol{\sigma}, \mathbf{q})}{\partial \boldsymbol{\sigma}}, \tag{17}$$

where $\partial F(\boldsymbol{\sigma}, \mathbf{q})/\partial \boldsymbol{\sigma}$ is the $\mathbb{S}_3$-valued gradient of F with respect to the components, σ_{ij}'s, of $\boldsymbol{\sigma}$.

The set of the hardening variables $\mathbf{q}$ as mentioned above evolves with the current plastic state of the material. The evolution of $\mathbf{q}$ is governed by the hardening law which is parameterized by a set of material parameters called the *hardening parameters*. Let us denote all the hardening parameters (say, M hardening parameters) together by $\mathbf{A} \in \mathbb{R}^M$. The $\mathbb{R}^M$-valued hardening parameter $\mathbf{A}$ is calibrated by fitting a functional form of the hardening law to a set of experimental data (Ohno and Wang 1993; Zhao and Lee 2001). At this stage, the hardening law is assumed to be a function of $\mathbf{q}$ and $\mathrm{d}\boldsymbol{\varepsilon}^p$, which is parameterized by $\mathbf{A}$. Given $\mathbf{A}$, this hardening law, $(\mathbf{q}, \mathrm{d}\boldsymbol{\varepsilon}^p \mid \mathbf{A}) \longmapsto \mathrm{d}\mathbf{q}$, is so chosen that it can be eventually cast in the following form,

$$\mathrm{d}\mathbf{q} = \mathrm{d}\lambda \; \mathbf{h}(\boldsymbol{\sigma}, \mathbf{q} \mid \mathbf{A}), \tag{18}$$

when $\mathrm{d}\boldsymbol{\varepsilon}^p$ is expressed in terms of $(\boldsymbol{\sigma}, \mathbf{q})$ using the flow rule (16).

Now that the flow rule and the hardening law are summarized, we can define the yield criterion. The yield criterion is defined by introducing a scalar-valued function, $(\boldsymbol{\sigma}, \mathbf{q}) \longmapsto f(\boldsymbol{\sigma}, \mathbf{q})$, called the *yield function*, as follows,

$$\begin{aligned} f(\boldsymbol{\sigma}, \mathbf{q}) < 0 \quad &\Rightarrow \quad \text{Elastic state}, \\ f(\boldsymbol{\sigma}, \mathbf{q}) = 0 \quad &\Rightarrow \quad \text{Plastic state}. \end{aligned} \tag{19}$$

The case $f(\boldsymbol{\sigma}, \mathbf{q}) > 0$ is not allowed. Staring from some elastic state $f(\boldsymbol{\sigma}, \mathbf{q}) < 0$, the material reaches the plastic state $f(\boldsymbol{\sigma}, \mathbf{q}) = 0$ thereby starting the plastic flow. If there is unloading, $f(\boldsymbol{\sigma}, \mathbf{q})$ again becomes negative indicating that the elastic constitutive property of the material should be invoked. In the plastic state, since $\boldsymbol{\sigma}$ and $\mathbf{q}$ are interrelated by the hardening law (18), the yield function f can be *implicitly* viewed as function of $\boldsymbol{\sigma}$ only. When the surface $f = 0$ is viewed in $\mathbb{R}^6$ whose principal cartesian coordinate system labels the six stress components σ_{ij}s, then the surface defined by $f = 0$ in this six-dimensional stress space is referred to as the *yield surface*. During the plastic flow, changes in $\boldsymbol{\sigma}$ and $\mathbf{q}$ are *assumed* to alter the shape or/and location of the yield surface in the six-dimensional stress space so that $\boldsymbol{\sigma} \in \mathbb{R}^6$ always remains on the yield surface satisfying the condition $f(\boldsymbol{\sigma}, \mathbf{q}) = 0$, which is known as the *consistency condition*. The transformation and translation of the yield surface in the six-dimensional stress space are governed by the hardening law. When the plastic potential F in (17) is also treated as an implicit function of $\boldsymbol{\sigma}$ and plotted in the six-dimensional stress space, then the gradient $\partial F(\boldsymbol{\sigma}, \mathbf{q})/\partial \boldsymbol{\sigma}$ is a $\mathbb{R}^6$-valued variable and defines the direction of $\mathrm{d}\boldsymbol{\varepsilon}^p$ as conjectured by the flow rule. If $F = f$, then the flow rule is termed as *associative* to indicate that $\mathrm{d}\boldsymbol{\varepsilon}^p$ is associated with the yield function. On the other hand, when F is different from f, then the flow rule is referred to as *non-associative*.

The associative flow rules are experimentally found to be suitable for ductile metals (Hill 1950). In this case, (16) and (17) imply that $\mathrm{d}\boldsymbol{\varepsilon}^p$ is co-directional with the outward normal to the yield surface (often referred to as *normality hypothesis* in the literature). The normality condition can be shown to follow from Il'yushin's work postulate (Hill 1968, 1972; Lubarda 2002). In the literature, it is often taken granted that the associative flow rule and the normality hypothesis are equivalent. The normality condition, however, may not follow from the associative flow rules in certain applications without special care in defining the elastic and inelastic strain parts of the total strain (Houlsby 1981, Chap. 5). In the context of the present work, the transition of the normality hypothesis from the microlevel to the macrolevel is perhaps one of the concerns. Based on the numerical study, it is found that if the normality hypothesis applies at the microlevel, then it may not hold at the macrolevel (Ostoja-Starzewski 2002, 2008). However, it is theoretically proven in the literature (Hill 1972; Hill and Rice 1973; Lubarda 2002) that if the associative flow rule involving a work-conjugate pair of stress and strain tensors is asserted at the microlevel, then it is transmitted *unchanged* to the macrolevel provided the macrolevel variables are appropriately defined based on a few basic volume averaged microlevel variables (Nemat-Nasser 1999). Based on this theoretical result, we will assume in the following discussion that the normality hypothesis at the microlevel is appropriately transferred to the macrolevel. We also note in passing that while the non-associative flow rules have been demonstrated to be useful in certain applications (Miller and Cheatham 1972; Rudnicki and Rice 1975; Senseny and Pfeifle 1983), their mathematical validity has been disputed in the literature on the physical grounds (Molenkamp and Van Ommen 1987; Sandler and Pucik 2001).

In Sect. 2, we have commented on the symmetry of the constitutive elasticity matrix $\mathbf{C}$ from a macroscopic viewpoint. We now elaborate its transmission from the microlevel to the macrolevel perspective since the focus of the present work is to characterize the coarse scale constitutive matrix from the fine scale material features. If the constitutive elasticity matrices at the microlevel or constituent level are symmetric, then it can be shown that the resulting macrolevel or coarse scale constitutive elasticity matrix **C** is also symmetric (Hill 1967; Lubarda 2002). Based on this symmetry of **C** and the unchanged transmissible capability of the associative flow rule from the microlevel to the macrolevel, it can be further shown, under fairly general conditions, that the constitutive tangential elastoplastic matrix $\mathbf{C}_{\text{ep}}$ is also symmetric (Simo and Hughes 1998; Lubarda 2002). By assuming Drucker's postulate (Drucker 1988) and following the literature, it can also be concluded that $\mathbf{C}_{\text{ep}}$ is a positive-definite matrix in the strain-hardening phase from the consideration of the *incremental stability criteria* provided the stress magnitudes, that the material is subjected to, are not sufficiently high (Hill 1967, 1968). This is practically valid for almost all applications involving metals, which is the focus of the present work, provided the portion of the stress–strain curve beyond the point of maximum load, which is followed by the phenomenon of necking instability, is excluded. A few rare applications involving, for instance, the material being subjected to the tensile test under sufficiently high fluid or superposed pressure (as might happen for a submarine under deep water being subjected to tensile stress caused by underwater detonation), may not guarantee the positive-definiteness of $\mathbf{C}_{\text{ep}}$ (Hill 1968; Lubarda 2002). Such problems are, therefore, naturally precluded from the domain of applicability of our approach. We will rely on the two properties (symmetry and positive-definiteness) of $\mathbf{C}_{\text{ep}}$ to construct its probability measure in Sect. 4.

Let us now derive the expression of a constitutive tangential elastoplastic matrix by considering specific model-forms of the hardening law and the yield surface to explicate the above abstract discussion in rather simplified manner. Consider, for this purpose, the combined isotropic and kinematic hardening law. The isotropic hardening reflects the experimentally observed fact that many metals, when deformed plastically, harden. The kinematic hardening implies the Bauschinger effect typically observed under cyclic loading in polycrystalline aggregate (compact aggregate of several crystal grains of varying shapes and orientations) as metals are generally found and used in practical applications. The Bauschinger effect alludes to the phenomenon of translation of the center of the yield surface in the direction of the plastic flow.

One of the popularly used models that governs the isotropic hardening variable, r, is given by,

$$\mathrm{d}r(p) = b\,(Q - r(p))\,\mathrm{d}p, \tag{20}$$

in which b and Q are scalar-valued material parameters, p represents the accumulated effective plastic strain, and the increment $\mathrm{d}p$ in the effective plastic strain is defined by,

$$\mathrm{d}p := \left(\frac{2}{3}\,\mathrm{d}\boldsymbol{\varepsilon}^{p} : \mathrm{d}\boldsymbol{\varepsilon}^{p}\right)^{1/2}. \tag{21}$$

It should be noted that b and Q are hardening parameters that go to form a part of **A** as defined earlier, while r is a hardening variable that goes to form a part of **q**.

The isotropic hardening in (20) can be expressed in a form analogous to (18) if one uses the flow rule in (16) and the definition of $\mathrm{d}p$ in (21),

$$\mathrm{d}r(p) = \mathrm{d}\lambda\, h_1(\boldsymbol{\sigma}, r(p) \mid b, Q), \tag{22}$$

where $h_1(\boldsymbol{\sigma}, r(p) \mid b, Q) = [b\,(Q - r(p))\,\{(2/3)\,\mathbf{r}(\boldsymbol{\sigma}, r(p)) : \mathbf{r}(\boldsymbol{\sigma}, r(p))\}^{1/2}]$ and the choice of the flow rule defines $\mathbf{r}(\boldsymbol{\sigma}, r(p))$. The function $h_1(\cdot)$ constitutes a part of $\mathbf{h}$ in (18).

The following special case of the Armstrong and Frederick kinematic hardening rule is often used in practice,

$$\mathrm{d}\boldsymbol{\alpha} = \frac{2}{3} c\, \mathrm{d}\boldsymbol{\varepsilon}^p - \gamma\, \boldsymbol{\alpha}\, \mathrm{d}p, \tag{23}$$

where c and γ are further material parameters, and $\boldsymbol{\alpha}$ is the kinematic hardening variable, often called *back stress*, that determines the magnitude of translation of the center of the yield surface resulting from the Bauschinger effect. The material parameters c and γ comprise a part of $\mathbf{A}$, and $\boldsymbol{\alpha}$ constitutes a part of $\mathbf{q}$. We can also form $h_2(\boldsymbol{\sigma}, \boldsymbol{\alpha} \mid c, \gamma)$ which, along with h_1, eventually composes $\mathbf{h}$.

The associative flow rule implies that the plastic strain $\boldsymbol{\varepsilon}^p$ in (14) is governed by,

$$\mathrm{d}\boldsymbol{\varepsilon}^p = \mathrm{d}\lambda\, \frac{\partial f}{\partial \boldsymbol{\sigma}}, \tag{24}$$

where $\mathrm{d}\lambda$ is the plastic multiplier and f represents the scalar-valued yield function explicitly defined later in this section. The second-order tensor $\partial f / \partial \boldsymbol{\sigma}$ in (24), whose (i, j)th element is given by $\partial f / \partial \sigma_{ij}$, represents the direction of $\mathrm{d}\boldsymbol{\varepsilon}^p$ according to the associative flow rule.

The yield function f can be defined now as,

$$f = \sigma_e - r(p) - \sigma_y, \tag{25}$$

in which σ_y represents the initial yield strength which is a material parameter and σ_e represents the effective stress, which is a function of the deviatoric stress parts of $\boldsymbol{\sigma}$ and $\boldsymbol{\alpha}$, as given below,

$$\sigma_e := \left[\frac{3}{2}(\boldsymbol{\sigma}' - \boldsymbol{\alpha}') : (\boldsymbol{\sigma}' - \boldsymbol{\alpha}')\right]^{\frac{1}{2}}. \tag{26}$$

The deviatoric stress $\boldsymbol{\sigma}'$ is defined by,

$$\boldsymbol{\sigma}' = \boldsymbol{\sigma} - \frac{1}{3}\mathrm{Tr}(\boldsymbol{\sigma})\mathbf{I}, \tag{27}$$

in which $\mathrm{Tr}(\cdot)$ represents the trace operator of a second-order tensor (i.e., a matrix) and $\mathbf{I}$ is the second-order identity tensor or identity matrix. The deviatoric part of $\boldsymbol{\alpha}$ can be similarly defined.

For von Mises material, the plastic multiplier $\mathrm{d}\lambda$ in (24) turns out to be the increment in effective plastic strain $\mathrm{d}p$ implying (Dunne and Petrinic 2005),

$$\mathrm{d}p = \mathrm{d}\lambda. \tag{28}$$

For von Mises material and using (15), (20)–(25) along with the consistency condition and noting that $\partial f/\partial \boldsymbol{\alpha} = -(\partial f/\partial \boldsymbol{\sigma})$ for yield function (25), it can be shown for the combined isotropic and kinematic hardening in matrix–vector notation (not tensorial notation) that (Dunne and Petrinic 2005)

$$\mathrm{d}\boldsymbol{\sigma} = \mathbf{C}_{\mathrm{ep}}\, \mathrm{d}\boldsymbol{\varepsilon}, \tag{29}$$

in which the constitutive tangential elastoplastic matrix $\mathbf{C}_{\mathrm{ep}}$ is given by,

$$\mathbf{C}_{\mathrm{ep}} = \mathbf{C}\left(\mathbf{I} - \frac{\partial f}{\partial \boldsymbol{\sigma}} \frac{\dfrac{\partial f}{\partial \boldsymbol{\sigma}}}{\dfrac{\partial f}{\partial \boldsymbol{\sigma}}\mathbf{C}\dfrac{\partial f}{\partial \boldsymbol{\sigma}} - \gamma \dfrac{\partial f}{\partial \boldsymbol{\sigma}}\boldsymbol{\alpha} + \dfrac{2}{3} c \dfrac{\partial f}{\partial \boldsymbol{\sigma}}\dfrac{\partial f}{\partial \boldsymbol{\sigma}} + b(Q - r(p))}\right). \tag{30}$$

Here $\mathbf{I}$ is an identity matrix and $(\partial f/\partial \boldsymbol{\sigma})$ must be interpreted in Voigt notation as explained for $\boldsymbol{\sigma}$ in (4). The constitutive matrix $\mathbf{C}_{\mathrm{ep}}$ can be regarded as a generalized tangent modulus. It is symmetric and positive-definite matrix for the reasons as explained earlier. It should be noted that (30) is derived for von Mises material because (28) is used. For unloading from a plastic state ($f = 0$ and $\mathrm{d}f < 0$) or when yielding is yet to take place ($f < 0$), the second term within the parentheses in (30) is set to the zero matrix so that $\mathbf{C}_{\mathrm{ep}} = \mathbf{C}$.

4 Nonparametric Modeling of $\mathbf{C}_{\mathrm{ep}}$

It should be noted that the expression of $\mathbf{C}_{\mathrm{ep}}$ as shown in (30), in general, is not analytically available except for few simple hardening models. Within a FE formulation, it is often required to invoke some suitable nonlinear optimization technique and differential equation solver to construct $\mathbf{C}_{\mathrm{ep}}$ at the integration or quadrature points. Similar to the context of $\mathbf{C}$ as explained below (3), we also emphasize here that $\mathbf{C}_{\mathrm{ep}}$ depends on the underlying microstructural fields at the macroscopic point $\mathbf{x} \in \mathscr{B}$. In other words, $\mathbf{C}_{\mathrm{ep}}(\mathbf{x}_i)$, at each integration point $\mathbf{x}_i \in \mathscr{B}$, should ideally be constructed based on the material properties and the morphological features of the microconstituents included in a very small material volume element $V(\mathbf{x}_i)$ around each $\mathbf{x}_i$, where the actual size of $V(\mathbf{x}_i)$ is application-specific and depends on many other practical constraints as briefly mentioned in Sect. 1. This ideal approach would be equivalent to tackling the problem at the microscale level rather than at the macroscale level. Such an attempt is clearly a formidable undertaking due to the enormous computational overhead, experimental constraints, and the limited financial resource. It demands an absolute access to the accurate fine scale details of the heterogeneous materials over the entire domain $\mathscr{B}$ or at least around each and every integration points. This amount of information is seldom available in practice. In addition, such fine scale details would be too intricate to be useful for the intended purpose of characterizing the coarse scale material properties. Motivated by this observation, a few stochastic upscaling schemes have been proposed in the recent past to compute the coarse scale representation of a *specified* fine scale model. Specifically, approaches based on spectral decomposition of the microstructural random field (Jardak and Ghanem 2004) and minimization of the relative entropy between the specified fine scale model and its coarse scale representation (Koutsourelakis 2007; Arnst and Ghanem 2008) have been presented. Both the approaches are hinged on the assumption that complete probabilistic descriptions of the fine scale features are available over the entire spatial domain $\mathscr{B}$.

The difficulty of having limited or partial information at the fine scale level can be tackled by invoking the MaxEnt principle. The MaxEnt principle yields an *unique* probability model that is consistent with the limited information, while it concurrently remains least committal to all the unavailable information. Within the conventional parametric formulation, a few MaxEnt-based statistical approaches have been proposed in the literature to estimate the probability models of the fine scale morphological features based on partial information (Zhu et al. 1998; Koutsourelakis 2006; Sankaran and Zabaras 2007). However, the macroscopic material properties derived from the digital samples of the fine scale morphological random fields are likely to violate the energy-based constitutive bounds that encapsulate the underlying physics of the problem under consideration. Standard probabilistic models (e.g., Gaussian distribution and log-normal distribution) are also calibrated to fit the probability models of the constitutive material properties based on simulated results that are obtained by carrying out micromechanical analyses on several realizations of the small material volume element V (Yin et al. 2008). These calibrated probability models certainly violate the corresponding energy-based theoretical bounds. Violation of these physics-based constitutive bounds induces errors, the impact of which is difficult to quantify on the simulated behavior.

To respect these physics-based bounds, we propose to construct the probability density function (pdf) of the random $\mathbf{C}_{\text{ep}}$ that is supported only over the energy-based theoretical bounds. It should be noted that $\mathbf{C}_{\text{ep}}$ as shown in (30), or in a general case, requires the knowledge of the conventional local system parameters, e.g., Young's modulus, Poisson ratio, Q, c and γ. In the present work, we characterize the random $\mathbf{C}_{\text{ep}}$ within the RMT-based nonparametric formulation because of the reasons already highlighted in Sect. 1. In the stochastic mechanics literature, the probabilistic descriptions of the conventional local parameters are often characterized independently rather than jointly. This is hardly true in a practical context. The RMT-based characterization of $\mathbf{C}_{\text{ep}}$ as presented in this work will alleviate this shortcoming to quite an extent, since the resulting pdf prescribes a joint probabilistic description of the elements of $\mathbf{C}_{\text{ep}}$ by appropriately preserving the essential

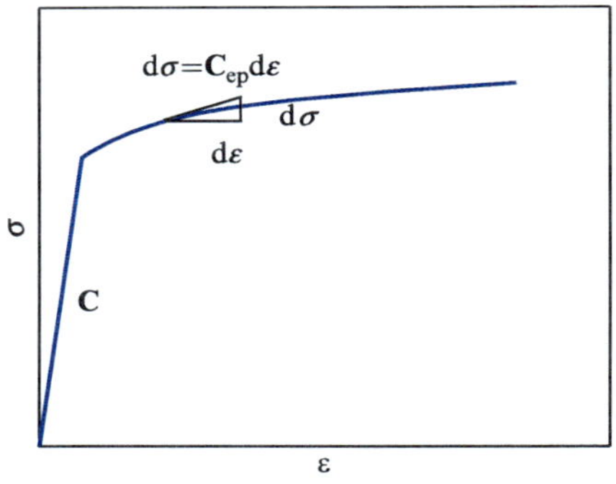

Fig. 1 Schematic illustration of active regions for **C** and $\mathbf{C}_{\rm ep}$

symmetry and positive-definiteness properties of $\mathbf{C}_{\rm ep}$. Characterization of $\mathbf{C}_{\rm ep}$ within this RMT-based formalism entirely bypasses the need for modeling several conventional parameters.

In the nonlinear inelastic stress–strain regime (Fig. 1), $\mathbf{C}_{\rm ep}$ evolves with the current state of the material. We propose in the present work to construct $\mathbf{C}_{\rm ep}$ as a function of effective total strain, ε_e, which we introduce below,

$$\varepsilon_e := \left(\frac{2}{3}\boldsymbol{\varepsilon} : \boldsymbol{\varepsilon}\right)^{1/2}. \tag{31}$$

Once the effective stress σ_e at a material point reaches the yield strength σ_y, yielding commences and $\mathbf{C}_{\rm ep}$ starts playing the role of $\mathbf{C}$. The factor $(2/3)$ and square-root in (31) are simply used by following the definition of the effective plastic strain p in (21). Since $\boldsymbol{\varepsilon}$ is tensor-valued variable, (31) is introduced only to associate the current tensor-valued strain state with a scalar-valued equivalent state. Clearly, this is based on an assumption that different strain states, which have the same effective total strain ε_e, will yield the same $\mathbf{C}_{\rm ep}$.

Following the literature (Jiang et al. 2001; Ostoja-Starzewski 2008), if it is presumed that the material behaves *locally* elastic over an *infinitesimal* region around each and every point on the strain-hardening elastoplastic nonlinear stress-strain surface, then it can be shown that $\mathbf{C}_{\rm ep}$ is bounded by two symmetric and positive-definite deterministic matrices C_l and C_u that also evolve concurrently with the current state of the material,

$$\mathbf{0} < C_l(\varepsilon_e) < \mathbf{C}_{\rm ep}(\varepsilon_e) < C_u(\varepsilon_e) \quad \text{a.s.}, \tag{32}$$

where the effective total strain ε_e reflects the current state of the material, $\mathbf{0}$ is a zero matrix, and the inequalities should be interpreted in the positive-definite sense (for instance, $C_l(\varepsilon_e) < \mathbf{C}_{\rm ep}(\varepsilon_e)$ implies that $(\mathbf{C}_{\rm ep}(\varepsilon_e) - C_l(\varepsilon_e))$ is positive-definite matrix a.s.). Finally, a.s. (almost surely, i.e., with probability one) should be interpreted with respect to (w.r.t.) the joint probability measure of all the underlying microstructural random fields. The lower and upper bounds are computed as follows.

The small material volume element V is uniformly partitioned into a set of smaller specimens of size $V_{\text{sub}}^{(i)} < V$ such that $\bigcup_i V_{\text{sub}}^{(i)} = V$. Let us first consider the procedure to compute the lower bound. Starting from the zero strain and stress states, each smaller specimen, $V_{\text{sub}}^{(i)}$, is subjected to incremental static uniform boundary condition (SUBC) or traction control boundary condition of the following form: $\mathbf{dt}(\mathbf{y}) = \mathbf{d}\boldsymbol{\sigma}_o\, \mathbf{n}(\mathbf{y}), \forall \mathbf{y} \in \partial V_{\text{sub}}^{(i)}$. Here, $\mathbf{dt}(\mathbf{y})$ is the increment in the applied traction vector surface density, $\mathbf{d}\boldsymbol{\sigma}_o$ is the increment in the stress tensor that does not depend on $\mathbf{y}$, and $\mathbf{n}(\mathbf{y})$ denotes the unit vector normal to the boundary $\partial V_{\text{sub}}^{(i)}$ of $V_{\text{sub}}^{(i)}$ at $\mathbf{y}$. Since the microstructural properties of the microconstituents enclosed within V are known (experimentally identified), a micromechanical analysis is subsequently carried out to obtain the *volume averaged* (over $V_{\text{sub}}^{(i)}$) incremental stress tensor $\langle \mathbf{d}\boldsymbol{\sigma} \rangle_{V_{\text{sub}}^{(i)}}$ and strain tensor $\langle \mathbf{d}\boldsymbol{\varepsilon} \rangle_{V_{\text{sub}}^{(i)}}$, respectively, given by,

$$\langle \mathbf{d}\boldsymbol{\sigma} \rangle_{V_{\text{sub}}^{(i)}} = \frac{1}{V_{\text{sub}}^{(i)}} \int_{V_{\text{sub}}^{(i)}} \mathbf{d}\boldsymbol{\sigma}(\mathbf{y})\, \mathbf{dy}, \quad \text{and} \quad \langle \mathbf{d}\boldsymbol{\varepsilon} \rangle_{V_{\text{sub}}^{(i)}} = \frac{1}{V_{\text{sub}}^{(i)}} \int_{V_{\text{sub}}^{(i)}} \mathbf{d}\boldsymbol{\varepsilon}(\mathbf{y})\, \mathbf{dy}. \tag{33}$$

Based on the assumption that the perfect bonding exists between the different constituents of the materials, it can be shown that $\langle \mathbf{d}\boldsymbol{\sigma} \rangle_{V_{\text{sub}}^{(i)}} = \mathbf{d}\boldsymbol{\sigma}_o$ (Huet 1990). Since $\langle \mathbf{d}\boldsymbol{\varepsilon} \rangle_{V_{\text{sub}}^{(i)}}$ and $\mathbf{d}\boldsymbol{\sigma}_o$ are known, an *apparent* constitutive tangential modulus $C_l^{(i)}(\varepsilon_e)$ associated with the current level of the effective total strain ε_e is then determined from: $\mathbf{d}\boldsymbol{\sigma}_o = C_l^{(i)}(\varepsilon_e) \langle \mathbf{d}\boldsymbol{\varepsilon} \rangle_{V_{\text{sub}}^{(i)}}$. Standard techniques within the parametric formulation (Ostoja-Starzewski 2008) or nonparametric formulation (Das 2008; Das and Ghanem 2009) can be used here to determine $C_l^{(i)}(\varepsilon_e)$ [1]. At every step of the incremental approach, the effective total strain ε_e is computed from (31) by taking $\boldsymbol{\varepsilon} = \langle \boldsymbol{\varepsilon} \rangle_{V_{\text{sub}}^{(i)}}$. Here, $\langle \boldsymbol{\varepsilon} \rangle_{V_{\text{sub}}^{(i)}}$ is obtained by updating the previous state of the strain as follows: $\langle \boldsymbol{\varepsilon} \rangle_{V_{\text{sub}}^{(i)}}^{(k+1)} = \langle \boldsymbol{\varepsilon} \rangle_{V_{\text{sub}}^{(i)}}^{(k)} + \langle \mathbf{d}\boldsymbol{\varepsilon} \rangle_{V_{\text{sub}}^{(i)}}^{(k)}$, where the $k = 0$ corresponds to the initial zero state, $\langle \boldsymbol{\varepsilon} \rangle_{V_{\text{sub}}^{(i)}}^{(k)}$ is the volume averaged strain tensor at the kth step, and $\langle \mathbf{d}\boldsymbol{\varepsilon} \rangle_{V_{\text{sub}}^{(i)}}^{(k)}$ is the volume averaged incremental strain tensor obtained from the micromechanical analysis at the kth step. Because of the variations of the microstructural features across different smaller specimens $V_{\text{sub}}^{(i)}$'s, the values of $\langle \boldsymbol{\varepsilon} \rangle_{V_{\text{sub}}^{(i)}}$'s will be different from each other even though the corresponding volume averaged stress tensors, $\langle \boldsymbol{\sigma} \rangle_{V_{\text{sub}}^{(i)}}^{(k+1)} = \boldsymbol{\sigma}_o^{(k)} + \mathbf{d}\boldsymbol{\sigma}_o^{(k)}$, remain the same for all $V_{\text{sub}}^{(i)}$'s. Since C_l is assumed to model as a function of ε_e and $\langle \boldsymbol{\varepsilon} \rangle_{V_{\text{sub}}^{(i)}}$'s differ from each other, an interpolation scheme can be invoked to obtain $C_l^{(i)}$'s at any given ε_e. Subsequently, the lower bound C_l is computed as $C_l(\varepsilon_e) = \left(\overline{(C_l^{(i)}(\varepsilon_e))^{-1}}\right)^{-1}$, where the overbar symbol

[1] We can determine $C_l^{(i)}$ both parametrically or nonparametrically even though our goal is to characterize $\mathbf{C}_{\text{ep}}$ within a nonparametric formulation.

and $(\cdot)^{-1}$, represent, respectively the average (over i) and inverse operator (Huet 1990). As ε_e increases, $C_l(\varepsilon_e)$ decreases reflecting the effect of the continuous yielding of the inelastic material behavior of the microconstituents.

The upper bound $C_u(\varepsilon_e)$ can be similarly computed by following the above incremental approach with the following exceptions. Each sub-specimen $V_{\text{sub}}^{(i)}$ is now subjected to incremental kinematic uniform boundary condition (KUBC) or displacement-controlled boundary condition of the following form: $\mathbf{du(y)} = \mathrm{d}\boldsymbol{\varepsilon}_o\,\mathbf{y}$ and $\forall \mathbf{y} \in \partial V_{\text{sub}}^{(i)}$. Here, $\mathbf{du(y)}$ represents the increment in the prescribed displacement vector, and $\mathrm{d}\boldsymbol{\varepsilon}_o$ is the increment in the strain tensor that does not depend on $\mathbf{y}$. In this case, it can be shown that $\langle \mathrm{d}\boldsymbol{\varepsilon}\rangle_{V_{\text{sub}}^{(i)}} = \mathrm{d}\boldsymbol{\varepsilon}_o$ (Huet 1990) and another *apparent* constitutive tangential modulus $C_u^{(i)}(\varepsilon_e)$ associated with the current ε_e is determined from: $\langle \mathrm{d}\boldsymbol{\sigma}\rangle_{V_{\text{sub}}^{(i)}} = C_u^{(i)}(\varepsilon_e)\mathrm{d}\boldsymbol{\varepsilon}_o$. The values of $\langle \boldsymbol{\sigma}\rangle_{V_{\text{sub}}^{(i)}}$'s will be different from each other, while the corresponding volume averaged strain tensors, $\langle \boldsymbol{\varepsilon}\rangle_{V_{\text{sub}}^{(i)}}^{(k+1)} = \boldsymbol{\varepsilon}_o^{(k)} + \mathrm{d}\boldsymbol{\varepsilon}_o^{(k)}$, remain same for all $V_{\text{sub}}^{(i)}$'s. Finally, the upper bound $C_u(\varepsilon_e)$ is determined from $C_u(\varepsilon_e) = \overline{C_u^{(i)}(\varepsilon_e)}$ (Huet 1990). As ε_e increases, $C_u(\varepsilon_e)$ decreases reflecting the effect of the continuous yielding of the inelastic material behavior of the microconstituents and simultaneously maintains that $(C_u(\varepsilon_e) - C_l(\varepsilon_e))$ is a positive-definite matrix at any given ε_e during strain-hardening.

Further constitutive bounds for both the linear and nonlinear material behaviors across a wide range of different applications have also been developed, and we refer the readers to the literature as briefly mentioned in Sect. 1. The objective of the present work is not to discuss how further sophisticated constitutive bounds can be developed. Rather our aim is to use the already available bounds in capturing the effects of the microstructural fields while characterizing the macroscopic constitutive material properties. The microstructural features vary spatially over a given heterogeneous material specimen as well as across different heterogeneous material specimens and influence the macroscopic material properties. Our proposition is that these variabilities in the microstructural fields can be captured by modeling $\mathbf{C}_{\text{ep}}$ as bounded random matrix, where the associated bounds reflect the variabilities of the microstructural properties in some appropriate sense. Following the MaxEnt-based random matrix formulation as developed for $\mathbf{C}$ in the elastic region (Das 2008; Das and Ghanem 2009), we propose a similar formulation to construct the probabilistic description of $\mathbf{C}_{\text{ep}}$. Unlike the probability model of $\mathbf{C}$, the probability model of the matrix-valued random variable $\mathbf{C}_{\text{ep}}$ now changes depending on the current state of the material, expressed summarily via ε_e, in the region of the nonlinear material behavior. We explain how the probability model of the random $\mathbf{C}_{\text{ep}}$ can be constructed below.

Once $C_l(\varepsilon_e)$ and $C_u(\varepsilon_e)$ are determined, we postulate that the $N \times N$ random matrix $\mathbf{C}_{\text{ep}}(\varepsilon_e)$ would be uniformly distributed over the set $\mathscr{C}(\varepsilon_e) = \{C \in \mathbb{M}_N^+(\mathbb{R}) : C_l(\varepsilon_e) < C < C_u(\varepsilon_e)\}$ in the absence of any further statistical and physical information, where $\mathbb{M}_N^+(\mathbb{R})$ is the set of symmetric positive-definite real matrices of size $N \times N$. This would also be perfectly consistent with the original proposition of the MaxEnt principle in the absence of sufficient experimental data. In fact, this would be practically appealing specially for those cases when the 'gap' between $C_l(\varepsilon_e)$ and $C_u(\varepsilon_e)$ is small so that imposing any additional constraint is not useful in constructing the probabilistic model of $\mathbf{C}_{\text{ep}}$ at the cost of additional computational resources. The proposed matrix-variate uniform distribution is a special case of the matrix-variate Beta type I distribution (Gupta and Nagar 2000) and the associated pdf is given by,

$$p_{\mathbf{C}_{\text{ep}}}(C) = \frac{1}{\beta_N\left(\frac{1}{2}(N+1), \frac{1}{2}(N+1)\right) \det[C_u(\varepsilon_e) - C_l(\varepsilon_e)]^{\frac{1}{2}(N+1)}} \mathbb{I}_{\mathscr{C}(\varepsilon_e)}(C). \tag{34}$$

Here, $\beta_N(\cdot)$ is the multivariate Beta function given by (Gupta and Nagar 2000),

$$\beta_N(x, y) \equiv \int_{\mathscr{I}} \det(U)^{x-\frac{1}{2}(N+1)} \det(\mathbf{I} - U)^{y-\frac{1}{2}(N+1)} \, \mathrm{d}U = \frac{\Gamma_N(x)\Gamma_N(y)}{\Gamma_N(x+y)}, \tag{35}$$

in which $\Re(x) > (1/2)(N-1)$, $\Re(y) > (1/2)(N-1)$, $\mathbf{I}$ is the $N \times N$ identity matrix, $\mathscr{I} = \{U \in \mathbb{M}_N^+(\mathbb{R}) : \mathbf{0} < U < \mathbf{I}\}$, and $\Gamma_N(\cdot)$ represents the multivariate Gamma function given by (Gupta and Nagar 2000),

$$\Gamma_N(z) = \pi^{\frac{1}{4}N(N-1)} \prod_{i=1}^{N} \Gamma\left[z - \frac{1}{2}(i-1)\right], \quad \Re(z) > \frac{1}{2}(N-1), \tag{36}$$

with $\Gamma(\cdot)$ being the usual Gamma function (Abramowitz and Stegun 1970). Let the probability distribution function associated with the pdf in (34) be denoted by $\mathscr{U}_N(C_l(\varepsilon_e), C_u(\varepsilon_e))$.

Sampling scheme to generate the realizations of $\mathbf{C}_{\text{ep}}$ is already discussed, and we refer the readers to the literatures (Das 2008; Das and Ghanem 2009). Referring to these literatures, we simply summarize the sampling scheme (see, e.g., Algorithm 3.6 of (Das and Ghanem 2009)). Generate matrix-valued random variable $\mathbf{U}$ from the *standard matrix-variate Beta type I distribution* with distribution parameters $a = (1/2)(N+1)$ and $b = (1/2)(N+1)$, i.e., $\mathbf{U} \sim B_N^I((1/2)(N+1), (1/2)(N+1))$. Then, it can be shown that $\mathbf{C}_{\text{ep}}$ as defined below,

$$\mathbf{C}_{\text{ep}}(\varepsilon_e) = [C_u(\varepsilon_e) - C_l(\varepsilon_e)]^{\frac{1}{2}} \, \mathbf{U} \, [C_u(\varepsilon_e) - C_l(\varepsilon_e)]^{\frac{1}{2}} + C_l(\varepsilon_e), \tag{37}$$

is distributed as $\mathscr{U}_N(C_l(\varepsilon_e), C_u(\varepsilon_e))$.

Equation (37) reveals one appealing computational advantage of the proposed sampling scheme and furnishes a salient physical interpretation of the proposed RMT-based characterization of $\mathbf{C}_{\text{ep}}$. Since $C_l(\cdot)$ and $C_u(\cdot)$ are already determined as functions of ε_e, we need to sample $\mathbf{U}$ only once for any given integration point within the FE formulation for each realization of the (macroscopic) mechanical system being analyzed. As ε_e changes at the integration point depending on the applied loading condition, $C_l(\cdot)$ and $C_u(\cdot)$ also evolve with ε_e. Clearly, use of the *same* realization of $\mathbf{U}$ at the integration point produces different values of $\mathbf{C}_{\text{ep}}(\cdot)$

depending on the current state of ε_e. This feature of the proposed approach can be interpreted in the following sense. All the conventional local system parameters (e.g., Young's modulus, Poisson ratio, Q, c, and γ) associated with the microconstituents are collectively and alternatively characterized by a single macroscopic matrix-valued material parameter $\mathbf{U}$ within the nonparametric formulation.

The next section numerically illustrates the proposed approach on a simple 2-D cantilever beam.

5 Numerical Illustration

To determine the lower and upper bounds, $C_l(\cdot)$ and $C_u(\cdot)$, we consider a two-phase heterogeneous material with a dominant matrix phase and a secondary phase (inclusions). A small material volume element V showing the microconstituents is presented in Fig. 2. We assume that such a microstructure can be associated with each and every macroscopic point $\mathbf{x} \in \mathscr{B}$. If the microstructural features change considerably over $\mathscr{B}$, then the evolution of these microstructural features must be experimentally identified, and a stochastic field representation of the underlying microstructural field should be appropriately constructed to use in the subsequent analysis for the prediction purpose. This involves both additional experimental and computational resources. For the purpose of illustration, we assume in the present work that the inclusions are randomly distributed over V. It is not required to explicitly specify the size of V, provided it is ensured that the volume fractions of the different microconstituents are consistent with the experimentally identified ones. In the present case, it is simply assumed that the volume fraction, v_i, of inclusions is 0.35. Of course, the volume element V as considered in the present work is a grossly simplified version of the heterogeneous material encountered in practice. All the

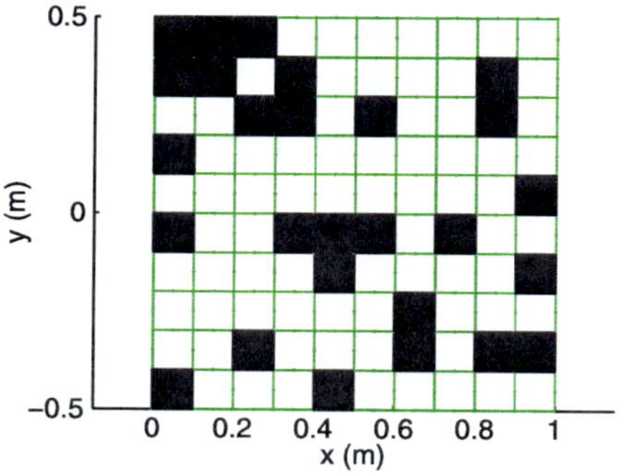

Fig. 2 A small material volume element V of the two-phase heterogeneous material; the black phase represents the inclusions, and the spatial regions are randomly selected by following the scheme as already discussed in the previous work (Das 2008; Das and Ghanem 2009); micromechanical FE analysis carried out with nine-node quadrilateral plane stress elements

simplifications nevertheless are general enough within the context of the present work and are used only to highlight the primary contributions of the proposed approach. For inquisitive readers, we comment that each small square in Fig. 2 can be viewed as a material grain.

We consider only one realization of V as shown in Fig. 2 to determine the bounds, $C_l(\cdot)$ and $C_u(\cdot)$. The Sachs bound and Taylor bound are used, respectively, for $C_l(\cdot)$ and $C_u(\cdot)$. They are straightforward generalizations of the Reuss and Voigt bounds in the linear elastic material behavior (Hill 1963; Nemat-Nasser and Hori 1999) to the nonlinear material behavior, and under the assumptions indicated earlier, they can be shown to follow from the application of the classical principles of minimum complementary energy and minimum potential energy, respectively (Huet 1990; Jiang et al. 2001). Computation of the Taylor and Sachs bounds requires determining the volume averages (over the microstructural domain shown in Fig. 2) of two matrix-valued fields – one associated with the constitutive tangential elastoplastic matrix and the other its inverse – at each step of the incremental procedures involving two different incremental boundary conditions (SUBC and KUBC). In the inelastic regime, these matrix fields at the fine scale level spatially vary over the microstructural domain V. A numerical integration scheme is, therefore, required to compute the corresponding matrix-valued volume averages. For this purpose, we used a standard FE-type Gauss quadrature integration rule to compute the following integrals representing, respectively, the Sachs and Taylor bounds,

$$C_l(\varepsilon_e) = \left\{ \frac{1}{V} \sum_i \int_{V_i} \left(\mathbf{C}_{\text{ep}}^{\text{SUBC}}(\varepsilon_e, \mathbf{y}) \right)^{-1} d\mathbf{y} \right\}^{-1}, \text{ and } C_u(\varepsilon_e) = \frac{1}{V} \sum_i \int_{V_e} \mathbf{C}_{\text{ep}}^{\text{KUBC}}(\varepsilon_e, \mathbf{y})\, d\mathbf{y}, \tag{38}$$

where $V_i = \int_{V_i} d\mathbf{y}$ is the volume of the ith finite element such that $\sum_i V_i = V$, ε_e is the associated *macroscopic* effective total strain computed from (31) by taking $\boldsymbol{\varepsilon} = \langle \boldsymbol{\varepsilon} \rangle_V$, and $\mathbf{C}_{\text{ep}}^{\text{SUBC}}(\varepsilon_e, \mathbf{y})$ associated with ε_e represents the microscopic constitutive tangential elastoplastic matrix at $\mathbf{y} \in V$ when the micromechanical analysis involving the incremental SUBC is carried out. It should be noted that the value of the *microscopic* strain tensor $\boldsymbol{\varepsilon}(\mathbf{y})$ at $\mathbf{y}$ in conjunction with (31) does not yield the *macroscopic* effective total strain ε_e. At any given step of the incremental procedure, ε_e remains same for all $\mathbf{y} \in V$. The other microscopic constitutive tangential elastoplastic matrix $\mathbf{C}_{\text{ep}}^{\text{KUBC}}(\varepsilon_e, \mathbf{y})$ can be similarly defined. The tangential elastoplastic matrices, $\mathbf{C}_{\text{ep}}^{\text{SUBC}}(\varepsilon_e, \mathbf{y})$ and $\mathbf{C}_{\text{ep}}^{\text{KUBC}}(\varepsilon_e, \mathbf{y})$, are computed from (30). Here, the inelastic material behavior of the matrix phase is thus assumed to be governed by the combined isotropic and kinematic hardening law at the fine scale level. The material properties of the matrix phase are tabulated in Table 1. The inclusion phase is assumed to remain elastic throughout the micromechanical analyses and the corresponding elastic material properties are same as those of the matrix phase. It implies that material is homogeneous in the elastic regime. Once yielding sets out, the material displays the heterogeneous characteristics. This is done to restrict the material heterogeneity only in the nonlinear inelastic regime which is the focus of the present work. The evolution of the traces of $C_l(\cdot)$ and $C_u(\cdot)$ is reported in Fig. 3.

Table 1 Material parameters of the matrix phase for the micromechanical FE analyses for computation of the Sachs and Taylor bounds

Young's modulus, E	70 GPa
Poisson's ratio, ν	0.35
Initial yield strength, σ_y	0.4 GPa
Q	0.14 GPa
b	7.094
c	7.019 GPa
γ	118.6

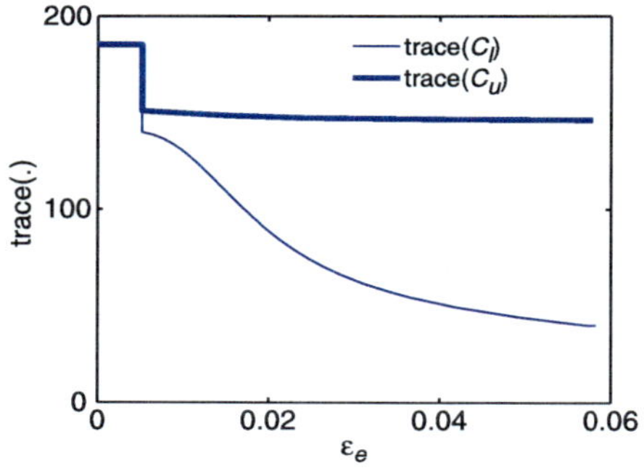

Fig. 3 Evolution of the traces of $C_l(\cdot)$ and $C_u(\cdot)$

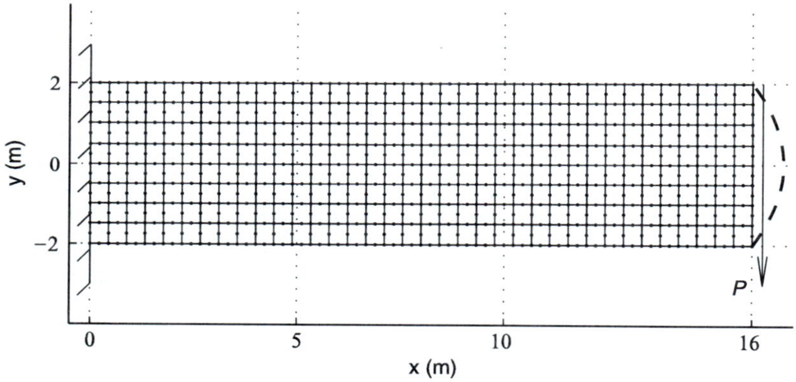

Fig. 4 A 2-D cantilever beam modeled with nine-node quadrilateral nonparametric plane elements; the total load P is distributed parabolically as shown with a dashed line at $x = L$

A cantilever beam as shown in Fig. 4 is analyzed next using the nonparametric $\mathbf{C}_{\text{ep}}$ which is probabilistically characterized using the scheme as discussed in Sect. 4. In a Monte Carlo fashion, we can compute the usual response quantities (stress, strain, etc.) of the cantilever beam using the standard FE technique. Several samples (144 samples) of $\mathbf{C}_{\text{ep}}$ are first simulated by following the sampling scheme as suggested in Sect. 4. As indicated at the beginning of this section, it is assumed that each sample of $\mathbf{C}_{\text{ep}}$ characterizes the material property over the entire spatial domain of the corresponding sample of the beam. In Fig. 5, plots of the trace of four different realizations of $\mathbf{C}_{\text{ep}}$ at a Gauss point near the left lower corner of the

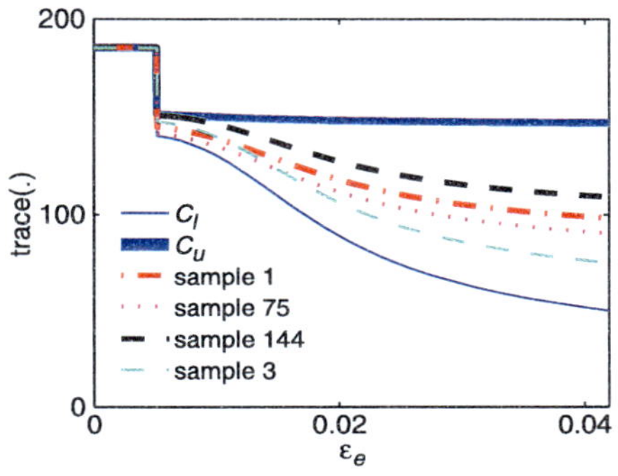

Fig. 5 Traces of four realizations of $\mathbf{C}_{ep}$ at a Gauss point near the left lower corner of the cantilever beam

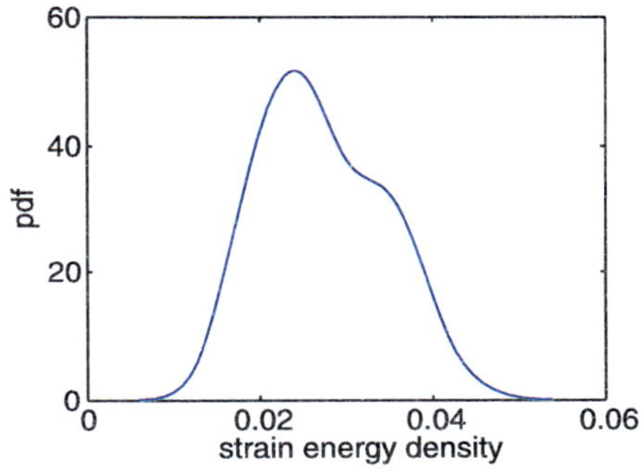

Fig. 6 Strain energy density at a Gauss point near the left lower corner of the cantilever beam

cantilever beam are shown. Four different curves as shown for the plastic region illustrate the effects of the variability in the material properties associated with the nonlinear plastic material behavior characterized in the present work by the random $\mathbf{C}_{ep}$. The pdf of the associated random strain energy density at the same Gauss point is also reported in Fig. 6.

6 Conclusion

The MaxEnt-based random matrix formulation can provide meaningful packaging of information, where general knowledge-based constraints can be imposed to construct the probability model of the associated random matrix. In the present work, the constitutive tangential elastoplastic matrix $\mathbf{C}_{ep}$ is probabilistically modeled within the random matrix formalism to characterize the nonlinear inelastic material behavior. The realizations of the random $\mathbf{C}_{ep}$ are always bounded (in the positive-definite sense) by two physics-based matrix-valued lower and upper bounds that capture the variability of the underlying microstructural features in some sense. The proposed formulation can be readily adapted to other nonlinear behaviors, provided the associated random matrix is symmetric, positive-definite, and bounded from below and above, in the positive-definite sense, by two symmetric and positive-definite matrices.

References

Abramowitz M, Stegun IA (1970) Handbook of Mathematical Functions: with Formulas, Graphs, and Mathematical Tables. Dover, New York

Arnst M, Ghanem R (2008) Probabilistic equivalence and stochastic model reduction in multiscale analysis. Computer Methods in Applied Mechanics and Engineering 197(43-44):3584–3592

Das S (2008) Model, identification & analysis of complex stochastic systems: Applications in stochastic partial differential equations and multiscale mechanics. PhD thesis, Department of Civil and Environmental Engineering, University of Southern California, Los Angeles, USA, http://digarc.usc.edu/assetserver/controller/view/search/etd-Das-20080513

Das S, Ghanem R (2009) A bounded random matrix approach for stochastic upscaling. SIAM Multiscale Modeling and Simulation 8(1):296–325

Drucker DC (1988) Conventional and unconventional plastic response and representation. Applied Mechanics Reviews 41(4):151–167

Drugan WJ, Willis JR (1996) A micromechanics-based nonlocal constitutive equation and estimates of representative volume element size for elastic composites. Journal of the Mechanics and Physics of Solids 44:497–524

Dunne F, Petrinic N (2005) Introduction to Computational Plasticity. Oxford University Press, New York (Reprinted with corrections in 2007)

Ghanem R, Das S (2009) Hybrid representations of coupled nonparametric and parametric models for dynamic systems. AIAA Journal 47(4):1035–1044

Gupta A, Nagar D (2000) Matrix Variate Distribution. Chapman & Hall/CRC, Boca Raton

Gusev AA (1997) Representative volume element size for elastic composites : A numerical study. Journal of the Mechanics and Physics of Solids 45(9):1449–1459

Hashin Z, Shtrikman S (1963) A variational approach to the theory of the elastic behavior of multiphase materials. Journal of the Mechanics and Physics of Solids 11:127–140

Hazanov S, Huet C (1994) Order relationship for boundary conditions effect in heterogenous bodies smaller than the representative volume. Journal of the Mechanics and Physics of Solids 42(12):1995–2011

Hill R (1950) The Matthematcal Theory of Plasticity. Oxford University Press, New York

Hill R (1963) Elastic properties of reinforced solids: Some theoretical principles. Journal of the Mechanics and Physics of Solids 11(5):357–372

Hill R (1967) The essential structure of constitutive laws for metal composites and polycrystals. Journal of the Mechanics and Physics of Solids 15(2):79–95

Hill R (1968) On constitutive inequalities for simple materials – II. Journal of the Mechanics and Physics of Solids 16(5):315–322

Hill R (1972) On constitutive macro-variables for heterogeneous solids at finite strain. Proceedings of the Royal Society of London Series A, Mathematical and Physical Sciences, 326(1565): 131–147

Hill R, Rice JR (1973) Elastic potentials and the structure of inelastic constitutive laws. SIAM Journal on Applied Mathematics 25(3):448–461

Houlsby GT (1981) A study of plasticity theories and their applicability to soils. PhD thesis, St. John's College, Cambridge University, Los Angeles, USA, http://www-civil.eng.ox.ac.uk/people/gth/thesis/thesis.htm

Huet C (1990) Application of variational concepts to size effects in elastic heterogeneous bodies. Journal of the Mechanics and Physics of Solids 38(6):813–841

Jardak M, Ghanem R (2004) Spectral stochastic homogenization for of divergence-type pdes. Computer Methods in Applied Mechanics and Engineering 193(6-8):429–447

Jaynes E (1957) Information theory and statistical mechanics. Physical Review 106(4):620–630

Jaynes ET (2003) Probability Theory: The Logic of Science. Cambridge University Press

Jiang M, Ostoja-Starzewski M, Jasiuk I (2001) Scale-dependent bounds on effective elastoplastic response of random composites. Journal of the Mechanics and Physics of Solids 49:655–673

Kanit T, Forest S, Galliet I, Mounoury V, Jeulin D (2003) Determination of the size of the representative volume element for random composites: statistical and numerical approach. International Journal of Solids and Structures 40(13-14):3647–3679

Kapur J, Kesavan H (1992) Entropy Optimization Principles with Applications. Academic, Boston, USA

Koutsourelakis PS (2006) Probabilistic characterization and simulation of multi-phase random media. Probabilistic Engineering Mechanics 21(3):227–234

Koutsourelakis PS (2007) Stochastic upscaling in solid mechanics: An excercise in machine learning. Journal of Computational Physics 226(1):301–325

Liu WK, Siad L, Tian R, Lee S, Lee D, Yin X, Chen W, Chan S, Olson GB, Lindgen HMF Lars-Erik, Chang YS, Choi JB, Kim YJ (2009) Complexity science of multiscale materials via stochastic computations. International Journal for Numerical Methods in Engineering 80(6-7):932–978

Lubarda VA (2002) Elastoplasticity Theory. CRC, Boca Raton

Miller TW, Cheatham JB (1972) A new yield condition and hardening rule for rocks. International journal of rock mechanics and mining sciences 9:453–474

Molenkamp F, Van Ommen A (1987) Peculiarity of non-associativity in plasticity of soil mechanics. International Journal for Numerical and Analytical Methods in Geomechanics 11:659–661

Murihead R (1982) Aspects of Multivariate Statistical Theory. Wiley, revised printing in 2005

Nemat-Nasser S (1999) Averaging theorems in finite deformation plasticity. Mechanics of Materials 31(8):493–523, (Erratum in *Mechanics of Materials*, vol. 32, issue 5, 2000, page 327)

Nemat-Nasser S, Hori M (1999) Micromechanics: Overall properties of heterogeneous materials, 2nd edn. Elsevier, Amsterdam

Ogden RW (1978) Extremum principles in non-linear elasticity and their application to composites–I : Theory. International Journal of Solids and Structures 14:265–282

Ogden RW (1984) Non-Linear Elastic Deformations. Ellis Horwood Limited, Chichester, re-published by Dover in 1997

Ohno N, Wang JD (1993) Kinematic hardening rules with critical state of dynamic recovery, Part II: Application to experiments of ratchetting behavior. International Journal of Plasticity 9: 391–403

Ostoja-Starzewski M (1999) Microstructural disorder, mesoscale finite elements and macroscopic response. Proceedings of the Royal Society of London Series A, Mathematical, Physical and Engineering Sciences 455(1989):3189–3199

Ostoja-Starzewski M (2002) Scale effects in plasticity of random media: status and challenges. International Journal of Plasticity 21(6):1119–1160

Ostoja-Starzewski M (2008) Microstructural Randomness and Scaling in Mechanics of Materials. Chapman & Hall/CRC

Ostoja-Starzewski M, Wang X (1999) Stochastic finite elements as a bridge between random material microstructure and global response. Computer Methods in Applied Mechanics and Engineering 168(1-4):35–49

Rudnicki JW, Rice JR (1975) Conditions for the localization of deformation in pressure-sensitive dilatant materials. Journal of the Mechanics and Physics of Solids 23(6):371–394

Sandler IS, Pucik TA (2001) Non-uniqueness in dynamic rate-independent non-associated plasticity. In: Voyiadjis G (ed) Mechanics of Materials and Structures, Elsevier Science, New York, pp 221–240

Sankaran S, Zabaras N (2007) Computing property variability of polycrystals induced by grain size and orientation uncertainties. Acta Materialia 55(7):2279–2290

Senseny AF Paul E abd Fossum, Pfeifle TW (1983) Non-associative constitutive laws for low porosity rocks. International Journal for Numerical and Analytical Methods in Geomechanics 7(1):101–115

Simo JC, Hughes TJR (1998) Computational Inelasticity. Springer, New York

Soize C (2001) Maximum entropy approach for modeling random uncertainties in transient elastodynamics. Journal of the Acoustical Society of America 109(5):1979–1996, pt. 1

Soize C (2006) Non-Gaussian positive-definite matrix-valued random fields for elliptic stochastic partial differential operators. Computer Methods in Applied Mechanics and Engineering 195(1-3):26–64

Soize C (2008) Tensor-valued random fields for meso-scale stochastic model of anisotropic elastic microstructure and probabilistic analysis of representative volume element size. Probabilistic Engineering Mechanics 23(2-3):307–323

To AC, Liu WK, Olson GB, Belytschko T, Chen W, Shephard MS, Chung YW, Ghanem R, Voorhees PW, Seidman DN, Wolverton C, Chen JS, Moran B, Freeman AJ, Tian R, Luo X, Lautenschlager E, Challoner AD (2008) Materials integrity in microsystems: a framework for a petascale predictive-science-based multiscale modeling and simulation system. Computational Mechanics 42(4):485–510

Torquato S (2002) Ranom Heterogeneous Materials: Microstructure and Macroscopic Properties. Springer

de Veubeke BF (1964) Upper and lower bounds in matrix structural analysis. In: de Veubeke BF (ed) Matrix Methods of Structural Analysis, The McMillan Company, New York, pp 165–201

Yin X, Chen W, To A, McVeigh C, Liu WK (2008) Statistical volume element method for predicting microstructure-constitutive property relations. Computer Methods in Applied Mechanics and Engineering 197(43-44):3516–3529

Yoder PJ (1980) A strain-space plasticity theory and numerical implementation. PhD thesis, Earthquake Engineering Research Laboratoty, Calfornia Institute of Technology, USA, http://caltecheerl.library.caltech.edu/146/00/8007.pdf

Zhao KM, Lee JK (2001) Material properties of aluminum alloy for accurate draw-bend simulation. Journal of Engineering Materials and Technology 123:287–292

Zhu SC, Wu Y, Mumford D (1998) Filters, random fields and maximum entropy (FRAME): Towards a unified theory for texture modeling. International Journal of Computer Vision 27(2):107–126

Zienkiewicz OC (2001) Displacement and equilibrium models in the finite element method by B. Fraeijs de Veubeke, Chapter 9, Pages 145–197 of *Stress Analysis*, Edited by O. C. Zienkiewicz and G. S. Holister, Published by Wiley, 1965. International Journal for Numerical Methods in Engineering 52(3):287–342, (Classic Reprint)

DDSim: Framework for Multiscale Structural Prognosis

John M. Emery and Anthony R. Ingraffea

1 Prologue: 2025

On June 1, 2025, Boeing delivers the first B-17 HyperFortress, tail number 20–0001, a Mach 6 aircraft. Along with this physical instantiation of the aircraft, Boeing also delivers an as-built *digital instantiation* of this tail number, 20–0001D/I. 20–0001D/I is a 1,000 billion DOF, hierarchical, computational structures model of 20–0001. This "Digital Fortress" is ultra-realistic in geometric detail, including manufacturing anomalies, and in material detail, including the statistical microstructure level. 20–0001D/I accepts probabilistic input of loads, environmental, and usage factors, and it also tightly couples to an outer-mold-line, as-built, CFD model of 20–0001.

20–0001D/I can be virtually flown over a 1-h, design-point flight in 1 h on an exaflop-scale high-performance computer. During each such virtual flight, 20–0001D/I accumulates usage damage according to best-physics-based, probabilistic simulations, and outputs about 1 petabyte of material, structural performance, and damage data. 20–0001D/I is "flown" for 1,000 h during ground testing of 20–0001. During this accelerated, preliminary lifing, a number of unexpected limit states are encountered leading to loss of primary structural elements, with two incidents likely leading to loss of aircraft forecast. Appropriate repairs, redesigns, and retrofits are planned and implemented on 20–0001 before its first flight to preclude such events from actually occurring. The HyperFortress becomes the first Air Force flight vehicle to be certified mostly through simulation.

It is recognized, however, that design-point usage is always trumped by actual usage, involving unplanned mission types and payloads. Therefore, a second digital instantiation, 20-0001D/A, is linked to the structural sensing system deployed on 20-0001. This structural health monitoring system records, at high frequency, actual,

A.R. Ingraffea (✉)
Dwight C. Baum Professor of Engineering, Cornell Fracture Group, 643 Rhodes Hall, Cornell University, Ithaca, NY 14853, USA
e-mail: ari1@cornell.edu

S. Ghosh and D. Dimiduk (eds.), *Computational Methods for Microstructure-Property Relationships*, DOI 10.1007/978-1-4419-0643-4_13,

six-DOF accelerations, as well as surface temperature/pressure readings during each actual flight of 20-0001. Each hour of real flight produces about 1 petabyte of such real data. These data are input into the 20-0001D/A structural model, and this model itself becomes a virtual sensor, interpolating sparse acquired data over the entire airframe. Using Bayesian statistical techniques, 20-0001D/I is periodically updated to reflect actual usage recorded by 20-0001 and by 20-0001D/A, and is rerun for prognosing the remaining life of 20-0001, and for updating reliability estimates for all primary structural components. This prognosis leads to time-and-budget-appropriate execution of rehab plans resulting from such updated lifing and reliability estimates. This process is being executed for all 12 of the HyperFortress aircraft.

2 Introduction

From the perspective of the air forces of the US military, the capabilities envisioned in the prologue to this chapter are what this book is all about. What science and technology, what simulation systems, need to be discovered and developed to make this vision possible? Can complex structural airframe and engine systems, to be operated in ever more extreme conditions, be designed and certified mostly through computer simulation? Can tail-number-specific, reliability-informed, condition-based maintenance replace fleet-wide damage tolerance procedures with their attendant high inspection costs? To answer "yes" to these questions does not require a fundamentally new approach to thinking about how to design, build, and maintain predictably safe structures: make sure driving "forces" determined from response analyses are always acceptably less than resisting "forces" defined through test and theory, lest one exceed a structural limit state. *Rather, in the authors' opinions, what is required is confidence in the belief that one can significantly reduce the testing needed to uncover and quantify material and structural limit states with reliability predictions based on understanding of the rules of physics and mechanics, expressed in computer code, and executed on our most advantageous tool – nearly infinite computing power.*

A graphical depiction of the hierarchical computational simulation system of a "digital aircraft" is shown in Fig. 1. The OEM would deliver as-built geometry and numerical models of each primary structural component (Fig. 1a). These models can be analyzed at any time during acceptance trials or during actual aircraft life to determine current field responses. These analyses would flag "hot spots," likely to lead to a local limit state, such as fatigue crack initiation (Fig. 1b). At each such indication, a lower length-scale model is introduced, involving all the microstructural modeling, constitutive behavior, and damage modes discussed in earlier chapters of this book (Fig. 1c). Multiscaling techniques, also addressed in this book, inform these models of evolving fields at the structural scale (Fig. 1d), and damage evolution is simulated at the microstructural scale (Fig. 1e). Information concerning stiffness

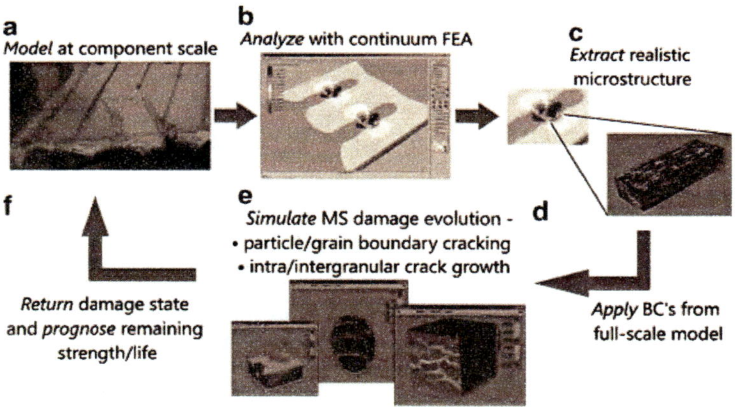

Fig. 1 Graphical depiction of information flow in the hierarchical computational simulation system of the "digital aircraft" of the future. This figure shows the probabilistic prognosis of fatigue life as the structural limit state. The sub-figures show: (**a**) as-built geometry and numerical model; (**b**) analysis to flag hot-spots; (**c**) high fidelity lower-length scale model; (**d**) multiscale techniques; (**e**) microstructural damage evolution; and (**f**) update as-built model to account for microstructural damage

and strength changes resulting from such evolution is periodically uplinked to the structural scale model, and its remaining life and residual strength are probabilistically estimated (Fig. 1f).

For this fleet management paradigm to be completely viable, many scientific advances still need to be made. The purpose of this chapter is to describe a prototype hierarchical computational simulation system, a Damage and Durability Simulator (DDSim), as it would apply to prognosis of fatigue life of a major airframe component and to highlight areas of research that must see commensurate progress in the coming years. The prototype focuses on fatigue cracking; however, the general framework can readily be extended to other modes of damage and other material systems. In the next section, the architecture of DDSim is described. Each level of its hierarchy is described in the following sections, and these are interconnected with the thread of a consistent example problem. In each section, the relevance of a DDSim-like environment to the vision in the prologue is highlighted, and the major shortcomings in our present capabilities to reach this vision are identified.

3 DDSim Architecture

Those 1,000 billion degrees-of-freedom (DOF) and petabyte databases associated with the "Digital Fortress" have to be hierarchical and tightly integrated. Many computational models, at different length/time scales, and with multi-physics coupling, will consume those DOF over an entire air vehicle. Data associated with response analyses from these models need to be mined for indications of impending problems, and health prognoses will then need to be performed by projecting damage evolution forward with even more computational models. DDSim is a rudimentary prototype of a system that embodies these needs in the context of fatigue cracking.

DDSim uses a hierarchical "search and simulate" strategy consisting of three main levels. The strategy assumes that the total fatigue life of a structure can be decomposed as:

$$N^{\mathrm{T}} = N^{\mathrm{MLC}} \oplus^{a^{\mathrm{MLC}}} N^{\mathrm{MSC}} \tag{1}$$

where N^{T} is the total fatigue life of the structure, N^{MLC} is the number of loading cycles consumed by microstructurally large crack (MLC) growth processes, N^{MSC} is the number of loading cycles consumed by microstructurally small crack (MSC) growth processes, and a^{MLC} is the characteristic length of a crack when it can be considered microstructurally large. Here, microstructurally small is used to describe all cracks whose growth rate and shape are dominated by local microstructural effects. All other crack sizes are collectively referred to as microstructurally large.

The operator $\oplus^{a^{\mathrm{MLC}}}$ implies summation that is dependent on the definition of a^{MLC} which can be arbitrarily chosen by the engineer. The notation here will use capital letters to denote random variables and lowercase letters to represent deterministic variables or specific samples of a random variable. Hence, while N^{T} is the random variable representing total fatigue life of the structure described by a probabilistic distribution, n^{T} is one realization of that total life and is given by an integer value of cycles.

For a structural component subject to a given loading program, the functional dependence of N^{T} can be expressed by

$$N^{\mathrm{T}} = N^{\mathrm{T}}(\boldsymbol{X}, \boldsymbol{A}, \boldsymbol{E}) \tag{2}$$

where $\boldsymbol{X} \in \mathfrak{R}^3$ is the dominant flaw location, $\boldsymbol{A} \in \mathfrak{R}^{\mathrm{p}}$ is a p-dimensional array describing flaw geometry, and $\boldsymbol{E} \in \mathfrak{R}^{\mathrm{s}}$ is a s-dimensional array describing material response including crack growth resistance and environmental effects. Because N^{T} is a function of the random vectors ***X***, ***A***, and ***E***, N^{T} is a random variable. Consequently, the total fatigue life of a structural component has the cumulative distribution function (CDF)

$$F_{N^{\mathrm{T}}}(n^{\mathrm{T}}) = P\left[N^{\mathrm{T}}(\boldsymbol{X}, \boldsymbol{A}, \boldsymbol{E}) \leq n^{\mathrm{T}}\right] \tag{3}$$

Furthermore, the probability of failure at a specified critical number of cycles is defined as:

$$P_f = P\left[N^{\mathrm{T}}(\boldsymbol{X}, \boldsymbol{A}, \boldsymbol{E}) \leq n^{\mathrm{T}}_{\mathrm{cr}}\right] = F_{N^{\mathrm{T}}}\left(n^{\mathrm{T}}_{\mathrm{cr}}\right) \tag{4}$$

where $n^{\mathrm{T}}_{\mathrm{cr}}$ is the critical number of loading cycles. The reliability of the structure is the probability of non-failure and can be readily computed as:

$$P\left[N^{\mathrm{T}}(\boldsymbol{X}, \boldsymbol{A}, \boldsymbol{E}) > n^{\mathrm{T}}_{\mathrm{cr}}\right] = 1.0 - P_f \tag{5}$$

A flowchart showing the architecture of DDSim is shown in Fig. 2. The components in the flowchart bear the following significance:

- Dashed boxes indicate user input/control into the system.
- Solid rectangular boxes indicate operations within the system (bold for primary hierarchical components).
- Diamonds indicate decision points.
- Ovals indicate probabilistic life predictions (output).
- Arrows indicate the direction of data flow.

The components above the dashed line are collectively labeled as the "predictor." Shown as input to the predictor are the, generally, stochastic data supporting the simulation.

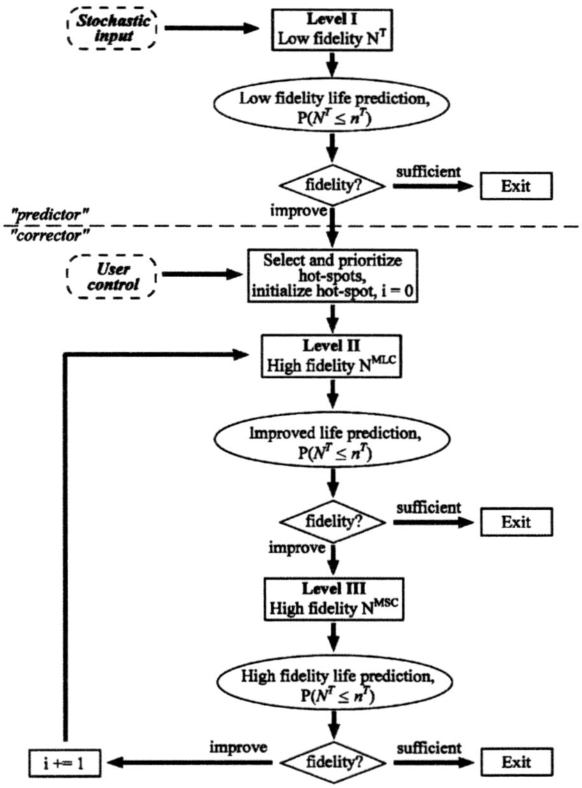

Fig. 2 Flowchart showing the three-level architecture of DDSim

Level I operates on this data and the output is a low fidelity estimate of the probability of failure. Then, the first diamond in the data flow asks whether refined fidelity is necessary and passes data along or exits, accordingly. If refinement is necessary, the low fidelity estimate becomes the prediction of reliability to be corrected by subsequent analyses, and the data enter the "corrector," actions below the dashed line. The first step in the corrector requires intelligent control to select hotspots (criterion for hotspot selection are discussed in detail below). The corrector improves fidelity by operating on a subregion of the original domain. The volume of this subdomain is appropriately chosen surrounding a hotspot. Recalling (1), there are obvious ways to improve the Level I prediction of reliability: improve the estimate of N^{MLC}, and/or improve the estimate of N^{MSC}. DDSim Levels II and III address these opportunities.

Level II operates at the selected hotspots to improve the estimate of N^{MLC} which can be combined through (1) to provide an improved estimate of the total structure reliability. Upon exiting Level II, the level of fidelity is again assessed and the decision to continue refinement or exit is made.

If continued refinement is necessary, Level III operates at the selected hotspots to improve the estimate of N^{MSC}. The Level III tool is generally the most computationally expensive, as it is the bridge between length scales. Consequently, Level III requires intelligent control to determine size and number of microstructural simulations. The final result of the simulation will be a higher fidelity structural reliability estimate than that obtained using only Level I. If the fidelity of this reliability estimate is insufficient, the engineer must refine the model. Obviously, there are three ways to control the quality of the final estimate of reliability. First, one could provide higher resolution input data and run the entire system again. Second, one could include a larger subset of the original domain to be analyzed in the corrector by selecting additional hotspots. Finally, one could include more or larger microstructural simulations. It should be emphasized that the decomposition indicated in (1) is not intended to suggest that N^{T} is evenly divided into two parts. In fact, N^{MSC} can consume most of N^{T}. If that is the case, it may not be cost-effective to use Level II, and one can choose to skip directly to Level III. Conversely, for a vehicle already in service, the average MLC length may be known by NDE. If so, the Level III simulation might be too costly and unnecessary, and one could stop the simulation after Level II.

In the following sections, each of these levels is described and its usage is exemplified through a consistent, simple example problem, the stiffened wing panel shown in Fig. 3. Proprietary physical testing on this example has been performed (Papazian et al. 2007a,b). Although specific geometrical data associated with this component cannot be presented here, results from this testing will help to identify the validity and shortcomings of reliability predictions based on current capabilities.

Space limitations here preclude detailed descriptions of all the functions of DDSim, all levels; the reader can refer to Emery (2007) and Emery et al. (2009) for more details.

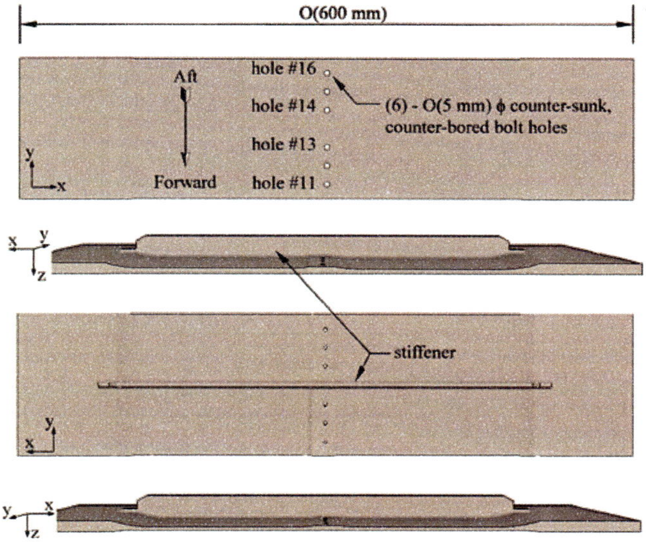

Fig. 3 Example application problem for three levels of DDSim. Stiffened wing panel is loaded in tension in the x-direction

4 DDSim Level I: Reduced-Order, Probabilistic, Low-Fidelity Life Prediction and Initial Screening

Essentially, Level I in DDSim is an automated way of performing state-of-practice, damage tolerance type assessments at every possible flaw location in a structure (Fig. 4). DDSim Level I is a generalization and automation of the approach taken in such familiar codes as NASGRO (2008) or AFGROW (2008). Level I idealizes the initial crack geometry as penny-shaped, semicircular, or quarter-circular, and uses analytically computed stress intensity factor (SIF) versus crack size solutions to compute crack driving forces. These solutions are then related to crack growth rate through standard relationships, and these are integrated to predict crack size versus number of cycles. The result is a low fidelity reliability estimate that can be used for preliminary design, selection of subdomains to perform the higher level simulations, and as the prediction to be corrected by the higher-order analyses of Levels II and III. Level I produces a highly automated, probabilistic, conservative, and fast prediction of fatigue life.

Inputs to Level I are distributions of initial crack radii, a finite element (FE) mesh and boundary conditions for the component, the finite-element-generated stress analysis results, and required material parameters for the requested crack growth rate model. The user-defined initial crack sizes can be described deterministically, randomly by distribution, or randomly through the use of a particle cracking filter for materials where intrinsic flaw size is correlated with second phase particle size.

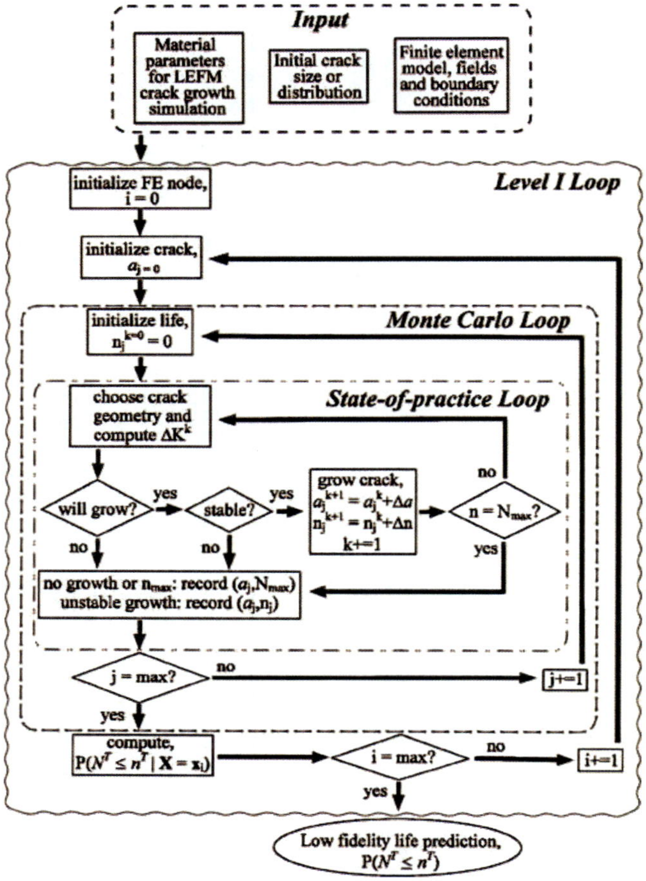

Fig. 4 Flowchart of DDSim Level I operations

For the example component described herein, this last option uses a particle cracking filter described by Bozek et al. (2008) specific to AA 7075-T651, the material used in the component shown in Fig. 3.

The input FE model can be of arbitrary geometrical complexity, and many such models would be expected to represent possibly hundreds of primary structural components in a digital aircraft. This FE model does not include damage; rather, it represents structure-scale behavior using standard continuum behavior models. The results from the stress analysis are input as nodal stresses, but could be any other field variable specific to the component and its material systems.

For a deterministic analysis, Level I uses the nodal stresses in conjunction with the initial crack size and analytically computed SIF solutions to compute SIFs for the assumed crack at each FE node. The orientation of the initial crack is assumed to be perpendicular to the local maximum principal tensile stress at the node, and no crack interaction is currently allowed. The use of analytically computed SIF solutions is the main source of speed and, subsequently, reduced order accuracy. Level I combines the SIFs with input material parameters and empirical models to calculate the crack growth rate. With the growth rate, a new crack size is computed and the procedure is repeated for each load cycle. Following this procedure, Level I performs an automated interrogation resulting in a life prediction at every FE node. This produces a deterministic scalar field of life prediction computed over the entire domain of the component.

For probabilistic life assessment, Level I performs Monte Carlo simulation using the procedure outlined above for each initial crack radius from its distribution. Assuming the initial crack radii are less than a^{MLC}, the result at one node is a list of life predictions that corresponds to the list of initial crack sizes, or

$$\boldsymbol{n}_I^{\mathrm{T}} = \{n_1^{\mathrm{T}}, n_2^{\mathrm{T}}, \ldots, n_l^{\mathrm{T}}\} \tag{6}$$

where the subscript I indicates that this is the Level I estimate, the n_j^{T} are the number of load cycles computed corresponding to initial crack size j, and l is the total number of initial crack sizes. $\boldsymbol{n}_I^{\mathrm{T}}$ represents a list of samples from the conditional distribution of total fatigue life, given the event that a crack originates at node i:

$$F_{N^{\mathrm{T}}|x}(n^{\mathrm{T}}) = P[N^{\mathrm{T}}(\boldsymbol{X},A) \leq n^{\mathrm{T}} | \boldsymbol{X} = \boldsymbol{x}_i] \tag{7}$$

where $A \in \boldsymbol{R}$ is the random initial crack radius, and $\boldsymbol{x}_i \in \boldsymbol{R}^3$ is the position of node i. The conditional probabilities for each node are combined, via the theorem of total probability, to give the cumulative probability of failure of the component as

$$F_{N^{\mathrm{T}}}(n^{\mathrm{T}}) = \sum_{i=1}^{m} P[N^{\mathrm{T}}(\boldsymbol{X},A) \leq n^{\mathrm{T}} | \boldsymbol{X} = \boldsymbol{x}_i] \, P(\boldsymbol{X} = \boldsymbol{x}_i) \tag{8}$$

where m is the total number of nodes in the FE model, and $P(\boldsymbol{X} = \boldsymbol{x}_i)$ is the probability that the flaw originates at node i. The probabilistic distribution of flaw origin is assumed to be the uniform distribution or, using Bozek's particle cracking criteria (2008) for a more physics-based estimate, explicitly given by the ratio of broken particles at a damage origin divided by the total number of broken particles in the component. Finally, the probability of failure at some critical number of load cycles, $n_{\mathrm{cr}}^{\mathrm{T}}$, is:

$$P_f = F_{N^{\mathrm{T}}}(n_{\mathrm{cr}}^{\mathrm{T}}) = P\left[N^{\mathrm{T}}(\boldsymbol{X},A) \leq n_{\mathrm{cr}}^{\mathrm{T}}\right] \tag{9}$$

and the reliability, R, is computed as the probability of non-failure,

$$R = P\left[N^{\mathrm{T}}(\boldsymbol{X},A) > n_{\mathrm{cr}}^{\mathrm{T}}\right] = 1.0 - P_f \tag{10}$$

where $n_{\mathrm{cr}}^{\mathrm{T}}$ is some critical life measure.

Since the Level I simulation involves l life predictions at m FE nodes, there are $l \times m$ fatigue life predictions required. For reasonable FE meshes with a reasonable sample of initial crack radii, the number of fatigue life predictions can readily climb into the millions. This can be reduced, for example, by only considering surface nodes. However, with the domain decomposition and parallelization of tasks currently implemented in DDSim, nearly linear scaling is achieved and large simulations can be completed in relatively short amounts of time. For the example considered here, DDSim Level I processed 10,000 initial flaw sizes at about 64,000 surface nodes, in about 15 min on 340 processors (single core, 3.6 GHz, about 10^{12} FLOPS). For the "Digital Fortress," parallel, exascale (10^{18} FLOPS, about 10,000 multi-core processors, about a billion times faster than current desktop computers) computers would make very short work of a Level I filter on all its primary structural components.

Since the result of the Level I simulation is a scalar field, the conditional probability, or mean life prediction, or any other meaningful statistic, can be easily color-contoured and plotted on the surface of a component for effective visualization of results, as will be shown below in the example problem. This facilitates post-processing and provides visual guide to regions that are particularly susceptible to fatigue damage. No doubt an immersive stereo visualization of the "Digital Fortress," with all its hotspots color coded for priority, could be available at maintenance depots and would be an invaluable tool for fleet management.

Finally, the hotspots requiring higher-order evaluation are selected from the FE node locations using a criterion based on the Level I life prediction. Three criteria easily identified are: conditional probability of failure, absolute lowest life, and mean life. In this chapter, the minimum mean life prediction is used on the example component. The resulting hotspot is the *one* FE node location which has the lowest average Level I life prediction; however, choosing one location is for illustrative purposes only. In application, many locations would be chosen and prioritized obviously requiring high-performance computing resources.

4.1 Application of DDSim Level I to Example Problem

Figure 3 shows the geometry of the example problem. DDSim Level I was applied to this problem under constant amplitude and spectrum loading conditions. The component has six countersunk, counter-bored bolt holes at midlength which cause large, but differing, stress concentrations. The component was machined from rolled AA 7075-T651 plate. The specimen is modeled such that the x-direction corresponds to the material rolling direction (RD), the y-axis corresponds to the transverse direction (TD), and the z-axis corresponds with the normal direction (ND). This geometry is purposely *not symmetric* about the stiffener. There are two sources of asymmetry. First, the component is thicker for holes 11–13 than for holes 14–16. Second, the space between hole 16 and the aft edge of the specimen is less than the space between hole 11 and the forward edge of the specimen.

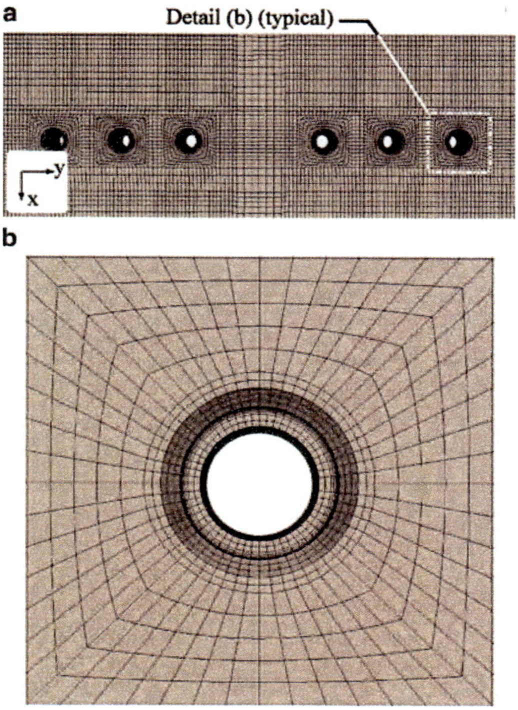

Fig. 5 Finite element model of example problem. (**a**) Surface mesh on un-stiffened side. (**b**) Typical detail around fastener hole

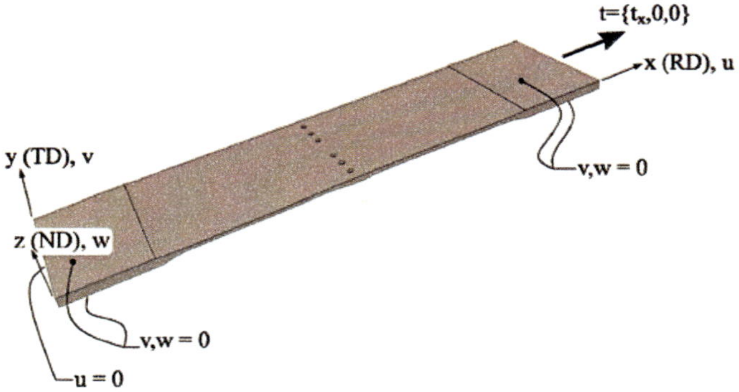

Fig. 6 Boundary conditions applied to FE model of example problem

Figure 5a shows the surface FE mesh on the model and 5b a close-up at one of the bolt holes. There were 140,024 eight-noded brick elements and 184 six-noded wedge elements in the input to DDSim. There are 63,974 nodes on the surface of the model.

Figure 6 shows the boundary conditions used in the model. The applied load for the spectrum loading corresponds to the maximum load in the spectrum. This spectrum, which contains many over- and underload excursions, was developed at the

Northrop Grumman Corporation (NGC) to model the service loads experienced by military aircraft. The applied load for the constant amplitude simulation corresponds to the root mean square (RMS) of the spectrum. The maximum load in the spectrum was 316.8 kN. In the constant amplitude loading analysis, the RMS load, 177.9 kN, was applied with $R = 0$. The constitutive model was linear elastic and isotropic.

Crack growth rate was computed using the familiar NASGRO equation (Forman and Mettu 1992). The material properties used with this equation were taken from the NASGRO (2008) material database for AA 7075-T651.

The distribution of a set of 10,000 potential flaw radii is shown in Fig. 7. This distribution was then passed through the filter developed by Bozek et al. (2008) to produce a subset of final flaw radii, a function of the state (stress, grain orientation, etc.) at each node location.

Figure 8 shows the mean life prediction contour plot for the component under the spectrum loading program with initial flaw sizes from the particle cracking filter. The lowest mean life prediction is 29,058 cycles, which occurs at the aft side of hole 16, near the intersection of the counter-bore with the main-bore. Figure 9 shows the predicted reliability of the component. This figure includes the results from the constant amplitude and spectrum loading programs.

The fatigue surfaces in hole 16 which showed large crack growth increments toward the outward edge of the fatigue crack in the aft side of hole 16 had the largest growth increments and the greatest surface area and, therefore, appeared to be the dominant crack. From these surfaces, it is evident that the origin of the fatigue cracks, in particular the dominant crack, is near the corner created by the intersection of the counter-bore with the main-bore. The Level I simulation matches this observation.

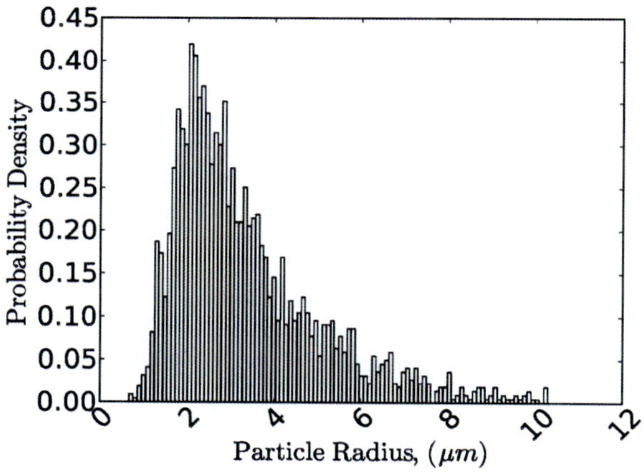

Fig. 7 Distribution of potential flaw size used in the particle cracking filter (Bozek et al. 2008) for DDSim Level I

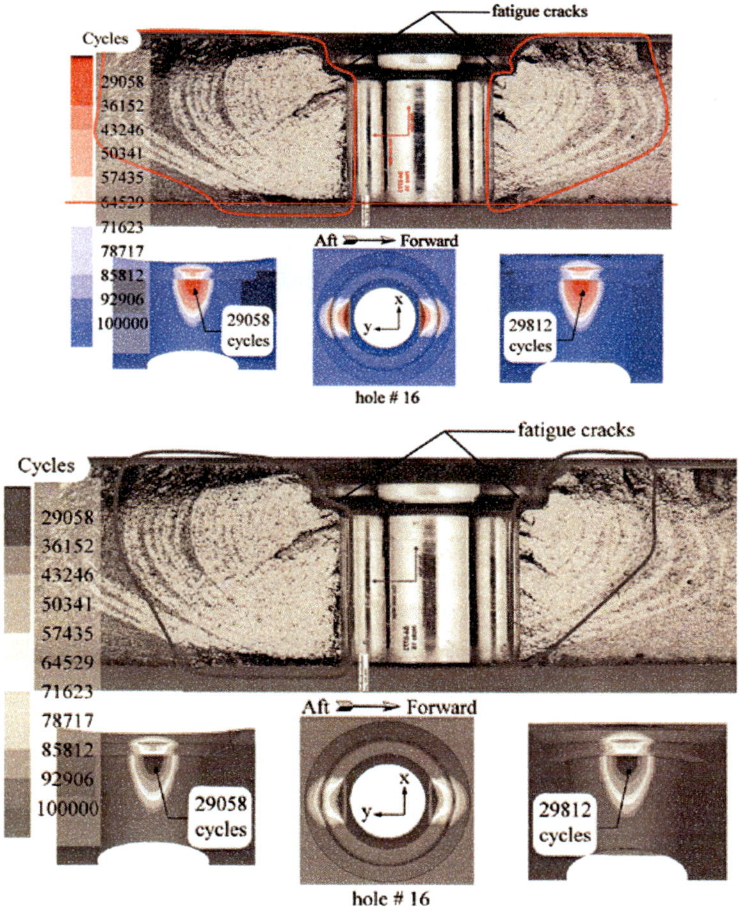

Fig. 8 *Above*, fracture surface at hole 16 in actual component, showing origins of fatigue cracks at shoulder of counterbore. *Below*, Level I prediction of location of minimum life flaw, and the average life prediction for the example problem under variable amplitude loading with 10,000 initial particles

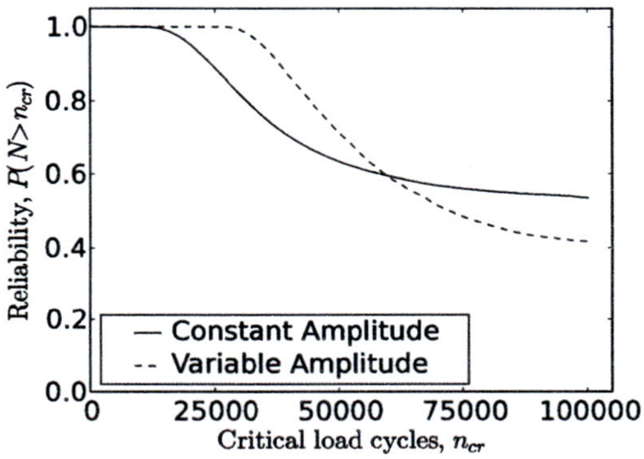

Fig. 9 Reliability prediction for the example component under both constant amplitude and variable amplitude loading with initial flaws generated with the particle cracking filter

5 DDSim Level II: High-Fidelity, MLC Growth Simulation

DDSim Level II is the second tier in the hierarchical fatigue life simulation. Level II performs high fidelity computations for the number of cycles consumed by MLC growth, n^{MLC}. The primary technical advances offered by Level II are automated crack insertion at hotspot locations determined from the Level I results, and support for cracks of arbitrary shape and orientation in a component of complex geometry and boundary conditions. The hotspot selection is based on the mean Level I life prediction. The three-dimensional crack growth simulations are conducted with the fracture analysis code FRANC3D/NG (2009). FRANC3D/NG performs its fracture mechanics computations based on field data obtained through any suitable FE analysis code.

Carter et al. (2000) describe crack growth simulation as an incremental process, where a series of steps are repeated for a progression of models. Each increment of the simulation relies on previously computed results and represents one crack configuration. There are four primary databases required for each increment. The first is the representational database, denoted by $\boldsymbol{R}_i$, where the subscript identifies the increment number. The representational database contains a description of the solid model geometry, including the cracks, boundary conditions, and material properties. The representational database is transformed by a discretization process $\boldsymbol{D}$ to a stress analysis database $\boldsymbol{\Sigma}_i$. The discretization process includes a meshing function $\boldsymbol{M}$:

$$\boldsymbol{D}[\boldsymbol{R}_i, \boldsymbol{M}(\boldsymbol{R}_i)] \Rightarrow \boldsymbol{\Sigma}_i \tag{11}$$

Level II automatically modifies the structural scale FE model from Level I to include the MLC with characteristic dimension a^{MLC}. The uncracked FE model that was read into Level I as input is first converted to the representational database. Then, the database is altered to account for the new geometry of the crack surface. The automatic crack insertion orients the crack perpendicular to the local maximum principal stress direction. At this point, the original discretization is no longer valid because the background geometry has changed. Hence, a new mesh must be created using the discretization process $\boldsymbol{D}$. The meshing function in DDSim uses an advancing front algorithm that originated with the work of Neto et al. (2001, 2008). FRANC3D/NG (2009) surrounds the crack front with a template of well-formed, singular, crack-front elements.

The analysis database contains a complete, but approximate, description of the body suitable for input to a solution procedure, $\boldsymbol{S}$, often a finite or boundary element stress analysis program. Any suitable analysis program is sufficient; however, in the example problem in this chapter, the commercial FE code, ANSYS (2006), was used. The analysis database, $\boldsymbol{\Sigma}_i$, is exported to the analysis program where the solution procedure, $\boldsymbol{S}$, is used to transform $\boldsymbol{\Sigma}_i$ to an equilibrium database, $\boldsymbol{Q}_i$. The equilibrium database consists of field variables, such as displacements and stresses, that define the equilibrium solution and contains appropriate material state information:

$$\boldsymbol{S}(\boldsymbol{\Sigma}_i) \Rightarrow \boldsymbol{Q}_i \tag{12}$$

The equilibrium database is then read back into FRANC3D/NG, where it is converted to the crack driving forces database, $\boldsymbol{K}_i$, with the fracture mechanics procedure, $\boldsymbol{F}$:

$$\boldsymbol{F}(\boldsymbol{Q}_i) \Rightarrow \boldsymbol{K}_i \tag{13}$$

Mixed-mode SIFs are automatically computed in FRANC3D/NG using the M-integral approach (Banks-Sills et al. 2005, 2007). By means of an update function, $\boldsymbol{U}$, $\boldsymbol{K}_i$, in conjunction with $\boldsymbol{R}_i$, is used to create a new representational model $\boldsymbol{R}_{i+1}$, which includes the crack growth increment. The crack growth function, $\boldsymbol{C}$, which is part of $\boldsymbol{U}$, determines the shape and extent of the crack growth increment:

$$\boldsymbol{U}[\boldsymbol{R}_i, \boldsymbol{C}(\boldsymbol{K}_i)] \Rightarrow \boldsymbol{R}_{i+1} \tag{14}$$

In the present example, the update function performs an automatic geometry and mesh update. Other techniques, e.g., partition of unity methods, could be implemented. This sequence of operations is repeated until a suitable termination condition is reached. The flowchart presented in Fig. 10 illustrates the path of data flow in the context of the Level II simulation. Of course, other approaches for simulating growth of cracks in a computational model are available. The advantages and disadvantages of these are discussed in Ingraffea (2008).

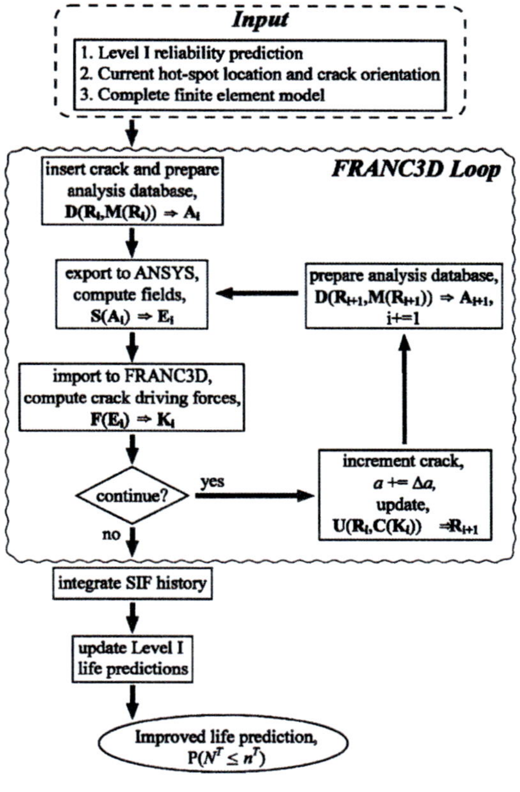

Fig. 10 Flowchart of DDSim Level II operations

The representation, analysis, and equilibrium databases would consume much of the petabytes of data used in tracking the life of the "Digital Fortress." Results of such a simulation might include one or more of the following: a final crack geometry, a loading versus crack size history, a crack opening profile, or a history of the crack-front fracture parameters. When a suitable stopping criterion is met, the SIF history is integrated to produce an estimate for the number of cycles consumed by MLC growth. Following the integration, the Level I life prediction is updated. The following section elaborates on these processes.

5.1 Input and the FRANC3D/NG Loop

The input required for a Level II analysis is as follows:

- Level I life prediction
- A hotspot location and crack orientation
- The complete structural scale FE model

Level II uses the complete, structural scale FE model to create its representational database. This does not require any additional user effort because the entire model, including boundary conditions, was input to Level I. In the case of variable amplitude loading, the analysis database is computed for a single loading with the assumption that linear superposition is valid. The choice is arbitrary and user defined, but in the example shown herein the maximum load in the spectrum was used. During the SIF history integration phase, the driving force is scaled appropriately to correspond with the load at the current cycle in the spectrum.

The output from FRANC3D/NG is the history of the crack driving force database $\{\boldsymbol{K}_0, \ldots, \boldsymbol{K}_n\}$, where the subscript n is the total number of increments in the loop. This history relates crack length to the corresponding SIF. The integration routine treats discrete points along the crack front as separate, non-interacting cracks in a two-dimensional body.

Crack *length* is ill-defined for a crack in a three-dimensional body because a crack is actually a two-dimensional manifold; so a convenient convention is adopted. The crack length for Level II is defined as the distance along the set of straight lines connecting a discrete crack front point. The crack front point is given by the normalized crack front position. The dash-dot line in Fig. 11 illustrates the crack length for the midpoint, B_i, along a crack front. This illustration is for a very large crack increment; smaller increments generally lead to a more accurate representation.

The SIF history is extracted from the $\{\boldsymbol{K}_0, \ldots, \boldsymbol{K}_n\}$ database with a utility which requires the normalized crack front point to be specified. The integration routine uses a modified version of the state-of-practice loop from Level I. However, in place of the analytical SIF solutions used in Level I, the FE-computed SIF history is interpolated to compute the cyclic crack driving force. The integration is conducted at each crack front point considered, and the minimum is taken as the high-fidelity estimate of the number of cycles consumed by MLC growth, N^{MLC}. In the example

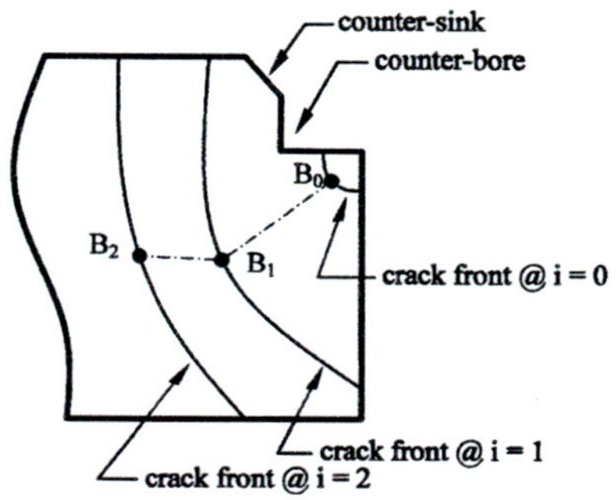

Fig. 11 The crack length for point B_i at crack increment $i = 0, 1, 2$

in this chapter, three crack front points are used. One point is used at each surface-breaking location and one is used at the midpoint of the crack front. The final step in the Level II process is to update the prediction from Level I.

5.2 *Application of DDSim Level II to Example Problem*

In this section, one hotspot in the example component is evaluated with DDSim Level II. A crack is inserted at the hotspot with the lowest mean life prediction computed in the Level I simulation under variable amplitude loading with the initial flaws determined by the particle cracking filter.

The lowest mean total life prediction from that dataset was 29,058 cycles with a crack originating from the intersection between the counter-bore and main-bore at the aft side of bolt hole 16. The Level II result should be an improvement of the Level I fatigue life estimate. The following subsections describe the cracked model, the updated Level I prediction, and discussion.

To facilitate the crack insertion process, a subregion of the structural-scale mesh was used to prepare the initial analysis database, $\mathbf{\Sigma}_o$, around bolt hole 16. Figure 12a shows a portion of this subregion at the aft edge of bolt hole 16 and the surface mesh. Figure 12b is a cross-section in the plane of the aft edge crack showing the quarter-penny-shaped initial crack of size $a^{\mathrm{MLC}} = 0.381$ mm, the default initial crack size for MLC.

There were a total of 13 crack increments used in the Level II simulation (Fig. 13). Although the predicted cracks could have taken any shape, unrestricted by the mesh model, they remained essentially in their original plane, perpendicular to the local maximum principal stress direction at the crack origin. Figure 13 shows the fracture surface of the fatigue specimen at hole 16 (top) and the crack growth steps of the Level II simulation (bottom). This allows qualitative visual validation. Generally, the simulated crack fronts compare well with the actual fracture surfaces, especially during the early stages of MSC growth. Thereafter, the observed crack front has more curvature, but comparison of what appear to be fatigue striations with the last five or so predicted shapes is not valid. These are not fatigue

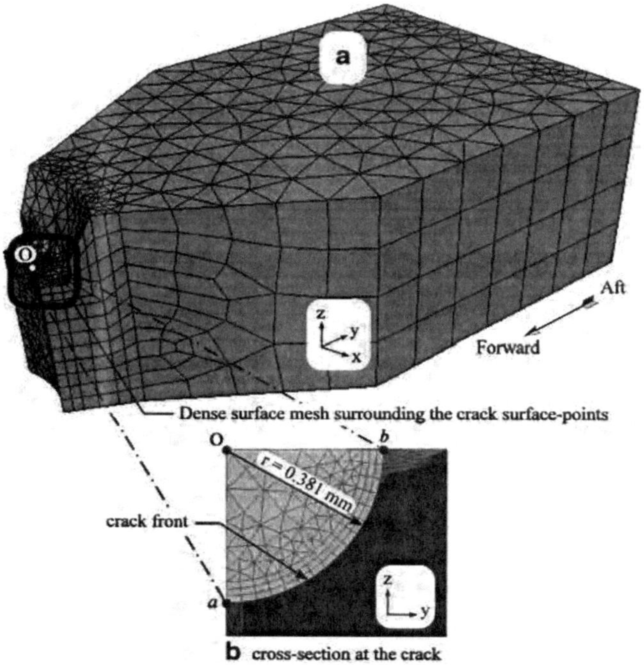

Fig. 12 Portion of analysis database, Σ_o, at initial step showing FE model of one side of bolt hole: (**a**) portion of subregion; and (**b**) one of the crack fronts. The point labeled "O" is the crack origin. The points labeled *a* and *b* are the crack front surface points

striations, but rather are pop-in markings, each associated with a few of the final spectrum cycles. Nevertheless, this comparison highlights some as yet unresolved issues in simulating fully 3D, arbitrary fatigue crack growth. Among these is that the K-range is computed assuming plane strain conditions at all locations along the crack front. This over-predicts the K-range at the plane stress, surface points on the crack front. Next, the algorithm to advance the crack front for each increment uses only the ratio of the SIFs and cannot account for varying fatigue resistance with material direction: in general, no point along the crack front is always moving in a constant material direction. Finally, the specimen was machined from a thick plate. Clearly, the thickening-pad on the stiffener-side (inner-wing surface) of the plate is deeper into the original plate than the outer-wing surface. This inevitably results in asymmetry in the material toughness, producing faster growth rates toward the outer-wing surface. Overall, it is inconclusive how these effects impact the Level II prediction of n^{MLC}.

The Level II life prediction is the result of cycle-by-cycle integration of the SIF histories conducted at three crack front points. The minimum life prediction is taken to be the high fidelity estimate of number of cycles consumed by MLC growth, n^{MLC}. The life prediction was governed by surface point b (see figure) which had

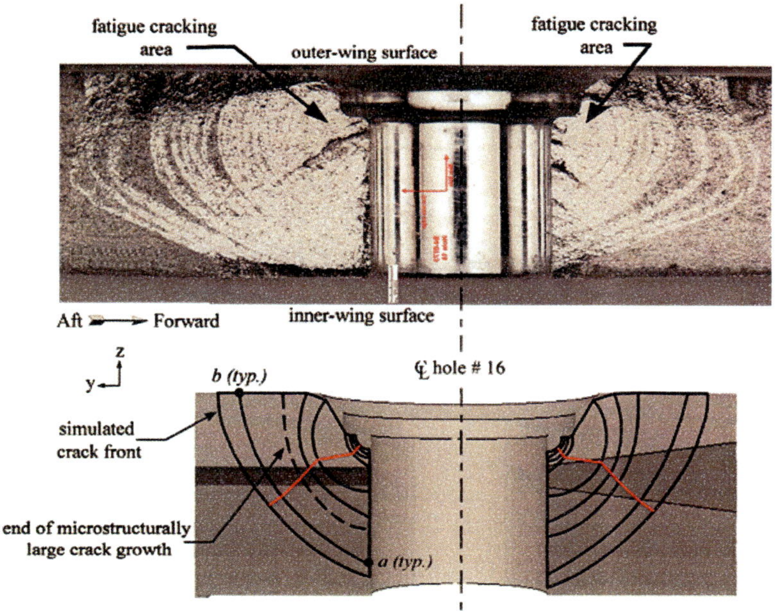

Fig. 13 Image of fracture surface at the aft side of bolt hole 16 (*top*) compared with the simulated fracture surface (*bottom*). Dotted, simulated crack front marks end of microstructurally large fatigue crack growth region, at $n^{\mathrm{MLC}} = 4070$ cycles

the minimum number of cycles of 4,070. The Level I conservative estimate of n^{MLC} was only 147 cycles. With the update from Level II, the mean life prediction at the hotspot, for spectrum loading with initial flaws from the particle filter, shifts from 29,058 cycles to 32,981 cycles of spectrum load. Applying the update at only one hotspot has no visible effect on the total reliability of the structure. This is the expected result considering the summation in (8) is over all surface nodes.

From the experiments conducted at NGC, the component failed after 53,485 cycles of fatigue loading. The minimum average Level I fatigue life prediction was 29,058, or, conservatively as desired, 46% less than the test. Including the update from the Level II simulation, that average life prediction changed to 32,981, or 38% less than the test. However, there is no way to guarantee that the current hotspot remains the location of minimum average life prediction without performing updates at all damage origins having Level I life predictions $\leq$32,981 cycles. Also, it is of interest to note that the predicted 4,070 cycles of MLC growth are approximately 8% of the observed total fatigue life. This estimate agrees well with the literature that suggests microstructural growth processes can consume most of the total fatigue life (Suresh 1998; Fan et al. 2001). Finally, the predicted probability of failure at 53,485 cycles is not affected by Level II updating of only one hotspot.

6 DDSim Level III: High-Fidelity, MSC Growth Simulation

DDSim Level III is the third tier in the hierarchical fatigue life simulation. Level III directly couples FE models of the material microstructure with FE models of the structural length scale. The microstructural models can include as much detail about phase, morphology, and texture as necessary to capture crack incubation and nucleation processes. Hence, the hotspots selected from the Level I simulation are used as the focal points of the microstructural models. Much of the work supporting Level III is ongoing, as described in other chapters of this book. Nevertheless, the discussion here will include a vision for the final product which will result in a high-fidelity fatigue life prediction by accurately computing the number of cycles consumed by microstructurally small cracking processes. The processes to be included are crack incubation and nucleation – the nano- and microscale processes preceding the appearance of new surface area, and the appearance of new surface area, respectively – and MSC propagation.

The work presented here is intended to provide the foundation for further advancement of the DDSim methodology. Continued work is being done in the following arenas:

- development of FE models that are accurate statistical realizations of the microstructure of many important aerospace alloys, such as those discussed in other chapters of this book
- development of models for the mechanics of MSC propagation, and the damage processes discussed in other chapters
- development of numerical methods to couple length scales
- development of statistical methods to maximize the efficiency of the DDSim methodology

Many multiscale simulation approaches in the literature allow unilateral data flow either downward from the structural length scale or upward from the microstructural length scale, by way of micromechanically informed constitutive modifications. Often, it is assumed that the local fields in the structural length-scale model have negligible gradient over the relatively short dimensions of the microstructural model (Fish and Shek 2000; Fish and Belsky 1997). DDSim Level III is designed to allow two-way data flow. The FE models of the two length scales are directly coupled either with multipoint constraints or with a modified multigrid approach that does not require homogenization (Bozek 2007; Datta et al. 2004). The microstructural damage is allowed to accumulate in situ, as it would in service, and the response is directly palpable by the structural scale model as envisioned within the evolution of the digital HyperFortress. That is, the fields resulting from the boundary conditions applied to the structural model follow their preferred paths – trickling down to the crystal lattice length scale – and the material's response is propagated back upward.

6.1 Generation of a Microstructural Model

The microstructural model should be statistically realistic. Such a representation can begin with a grain morphology created by a simulated annealing process (Brahme et al. 2006; Saylor et al. 2004). This process itself begins by creating a Voronoi tessellation. Subsequently, the volume is densely packed with ellipsoids whose dimensions are generated from the observed statistics of grain dimensions. The grain geometry is created by grouping Voronoi cells whose centroids are contained within a common ellipsoid. After the geometry is generated, meshing is automatically completed using the same advancing front algorithm as in Level II. Both representations can include cracked second-phase particles, as described by Bozek et al. (2008) and Veilleux (2007). Figure 14a shows the geometry of a 128 grain polycrystal which includes 28 second-phase surface particles and 14(b) shows the surface mesh. The volume mesh of the polycrystal model shown in the figure consists of over 3.2 M quadratic tetrahedra with over 13 M DOF.

Finally, the mechanics of the crystallographic response are approximated with a crystal plasticity model (Matous and Maniatty 2004). This model includes the anisotropic elasto-plastic behavior of a grain, and includes parameters that describe the lattice orientation of the grain. The polycrystal model then accounts for texture by assigning orientations to grains based on observed statistics.

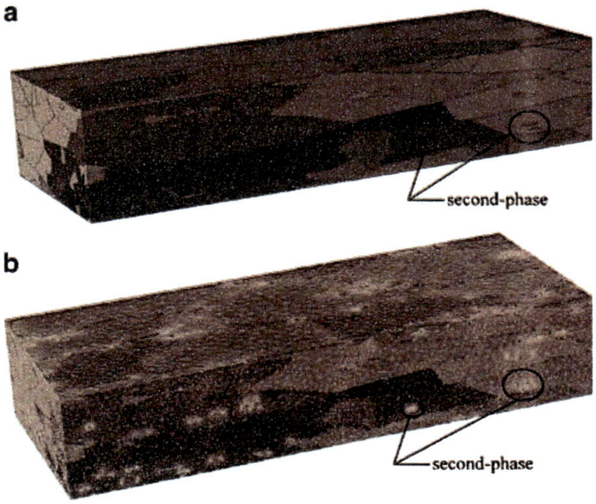

Fig. 14 A digital replication of a statistically accurate 128 grain polycrystal of 7075-T651 which includes 28 second-phase particles: (**a**) the geometry; and (**b**) the surface mesh

6.2 Level III Input and Operations

Figure 15 is the flowchart for the Level III simulation using the multipoint constraint approach. The input to Level III comes from Levels I and II and includes the complete FE model of the structural length scale. The geometry model of this scale is modified to accommodate the aforementioned polycrystal models. The polycrystal model is generated, the multipoint constraint equations are written, or the multigrid procedure is executed, and the multiscale model is analyzed. Following the analysis, the conditional probability of failure is computed and the loop is repeated as necessary. The following discussion elaborates on the process.

The input to Level III includes the structural length-scale FE model that was used to make the Level I predictions. The other essential inputs to Level III are the current hotspot location and the conditional reliability at that hotspot. The former is required so that the structural model can be modified. The latter is required because it is conceivable that a Bayesian analysis could be developed, whereby the Level I reliability prediction is used as the prior distribution and is improved by only a few multiscale analyses.

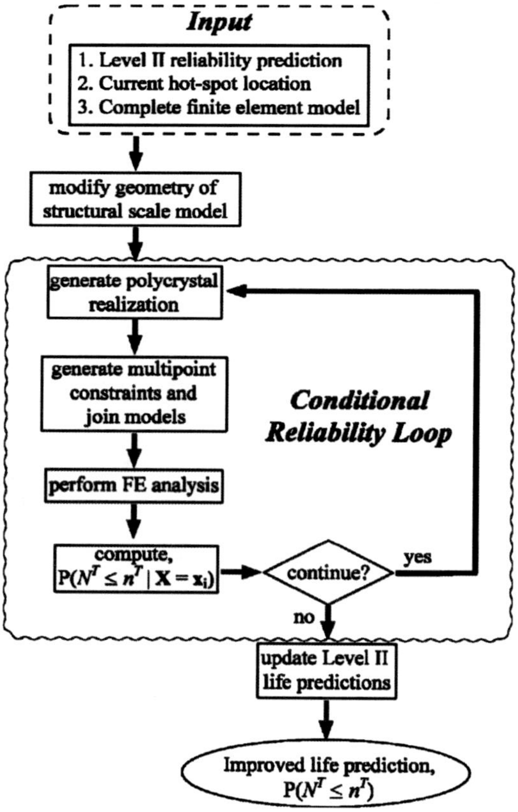

Fig. 15 DDSim Level III flowchart for the multipoint constraint approach

Interactive tools have been created to modify automatically the geometry and mesh of the structural length-scale model. The first tool allows entry of the coordinate position of the centroid of the polycrystal model, information obtained from Level I. The next tool defines the region to be removed from the structural scale model. This tool automates this process and allows the user to control the mesh density on the new surfaces created during the process. Figure 16a shows the region definition and mesh density control panel. The mesh density control panel allows the user to control the number of elements along the edges of the void. Subsequently, the region can be rotated or translated, as shown in Fig. 16b, to allow the microstructural model to be aligned with the rolling, transverse, and normal direction of the structure.

Figure 17 shows the renovated example model. The void to accommodate the polycrystal is circled in black. The automatically remeshed region, between the undisturbed structured mesh and the inserted polycrystal FE model, is obviously distinguishable.

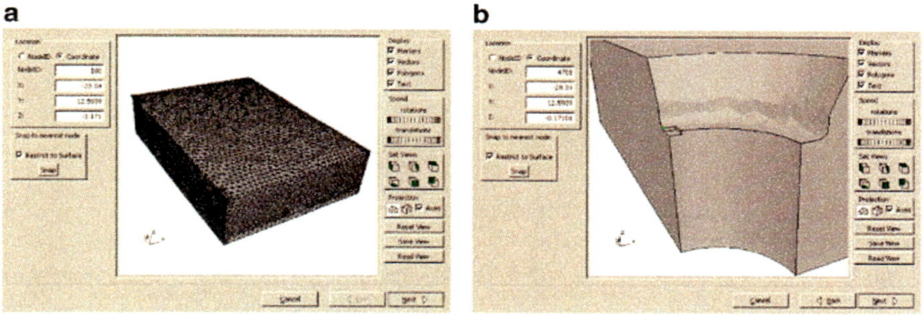

Fig. 16 (**a**) The microstructure region definition and mesh control panel. (**b**) The region rotation and translation panel. Size and location of microstructural model shown

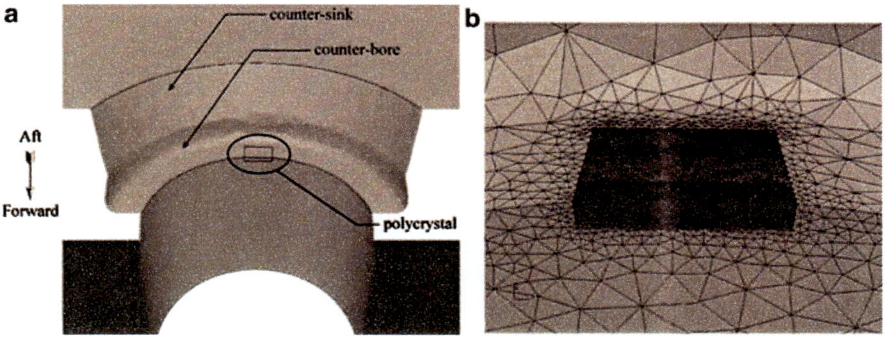

Fig. 17 The renovated example (**a**) with the structure-scale mesh locally modified, and (**b**) a polycrystal FE model inserted

The conditional reliability loop is where the high fidelity prediction of conditional reliability for the given hotspot is computed. The necessary steps for the multipoint constraint option are to:

- generate the polycrystal realization
- generate the multipoint constraints and merge the models
- perform the FE analysis
- compute the conditional reliability

After the multiscale analysis, the conditional reliability is computed. The conditional reliability loop is repeated until the required fidelity at the current hotspot is achieved. After achieving adequate fidelity in the conditional reliability, the conditional reliability loop is exited and the total structural reliability is updated. With the total reliability updated at the current hotspot, the simulation continues at the next hotspot as shown in Fig. 15. The details regarding the requisite number of iterations through the conditional reliability loop are beyond the scope of this chapter and are an open topic of research. However, if computing resources were not a limitation, the conditional reliability loop of Level III could be used in a Monte Carlo simulation with a large number of microstructural realizations.

6.3 Application of DDSim Level III to Example Problem

As noted earlier in this section, the rules of behavior for all the MSC processes are still being discovered and encoded, as evidenced by other chapters in this book. Therefore, a Level III update to N^{MSC} is not yet possible. However, it is possible to show how a microstructural model can be analyzed within the DDSim framework. The one-way example shown here is the two-phase polycrystal shown in Fig. 14, under a single cycle of uniaxial straining. A result is depicted in Fig. 18a, which clearly shows the highly heterogeneous RD stress distribution resulting from crystal-plastic deformation operating on statistically accurate texture. Figure 18b shows local fields around, and stress concentrations in, some of the stiffer surface particles, here idealized as semi-ellipsoids. This state would be the starting point for simulation of incubation in this alloy: particle cracking. This would be followed by the nucleation stage, during which some of the cracked particles spawn intragranular cracks, the MSC propagation stage leading to intergranular cracking and coalescence, and, ultimately, the MLC stage leading to component failure. The physics, mechanics, and statistics of all these stages need to be discovered or tested for validity in this new era of combined physical and computational simulation.

This single elasto-crystal plastic analysis required about 64 h to execute on a 240 (single core, 3.6 GHz) processor cluster, with a parallel FE code developed by the Cornell Fracture Group (FRANC3D/NG 2009). The code uses PETSc (Balay et al. 2006) for parallel equation solving, and an in-house FE library, FemLib, for element and constitutive formulations. Again, the need for exascale computing power for reliable prognosis for "digital aircraft" is evident.

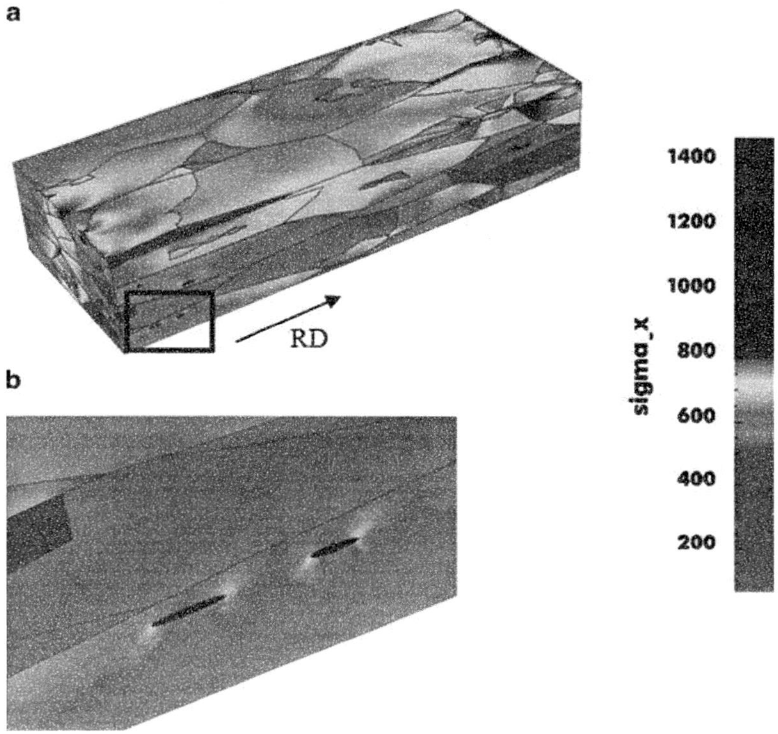

Fig. 18 (**a**) Contours of RD normal stress (MPa) on microstructural model under a 1% RD applied strain. (**b**) Detail showing large stress concentration in uncracked, stiffer particles

7 Conclusions

Highly detailed digital instantiations of future air vehicles will co-exist with their tangible counterparts. With such "digital aircraft" as both database and virtual sensor, computational simulations of material degradation and structural performance will be done during design, development, testing, and service. This chapter described a humble prototype of a system within which multiscale simulations of fatigue cracking could be coherently performed.

DDSim's hierarchical design allows reduced-order, fast searches for likely trouble spots across the many FE models which will exist for the primary structural components of airframe and engine. It predicts a conservative reliability estimate for each and then offers opportunities to improve the fidelity of the reliability prediction. Rigorous component-scale FE simulations can then be performed to improve the reliability of the MLC portion of total life. In addition, material-scale simulations can be performed at these trouble spots, involving all the developments in understanding of nano- and microstructural damage processes described in the other chapters of this book, to improve reliability of the MSC portion of life.

Much work remains to be done to reach the vision offered in the prologue to this chapter. To "...uncover and quantify material and structural limit states with reliability predictions..." the rules of physics mechanics, and chemistry governing all aspects of fully 3D MSC processes, still need to be discovered. The stochastics of these processes will then have rational explanation. These rules must then be properly encoded, within software not yet invented, that can be executed efficiently, on computers not yet built, *so we can use our "... most advantageous tool – nearly infinite computing power."*

Acknowledgments The authors gratefully acknowledge partial sponsorship for the research reported here by the Defense Advanced Research Projects Agency under contract HR0011–04-C-0003, Dr. Leo Christodoulou, Program Manager, and by NASA through the Constellation University Institutes Program, grant number NCC3–994. This manuscript has been co-authored by a contractor of the US Government at Sandia National Laboratories. Sandia is a multi-program laboratory operated by Sandia Corporation, a Lockheed Martin Company, for the United States Department of Energy's National Nuclear Security Administration under Contract DE-AC04–94AL85000. The authors also thank Drs. Wash Wawrzynek and Bruce Carter for their development of FRANC3D/NG. Finally, the authors wish to dedicate this chapter to Dr. John Papazian: he has led from wisdom and with collegiality.

References

AFGROW (2008) Users' Guide and Technical Manual. Technical Report AFRL-VA-WP-TR-2008-XXXX. Air Force Research Laboratory, WPAFB, OH

ANSYS (2006) ANSYS, Release 10.0. ANSYS, Inc., Canonsburg, PA

Balay S, Buschelman K, Eijkhout V, Gropp W, Kaushik D, Knepley M (2006) PETSc Users Manual. Argonne National Laboratory report anl-95/11-rev. 2.3.2 edn

Banks-Sills L, Hershkovitz I, Wawrzynek PA, Eliasi R, Ingraffea AR (2005) Methods for calculating stress intensity factors in anisotropic materials: Part I – $z = 0$ is a symmetric plane. Eng Fract Mech 72:2328–2358

Banks-Sills L, Wawrzynek PA, Carter BJ, Ingraffea AR, Hershkovitz I (2007) Methods for calculating stress intensity factors in anisotropic materials: Part II – arbitrary geometry. Eng Fract Mech 74:1293–1307

Bozek JE (2007) A 2D Multiscale Procedure for Fatigue Crack Nucleation. M.S. Thesis, Cornell University

Bozek JE, Hochhalter JD, Veilleux MG, Liu M, Heber G, Sintay SD, Rollett AD, Littlewood DJ, Maniatty AM, Weiland H, Christ Jr. RJ, Payne J, Welsh G, Harlow DG, Wawrzynek PA, Ingraffea AR (2008) A geometric approach to modeling microstructurally small fatigue crack formation, part I: probabilistic simulation of constituent particle cracking in AA 7075-T651. Modell Simul Mater Sci Eng 16: article number 065007

Brahme A, Alvi MH, Saylor D, Fridy J, Rollett AD (2006) 3D reconstruction of microstructure in a commercial purity aluminum. Scripta Mater 55:75–80

Carter BJ, Wawrzynek PA, Ingraffea AR (2000) Automated 3-D crack growth simulation. Int J Numer Methods Eng 47:229–253

Datta DK, Picu RC, Shepard MS (2004) Composite grid atomistic continuum method: an adaptive approach to bridge continuum with atomistic analysis. Int J Multiscale Comp Eng 2(3):401–419

Emery JM (2007) Hierarchical, Probabilistic, Damage and Durability Simulation Methodology. Ph.D. Thesis, Cornell University

Emery JM, Hochhalter JD, Wawrzynek PA, Ingraffea AR (2009) DDSim: a hierarchical, probabilistic, multiscale damage and durability simulation methodology – Part I: methodology and Level I. Eng Fract Mech (in press)
Fan J, McDowell DL, Horstemeyer MF, Gall K (2001) Computational micromechanics analysis of cyclic crack-tip behavior for microstructurally small cracks in dual-phase al-si alloys. Eng Fract Mech 68:1687–1706
Fish J, Belsky V (1997) Generalized aggregation multilevel solver. Int J Numer Methods Eng 40(23):4341–4361
Fish J, Shek KL (2000) Multiscale analysis of large scale nonlinear structures and materials. Int J Comp Civil Struct Eng 1(1):79–90
Forman RG, Mettu SR (1992) Behavior of surface and corner cracks subjected to tensile and bending loads in Ti-6Al-4V alloy. In: Fracture Mechanics: Twenty second Symposium, ASTM STP-1131. American Society for Testing and Materials, Philadelphia, pp. 519–546
FRANC3D/NG (2009) Three-dimensional fracture analysis code. http://www.cfg.cornell.edu/software/software.htm. The Cornell Fracture Group, Cornell University, Ithaca, NY
Ingraffea AR (2008) Computational fracture mechanics. In: Stein E, de Borst R, Hughes T (eds) Encyclopedia of Computational Mechanics, 2nd edn, John Wiley and Sons, New York, Volume 2, Chapter 11
NASGRO (2008) Fatigue crack growth analysis software, version 5.2. Southwest Research Institute and National Aeronautics and Space Administration, San Antonio, TX
Matous K, Maniatty A (2004) Finite element formulation for modelling large deformations in elasto-viscoplastic polycrystals Int J Numer Methods Eng 60:2313–33
Neto JCB, Wawrzynek PA, Carvalho MTM, Martha LF, Ingraffea AR (2001) An algorithm for three-dimensional mesh generation for arbitrary regions with cracks. Eng Comput 17:75–91
Neto JCB, Miranda A, Martha L, Wawrzynek PA, Ingraffea AR (2008) Surface mesh regeneration considering curvatures. Eng Comput (in press)
Papazian JM, Anagnostou EL, Engel SJ, Fridline DR, Hoitsma DH, Madsen JS, Silberstein RP, Whiteside JB (2007a) Structural integrity prognosis. In: Lazzeri L, Salvetti A (eds) Proceedings 24th Symposium of the International Committee on Aeronautical Fatigue. pp. 109–125
Papazian JM, Anagnostou EL, Engel SJ, Fridline DR, Hoitsma DH, Madsen JS, Nardiello J, Silberstein RP, Welsh G, Whiteside JB (2007b) SIPS, a structural integrity prognosis system. In: Proceedings 2007 IEEE Aerospace Conference, Big Sky MT, paper no. 11.0901 (T11/Z11_0901)
Saylor DM, Fridy J, El-Dasher BS, Jung KY, Rollett AD (2004) Statistically representative three-dimensional microstructures based on orthogonal observation sections. Metall Mater Trans 25A(7):1969–1979
Suresh S (1998) Fatigue of Materials. Cambridge University Press, Cambridge
Veilleux MG (2007) Finite element model generation of statistically accurate 7075-T651 aluminum alloy microstructures. M.S. Thesis, Cornell University

List of Symbols and Abbreviations

$\boldsymbol{\Sigma}_i$	stress analysis database
$\boldsymbol{A}$	random vector, p-dimensional array describing flaw geometry
a^{MLC}	the characteristic length of a crack when it can be considered microstructurally large
$\boldsymbol{C}$	crack growth function
CDF	cumulative distribution function
$\boldsymbol{D}$	discretization process

$\boldsymbol{E}$	random vector, s-dimensional array describing material resistance to crack growth
$\boldsymbol{F}$	fracture mechanics process
FE	finite element
FLOPS	floating-point operations per second
$\boldsymbol{K}_i$	crack driving force database
MLC	microstructurally large crack
MSC	microstructurally small crack
ND	normal direction
N^{MLC}	a random variable, the number of cycles consumed by microstructurally large crack growth processes
n^{MLC}	a realization of N^{MLC}
N^{MSC}	a random variable, the number of cycles consumed by microstructurally small crack growth processes
n^{MSC}	a realization of N^{MSC}
N^{T}	a random variable, the total fatigue life of a structure, integer, cycles
n^{T}	a realization of N^{T}
$n^{\mathrm{T}}_{\mathrm{cr}}$	a critical realization of N^{T}
$\boldsymbol{P}_f$	probability of failure
$\boldsymbol{Q}_i$	equilibrium database
R	load ratio
R	reliability
RD	rolling direction
$\boldsymbol{R}_i$	representational database
RMS	root mean square
$\boldsymbol{S}$	solution procedure
TD	transverse direction
$\boldsymbol{U}$	update function
$\boldsymbol{X}$	random vector, dominant flaw location

Modeling Fatigue Crack Nucleation Using Crystal Plasticity Finite Element Simulations and Multi-time Scaling

Somnath Ghosh, Masoud Anahid, and Pritam Chakraborty

Abstract This chapter addresses two important aspects of predicting fatigue crack nucleation in polycrystalline alloys under dwell cyclic loading. The first is a microstructure sensitive criterion for dwell fatigue crack initiation in polycrystalline titanium alloys. Local stress peaks due to load shedding from time-dependent plastic deformation fields in neighboring grains are responsible for crack initiation. Crystal plasticity finite element simulation results are post-processed to provide inputs to the fatigue crack nucleation model. The second part of this chapter discusses a wavelet transformation based multi-time scaling (WATMUS) algorithm for accelerated crystal plasticity finite element simulations. The WATMUS algorithm does not require any scale-separation and naturally transforms the coarse time scale response into a "monotonic cycle scale" without the requirement of subcycle resolution. The method significantly enhances the computational efficiency in comparison with conventional single timescale integration methods. Adaptivity conditions are also developed for this algorithm to improve accuracy and efficiency.

1 Introduction

Many metals and alloys, such as titanium alloys and Ni-base superalloys, find widespread utilization in various high performance applications, such as in the automotive and aerospace sectors. Developments in advanced materials have contributed tremendously to the design and implementation of components with improved properties and reliability. During service, many of these components are exposed to cyclic loading conditions due to start up and shut down processes or load reversals. In many cases, this results in their fatigue or time-delayed fracture. Fatigue failure in

S. Ghosh, John B. Nordholt Professor (✉)
Department of Mechanical Engineering, The Ohio State University, W496 Scott Laboratory, 201 West 19th Avenue, Columbus, OH 43210, USA
e-mail: ghosh.5@osu.edu

S. Ghosh and D. Dimiduk (eds.), *Computational Methods for Microstructure-Property Relationships*, DOI 10.1007/978-1-4419-0643-4_14,

the microstructure evolves in three stages (Suresh 1998): (a) crack nucleation due to inhomogeneous plastic flow or grain boundary failure, (b) subsequent crack growth by cyclic stresses, and (c) coalescence of cracks to cause fast crack propagation.

A large body of literature exists in the field of fatigue of metals. The phenomena of high cycle and low cycle fatigue have been traditionally characterized using macroscopic parameters such as applied stresses, cyclic frequency, loading waveform, hold time, as well as statistical distributions of fatigue life and fatigue strength (Suresh 1998; Coffin 1973; Laird 1976; Fleck et al. 1994; Hashimoto and Pereira 1996). Fatigue analysis by total life approaches includes (a) the stress-life or S–N approach, where the stress amplitude versus life is determined, and (b) the strain-life approach, e.g., the Coffin–Manson rule, where the number of cycles to failure is determined as a function of plastic strain. In the stress-life approach, the applied stresses are nominally elastic and the number of cycles to failure is large as in high cycle fatigue. On the other hand, in the strain-life approach components undergo significant plastic straining and crack propagation. The total life approaches have been adjusted for notch effects using fatigue strength reduction and for variable amplitude fatigue, e.g., in the Palmgren–Miner rule of cumulative damage. In all the cases, microstructural effects are represented by shifts in such data curves after extensive testing. Alternatively, the defect or damage tolerance approaches determine fatigue life as the number of cycles to propagate a crack from a certain initial size to a critical size. These are determined from threshold stress intensity, fracture toughness, limit load, allowable strain or allowable compliance. The models assume the presence of a flaw in the structure, and predict life using laws such as the Paris law (Paris 1964). Fatigue crack advance has been conventionally based on linear elastic fracture mechanics analysis related to the concepts of similitude, in which, stress intensity factors uniquely characterize fatigue crack growth. Two design criteria are conventionally employed in current practice (Fredell 2008). In the low-cycle-fatigue design (*safe life*), which is based on statistical distribution of in-service stress levels versus fatigue life, the "safe level" low bound is selected as when 1 in 1,000 components is predicted to initiate a 0.8 mm crack. To eliminate this one possibly-defective component, all the 1,000 are removed from service (rejected) at this service life. As a result of this practice, a great number of expensive parts are retired prematurely before damage initiation, thus shortening their full useful life. Damage-tolerant design, on the other hand, is based on deterministic fracture mechanics and requires 1 or 2 safety inspections during service life. This also incurs additional costs.

However, predictions of these widely used models can suffer from significant scatter. This is primarily due to the absence of robust underlying physical mechanisms and information on the material microstructure in their representation. Morphological and crystallographic characteristics of the microstructure, e.g., crystal orientations, misorientations, and grain size distribution, play significant roles in the mechanical behavior and fatigue failure response. Fatigue studies in (Lord and Coffin 1973; Tsuji and Kondo 1987; Antolovich et al. 1981) have demonstrated the influence of deformation and damage mechanisms, creep, oxidation, and microstructural instabilities on cyclic life. Accurate modeling of fatigue failure inherently involves coupling of multiple spatial scales ranging from those of individual grains or polycrystalline aggregates to that of the structural component. Detailed modeling of real microstructures with commercial FEM codes are computationally intense, and can suffer from limitations in computational efficiency, accuracy, resolution, and numerical stability in regions of localized deformation and damage. To overcome limitations incurred in oversimplified models, probabilistic methods have been developed to provide uncertainty quantification for life prediction. The DARWIN (Design Assessment of Reliability with Inspections) code (McClung et al. 2004) has been developed to calculate the risk of fracture caused by specific damage sources and the total accumulated risk over the projected lifetime of the part. It uses risk assessment methods to determine the probability of fracture by integrating finite element stress analysis results, fracture mechanics models, material anomaly data, probability of crack detection, and inspection schedules within a graphical user interface. A severe limitation of this analysis arises from the quality of stress analysis and damage data, conventionally derived from isotropic elastic finite element simulations. Much can be gained in reducing uncertainties in life prediction if better simulations, accounting for the realistic deformation and failure mechanics can be incorporated into the probabilistic analysis.

The recent years have seen a paradigm shift towards the use of material microstructure based detailed mechanistic models for predicting fatigue crack nucleation and propagation, as a promising alternative to the empirical models. Many of these approaches seek accurate description of material behavior through crystal plasticity-based finite element models. Crystal plasticity theories with explicit grain structures are effective in predicting localized cyclic plastic strains (Mineur et al. 2000; Bennett and McDowell 2003; Chu et al. 2001). The mechanical response of polycrystalline aggregates are deduced from the behavior of constituent crystal grains with specific assumptions about their interaction. Various computational studies have modeled anisotropy and its evolution in large deformation processes with this approach (Harren and Asaro 1989; Mathur et al. 1990; Kalidindi et al. 1991, 1992; Beaudoin et al. 1995). Finite element calculations have shown that depending on the loading conditions, significant gradients of stresses or strains can evolve, even within a single slip system. Dawson and coworkers (Turkmen et al. 2003; Dawson 2000) have investigated the mechanical behavior of aluminum alloys in cyclic loading using crystal plasticity-based FEM simulations of crystalline aggregates. McDowell and coworkers have incorporated a crystal plasticity model with kinematic hardening to model cyclic plasticity in high cycle fatigue in Ti-6Al-4V (Bennett and McDowell 2003). Ghosh et al. have developed crystal plasticity models for deformation and creep in titanium alloys in (Hasija et al. 2003; Deka et al. 2006; Venkatramani et al. 2006; Venkataramani et al. 2007, 2008) and have modeled deformation and ratcheting fatigue of HSLA steels in (Xie et al. 2004; Sinha and Ghosh 2006). These calculations provide a platform for the implementation of physics-based crack nucleation and propagation criterion that accounts for the effects of microstructural inhomogeneity. In Ghosh et al. (Kirane and Ghosh 2008; Kirane et al. 2009; Anahid et al. 2009), a crystal plasticity simulation-based crack nucleation model has been developed incorporating nonlocal effects of dislocation pile-up in adjacent grains.

A major bottleneck with 3D crystal plasticity finite element (CPFE) simulations for fatigue life prediction is the accommodation of large number of cycles to failure, often observed in experiments. In single time scale CPFE solutions using conventional time integration algorithms, each cycle is resolved into a large number of time steps. A high time step resolution is required for each cycle throughout the loading process, often leading to prohibitively large computational requirements. Consequently, a number of 3D cyclic crystal plasticity studies, e.g., (Bennett and McDowell 2003; Turkmen et al. 2003; Dawson 2000; Sinha and Ghosh 2006), have simulated a small number of cycles ($\approx$100) and subsequently extrapolated the results over thousands of cycles. Such extrapolation can lead to considerable error in the prediction of microstructural variables pertinent to fatigue life. It is desirable to conduct simulations for a high number of cycles, so as to reach local states of damage nucleation and growth in the microstructure. A few methods of multi-time scaling (Oskay and Fish 2004; Yu and Fish 2002; Manchiraju et al. 2007, 2008) have been introduced to avert these challenges. The time scales range from the loading frequency dependent scale of each cycle to the material dependent scale of relaxation times or the overall life of the component. In general, the methods have had very limited success in crystal plasticity solutions due to considerable localization and nonperiodic response with evolving plastic variables. In addition, some of these methods invokes two-way coupling between the time scales that requires having to solve initial value problems in each step at both time scales. This can result in very high computational time and may not provide any advantage over single time scale computations.

This chapter addresses two important aspects of predicting fatigue crack nucleation in polycrystalline alloys. Section 2 discusses the systematic development of a fatigue crack nucleation model for titanium alloys under dwell loading. Crystal plasticity finite element (CPFE) simulation results are post-processed to provide inputs to the fatigue crack nucleation model. The second part of this chapter discusses a wavelet transformation based multi-time scaling (WATMUS) algorithm for accelerated crystal plasticity finite element simulations in Sect. 3. The WATMUS algorithm does not require any scale-separation and naturally transforms the coarse time scale response into a *monotonic cycle scale* without the requirement of subcycle resolution (Anahid et al. 2009). Numerical studies with 1D and 3D crystal plasticity are conducted to establish the effectiveness of the WATMUS methodology.

2 Grain Level Dwell Fatigue Crack Nucleation Model based on Crystal Plasticity Finite Element Simulations

Premature fatigue failure in polycrystalline alloys, e.g., Ti alloys, under dwell fatigue loading conditions is attributed to room temperature creep (Inman and Gilmore 1979). During the hold period in each dwell cycle, grains in the microstructure with favorably oriented slip systems can undergo significant plastic straining due to slip. Compatibility requirements cause substantial increase in the local stress in adjacent unfavorably oriented grains. This phenomenon is known as load shedding (Hasija et al. 2003), in which time-dependent local stress concentration near grain boundaries are caused by dislocation pileup in neighboring grains. This local stress rise causes early crack initiation under dwell loading conditions (Bache 2003). Several microstructural and macroscopic factors affect stress evolution due to load shedding. A significant reduction in dwell fatigue life of Ti-6242 has been reported in (Woodfield et al. 1995) for a high microtexture, while shorter hold times have been seen to improve life in (Rokhlin et al. 2005). Material microstructure-based detailed mechanistic models for fatigue crack nucleation are seen as promising alternatives to empiricism with a higher probability of accurate fatigue failure prediction. An experimentally validated computational model has been developed for cyclic deformation-induced crack nucleation in polycrystals (Kirane and Ghosh 2008; Kirane et al. 2009; Anahid et al. 2009). In this section, the model in (Kirane and Ghosh 2008; Kirane et al. 2009) is modified to accurately account for crack evolution ahead of a dislocation pile-up.

The model developed in this section is for the dual phase alloy Ti-6242, used in aerospace engine components. The material response is modeled by an experimentally validated rate and size-dependent anisotropic elastic-crystal plasticity constitutive model (Hasija et al. 2003; Deka et al. 2006; Venkatramani et al. 2006; Venkataramani et al. 2007). Features of the real material microstructure, e.g., grain size and orientation, grain neighborhood distributions, as well as correlation between different characterization functions are accounted for in a statistically equivalent sense in the model. A method of simulating statistically equivalent 3D microstructural models from projected 2D orientation microscopy images has been developed in (Ghosh et al. 2008a; Groeber et al. 2008b; Groeber 2008) and in Chap. 3 of this book. This method is used in this work as a microstructure and FE model builder. A microstructural based nonlocal crack nucleation criteria is proposed for fatigue failure in a polycrystalline microstructure.

2.1 Experimental Observations on Crack Evolution

Experimental studies on crack evolution and crystallographic orientations in Ti-6242 samples have been conducted in (Rokhlin et al. 2005) using quantitative tilt fractography and Electron Back Scattered Diffraction (EBSD) techniques in SEM. Figure 1a shows a fractograph of a small region of crack initiation site for a failed Ti-6242 sample in dwell fatigue. The failure site is found to consist of facets that form on the basal plane of the primary α grains with a *hcp* lattice structure. The facets predominantly lie on a plane perpendicular to the principal tensile loading direction (Sinha et al. 2006). It has been observed in (Sinha et al. 2006) that the angle θ_c between the loading axis and the 'c' axis of grains at the failure site is quite small ($\sim 0^o$–30°). Furthermore, the failure site shows a low prism activity with Schmid factor (*SF*) $\sim$0–0.1 and a moderate basal activity with an *SF* $\sim$0.3–0.45. However, the region surrounding the failure site has a high prismatic and basal activity with

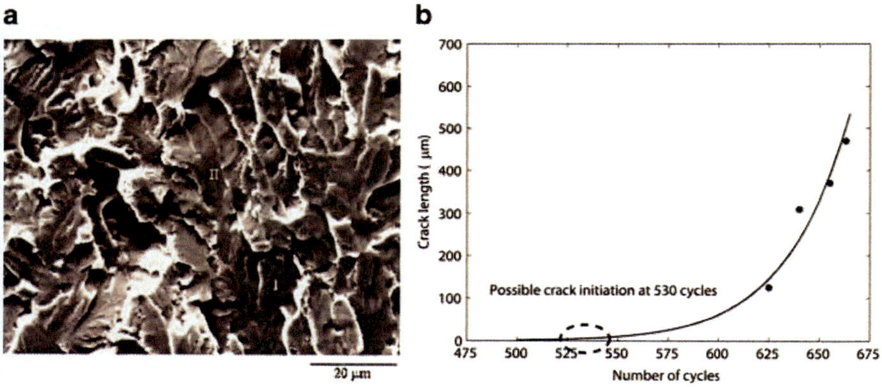

Fig. 1 (**a**) Fractograph of a faceted initiation site for a failed Ti-6242 dwell fatigue sample, (**b**) length as a function of number of cycles for a secondary crack in the microstructure MS2

a *SF* ~0.5. Thus, it may be inferred that while crack initiation occurs in a region that is unfavorably oriented for slip, it is surrounded by grains that are favorably oriented for slip. In other words, crack initiates in a hard-orientation grain surrounded by soft-orientation grains. These observations suggest time-dependent accumulation of stress in hard oriented grains due to load shedding with increasing plastic deformation in the surrounding soft grains.

2.1.1 Crack Detection and Monitoring in Tests on α/β Forged Ti-6242

Ultrasonic techniques, such as in-situ surface acoustic wave techniques have been developed for monitoring subsurface crack initiation in high micro-texture α/β forged Ti-6242 samples for dwell fatigue and creep experiments in (Rokhlin et al. 2005). The experiments monitor crack initiation and growth in real time, making estimation of the time for crack initiation possible. Dwell fatigue experiments are conducted with three microstructural samples labeled as *MS1*, *MS2*, and *MS3* that loaded with trapezoidal load cycles. Each load cycle has a maximum applied traction of 869 MPa (95% of the macroscopic yield stress) at a hold time of 2 min, and a loading/unloading time of 1 s. The stress ratio, measured as the ratio of the minimum to maximum load, is zero. In (Rokhlin et al. 2005; Williams et al. 2006), crack growth in samples *MS2* and *MS3* is monitored through micro-radiographic images taken by interrupting the experiment every 15 cycles. Figure 1b is a sample plot of the observed crack length as a function of the number of cycles for a secondary crack in the *MS2* sample. This crack is of length 125 μm at 625 cycles, while at 663 cycles, it is of length 470 μm. Extrapolating backwards to zero length as shown in Fig. 1b, the number of cycles to crack initiation for this crack is estimated to be approximately 530. The crack initiation cycles, extrapolated from plots for primary cracks that grew to cause final failure, are given in the Table 1. The primary crack

Table 1 Primary crack initiation data in dwell fatigue experiments on Ti-6242 by ultrasonic monitoring

Sample microstructure label	Test type	Sample life	Time to crack initiation	% Life at primary crack initiation
MS1	2-min dwell load	352 cycles	–	–
MS2	2-min dwell load (with modulation)	663 cycles	550 cycles	83%
MS3	2-min dwell load (with modulation)	447 cycles	380 cycles	85%

initiated at 83% life (550 cycles) for the *MS2* sample, while it initiated at 85% life (380 cycles) for the *MS3* sample. The results generally suggest that primary crack initiation in dwell fatigue occurs in the range 80–90% of the total number of cycles to failure.

2.2 *The Crystal Plasticity Finite Element Model (CPFEM)*

Ti alloys are often characterized by time-dependent deformation characteristics at low temperatures (Hasija et al. 2003; Inman and Gilmore 1979; Neeraj et al. 2000). This "cold" creep phenomenon occurs at temperatures lower than that at which diffusion-mediated deformation is expected. The creep process is not expected to be associated with diffusion-mediated mechanisms, such as dislocation climb. TEM studies, e.g., in (Neeraj et al. 2000), have shown that deformation actually proceeds via dislocation glide, where the dislocations are inhomogeneously distributed into planar arrays. Significant creep strains can accumulate at applied stresses, even as low as 60% of the yield strength. This characteristic has been attributed to rate sensitivity effects in (Inman and Gilmore 1979).

2.2.1 Crystal Plasticity Constitutive Model

The α/β forged Ti-6242 is a dual-phase polycrystalline alloy, which consists of colonies of transformed β phase in a matrix of the primary α phase. The primary α phase consists of equiaxed grains with a *hcp* crystalline structure, whereas the transformed β colonies have alternating α (hcp) and β (bcc) laths. The hcp crystals consist of 5 different families of slip systems, namely the basal$< a >$, prismatic $< a >$, pyramidal $< a >$, first-order pyramidal $< c + a >$, and second-order pyramidal $< c + a >$ with a total of 30 slip systems while the bcc crystal system consists of different slip families, $< 111 >$ 110, $< 111 >$ 112 and $< 111 >$ 123 with a total of 48 slip systems. The alloy considered in this study consists of 70% primary α and 30% transformed β grains. To incorporate the effect of various microstructural parameters, a size and time-dependent, finite strain crystal plasticity-based FE models have been developed by Ghosh et al. in (Hasija et al. 2003; Deka et al. 2006; Venkatramani et al. 2006; Venkataramani et al. 2007, 2008). For the transformed β-phase colony regions, a homogenized equivalent crystal model is developed in (Deka et al. 2006). The homogenized model consists of 78 slip systems, of which 30 correspond to hcp (secondary α) and 48 correspond to bcc slip systems.

The stress-strain relation is written in terms of the second Piola–Kirchoff stress $\mathbf{S}$ and the work conjugate Lagrange–Green strain tensor $\mathbf{E}$ as

$$\mathbf{S} = \mathbf{C} : \mathbf{E}^{\mathrm{e}}, \quad \text{where } \mathbf{E}^{\mathrm{e}} = \frac{1}{2}\left(\mathbf{F}^{\mathrm{e}^{\mathrm{T}}}\mathbf{F}^{\mathrm{e}} - \mathbf{I}\right) \tag{1}$$

Here, $\mathbf{C}$ is a fourth-order anisotropic elasticity tensor and $\mathbf{F}^e$ is the elastic component of the deformation gradient which is obtained by multiplicative decomposition

$$\mathbf{F} = \mathbf{F}^e\mathbf{F}^p, \quad \det(\mathbf{F}^e) > 0, \tag{2}$$

where $\mathbf{F}$ is the deformation gradient tensor and $\mathbf{F}^p$ is its incompressible plastic component, i.e., $\det \mathbf{F}^p = 1$. The flow rule, governing evolution of plastic deformation, is expressed in terms of the plastic velocity gradient $\mathbf{L}^p$ as:

$$\mathbf{L}^p = \dot{\mathbf{F}}^p\mathbf{F}^{p-1} = \sum_{\alpha}^{\mathrm{nslip}} \dot{\gamma}^\alpha \mathbf{s}^\alpha, \tag{3}$$

where the Schmid tensor associated with α-th slip system $\mathbf{s}^\alpha$ is expressed in terms of the slip direction $\mathbf{m}_0^\alpha$ and slip plane normal $\mathbf{n}_0^\alpha$ in the reference configuration as $\mathbf{s}^\alpha = \mathbf{m}_0^\alpha \otimes \mathbf{n}_0^\alpha$. For the plastic slip rate $\dot{\gamma}^\alpha$ on the slip system α, a power law dependence on the resolved shear stress τ^α and the slip system deformation resistance g^α has been described in crystal plasticity models (Harren and Asaro 1989; Kalidindi et al. 1991; Asaro and Rice 1977) as:

$$\dot{\gamma}^\alpha = \dot{\bar{\gamma}} \left|\frac{\tau^\alpha - \chi^\alpha}{g^\alpha}\right|^{1/m} \operatorname{sign}(\tau^\alpha - \chi^\alpha), \quad \tau^\alpha = \left(\mathbf{F}^{e^{\mathrm{T}}}\mathbf{F}^{\mathrm{e}}\mathbf{S}\right) : \mathbf{s}^\alpha. \tag{4}$$

Here, m is the material rate sensitivity parameter, $\dot{\bar{\gamma}}$ is the reference plastic shearing rate and χ^α is the back stress that accounts for kinematic hardening in cyclic deformation.

Slip System Deformation Resistance: The evolution of slip system deformation resistance is assumed to be controlled by two classes of dislocations, viz. statistically stored dislocations (SSDs) and geometrically necessary dislocations (GNDs). SSDs correspond to homogeneous plastic deformation, while GNDs accommodate incompatibility of the plastic strain field due to lattice curvature, especially near grain boundaries. Thus, the deformation resistance rate is expressed as:

$$\dot{g}^\alpha = \sum_\beta h^{\alpha\beta}|\dot{\gamma}^\beta| + \frac{k_0\hat{\alpha}^2G^2b}{2(g^\alpha - g_0^\alpha)} \sum_\beta \lambda^\beta |\dot{\gamma}^\beta| \tag{5}$$

Statistically Stored Dislocations: The first term in (5) corresponds to SSDs, where the modulus $h^{\alpha\beta} = q^{\alpha\beta} h^{\beta}$ (no sum on β) is the strain hardening rate due to self- and latent hardening on the α-th slip system by slip on the β-th slip system. Here, h^{β} is the self-hardening coefficient and $q^{\alpha\beta}$ is a matrix describing latent hardening. The evolution of self hardening for the hcp phase, used in (Deka et al. 2006; Venkatramani et al. 2006; Balasubramanian and Anand 2002), is of the form:

$$h^{\beta} = h_0 \left|1 - \frac{g^{\beta}}{g_s^{\beta}}\right|^{r} \operatorname{sign}\left(1 - \frac{g^{\beta}}{g_s^{\beta}}\right), \quad g_s^{\beta} = g_{s0}\left(\left|\frac{\dot{\gamma}^{\beta}}{\dot{\gamma}_0}\right|\right)^{c}, \tag{6}$$

where h_0 is the initial hardening rate, g_s^{β} is the saturation slip deformation resistance, and r, g_{s0}, and c are slip system hardening parameters. For the bcc phase, the self-hardening evolution law is of the form (Deka et al. 2006; Venkatramani et al. 2006; Venkataramani et al. 2007):

$$h^{\beta} = h_s^{\beta} + \operatorname{sech}^2\left[\left(\frac{h_0^{\beta} - h_s^{\beta}}{\tau_0^{\beta} - \tau_s^{\beta}}\right)\gamma_a\right]\left(h_0^{\beta} - h_s^{\beta}\right)\gamma_a = \int_0^t \sum_{\beta}^{\text{nslip}} |\dot{\gamma}^{\beta}| \mathrm{d}t, \tag{7}$$

where h_0^{α} and h_s^{β} are the initial and asymptotic hardening and τ_s^{β} represents the saturation value of the shear stress when $h_s^{\beta} = 0$.

Geometrically Necessary Dislocations: The second term in (5) accounts for the effect of GNDs on work hardening (Acharya and Beaudoin 2000). Here, k_0 is a dimensionless material constant, G is the elastic shear modulus, b is the Burgers vector, g_0^{α} is the initial deformation resistance, and $\hat{\alpha}$ is a nondimensional constant. $\hat{\alpha}$ is taken to be $\frac{1}{3}$ in this work following (Ashby 1970). λ^{β} is a measure of slip plane lattice incompatibility, which can be expressed for each slip system as a function of slip plane normal $\mathbf{n}^{\beta}$ and an incompatibility tensor Λ as:

$$\lambda^{\beta} = \left(\mathbf{\Lambda n}^{\beta} : \mathbf{\Lambda n}^{\beta}\right)^{\frac{1}{2}}. \tag{8}$$

The dislocation density tensor $\mathbf{\Lambda}$, introduced in Nye (1953), is a direct measure of the GND density. It can be expressed using the curl of plastic part of the deformation gradient tensor $\mathbf{F}^{P}$. Since this crystal plasticity formulation does not explicitly incorporate a dislocation density tensor, it can be indirectly extracted from the CPFEM output data as:

$$\mathbf{\Lambda} = \nabla^{T} \times \mathbf{F}^{P}. \tag{9}$$

Back Stress Evolution: The evolution of slip system back stress χ^{α} in (4) is given by the expression (Hasija et al. 2003; Xie et al. 2004; Harder 1999; Morrissey et al. 2001):

$$\dot{\chi}^{\alpha} = c\,\dot{\gamma}^{\alpha} - d\,\chi^{\alpha}|\dot{\gamma}^{\alpha}|, \tag{10}$$

where c and d are the direct hardening and dynamic recovery coefficients, respectively.

Size Dependence: Polycrystalline Ti-6242 exhibits a strong grain size dependence of the slip system deformation resistance g^α. In (Venkataramani et al. 2007, 2008), a Hall–Petch type relationship that relates g^α to various characteristic length scales depending on the location in the $\alpha + \beta$ microstructure has been incorporated to account for size effects. The relation is expressed as:

$$g^\alpha = g_0^\alpha + \frac{K^\alpha}{\sqrt{D^\alpha}}, \tag{11}$$

where g_0^α and K^α are slip system constants and D^α is a characteristic length scale. For multiphase materials such as Ti-6242, the overall grain size, colony size, α lath thicknesses, or β lath thicknesses are candidate characteristic length scales. For example, in the primary α region, the grain boundary impedes the transmission of slip for all systems termed as soft slip modes. Correspondingly, the grain size D_g is the characteristic length for primary α grains. In the transformed β region, the plastic deformation is activated through both the hard and the soft slip modes. The orientations of α and β lamellae follow the Burger's orientation relationship (see Deka et al. 2006; Venkatramani et al. 2006), which brings the hcp a_1 slip direction into coincidence with the bcc b_1 slip. This results in a relatively easy slip transmission across the interface and hence it is classified as soft slip mode as the dislocations glide freely across the α/β interface. The resistance to slip is only from the colony boundary and hence the colony size D_c is the characteristic length. On the other hand, there is considerable misalignment between the α phase a_2 and β phase b_2 slip directions, and also between the α a_3 and all $< 111 >_\beta$ directions in the β phase. The slip transmission in these cases is impeded by the α/β interface and hence these systems are classified as hard slip modes. Thus, the characteristic lengths for systems with hard slip modes are lath thickness l_α and l_β for the hcp and bcc phases, respectively.

Material properties for each of the constituent phases and individual slip systems in the crystal plasticity model have been calibrated in (Deka et al. 2006) with single colony and single crystal experiments. Calibrated values of important material constants are listed in Tables 2 and 3. Other parameters used in (5) are listed in Table 4. The computational model is validated by comparing the results of simulations of constant strain rate and creep tests in (Deka et al. 2006; Venkatramani et al. 2006; Venkataramani et al. 2007, 2008).

Table 2 Material flow and hardness parameters for *hcp* Ti-6242 slip systems in tension

	a_1 basal	a_2 basal	a_3 basal	a_1 prism	a_2 prism	a_3 prism
g_0^α (Primary α)	357.6	357.6	357.6	355.8	355.8	355.8
g_0^α (Transformed β)	349.9	382.0	382.0	305.9	255.8	287.4
m	0.02	0.02	0.02	0.02	0.02	0.02
$\dot{\gamma}$ (sec^{-1})	0.0023	0.0023	0.0023	0.0023	0.0023	0.0023

Table 3 Hardness parameters for *hcp* Ti-6242 slip systems in compression (m, $\dot{\gamma}$ are same as for tension)

	a_1 basal	a_2 basal	a_3 basal	a_1 prism	a_2 prism	a_3 prism
g_0^α (Primary α)	395.6	395.6	395.6	393.6	393.6	393.6
g_0^α (Transformed β)	450.9	574.7	590.6	448.5	571.1	586.9

Table 4 Parameters used in the hardening evolution (5)

Shear modulus G	48 GPa
Magnitude of Burgers vector b	0.30 nm
Material constant k_0	2

2.3 *Representation of Microstructural Images in CPFEM*

Accurate representation of morphological and crystallographic features of the microstructure, at least in a statistical sense, are important for meaningful stress-strains predictions and associated localization or crack evolution. Significant advances have been made in the reconstruction and simulation of 3D polycrystalline microstructures in (Ghosh et al. 2008a; Groeber et al. 2008b; Groeber 2008; Bhandari et al. 2007), based on data obtained from dual beam focused ion-beam scanning electron microscope (FIB-SEM) systems. The system acquires 3D orientation or electron backscatter diffraction (EBSD) data for a series of material cross sections. This information has been used in (Ghosh et al. 2008a; Bhandari et al. 2007) for automatic segmentation of individual grains from the image and subsequently translated into a 3D mesh for finite element analysis. Through a multitude of data sets, intrinsic distributions of microstructural parameters can be captured and accurately represented through 3D microstructure reconstruction. Computational tools have been developed in (Groeber et al. 2008b; Groeber 2008) to create synthetic microstructures that are statistically equivalent to the measured structure with respect to certain microstructural features. These methods are used in this work for computational simulations leading to the crack initiation model.

Microstructures are created from orientation imaging microscopy or OIM images at two sites in the material samples, viz. (a) a critical region in the vicinity of a dwell fatigue crack tip (corresponding to the sample *MS1* in Sect. 2.1.1) as shown in Fig. 2, and (b) a noncritical region (*MS0*) away from it. The following steps are performed for the construction of 3D microstructures from 2D OIM scans at the critical and noncritical sites. As discussed in (Groeber et al. 2008b; Groeber 2008), statistical distribution functions of various microstructural parameters in the 2D OIM scan are generated and stereologically projected in the third dimension for creating 3D statistics. The assumptions and process implemented to extract 3D statistics from the 2D distributions of morphological and crystallographic features are briefly outlined.

1. *Distribution functions of grain size and shape:* An assumption made is that the size and shape correlation of 2D OIM scans of grain sections to their parent 3D structures is similar to that of elliptical sections to their parent 3D ellipsoids. For determining the size and shape distributions of 3D grains in the microstructure,

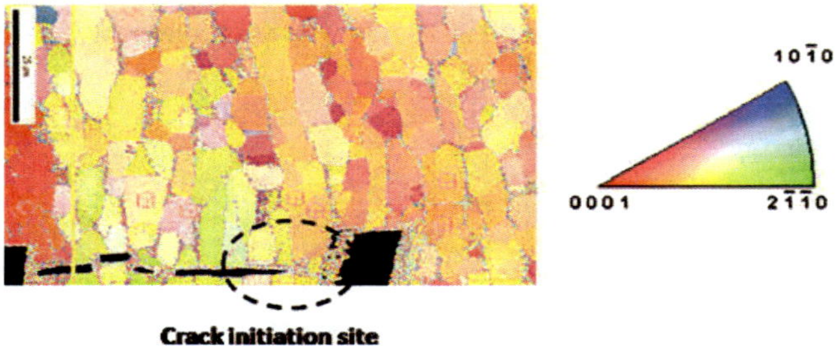

Fig. 2 OIM scan of the critical primary crack initiation site in the MS1 microstructure

a large number of ellipsoids of different sizes and shapes are randomly sectioned and the resulting elliptical sections are recorded. Probabilistic weighting functions are created for the grain reconstruction process. The 3D ellipsoid that produces an elliptical section closest in shape and size to a 2D OIM grain scan is assumed to have a high probability in the representing corresponding 3D grain. An assumption is needed for the orientation distribution of the ellipsoids relative to the sectioning plane of the OIM scan. While in (Groeber 2007), three orthogonal sections have been taken, only one section of the surface scan is available in this work. The orientation distribution is assumed to be random. A constrained Voronoi tessellation, with initial seed points at the centroid of the ellipsoid, is executed for generating the grain shapes.

2. *Distribution of number of neighbors:* The reconstructed 3D grains are placed in a representative cubic volume with a constraint that each grain has appropriate number of neighbors, as determined by 3D projection of the OIM scan. In the 2D OIM scan, each grain has approximately 3–12 neighbors, while grains in the 3D representation have 8–25 neighbors. The representative cube of dimensions $65\times65\times65$ m for the sample critical microstructure consists of 949 grains.
3. *Distribution of crystallographic orientations:* The crystallographic orientation assignment to the grains in the cubic volume is executed by the 3 major steps, described in (Deka et al. 2006; Groeber 2008). These are delineated as: (a) Orientation probability assignment method; (b) Misorientation probability assignment method; and (c) Micro-texture probability assignment method. Figure 3 shows the microtexture distribution in the 2D scan and in the 3D model, with satisfactory agreement.

The reconstructed 3D model has distributions of orientation, misorientation, microtexture, grain size, and number of neighbors that are statistically equivalent to those observed experimentally in the OIM scan. The model of 949 grains is subsequently discretized into a finite element mesh of 78,540 tetrahedron elements as shown in Fig. 4. Local variables in the finite element simulations are pivotal to the development of the nucleation criterion. Consequently, a mesh sensitivity study of the local variables with respect to mesh density is done prior to the dwell fatigue analysis. The critical microstructure (*MS1*) is used with two different mesh

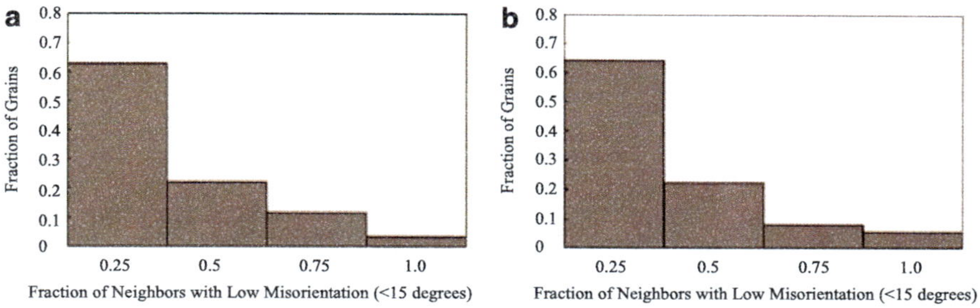

Fig. 3 Microtexture distributions (**a**) from OIM scan of critical region (**b**) in the critical FE model of MS1

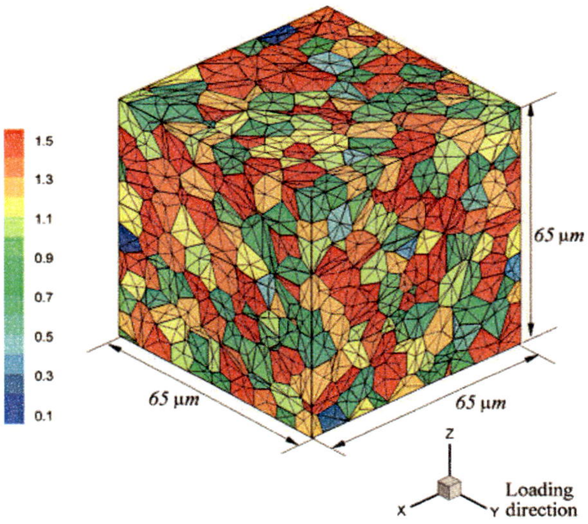

Fig. 4 FE model for polycrystalline Ti-6242, which is statistically equivalent to the OIM scan of the critical region of microstructure *MS1*. Also shown is the contour of '*c*' axis orientation distribution (radians)

densities. The first model consists of 78,540 elements while the second has 116,040 elements, which is approximately 150% higher in mesh density. A creep simulation is performed for both these models for 1,000 s at an applied load of 869 MPa in the Y-direction. The local stress component in the loading direction and the local plastic strain at the end of 1,000 s are compared for various sections in the FE models. Plots comparing the distribution of these variables along a section parallel to the X-axis in Fig. 5a, b show excellent agreement between the two mesh densities. Thus, the 78,540-element mesh is a converged model for the loading considered and is used for the development of the crack initiation criterion.

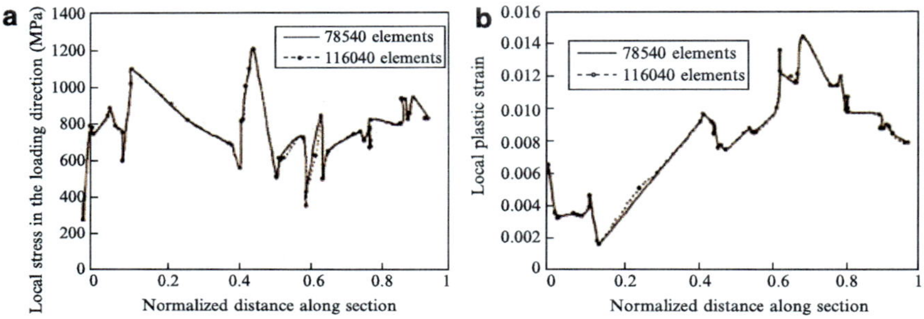

Fig. 5 Distribution of local variables: (**a**) loading direction stress (σ_{22}), (**b**) local plastic strain along a section parallel to the x-axis at the end of 1,000 s for a creep simulation on the two models of microstructure *MS1* with two different mesh densities

2.4 A Nonlocal Crack Nucleation Criterion from CPFE Variables

Stress concentration induced mode-I crack nucleation at the grain boundary of a crystalline solid has been modeled in Stroh (1954) using a dislocation pileup model. The model proposes that a crack is initiated under the condition that $n\sigma_0 \geq 12\alpha G$. Here, n is the number of dislocations in the pile up, σ_0 is the applied stress on the slip plane, G is the shear modulus and $\alpha = \frac{\gamma_s}{bG}$ is a material constant, in which γ_s is the surface energy and b is the Burgers vector. Bache et al. (2003) have argued that a combined effect of normal and shear stresses is responsible for dwell fatigue of Ti alloys. Crack nucleation by dislocation pileups has also been studied in (Smith 1979, 1966), where it was proposed that a dislocation pile up leads to a mode-II crack by dislocation coalescence. Dislocation pileup is represented from equilibrium consideration in a grain of size d, yielding a cleavage fracture criterion as:

$$\sigma_E \geq \left(\frac{2\gamma_s G}{\pi(1-\nu)d}\right)^{0.5}. \tag{12}$$

The stress σ_E required to fracture a grain is inversely related to the square root of the grain size. While the functional forms are different, the models concur on the fact that the crack nucleates at the hard–soft grain boundary as a consequence of stress concentration caused by dislocation pileup in the soft grain. Various dislocation level fatigue failure models, e.g., (Stroh 1954; Baker 1999; Tanaka and Mura 1981), have discussed the equivalence in the crack evolution ahead of dislocation pile-ups and at crack tips.

2.4.1 Limitations of a Purely Stress based Crack Nucleation Criterion

A purely stress-based criterion, where a crack is nucleated when the effective stress exceeds a critical value, i.e., $T_{\mathrm{eff}} \geq T_{\mathrm{crit}}$ has been tested in (Kirane and Ghosh 2008).

Results show that the rate of change of T_{eff} per cycle is very low at higher cycles (increases only by 3.7% in the last 250 cycles). If T_{crit} is a constant, even a small variation in T_{crit} might cause the predicted life to vary by hundreds of cycles. For example, a 3.5% change in T_{crit} from 1,660 to 1,724 MPa will result in a 150% change in the predicted life from 100 cycles to 352 cycles. This is not practical and should be amended by taking into account other evolving variables in the analysis. Consequently, a dislocation pile-up-based nonlocal nucleation criterion has been proposed in (Kirane and Ghosh 2008; Kirane et al. 2009). A mechanistically more accurate nucleation model is however proposed in the next section.

2.4.2 Nonlocal Nucleation Criterion Incorporating Soft Grain Dislocation Pile-Up

The crack nucleation model, proposed in this work, is at the length scale of individual grains and is set for each hard grain–soft grain interface in a polycrystalline aggregate. It is nonlocal in the sense that while crack nucleates in the hard grain due to load-shedding, it is affected by plasticity in the neighboring soft grains due to dislocation pile-ups at grain boundary barriers.

The nucleation model is founded upon the Stroh's fracture mechanism (Stroh 1957), in which a wedge micro-crack nucleates in the neighboring grain when a dislocation impinges upon the grain boundary. An edge dislocation is an extra half plane of atoms wedged between two complete planes. This is equivalent to a micro-crack with opening displacement of one atomic spacing, b. As more dislocations pile up together, the opening displacement gets bigger as shown in Fig. 6a. From this figure, the crack opening displacement is simply the closure failure along a circuit surrounding the piled up dislocations. If n edge dislocations with Burgers vector b

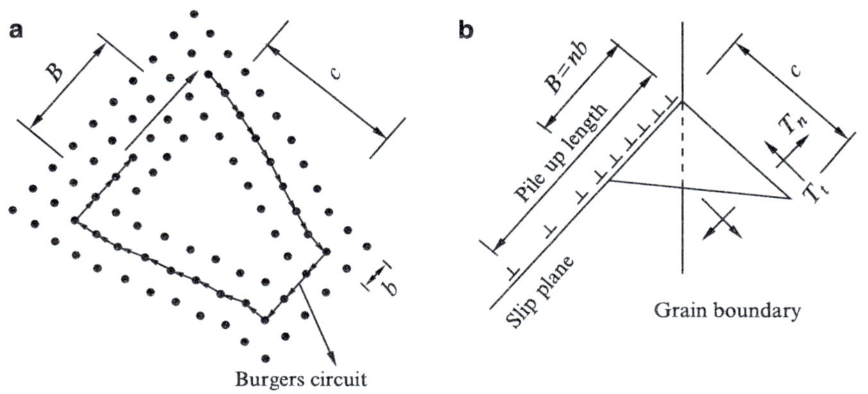

Fig. 6 (**a**) A wedge micro-crack with opening displacement of $4b$ produced from coalescence of 4 dislocations, (**b**) nucleation of a wedge crack in the hard grain resulting from a dislocation pile up in the soft grain

contribute to the formation of a micro-crack, a wedge with opening displacement of $B = nb$ is produced. It should be noted that while the dislocations are piled up at the grain boundary of a soft grain, the wedge crack initiates in the adjacent hard grain (see Fig. 6b).

As shown in Fig. 6a, the micro-crack length c can be considered as the length after which the disturbance in the lattice structure of the hard grain disappears. This disturbance is caused by extra half planes of atoms in the soft grain. This micro-crack length c can be related to the crack opening displacement B using a 90° intercept definition suggested by Rice (1968), where $c = B/2 = nb/2$. The wedge crack is initially stable. As more dislocations enter the crack, the crack opening size increases and therefore the crack length also increases. However, the applied stress across the micro-crack, which is associated with the hard grain, also helps open up the crack. The acting stress on the micro-crack surface is a combination of normal and shear stresses as shown in Fig. 6b. The micro-crack becomes unstable when the mixed mode stress intensity factor K_{mix} exceeds a critical value K_c. K_{mix} is expressed in terms of the normal stress intensity factor K_n and the shear stress intensity factor K_t as:

$$K_{\text{mix}} = K_n^2 + \beta K_t^2. \tag{13}$$

Here, β is a shear stress factor, which is used to assign different weights to the normal and shear traction components for mixed mode. It is defined as the ratio of the shear to normal fracture toughness , i.e., $\beta \approx K_{\text{nc}}/K_{\text{tc}}$ (Ruiz et al. 2001). The stress intensity factors are expressed in terms of the micro-crack length c as:

$$K_n =< T_n > \sqrt{\pi c} \text{ and } K_t = T_t \sqrt{\pi c}. \tag{14}$$

Noting that the micro-crack grows when $K_{\text{mix}} \geq K_c$, the hard grain crack nucleation criterion ahead of dislocation pile-ups in adjacent soft grain may be stated as:

$$T_{\text{eff}} = \sqrt{< T_n >^2 + \beta T_t^2} \geq \frac{K_c}{\sqrt{\pi c}} \tag{15}$$

Replacing c by $B/2$ and rearranging, the criteria is obtained as

$$R = T_{\text{eff}}.\sqrt{B} \geq R_c \qquad R_c = K_c \sqrt{\frac{2}{\pi}}, \tag{16}$$

where T_{eff} is an effective stress for mixed mode crack nucleation. The traction component normal to the crack surface is expressed as $T_n = n_i^b(\sigma_{ij} n_j^b)$, where σ_{ij} are components of the Cauchy stress tensor and n_i^b are components of the unit outward normal to the crack surface. Only the tensile normal stress $< T_n >$, represented by the McCauley bracket <> contributes to the effective stress. Compressive stresses do not contribute to crack opening. The shear stress component, T_t, is obtained by the vector subtraction of T_n from the stress vector on the crack surface, i.e., $T_t \mathbf{t}^b = \mathbf{T} - T_n \mathbf{n}^b$, where $\mathbf{t}^b$ is the tangential unit vector to the surface. The stress components in (15) correspond to remote applied stresses. Typical values of c, which give rise to an unstable cracking are of the order of nanometers, while the typical grain size is of the order of microns. Thus, it is reasonable to assume that the maximum stress at the hard grain boundary is the remote stress.

R_c is a parameter that depends on the material elastic properties, as well as on the critical strain energy release rate G_c. It has the units of stress intensity factor (MPa$\sqrt{\mu\text{m}}$). R_c is calibrated from crack initiation data extracted from experiments conducted in (Rokhlin et al. 2005). A value of $\beta = 0.7071$ has been suggested for Ti-64 alloys in (Parvatareddy and Dillard 1999), and is used in this study. Sensitivity analyses with different values of β indicate that T_{eff} is not very sensitive to β for $< c + a >$ oriented hard grains, since $T_n \gg T_t$. As more dislocations are added to the pile-up with time, the wedge crack opening displacement increases. This implies a smaller T_{eff} to initiate a crack with increasing plastic deformation and pile-up.

2.4.3 Implementation of the Nonlocal Nucleation Criterion

A method of calculating the micro-crack opening displacement B in (16) is described in this section. The most probable experimentally observed nucleated crack is along the basal plane. Accordingly, a hard grain basal plane crack nucleation criterion is modeled in this study. For estimating B, it is necessary to have the distribution of dislocations inside the soft grain. However, the CPFEM in (Hasija et al. 2003; Deka et al. 2006; Venkatramani et al. 2006; Venkataramani et al. 2007, 2008) does not explicitly have dislocation density as a state variable. Hence, the plastic strains and its gradients, available from the results of the crystal plasticity FE simulations, are used to estimate the micro-crack opening displacement B. This contributes to the nonlocality aspect of the crack nucleation criterion.

From the fracture mechanisms in Stroh (1954), the wedge opening displacement is equal to the closure failure along a circuit surrounding the piled up edge dislocations on one slip plane, shown in Fig. 6a. This can be extended to the general case of 3D representation of dislocations including edge and screw dislocations considering multiple slip systems. In this case, both closure failure and crack opening displacement are vector quantities. In the dislocation glide model, the existence of dislocations results in an incompatible, non-physical intermediate configuration. The lattice incompatibility can be measured by the closure failure of a line integral along a Burgers circuit $\bar{\Gamma}$ in the intermediate configuration. Closure failure is equivalent to the net Burgers vector $\mathbf{B}$ of all dislocations passing through the region $\bar{\Omega}$ bounded by the circuit. The Burgers vector can be mapped to a line integral along a referential circuit, Γ using $\mathbf{F}^p$. Then the closure failure is related to a surface integral of the curl of $\mathbf{F}^p$ over a referential surface Ω by application of the classical Stokes theorem

$$\mathbf{B} = \oint_{\bar{\Gamma}} d\bar{\mathbf{x}} = \oint_{\Gamma} \mathbf{F}^p d\mathbf{X} = \int_{\Omega} \mathbf{\Lambda} \cdot \mathbf{n} d\Omega, \tag{17}$$

where $\mathbf{n}$ is the unit normal to the surface Ω and $\mathbf{\Lambda}$ is the Nye's dislocation tensor given in (9). Components of $\mathbf{\Lambda}$ are evaluated at each quadrature point using shape function based interpolation of nodal values of $\mathbf{F}^p$, described in (Anahid et al. 2009).

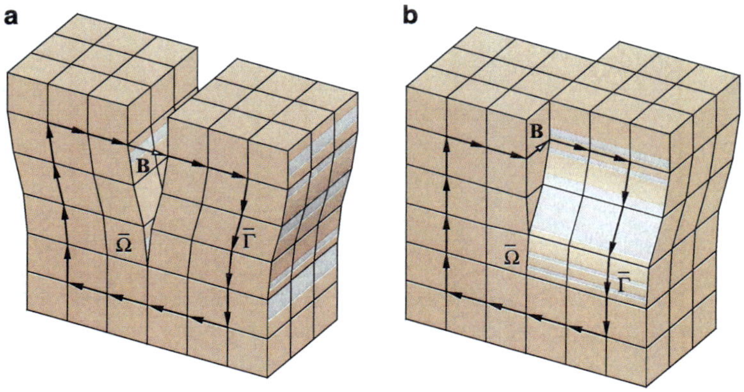

Fig. 7 Closure failure and crack opening displacement for (**a**) single pure edge dislocation, (**b**) single pure screw dislocation

The closure failure obtained from (17) can make any arbitrary angle with respect to the surface, depending on the type of dislocations passing through $\bar{\Omega}$. For a single pure edge dislocation with a dislocation line perpendicular to $\bar{\Omega}$, the closure failure **B** lies in the plane, see Fig. 7a. This is equivalent to a mode-I micro-crack with opening displacement of b. For a single pure screw dislocation with the same dislocation line, **B** is perpendicular to the plane as shown in Fig. 7b. This corresponds to a shear mode micro-crack. For a mixed type of dislocation with edge and screw components, **B** is neither perpendicular nor parallel to Ω, which is equivalent to a mixed-mode micro-crack.

Consider a material point A surrounded by dislocations. There are different planes with different normal vectors which contain point A. The closure failure (or equivalently the crack opening displacement) caused by the dislocations piercing each of these planes depends on the normal **n** in (17). Therefore, different micro-cracks with different crack opening displacements can be assumed at point A. Also, the stresses acting on each of these planes are different. Thus, it can be inferred that there are several micro-cracks with dissimilar stress intensity factors competing at point A. The micro-crack with the highest mixed-mode stress intensity factor can be considered as the critical one. However, experimental observations suggest that the fatigue crack happens on the basal plane of a hard grain. Hence, it is assumed that the critical closure failure at the hard–soft grain boundary is perpendicular to the hard grain basal plane, i.e., $\mathbf{B} = B\mathbf{n}^*$, with $\mathbf{n}^*$ denoting the normal to the basal plane (see Fig. 8a). Correspondingly, there is a unit area in the soft grain with normal vector $\mathbf{n}_{\mathrm{cr}}$, in which the piercing dislocations create a closure failure parallel to $\mathbf{n}^*$. $\mathbf{n}_{\mathrm{cr}}$ is obtained using (17) as:

$$\mathbf{n}_{\mathrm{cr}} = \mathbf{\Lambda}^{-1}.\mathbf{B} = B\mathbf{\Lambda}^{-1}.\mathbf{n}^*. \tag{18}$$

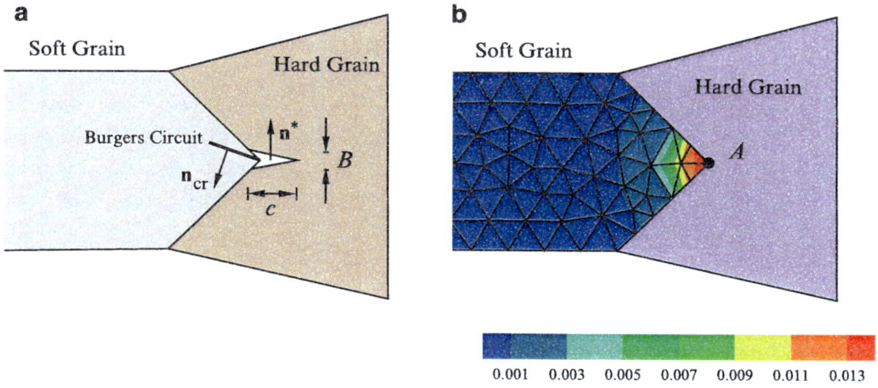

Fig. 8 (**a**) Basal plane crack nucleation in the hard grain, (**b**) distribution of the norm of Nye's dislocation tensor inside a representative soft grain

Noting that both $\mathbf{n}_{cr}$ and $\mathbf{n}^*$ are unit vectors, B is calculated as,

$$B = \frac{1}{\|\boldsymbol{\Lambda}^{-1}.\mathbf{n}^*\|} \tag{19}$$

with $\| \quad \|$ representing the magnitude of a vector. Equation (19) holds when all dislocations are lumped in a very small portion of the soft grain near boundary. However, the dislocations are distributed in the whole soft grain. Figure 8b shows the distribution of the norm of Nye's dislocation tensor inside a representative soft grain. The maximum value occurs at the hard–soft grain boundary (point A), and the values decrease with increasing distances from the grain boundary. Values of $\boldsymbol{\Lambda}$ are available at the Gauss points of all tetrahedron elements in the soft grain. Each element I contains its own dislocations quantified by Nye's dislocation tensor at that element, $\boldsymbol{\Lambda}_I$. Dislocations associated with the element I can produce a crack opening displacement normal to the hard grain basal plane, given as:

$$B_I = \frac{W_I A_I}{\|\boldsymbol{\Lambda}_I^{-1}.\mathbf{n}^*\|}, \tag{20}$$

where A_I is the surface area associated with an element I into which the dislocations penetrate. This is estimated by creating an equivalent spherical domain to eliminate the dependence of the surface area on the shape of the element. It only depends on the element volume, since the sphere is created from volume equivalence of the element. The center of this sphere coincides with the Gauss point. Assuming that the plane containing the Burgers circuit in element I passes through the Gauss point, A_I is equal to the area of the circular cross-section passing through the center of sphere. Assuming that V_I denotes the volume of element I, the sphere's radius R_I and the corresponding area A_I are obtained as

$$R_I = \sqrt[3]{\frac{3}{4\pi} V_I} \quad \text{and} \quad A_I = \pi R_I^2 = 1.77 (V_I)^{\frac{2}{3}}. \tag{21}$$

W_I is a weighting parameter applied to B, which accounts for the fact that the effect of a dislocation on the crack opening displacement decreases with distance. W_I is 1 when the distance is zero. The weighting function is expressed in (Engelen et al. 2003) as $W_I = \exp(-r_I^2/2l^2)$, in which r_I is the distance between the point with maximum dislocation density (point A in Fig. 8b) and the Ith element integration point. This weighting function decays to zero beyond a critical distance corresponding to l. Considering the contribution of all elements in the soft grain on the hard grain crack opening displacement, B is stated as

$$B = \sum_I B_I = \sum_I \frac{W_I A_I}{\|\mathbf{\Lambda}_I^{-1}.\mathbf{n}^*\|} = \sum_I \frac{1.77 W_I (V_I)^{2/3}}{\|\mathbf{\Lambda}_I^{-1}.\mathbf{n}^*\|}. \tag{22}$$

The effective basal traction T_{eff} at a point in the hard grain, as well as the crack opening displacement B evaluated from the dislocation density field in adjacent softer grain, can be used in (16) to calculate the effective nucleation variable R. R is checked for every grain pair in the crystal plasticity FE model in the post-processing stage. The condition posed in (16) is nonlocal, in that the stress required to initiate a crack at a point in the hard grain depends on the gradient of plastic strain in the neighboring soft grain as well.

2.5 *Calibration and Validation of the Nucleation Criterion*

Data from three experiments on α/β forged Ti-6242, discussed in Sect. 2.1.1, are considered for calibrating and validating the crack nucleation model. The three microstructural samples *MS1*, *MS2*, and *MS3* differ in microstructural orientation, misorientation, and micro-texture distribution. The corresponding cycles to crack initiation under dwell loading are provided in Tables 1 and 5. Results in (Rokhlin et al. 2005) and Table 1 suggest that generally primary crack initiation in dwell fatigue occurs in the range 80–90% of total number of cycles to failure. Based on observations made for the samples *MS2* and *MS3*, crack initiation in *MS1* is assumed at 80% of the total life, viz. 282 cycles. This data is used for calibrating the parameter R_c in (16) for sample *MS1*. The calibrated value of R_c is then used to predict the number of cycles to crack initiation in samples *MS2* and *MS3*.

CPFEM models of statistically equivalent simulated microstructures at the critical and noncritical regions described in Sect. 2.3, are developed for analysis. Representative 65×65×65 μm microstructural volume elements of the samples consist of 949 grains discretized into 78,540 tetrahedron elements, as shown in Fig. 4. In the development of the crack initiation model, it is expected that the initiation criterion will be met at some location in the critical FE model, but will not be satisfied in the noncritical FE model.

2.5.1 Calibration of R_c for α/β Forged Ti-6242

The material parameter R_c in (16) is calibrated from results of 2-minute dwell fatigue CPFEM simulations of the *MS1* microstructure. The CPFEM model is constructed at the critical region and is run for 352 cycles. Crack initiation is assumed at 80% (282 cycles) of the total life. The variable R corresponding to the LHS of (16) is determined for all grain pairs at the end of 282 cycles. The hard grain with the maximum value of R is located and the evolution of this maximum R with number of cycles is plotted in Fig. 9a. The value of R_c is determined from the value of R at 282 cycles. It is $R_{c(80\%)} = 137.55\,\mathrm{MPa}\,\mu\mathrm{m}^{1/2}$. This critical value is subsequently used for predicting crack initiation in other samples.

Crystal plasticity finite element simulation of the noncritical microstructure *MS0* is conducted for the 2-min dwell loading conditions and R-values are evaluated for all grain-pairs. Figure 9b plots the evolution of the maximum R as a function of cycles. The maximum R reached at the end of 352 cycles is only $115\,\mathrm{Mpa}\,\mu\mathrm{m}^{1/2}$, which is far less than the threshold value R_c. Thus, the criterion predicts no crack initiation for this noncritical region, which is consistent with the experimental observation.

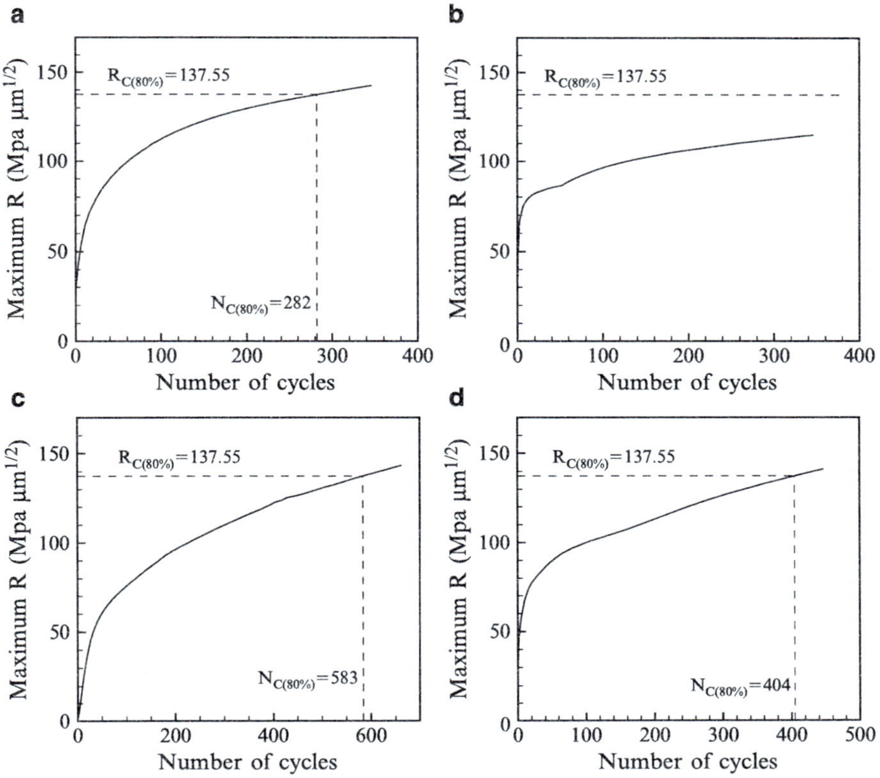

Fig. 9 Evolution of the maximum R over number of cycles for the FE models of microstructures (**a**) *MS1* critical region, (**b**) *MS0* noncritical region, (**c**) *MS2* and (**d**) *MS3*

2.5.2 Predictions of Crack Nucleation in *MS2* and *MS3*

For the *MS2* microstructure, a statistically equivalent FE model is generated from an OIM scan surrounding one of the secondary cracks in the failed sample. As discussed in Sect. 2.1.1, crack nucleation is determined to occur at 530 cycles. The 2-min FE simulation is performed for 663 cycles with loading conditions described in (Rokhlin et al. 2005). Figure 9c shows the evolution of the maximum R with cycles. The number of cycles to initiation $N_{c(80\%)}$ is predicted for *MS2*, from where the R curve intersects the critical value of $R_{c(80\%)}$. The corresponding number of cycles to initiation is found to be $N_{c(80\%)} = 583$. The difference with the experimentally determined value of 530 cycles is 10.13%. This agreement is considered to be very good, given the number of uncertainties in the developed model.

For the *MS3* microstructure, a similar 2-min dwell fatigue simulation is performed for 447 cycles. The evolution of maximum R is plotted in Fig. 9d. The number of cycles to initiation is predicted to be $N_{c(80\%)} = 404$. From Table 1, this crack initiates at 380 cycles. The difference with the experimentally determined value is 6.49%. The results are summarized in Table 5. The agreement is considered to be satisfactory. Alternatively, if R_c is calibrated from results on *MS2*, the calibrated value is found to be $R_c = 133.09\,\text{MPa}\,\mu\text{m}^{1/2}$. Correspondingly, the number of cycles to nucleation for *MS3* is found to be $N_c = 356$, which corresponds to a −6.2% difference with the experimentally determined value of 380 cycles.

For microstructural validation of the nucleation criterion, locations of the initiation sites in the microstructures are examined and compared with experimental observations in Table 6. In each case, the predictions for grains with prismatic and basal Schmid factors and c-axis orientations are consistent with the observations in (Sinha et al. 2006). These results prove convincingly the predictive capability of the crack nucleation criterion in the hard grain.

Table 5 Comparison of predicted cycles to crack initiation with experimentally observed life

Microstructure label	Cycles to crack initiation (experiment)	Cycles to crack initiation based on 80% of life (predicted)	% Relative error
MS2	530	583	+10.13
MS3	380	404	+6.49

Table 6 Microstructural features of predicted location of crack initiation site in dwell fatigue of Ti-6242

Microstructural parameters	Experiments	MS1	MS2	MS3
θ_c	$\sim 0 - 30°$	17.8°	23.0°	28.5°
Prismatic Schmid factor	$\sim 0.0 - 0.1$	0.04	0.07	0.11
Basal Schmid factor	$\sim 0.3 - 0.45$	0.27	0.36	0.40

3 A Novel Multi-time Scaling Method for Cyclic Crystal Plasticity FE Simulations

Typically, fatigue life in metallic materials could be of the order of thousands of cycles, depending on the material and loading conditions. A major challenge with the crystal plasticity finite element simulations for fatigue life prediction is accommodating the large number of cycles to failure, or even crack nucleation. In single time-scale finite element solutions using conventional time integration algorithms, each cycle is discretized into a number of time steps for time integration. A high time step resolution may be required for each cycle in crystal plasticity calculations, depending on the evolution pattern of the response variables throughout the loading process. In addition, it is often necessary to conduct simulations for a significantly high number of cycles to reach local states of damage initiation and growth. This presents significant challenges due to the presence of two distinct time scales, viz.:

1. The fine time scale τ of each cycle, dictated by the frequency of loading.
2. The coarse time scale t of material evolution, characterized by the material relaxation time or time to failure.

Typically, studies in 3D crystal plasticity (Bennett and McDowell 2003; Turkmen et al. 2003; Sinha and Ghosh 2006) simulate a small number of cycles ($\sim$100) and subsequently extrapolate the results to thousands of cycles. This can lead to considerable error in the evolution of variables at the microstructural level and consequently in fatigue life predictions. Methods of multi-scaling in the temporal domain can be introduced to avert some of these challenges. The method of direct separation of motions has been traditionally used to study the vibratory response under the application of high frequency loads (Blekhman 2000; Thomsen 2004). It involves defining two separate integro-differential equations, one each for the high- and low-frequency components of the response. These methods are based on the assumption that all variables are either locally periodic or nearly periodic in the temporal domain, e.g., in (Oskay and Fish 2004; Yu and Fish 2002), and implicitly assume time scale separation of the governing equations. However, these methods cannot be extended to crystal plasticity solutions due to the strong non-periodic response of evolving plastic variables and also due to localization in the spatial domain. In addition, these methods often invoke two-way coupling between the time scales that requires having to solve initial value problems in each step at both time scales. This can result in very high computational time and may not provide any advantage over single time scale calculations. In (Manchiraju et al. 2007), cyclic averaging in conjunction with asymptotic expansion of variables in the time domain has been proposed as a basis of multi-time scaling. However, asymptotic expansion methods are not suitable for crystal plasticity simulations at or near fully reversed loading for with the amplitude ratio $R \sim -1$.

This section proposes a novel wavelet transformation-based multi-time scaling (WATMUS) algorithm for accelerated crystal plasticity finite element simulations to overcome the above deficiencies. The wavelet decomposition naturally retains the high frequency response through the wavelet basis functions and transforms the low frequency material response into a "monotonic cycle scale" problem undergoing monotonic evolution. No assumption of scale separation is needed with this method. The section starts with a brief review of a few existing methods of time scale acceleration and then introduces the wavelet based multi-time scaling scheme. Numerical examples are executed for establishing its capabilities of the WATMUS algorithm.

3.1 *Review of Some Accelerated Time Integration Methods*

Various accelerated time integration schemes have been proposed with limited success. A few of these methods are discussed in the following sections.

3.1.1 Extrapolation based Methods

In extrapolation methods, the cyclic response of a representative microstructural volume is simulated up to a certain number of cycles and the evolution of state variables are extrapolated from these results thereafter. High cycle fatigue problems have been studied in (Bennett and McDowell 2003), where fatigue cracks usually initiate after thousands of cycles. However, only two complete strain cycles were simulated to obtain fatigue parameters with the assumption that the response stabilizes thereafter. Fretting fatigue in titanium alloys has been studied using a cyclic crystal plasticity model in (Goh et al. 2001, 2003), where the response is assumed to stabilize after three load cycles. An extrapolation-based approach has also been utilized in (Sinha and Ghosh 2006) for predicting fatigue life of HSLA steels from the results of crystal plasticity simulations. The stabilized plastic strain response is used to derive a functional relationship for the local ratcheting rate in terms of the number of cycles and the mean applied stress, of the form:

$$\frac{d\varepsilon^p}{dN} = \left[\frac{d\varepsilon^p}{dN}\right]_1 - C_1\left(1 - \frac{1}{N(1+\ln(N))^{C_2}}\right), \tag{23}$$

where C_1 and C_2 are parameters which depend on the applied mean stress and $\left[\frac{d\varepsilon^p}{dN}\right]_1$ is the ratcheting rate for the first cycle.

A major drawback of the extrapolation-based approaches is that they fail to adequately track the evolution of microstructural variables that are often highly localized, e.g., the phenomenon of load shedding from one grain to its neighbors. This can lead to significant errors in the predicted value of the microstructural state variables and can render any microstructure-based crack initiation criterion completely inaccurate. Another issue with these methods is that their accuracy is very sensitive to the points in time, from which values are being extrapolated. Reaching a stabilized state at all points in the polycrystalline microstructure may involve simulating a considerable number of cycles before the extrapolation is carried out. Even then, it does not exclude the possibility of inaccurate representation of local deformation states in the microstructure that are crucial for predicting fatigue life.

As a test of the extrapolation algorithms, a one-dimensional viscoplastic bar problem is solved with the following constitutive relations and boundary conditions.

$$\sigma = E(\varepsilon - \varepsilon^p), \quad \dot{\varepsilon}^p = \dot{a}\left|\frac{\sigma}{g}\right|^{\frac{1}{m}} \mathrm{sgn}(\sigma) \quad \dot{g} = h\,|\dot{\varepsilon}^p|$$

$$\text{with } \varepsilon^p(x, t=0) = 0 \text{ and } g(x, t=0) = \begin{cases} g_{o1} & \text{for } 0 < x < L \\ g_{o2} & \text{for } L < x < 2L \end{cases} \tag{24}$$

with σ, ε, ε^p, g, m, and $\dot{a}$ being the stress, strain, plastic strain, hardness, rate sensitivity exponent, and reference strain rate. The cyclic boundary conditions are:

$$u(x=0,t) = 0 \quad \text{and}$$
$$u(x=2L,t) = \bar{u}_o + \tilde{u}_o \sin\left(\frac{2\pi}{\tau_o}\frac{t}{\zeta}\right) \quad \text{with } \zeta \to 0. \tag{25}$$

Equations (24) and (25) are solved using the finite element method using two 1-D bar elements for up to 2,000 cycles of the applied displacement loading. Material parameters used are shown in Table 7 and the time period of the applied loading is 1s, i.e., $\zeta\tau_o = 1$.

For testing the extrapolation-based schemes, the simulation is continued till stabilization in plastic strain occurs corresponding to a tolerance of $\Delta\,\varepsilon^p(N) \leq 10^{-3} \times \varepsilon^{p0}(N)$. Here, $\varepsilon^{p0}(N)$ is the plastic strain at the beginning of cycle N and $\Delta\,\varepsilon^p(N)$ is the change in plastic strain over the Nth cycle. For the problem considered, stabilization occurs at around $N_o = 1{,}590$ cycles and the variable $\varepsilon^{p0}(N)$ is extrapolated from the corresponding state using the formula:

$$\varepsilon^{p0}(N) = \varepsilon^{p0}(N_o) + (\varepsilon^{p0}(N_o+1) - \varepsilon^{p0}(N_o))(N - N_o), \quad N > N_o. \tag{26}$$

Figure 10 shows a comparison of the extrapolated solution with the completely solved finite element solution for 2,000 cycles. The extrapolated solution accumulates error with increasing cycles, with an error of approximately 12% at around 5,000 cycles even in this macroscopic variable. In another scheme, a cubic polynomial is fitted with the initial values of the plastic strain and the corresponding extrapolation is shown. This also fails in capturing the response after the stabilized cycle. It is obvious that these methods are likely to produce significant local errors even when overall stabilization has been achieved in the solution.

Table 7 Properties for the 1-D viscoplastic problem

E (MPa)	$\dot{a}$ (s^{-1})	m	g_{o1} (MPa)	g_{o2} (MPa)	h (MPa)
200×10^3	0.0023	0.02	320	600	100

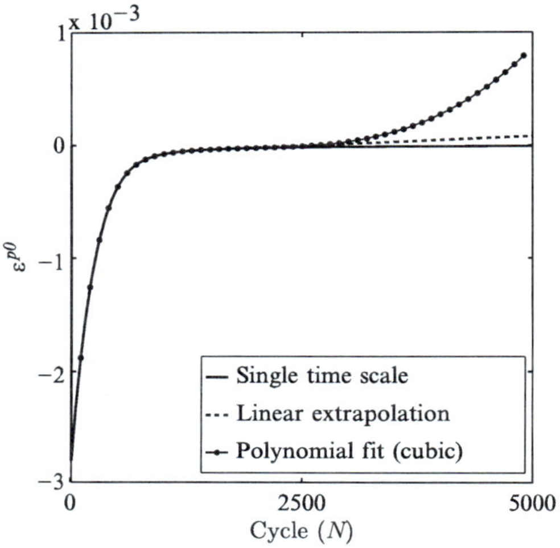

Fig. 10 Comparison between the single scale and extrapolated solution for the 1-D problem

3.1.2 Block Integration Methods

Block integration methods divide the entire loading history into a series of blocks and subsequently evaluate the evolution of state variables within a block by using their starting values within the block. Block integration schemes have been developed in the context of continuum damage mechanics in (Chow and Wei 1991; Paas et al. 1993; Fish and Yu 2002). In these methods, a constant amplitude loading history is subdivided into a series of load cycle blocks. Each block consists of a series of load reversals between two fixed amplitudes. As pointed out in (Fish and Yu 2002), limitations of these approaches are as follows:

- Solutions deviate from the equilibrium path caused by assumptions in the integration of the fatigue damage accumulation law (damage evolution independent of stress).
- It is difficult to estimate the adequate block size especially when the growth of the fatigue damage is high.
- The methods have limited applicability to heterogeneous materials.

As a remedy, the cyclic derivative of the damage parameter using an incremental finite element analysis has been proposed in (Fish and Yu 2002). This is then integrated over multiple cycles by a modified Euler approach with time step control. However, this approach violates the constitutive relations due to keeping state variables fixed while updating the damage variables, and a consistency adjustment procedure is required. Consequently, the number of cycles traversed in each step can turn out to be very small and the method will lose its advantage. Additionally, the consistency adjustment procedure can lead to nonuniqueness in the solution.

3.1.3 Asymptotic Expansion Based Methods

Dual time scale methods that assume existence of two time scales in the solution have been proposed in (Yu and Fish 2002; Manchiraju et al. 2007; Oskay and Fish 2004). The first corresponds to a coarse time scale t that characterizes the slow varying behavior of the solution, while the second is a fine time scale τ that characterizes the fast varying behavior. These methods make use of an asymptotic expansion of the primary state variables to decouple their coarse and fine scale evolution. Yu and Fish (2002) have proposed a temporal homogenization method in which they assumed all the variables to be locally periodic in time. This assumption was nominally relaxed in (Oskay and Fish 2004) to include variables, which are nearly periodic in time. Temporal periodicity or even near-periodicity are, however, not valid assumptions for evolving microstructural variables in crystal plasticity simulations under certain loading conditions. In addition, these method may invoke two-way coupling between the time scales that requires solution of initial value problems at both time scales in each step. This can result in very high computational cost and may not provide any advantage over single time scale calculations.

To avoid two-way coupling, Ghosh et al. (Manchiraju et al. 2007,2008) have developed a decoupled set of crystal plasticity-based governing equations. The problems are characterized by a cycle-averaged, low-frequency behavior and a short time scale (high frequency) problem for the remaining oscillatory portion. Effective constitutive equations are developed for the cycle-averaged problem by interpolating in a parameter space that is created from single time scale solutions of single slip-system problems. The resulting averaged constitutive equations do not assume periodicity and are decoupled from the short time scale oscillatory behavior. Consequently, they can be solved with time increments that are of the order of multiple time periods of the cyclic loading. This can yield significant computational gain over single time scale solution. Furthermore, asymptotic expansions of various field variables are used to decompose the oscillatory problem into various orders of oscillations. Each order of oscillatory solution can be solved locally in temporal domain with the knowledge of the averaged solution. While computational efficiency can be significantly enhanced with this method for certain problems, the asymptotic expansion-based methods can face serious deficiencies for some load cases.

The main idea of the asymptotic methods and its application in (Manchiraju et al. 2007,2008) can be understood through the 1-D viscoplastic problem introduced in Sect. 3.1.1. All state variables, e.g., plastic strain, stress, and hardness at a point x are assumed to depend on the two scales t and τ and may be expressed in terms of an asymptotic series as:

$$\begin{aligned}
\varepsilon^{p(\zeta)}(x,t,\tau) &= \varepsilon^{p(0)}(x,t,\tau) + \zeta\varepsilon^{p(1)}(x,t,\tau) + \mathcal{O}(\zeta^2) \\
\sigma^{\zeta}(x,t,\tau) &= \sigma^{(0)}(x,t,\tau) + \zeta\sigma^{(1)}(x,t,\tau) + \mathcal{O}(\zeta^2) \\
g^{\zeta}(x,t,\tau) &= g^{(0)}(x,t,\tau) + \zeta g^{(1)}(x,t,\tau) + \mathcal{O}(\zeta^2).
\end{aligned} \quad (27)$$

In addition, these evolving variables $\phi(x, t, \tau)$ can be decomposed into their average and oscillator parts as:

$$\bar{\phi}(x,t) = \frac{1}{\tau_o} \int_{\frac{t}{\zeta}}^{\frac{t}{\zeta}+\tau_o} \phi(x,t,\tau)\,\mathrm{d}\tau = \langle \phi \rangle \quad \phi_{\mathrm{osc}}(x,t,\tau) = \phi(x,t,\tau) - \bar{\phi}(x,t). \tag{28}$$

Substituting these decompositions into the constitutive equations and separating terms corresponding to various orders of $\varepsilon^{p(\zeta)}$ result in:

$$\mathcal{O}(\zeta^{-1}): \quad \frac{\partial \varepsilon_{\mathrm{osc}}^{p(0)}}{\partial \tau} = 0 \Rightarrow \varepsilon^{p(0)}(t,\tau) = \bar{\varepsilon}^{p(0)}(t) \text{ and } \varepsilon_{\mathrm{osc}}^{p(0)}(t,\tau) = 0 \tag{29a}$$

$$\frac{\partial g_{\mathrm{osc}}^{(0)}}{\partial \tau} = h\,\mathrm{sgn}(\sigma) \frac{\partial \varepsilon_{\mathrm{osc}}^{p(0)}}{\partial \tau} = 0 \tag{29b}$$

$$\mathcal{O}(\zeta^{0}): \quad \frac{\partial \varepsilon^{p(0)}}{\partial t} + \frac{\partial \varepsilon^{p(1)}}{\partial \tau} = \dot{a} \left| \frac{\sigma^{(0)}}{g^{(0)}} \right|^{\frac{1}{m}} \mathrm{sgn}(\sigma) \tag{29c}$$

$$\frac{\partial g_{\mathrm{osc}}^{(0)}}{\partial t} + \frac{\partial g^{(1)}}{\partial \tau} = h\,\mathrm{sgn}(\sigma) \left(\frac{\partial \varepsilon^{p(0)}}{\partial t} + \frac{\partial \varepsilon^{p(1)}}{\partial \tau} \right). \tag{29d}$$

Local temporal periodicity that has been assumed in (Yu and Fish 2002; Oskay and Fish 2004) is in general not valid for problems involving plastic accumulation due to evolving hardness and other parameters. From the asymptotic analysis, the plastic strain oscillator is derived to be of $\mathcal{O}(\zeta)$. This implies that the stress oscillator $\sigma_{\mathrm{osc}}^{\zeta}$ is of the form:

$$\sigma_{\mathrm{osc}}^{\zeta} = E\varepsilon_{\mathrm{osc}} - \zeta E \varepsilon_{\mathrm{osc}}^{p(1)} + \mathcal{O}(\zeta^2). \tag{30}$$

Considering only $\mathcal{O}(\zeta^0)$ contributions, the stress oscillator can be assumed to be elastic, i.e., $\sigma_{\mathrm{osc}}^{\zeta} \approx E\varepsilon_{\mathrm{osc}}$ as long as $\zeta \varepsilon^{p(1)} \to 0$. This result ensures that the stress oscillator can be obtained at every material point by simulating a single cycle. The approach used in (Manchiraju et al. 2007, 2008) makes use of this property and assumes a coarse scale evolution equation for the average plastic strain of the form:

$$\frac{\mathrm{d}\bar{\varepsilon}^p}{\mathrm{d}t} = f(\bar{\sigma}, \sigma_{\mathrm{osc}}^{\mathrm{char}}, \bar{g}), \tag{31}$$

thereby decoupling the average and the oscillatory parts of the problem. The function f is obtained through a calibration process. In (31), $\sigma_{\mathrm{osc}}^{\mathrm{char}}$ is selected as a characteristic function of the stress oscillator, e.g., the stress amplitude. From the results of the asymptotic analysis, the stress oscillator may be assumed to remain constant at a material point from cycle to cycle. Equation (31) can now be used to integrate over many cycles in one integration step over coarse time t.

Remark. Failure of the asymptotic methods for certain R ratios

Load reversal in cyclic loading is expressed in terms of the R ratio, defined as $R = \frac{\sigma_{\min}}{\sigma_{\max}}$. While the asymptotic expansion-based methods work well for loads that are not in the vicinity of fully reversed loading, i.e., for $R \neq -1$, they face severe limitations as the loads approach the fully reversed loading case, i.e., $R \rightarrow -1$. Large oscillations in the plastic variables are encountered near $R = -1$ that violate the assumption of decay of higher order terms in the asymptotic expansion. This implies that the stress oscillator is no longer elastic and has a strong inelastic component. Hence, it is not possible to characterize it at a point across multiple cycles from the results of just one cycle. As a result, equations of the form (31) are not usable in such situations as discussed next.

The 1-D bar is subjected to an applied harmonic strain of $\varepsilon(t,\tau) = \bar{\varepsilon}_o + \tilde{\varepsilon}_o \sin(\frac{2\pi}{\tau_o}\tau)$. Here, t represents the coarse or slow varying time scale and $\tau = \frac{t}{\zeta}$ with $\zeta \rightarrow 0$ represents the fast varying time scale. Application of chain rule together with a relation connecting derivatives in the two scales results in:

$$\frac{\partial \varepsilon^{p(\zeta)}}{\partial t} + \frac{1}{\zeta}\frac{\partial \varepsilon^{p(\zeta)}}{\partial \tau} = \dot{a}\left|\frac{E(\bar{\varepsilon}_o + \tilde{\varepsilon}_o \sin\left(\frac{2\pi}{\tau_o}\tau\right) - \varepsilon^p)}{g}\right|^{\frac{1}{m}} \operatorname{sgn}(\sigma). \tag{32}$$

Using the asymptotic expansion in (27) for a simplified case with $m = 1$ and $h = 0$, the following solution can be obtained analytically:

$$\begin{aligned} &\varepsilon^{p\zeta} = \bar{\varepsilon}^{p(0)}(t) + \zeta(\bar{\varepsilon}^{p(1)}(t) + \varepsilon^{p(1)}_{\text{osc}}(t,\tau)) + \mathcal{O}(\zeta^2)\cdots, \quad \text{with} \\ &\bar{\varepsilon}^{p(0)}(t) = \bar{\varepsilon}_o(1 - \mathrm{e}^{-\frac{t}{t_r}}), \;\; \bar{\varepsilon}^{p(1)}(0) = \frac{\tau_o}{2\pi t_r}, \;\; \varepsilon^{p(1)}_{\text{osc}} = -\tilde{\varepsilon}_o \frac{\tau_o}{2\pi t_r}\cos\left(\frac{2\pi}{\tau_o}\tau\right) \end{aligned} \tag{33}$$

Additionally, the stress expansion is

$$\begin{aligned} \sigma^{\zeta}(t,\tau) &= \bar{\sigma}^{\zeta}(t) + \sigma^{\zeta}_{\text{osc}}(t,\tau), \quad \text{where} \\ \bar{\sigma}^{\zeta} &= E(\bar{\varepsilon}_o - \bar{\varepsilon}^{p(0)}) - \zeta E \bar{\varepsilon}^{p(1)} + \mathcal{O}(\zeta^2), \quad \text{and} \\ \sigma^{\zeta}_{\text{osc}}(t,\tau) &= E\tilde{\varepsilon}_o \sin\left(\frac{2\pi}{\tau_o}\tau\right) + \zeta E \varepsilon^{p(1)}_{\text{osc}} + \mathcal{O}(\zeta^2) \end{aligned} \tag{34}$$

with $t_r = \frac{g}{\dot{a}E}$. This solution demonstrates that as long as $\zeta\varepsilon^{p(1)}_{\text{osc}} \rightarrow 0$, the stress oscillator is elastic and can be obtained from just one cycle of the simulation. This is a key assumption in (Yu and Fish 2002; Manchiraju et al. 2007), and it is true as long the first-order plastic strain $\varepsilon^{p(1)}_{\text{osc}}$ remains relatively small, i.e., the oscillations in the plastic strain are not excessive. For an applied reversible sinusoidal loading with amplitude 0.006 and time period of 1 s, the variation of different orders of the stress oscillator over one cycle is shown in Fig. 11.

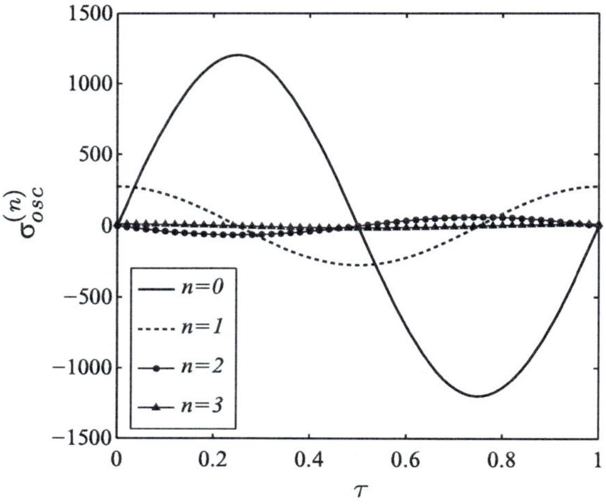

Fig. 11 Various orders of the stress oscillator in asymptotic expansion ($n = 0$ is the elastic stress oscillator)

The figure shows that the first-order stress oscillator contribution is around 22% of the elastic stress oscillator. Consequently, it cannot be neglected in the construction of the stress. From (33), the amplitude of the first order plastic strain oscillator is proportional to $\tilde{\varepsilon}_o$, i.e., the oscillatory part of the applied strain. Using the first-order terms in σ^ζ to calculate the R ratio, the following relationship is obtained:

$$\tilde{\varepsilon}_o \approx \begin{cases} \frac{1}{2}\{1 - R\}(|\bar{\varepsilon}_o| + |\tilde{\varepsilon}|_{\max}) & \text{for } |R| \leq 1 \\ \frac{1}{2}\{1 - \frac{1}{R}\}(|\bar{\varepsilon}_o| + |\tilde{\varepsilon}|_{\max}) & \text{for } |R| \geq 1. \end{cases} \tag{35}$$

The normalized amplitude of the plastic oscillator, denoted as (normalized $|\varepsilon_{\text{osc}}^{p(1)}|_{\max} = \frac{2\pi t_r}{|\bar{\varepsilon}_o|+|\tilde{\varepsilon}|_{\max}}|\varepsilon_{\text{osc}}^{p(1)}|_{\max}$) is plotted as a function of R in Fig. 12. Here, t_r is a material parameter and $|\bar{\varepsilon}_o| + \left|\tilde{\varepsilon}\right|_{\max}$ is the maximum absolute value of the applied strain. The maximum value of the plastic strain oscillator $\varepsilon_{\text{osc}}^{p(1)}$ occurs as $R \to -1$. Therefore, the possibility of large oscillations in the plastic strain increases as $R \to -1$ and hence the stress oscillator does not remain elastic.

The study is helpful for making the following conclusions.

1. Asymptotic expansion of state variables in the viscoplastic constitutive laws results in the condition that the plastic oscillations are of $\mathcal{O}(\zeta)$ or are very small.
2. As a consequence of negligible plastic oscillations, the stress oscillator remains elastic and shows very little variation across cycles. Decoupling between the scales is achieved by integrating only the slow-varying or low-frequency components like the average response with large time increments. On the other hand, the fast-varying or high-frequency components remain the same across the cycles (stress oscillator) or are negligible (plastic strain oscillations).

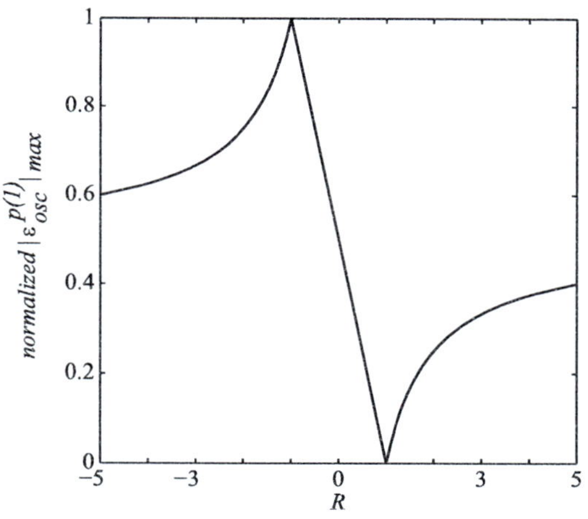

Fig. 12 Effect of R ratio on the amplitude of plastic oscillator using a normalizing factor $\frac{2\pi t_r}{|\bar{\varepsilon}_o|+|\tilde{\varepsilon}|_{\max}}$

3. In the case of reversible loading, the plastic oscillations are not negligible and the assumption of asymptotic decay of the successive terms in the series does not hold. The stress oscillator is now inelastic and needs to be explicitly solved from cycle to cycle. As a result, decoupling is no longer accurate, as the high frequency components have to be explicitly accounted for with progressing cycles.

3.1.4 Methods on Homogenization of Almost Periodic Functions

Oskay and Fish (2004) have proposed an a-periodic temporal homogenization (APTH) operator to track the evolution of almost-periodic variables. Such near-periodicity may arise in constitutive laws due to irreversibility or damage accumulation. The averaging APTH operator $\langle \cdots \rangle_{\text{APTH}}$ acting on the a-periodic function $\phi_{\text{ap}}(x, t, \tau)$ has been defined in (Oskay and Fish 2004) through a coarse scale differential equation as:

$$\frac{\mathrm{d}\langle \phi^{\xi}_{\text{ap}} \rangle_{\text{APTH}}}{\mathrm{d}t} = \langle \dot{\phi}^{\xi}_{\text{ap}} \rangle (t). \tag{36}$$

The fine time scale response of the function denoted by $\tilde{\phi}^{\xi}_{\text{ap}}$ is defined as:

$$\tilde{\phi}_{\text{ap}} = \phi_{\text{ap}}(t, \tau) - \langle \phi_{\text{ap}} \rangle_{\text{APTH}}(t). \tag{37}$$

The 1-D viscoplastic model of Sect. 3.1.1 can be resolved into a coupled set of coarse and fine time scale initial-boundary value problems by using the APTH operator. The equations are solved using a staggered, global–local, integration scheme. In this adaptive scheme, coarse scale variables $\langle \phi \rangle_{\text{APTH}}(t_{n+1})$ at time t_{n+1} are updated, while keeping the fine scale variables at t_n, i.e., $\tilde{\phi}_{n+1} = \tilde{\phi}_n$. The local, fine scale initial-boundary value problem is then solved with the updated values of the coarse variables over the time domain $[0, \tau]$.

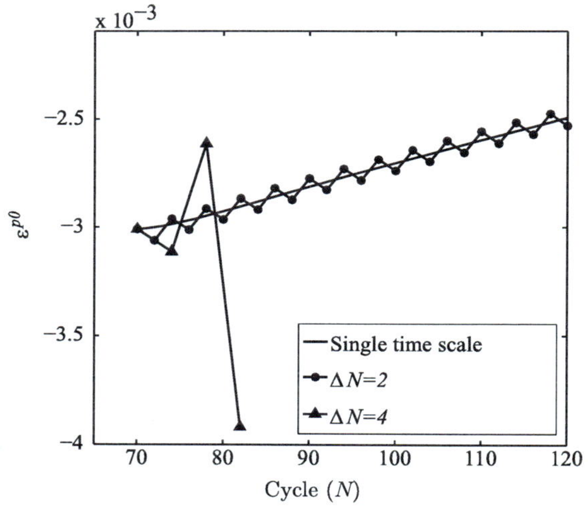

Fig. 13 Comparison of the fine-scale solution with the APTH operator-based solution with different step sizes, for the 1-D viscoplastic problem

This global–local method does not make use of the asymptotic expansion and hence does not suffer from the limitations pertaining to $R \to -1$. However if the coupling between the two scales is strong, it may lead to very small time steps in the coarse scale, thus affecting the computational efficiency adversely. This is illustrated for the 1-D viscoplastic problem in Fig. 13 for the fully reversible loading case $R = -1$. The results show increasing instability with higher step sizes (number of cycles traversed in each step) in the staggered approach. Figure 3 shows oscillations in the solution for step size of two cycles, i.e., $\Delta N = 2$, while for $\Delta N = 4$ it leads to complete instability. The method does give accurate result for $\Delta N = 1$, but this corresponds to the normal cycle by cycle or fine scale integration with no computational advantage. In summary, the global–local approach using the APTH operator suffers from the following limitations for simulating rate-dependent crystal plasticity problems.

1. Inability to solve the entire set of governing equations (equilibrium and constitutive) in a consistent manner leads to global–local or staggered approach, where one set of field variables are artificially kept constant while others are updated.
2. This approach may require unreasonably small time steps to preserve stability in crystal plasticity simulations, thus negating the proposed time advantage due to decoupling.

3.2 *Wavelet Transformation based Multi-time Scaling Methodology for Cyclic Plasticity*

In this section, a new wavelet transformation-based multi-time scaling (WATMUS) method is introduced as a promising alternative to the conventional time acceleration

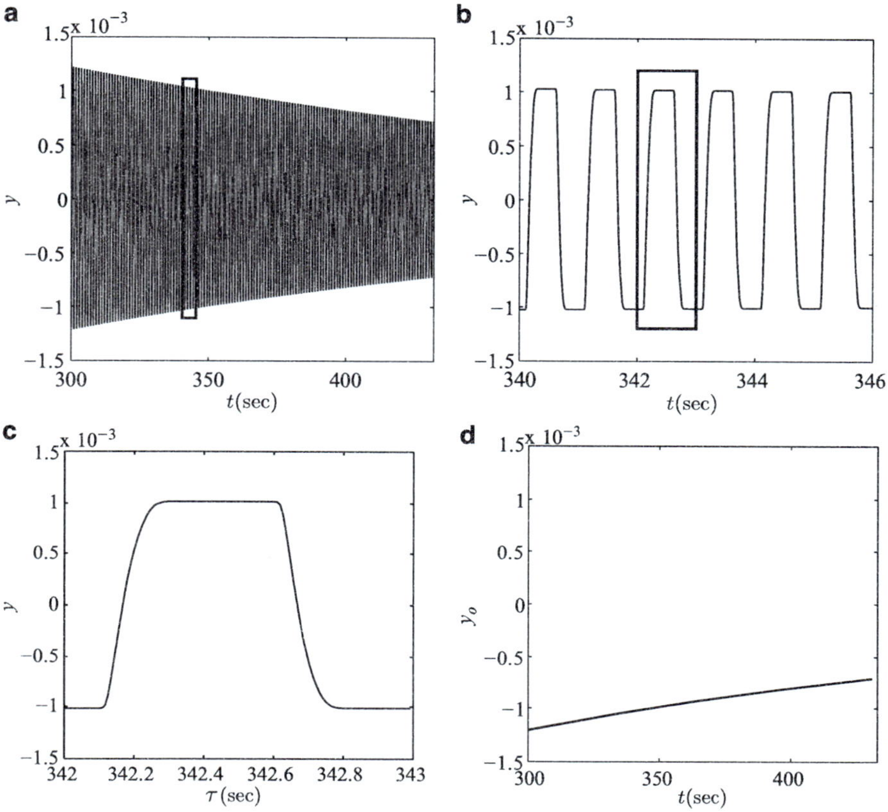

Fig. 14 Decoupling the fine and coarse time scale responses for a chosen viscoplastic state variable under cyclic loading: (**a**) fine scale solution, (**b**, **c**) zoomed in fine scale solution, (**d**) coarse scale solution corresponding to the value at the start of a cycle

methods. Unlike some conventional approaches, this method makes no assumption on the periodicity of the solution or asymptotic behavior of variables. The method is valid for all conditions of loading including the fully reversible case.

The WATMUS methodology for accelerated time integration in crystal plasticity finite element analyses may be motivated by the viscoplastic response of a material undergoing cyclic plastic deformation. Shown in Fig. 14a is the evolution of a rate-dependent crystal plasticity state variable (y), solved by the finite element method. Clearly, the material response can be resolved into two time scales, viz. (i) a rapidly oscillating response within each cycle corresponding to a fine time scale τ as shown in Fig. 14b,c, and (ii) a slowly varying monotonic response over the entire loading time span as shown in Fig. 14d, corresponding to a coarse time scale t. The coarse time scale is identified with the cycle scale N, and hence projects a monotonic behavior. Correspondingly, the value of any given state variable y_o at the beginning of a given cycle can be thought of as a coarse time scale variable. This state variable will not vary in the τ-scale within each cycle and hence, it can be considered to be purely a function of the cycle number N, i.e., $y^o = y^o(N)$. The objective then is to obtain a coarse time scale (cycle scale) evolution equation of the form:

$$\frac{\mathrm{d}y^o}{\mathrm{d}N} = f(y^o(N), \varepsilon^k(N)), \tag{38}$$

where $\varepsilon^k(N)$ are the wavelet decomposed strain coefficients over the cycle N that is resolved with respect to wavelet basis functions $\psi_k(\tau)$ in the τ-scale. Note that without the wavelet components $\varepsilon^k(N)$, it is not possible to obtain the right hand side in (38) and a global–local approach would have to be used along with its limitations. Thus, any variable, e.g., strain can be resolved in a wavelet basis as:

$$\varepsilon(x,t) = \varepsilon(x,N,\tau) = \sum_k \varepsilon^k(x,N)\psi_k(\tau). \tag{39}$$

The coefficients $\varepsilon^k(x,N)$ depends only on the cycle number N and the location x in the material microstructure. It is important to note that the τ-scale basis functions $\psi_k(\tau)$ do not change with cycles N. For any material variable, the coarse scale behavior is associated with the cycle number N and the fine scale behavior with the time scale $\tau \in [0,T]$, T being the loading period. Equations (38) and (39) can then be used to delineate the coarse and fine scale behavior of the constitutive equations. A single integration step of (38) can traverse many cycles ΔN, resulting in significant computational efficiency of the algorithm. The value of ΔN is expected to increase with response stabilization.

The WATMUS methodology work requires appropriate basis functions for temporally resolving the displacement vector, and the corresponding strain or deformation gradient fields as in (39). An ideally chosen set of basis functions should satisfy the following conditions:

- The functions should be orthogonal, i.e., form a linearly independent set.
- The functions should be able to represent all possible waveforms in the response variables to a pre-determined resolution.
- The number of coefficients, corresponding to the number of basis functions used in this representation must be optimally small. This should hold even as the number evolves with progressing cyclic deformation.

A spectral basis representation in terms of Fourier series functions suffer from the following shortcomings.

- Basis functions in a Fourier series have infinite support. The use of a finite set of Fourier coefficients, while truncating others in the infinite series, can lead to instabilities in the oscillatory response. Such instabilities, viz. the Gibbs phenomenon give spurious oscillations at regions, where the signal is cut-off. This in turn can lead to inaccuracies in the coarse scale solution.

- The dominant terms, needed to match a response signal, are not known a-priori in the selection of a finite subset of the infinite Fourier series. A trial-and-error process is needed to establish these terms.

It is deemed that a basis of wavelet functions avoids these shortcomings and hence is considered for the multi-time scale (WATMUS) approach developed here.

3.2.1 Brief Overview of Wavelet Basis Functions

Wavelet basis functions span the space of square integrable functions $L^2(R)$ through translation and dilation of the scaling function $\phi(\tau)$ (Walker 1999; Strang and Nguyen 1996), which satisfies the following refinement condition:

$$\phi(\tau) = \sum_{k=1}^{N_{\text{filt}}} h_k \phi(2\tau - k) \tag{40}$$

Parameters h_k and N_{filt} characterize the wavelet basis and correspond to the components of a low-pass filter. Any function in this space can be expressed as:

$$f(\tau) = \sum_{m} \sum_{n} C_{m,n} \phi_{m,n}(\tau), \tag{41}$$

where $\phi_{m,n} = 2^{\frac{m}{2}} \phi(2^m \tau - n)$ with $m, n \in Z$ corresponding to the dilatation and translation of ϕ, respectively. This means that at a certain resolution, the subspace is spanned by translations of the scaling function in (40), dilated to that resolution. The scaling function $\phi(\tau)$ function may be used to produce a multiresolution analysis in $L^2(R)$ using the property:

$$\{0\} \subset \cdots \subset V_{-1} \subset V_0 \subset V_1 \subset \cdots V_m \subset V_{m+1} \subset \cdots \subset L^2(R), \tag{42}$$

where V_m denotes the subspace of $L^2(R)$ at resolution m with the basis ϕ_{mn}. A complimentary space of interest in the multiresolution analysis is the detail space W_m, which contains the orthogonal difference between two consecutive resolutions V_m and V_{m+1}, i.e., $V_{m+1} = V_m \oplus W_m$. The basis functions for this space is generated in a similar manner to that of V_m, through the translation and dilation of a mother wavelet function $\psi(\tau)$, i.e., $W_m = span\{\psi_{m,n} = 2^{\frac{m}{2}} \psi(2^m \tau - n), \quad m, n \in Z\}$. The mother wavelet satisfies a condition similar to (40):

$$\psi(\tau) = \sum_{k=1}^{N_{\text{filt}}} g_k \phi(2\tau - k), \tag{43}$$

where g_k corresponds to the components of a high-pass filter and is another characteristic of the wavelet basis. In other words, the wavelet basis can be completely represented through the filter coefficients h_k and g_k using (40) and (43). Filter coefficients for the Daubechies-4 wavelet used are given as:

$$h_1 = \frac{1+\sqrt{3}}{4}, h_2 = \frac{3+\sqrt{3}}{4}, h_3 = \frac{3-\sqrt{3}}{4}, h_4 = \frac{1-\sqrt{3}}{4}$$
$$g_1 = \frac{1-\sqrt{3}}{4}, g_2 = \frac{-3+\sqrt{3}}{4}, g_3 = \frac{3+\sqrt{3}}{4}, g_4 = \frac{-1-\sqrt{3}}{4} \tag{44}$$

These coefficients are obtained from the following considerations:

- Compact support: The Daubechies-4 wavelet has four filter coefficients ($N_{\text{filt}} = 4$). This leads to a $\phi(\tau)$ and $\psi(\tau)$ with support [0, 3].
- Orthogonality of translation: Scaling functions obtained through the coefficients h_k satisfy $\int_{-\infty}^{\infty} \phi(\tau - k)\phi(\tau - l)\mathrm{d}\tau = \delta_{kl}$.
- Smoothness: Coefficients are selected so as to have the maximum number of vanishing moments for a given support (2 in the case of Daubechies-4 wavelet). This implies that the basis can capture a linear function exactly.

Using (40) at discrete points $\tau = 0, 1, 2, 3$ in the time scale and setting to zero the points outside the compact support [0, 3] results in the following equation:

$$\begin{Bmatrix} \phi(0) \\ \phi(1) \\ \phi(2) \\ \phi(3) \end{Bmatrix} = \begin{bmatrix} h_1 & 0 & 0 & 0 \\ h_3 & h_2 & h_1 & 0 \\ 0 & h_4 & h_3 & h_2 \\ 0 & 0 & 0 & h_4 \end{bmatrix} \begin{Bmatrix} \phi(0) \\ \phi(1) \\ \phi(2) \\ \phi(3) \end{Bmatrix} \tag{45}$$

One of the eigen-values of the square matrix on the right hand side of the above equation may be set to 1, through appropriate choice of h_k in (44). Correspondingly, the values of the scaling function at $\tau = 0, 1, 2, 3$ are obtained from the eigen-vector and this is used as starting values in the following recursive relationship once again obtained from (40):

$$\phi\left(\frac{n}{2^j}\right) = h_1\phi\left(\frac{n}{2^{j-1}}\right) + h_2\phi\left(\frac{n}{2^{j-1}} - 1\right) + h_3\phi\left(\frac{n}{2^{j-1}} - 2\right)$$
$$+ h_4\phi\left(\frac{n}{2^{j-1}} - 3\right) \forall \; n = 1, 2, \ldots, 3.2^j, \;\; j = 1, 2, 3, \ldots. \tag{46}$$

A similar procedure can be obtained for the mother wavelet $\psi(\tau)$. The values so obtained for the scaling function and the mother wavelet are shown in Fig. 15. A function, belonging to the space V_m, can thus be split into two orthogonal components. One of the components belongs to a lower resolution space V_{m-1}, while the other to the orthogonal space W_{m-1} that corresponds to the difference signal between the two successive resolutions. Mathematically,

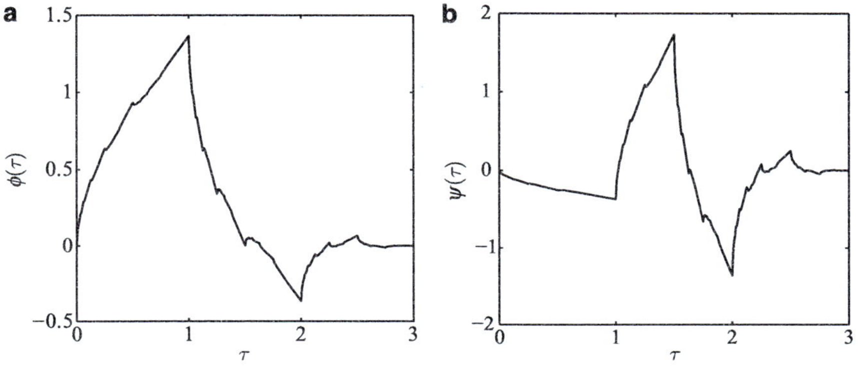

Fig. 15 Daubechies-4 wavelets: (**a**) Scaling function ϕ, and (**b**) Mother wavelet ψ

$$\begin{aligned} f^m(\tau) &= \sum_n \langle f^m, \phi_{m-1,n}\rangle \phi_{m-1,n} + \sum_n \langle f^m, \psi_{m-1,n}\rangle \psi_{m-1,n} \\ &= \sum_n a^{m-1,n}\phi_{m-1,n} + \sum_n d^{m-1,n}\psi_{m-1,n} \\ &= f^{m-1} + \sum_n d^{m-1,n}\psi_{m-1,n}. \end{aligned} \tag{47}$$

Coefficients $a^{m-1,n}$, $d^{m-1,n}$ are called the approximation and the detail coefficients respectively. The function f^{m-1} is an approximation of the function f^m at a lower resolution. The same procedure can be carried out on f^{m-1} and so on, each time reducing the resolution by half and generating an additional set of detail coefficients. The final decomposition, based on Daubechies-4 wavelets consists of only two approximation coefficients and remaining detail coefficients. The refinement (40) and (43) can be used to connect the basis functions at the two consecutive resolutions $m-1$ and m. For the Daubechies-4 wavelet, this yields:

$$\begin{aligned} \phi(2^{m-1}\tau - n) &= \sum_{k=1}^{4} h_k \phi(2^m\tau - 2n - k) \\ \psi(2^{m-1}\tau - n) &= \sum_{k=1}^{4} g_k \psi(2^m\tau - 2n - k) \end{aligned} \tag{48}$$

leading to the connection between the bases:

$$\begin{aligned} \phi_{m-1,n} &= \sum_{k=1}^{4} \frac{h_k}{\sqrt{2}}\phi_{m,2n+k} \\ \psi_{m-1,n} &= \sum_{k=1}^{4} \frac{g_k}{\sqrt{2}}\phi_{m,2n+k}. \end{aligned} \tag{49}$$

The approximation coefficients at resolution $m-1$ can now be written as:

$$\begin{aligned} a^{m-1,n} &= \langle f(\tau), \phi_{m-1,n} \rangle \\ &= \left\langle f(\tau), \sum_{k=1}^{4} \frac{h_k}{\sqrt{2}} \phi_{m,2n+k} \right\rangle \\ &= \sum_{k=1}^{4} \frac{h_k}{\sqrt{2}} \langle f(\tau), \phi_{m,2n+k} \rangle \\ &= \sum_{k=1}^{4} \frac{h_k}{\sqrt{2}} a^{m,2n+k}. \end{aligned} \tag{50}$$

Similarly for the detail coefficients at resolution $m-1$:

$$d^{m-1,n} = \sum_{k=1}^{4} \frac{g_k}{\sqrt{2}} a^{m,2n+k}. \tag{51}$$

The above two relationships can be written in matrix form, assuming periodic extensions for the coefficients at resolution m as:

$$\begin{Bmatrix} \mathbf{a}^{m-1} \\ \mathbf{d}^{m-1} \end{Bmatrix} = \frac{1}{\sqrt{2}} \begin{bmatrix} \mathbf{H} \\ \mathbf{G} \end{bmatrix} \mathbf{a}^m, \tag{52}$$

where:

$$\begin{bmatrix} \mathbf{H} \\ \mathbf{G} \end{bmatrix} = \begin{bmatrix} h_1 & h_2 & h_3 & h_4 & 0 & 0 & \cdots\cdots & 0 & 0 \\ 0 & 0 & h_1 & h_2 & h_3 & h_4 & \cdots\cdots & 0 & 0 \\ \vdots & \vdots & \vdots & \vdots & \vdots & \vdots & \vdots & \vdots & \vdots \\ h_3 & h_4 & 0 & 0 & 0 & 0 & \cdots\cdots & h_1 & h_2 \\ g_1 & g_2 & g_3 & g_4 & 0 & 0 & \cdots\cdots & 0 & 0 \\ 0 & 0 & g_1 & g_2 & g_3 & g_4 & \cdots\cdots & 0 & 0 \\ \vdots & \vdots & \vdots & \vdots & \vdots & \vdots & \vdots & \vdots & \vdots \\ g_3 & g_4 & 0 & 0 & 0 & 0 & \cdots\cdots & g_1 & g_2 \end{bmatrix}. \tag{53}$$

The operation in (52) is similar to the action of a low-pass **H** filter and a high-pass **G** filter on the original data, yielding the approximation and detail coefficients, respectively. The $m-1$ level approximation coefficients can be further decomposed into approximation and detail coefficients at resolution $m-2$, and can be successively continued till $m = 1$. The end results are approximation coefficients at $m = 1$ and detail coefficients at each filtering step. The coefficients are stored in a single array $\{\mathbf{c}\}$ and Table 8 shows the connection to the projection spaces for $m = 7$.

Table 8 Coefficient numbers and resolution space

Coefficient	$c^1 - c^2$	$c^3 - c^4$	$c^5 - c^8$	$c^9 - c^{16}$	$c^{17} - c^{32}$	$c^{33} - c^{64}$	$c^{65} - c^{128}$
Space	V_1	W_1	W_2	W_3	W_4	W_5	W_6

For all linear operations, the multiresolution decomposition may be represented as:

$$c^k = \sum_{l=1}^{N_{\text{wav}}} T_{kl} f(\tau_l) \quad k = 1 \cdots N_{\text{wav}}, \tag{54}$$

where c^k are the wavelet approximation and detail coefficients. The function is sampled at $N_{\text{wav}}(= 2^m)$ points in the fine time scale τ_l. $\mathbf{T}(= T_{ij}\mathbf{e}_i \otimes \mathbf{e}_j)$ is an orthogonal matrix containing the filter coefficients constructed by repeated application of (52).

Advantageous Wavelet Properties

Wavelet functions form excellent bases in the representation of fine time or τ-scale cyclic response patterns, on account of the following properties:

- *Compact Support:* Each wavelet basis has a compact support, i.e., spans a finite domain. As a result, the wavelet decomposition does not exhibit spurious instabilities such as the Gibbs phenomena, commonly encountered with Fourier series representations.
- *Multi-resolution:* For a given resolution of the fine scale response, the space of basis functions is well defined and finite, unlike in the Fourier basis. This implies that it is a-priori possible to identify known set of wavelet basis functions to represent the fine scale variables in crystal plasticity analysis under cyclic loading. For the Fourier series, this has to be found by trial and error.
- *Number of Coefficients:* The number of wavelet terms and coefficients can be minimized for a known response function. For example, in the case of dwell or triangular loading, response functions over a cycle might have segments that are almost linear. This makes the Daubechies-4 wavelets an ideal choice, since they are able to represent linear behavior exactly with only a few coefficients.

Choice of the optimal resolution in the wavelet representation is critical for both accuracy and efficiency of the problem solved. The maximum representative resolution can be obtained from the minimum time step required to integrate the single time scale problem to within a prescribed accuracy. In practice, this is determined from the first few cycles and the same resolution is retained throughout. This assumption is valid as long as there is no sudden change in the loading or boundary conditions in the problem course.

3.2.2 Coarse (Cycle) Scale Evolutionary Constitutive Relations

Coarse scale constitutive equations of the form in (38) are developed in this section. The evolution equation of a state variable y is assumed to be of the form:

$$\dot{y} = f(y, \varepsilon(t)) = f(y, \varepsilon^k(N), \tau), \tag{55}$$

where $\varepsilon^k(N)$ are known wavelet coefficients for the applied strain $\varepsilon(t)$. Given the initial value $y_o(N) = y(N, 0)$ for a cycle N and the wavelet decomposed strain components $\varepsilon^k(N)$, the value of y at any fine time scale point (τ) within the cycle can be expressed as:

$$\begin{aligned} y(N,\tau) &= y(N,0) + \int_0^\tau f(y, \varepsilon^k(N), \tau')\mathrm{d}\tau' \\ &= y_o(N) + \int_0^\tau f(y, \varepsilon^k(N), \tau')\mathrm{d}\tau'. \end{aligned} \tag{56}$$

Continuity of y across cycles with time period T yields the relation $y(N, T) = y(N + 1, 0) = y_o(N + 1)$. Consequently, the coarse time scale derivative of y can be expressed as:

$$\frac{\mathrm{d}y_o}{\mathrm{d}N} = y(N,T) - y(N,0) = \int_0^T f(y, \varepsilon^k(N), \tau')\mathrm{d}\tau' = Y_o(y_o, \varepsilon^k). \tag{57}$$

Equation (57) represents the rate of change of the initial value of y per cycle, for which the integral expression is evaluated numerically using the backward Euler method. A second-order implicit backward difference integration scheme is utilized for integration of this equation. A multi-time step formalism is adopted in this scheme, where the derivative is expressed as:

$$\frac{\mathrm{d}y_o}{\mathrm{d}N}(N + \Delta N) = \frac{a_1 y_o(N+\Delta N) - a_2 y_o(N) + a_3 y_o(N - \Delta N_p)}{\Delta N}. \tag{58}$$

ΔN and ΔN_p are cycle jumps corresponding to the current and previous steps. The parameters are expressed in terms of the cycle step ratio $r = \frac{\Delta N_p}{\Delta N}$, as:

$$a_1 = \frac{\{(r+1)^2 - 1\}}{\{(r+1)^2 - (r+1)\}}, \quad a_2 = \frac{(r+1)^2}{\{(r+1)^2 - (r+1)\}}, \quad a_3 = \frac{1}{\{(r+1)^2 - (r+1)\}}.$$

3.2.3 Coarse (Cycle) Scale Crystal Plasticity Finite Element Equations

The wavelet transform-based multi-time scale (WATMUS) algorithm is applied to the crystal plasticity constitutive models presented in Sect. 2. The evolving microstructural variables in the crystal plasticity constitutive relations (1)–(10) are:

$\mathbf{F}^p$, g^α, χ^α. Applying the methodology discussed in the previous section, the coarse scale evolution equations for these variables can be expressed as:

$$\begin{aligned}
\frac{\mathrm{d}F_{ij}^{po}}{\mathrm{d}N} &= f_{ij}(F_{ij}^k(N), F_{ij}^{po}, g^{\alpha o}, \chi^{\alpha o}) \\
\frac{\mathrm{d}g^{\alpha o}}{\mathrm{d}N} &= G^\alpha(F_{ij}^k(N), F_{ij}^{po}, g^{\alpha o}, \chi^{\alpha o}) \\
\frac{\mathrm{d}\chi^{\alpha o}}{\mathrm{d}N} &= B^{\alpha o}(F_{ij}^k(N), F_{ij}^{po}, g^{\alpha o}, \chi^{\alpha o})
\end{aligned} \tag{59}$$

Here, F_{ij}^{po} is the initial value of plastic deformation gradient, $g^{\alpha o}$ and $\chi^{\alpha o}$ are the initial values of hardness and back stress, respectively, for slip system α for the cycle N. The right-hand side in (59) is calculated using the formula in (57), i.e.:

$$\frac{\mathrm{d}F_{ij}^{po}}{\mathrm{d}N} = \int_0^T \dot{F}_{ij}^p(N,\tau)\mathrm{d}\tau = \int_0^T \sum_\alpha \dot{\gamma}^\alpha s_{oi}^\alpha m_{ok}^\alpha F_{kj}^p \mathrm{d}\tau. \tag{60}$$

The integral in the numerator of the RHS is evaluated numerically using the backward Euler method. Once the values of coarse scale variables are known, the increments of the Cauchy stress $\Delta\sigma(N,\tau)$ and other state variables can be computed over the cycle N.

The finite element formulation for the coarse scale equations introduces wavelet coefficients of nodal displacements as the primary solution variable, as opposed to nodal displacement components in conventional FEM. In this formulation, the element displacement field and the associated generalized nodal displacements of each element are expressed in terms of the wavelet basis expansion as:

$$u_i(\mathbf{X}, N, \tau) = \sum_\alpha N_\alpha(\mathbf{X}) q_{\alpha i}(N,\tau) = \sum_\alpha N_\alpha(\mathbf{X}) \sum_k^{N_{\mathrm{wav}}} q_{\alpha i}^k(N)\psi_k(\tau), \tag{61}$$

where $N_\alpha(\mathbf{X})$ is the shape functions corresponding to the node α, $q_{\alpha i}$ is the nodal displacement component i for the α-th node in an element, and $q_{\alpha i}^k(N); k = 1 \cdots N_{\mathrm{wav}}$ are the corresponding wavelet coefficients. The wavelet coefficients are functions of N and not of τ, i.e., they evolve in the cycle scale alone. The corresponding wavelet coefficients of the deformation gradient field is derived to be:

$$\begin{aligned}
F_{ij}^k(\mathbf{X}, N) &= \int_0^T \left(\delta_{ij} + \frac{\partial u_i}{\partial X_j}\right) \psi_k(\tau)\mathrm{d}\tau \\
&= \delta_{ij}\int_0^T \psi_k(\tau)\mathrm{d}\tau + \sum_\alpha \frac{\partial N_\alpha(\mathbf{X})}{\partial X_j} q_{\alpha i}^k(N).
\end{aligned} \tag{62}$$

Starting Procedure for Solving the Cycle Scale Problem

Initial conditions of the cycle scale governing equations are generated by conducting single (fine) scale analysis for the first few cycles ($N_o \leq 5$). This yields the initial displacement response and the initial values of the wavelet coefficients discussed in (54), i.e.,

$$q^k_{\alpha i}(N_o) = \sum_{l=1}^{N_{\text{wav}}} T_{kl} q_{\alpha i}(N_o, \tau_l) \quad k = 1 \cdots N_{\text{wav}}. \tag{63}$$

$q^k_{\alpha i}(N_o)$'s are wavelet coefficients of the nodal displacements resolved over the cycle N_o and N_{wav} corresponds to the number of wavelet basis functions or coefficients. The number of degrees of freedom for the corresponding finite element problem is then $N_d \times N_{\text{wav}}$, where N_d is the number of displacement degrees freedom in a conventional single time-scale FEM problem.

3.3 WATMUS Adaptivity for Accuracy and Efficiency

The accuracy and efficiency of the WATMUS methodology depends on two specific parameters, viz. (a) the number (N_{wav}) of wavelet bases or displacement coefficients $q^k_{\alpha i}(N)$ selected in (61) and (b) the step size ΔN or the number of cycles traversed in each increment of the numerical integration scheme. Optimally, N_{wav} should be low and ΔN as high as possible, while keeping the net errors due to waveform representation and series truncation, respectively, to under pre-determined bounds.

3.3.1 Evolving and Active Wavelet Basis Functions

Retaining all the wavelet basis functions at a given resolution for representing a response function in (61) may lead to a large number of degrees of freedom in the crystal plasticity FE model. The efficiency of the WATMUS method can benefit significantly from optimally reducing N_{wav}, by incorporating only the active and evolving wavelet bases required to represent the fine scale cyclic behavior of any state variable. The following scheme is developed for adaptively selecting those coefficients that change considerably in time. For a given degree of freedom, it divides the total set of wavelet coefficients $\mathscr{I}_{\alpha i} = \{q^k_{\alpha i} | k = 1 \cdots N_{\text{wav}}\}$ into a set of evolving ($\mathscr{I}^{\text{evol}}_{\alpha i}$) and nonevolving ($\mathscr{I}^{\text{non-evol}}_{\alpha i}$) coefficients respectively, i.e., $\mathscr{I}_{\alpha i} = \mathscr{I}^{\text{evol}}_{\alpha i} \bigcup \mathscr{I}^{\text{non-evol}}_{\alpha i}$. The division is based on the following criterion.

$$\begin{aligned} \mathscr{I}^{\text{evol}}_{\alpha i} &= \{q^k_{\alpha i} | q^{k+1}_{\alpha i} - 2q^k_{\alpha i} + q^{k-1}_{\alpha i} > \eta_1 C^{\text{tol}}_k, k = 1 \cdots N_{\text{wav}}\} \\ &= \{\hat{q}^k_{\alpha i} | k = 1 \cdots N^{\text{evol}}_{\alpha i}\} \\ \mathscr{I}^{\text{non-evol}}_{\alpha i} &= \mathscr{I}_{\alpha i} \setminus \mathscr{I}^{\text{evol}}_{\alpha i}. \end{aligned} \tag{64}$$

The coefficient $C_k^{\text{tol}} = \max_{2^{[log_2 k]}<l<2^{[log_2 k]+1}}(c_{l+1} - 2c_l + c_{l-1})$ is the maximum value of all coefficients at each resolution and η_1 is a prescribed tolerance. The variation of nodal displacements over a given cycle can be obtained from its wavelet coefficients using the orthogonality property of the transformation matrix (see (52)):

$$\begin{aligned} q_{\alpha i}(N, \tau_j) &= \sum_{k=1}^{N_{\text{wav}}} T_{kj}^{\alpha i} q_{\alpha i}^k(N) \qquad j = 1 \cdots N_{\text{wav}} \\ &= \sum_{q_{\alpha i}^k \in \mathscr{I}_{\alpha i}^{\text{evol}}} T_{kj}^{\alpha i} q_{\alpha i}^k(N) + \sum_{q_{\alpha i}^k \in \mathscr{I}_{\alpha i}^{\text{non-evol}}} T_{kj}^{\alpha i} q_{\alpha i}^k(N) \\ &= \sum_{k=1}^{N_{\alpha i}^{\text{evol}}} \hat{T}_{kj}^{\alpha i} \hat{q}_{\alpha i}^k(N) + q_{\alpha i}^{\text{add}}(\tau_j) \end{aligned} \tag{65}$$

The matrix $\hat{\mathbf{T}}^{\alpha i}$ is constructed from the wavelet transform matrix $\mathbf{T}$ by removing the rows corresponding to the nonevolving coefficients. The nonevolving coefficients remain constant during an integration step and is denoted by $q_{\alpha i}^{\text{add}}(\tau_j) = \sum_{q_{\alpha i}^k \in \mathscr{I}_{\alpha i}^{\text{non-evol}}} T_{kj}^{\alpha i} q_{\alpha i}^k(N)$.

The primary solution variables, corresponding to the displacement degrees of freedom, for the coarse-scale finite element model are $\hat{q}_{\alpha i}^k$. Accordingly, the wavelet coefficient degrees of freedom is $\sum_{\alpha=1}^{N_{\text{nodes}}} \sum_{i=1}^{3} N_{\alpha i}^{\text{evol}}$, corresponding to the sum of the number of evolving coefficients for all the displacement components at all nodes α. The unknown wavelet coefficients for a cycle are solved using the standard weak form of the equilibrium equation, decomposed into its wavelet components and expressed as:

$$\begin{aligned} f_i^k(N) &= \sum_{l=1}^{N_i^{\text{evol}}} \hat{T}_{kl}^i f_i(N, \tau_l) = 0, \\ \text{where} \quad f_i(N, \tau_l) &= \int_{V(\tau_l)} B_{ji} \sigma_j(N, \tau_l) \mathrm{d}V - \int_{S_t(\tau_l)} N_{ji} \bar{t}(N, \tau_l) \mathrm{d}S. \end{aligned} \tag{66}$$

Here, B_{ji} and N_{ji} are matrices that, respectively, relate the element strain and displacement components to the nodal displacements for the i-th degree of freedom in a conventional, single time scale FEM analysis. Furthermore, $\bar{t}(N, t)$ is the applied load over cycle N. The change in Cauchy stress σ_j in the Voigt form over the N-th cycle is obtained as a function of the nodal displacement components by solving the coarse cycle scale equations. Equation (66) are solved using the Newton family of nonlinear iterative solvers, for which the Jacobian matrix is approximated as:

$$\frac{\partial f_i^k}{\partial \hat{q}_j^l} = \sum_{m=1}^{N_{\text{wav}}} \hat{T}_{km}^i \frac{\partial f_i}{\cdot} \partial q_j(\tau_m) \hat{T}_{lm}^j \tag{67}$$

It is obtained by transforming the fine time scale Jacobian $\frac{\partial f_i}{\partial q_j}$, for geometrically nonlinear problems (see Crisfield 1996) at a point τ_m, by the wavelet transformation matrix **T**. Quasi-Newton nonlinear solvers (Bathe 1995) are used for solving the equations and the expression (67) is used as the initial estimate for the Jacobian matrix.

Procedure for Adding and Removing Wavelet Coefficients

Adding and removing wavelet bases for accurate representation of the response functions is important for optimal convergence of the WATMUS method. The procedure for adaptively adding and removing wavelet coefficients $q^k; k = 1 \cdots N_{\text{wav}}$ is illustrated with a 1-D problem (Sect. 3.4.1).

1. Start with an initial guess on the set of evolving coefficients $\mathscr{I}^{\text{evol}}$ based on a selection criterion. In the case of the 1-D problem, the criterion used is: $\mathscr{I}^{\text{evol}} = \{k|\frac{\text{d}q^k}{\text{d}N} \geq ctol \times \max_l |q^l|\}$ and $\mathscr{I}^{\text{non-evol}} = \mathscr{I} \setminus \mathscr{I}^{\text{evol}}$.
2. Solve the problem for the evolving coefficients:

$$f^k(q^l) = 0 \quad \forall (k,l) \in \mathscr{I}^{\text{evol}} \times \mathscr{I}^{\text{evol}} \tag{68}$$

3. Based on the solution, select the set of coefficients $\mathscr{I}^{\text{add}}$ to be added according to the criterion:

$$\mathscr{I}^{\text{add}} = \{q^k | q^k \in \mathscr{I}^{\text{non-evol}} \text{ and} |f_k| > \eta_F\} \tag{69}$$

   It implies all nonevolving coefficients, whose residual is greater than a tolerance η_F given by the maximum value of the converged residual corresponding to the evolving coefficients, are selected. If $\mathscr{I}^{\text{add}} \neq \emptyset$, then $\mathscr{I}^{\text{evol}} = \mathscr{I}^{\text{evol}} \cup \mathscr{I}^{\text{add}}$.
4. Go to Step 1.

This procedure ensures that an appropriate set of displacement coefficients are selected such that all the wavelet coefficients of the residual are within a certain tolerance, as opposed to just the ones corresponding to the evolving set. This in turn guarantees that the error in the displacement coefficients remain bounded. The procedure developed can be combined with a criterion for removing coefficients to obtain the optimal set of evolving coefficients. Coefficients whose cyclic rate of change is less than a pre-set tolerance (η_{del}) are removed from $\mathscr{I}^{\text{evol}}$, i.e.,

$$\begin{aligned} \mathscr{I}^{\text{non-evol}} &= \mathscr{I}^{\text{non-evol}} \bigcup \{q^k | \quad |q^k(N) - q^k(N - \Delta N)|/\Delta N \leq \eta_{\text{del}} \max_{\mathscr{I}^{\text{evol}}} |q^l|\} \\ \mathscr{I}^{\text{evol}} &= \mathscr{I} \setminus \mathscr{I}^{\text{non-evol}} \end{aligned} \tag{70}$$

This yields the new set of evolving coefficients.

3.3.2 Coarse (Cycle Scale) Integration Step Size Control

The step size in the integration of cycle scale equations corresponds to the number of cycles traversed in each increment of the numerical integration scheme. An optimal step size is estimated in the WATMUS method from a truncation error criterion in conjunction with the residual of the equilibrium equation. For the implicit second-order backward difference scheme in (58), the truncation error criterion is derived from the truncation of second-order terms in the Taylor's expansion of a variable y_o as:

$$\delta y_o = \frac{1}{6}\left|\frac{\mathrm{d}^3 y}{\mathrm{d}N^3}\frac{(r+1)^2-(r+1)^3}{(r+1)^2-1}\right|\Delta N^3, \tag{71}$$

where $r = \frac{\Delta N_p}{\Delta N}$ is the ratio of the previous to the current step size. The relative error, plotted as a function of r in Fig. 16, shows that the error decreases asymptotically as r increases.

The norm of the error that is propagated to the equilibrium residual δf from the constitutive level truncation error (71) is derived as:

$$\begin{aligned} \delta f &= \delta f_{\text{err}}\,\Delta N^3, \\ \text{where } \delta f_{\text{err}} &= \frac{1}{6}\max_{el}\left\|\int_{V_{el}} \mathbf{B}^T\frac{\partial\sigma}{\partial\mathbf{F}^{po}}\frac{(r+1)^2-(r+1)^3}{(r+1)^2-1}\frac{\mathrm{d}^3\mathbf{F}^{po}}{\mathrm{d}N^3}\mathrm{d}V\right\|. \end{aligned} \tag{72}$$

Equation (72) accounts for the effect of the local constitutive error on the global equilibrium equations in the coarse scale problem. For the error to be bounded by a prescribed relative tolerance η, the maximum allowed cycle step jump ΔN_{jump} is estimated as:

$$\Delta N_{\text{jump}} \le \left(\frac{6\eta f_{\text{err}}}{\delta f_{\text{err}}}\right)^{\frac{1}{3}}, \quad \text{where} \quad f_{\text{err}} = \sum_{el}\left\|\int_{V^{el}}\mathbf{B}^T\frac{\partial\sigma}{\partial\mathbf{F}^{po}}\mathrm{d}V\right\| \tag{73}$$

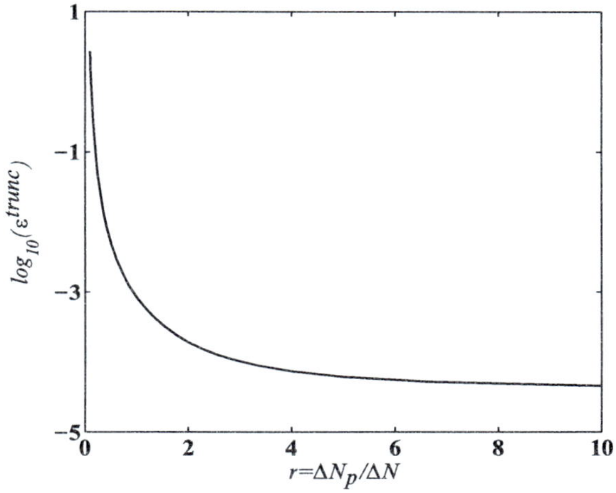

Fig. 16 Variation of truncation error $\varepsilon^{\text{trunc}} = \left|\frac{\varepsilon^{p0}_{\text{integ}}-\varepsilon^{p0}_{\text{exact}}}{\varepsilon^{p0}_{\text{exact}}}\right|$ with step size ratio r

3.4 Numerical Examples Solved with the WATMUS Algorithm

The wavelet transformation-based multi-time scaling (WATMUS) methodology is applied to study the cyclic response of a one-dimensional viscoplastic problem, and a three-dimensional rate dependent crystal plasticity problem is discussed next.

3.4.1 One-Dimensional Elastic-Viscoplastic Problem

The WATMUS methodology is applied to study cyclic response of a 1-D elastic-viscoplastic bar introduced in Sect. 3.1. The bar is fixed at the left end and is subjected to a sinusoidal loading at the right end with frequency $\omega = 2\pi$, i.e., period $T = 1s$. Variables in (24) that are integrated using the second-order backward difference algorithm in (58) are the initial values of the plastic strain (ε^{po}) and hardness (g^o). For the 1-D problem, the algorithm is written as:

$$\begin{aligned} \mathbf{f}^{\text{const}} &= a_1\boldsymbol{\alpha}_o(N+\Delta N) - a_2\boldsymbol{\alpha}_o(N) + a_3\boldsymbol{\alpha}_o(N-\Delta N_p) \\ &\quad - \Delta N \frac{\mathrm{d}\boldsymbol{\alpha}_o}{\mathrm{d}N}(N+\Delta N) = 0, \\ \text{where} \quad \boldsymbol{\alpha}_o &= \begin{Bmatrix} \varepsilon^{po} \\ g^o \end{Bmatrix} \end{aligned} \tag{74}$$

Derivatives in (74) are calculated using (57). The Newton–Raphson nonlinear solver is used to calculate $\boldsymbol{\alpha}_o(N+\Delta N)$ from (74). The only unknown displacement in the model is that of node 2 (between the two elements), which is decomposed into its wavelet coefficients. The equilibrium equation, resolved in the wavelet basis, over a cycle step at $N + \Delta N$ is:

$$\begin{aligned} f^k(N+\Delta N) = \sum_{l=1}^{N_{\text{wav}}} \hat{T}_{kl}\, [\sigma_1(N+\Delta N, \tau_l) \\ -\sigma_2(N+\Delta N.\tau_l)] = 0, \quad k = 1\cdots N_{\text{evol}}, \end{aligned} \tag{75}$$

where σ_i corresponds to the stress in element i and N_{evol} is the number of evolving displacement coefficients. The criterion for the maximum integration step-size is obtained similar to (73).

Results of Cyclic Loading Simulation of 1-D Viscoplastic Bar Problem

The 1-D viscoplastic bar model, described in Sect. 3.1 is subjected to a fully reversed ($R = -1$) sinusoidal loading for 2,000 cycles. The WATMUS method is used to integrate the problem in coarse cycle scale and the results are compared with those obtained from a single (fine) time-scale analysis, with respect to both accuracy and efficiency. The effects of (a) evolving wavelet coefficients of nodal displacements that are adaptively chosen for a tolerance *ctol*, and (b) the maximum allowed step size (number of cycles), on the solution are also investigated. Three displacement tolerances, viz. $ctol = 10^{-10}, 10^{-4}$, and10^{-3} are considered for selecting the number of coefficients that are expected to evolve during the loading process. The tolerances result in 128, 30, and 10 initial coefficients, respectively, which can evolve freely with deformation. For each of the three tolerances, four different bounds on the maximum cycle step size $\Delta N_{\max}$ in (73) are considered. The step size for a coarse scale integration increment is assessed as $\Delta N = \min(\Delta N_{\text{jump}}, \Delta N_{\max})$. The value of η in (73) is set to be 10^{-3}. Figure 17 shows the variation of the coarse scale variables (ε^{po} and g^{o}) as functions of the number of cycles N.

Once the coarse scale variables (ε^{po} and g^{o}), along with the wavelet displacement coefficients are known for a given cycle N, the fine scale response for that cycle can be calculated by superimposing the wavelet bases according to (56). This is shown in Fig. 18, in which the plastic strain response over the 2,000th cycle is obtained from the coarse scale solution. The fine scale response is plotted for different step bounds $\Delta N_{\max}$ and for different number of evolving coefficients $\hat{\mathbf{q}}$. Excellent agreement is observed between the coarse scale and the single (fine) time-scale results, except for the case with very few (~10) evolving wavelet coefficients, which suffers from a significant error regardless of the refinement in the time step as seen from Fig. 18(c). This is due to the fact that there are not enough basis functions available to represent the displacement response accurately, regardless of the time step. However, the accuracy increases rapidly, and the difference in response between reconstructions with 30 and 128 evolving coefficients is minimal. This justifies the development of the criterion in Sect. 3.3 that can yield an optimal number of evolving coefficients during the loading process while retaining accuracy. Table 9 shows the effects of the number of coefficients on the computational speedup and accuracy, compared with the single (fine) time-scale analysis. A computational speedup of approximately 80 times is obtained with comparable accuracy for this problem. Results in the table demonstrate that the number of coefficients used in the solution plays a crucial role in the accuracy of the solution. To ensure that all the required coefficients are included during the loading process, the criterion developed in Sect. 3.3.1 especially in (68) and (69), is utilized for the problem. The starting list of coefficients are obtained using $ctol = 10^{-3}$. Note that while coefficients chosen based on *ctol* alone can give large errors, substantial improvement in accuracy can be achieved through the implementation of the adding–removal algorithm in Sect. 3.3. The results of this adaptive procedure are illustrated in Figs. 19(a) and 19(b). Excellent agreement is observed with the fine scale results. The corresponding number of selected evolving coefficients is shown in Fig. 19(c).

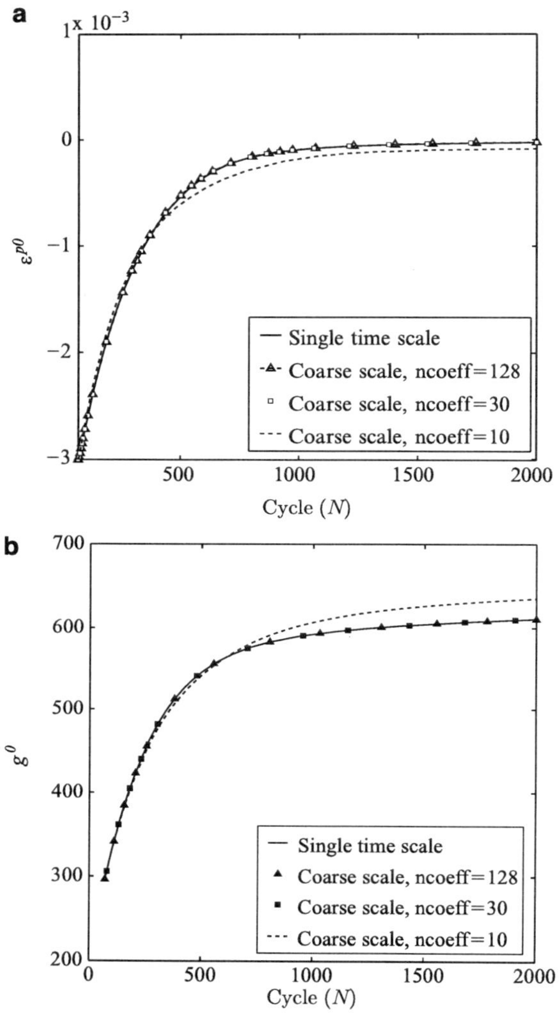

Fig. 17 Evolution of coarse scale variables as a function of cycles (N): (**a**) initial plastic strain for element 1 with $\Delta N_{\max} = 277$, (**b**) initial hardness for element 1 with $\Delta N_{\max} = 25$

3.4.2 3D Crystal Plasticity FE Simulation Under Cyclic Loading

The WATMUS methodology is used in this example to simulate cyclic deformation of the Ti-6Al alloy using the 3D crystal plasticity finite element (CPFE) model discussed in Sect. 2. A model problem of a polycrystalline microstructure is developed, consisting of 27 cubic grains with different crystallographic orientations. Each grain is subdivided into 27 brick elements resulting in a total of 729 elements for the CPFE model. Euler angles corresponding to the crystallographic orientation

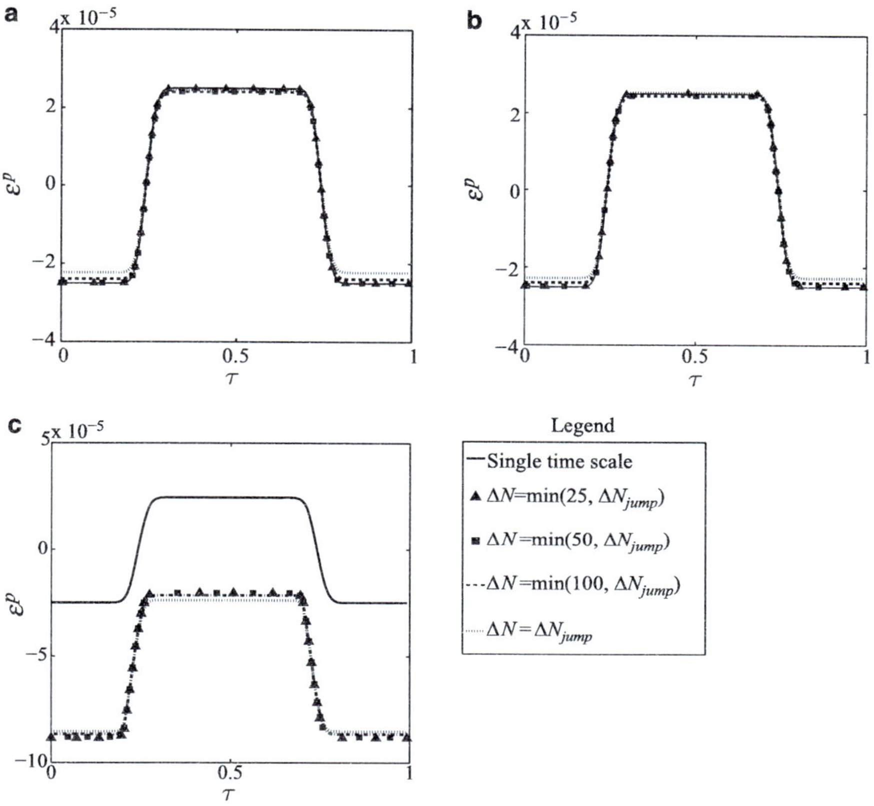

Fig. 18 Reconstructed profile of the fine-scale plastic strain response at the 2,000th cycle for different values of ΔN_{max}, constructed with: (**a**) 128 wavelet coefficients, (**b**) 30 wavelet coefficients, and (**c**) 10 wavelet coefficients. ΔN_{jump} obtained from (73)

distribution for the model are depicted in Fig. 20. Material parameters for the crystal plasticity model are provided in Sect. 2. The model is subjected to cyclic loading with a mean stress of 500 MPa, together with a superposed oscillatory portion with a peak of 350 MPa and time period of $T = 1$ s. To start the problem with appropriate values of cycle-derivatives, a fine-scale simulation of the CPFE model is carried out for the first 6 cycles. Thereafter, the coarse-scale WATMUS method is commenced. The starting value of cycle jump is taken as $\Delta N_{\mathrm{jump}} = 2$, while the subsequent cycle steps with loading is obtained from the criterion in (73) with $\eta = 10^{-2}$. The Daubechies-4 wavelet basis is used to decompose the nodal displacement components in each cycle according to (61). The value of η_1 in (64) is set to 10^{-3} in the present simulations. This criterion selects coefficients from regions having rapid fluctuations in values of the response functions and those that are most likely to change as the problem progresses leading to a total of 116,521 coefficients with an

Table 9 Effect of step size ΔN and number of coefficients $\hat{\mathbf{q}}$ on error and efficiency at $N_f = 2{,}000$

Coefficient Tolerance	ncoeff	$\Delta N_{\max}$	$\frac{\|\varepsilon^p_{\text{coarse}}(N_{\text{wav}},\tau)-\varepsilon^p_{\text{fine}}(N_{\text{wav}},\tau)\|}{\|\varepsilon^p_{\text{fine}}(N_{\text{wav}},\tau)\|}$ (%)	Speedup
10^{-10}	128	25	0.336	29
10^{-10}	128	50	2.544	43
10^{-10}	128	100	4.11	58
10^{-10}	128	255	8.11	67
10^{-4}	30	25	1.05	35
10^{-4}	30	50	1.89	53
10^{-4}	30	100	3.73	72
10^{-4}	30	277	6.87	84
10^{-3}	10	25	232.4	36
10^{-3}	10	50	230.8	57
10^{-3}	10	100	229.8	74
10^{-3}	10	214	230.8	83

average of 39 coefficients per degree of freedom for the present problem. Due to the computational cost, the Jacobian matrix is factored at the start of the simulation and is used as the initial Jacobian estimate for every iteration cycle of the quasi-Newton method solution of the equilibrium equations. This is observed to be a good estimate due to the monotonic response of the evolving displacement coefficients. The simulation is run for only 1,000 cycles, so that results can be compared with results from computationally exhaustive single (fine scale) simulations. Figure 21 shows a comparison of the cycle-scale plastic deformation gradient F^{po} by coarse and fine time scale simulations at a typical material point in a grain as a function of the number of cycles. The results of the two methods are indistinguishable for the range considered. The state of stress σ_{zz} in the microstructure along the loading direction at the 1,000th cycle is shown in Figs. 22 and 23 for both coarse and fine time-scale simulations. As is observed, the coarse time scale and fine time-scale results are in excellent agreement with each other.

Once the coarse scale variables for a given cycle are known, the fine scale response over that cycle can be reconstructed as shown in Fig. 24. A computational time advantage of approximately 7 times is observed for the current problem. The predicted cycle jump ΔN_{jump} from (73) is expected to increase as the problem progresses due to stabilization of the response, resulting in even higher computational savings.

4 Conclusions

This chapter presents important ingredients for a crystal plasticity finite element simulation based prediction of fatigue crack nucleation in polycrystalline alloys. The novel framework introduced in this chapter paves the way for a paradigm

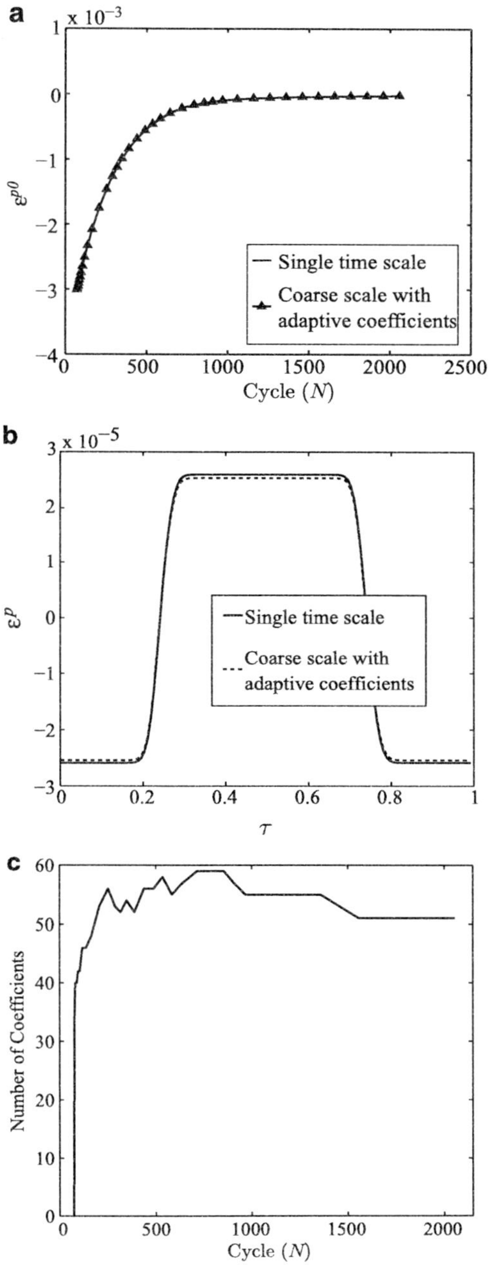

Fig. 19 Effect of adaptive wavelet coefficients: (**a**) coarse scale plastic strain as a function of cycle number N, (**b**) reconstructed plastic strain response with $\Delta N = \min(100, \Delta N_{\text{jump}})$ at the 2,000th cycle, and (**c**) number of evolving coefficients with increasing cycles N

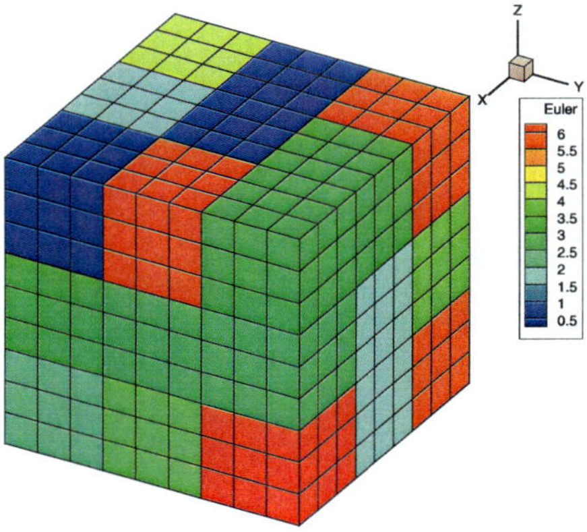

Fig. 20 Euler angle distribution for the 729 element crystal plasticity FE model

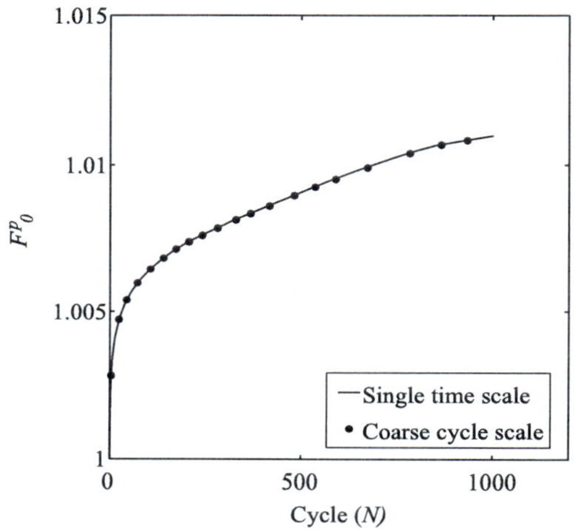

Fig. 21 Coarse scale evolution of plastic deformation gradient $\mathbf{F}^{po}$ as a function of cycles N

change in fatigue modeling of metals and alloys. Two specific topics are at the core of this development. The first is a nonlocal fatigue crack nucleation model at the scale of grains in a polycrystalline microstructure based on post-processing results of crystal plasticity finite element simulations under cyclic loading. The second development is a unique wavelet transformation based multi-time scaling algorithm for accelerated crystal plasticity finite element simulations. It is motivated by the large number of cycles that may be required to initiate a fatigue crack in a polycrystalline sample. Simulating such large number of cycles remains intractable to conventional single time scale finite element analysis. The unique aspect of the proposed method is that the algorithm does not require inherent scale separation as with other conventional methods that assume averaging, periodicity, or near periodicity.

The microstructure-based crack nucleation criterion in polycrystalline alloys uses the results of an image-based rate dependent, crystal plasticity finite element model. It is postulated that dwell fatigue crack initiates due to stress concentration caused

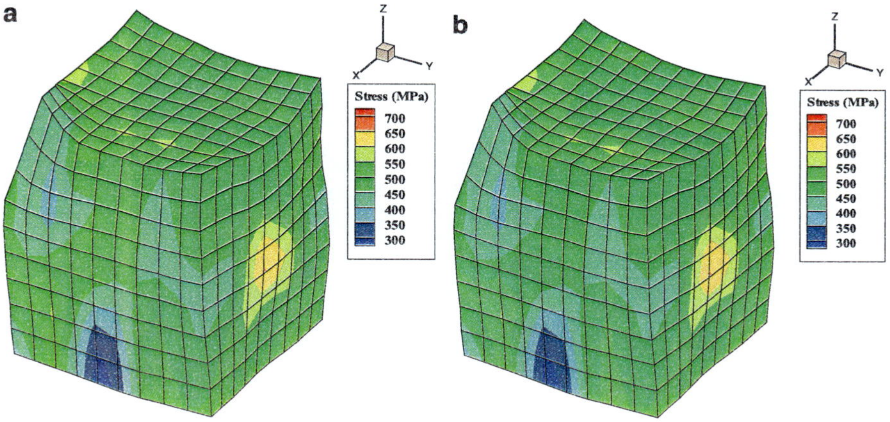

Fig. 22 Stress (σ_{33}) contour for 729 element crystal plasticity by: (**a**) multi time-scale simulations and (**b**) fine time-scale simulations

Fig. 23 Variation of stress along the diagonal of the model cube (729 element case)

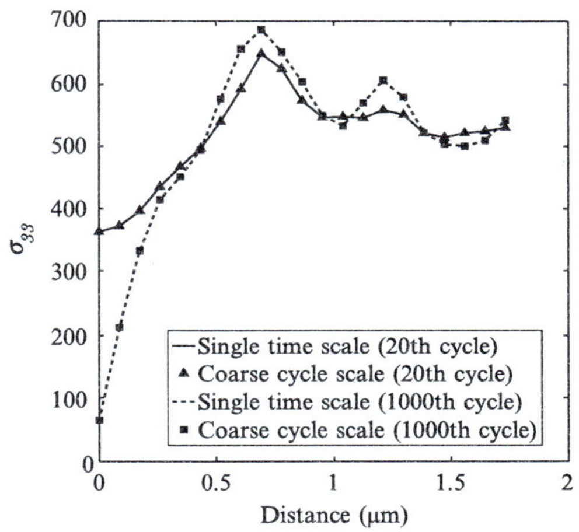

by the load shedding between adjacent hard and soft grains. To predict load shedding induced crack initiation in the hard grain, a nonlocal criterion that depends on energy release rate in the hard grain, as well as the nonlocal plastic dissipation rate in an adjacent soft grain is proposed. A numerical approach is proposed to evaluate the energy release and dissipation rates in a hard–soft grain combination. Correspondingly, the surface energy density, a material parameter in this crack nucleation model, is calibrated from the experimental results of ultrasonic monitoring of crack evolution in dwell fatigue experiments. The model is validated through accurate predictions of the cycles to failure, as well as from critical features of the failure site in dwell fatigue experiments.

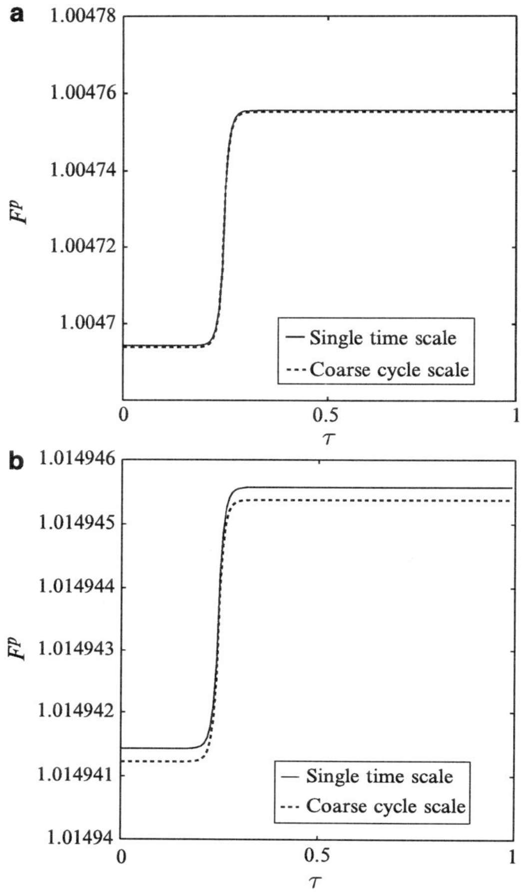

Fig. 24 Reconstructed value of the initial plastic deformation gradient (F^p_{33}) at the (**a**) 20th Cycle and (**b**) 1,000th Cycle

In the second part of this chapter, a novel wavelet transformation-based multi-time scaling (WATMUS) method is developed for crystal plasticity finite element simulations of cyclic deformation in polycrystalline materials leading to fatigue failure. The WATMUS methodology introduces wavelet decomposition of nodal displacements and all associated variables in the finite element formulation to decouple the response into a monotonic coarse cycle-scale behavior and an oscillatory fine time scale behavior within each cycle. Multiresolution wavelet bases functions are effectively able to capture the rapidly varying fine scale response, which necessitates very small time steps in conventional single time scale FEM simulations. An effective criterion is developed for selecting an optimal number of wavelet coefficients in the representation of all response functions in the crystal plasticity simulations. The wavelet transformation is utilized to obtain the coarse scale evolution equations for the microstructural state variables. The coarse scale variables exhibit monotonic behavior that stabilizes with saturating hardness at higher levels of deformation. Relatively large increments, traversing several cycles at a time, can therefore be utilized in the numerical integration scheme with significantly enhanced efficiency. The numerical simulations exhibit approximately 80–100 times speedup, even with relatively low number of cycles. Subsequently, fine scale variations in temporal response at any point in a microstructural point can be recovered from values of the nodal displacement wavelet coefficients and the coarse scale state variables at that point.

Acknowledgements This work has been supported by the US National Science Foundation (Grant # CMMI-0800587, Program Manager: Dr. Clark Cooper), Federal Aviation Administration (Grant # DTFA03-01-C-0019, Program Manager: Dr. Joe Wilson), the Air Force Office of Scientific Research (Grant # FA9550-05-1-0067, Program Manager: Dr. David Stargel) and the Office of Naval Research (Grant # N00014-05-1-0504, Program Manager: Dr. Julie Christodolou). This sponsorship is gratefully acknowledged. Computer support by the Ohio Supercomputer Center through grant PAS813-2 is also gratefully acknowledged.

References

Acharya, A. and Beaudoin, A.J.: Grain-size effect in viscoplastic polycrystals at moderate strains. J. Mech. Phys. Solids **48**, 2213–2230 (2000)

Anahid, M., Chakraborty, P., Joseph, D.S. and Ghosh, S.: Wavelet decomposed dual-time scale crystal plasticity FE model for analyzing cyclic deformation induced crack nucleation in polycrystals. Model. Simul. Mater. Sci. Eng. **37**, 064009 (2009)

Antolovich, S.D., Liu, S. and Baur, R.: Low cycle fatigue behavior of Rene 60 at elevated temperature. Metal. Trans. **12A**, 473–481 (1981)

Asaro, R.J., and Rice, J.R.: Strain localization in ductile single crystals. J. Mech. Phys. Solids **25**, 309–338 (1977)

Ashby, M.F.: The deformation of plastically non-homogeneous materials. Phil Mag. **21**, 399–424 (1970)

Bache, M.R.: A review of dwell sensitive fatigue in titanium alloys: the role of microstructure, Texture and Operating Conditions. Int. J. Fatigue. **25**, 1079–1087 (2003)

Baker, I.: Improving the ductility of intermetallic compounds by particle-induced slip homogenization. Scripta Materialia **41(4)**, 409–414 (1999)

Balasubramanian, S., Anand, L.: Elasto-viscoplastic constitutive equations for polycrystalline fcc materials at low homogeneous temperatures, J. Mech. Phys. Solids. **50**, 101–126 (2002)

Bathe, K.J.: Finite element procedures, Prentice Hall (1995)

Beaudoin, A.J., Mecking, H. and Kocks, U.F.: Development of local shear bands and orientation gradients in fcc polycrystals. In Shen, S.F. and Dawson, P.R. (eds.), Simulation of Materials Processing; Theory, Methods and Applications, NUMIFORM'95, pp. 225–230, A.A. Balkema, Rotterdam (1995)

Bennett, V.P. and McDowell, D.L.: Polycrystal orientation distribution effects on microslip in high cycle fatigue. Int. J. Fatigue. **25**, 27–39 (2003)

Bhandari, Y., Sarkar, S., Groeber, M., Uchic, M., Dimiduk D. and Ghosh, S.: 3D polycrystalline microstructure reconstruction from FIB generated serial sections for FE Analysis. Comp. Mater. Sci. **41**, 222–235 (2007)

Blekhman, I.I.: Vibrational Mechanics, World Scientific (2000)

Chow, C.L. and Wei, Y.: A model of continuum damage mechanics for fatigue failure. Int. J. Fracture **50**, 301–316 (1991)

Chu, R.Q., Cai, Z., Li, S.X., and Wang, Z.G.: Fatigue crack initiation and propagation in an a-iron polycrystals. Mater. Sci. Eng. **313**, 61–68 (2001)

Coffin, L.F.: Fatigue. Ann. Rev. Matls. Sci. **2**, 313–348 (1973)

Crisfield, M.A.: Non-linear finite element analysis of solids and continuam, Academic (1996)

Dawson, P. R.: Computational crystal plasticity. Int. J. Sol. Struct. **37**, 115–130 (2000)

Deka, D., Joseph, D.S., Ghosh, S. and Mills, M.J.: Crystal plasticity modeling of deformation and creep in polycrystalline Ti-6242. Metall. Trans. A. **17A(5)**, 1371–1388 (2006)

Engelen, R.A.B., Geers, M.G.D. and Baaijens F.P.T.: Nonlocal implicit gradient-enhanced elasto-plasticity for the modeling of softening behavior. Int. J. Plasticity **19**, 403–433 (2003)

Fish, J. and Yu, Q.: Computational mechanics of fatigue and life predictionsfor composite materials and structures. Comp. Meth. Appl. Mech. Eng. **191**, 4827–4849 (2002)

Fleck, N.A., Kang, K.J. and Ashby, M.F.: The cyclic properties of engineering materials. Acta Metall. Mater. **42**, 365–381 (1994)

Fredell, R.S.: Operational needs for a paradigm shift in life prognosis. Workshop on Prognosis of Aircraft and Space Devices, Components, and Systems, University of Cincinnati, Cincinnati, Ohio (2008)

Ghosh, S., Bhandari, Y. and Groeber, M.: CAD based reconstruction of three dimensional polycrystalline microstructures from FIB generated serial sections. J. Comput. Aid. Des. **40(3)**, 293–310 (2008)

Goh, C.H., Neu, R.W. and McDowell, D.L.: Crystallographic plasticity in fretting of Ti-6Al-4V. Int. J. Plasticity **19**, 1627–1650 (2003)

Goh, C.H., Wallace, J.M., Neu, R.W. and McDowell, D.L.: Polycrystal plasticity simulations of fretting fatigue. Int. J. Fatigue **23**, 5423–5435 (2001)

Groeber, M.: Development of an automated characterization-representation framework for the modeling of polycrystalline materials in 3D. Ph.D. dissertation. The Ohio State University, Columbus, OH (2007)

Groeber, M., Ghosh, S., Uchic, M.D. and Dimiduk, D.M.: A framework for automated analysis and representation of 3D polycrystalline microstructures, part 1: statistical characterization. Acta. Mater. **56(6)**, 1257–1273 (2008)

Groeber, M., Ghosh, S., Uchic, M.D. and Dimiduk, D.M.: A framework for automated analysis and representation of 3D polycrystalline microstructures, part 2: synthetic structure generation. Acta. Mater. **56(6)**, 1274–1287 (2008)

Harren, S.V., and Asaro, R.J.: Nonuniform deformations in polycrystals and aspects of the validity of the Taylor model. J. Mech. Phys. Solids. **37(2)**, 191–232 (1989)

Hashimoto, T.M. and Pereira, M.S.: Fatigue life studies in carbon dual-phase steels. Int. J. Fatigue. **18(8)**, 529–533 (1996)

Hasija, V., Ghosh, S., Mills, M.J. and Joseph, D.S.: Modeling deformation and creep in Ti-6Al alloys with experimental validation. Acta Mater. **51**, 4533–4549 (2003)

Harder, J.: Crystallographic model for the study of local deformation processes in polycrystals. Int. J. Plastic. **15**, 605–624 (1999)

Inman, M.A. and Gilmore, C.M.: Room temperature creep of Ti-6Al-4V. Metall. Trans. A. **10A**, 419–425 (1979)

Kalidindi, S.R., Bronkhorst, C.A. and Anand, L.: Crystallographic texture evolution in bulk deformation processing of fcc metals. J. Mech. Phys. Solids. **40**, 537–569 (1992)

Kalidindi, S.R., Bronkhorst, C.A. and Anand, L.: On the accuracy of the Taylor assumption in polycrystalline plasticity. In Boehler, J.P. and Khan, A.S. (eds.), Anisotropy and Localization of Plastic Deformation, pp. 139–142, Elsevier Applied Science, London and New York (1991)

Kirane, K. and Ghosh, S.: A cold dwell fatigue crack nucleation criterion for polycrystalline Ti-6242 using grain-level crystal plasticity FE model. Int. J. Fatigue **30**, 2127–2139 (2008)

Kirane, K., Ghosh, S., Groeber, M. and Bhattacharjee, A.: Crystal plasticity finite element based grain level crack nucleation criterion for Ti-6242 alloys under dwell loading. J. Eng. Mater. Tech. ASME **131**, 27–37 (2009)

Laird, C.M.: The fatigue limit of metals. Mater. Sci. Eng. **22**, 231–236 (1976)

Lord, D.C. and Coffin, L.F.: Low cycle fatigue hold time behavior of cast Rene 80. Metal. Trans. **4**, 1647–1653 (1973)

Manchiraju, S., Asai, M. and Ghosh, S.: A dual time scale finite element model for simulating cyclic deformation of polycrystalline alloys. J. Strain Anal. Engrg. Des. **42**, 183–200 (2007)

Manchiraju, S., Kirane, K. and Ghosh, S.: Dual-time scale crystal plasticity FE model for cyclic deformation of Ti alloys. J. Comp. Aid. Des. **14**, 47–61 (2008)

Mathur, K.K., Dawson, P.R. and Kocks, U.F.: On modeling anisotropy in deformation processes involving textured polycrystals with distorted grain shape. Mech. Mater. **10**, 183–202 (1990)

McClung, R.C., Enright, M.P., Millwater, H.R., Leverant, G.R., and Hudak, S.J.: A software framework for probabilistic fatigue life assessment of gas turbine engine rotors. J. ASTM Int. **1(8)**, JAI19025-12 (2004)

Mineur, M., Villechaise, P., and Mendez, J.: Influence of the crystalline texture on the fatigue behavior of a 316L austenitic stainless steel, Mater. Sci. Eng. **A286**, 257–268 (2000)

Morrissey, R.J., McDowell, D.L., and Nicholas, T.: Microplasticity in HCF of Ti-6Al-4V. Int. J. Fatigue. **23**, S55–S64 (2001)

Neeraj, T., Hou, D.H., Daehn, G.S. and Mills, M.J.: Phenomenological and microstructural analysis of room temperature creep in Titanium alloys. Acta Mater. **48**, 1225–1238 (2000)

Nye, J.F.: Some geometrical relations in dislocated crystals. Acta. Metall. **1**, 153–162 (1953)

Oskay, C. and Fish, J.: Fatigue Life Prediction using 2-scale Temporal Asymptotic Homogenization. Int. J. Numer. Meth. Eng. **61**, 329–359 (2004)

Oskay, C. and Fish, J.: Multiscale modeling of fatigue for ductile materials. Int. J. Multiscale Comput. Eng. **2**, 1–25 (2004)

Paas, M.H.J.W., Schreurs, P.J.G. and Berkelmans, W.A.M.: A continuum approach to brittle and fatigue Damage: theory and numerical Procedures. Int. J. Solids Struct. **30**, 579–599 (1993)

Paris, P.C.: The fracture mechanics approach to fatigue, Fatigue-An Interdisciplinary Approach, Syracuse University Press, Syracuse (1964)

Parvatareddy, H. and Dillard, D.A.: Effect of mode-mixity on the fracture toughness of Ti-6Al-4V/FM-5 adhesive joints. Int. J. Fracture **96**, 215–228 (1999)

Rice, J.R.: A path independent integral and the approximate analysis of strain concentration by notches and cracks. J. Appl. Mech. **35**, 379–386 (1968)

Rokhlin, S., Kim, J.Y. and Zoofan, B.: The Ohio State University, Columbus, OH., unpublished research (2005)

Ruiz, G., Pandolfi, A. and Oritz, M.: Three-dimensional cohesive modeling of dynamic mixed mode fracture. Int. J. Numer. Meth. Eng. **52**, 97–120 (2001)

Sinha, S. and Ghosh, S.: Modeling cyclic ratcheting based fatigue life of HSLA steels using crystal plasticity FEM simulations and experiments. Int. J. Fatigue. **28**, 1690–1704 (2006)

Sinha, V., Mills, M.J. and Williams, J.C.: Crystallography of fracture facets in a near alpha titanium alloy. Metall. Trans. A. **37A**, 2015–2026 (2006)

Sinha, V., Spowart, J.E., Mills, M.J. and Williams, J.C.: Observations on the faceted initiation site in the dwell-fatigue tested Ti-6242 alloy: Crystallographic orientation and size effects. Metall. Trans. A. **37A**, 1507–1518 (2006)

Smith, E.: Cleavage fracture in mild steel. Acta Metal. **14**, 985–989 (1966)

Smith, E.: Disloclations and cracks. In Nabarro, F.R.N. (ed.), Dislocations in Solids **4**, North Holland Amsterdam, The Netherlands, 363 (1979)

Strang, G. and Nguyen, T.: Wavelets and filter banks, Wellessey College (1996)

Stroh, A.N.: A theory of the fracture of metals. Adv. Phys. **6**, 418–465 (1957)

Stroh, A.N.: The formation of cracks as a result of plastic flow. Proc. R. Soc. London, Ser. A. **223**, 404–414 (1954)

Suresh, S.: Fatigue of Materials, Cambridge University Press, Cambridge (1998)

Tanaka, K. and Mura, T.: A dislocation model for fatigue crack initiation. J. Appl. Mech. **48**, 97–103 (1981)

Thomsen, J.J.: Vibrations and Stability: Theory, Analysis and Tools, Springer, Berlin (2004)

Tsuji, H. and Kondo, T.: Strain-time effects in low cycle fatigue of Nickel-based heat resistant alloys at high temperature. J. Nucl. Mater. **190**, 259–265 (1987)

Turkmen, H.S., Loge, R.E., Dawson, P.R. and Miller, M.: On the mechanical behavior of AA 7075-T6 during cyclic loading. Int. J. Fatigue. **25**, 267–281 (2003)

Venkatramani, G., Deka, D. and Ghosh, S.: Crystal plasticity based FE model for understanding microstructural effects on creep and dwell fatigue in Ti-6242, ASME J. Eng. Mater. Tech. **128(3)**, 356–365 (2006)

Venkataramani, G., Ghosh, S. and Mills, M.J.: A size dependent crystal plasticity finite element model for creep and load-shedding in polycrystalline Titanium alloys. Acta Mater. **55**, 3971–3986 (2007)

Venkataramani, G., Kirane, K. and Ghosh, S.: Microstructural parameters affecting creep induced load shedding in Ti-6242 by a size dependent crystal plasticity FE model. Int. J. Plas. **24**, 428–454 (2008)

Walker, J.S.: A primer on wavelets and their scientific applications, CRC (1999)

Williams et al.: The evaluation of cold dwell fatigue in Ti-6242. FAA report summary, The Ohio State University (2006)

Woodfield, A.P., Gorman, M.D., Corderman, R.R., Sutliff, J.A., and Yamron, B.: Effect of microstructure on dwell fatigue behaviour of Ti-6242. Titanium '95 Science and Technology, 1116–1124 (1995)

Xie, C.L., Ghosh, S. and Groeber, M.: Modeling cyclic deformation of HSLA steels using crystal plasticity. ASME J. Eng. Mater. Tech. **126**, 339–352 (2004)

Yu, Q. and Fish, J.: Temporal homogenization of viscoelastic and viscoplastic solids subjected to locally periodic loading. Comput. Mech. **29**, 199–211 (2002)

Challenges Below the Grain Scale and Multiscale Models

Hussein M. Zbib and David F. Bahr

Abstract The main point in this chapter is the multiscale aspect of plastic deformation and strength in crystalline materials. Particular emphasis is placed on models and experiments below the grain level where the length scale is too small for classical continuum constitutive equations to be meaningful but yet is not small enough to be treated within an atomistic framework. More specifically, this chapter deals with experimental advances and theoretical models that have been developed recently to characterize dislocations at the subgrain level. Emphases is placed on recent developments in the nanoindentation technique to measure mechanical properties in small volumes, and on most recent experiments which attempt to measure mechanical properties using more traditional experiments such as tension and compression of microscale specimens. A discussion of a multiscale modeling framework is provided illuminating the important role discrete dislocation dynamics models play in this area.

1 Introduction

With so much interest in nanotechnology which relies on the use of materials whose dimensions are in the submicron scale, it is no surprise that there has been abundance of activities in the area of dislocation mechanics over the past few years. It is now well established by experiments and theory that strength in metals at all length scales is strongly dependent on unit dislocation processes and on the manner in which dislocations interact with grain boundaries, interfaces, and various defects that may be present in the crystal. This dependence becomes more pronounced in small volumes ranging from the micrometer level down to a few nanometers. However, this fundamental understanding has not been transformed into a continuum theory of crystal

---

H.M. Zbib (✉)
School of Mechanical and Materials Engineering, Washington State University, Pullman, WA, USA
e-mail: zbib@wsu.edu

S. Ghosh and D. Dimiduk (eds.), *Computational Methods for Microstructure-Property Relationships*, DOI 10.1007/978-1-4419-0643-4_15,

plasticity based on dislocation mechanisms. While classical plasticity provides a physical framework for modeling dislocation-dependent plasticity phenomena at large scales, it fails to predict size effect and related phenomena at small scales. The major difficulty in developing such a theory is the multiplicity and complexity of the mechanisms of dislocation motion and interactions that make it impossible to develop a quantitative analytical approach. The problem becomes more complicated when trying to predict the evolution of a very large number of dislocations over very long periods of time, as required for the calculation of plastic response in a representative volume element. Such a difficulty of the dislocation-based approaches, on one hand, and the developing needs of material engineering at the nano- and microlength scales, on the other hand, have created a situation where equations of crystal plasticity used for continuum modeling are phenomenological and largely disconnected from the physics of the underlying dislocation behavior. Recently, however, bridging the gap between crystal plasticity and dislocation physics has become possible through the emergence of multiscale models that attempt the appropriate coupling of models for each of the scales involved. In the case of deformation in small volumes, such coupling involves the continuum and dislocation dynamics models, although a further coupling to the atomic scale is theoretically possible but practically very complex.

In the following sections, we provide a brief review of recent developments in the area of plastic deformation in small volumes. First, in Sect. 2, we review some key experimental methods that are used to characterize deformation at small scales. We specifically discuss the nanoindentation experiment and recent micropillars and microtensile experiments. In Sect. 3, we discuss a multiscale frame work; in Sect. 4, we discuss the dislocation dynamics method. In Sect. 5, we focus on the coupling of dislocation dynamics with continuum mechanics. Finally in Sect. 6, we discuss a few problems of small scale plasticity phenomena.

2 Advances in Experimental Methods

This section provides a brief review of the developments in the area of experimental determination of mechanical properties at the microscale and nanoscale. Emphasis is placed on experimental methods using nanoindentation; readers who wish for a more extensive treatment of the technique or specific aspects are referred to several reviews (Li and Bhushan 2002; VanLandingham 2003; Fischer-Cripps 2004; Oliver and Pharr 2004). We also briefly discuss recent experiments that attempt to determine mechanical properties from microscale tensile, compression, or bending experiments.

To perform an indentation test, a tip is pressed into a solid and the resulting load to the sample and penetration of the tip relative to the initial surface can be recorded during the process. This process, depth-sensing indentation, typically produces a load–depth curve for a particular tip–solid system, shown schematically in Fig. 1. The load–depth curve has several important features used to interpret

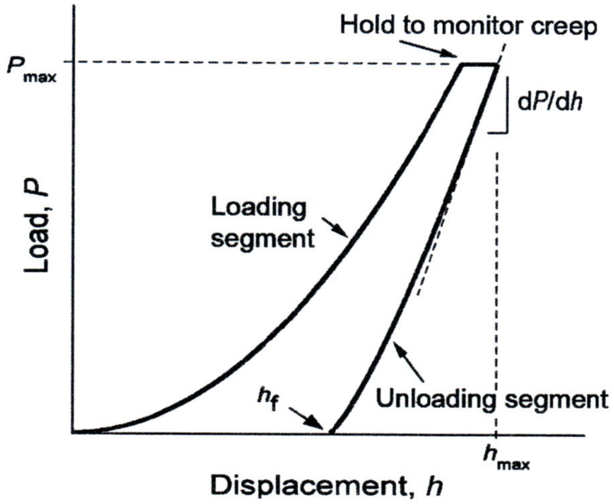

Fig. 1 Schematic load–depth curve for the relationship during penetration of a conical-equivalent tip into a flat solid

materials properties. The maximum load, P_{max}; the maximum penetration depth, h_{max}; and the final depth, h_f, are often used to describe the indentation process. In addition, the load–depth relationship can be monitored throughout the loading process, and the stiffness would be given by dP/dh. As shown by Doerner and Nix (1986) and further developed by Oliver and Pharr (1992), an analysis of the unloading portion of the load–depth curve allows the modulus and hardness of the material to be determined; the following treatment is based on that work and the work of Sneddon (1965) for the indentation of elastic half space with a rigid shape or column. Under load, it is often assumed that the profile of the indentation is given by Fig. 2, where the contact radius (if a conical indenter were used) would be defined by the contact radius a, and the maximum penetration, h_{max}.

Contact mechanics since has long recognized that the elastic deformation of two solids can be described by composite terms. Extensive treatments of contact mechanics can be found in Johnson's text (Johnson 1985). For the elastic compression of two solids in contact, the compliances add in series, and it is convenient to define a reduced elastic modulus, E^*,

$$\frac{1}{E^*} = \frac{1-\nu_i^2}{E_i} + \frac{1-\nu_s^2}{E_s}, \tag{1}$$

where E and ν are the elastic modulus and Poisson's ratio, respectively, and the subscripts i and s refer to indenter and sample, respectively. For standard indentation techniques, diamond is selected for the indenter tip; hence, 1,141 GPa is commonly used for E_i, while ν_i is 0.07.

The standard assumption during depth-sensing indentation of metals is that both elastic and plastic deformations occur during loading, beginning at the lowest

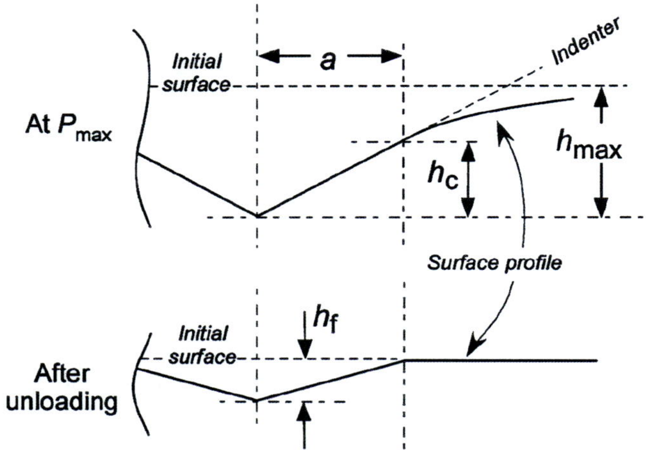

Fig. 2 Schematic of the surface of a solid while under load with an indentation tip and after removing the loaded tip

measurable loads. Section 2.1 will address situations where this is not the case; but for materials with reasonable dislocation densities, this is assumption is valid. With a shallow indenter (i.e., one with an included angle similar to that of the Berkovich geometry), the unloading of the indentation is used to determine the elastic properties of the solid, assuming that the unloading is purely described by the elastic response of the system. While this may not be true for materials at high homologous temperatures, in many metallic and ceramic systems this is experimentally observed (i.e., after partially unloading, subsequent reloading follows the load–depth curve until the previous maximum load is achieved). In this case, the unloading of the indenter from the solid can be described by

$$\frac{\mathrm{d}P}{\mathrm{d}h} = \beta \frac{2}{\sqrt{\pi}} E^* \sqrt{A}, \tag{2}$$

where A is the projected contact area of the solid ($A = \pi a^2$) and β is an emperical constant which accounts for the differences between the axisymmetric contact models and the experimental variations in using pyramidal geometries. This constant also accounts for possible variations due to plastic deformation and the fact that the contact area is beyond the small strain conditions assumed in many contact mechanics problems. This correction factor β is still a subject of debate and research Oliver and Pharr (1992) and Fischer-Cripps (2006); reported values range from the 1.034 to 1.1 (King 1987). A recent study by Strader et al. (2006) describes these effects in detail, but for the purposes of this work the specific value is not of great importance.

If the projected contact area can be related to the penetration depth through a calibration routine, it is possible to then define the hardness of the material. The concept that hardness was a materials property, and not dependent on test method, led to the observations that the parameter which remains constant for large indentations appears to be the mean pressure, defined as the load divided by the *projected* contact area of the surface. Meyer (1908) suggested that the hardness of a spherical impression with projected diameter d should therefore be defined as:

$$H = \frac{4P}{\pi d^2}, \tag{3}$$

where H is the hardness. Confirmation of Meyer's work (in English) was shown by Hoyt (1924). This type of hardness measurement is alternately referred to as the mean pressure of an indentation, and noted as either p_o or H in the literature, and differs from Vickers hardness or Brinnel hardness, which are based on load to surface, not projected, area ratios. Tabor describes in detail one of the direct benefits of using spherical indentation for determining the hardness of a material (Tabor 1951). As the indentation of a spherical indentation progresses, the angle of contact between the tip and solid changes. From extensive experiments in the 1920s to 1950s, the effective strain of the indentation, ε_i, imposed by a spherical tip was shown to be approximated by:

$$\varepsilon_i \approx 0.2\frac{d}{D}. \tag{4}$$

For materials that exhibit substantial plastic deformation, such as copper and steel, H is approximately 2.8 times the yield stress. These relationships allowed the creation of "indentation stress–strain curves" by applying different loads to spherical tips and measuring the impression diameters. In this case, the hardness of a material which work hardens increases with increasing indentation area for spherical tips. Current methods of partial unloading using depth-sensing indentation with spherical tips use this methodology (Field and Swain 1993).

The spherical cavity model of indentation, which considers the radial and tangential deformation fields around an indenter, has been developed extensively by Johnson (1970). However, the spherical cavity model does not address the amount of upheaval or "pile up" around an indentation. It has been shown (Samuels and Mulheam 1957) that materials that perfectly exhibit plastic behavior tend to pile up around an indenter, while annealed materials that have high work hardening exponents tend to sink in around an indenter. This behavior has been modeled using finite element methods by Bower et al. (1993). Previous work on pile up around indentations has shown the amount of pile up to vary with the indenter angle for a given load, with sharper indenters providing more pile up (Dugdale 1954). The following discussion is only strictly valid for cones between the angles of 105° and 137°, where the spherical cavity model has been used to successfully determine plastic zones around indentations (Stelmashenko et al. 1993). Cones sharper than this may begin to approach cutting mechanisms modeled by slip line fields described by Hill.

In addition to the upheaval around the indenter being used to define the extent of deformation around a solid, there is a region centered on the indenter tip which experiences radial compression from the indenter tip. The ratio of the plastic zone size to the contact radius, c/a, will quantify the extent of plastic deformation. Plastic zones under materials have been successfully mapped by several researchers, including a particularly thorough study by Tromas and co-workers (Tromas et al. 2000). According to Johnson's conical spherical cavity model (Johnson 1970) of an elastic–plastic material, the plastic zone is determined by:

$$\frac{c}{a} = \left[\frac{E^* \tan\alpha}{6\sigma_y(1-\nu)} + \frac{2}{3}\left(\frac{1-2\nu}{1-\nu}\right)\right]^{1/3}, \tag{5}$$

where α is the angle between the face of the cone and the indented surface, and σ_y is the yield strength of an elastic–plastic material. When the first term in (5) is greater than 40, Johnson suggests that the analysis of elastic–plastic indentation is not valid, and that the rigid-plastic case has been reached. Once the fully plastic state has been reached, the ratio c/a becomes about 2.33. Another model of the plastic zone size (Harvey et al. 1993) suggests that the plastic zone boundary, c, is determined by:

$$c = \sqrt{\frac{3P}{2\pi\sigma_y}}. \tag{6}$$

With this model, and as noted by Tabor (1951), the hardness is approximately three times the yield strength of a fully plastic indentation, and the ratio c/a should be 2.12 (similar to the suggested values of Johnson). In either case, this justifies the conventional lateral positioning of indentations which require a spacing of at least five times the indenter diameter to ensure no influence of a previous indentation interferes with the hardness measurement. Experimental measurements of pile up around macroscopic indentations (Bahr and Gerberich 1996) have shown the pile up models described in this section are reasonable, and both on the macroscopic and nanoscale (Kramer et al. 1998) the plastic zone size approximation for the yield strength is reasonable. Therefore, it is reasonable to assume that subgrain size measurements of plasticity can be obtained using both direct observations of impressions with atomic force microscopy (Sect. 2.3) or by measuring the load–displacement relationships.

2.1 Onset of Plasticity Modeling

Small volumes of materials regularly exhibit experimentally measured mechanical strengths greatly in excess of macroscopic volumes of the same materials. Observations of the deformation in micron-scale whiskers were among the first studies to demonstrate the ability of metals to sustain near theoretical shear stresses of the material. The "size effect" in metals (Fleck et al. 1994; Ma and Clarke 1995; Nix and Gao 1998; Elmustafa et al. 2000) has been the subject of many studies with nanoindentation due to the ability to neglect elastic recovery (i.e., the indentation hardness is measured under load) and to measure very small sizes. This differs from the effect in ceramics, which has been explained by variations in elastic recovery (Bull et al. 1989). Mechanical testing of nanocrystalline metals have continued this trend; the current literature has reached a point where models suggest that grain sizes too small to support the nucleation and growth of stable dislocation loops will likely deform via other mechanisms such as grain boundary sliding. Nanoscale metallic laminates have reached a point where individual dislocation motion may dominate deformation (Misra et al. 2005). Recent experiments in which a flat punch in a nanoindenter has been used to carry out compression tests on machined structures fabricated via focus ion beam machining have demonstrated that just having a smaller volume of material, with the concomitant increase in surface area and decrease in sample size, may indeed impact plasticity in ways not immediately obvious through scaling arguments (Uchic et al. 2004; Volkert and Lilleodden 2006). This is further discussed in Sect. 6.2. Indentation testing of relatively dislocation free solids (Sayed-Asif and Pethica 1997; Bahr et al. 1998; Minor et al. 2004) show shear stresses under the indenter tip approach the theoretical shear strength of the material. Testing defect-free metals is only possible using small-scale mechanical testing; large volume mechanical testing will generally measure the motion of pre-existing dislocations under applied stresses. Nanoindentation couples well with small-scale mechanical modeling in size space if not time scales, as it approaches volumes which can be simulated using molecular dynamics (Kelchner et al. 1998; Choi et al. 2003); Gerberich et al. 2002); however, the time scales between the models and experiments are orders of magnitude different.

In materials with low dislocation densities, the sudden onset of plasticity occurs at stresses approaching the theoretical strength of a material. Three primary models are often used to account for this behavior: homogenous nucleation, films that resist deformation on the underlying metal, and thermally activated processes. Gane and Bowden (1968) were the first to observe the extreme strengths during indentation. They observed the indentation of an electropolished gold crystal in a scanning electron microscope. No permanent penetration was initially observed; however, at some point during the process a sudden jump in displacement occurred, which corresponded to the onset of permanent deformation. Pethica and Tabor (1979) made similar observations during indentation using a tungsten tip and a nickel (111) surface in an ultra high vacuum environment while monitoring the electrical resistance between the tip and surface. The nickel had been annealed, sputtered to remove contaminants and oxides, and then annealed again to remove surface damage from the sputtering process, leaving behind an assumed pristine surface. When indentation was performed on the sample after a 50 Å oxide was grown on the nickel, the loading was largely elastic while the electrical resistance remained extremely high, with only minimal evidence of plastic deformation. However, at some critical load the resistance between the tip and sample dropped dramatically; continued loading was suggestive of primarily plastic deformation, implying that an oxide film on metals may be responsible for high strengths during nanoscale contacts.

With instrumented indentation testing on the nanometer scale, it is possible to monitor the "pop in" or "excursion" in depth as represented in Fig. 3. The tests use experimental methods that are either depth-controlled (Kiely and Houston 1998;

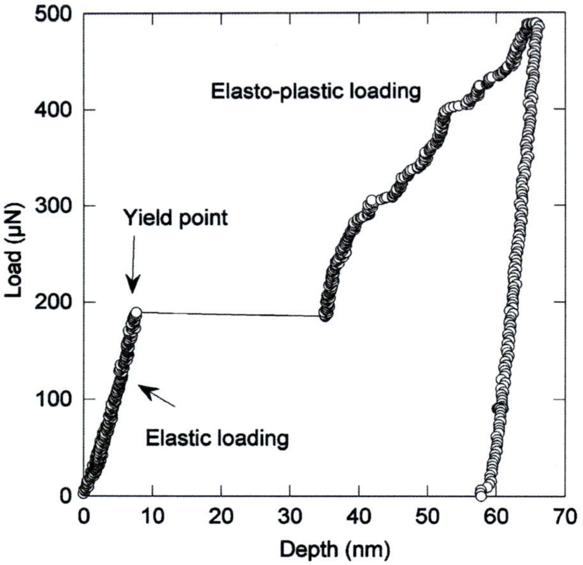

Fig. 3 Nanoindentation into electropolished nickel, with an initial elastic loading followed by plastic deformation. Verification of elastic loading was done by unloading from 100 to 50 µN and reloading, showing complete reversibility in loading

Michalske and Houston 1998) or a displacement (and not penetration) (Bahr et al. 1999) or load-controlled (Sayed-Asif and Pethica 1997). Post-indentation microscopy demonstrated that after these types of discontinuous events during loading, dislocation structures were present (Page et al. 1992; Gaillard et al. 2006). Using the Hertzian elastic model,

$$P = \frac{4}{3}E^*\sqrt{R\delta^3} \tag{7}$$

with δ being the indentation depth, the maximum shear stress under the indenter tip, $\tau_{\max}$, is related to the applied load, P, and by

$$\tau_{\max} = 0.31\left(\frac{6PE^{*2}}{\pi^3R^2}\right)^{1/3} \tag{8}$$

at a position underneath the indenter tip at a depth of 0.48 a, where a is the radius of the contact area and is determined by solving

$$\frac{3P}{2\pi a^2} = \left(\frac{6PE^{*2}}{\pi^3R^2}\right)^{1/3}. \tag{9}$$

The shear stress field does not drop off rapidly with position; there is a region under the indenter tip extending to approximately $2a$ which has an applied shear stress of approximately $P/2\pi a^2$. At the yield point in Fig. 3, this stress approaches the theoretical shear stress in a crystal. The similarity between the applied shear stress and shear modulus is close to the classical models of homogeneous dislocation nucleation. For instance, Cotrell (1953) reported that the classical theoretical strength of a material is often approximated by $E/30$ to $E/100$, where E is the elastic modulus. Recent work (Li et al. in press) has shown that given an exact model of the surface shape of the probe, the shear stress may be underestimated by up to 30% using this basic Hertzian model as the small strain condition for spherical indentation is exceeded at these depths.

New experimental tools, using in situ nanoindentation in a transmission electron microscope, have been used to support the homogeneous dislocation nucleation model. Minor et al. (2004) observed dislocations to pop in underneath an indenter tip after an initial elastic loading during contact between a tip and an aluminum sample. This direct observation is a strong evidence that dislocations nucleate underneath the indenter tip. However, recent studies from in situ microscopy have shown that even the very initial contact can generate dislocations that nucleate and propagate to grain boundaries, after which an elastic loading behavior is again demonstrated followed by a second strain burst (Minor et al. 2006). Multiple yield points during nanoindentation or contact loading (Corcoran et al. 1997; Kiely and Houston 1998; Uchic et al. 2004) are often referred to as "staircase" yielding.

Schuh and co-workers (Schuh et al. 2005) have carried out indentation experiments that probe the onset of plastic deformation at elevated temperatures to determine an effective activation energy and volume for the initiation of plasticity. By describing the probability of an excursion occurring underneath the indenter tip using an Arrhenius model for dislocation nucleation, they suggest that the activation energy is similar to that of vacancy motion, and that the activation volume is on the order of the volume of a cubic Burgers vector. This suggests that point defects have a direct relationship on dislocation nucleation, as no other atomistic feature sizes would have that dimension. Studies (Ngan and Wo 2006) of incipient plasticity in Ni_3Al suggests that the growth of a Frank dislocation source at subcritical stresses controls plasticity at lower stresses during indentation, and that self diffusion along the indenter–sample surface dominates the onset of plasticity.

2.2 *Analysis of Slip Around Indentation as a tie to Dislocation Mechanisms*

Indentation can also be used with post-indent microscopy to characterize specific dislocation structures responsible for deformation, providing a strong tie to simulations using discreet dislocation dynamics. Chang and Chen (1995) described a method for determining the surface orientation of a particular grain in an FCC material by measuring the angles of the slip step lines on the surface around indentations and calculating the orientation from the combination of angles. As availability of orientation imaging microscopy (OIM) has increased, it is possible to use the known orientation of a grain to determine the slip plane responsible for each slip step. Each slip plane can be indexed with respect to a reference direction taken from the OIM. Studies of the plastic zone around materials by analyzing the surface topography to determine the effects of tip geometry and crystal orientation on slip can be related directly to dislocation mechanisms (Nibur and Bahr 2003; Nibur et al. 2007). Figure 4 demonstrates the extent of surface deformation in a stainless steel alloy with an orientation of (1 2 20) normal to the plane of the indentation [effectively an (001) grain] indented with a 90° conical indenter with a tip radius of approximately 1 μm to a load of 100 mN. The figure is a derivative scanning probe microscopy image with the resulting true height profile shown below. The extent of plastic deformation surpasses the expected c/a ratio of 2.33 in the "x" direction in the image, but in the "y" direction noted in the figure the expected c/a ratio is found.

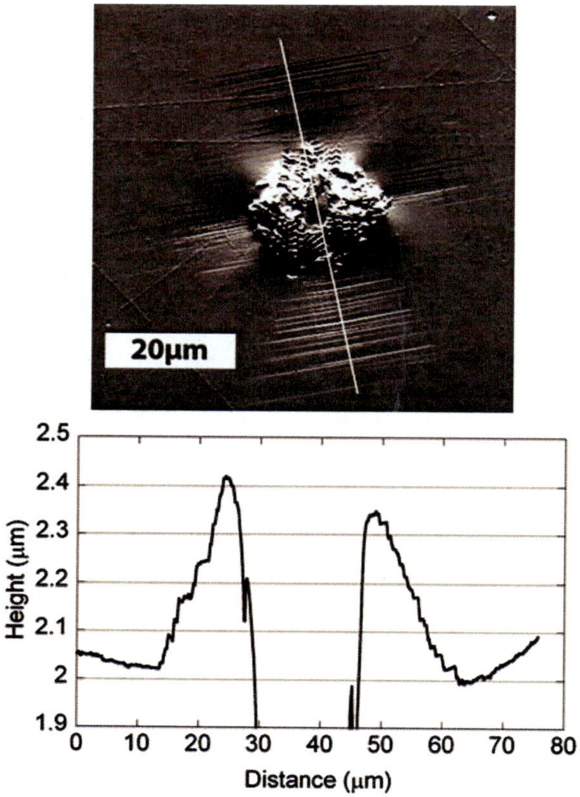

Fig. 4 Slip steps around a conical indentation in a near (001) grain of stainless steel showing differences in slip extent on the free surface as measured by scanning probe microscopy, and the ability to identify individual slip step features

These results are not unique to surface measurements; dislocation arrays identified by etching after indentation in MgO have demonstrated similar effects (Tromas et al. 1999) of crystallography and nonuniform dislocation structures. The extent of deformation in the lateral direction which is transferred to the surface is influenced by the effective strain of the indentation. Of particular interest in recent studies using dislocation etch pitting in conjunction with nanoindentation experiments (Gaillard et al. 2006) is that the lattice friction stress for individual screw and edge dislocation components can be measured; Gaillard and co-workers have determined lattice friction stresses of 65 MPa for edge dislocations and 86 MPa for screw dislocations in MgO.

2.3 *Micro-Uniaxial Tests*

The nanoindentation method discussed above is the most frequently used technique to determine mechanical properties at small scale. The method can be easily applied to a wide range of materials, and a number of mechanical properties can be deduced from the experimental data. Because of the complexity of the stress–strain field under the indenter tip, extraction of the mechanical properties that exactly correspond to more traditional features found on uniaxial stress–strain curves, such as yield strength and strain hardening behavior, relies on models that may include certain assumptions and approximations with a direct dependence on the shape and size of the indenter tip. However, the exact shape of the indenter tip can vary from one experiment to another, and even if this variation is small, it may impact the actual contact area and the calculated values. To address these issues, some researchers have worked to adapt conventional material test techniques, such as torsion, bending, and uniaxial tests, to microscale specimens. Fleck et al. (1994) performed torsional experiments on microscale copper wires to determine mechanical properties at small scales, and Stolken and Evans (1998) carried out bending tests on nickel foils and showed strong dependence of strength on foil thickness. In both cases, the size dependence in mechanical properties was attributed to strain gradients that result from the formation of dislocation walls that form to accommodate curvature distortion of the lattice, rather than as an effect of dislocation density as noted by some of the indentation work in the previous section. However work by Horstemeyer et al. (2001) showed through molecular simulations that size-dependent behavior can exist even in the absence of strain gradients. One of their main discoveries is the presence of a strong inverse–power relationship between the yield strength and the volume-to-surface-area ratio of the computational cell. Subsequently, a number of experiments have been performed confirming that size-dependent behavior exists in homogenously deformed small-scale specimens with no strain gradients. For example, unlike the size-independent behavior showed by Fleck et al. (1994) of 15–170-μm-thick Cu polycrystalline specimens under tension, experiments on compression of single crystal micropillars (Dimiduk et al. 2005; Greer and Nix 2006; Volkert and Lilleodden 2006; Frick et al. 2008) show a significant size effect in Cu specimens of similar sizes. Therefore, the argument that a size effect is inherently due to the presence of strain gradients resulting from inhomogeneous loading does not explain the observed size effect in compression experiments of micropillars.

Small-scale compression experiments have also been carried out on nanoscale gold pillars with diameters ranging from 200 nm to several micrometers along <001> loading direction (Dimiduk et al. 2005). Even though the orientation is one of high symmetry and has a low Schmid factor, the lack of stage II hardening in the stress–strain response lead to an explanation of size effects based on dislocation starvation hypothesis (Greer and Nix 2006). According to the dislocation starvation hypothesis, as the specimen dimensions are small and unconstrained, all the mobile dislocations glide to the surface and annihilate causing a strain burst in the macroscopic response resulting in a dislocation free crystal; further plastic deformation requires dislocations to nucleate under very high stresses. The elastic loading between strain bursts is associated with the lack of dislocations in the crystal. Volkert and Lilleodden (2006) and Frick et al. (2008) have performed compression experiments on Au and Ni micropillars, respectively, oriented effectively for single slip and showed significant hardening. Frick et al. (2008) also showed the TEM micrographs of dislocation structure in the compressed micropillars, while Shaw et al. (2008) show evidence of dislocation free pillars in very small pillars. Tensile testing that eliminates some stress concentrations present at the base of micropillar have been used to suggest that there are also hardening effects due to nonuniform stresses in the samples (Kiener et al. 2008).

In an attempt to understand the underlying dislocation mechanisms responsible for the macroscopic response of micropillars, many discrete dislocation dynamics (DDD) simulations have been performed and are discussed in Sect. 5.

3 Multiscale Framework

This section gives a brief description of how one can break down the different plasticity formulations related to the different length and time scales. Mesoscale analyses typically start at the scale of the grain or crystal, whereas macroscale analyses start at the polycrystalline level. Mesoscale analyses focus on just activities within a single crystal. In a sense, mesoscale analyses are ascribed as discrete methods but can address polycrystalline materials as averaging schemes over the single crystals are performed. Hence, mesoscale analyses use continuum mechanics as well. Several types of mesoscale analyses can be performed. In this chapter, dislocation dynamics occurring at the scale of the grain is presented, which is a discrete method.

Because the mesoscale dislocation dynamics and crystal plasticity formulations require a large capacity of computing power, they are not generally used to solve large-scale boundary value problems of structural components or systems. It is the macroscale internal state variable continuum theory that is often used in solving practical engineering problems. The use of the mesoscale dislocation dynamics and crystal plasticity formulations arises in material analyses studies, which in turn can play a role in the macroscale efforts. We also note that the details of dislocation nucleation, motion, and interaction can be captured in the dislocation dynamics formulation. Now the crystal plasticity formulation uses discrete crystals but treats the dislocation effects in a phenomenological manner. Hence, it loses some of the details but can capture larger scale boundary value problems. The macroscale internal state variable captures even less detail than the crystal plasticity formulation but can address even larger scale boundary value problems.

When considering the multiscale modeling and simulation methods described in this section, one can consider that the ab initio and atomistic method simulation results can be imported into the dislocation dynamics formulation. The results from the dislocation dynamics formulation can be imported into the crystal plasticity formulation. And the crystal plasticity results can be imported into the macroscale internal continuum formulation. This bridging of the length scales is a relatively recent area of research in the areas of plasticity, damage/fracture, and fatigue. Examples and details of models for each of the scales can be found in a recent book edited by Yip (2005). Here, we focus on the discussion on the microscale where one can analyze the dynamics of a collection of discrete dislocations using the so-called dislocation dynamics method.

Dislocation dynamics (DD) is one of the most recent and powerful tools developed to model the behavior of metallic materials at the microscale in a more physical manner than existing plasticity models. The DD approach has been advanced considerably over the past two decades. The early DD models were two-dimensional (2D) and consisted of periodic cells containing multiple dislocations whose behavior was governed by a set of simplified rules (Lepinoux and Kubin 1987; Van der Giessen and Needleman 1995). These simulations, although serving as a useful conceptual framework, were limited to 2D and, consequently, could not directly account for such important features in dislocation dynamics as slip geometry, line tension effects, multiplication, certain dislocation intersections, and cross-slip, all of which are crucial for the formation of dislocation patterns. In the 1990s, development of new computational approaches of DD in three-dimensional space (3D) generated hope for a principal breakthrough in our current understanding of dislocation mechanisms and their connection to crystal plasticity (Canova et al. 1992; Hirth et al. 1996; Zbib et al. 1996, 1998). This progress has been further magnified by the coupling of DD with continuum mechanics analysis in association with computational algorithms such as finite elements. This coupling has paved the way for researchers to investigate many complicated small-scale crystal plasticity phenomena that occur under a wide range of loading and boundary conditions, and cover a wide spectrum of strain rates, increasing the potential for future applications of this method in material, mechanical, structural, and process engineering analyses.

In the following section, we present the principles of DD analysis, followed by the procedure for the measurement of local quantities such as plastic distortion and internal stresses. The incorporation of the DD technique into the three-dimensional plastic continuum mechanics-based finite elements modeling is then described.

This method has been used to investigate a number of small-scale deformation phenomena including size effects in metal–matrix composites, size effects in models for free surface effects, nanoindentation, size effects in models for grain boundaries, and size effects in the bending of microbeams, shock waves, etc. Here, in Sect. 5, we present recent results pertaining to nanolaminate metallic composites and compression of micropillars.

4 The Dislocation Dynamics (DD) Method

To better describe the mathematical and numerical aspects of the DD methodology, first we identify the basic geometric conditions and kinetics that control the dynamics of dislocations. This is followed by discussions of the dislocation equation of motion, elastic interaction equations, and the descretization of these equations for numerical implementation, a thorough explanation of the method can be found in Zbib and Diaz de la Rubia (2002) and Zbib et al. (2009).

4.1 Kinematics and Geometric Aspects

A dislocation is a line defect in an otherwise perfect crystal described by its line sense vector $\boldsymbol{\xi}$ and Burgers vector **b**. The Burgers vector has two distinct components: edge, perpendicular to its line sense vector, and screw, parallel to its line sense vector. Under loading, dislocations glide and propagate on slip planes causing deformation and change of shape. When the local line direction becomes parallel to the Burgers vector, i.e., screw character, the dislocation may propagate into other slip planes. This switching of the slip plane, which makes the motion of dislocations three-dimensional, and is better known as cross slip, is an important recovery phenomena to be dealt with in dislocation dynamics. In addition to glide and cross slip, dislocations can also climb in a nonconservative three-dimensional motion by absorbing and/or emitting intrinsic point defects, vacancies, and interstitials. Some of these phenomena become important at high strain rate levels or temperatures when point defects become more mobile. In summary, the 3D dislocation dynamics accounts for the following geometric aspects:

- Identification of all possible slip planes for each dislocation
- Dislocation topology; 3D geometry, Burgers vector, and line sense
- Multiplication and annihilation of dislocation segments
- Changes in the dislocation topology when part of it cross-slips and or climbs to another plane
- Formation of complex connections and intersections such as junctions, jogs, and branching of the dislocation in multiple directions

4.2 Kinetics and Interaction Forces

The dynamics of the dislocation is governed by a *Newtonian* equation of motion, consisting of an inertia term, a damping term, and a driving force arising from short-range and long-range interactions. Since the strain field of the dislocation varies as the inverse of the distance from the dislocation core, dislocations interact among themselves over long distances. A moving dislocation has to overcome internal drag and local constraints such as the Peierls stresses (i.e., lattice friction). The dislocation may encounter local obstacles such as stacking fault tetrahedra, defect clusters, and vacancies that interact with the dislocation at short ranges and affect its local dynamics. Furthermore, the internal strain field of randomly distributed local obstacles gives rise to stochastic perturbations to the encountered dislocations, as compared with deterministic forces such as the applied load. This stochastic stress field also contributes to the spatial dislocation patterning in the later deformation stages. Therefore, the strain field of local obstacles adds spatially irregular uncorrelated noise to the equation of motion. In addition to the random strain fields of dislocations or local obstacles, thermal fluctuations also provide a stochastic source in dislocation dynamics. Dislocations also interact with free surfaces, cracks, and interfaces, giving rise to what is termed as image stresses or forces. In summary, the dislocation may encounter the following set of forces:

- Peierls stress $\boldsymbol{F}_{\text{Peierls}}$
- Drag force, $B\boldsymbol{v}$, where B is the drag coefficient and $\boldsymbol{v}$ is the dislocation velocity
- Force due to externally applied loads, $\boldsymbol{F}_{\text{External}}$
- Dislocation–dislocation interaction force $\boldsymbol{F}_D$
- Dislocation self-force $\boldsymbol{F}_{\text{Self}}$
- Dislocation–obstacle interaction force $\boldsymbol{F}_{\text{Obstacle}}$
- Image force $\boldsymbol{F}_{\text{Image}}$
- Thermal force $\boldsymbol{F}_{\text{Thermal}}$ arising from thermal fluctuations
- Osmotic force $\boldsymbol{F}_{\text{Osmotic}}$ resulting from nonconservative motion of dislocation (climb) and results in the absorption or emission of intrinsic point defects

The DD approach attempts to incorporate all of the aforementioned kinematics and kinetics aspects into a computational traceable framework. In the numerical implementation, three-dimensional curved dislocations are treated as a set of connected segments. One can represent smooth dislocations with any desired degree of realism, provided that the discretization resolution is high enough for a specified accuracy (limited by the size of the dislocation core radius r_0, typically the size of one Burgers vector $\mathbf{b}$). In such a representation, the dynamics of dislocation lines is reduced to the dynamics of discrete degrees of freedom of the dislocation nodes connecting the dislocation segments.

4.3 Dislocation Equation of Motion

The velocity v of a dislocation segment s is governed by a first-order differential equation consisting of an inertia term, a drag term, and a driving force vector (Hirth 1992; Hirth et al. 1998; Huang et al. 1999), such that:

$$m_s \dot{v} + \frac{1}{M_s(T, p)} v = F_s \tag{10}$$

$$F_s = F_{\text{Peirels}} + F_D + F_{\text{Self}} + F_{\text{External}} + F_{\text{Obstacle}} + F_{\text{Image}} + F_{\text{Osmotic}} + F_{\text{Thermal}}. \tag{11}$$

In the above equation, the subscript s stands for the segment, m_s is defined as the effective dislocation segment mass density (Hirth et al. 1998), M_s is the dislocation mobility which could depend both on the temperature T and the pressure p, and W is the total energy per unit length of a moving dislocation (elastic energy plus kinetic energy). As implied by (10), the glide force vector F_s per unit length arises from a variety of sources described in the previous section. Relations for the mass per unit dislocation length are given by Hirth et al. (1998).

4.4 Dislocation Mobility Function

The reliability of the numerical simulation depends critically on the accuracy of the dislocation drag coefficient $B(= 1/M)$ which is material dependent. There are a number of phenomenological relations for the dislocation glide velocity v_g, including relations of power law forms and forms with an activation term in an exponential or as the argument of a sinh form. Often, however, the simple power law form is adopted, resulting in nonlinear dependence of M on the stress. In some cases of pure phonon/electron damping control or of glide over the Peierls barrier, a constant mobility (with $m = 1$) predicts the results very well. This linear form has been theoretically predicted for a number of cases (Hirth and Lothe 1982).

Dislocation drag had been studied for some time and the drag coefficients have been estimated in numerous experimental and theoretical works by atomistic simulations or quantum mechanical calculations (see, e.g., the review by Al'shitz 1992). The determination of each of the two components (phonon and electron drag) that constitute the drag coefficient for a specific material is not trivial, and various simplifications have been made. For example, the Debye model neglects Van Hove singularities in the phonon spectrum, deformation potentials are approximated as isotropic, and so on. Also the values are sensitive to various parameters such as the mean free path or core radius. Nevertheless, in typical metals, the phonon drag B_{ph} range is 30–80 μPa s at room temperature and less than 0.1 μPa s at very low temperatures around 10 K, while for the electron drag B_{e}, the range is a few μPa s and expected to be temperature independent. Under strong magnetic fields at low temperature, macroscopic dislocation behavior can be highly sensitive to orientation relative to the field within accuracy of 1%. Except for special cases such as deformation under high strain rate, weak dependences of drag on dislocation velocity are usually neglected.

The dislocation mobility could be, among other things, a function of the angle between the Burgers vector and the dislocation line sense, i.e., dislocation character, especially at low temperatures. For example, in Wasserbäch (1986) it is observed that at low deformation temperatures (77–195 K) the dislocation structure in Ta single crystals consisted of primary and secondary screw dislocations and of tangles of dislocations of mixed characters, while at high temperatures (295–470 K) the behavior was similar to that of fcc crystals. In the work of Mason and MacDonald (1971), they measured the mobility of dislocation of an unidentified type in Nb as 4.2×10^4 (Pa s)$^{-1}$ near room temperature. A smaller value of 3.3×10^3 (Pa s)$^{-1}$ was obtained by Urabe and Weertman (1975) for the mobility of edge dislocation in Fe. The mobility for screw dislocations in Fe was found to be about a factor of two smaller than that of edge dislocations near room temperature. A theoretical model to explain this large difference in behavior is given in Hirth and Lothe (1982) and is based on the observation that in bcc materials the screw dislocation has a rather complex three-dimensional core structure, resulting in a high Peierls stress, leading to a relatively low mobility for screw dislocations while the mobility of mixed dislocations is higher.

4.5 Dislocation Collisions

When two dislocations collide, their response is dominated by their mutual interactions and becomes much less sensitive to the long-range elastic stress associated with external loads, boundary conditions, and all other dislocations present in the system. Depending on the shapes of the colliding dislocations, their approach trajectories and their Burgers vectors, two dislocations may form a dipole, or react to annihilate, or combine to form a junction, or intersect and form a jog. In the DD analysis, the dynamics of two colliding dislocations is determined by the mutual interaction force acting between them. In the case that the two dislocation segments are parallel (on the same plane and or intersecting planes) and have the same Burgers vector with opposite sign, they would annihilate if the distance between them is equal to the core size. Otherwise, the colliding dislocations would align themselves to form a dipole, a jog, or a junction depending on their relative position. A comprehensive review of short-range interaction rules can be found in Rhee et al. (1998).

4.6 Discretization of Dislocation Equation of Motion

Equation (10) applies to every infinitesimal length along the dislocation line. To solve this equation for any arbitrary shape, the dislocation curve may be discretized into a set of dislocation segments. Then the velocity vector field over each segment may be assumed to be linear, and, therefore, the problem is reduced to finding the velocity of the nodes connecting these segments. There are many numerical techniques to solve such a problem. Consider, for example, a straight dislocation segment s bounded by two nodes j and $j+1$. Then within the finite element formulation, the velocity vector field can be assumed to be linear over the dislocation segment length. Then this linear vector field v can be expressed in terms of the velocities of the nodes such that $v = \left[\mathbf{N}^D\right]^{\mathrm{T}} \mathbf{V}^D$ where $\mathbf{V}^D$ is the nodal velocity vector and $[\mathbf{N}^D]$ is the linear shape function vector. Upon using the Galerkin method, (10) for each segment can be reduced to a set of six equations for the two discrete nodes (each node has three degrees of freedom). The result can be written in the following matrix–vector form.

$$\left[M^D\right]\dot{\mathbf{V}}^D + \left[C^D\right]\mathbf{V}^D = F^D, \tag{12}$$

where

$$\left[M^D\right] = m_s \int \left[\mathbf{N}^D\right]\left[\mathbf{N}^D\right]^{\mathrm{T}} \mathrm{d}l$$

is the dislocation segment 6 × 6 mass matrix,

$$\left[C^D\right] = (1/\mathrm{M}_s) \int \left[\mathbf{N}^D\right]\left[\mathbf{N}^D\right]^{\mathrm{T}} \mathrm{d}l$$

is the dislocation segment 6 × 6-damping matrix, and

$$F^D = \int \left[\mathbf{N}^D\right] F_s \mathrm{d}l$$

is the nodal force vector. Then, following the standard element assemblage procedure, one obtains a set of discrete equations, which can be cast in terms of a global dislocation mass matrix, a global dislocation damping matrix, and a global dislocation force vector. In the case of one dislocation loop and with ordered numbering of the nodes around the loop, one can easily show that the global matrices are banded with a half-bandwidth equal to one. However, when the system contains many loops that interact among themselves and new nodes are generated and/or annihilated continuously, the numbering of the nodes becomes random and the matrices become unbanded. To simplify the computational effort, one can use the lumped mass and damping matrix method. In this method, the mass matrix $[\boldsymbol{M}^D]$ and damping matrix $[\boldsymbol{C}^D]$ become diagonal matrices (half-bandwidth equal to zero), and therefore the only coupling between the equations is through the nodal force vector $\mathbf{F}^D$. The computation of each component of the force vector is described below.

4.7 The Dislocation Stress and Force Fields

The stress induced by any arbitrary dislocation loop at an arbitrary field point p can be computed by the Peach–Koehler integral equation given in (Hirth and Lothe 1982). This integral equation, in turn, can be evaluated numerically over many loops of dislocations by discretizing each loop into a series of line segments. Let

r = distance from point p to the segment s
N_l = total number of dislocation loops
$N_s{}^{(l)}$ = number of segments of loop l
$N_n{}^{(l)}$ = number of nodes associated with the segments of loop l, i.e., $N_n{}^{(l)} = N_s{}^{(l)} + 1$
N_s = total number of segments = $N_s{}^{(l)} \times N_l$, where summation over l is implied
N_n = total number of nodes = $N_n{}^{(l)} \times N_l$, where summation over l is implied
l_s = length of segment s

Then the discretized form of the Peach–Koehler integral equation for the stress at any arbitrary field point p becomes

$$\overset{d}{\sigma}_{ij}(p) = \sum_{l=1}^{N_\ell} \sum_{s=1}^{N_s^{(\ell)}} \left\{ -\frac{G}{8\pi} \int_{l_s} b_p \in_{mpi} \frac{\partial}{\partial x'_m} \nabla'^2 R \, dx'_j - \frac{G}{8\pi} \int_{l_s} b_p \in_{mpi} \frac{\partial}{\partial x'_m} \nabla'^2 R \, dx'_j \right.$$
$$\left. - \frac{G}{4\pi(1-\nu)} \int_{l_s} b_p \in_{mpk} \left(\frac{\partial^3 R}{\partial x'_m \partial x'_i \partial x'_j} - \delta_{ij} \frac{\partial}{\partial x'_m} \nabla'^2 R \right) dx'_k \right\} \tag{13}$$

where $\in_{ijk}$ is the permutation symbol, and R is the magnitude of the $\mathbf{R} = \mathbf{r}' - \mathbf{r}$ (where $\mathbf{r}$ is the position vector of point p and $\mathbf{r}'$ is the position vector of a differential line segment of the dislocation loop or curve). The integral over each segment can be explicitly carried out using the linear element approximation. Exact solution of (4) for a straight dislocation segment can be found in DeWit (1960) and Hirth and Lothe (1982). However, evaluation of the above integral requires careful consideration as the integrand becomes singular in cases where point p coincides with one of the nodes of the segment that integration is taken over, i.e., self-segment integration. Thus,

- If p is not part of the segment s, there is no singularity since $R \neq 0$, and the ordinary integration procedure may be performed.
- If p coincides with a node of the segment s where the integration should be carried out, special treatment is required due to the singular nature of the stress field as $R \to 0$. Here, the regularization scheme developed by Zbib and co-workers has been used.

In general, the dislocation stresses can be decomposed into the following form:

$$\overset{d}{\sigma}(p) = \sum_{s=1}^{N_s-2} \overset{d}{\sigma}{}^{(s)} + \overset{d}{\sigma}{}^{(p+)} + \overset{d}{\sigma}{}^{(p-)}, \tag{14}$$

where $\overset{d}{\sigma}{}^{(s)}$ is the contribution to the stress at point p from a segment s, and $\overset{d}{\sigma}{}^{(p+)}$, $\overset{d}{\sigma}{}^{(p-)}$ are the contributions to the stress from the two segments that are shared by a node coinciding with p which will be further discussed below.

Then from the dislocation stress field, one can compute the forces on each dislocation segment by summing the stresses along the length of the segment. The stresses can be divided into those coming from the dislocations as formulated above and also from any other externally applied stresses plus the internal friction (if any) and the stresses induced by any other defects or micro-constituents. A model for the osmotic force $\boldsymbol{F}_{\text{Osmotic}}$ is given in Raabe (1998), and its inclusion in the total force is straightforward since it is a deterministic force. However, the thermal force $\boldsymbol{F}_{\text{Thermal}}$ is stochastic and its treatment is not trivial. This requires a special consideration and algorithm leading to what is called *stochastic dislocation dynamics* (SDD) as developed by Hiratani and Zbib (2002). Thus, the force acting on each segment can be written as:

$$F_s = \sum_{m=1}^{N_s} (\overset{d}{\sigma}{}^{(m)} + \overset{a}{\sigma}{}^{(m)} + \tau_s)\, b_s \times \xi_s = \overset{d}{F}_s + \overset{a}{F}_s + F_{\text{Thermal}}, \tag{15}$$

where $\overset{d}{\sigma}{}^{(m)}$, is the contribution to the stresses along segment s from another segment m (dislocation–dislocation interaction), $\overset{a}{\sigma}{}^{(s)}$ is the sum of all externally applied stresses, internal friction (if any) and the stresses induced by any other defects, and τ_s is the thermal stress; $\overset{d}{F}_s$, $\overset{a}{F}_s$, and F_{Thermal} are the corresponding total Peach–Koehler (PK) forces.

Also, the force $\overset{d}{F}_s$ can be decomposed into two parts, one arising from all dislocation segments and the other from the self-segment, which is better known as the self-force, that is:

$$\overset{d}{F}_s = \sum_{m=1}^{N_s-1} \overset{d}{F}_s{}^{(m)} + \overset{d}{F}_s{}^{(\text{self})}, \tag{16}$$

where $\overset{d}{F}_s{}^{(m)}$ and $\overset{d}{F}_s{}^{(\text{self})}$ are, respectively, the contribution to the force on segment s from segment m and the self-force. To evaluate the self-force, one should use a special numerical treatment, as given by Zbib et al. (1996, 2002) and Zbib and Diaz de la Rubia (2001, 2002), in which exact expressions for the self-force are given.

It is noted that the direct computation of the dislocation forces discussed above requires the use of a very fine mesh, especially when dealing with problems involving dislocation–defect interaction. Therefore, to capture the effect of the very small defects, the dislocation segment size must be comparable to the size of the defect. Otherwise, one can use large dislocation segments compared to the smallest defect size, provided that the force interaction is integrated over the segment length, or over a set number of Gauss points. The summation as in (6) would then follow according to:

$$F_s = F_s^{\mathrm{self}} + \sum_{m=1}^{N_s-1} \frac{1}{n_g} \sum_{g=1}^{n_g} \left[\overset{d}{\sigma}{}^{(m)}(p_g) + \overset{d}{\sigma}{}^{(m)}(p_g) + \dots \right] b_s \times \xi_s, \tag{17}$$

where p_g is the Gauss point g and n_g is the number of Gauss points along segment s. The number of Gauss points depends on the length of the segment. As a rule, the shortest distance between two Gauss points should be larger or equal to $2r_0$, i.e., twice the core size.

4.8 Evaluation of Plastic Strains

The motion of each dislocation segment gives rise to plastic distortion, which is related to the macroscopic plastic strain rate tensor $\dot{\varepsilon}^{\mathrm{p}}$, and the plastic spin tensor W^{p} via the relations

$$\dot{\varepsilon}^{\mathrm{p}} = \sum_{s=1}^{N_s} \frac{l_s v_{gs}}{2V} (n_s \otimes b_s + b_s \otimes n_s), \tag{18}$$

$$W^{\mathrm{p}} = \sum_{s=1}^{N_s} \frac{l_s v_{gs}}{2V} (n_s \otimes b_s - b_s \otimes n_s), \tag{19}$$

where n_s is a unit normal to the slip plane, v_{gs} is the magnitude of the glide velocity of segment s, and V is the volume of the representative volume element. The above relations provide the most rigorous connection between the dislocation motion (the fundamental mechanism of plastic deformation in crystalline materials) and the macroscopic plastic strain, with its dependence on strength and applied stress being explicitly embedded in the calculation of the velocity of each dislocation. Nonlocal effects are explicitly included into the calculation through long-range interactions. Another microstructure quantity, the dislocation density tensor α, can also be calculated according to

$$\alpha = \sum_{s=1}^{N} \frac{l_s}{V} b_s \otimes \xi_s. \tag{20}$$

This quantity provides a direct measure for the net Burgers vector that gives rise to strain gradient relief (bending of a crystal).

4.9 The DD Numerical Solution

The equation of motion (10) is nonlinear and coupled. Therefore, and to achieve fast convergence, it should be solved using an implicit algorithm with a backward integration scheme, i.e.

$$v^{t+\delta t}\left(1+\frac{\Delta t}{m_s M_s}\right)^{t+\delta t} = v^t + \frac{\Delta t}{m_s} F_s^{t+\delta t}. \tag{21}$$

As discussed in Zbib and Diaz de la Rubia (2002), this integration scheme is unconditionally stable for any time step size. However, in DD the time step is constrained by two factors: (1) the shortest flight distance for short-range interactions, and (2) the time step used in the dynamic finite element modeling to be described later. As suggested in Zbib and Diaz de la Rubia (2002), to ensure convergence and stable solution, the critical time t_c and the time step for both the DD and the FE should be $t_c = l_c/C_l$, and $\Delta t = t_c/20$, respectively, where l_c is the characteristic length scale.

The system of equations given above summarizes the basic elements that a dislocation dynamics simulation model should include. There are a number of variations in the manner in which the dislocation curves may be discretized, for example zero-order element (pure screw and pure edge), first-order element (or piecewise linear segment with mixed character), or higher order nonlinear elements, but this is purely a numerical issue. However, the DD model should have the minimum number of parameters and all of them are basic physical and material parameters and not phenomenological ones for the DD result to be predictive.

In summary, the DD model has the following set of physical and material parameters:

- Burgers vectors
- Dislocation mobility
- Elastic properties
- Core size (selected here as equal to one Burgers vector)
- Thermal conductivity and specific heat
- Mass density
- Stacking fault energy

In addition, there are two numerical parameters: the segment length and the time step. Both of these parameters should be determined in such a way to ensure convergence of the result. It should be emphasized that in general the dislocation mobility is an intrinsic material property that reflects the local drag mechanisms as discussed above. One can use an "effective" mobility that accounts for additional drag from dislocation–point defect interaction and thermal activation processes if the defects/obstacles are not explicitly impeded in the DD simulations. However, there is no reason not to include these effects explicitly in the DD simulations, i.e., dislocation–defect interaction, stochastic processes, and inertia effects, which actually permits the prediction of the "effective" mobility from the DD analysis.

5 Integration of DD and Continuum Plasticity

In the above, we have reviewed the basic ingredients needed to perform dislocation dynamics (DD) modeling. All of the essential parameters and quantities to perform the simulations have been discussed. One can enhance the power of DD simulations of the underlying dislocation microstructure and its evolution by coupling DD to finite-element modeling in real time. This coupling is now briefly described.

The discrete dislocation model can be coupled with continuum elasto-visoplasticity models, making it possible to correct for *dislocation image stress* and to address a wide range of complex boundary value problems at the microscopic level. In the following, a brief description of this coupling is provided. The coupling is based on a framework in which the material obeys the basic laws of continuum mechanics, i.e. the linear momentum balance.

$$\operatorname{div} \sigma = \rho \dot{v}, \tag{22}$$

and the energy equation

$$\rho c_v \dot{T} = k \nabla^2 T + \sigma \dot{\varepsilon}^{\mathrm{p}}, \tag{23}$$

where $v = \dot{u}$ is the particle velocity, $\mathbf{u}$, ρ, c_v, and k are the displacement vector field, mass density, specific heat, and thermal conductivity, respectively. In DD, the representative volume cell analyzed can be further discretized into subcells or finite elements, each representing a representative volume element (RVE). Then the internal stresses field induced by the dislocations (and other defects) $\boldsymbol{\sigma}^D$ and the plastic strain field within each RVE can be calculated at any point within each element. However, and to be consistent with the definition of a RVE, the heterogeneous internal stress field can be homogenized over each RVE, resulting into an equivalent internal stress S^D which is homogenous within the RVE, i.e.

$$S^D = \left\langle \sigma^D \right\rangle = \frac{1}{V_{\text{element}}} \int\limits_{\text{element}} \sigma^D(x) \mathrm{d}v. \tag{24}$$

Furthermore, the plastic strain increment results from only the mobile dislocations that exit the RVE (or subcell) and is computed as in (18) and (19) but with V being the volume of the element (or subcell). The dislocations that are immobile (zero velocity) do not contribute to the plastic strain increment. Therefore, the dislocations within an element induce an internal stress due to their elastic distortion. When some of these dislocations (or all) move and exit the element, they leave behind plastic distortion in the element, and the internal stress field should be recomputed by summing the stress from the remaining dislocations in the element. With this homogenization procedure, Hooke's law for the RVE becomes:

$$\sigma + S^D = [C^{\mathrm{e}}]\ [\varepsilon - \varepsilon^{\mathrm{p}}], \tag{25}$$

where C^e is the fourth-order elastic tensor. In the continuum plasticity theory, one would need to develop a phenomenological constitutive law for plastic stress–strain. Here, we avoid this ambiguity using the explicit expressions given by (18) for the plastic strain tensor as computed in the dislocation dynamics.

5.1 *Modifications for Finite Domains*

The solution for the stress field of a dislocation segment (13) is true for a dislocation in an infinite domain and for homogeneous materials. To account for finite domain boundary conditions, Van der Giessen and Needleman (1995) developed a 2D model based on the principle of superposition. The method has been extended by Yasin et al. (2001) and Zbib and Diaz de la Rubia (2002) to three-dimensional problems involving free surfaces and interfaces as summarized below.

5.1.1 Interactions with External Free Surfaces

In the superposition principle, the two solutions from the infinite domain and finite domain are superimposed. We assume that the dislocation loops and any other internal defects with self-induced stress are situated in the finite domain V bounded by the surface S and subjected to arbitrary external tractions and constraints. Then the stress, displacement, and strain fields are given by the superposition of the solutions for the infinite domain and the actual domain subjected to

$$\sigma = \sigma^{\infty} + \sigma^{*}, \quad u = u^{\infty} + u^{*}, \quad \varepsilon = \varepsilon^{\infty} + \varepsilon^{*}, \tag{26}$$

where σ^{∞}, ε^{∞}, and u^{∞} are the fields arising from the internal defects as if they were in an infinite domain, whereas σ^{*}, ε^{*}, and u^{*} are the field solutions corresponding to the auxiliary problem satisfying the following boundary conditions:

$$\begin{aligned} t &= t^{\mathrm{a}} - t^{\infty} \quad \text{on } S \\ u &= u^{\mathrm{a}} \quad \text{on part of the boundary } S, \end{aligned} \tag{27}$$

where t^{a} is the externally applied traction, and t^{∞} is the traction induced on S by the defects (dislocations) in the infinite domain problem. The traction $-t^{\infty} = \sigma\, n$ on the surface boundary S results in an *image stress* field, which is superimposed onto the dislocations segments and, thus, accounts for the surface–dislocation interaction.

5.1.2 Interactions with Interfaces

The framework described above for dislocations in homogenous materials can be implemented into a finite element code. One can also extend the model to the case of dislocations in heterogeneous materials using the concept of superposition as outlined by Zbib and Diaz de la Rubia (2002). For bi-materials, suppose that domain V is divided into two subdomains V_1 and V_2 with domain V_1 containing a set of dislocations. The stress field induced by the dislocations and any externally applied stresses in both domains can be constructed in terms of two solutions such that:

$$\sigma = \sigma^{\infty 1} + \sigma^*, \quad \varepsilon = \varepsilon^{\infty 1} + \varepsilon^*, \tag{28}$$

where $\sigma^{\infty 1}$ and $\varepsilon^{\infty 1}$ are the stress and strain fields, respectively, induced by the dislocations (the infinite solution) with the entire domain V having the same material properties of domain V_1 (*homogenous solution*). Applying Hooke's law for each of the subdomains, and using (18), one obtains the elastic constitutive equations for each of the materials in each of the subdomains as:

$$\begin{aligned} \sigma^* &= \left[C_1^{\mathrm{e}}\right] \varepsilon^*, \quad \text{in } V_1 \\ \sigma^* &= \left[C_2^{\mathrm{e}}\right] \varepsilon^* + \sigma^{\infty 21}; \sigma^{\infty 21} = \left[C_2^{\mathrm{e}} - C_1^{\mathrm{e}}\right] \varepsilon^{\infty 1}, \quad \text{in } V_2, \end{aligned} \tag{29}$$

where C_1^{e} and C_2^{e} are the elastic stiffness tensors in V_1 and V_2, respectively. The boundary conditions are:

$$t = t^{\mathrm{a}} - t_1^{\infty} \text{ on } S; \; u = u^{\mathrm{a}} \quad \text{on part of } S, \tag{30}$$

where t^{a} is the externally applied traction and t_1^{∞} is the traction induced on all of S by the dislocations in V_1 in the infinite-homogenous domain problem. The *eigenstress* $\sigma^{\infty 21}$ is due to the difference in material properties. The method described above can be extended to the case of heterogeneous materials with N subdomains.

The above equations are solved numerically using the finite element method as described by Zbib and Diaz de la Rubia (2002), resulting in a model, which they call multiscale discrete dislocation dynamics plasticity (MDDP). The MDDP consists of two main modules, the DD module and the continuum finite element module. The DD module computes the dynamics of the dislocations, the plastic strain field they produce, and the corresponding internal stresses field. These field values are passed to the continuum finite element module, in which the stress-displacement-temperature field is determined based on the boundary value problem at hand. The resulting stress field, in turn, is passed to the DD module and the cycle is repeated.

We have now described the DD method and its coupling to continuum mechanics concepts. In what follows, we present some examples of the problems that have been treated by DD to provide both solid mechanists and materials scientists with the data and insight that might be needed. In particular, DD (as will be illustrated) is very good at providing constitutive equations or quantification to relate microstructure to mechanical properties (e.g., flow stress). In particular, we treat two examples of size effects in small volumes.

6 Problems with Size Effects and the DD Approach

The discrete dislocation dynamics method has been used by many to investigate complicated small-scale crystal plasticity phenomena that occur under a wide range of loading and boundary conditions (Bulatov et al. 2001). Some of the phenomena that have been addressed include:

- Dislocation mechanisms in strain hardening (Devincre and Kubin 1997; Zbib et al. 2000)
- Dislocation–defect interaction problems; dislocation–void interaction, dislocation–SFT/void-clusters interaction in irradiated materials and the role of dislocation mechanisms in the formation of localized shear bands (Diaz de la Rubia et al. 2000; Khraishi et al. 2000, 2002)
- Size effect in metal–matrix composites (Rhee et al. 1994a, b; Khraishi and Zbib 2002a)
- Size effect in inclusions of different coefficient of thermal expansion from the metallic matrix on the composite strengthening (Khraishi and Shen 2004)
- Surface effects on the plasticity in a small material volume (Khraishi and Zbib 2002b)
- Crack tip plasticity and dislocation–crack interaction (Zbib et al. 2004; Zeng and Hatmaier 2010)
- Dislocation patterns such geometrically necessary boundaries (GNBs) (Khan et al. 2001, 2004)
- Plastic zone and hardening in nanoindentation tests (Zbib and Diaz de la Rubia 2002)
- High strain rate phenomena and shock wave interaction with dislocations (Shehadeh et al. 2005a, b, 2006)
- Size effect and dislocation mechanisms in thin films (Weinberger et al. 2009) and nanolaminate structures (von Blanckenhagen et al. 2003; Akasheh et al. 2007a, b; Nibur et al. 2007)

The DD method is typically used for solving problems where size effects and interfaces are important. Size effects in materials can be addressed by means of a strain-gradient constitutive phenomenological model for material behavior (Zbib and Aifantis 1988a, b, c, d). However, these theories have a number of problems, including identification of boundary conditions, non-uniqueness of geometrically necessary configurations associated with the same Nye's tensor, and statistically stored dislocation configurations with the long range stress. Moreover, the nature of the phenomenological length scale – the value of the length scale in these theories – is not intrinsic and varies from one experiment to the other for the same material. The complexities associated with gradient theories multiply when interface modeling is attempted. Upon interacting with dislocations, interfaces can block, transmit, emit, and/or absorb dislocations. Within the framework of FE Ashmawi and Zikry (2002), developed a set of constitutive equations, based on the relative orientation of slip planes across an interface, that allow for transmission, blockage, or partial transmission across the interface. The DD method is better suited for such problems because, as mentioned above, the constitutive behavior of a RVE, a material element, or a small material volume, is captured naturally within the DD simulations. Moreover, in DD, the evolution of plasticity and the constitutive behavior is *very sensitive* to *any changes* in the *scales* describing the problem at hand, and these changes are directly determined in DD.

The nanoindentation problem discussed in Sect. 2 has been addressed within the DD framework by a number of investigators over the past few years (see, e.g., Zbib and Diaz de la Rubia 2002). Here, we briefly address two problems that have been gained particular interest in the community: (1) size effects in nanolaminate metallic composites and (2) size effects in compression of micropillars.

6.1 Size Effect in Nanolaminate Metallic Composites

6.1.1 Dislocation near an Interface

When considering problems consisting of multiple media with different elastic properties, like is the case of nanolaminate composite,*image forces* on dislocations resulting from differences in elastic moduli becomes important. Here, we first show that the method based on the superposition principle discussed in Sect. 5 can accurately capture the image stress on a dislocation near an interface. Because of the availability of an exact solution, we consider the problem of an edge dislocation situated near a Cu–Ni interface as depicted on Fig. 5. The dislocation is placed in the Cu layer, and the Ni and Cu domains are divided into a finite element mesh as shown in Fig. 6a. The mesh size is made finest around the dislocation core with the smallest element size equal to $2b$ (b is the magnitude of the Burgers vector).

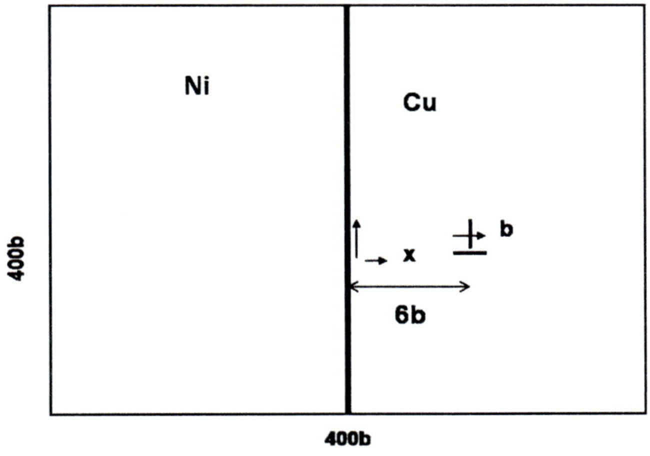

Fig. 5 Edge dislocation near a Cu–Ni interface

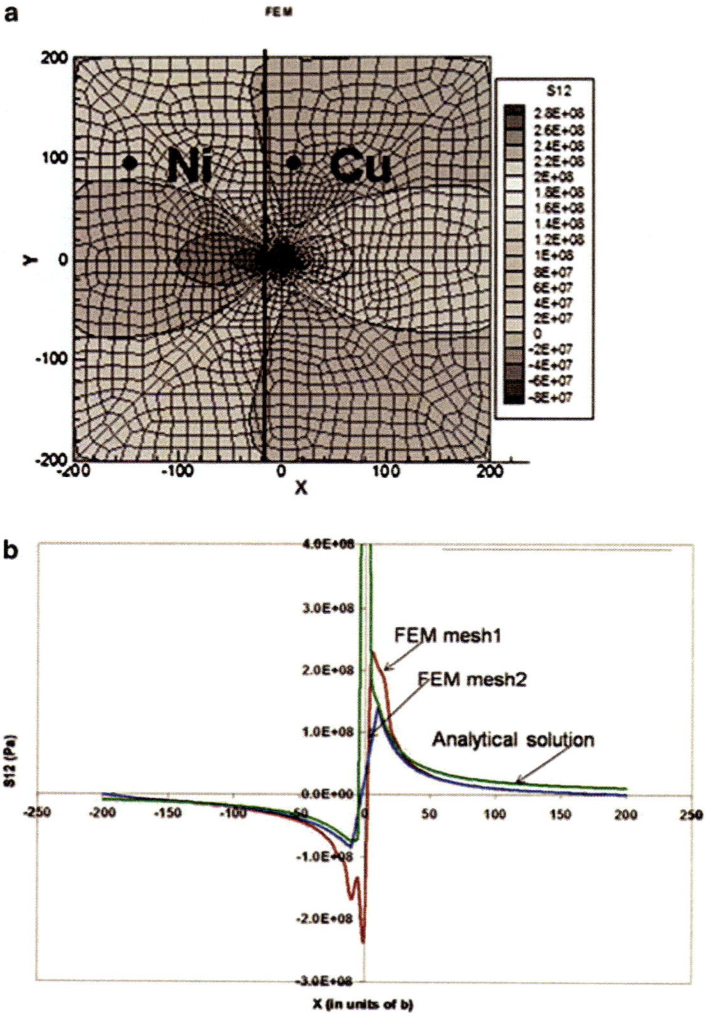

Fig. 6 Distribution of image stress of a dislocation near a Cu–Ni interface, (**a**) stress contour plot, (**b**) stress distribution along a line intersecting the dislocation core

Figure 6a, b shows the distribution of *image* stress that arises from the difference in elastic properties between the layers and compares the numerical results to analytical solution. As it can be deduced from Fig. 6a, as the mesh is refined, the results obtained using the superposition converges to the exact solution. The image stress results in a repulsive Peach–Kohler force which acts on the dislocation. The magnitude of the force is inversely proportional to the distance from the interface as shown in Fig. 7.

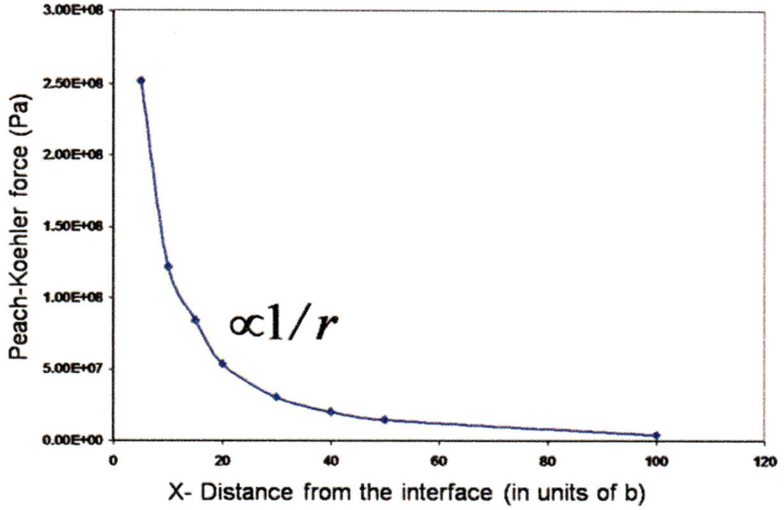

Fig. 7 The Peach Koehler force the image stress exerts on the dislocation core

6.1.2 Strengthening in Nanolaminate Metallic Composites

Next, we consider a three-layer {001} Cu/Ni cube-on-cube system (insert in Fig. 8) to analyze layer-confined slip in a coherent multilayer system with small lattice parameter mismatch and moderate elastic properties difference. We present results for a case with the initial configuration consisting of a single a/2<011> threading dislocation residing on a (111) slip plane in the Cu layer and sandwiched between two Ni layers. The results shown in the Fig. 8 indicate that the effect of the Koehler force (image stress) on the threading stress is significant and increases as the layer thickness decreases. This effect is linearly related to the difference in the elastic moduli between the two materials, and inversely proportional to the layer thickness. This is expected since the main effect of the Koehler force is most significant, as far as the dynamics of the threading dislocation is concerned, at the leading edge of the threading dislocation.

6.2 Size Effect in Micropillars

In an effort to identify the underlying dislocation mechanisms responsible for the macroscopic response of micropillars, two dimensional (2D) discrete dislocation dynamics (DD) simulations have been done on a planar single crystal both under tension and compression (Deshphande et al. 2005). These authors and others (Benzerga and Shaver 2006) showed the size dependence with and without constraining the loading axis. They concluded that the latter case showed more size effect than the former. Although the two dimensional discrete dislocation analyses

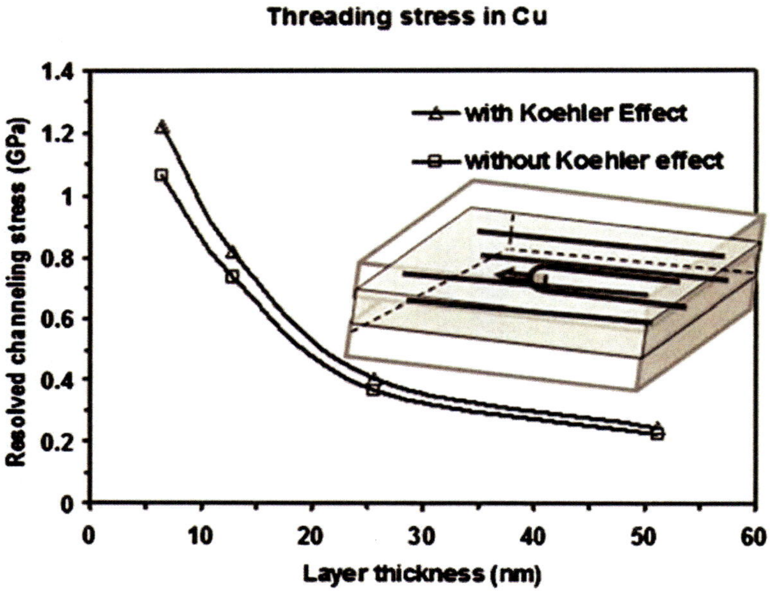

Fig. 8 Effect of Kohler force on flow stress in noanolaminate metallic composites

provided useful insights, they lack many key three-dimensional dislocation interactions. Most recently, other researchers (Tang et al. 2007, 2008; Motz et al. 2009) performed three-dimensional (3D) DD simulations of micropillar compression and showed the size dependence as the specimen size is decreased. However, in all these DD simulations, the stress and strain fields were assumed to be homogeneous and no surface effects were neglected. However, as pointed out in Akarapu et al. (2010) the deformation field in submicron scale specimens can be very heterogeneous and can become localized from the onset of plasticity. Moreover, surface effects in such small dimensions are shown to be important and cannot be neglected. To capture the heterogeneity of the macroscopic deformation and its influence on the microscopic mechanisms, Akarapu et al. (2010) used the MDDP model described in this chapter. The simulations consisted of cuboid-shaped Cu specimens with thickness ranging from 200 to 2,500 nm. Typical results are shown in Figs. 9 and 10. The stress–strain curves for various thicknesses are shown in Fig. 10. The result shows that there is a significant size effect in these micropillars. These results are similar to those reported in Akarapu et al. (2010) where it is shown that the intermittent operation of discrete dislocation arms is the prominent mechanism causing the heterogeneous deformation. One important aspect of the these simulations is the ability of the MDDP model to accurately capture surface effects (image stresses, in this case), which have important implications on the results and therefore, cannot be neglected as has been suggested in recent studies (Tang et al. 2007). To show this effect, Akarapu et al. (2010) considered two sets of simulation: (1) using only dislocation dynamics where no image forces from the free surfaces are accounted for, and (2) where they used

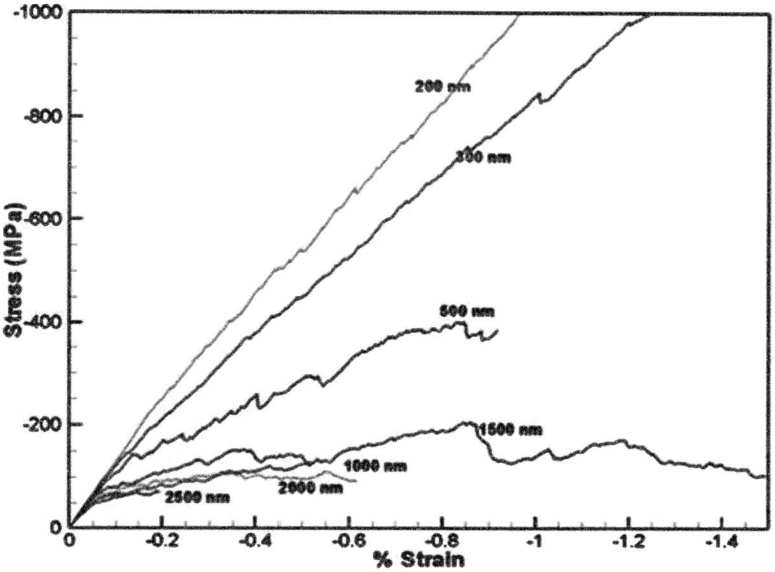

Fig. 9 MDDP results of size effect in micropillars: stress–strain curves

Fig. 10 MDDP simulation of micropillars with initial random distribution of dislocation arms and FR sources; here the initial dislocation density is 10^{14} $1/m^2$. (**a**) Dislocation structure after deformation, (**b**) deformed configuration (the nodal displacement is magnified by a factor of 5 and contour plot of shear strain)

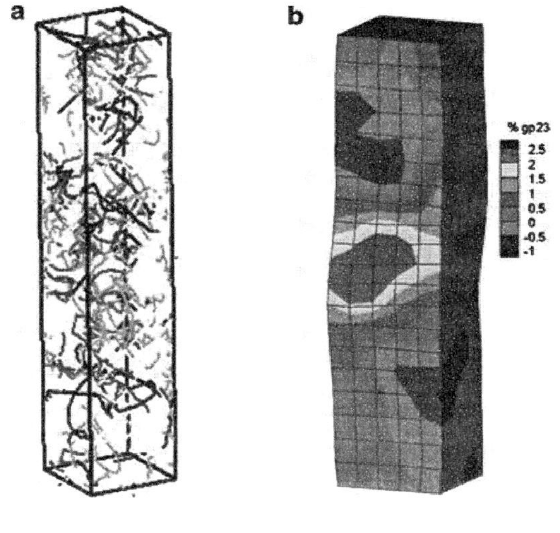

the full MDDP model. The MMDP results showed that because of these conditions and the manner in which the underlying dislocations get activated, the specimen deforms nonuniformly. When using only DD simulations, the box remains undistorted and the stress and strain are uniform throughout the specimen. Furthermore, the predicted stress–strain curves and dislocation density curves for both cases are quite different.

Acknowledgment The support from the Office of Basic Energy Science at the DOE under grant number DE-FG02–07ER46435 is gratefully acknowledged.

References

Akarapu, A., H. M. Zbib and D. F. Bahr (2010). "Analysis of heterogeneous defromation and dislocation dynamics in single crystal micropillars under compression." Int. J. Plast. 26: 239–257.

Akasheh, F., H. M. Zbib, J. P. Hirth, R. G. Hoagland and A. Misra (2007a). "Dislocation dynamics analysis of dislocation intersections in nanoscale multilayer metallic composites." J. Appl. Phys. 101: 084314.

Akasheh, F., H. M. Zbib, J. P. Hirth, R. G. Hoagland and A. Misra (2007b). "Interactions between glide dislocations and parallel interfacial dislocations in nanoscale strained layers." J. Appl. Phys. 102: 034324.

Al'shitz, V. I. (1992). "The phonon-dislocation interaction and its role in dislocation dragging and thermal resistivity." Elastic Strain and Dislocation Mobility (ed. V. L. Indenbom and J. Lother), Chapter 11 Elsevier Science Publishers B.V., Amsterdam.

Ashmawi, W. M. and M. A. Zikry (2002). "Prediction of grain-boundary interfacial mechanisms in polycrystalline materials." J. Eng. Mater. Tech. 124(1): 88–96.

Bahr, D. F. and W. W. Gerberich (1996). "Pile up and plastic zone size around indentations." Met. Mater. Trans. 27A: 3793–3800.

Bahr, D. F., D. E. Kramer and W. W. Gerberich (1998). "Non-linear deformation mechanisms during nanoindentation." Acta Mater. 46: 176–182.

Bahr, D. F., D. E. Wilson and D. A. Crowson (1999). "Energy considerations regarding yield points during indentation." J. Mater. Res. 14: 2269–2275.

Benzerga, A. A. and N. F. Shaver (2006). "Size dependence of mechanical properties of single crystals under uniform deformation." Scripta Mater. 54: 1937.

Bower, A. F., N. A. Fleck, A. Needleman and N. Ogboma (1993). "Indentation of a power law creeping solid." Proc. R. Soc. 441A: 97–124.

Bulatov, V., M. Tang and H. M. Zbib (2001). "Crystal plasticity from dislocation dynamics." MRS Bull. 26 (191–195).

Bull, S. J., T. F. Page and E. H. Yoffe (1989). "An explanation of the indentation size effect in ceramics." Philos. Mag. Lett. 59: 281–288.

Canova, G. R., Y. Brechet and L. P. Kubin (1992). "3D dislocation simulation of plastic instabilities by work?softening in alloys." Modelling of Plastic Deformation and Its Engineering Applications (ed. S.I. Anderson et al), Riso National Laboratory, Roskilde, Denmark.

Chang, S. C. and H. C. Chen (1995). "The determination of F.C.C. crystal orientation by indentation." Acta Metall. Mater. 43: 2501–2505.

Choi, Y., K. J. V. Vliet, J. Li and S. Suresh (2003). "Size effects on the onset of plastic deformation during nanoindentation of thin films and patterned lines." J. Appl. Phys. 94: 6050–6058.

Corcoran, S. G., R. J. Colton, E. T. Lilleodden and W. W. Gerberich (1997). "Anomalous plastic deformation at surfaces: Nanoindentation of gold single crystals" Phys. Rev. B 55: 16057–16060.

Cotrell, A. H. (1953). Dislocations and Plastic Flow in Crystals. Oxford, Oxford Press.

Deshphande, V. S., A. Needleman and E. Van der Giessen (2005). "Plasticity size effects in tension and compression of single crystals." J. Mech. Phys. Solids 53: 2661.

Devincre, B. and L. P. Kubin (1997). Mater. Sci. Eng. A234–236: 8.

DeWit, R. (1960). "The continuum theory of stationary dislocations." Solid State Phys 10: 249–292.

Diaz de la Rubia, T., H. M. Zbib, M. Victoria, A. Wright, T. Khraishi and M. Caturla (2000). "Flow localization in irradiated materials: a multiscale modeling approach." Nature 406: 871–874.

Dimiduk, D. M., M. D. Uchic and T. A. Parthasarathy (2005). "Size-affected single slip behavior of pure nickel microcrystals." Acta Mater. 53: 4065.

Doerner, M. F. and W. D. Nix (1986). "A method for interpreting the data from depth-sensing indentation instruments." J. Mater. Res. 1: 601–609.
Dugdale, D. S. (1954). "Cone indentation experiments." J. Mech. Phys. Solids 2: 265–277.
Elmustafa, A. A., J. A. Eastman, M. N. Ritter, J. R. Weertman and D. S. Stone (2000). "Indentation size effect: large grained aluminum versus nanocrystalline aluminum-zirconium alloys." Scripta Mater. 43: 951–955.
Field, J.S., Swain, M.V. (1993). "Simple predictive model for spherical indentation" J. Mater. Res. 8:297–306
Fischer-Cripps, A. C. (2004). Nanoindentation, Spriner, New York.
Fischer-Cripps, A. C. (2006). "Critical review of analysis and interpretation of nanoindentation test data." Surf. Coatings Tech 200: 4153–4165.
Fleck, N. A., G. M. Muller, M. F. Ashby and J. W. Hutchinson (1994). "Strain gradient plasticity: theory and experiment." Acta Metall. Mater. 42: 475–487.
Frick, C. P., B. G. Clark, S. Orso, A. S. Schneider and E. Artz (2008). "Size effect on strength and strain hardening of small-scale [111] nickel compression pillars." Mater. Sci. Eng. A.
Gaillard, Y., C. Tromas and J. Woirgrad (2006). "Quantitative analysis of dislocation pile-ups nucleated during nanoindentation in MgO." Acta Mater. 54: 1409–1417.
Gane, N. and F. P. Bowden (1968). "Microdeformation of solids." J. Appl. Phys. 39: 1432–1435.
Gerberich, W. W., N. I. Tymiak, J. C. Grunlan, M. F. Horstemeyer and M. I. Baskes (2002). "Interpretations of indentation size effects." J. Appl. Phys. 69: 433–442.
Greer, J. R. and W. D. Nix (2006). "Nanoscale gold pillars strengthened through dislocation starvation." Phys. Rev. B 73: 245210.
Harvey, S., H. Huang, Vannkataraman, and W. W. Gerberich (1993). "Microscopy and microindentation mechanics of single crystal Fe-3 wt. %Si: Part I. Atomic force microscopy of a small indentation." J. Mater. Res. 8: 1291–1299.
Hiratani, M. and H. M. Zbib (2002). "Stochastic dislocation dynamics for dislocation-defects interaction." J. Eng. Mater. Technol. 124: 335–341.
Hirth, J. P. (1992). Injection of Dislocations into Strained Multilayer Structures. Semiconductors and Semimetals, Academic, New York. 37: 267–292.
Hirth, J. P. and J. Lothe (1982). Theory of Dislocations. New York, Wiley.
Hirth, J. P., M. Rhee and H. M. Zbib (1996). "Modeling of deformation by a 3d simulation of multipole, curved dislocations." J. Comput. Aided Mater Des. 3: 164–166.
Hirth, J. P., H. M. Zbib and J. Lothe (1998). "Forces on high velocity dislocations." Model. Simul. Mater. Sci. Eng.. 6: 165–169.
Horstemeyer, M. F., M. I. Baskes and S. J. Plimpton (2001). Acta Mater. 49: 4363.
Hoyt, S. L. (1924). "The ball indentation hardness test." Trans. Am. Soc. Steel Treating 6: 396.
Huang, H., N. Ghoniem, T. Diaz de la Rubia, Rhee, Z. H.M. and J. P. Hirth (1999). "Development of physical rules for short range interactions in BCC crystals." ASME-JEMT 121: 143–150.
Johnson, K. L. (1970). "The correlation of indentation experiments." J. Mech. Phys. Solids 18: 115–126.
Johnson, K. L. (1985). Contact Mechanics, Cambridge University Press, Cambridge.
Kelchner, C. L., S. J. Plimpton and J. C. Hamilton (1998). "Dislocation nucleation and defect structure during surface indentation" Phys. Rev. B 58: 11085–11088.
Khan, A., H. M. Zbib and D. A. Hughes (2001). Stress Patterns of Deformation Induced Planar Dislocation Boundaries. MRS, San Francisco.
Khan, A., H. M. Zbib and D. A. Hughes (2004). "Modeling planar dislocation boundaries using a multi-scale approach." Int. J. Plast. 20: 1059–1092.
Khraishi, T. A. and H. M. Zbib (2002a). "Dislocation dynamics simulations of the interaction between a short rigid fiber and a glide dislocation pile-up." Comput. Mater. Sci. 24: 310–322.
Khraishi, T. A. and H. M. Zbib (2002b). "Free-surface effects in 3d dislocation dynamics: formulation and modeling." J. Eng. Mater. Tech. 124: 342–351.
Khraishi, T., H. M. Zbib, T. Diaz de la Rubia and M. Victoria (2000). "Modeling of flow localization and hardening in irradiated materials using discrete dislocation dynamics (DD)." Acta. Metall.

Khraishi, T., H. M. Zbib, T. Diaz de la Rubia and M. Victoria (2002). "Localized deformation and hardening in irradiated metals: three-dimensional discrete dislocation dynamics simulations." Metall. Mater. Trans. 33B: 285–296.
Khraishi, T. A. and Y.-L. Shen (2004). Int. J. Plast. 20: 1039.
Kiely, J. D. and J. E. Houston (1998). "Nanomechanical properties of Au(111), (001), and (110) surfaces." Phys. Rev. B 57.
Kiener, D., W. Grosinger, G. Dehm and R. Pippan (2008). "A further step towards an understanding of size-dependent crystal plasticity: In situ tension experiments of miniaturized single-crystal copper samples." Acta Mater. 56: 580.
King, R. B. (1987). "Elastic analysis of some punch problems for a layered medium." Int. J. Solids Struct. 23: 1657–1664.
Kramer, D., H. Huang, M. Kriese, J. Robach, J. Nelson, A. Wright, D. F. Bahr and W. W. Gerberich (1998). "Yield strength predictions from plastic zone around nanocontacts." Acta Mater. 47: 333–343.
Lepinoux, J. and L. P. Kubin (1987). "The dynamic organization of dislocation structures: a simulation." Scripta Metall. 21: 833–838.
Li, J., J. W. Morris, Jennerjohn, D. F. Bahr and Levin "in press." J. Mater. Res.
Li, X. and B. Bhushan (2002). "A review of nanoindentation continuous stiffness measurement technique and its applications." Mater. Char. 48: 11–36.
Ma, Q. and D. R. Clarke (1995). "Size dependent hardness of silver single crystals " J. Mater. Res. 10: 853–863.
Mason, W. and D. MacDonald (1971). "Damping of dislocations in niobium by phonon viscosity." J. Appl. Phys. 42: 1836.
Meyer, E. (1908). Ziets d. Vereines Deustscher Ingenieure 52: 645.
Michalske, T. A. and J. E. Houston (1998). "Dislocation nucleation at nano-scale mechanical contacts." Acta Mater. 46: 391–396.
Minor, A. M., E. T. Lilleodden, E. A. Stach and J. W. Morris (2004). "Direct observations of incipient plasticity during nanoindentation of Al." J. Mater. Res. 19: 176–182.
Minor, A. M., S. A. Syed-Asif, Z. Shan, E. A. Stach, E. Cyrankowski, T. J. Wyrobek and O. L. Warren (2006). "A new view of the onset of plasticity during the nanoindentation of aluminium." Nat. Mater. 5.
Misra, A., Hirth, J.P., Hoagland, R.G. (2005). "Length scale dependent deformation mechanisms in incoherent metallic multilayered composites." Acta Mater. 53:4817–4824.
Motz, C., D. Weygand, J. Senger and P. Gumbsch (2009). "Initial dislocation structures in 3-D discrete dislocation dynamics and their influence on microscale plasticity." Acta Mater. 57: 1744–1754.
Ngan, A. H. W. and P. C. Wo (2006). "Delayed plasticity in nanoindentation of annealed crystals." Phil. Mag. 86: 1287–1304.
Nibur, K. A., F. Akasheh and D. F. Bahr (2007). "Analysis of dislocation mechanisms around indentations through slip step observations." J. Mater. Res. 42: 889–900.
Nibur, K. A. and D. F. Bahr (2003). "Identifying slip systems around indentations in FCC metals." Scripta Mater. 49: 1055–1060.
Nix, W. D. and H. Gao (1998). "Indentation size effects in crystalline materials: a law for strain gradient plasticity." J. Mech. Phys. Solids 46: 411–425.
Oliver, W. C. and G. M. Pharr (1992). "An improved technique for determining hardness and elastic modulus using load and displacement sensing indentation experiments." J. Mater. Res. 7: 1564–1583.
Oliver, W. C. and G. M. Pharr (2004). "Review: Measurement of hardness and elastic modulus by instrumented indentation: Advances in understanding and refinements to methodology." J. Mater. Res. 19: 3–20.
Page, T. F., W. C. Oliver and C. J. McHargue (1992). "Deformation behavior of ceramic crystals subjected to very low load (nano)indentations." J. Mater. Res. 7: 450–473.
Pethica, J. B. and D. Tabor (1979). "Contact of characterized metal surfaces at very low loads: Deformation and adhesion." Surf. Sci. 89: 182–190.

Raabe, D. (1998). "Introduction of a hybrid model for the discrete 3d simulation of dislocation dynamics." Comput. Mater. Sci. 11: 1–15.
Rhee, M., J. P. Hirth and H. M. Zbib (1994a). "On the bowed out tilt wall model of flow stress and size effects in metal matrix composites." Scripta Metall. Mater. 31: 1321–1324.
Rhee, M., J. P. Hirth and H. M. Zbib (1994b). "A superdislocation model for the strengthening of metal matrix composites and the initiation and propagation of shear bands." Acta Metall. Mater. 42: 2645–2655.
Rhee, M., H. M. Zbib, J. P. Hirth, H. Huang and T. D. d. L. Rubia (1998). "Models for long/short range interactions in 3D dislocatoin simulation." Model. Simul. Mater. Sci. Eng. 6: 467–492.
Samuels, L. E. and T. O. Mulheam (1957). "An experimental investigation of the deformed zone associated with indentation hardness impressions " J. Mech. Phys. Solids 5: 125–134.
Sayed-Asif, S. A. and J. B. Pethica (1997). "Nanoindentation creep of single-crystal tungsten and gallium arsenide." Philos. Mag. A 76(6): 1105–1118.
Schuh, C. A., J. K. Mason and A. C. Lund (2005). "Quantitative insight into dislocation nucleation from high-temperature nanoindentation experiments." Nat. Mater. 4: 617–621.
Shaw, Z. N., R. K. Mishra, S. A. Syed-Asif, O. L. Warren and A. M. Minor (2008). "Mechanical annealing and source-limited deformation in submicrometre-diameter Ni crystals." Nature 7: 115.
Shehadeh, M., E. M. Bringa, H. M. Zbib, J. M. McNaney and B. A. Remington (2006). "Simulation of shock-induced plasticity including homogeneous and Heterogeneous dislocation nucleation." Appl. Phys. Lett. 89: 171918.
Shehadeh, M. A., H. M. Zbib and T. D. de la Rubia (2005a). "Modeling the dynamic deformation and patterning in FCC single crystals at high strain rates: dislocation dynamic plasticity analysis" Philos. Mag. A 85 1667–1684.
Shehadeh, M. A., H. M. Zbib and T. D. de la Rubia (2005b). "Multiscale dislocation dynamics simulations of shock compressions in copper single crystal." Int. J. Plast. 21: 2396–2390.
Sneddon (1965). "The relation between load and penetration in the axisymmetric Boussinesq problem for a punch of Arbitrary Profile." Int. J. Eng. Sci. 3: 47–56.
Stelmashenko, N.A., Walls, M.G., Brown, L.M., Milman, Y.V. (1993) "Microindentations on W and Mo oriented single crystals: An STM study." Acta Metall. Mater. 41:2855–2865.
Stolken, J. and A. G. Evans (1998). "A microbend test method for measuring the plasticity length scale." Acta Mater. 46: 5109–5115.
Strader, J. H., S. Shim, B. H., W. C. Oliver and G. M. Pharr (2006). "An experimental evaluation of the constant β relating the contact stiffness to the contact area in nanoindentation." Philos. Mag. 86: 5285–5298.
Tabor, D. (1951). The Hardnessof Metals, Oxford Press, Oxford.
Tang, H., K. W. Schwartz and H. D. Espinosa (2007). "Dislocation escape related size effects in single-crystal micropillars under uniaxial compression." Acta Mater. 55: 1607.
Tang, H., K. W. Schwartz and H. D. Espinosa (2008). "Dislocation-source shutdown and the plastic behavior of single-crystal micropillars." Phys. Rev. Lett. 100: 185503.
Tromas, C., Girard, J.C., Woigard, J. (2000). "Study by atomic force microscopy of elementary deformation mechanisms involved in low load indentations in MgO crystals." Philos. Mag. A 80:2325–2335.
Tromas, C., J. C. Girard, V. Audurier and J. Woirgrad (1999). "Study of the low stress plasticity in single-crystal MgO by nanoindentation and atomic force microscope." J. Mater. Sci. 34: 5337–5342.
Uchic, M. D., D. M. Dimiduk, J. N. Florando and W. D. Vix (2004). "Sample dimensions influence strength and crystal plasticity." Science 305: 986–989.
Urabe, N. and J. Weertman (1975). "Dislocation mobility in potassium and iron single crystals." Mater. Sci. Eng. 18: 41.
Van der Giessen, E. and A. Needleman (1995). "Discrete dislocation plasticity: a simple planar model." Mater. Sci. Eng. 3: 689–735.
VanLandingham, M. R. (2003). "Review of instrumented indentation." J. Res. Natl. Inst. Stand. Technol. 108: 249–265.

Volkert, C. A. and E. T. Lilleodden (2006). "Size effects in the deformation of sub-micron Au columns." Philos. Mag. 86: 5567–5579.
von Blanckenhagen, B., P. Gumbsch and E. Artz (2003). "Dislocation sources and the flow stress of polycrystalline thin metal films." Philos. Mag. Lett. 83: 1–8.
Wasserbäch, W. (1986). "Plastic deformation and dislocation arrangemnet of Nb-34 at. % TA alloy crystals,." Philos. Mag. A 53: 335–356.
Weinberger, C. R., S. Aubry, S. W. Lee, W. D. Nix and W. Cai (2009). "Modelling dislocations in a free-standing thin film." Model. Simul. Mater. Sci. Eng. 17: 1–26.
Yasin, H., H. M. Zbib and M. A. Khaleel (2001). "Size and boundary effects in discrete dislocation dynamics: coupling with continuum finite element." Mater. Sci. Eng. A309–310: 294–299.
Yip, S., Ed. (2005). Handbook of Materials Modeling. Springer, New York.
Zbib, H. M. and E. C. Aifantis (1988a). "A gradient-dependent model for the portevin-le chatelier effect." Scripta Metall 22(8): 1331–1336.
Zbib, H. M. and E. C. Aifantis (1988b). "On the localization and post localization behavior of plastic deformation-i. on the initiation of shear bands." Res. Mech., Int. J. Struct. Mech. Mater. Sci. 23: 261–277.
Zbib, H. M. and E. C. Aifantis (1988c). "On the localization and post localization behavior of plastic deformation-ii. on the evolution and thickness of shear bands." Res. Mech., Int. J. Struct. Mech. Mater. Sci. 23: 279–292.
Zbib, H. M. and E. C. Aifantis (1988d). "On the structure and width of shear bands." Scripta Metall 22(5): 703–708.
Zbib, H. M., S. Akarupa, F. Akasheh, C. Overman and D. F. Bahr (2009). "Deformation and size effects in small scale structures." Plasticity 2009: Macro to nano Scale Inelastic behavior of Materials: Plasticity, Fatigue and Fracture St. Thomas.
Zbib, H. M., T. D. de La Rubia, M. Rhee and J. P. Hirth (2000). "3D Dislocation dynamics: stress-strain behavior and hardening mechanisms in FCC and BCC metals." J. Nucl. Mater. 276: 154–165.
Zbib, H. M. and T. Diaz de la Rubia (2002). "A multiscale model of plasticity." Int. J. Plast. 18(9): 1133–1163.
Zbib, H. M., T. Diaz de la Rubia and V. A. Bulatov (2002). "A multiscale model of plasticity based on discrete dislocation dynamics." ASME J. Eng.. Mater. Tech. 124: 78–87.
Zbib, H. M. and T. A. Diaz de la Rubia (2001). "Multiscale Model of Plasticity: Patterning and Localization." Material Science For the 21st Century, The Society of Materials Science, Japan Vol A: 341–347.
Zbib, H. M., M. Hiratani and M. Shehadeh (2004). "Multiscale discrete dislocation dynamics plasticity. continuum scale simulation of engineering materials fundamentals - microstructures - process applications." D. Raabe, D. Roters, F. Baralt and L.-Q. Chen, Wiley-VCH, Weinheim: 202–229.
Zbib, H. M., M. Rhee and J. P. Hirth (1996). "3D simulation of curved dislocations: discretization and long range interactions." Advances in Engineering Plasticity and its Applications (eds. T. Abe and T. Tsuta) Pergamon, NY: 15–20.
Zbib, H. M., M. Rhee and J. P. Hirth (1998). "On plastic deformation and the dynamcis of 3d dislocations." Int. J. Mech. Sci. 40: 113–127.
Zeng, X. H. and H. Hatmaier (2010). "Modeling size effects on fracture toughness by dislocation dynamics." Acta Mater. 58: 301–310.

Emerging Methods for Matching Simulation and Experimental Scales

Andrew H. Rosenberger

Abstract Structures are on the scale of meters, yet the material deforms on the scale of the microstructure – micrometers or smaller. This chapter examines the experimental methods that are emerging at the different length scales that are important tools in building models for location-specific design. Current design practice is discussed to provide a baseline understanding of today's design methodology. Many of the design decisions are based on the stress or constitutive response of the structure to loading. This has been sufficient, but the idea behind location-specific design is that the material at each location can be tailored to the design requirements. Location-specific design will request a variation of material at each location, wherein each material will have a distinct constitutive response and specific damage accumulation. While special experimental techniques are needed to probe the material at finer scales to assess the local behaviors, testing methods at all scales are discussed to demonstrate the breadth of experimental capability available at each scale of the material. The chapter concludes with the thoughts on the experimental capabilities and material understanding that are still elusive and need to be developed.

Keywords Design methodology · Digital image correlation · Experiments · Material behavior · Micropillar · Microsample · Test techniques

1 Current Design Practice

The current design methodology used for safety and fracture critical components is largely based on experience and experimental mechanical property data. The geometry and loading conditions are analyzed using, typically, finite element tools

---

A.H. Rosenberger (✉)
Materials & Manufacturing Directorate, Air Force Research Laboratory,
Wright-Patterson Air Force Base, OH, USA
e-mail: Andrew.Rosenberger@wpafb.af.mil

S. Ghosh and D. Dimiduk (eds.), *Computational Methods for Microstructure-Property Relationships*, DOI 10.1007/978-1-4419-0643-4_16,

like ANSYS (http://www.ansys.com/default.asp) or ABAQUS (http://www.simulia.com/products/abaqus_fea.html) and continuum-based material property models. The input material property data form the basis of the structural stress analysis that ranges from a simple linear elastic analysis to a substantially more complex nonlinear analysis (see an example of the range of modeling options at ANSYS material models; http://www.ansys.com/solutions/solid-mechanics-materials.asp). Constitutive models start with Hooke's law and add complexity including multilinear and nonlinear behaviors, rate effects, hardening, creep, and viscoelasticity. Most models are isotropic, but anisotropic behavior can be incorporated. Cracks can be added using finite element or boundary element-based tools that can compute the stress intensity factors and shape of cracks in complex geometric structures subject to complex, multiaxial loading. In the early 2000s, these model approaches were 2D, but now fully 3D methods are available, e.g., FRANC3D (http://www.cfg.cornell.edu/index.htm) and ZenCrack (http://www.zentech.co.uk/zencrack_overview_what.htm).

In general, the goal of the current design modeling approach is to compute the maximum stress or strain in the component for the presumed loading. This would also be the maximum stress or strain range for cyclic loading. Using this information, the design engineer will see how this value compares to the design requirements. For example, if the structure is creep limited, the design engineer might compare this maximum stress with an allowable stress from a Larsen–Miller curve. The Larsen–Miller parameter (LMP) is calculated as:

$$\mathrm{LMP} = T[C + \log(t)], \tag{1}$$

where $\boldsymbol{C}$ is a material constant (usually 20), $\boldsymbol{t}$ is the time to failure or a prescribed level of strain (in hours), and $\boldsymbol{T}$ is the temperature in Kelvin. Using the temperature at the critical location, the required life, and the LMP material curve, the design engineer can then determine whether the calculated stress is safe or not. If the stress is too high, the design engineer will have to modify the design to reduce the stress, e.g., increase the section size or increase the notch radius. Ultimately all life prediction paths follow this approach.

If fatigue failure is the limit state, the stress range and stress ratio will be compared to a fatigue allowable curve for the appropriate material and temperature. One may fit the fatigue behavior using the Manson–Coffin relationship,

$$\Delta\varepsilon_{\mathrm{p}} = C(N_{\mathrm{f}})^{n}, \tag{2}$$

where $\Delta\varepsilon_{\mathrm{p}}$ is the plastic strain range, $\boldsymbol{C}$ is the Manson–Coffin coefficient, $\boldsymbol{N}_{\mathrm{f}}$ is the number of cycles to failure, and $\boldsymbol{n}$ is the Manson–Coffin exponent. Figure 1 shows a typical fatigue curve for Ti-6Al-4V at room temperature (Gallagher et al. 2004). This curve shows mean behavior as well as the $\pm 75\%$ and $\pm 99\%$ probability of failure curves indicating the degree of confidence in the fatigue results. Usually the design engineer will not use a mean fatigue curve but a minimum fatigue curve of -3σ or similar. The Manson–Coffin low-cycle fatigue law can also be used in

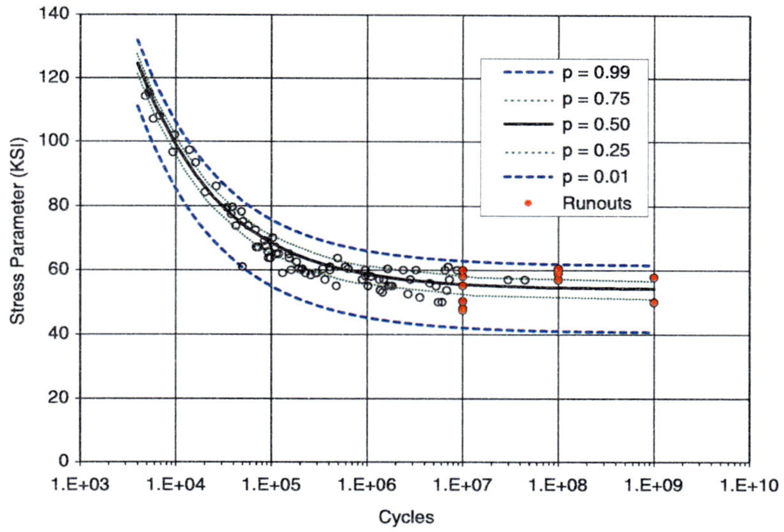

Fig. 1 Example of stress-life curve for Ti-6Al-4V showing ±75% and ±99% probability of failure curves for the data, from Gallagher et al. (2004)

high cycle fatigue like that in Fig. 1 using the elastic strain range or stress range divided by the modulus of elasticity. Again, the design requires the cyclic strain or stress range be known to assess the design.

Many aerospace structures require a damage tolerance assessment to ensure that, for example, a rogue material or process defect will not initiate a crack and grow to failure within the useful life of the component or safety inspection interval. This requires a crack growth assessment where the critical locations of a design are identified and the crack propagation life is calculated. The driving force for the crack is based on the stress range at the critical location and the crack size using,

$$\Delta K = \beta \Delta \sigma \sqrt{\pi a} \tag{3}$$

where $\Delta \boldsymbol{K}$ is the stress intensity factor range, β is the geometry factor that takes into account the crack shape and local geometry, $\Delta\sigma$ is the cyclic stress range, and $\boldsymbol{a}$ is the crack length. The geometry factor β for a simple crack can be found in handbooks, but cracking in complex geometries requires the use of more advanced, analytical tools such as FRANC3D (http://www.cfg.cornell.edu/index.htm) and ZenCrack (http://www.zentech.co.uk/zencrack_overview_what.htm) to compute the $\boldsymbol{K}$ solutions. The damage tolerant life (crack growth life) uses these component $\boldsymbol{K}$ solutions and is based on an integration of the crack growth rate ($\mathrm{d}\boldsymbol{a}/\mathrm{d}\boldsymbol{N}$) data. These data are generated in the laboratory on standard fracture mechanics specimens and fit to a crack growth law such as the simple Paris law (Paris 1962):

$$\frac{\mathrm{d}a}{\mathrm{d}N} = C(\Delta K)^n, \tag{4}$$

where $\boldsymbol{C}$ and $\boldsymbol{n}$ are the Paris law coefficient and exponent, or the Sinh law (Wallace et al. 1976):

$$\log\left(\frac{\mathrm{d}a}{\mathrm{d}N}\right) = C_1 \sinh\{C_2[\log(\Delta K) + C_3]\} + C_4, \tag{5}$$

where $\boldsymbol{C_1}$–$\boldsymbol{C_4}$ are material, loading, and environment-dependent constants. Equations 4 and 5 are then integrated from an assumed initial crack size to a final size that would be subcritical. If this cyclic life is insufficient, the design engineer would modify the geometry to reduce the stress intensity factor typically by reducing the stress.

These design decisions all depend on determining the local stress–strain response of the material to calculate the driving force for failure. The design engineer's primary response to the failure to meet the design intent (or the design allowable) is to change the geometry and reduce the stress or strain at the critical location in the component. These decisions are driven by the assessments of the material's damage progression in terms of its creep, fatigue, and fracture characteristics. The reaction of an individual material to applied loads is represented mathematically using constitutive equations. The deformation response of solid-state materials is complex, considering the influence of temperature and deformation rate, such that a single equation or set of equations can rarely be formulated to capture the range of behaviors. Idealized equations such as the generalized Hooke's law for elastic materials,

$$\sigma_{ij} = C_{ijrs}\varepsilon_{rs} \tag{6}$$

or the Prandtl and Reuss (1930) plasticity equation,

$$\mathrm{d}\varepsilon_{ij}^{\mathrm{p}} = \sigma_{ij}'\mathrm{d}\lambda \tag{7}$$

where for the Mises yield condition,

$$\mathrm{d}\lambda = \frac{1}{k}\sqrt{\Pi_{\mathrm{d}\varepsilon\mathrm{p}}}, \quad \text{where} \quad \Pi_{\mathrm{d}\varepsilon\mathrm{p}} = \frac{1}{2}\mathrm{d}\varepsilon_{ij}^{\mathrm{p}}\mathrm{d}\varepsilon_{ij}^{\mathrm{p}} \quad (\text{for } \mathrm{d}\varepsilon_{kk}^{\mathrm{p}} = 0) \tag{8}$$

represent only the most simple of continuum behaviors. Beyond this, the formulation of the constitutive behavior should move beyond a continuum, mean-field approach to predict local behavior. While a detailed discussion of the formulation of constitutive equations for the range of behaviors including anisotropy, viscoelasticity, viscoplasticity, etc. is beyond the scope of this chapter (see chapters titled "Representation of Materials Constitutive Responses in Finite Element-Based Design Codes," "Accounting for Microstructure in Large Deformation Models of Polycrystalline Metallic Materials," "Dislocation Mediated Continuum Plasticity: Case Studies on Modeling Scale Dependence, Scale-Invariance, and Directionality of Sharp Yield-Point," "Dislocation-Mediated Time-Dependent Deformation in Crystalline Solids"), it is nonetheless important to accurately represent the local stress–strain behavior. Methods to determine the constitutive behavior and damage progression of materials are discussed in the next section.

2 Physical Constitutive Behavior

The constitutive equations for a material are calibrated using careful experimental techniques that start simple and add complexity in loading and state of stress as required by the design application and the model formulation. This section will only address the experimental approaches and not the myriad of mathematical forms that are used to describe the behavior. The constitutive behavior of a material is typically determined by first conducting tests in simple uniaxial tension or tension/compression conducted at constant temperature. The specimen is subject to an axial force/force rate or strain/strain rate to assess the response of a relatively large volume of the material. A uniform state of stress is assumed over the volume (for isotropic materials) and a one-dimensional state of stress and a two-dimensional state of strain results.

The typical experimental apparatus is designed to apply a known, uniform stress, strain, and temperature over the gage length of a suitable specimen. The most simple and common test machine applies a uniaxial monotonic or cyclic load to a specimen. As shown in Fig. 2, the test system must: (1) grip the specimen for the application of the applied load or strain, (2) measure the dynamic force in the specimen, (3) apply the force or strain to the sample in a controlled manner, (4) measure the strain or displacement over a portion of the uniform gage length, and (5) heat (or cool) the sample uniformly. All test systems have these essential elements with differences in their actual construction, capabilities, and accuracy. A typical "button head" test sample is shown in Fig. 3 having a gauge length of $\boldsymbol{L}$ and a gauge diameter of $\boldsymbol{D}_{\text{gauge}}$. The load is applied through the large diameter ends of the specimen using a special specimen holder that maintains rigid alignment and allows both tensile

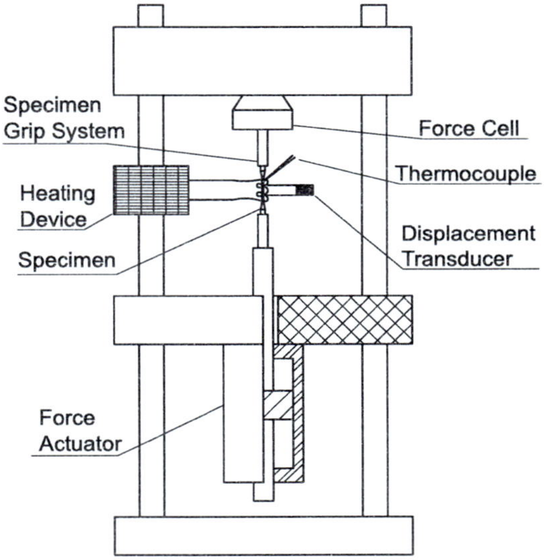

Fig. 2 Schematic of a uniaxial, tension/compression test system showing different components

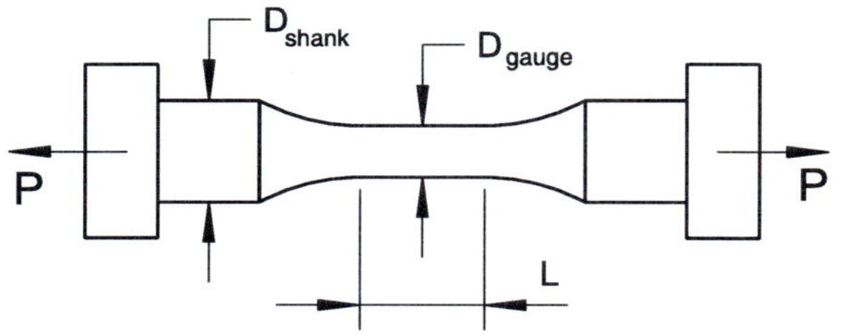

Fig. 3 Schematic of a uniaxial, tension/compression test specimen

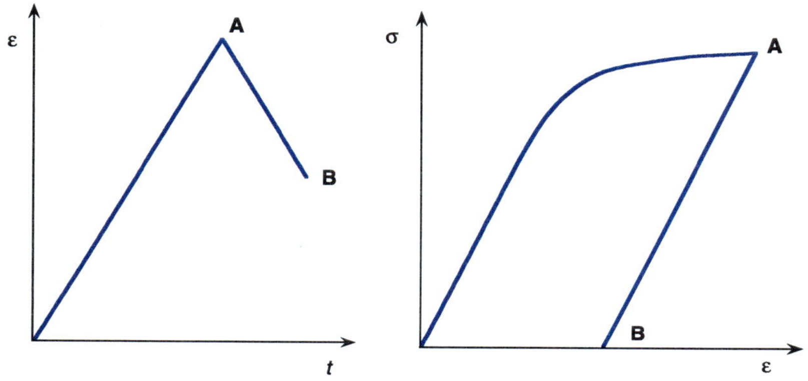

Fig. 4 Strain-controlled hardening test in tension; control input on left, resultant stress–strain response on the right

and compressive loading. Threaded and pin-loaded end conditions are also used. The gauge length and diameter are chosen to match the material availability and are selected based on the loading condition (compressive loading requires shorter, fatter specimens to resist buckling). Typical specimen diameters are 5 mm at minimum, though smaller and larger specimens are used as required by the component design system.

The most common test is a hardening test (or tension test) in tension or compression. This test is conducted in strain control (usually) and measures the monotonic deformation of the material – measuring the material modulus, yield point (usually at a particular offset strain, e.g., 0.2%), and flow or hardening behavior. Figure 4 shows the input control parameters and the resultant output. In this case, the test is controlled in strain and the resultant stress–strain response is measured.

The time-dependent behavior of a material is characterized by a creep test in tension or compression. This is a stress- or load-controlled test wherein the strain–time response is measured. Figure 5 shows the input control parameters and the resultant output. If the strain is measured after the removal of the load, point B, the strain decreases and some of the time-dependent deformation is recovered.

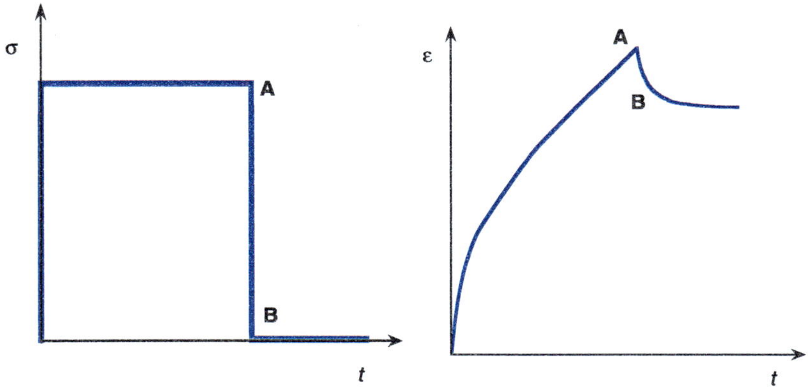

Fig. 5 Stress-controlled creep test in tension; control input on left, resultant strain-time response on the right. There is partial recovery of the strain after the removal of the stress, point B

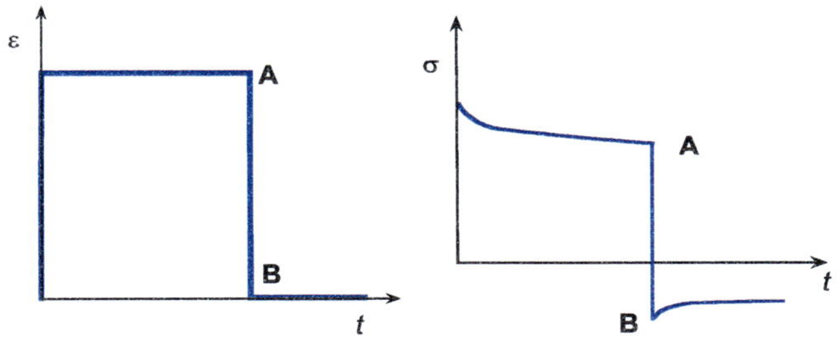

Fig. 6 Stress relaxation test in tension; control input on left, resultant stress-time response on the right

Similar to a creep test, a relaxation test subjects a test specimen to a constant strain and monitors the stress relaxation as a function of time (Fig. 6). This has been shown to be directly related to the steady-state creep behavior of a material within some limits of imposed strain (Woodford et al. 1992).

The cyclic behavior of a material is characterized by subjecting a specimen to a controlled variation of stress or strain. Figure 7 shows a strain-controlled test for a material exhibiting strain hardening (the stress at point A′ is greater than at A). Generally, the cyclic stress–strain stabilizes after a certain number of cycles but will show continued changes after many cycles, if damage, e.g., cracking or voiding, occurs.

It is cogent to discuss realistic loading here as it relates to stress determination as well as the ultimate failure prediction. Simply put, the load spectrum that is seen by a structure is usually not simple. Aircraft and automotive structures, for example, rarely see simple loading due to wind gusts and pot holes. In real structural applications, the simple fatigue loading in Fig. 7 would have various subcycles and

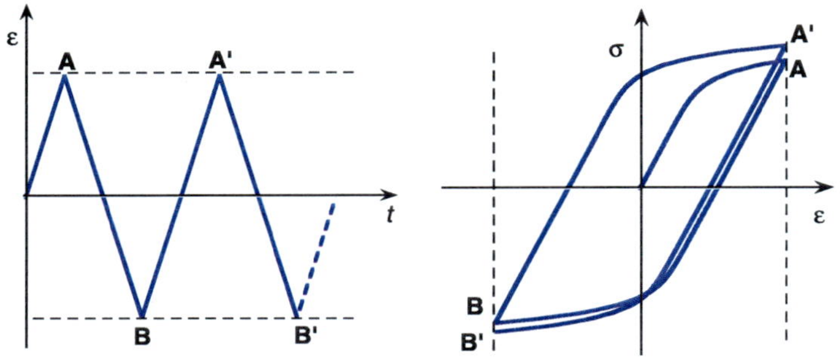

Fig. 7 Strain-controlled fatigue test; control input on left, resultant stress–strain response on the right

dwells that would substantially change the stress–strain response. This could result in significant shifts in the mean stress or strain that could change the damage mode from, for example, a pure fatigue failure to one that includes static modes (creep). The level of complexity in loading has to be balanced with the computational capabilities that are available for the analysis.

The above characterizations are 1D and are appropriate for isotropic and anisotropic materials. The principal directions of the material should generally align with the loading axis for the easiest interpretation of the results. Multiaxial tests can be conducted on special machines (e.g., biaxial or tension-torsion) but attention must be paid to anisotropic materials. For example, torsion testing of a anisotropic material in the form of a tubular specimen will not result in uniform shear along the circumference of the sample.

The tension, creep, and cyclic tests can also provide information on the fracture or damage accumulation of the material. For example, the number of load cycles to failure during a cyclic test (Fig. 7) can be used to generate a strain–life curve for a material. The width of the hysteresis loop at half-life would be the $\Delta\varepsilon_p$ for the Manson–Coffin fit (2). This information is critical to defining the durability of a component or structure and also provides a materials engineer with important information to design a better material or process for producing a material.

2.1 *Limitations of the Current Practices*

Current modeling approaches depend on material property data that are derived from the continuum scale and are easy for the computational tools to use. The property data are the average response of the gauge volume of the test specimen, and any damage accumulation will only be representative of that volume of material. Errors in the material response or damage accumulation can arise by choosing a specimen size that is too large or small. For example, a powder metallurgy component usually has included pores and/or nonmetallic particles from powder processing that are relatively rare. If the component to be designed is large and the fatigue data were generated on only small specimens, the true nature of the pores or foreign particles on fatigue failure may not be represented in the total volume of the test sample program (Jha et al. 2008). Alternatively, large complex forgings may have variations in the material property due to differences in forging strain or heat treatment response. Samples that are too large may average the material's response that could hide local property variations that are critical for highly stressed locations (e.g., bolt holes and fillets) in the finished component.

Many material properties as measured today have scatter in their experimental results. Much of this scatter is hidden in the original equipment manufacturers proprietary databases, but some data do exist in the open literature. The creep behavior does have considerable scatter. Igarashi et al. (2009) found creep rupture scatter up to 9× in low C 2.25Cr-1.6W-W-Nb pipe steel, though "average" scatter is typically 4–5×. Smaller sample sets by Dlouhý et al. (2009) on a TiAl alloy (Ti–46Al–2W–0.5Si–0.7B at%) and Liaw et al. (1989) on Grade 2-1/4Cr-1Mo steel showed a maximum variation in creep rupture life of 2×. The variation in minimum creep rate was approximately the same as the variation in time to creep rupture for these studies.

Fatigue behavior also has considerable scatter that varies for different materials (Cashman 2010). Casellas et al. (2005) studied the fatigue variability in a cast Al–Si (ANSI A360) alloy at room temperature and found that the variability ranged from 2.7X to 7.2X (calculated as the ratio of the longest life over the shortest life for a single fatigue condition) based on the processing route. Several materials were studied under various conditions by Jha et al. (2007). They found that the fatigue scatter in a nickel-base superalloy ranged from 4.5X to 200X at 650°C. The higher scatter was found at lower applied stress range which tends to reflect the change in slope of the stress–life curve as is apparent in Fig. 1. The scatter in $\alpha + \beta$ processed Ti-6246 was found to increase from 1.4X at high stresses near yield to 300X at stresses near the fatigue limit. The extreme in fatigue scatter was found in a titanium aluminide intermetallic at 593°C (Jha et al. 2005). This material has an extremely flat stress–life curve, and the scatter in fatigue life was found to be nearly 4,000X! Some of the scatter in fatigue could be caused by grain size variations and their extreme statistics but a fatigue variability study on the single crystal turbine blade material, PWA 1484 (Morrissey et al. 2009), found scatter of 280X at 593°C. It is clear that there is a need for quantitative methods to study the effects of microstructure and defect (extreme value) interactions within the mean-field microstructure.

Current design methods do not well manage the variability in material properties and tend to hide the causes in the fits such as equations 1, 2, 4, and 5. The simplicity of the fitting methods is not amenable to incorporating the details of the microstructure. Therefore, these models are not very useful in location-specific design. Finally, the current modeling and property representations do not offer insight into the causes of failure – tensile, creep, or fatigue. They offer no insight for the material producer or the design engineer to help produce a better material or a more suitable application of a material, respectively.

3 Need for New Design Tools

The design tools of today have enabled clear advances in technology, developed new products and devices, and helped to expand the realm of the possible. Since the development of the finite element method in the 1960s (Huebner et al. 2001), the method has become more powerful in terms of model complexity, size, range of loading conditions, and material responses. Yet, the mechanisms of failure are not integrated in the analysis methods, and usually only simple representations of the material response are used in the design process. Because of this, the design engineer still must use the minimum material properties which will generally result in overly conservative designs.

It is known that material properties depend on the material processing route. Of course, the processing affects the structure of the material that controls the mechanical behavior. A search of the current literature will show that there is a wealth of information linking processing and properties in a range of materials for a range of different applications (Sharma et al. 2009; Ahmad et al. 2009; Guitar et al. 2009). The design engineer has to pick a single material and process for component design – if the data are available. The variations in process-induced total strain or heat treatment response, for example, are not taken into account. The design optimization is hindered by the singular material property in the component – the minimum properties set the design. The cost of this approach is that a complex forging, i.e., bulkhead for a fighter aircraft, which does not require the same high material performance at all locations, is overly conservative in some locations. Critical locations such as attachment points and highly stressed ribs require higher strength and fatigue performance than other locations that simply impart stiffness to the structure. Another example of a simple structure in a complex environment is a high pressure turbine disk in a modern gas turbine engine. The outer diameter of the disk, the "rim," is close to the high temperature working fluid and therefore sees higher temperatures where creep resistance is important. The complex attachment geometry for the turbine blade has high contact stresses, so fretting and wear are also important at the rim location. The inner diameter, the "bore," is cooler and supports the centrifugal loading of the disk rim and pull of the blades. Therefore, it must be strong and resist low cycle fatigue damage accumulation. When a single material and microstructure are used in this disk, there is compromise due to the divergent requirements. For example, considering only the influence of grain size as a variable, the creep resistance improves with an increase in the grain size, but the strength decreases as the grain size is increased.

The design engineer would have considerably more flexibility if location-specific properties could be specified for the component. This could be achieved by localized heat treatment (Gayda et al. 2004), one of the additive manufacturing techniques (Dinda et al. 2009), surface enhancement through application of advantageous residual stress treatments (Prevéy and Cammett 2004), or microstructural refinement (Ma et al. 2008). Another benefit derived by the incorporation of location-specific properties in the design is that the microstructure in a finished part may not be the optimum because the part has a complex geometry or is difficult to process. If the design engineer understands the local material limitations, the design could work around this and result in a more robust component. The design engineer would have additional flexibility – in addition to changing the geometry of the design to reduce stress, local material properties could be specified at the critical location to reduce creep or cracking, for example.

Certainly, some of the variability in material behavior discussed above is due to variation in material. For example, some of the variability in the creep response of cast single crystal turbine airfoil materials is due to the crystal orientation of the test bar. These bars are typically cast to an orientation tolerance, i.e., $\pm 10°$, and not tracked further. By better understanding the relationship between specimen orientation and the resultant creep response, a better turbine blade design could result through specifying tighter blade orientation tolerance or taking the variation in crystal orientation into account for the complex geometric design.

Another strong example where microstructure plays an important role in material behavior has been recently highlighted by Telesman et al. (2008). The dwell crack growth behavior in an advanced nickel-base superalloy has been tied to the size of tertiary gamma prime. The larger the tertiary gamma prime size (at approximately constant gamma prime volume fraction), the slower the dwell crack growth rate. This is significant for two reasons: (1) by accounting for the size of the tertiary gamma prime size, the 20X variation in dwell crack growth rate can be understood, and (2) the size of the tertiary gamma prime is a result of subtle heat treatment steps that could be easily overlooked. Furthermore, the tertiary gamma prime is very small and must be characterized by TEM – not a normal tool used in material process development or verification.

An example of the need for exacting characterization of the mechanical behavior of a material is to improve the design practice for welded joints. These joints have a distinct variation in microstructure from the base material to the weld material and the heat affected zone (HAZ) in between the two. Molak et al. (2009) developed a tiny, 5.5-mm-long, dog-bone specimen for use in characterizing these three zones in 316L stainless steel plate. These small samples offered a means to determine the location-specific yield strength, ultimate strength, and elongation to failure. They found that the best properties were in the HAZ (339 and 580 MPa yield and ultimate strength), and the worst properties were found in the weld material (291 and 526 MPa yield and ultimate strength). Typically, the properties of a weld are qualitatively assessed with micro-hardness profiles in a transverse section, but this new method offers quantitative results. With quantitative data, more durable welded structures can be designed.

These are just a few examples where careful experimental characterization of the microstructure and the local mechanical behavior is critical to optimize the design process. Bringing the material into the engineering design task would have been unthinkable 10 years ago, but the improvement in computational efficiency now makes it possible to insert location-specific material properties into FE models. To do this, of course, one needs to assess the material behavior at a local level. That is the subject of the next section.

4 Multiscale Testing Methodology: From Coarse to Fine Scales

4.1 Advances and Needs

The other chapters of this book address how the models and analysis will be linking the length scales, determining the de-correlation length scales, etc. It might be misunderstood that an integrated computational materials engineering (ICME) (NMAB 2008) approach may not need experimentation. In fact, experimentation is even more important in ICME. In order to develop multiscale materials models, experiments are required at various length scales. The experimental approach is not to develop so-called design data, but to conduct experiments at the appropriate length scale to develop the material behavior models. In the end, the design engineer will not compare his critical stress to a minimum design allowable for a property assigned to the whole part, but to the validated property model that will set the allowable based on the location-specific material behavior and associated statistics. The experimental efforts (testing) will offer key insight into model formulation and validate the modeling approaches. As discussed by Dimiduk in chapter titled "Microstructure–Property–Design Relationships in the Simulation Era: An Introduction", it is not always clear what belongs in each level (scale) of analysis – careful experimentation has to be conducted at all scales of the material continuum. The experiments help to identify the roles of the different microstructural features and where they fit into the multiscale materials models. This chapter is focused on mechanical characterization which contrasts with the number of purely materials characterization tools discussed in this book (chapters titled "Serial Sectioning Methods for Generating 3D Characterization Data of Grain- and Precipitate-Scale Microstructures," "Digital Representation of Materials Grain Structure," "Multiscale Characterization and Domain Partitioning for Multiscale Analysis of Heterogeneous Materials," "Coupling Microstructure Characterization with Microstructure Evolution"). To demonstrate how mechanical and microstructural characterization can work together and that the size of a mechanical test specimen is important, consider the early fatigue work of Gabb et al. (1986) on nickel superalloy single crystal Rene N4. The material was characterized as having a pore density of 0.38% volume fraction with a length of $9.9 \pm 6.7\,\mu m$. These measurements were based on over 50 micrographs and represented the "wide pore size distribution." However, the paper shows a fracture surface containing a pore having a maximum length of $\sim 63\,\mu m$! This is especially important since pores were found to be one of the main locations for fatigue crack initiation. This shows that some of the rare (rogue) materials statistics cannot easily be determined metallographically, and that large-scale mechanical tests will always have a place even in ICME.

The tests need to be scaled to the model development – hierarchical test methods are needed to match the scale of the simulation. From a constitutive point of view, the measurements discussed earlier do span the scale of the microstructure. However, there are changes in the behavior of materials when the specimen size is changed. This is largely due to either a change in the state of stress – plane strain to plane stress – or a minimum sampling of the material, i.e., changing from polycrystalline behavior to monocrystalline behavior.

4.2 Advanced Test Techniques at the Conventional Scale

Conventional tests are important and stronger today than in the past. The mechanical test systems are now typically computer–controlled, which promotes better data collection and more precise force or displacement control during the test. The test specimens are also usually fabricated by dedicated vendors so that precision is improved and surface finish (both RMS roughness and residual stress) is better controlled and reproducible. More care is taken to align the test machine, and special alignment devices are in greater use (MTS Service Notes; http://www.mts.com/stellent/groups/public/documents/library/dev_002347.pdf). There are advances in testing that largely stem from two areas – measurement and load simulation.

4.2.1 Measurements

Typically, mechanical tests use average displacements over the gauge length of the specimen to assess the average deformation of the material. To develop a basic crystal plasticity model, for a TiAl material, the deformation of specially oriented, large single crystal samples was used to calibrate and validate the model (Brockman 2003). This approach is sound but costly – special specimens have to be prepared and tested individually. This approach also does not offer any insight into slip transfer between the individual grains or crystals since the material is used in a polycrystalline form, not in a single crystal. One powerful addition to conventional testing that can help to understand the material behavior is to measure the deformation field during the test in two- or three-dimensions. Using digital image correlation (DIC), the full field deformation can be measured in 2D using a single camera (Johnson 2004; Chiang 2009) or in 3D using a stereo pair of cameras (Avril et al. 2008; Tiwari et al. 2009). A comprehensive review of the current state of the art of DIC was recently published by Sutton et al. (2009). The displacement field is measured either directly on the material by tracking the movement of surface features (i.e., topography, grain structure, and microstructure) or by tracking the movement of a speckle pattern or grid that is applied and sized to compliment the measurements of the material behavior. The benefit of these approaches is that deformation inhomogeneities can be measured and incorporated in the development of models. The slip transfer in lamellar TiAl has been studied by Johnson (2004) to assess the applicability of continuum-based crystal plasticity at the local level. Under high strain conditions, Avril et al. (2008) have explored the development of Lüders bands in steel. Single point displacements cannot show this level of deformation in detail, which is important for model development. The power of DIC can be enhanced for the development of multiscale materials models by combining the material deformation field with the details of the microstructure and its crystal orientation (Héripré et al. 2007). In this paper, Héripré et al. apply DIC with a map of the grain structure and orientation determined using orientation imaging microscopy to calibrate and validate a local crystal plasticity model for zirconium and TiAl alloys. They constructed a finite element model of the actual 2D grain structure extruded in the 3D. The constitutive model was then tuned using the results from the DIC to activate the correct deformation systems in each grain at the right stress/strain condition.

The application of these optical techniques to high temperatures (Lyons et al. 1996) is difficult due to turbulence of heated air, oxidation, or modification of the specimen surface and background radiation. Some success has been made using optical methods at temperatures up to 650°C (Lyons et al. 1996), but laser speckle techniques are showing promise for even higher temperatures, up to 1,200°C (Anwander et al. 2000). These high temperature techniques are not commercially available yet, but should be coming soon.

4.2.2 Load Simulation

A number of specimen designs have been developed that can simulate a load environment other than uniform stress (Mayer et al. 1995, Touratier et al. 2009). These techniques strongly tie modeling to the experiments, as the interpretation of the results is not intuitively as clear as for the uniform stress case. The specimens shown in Fig. 8 have been used to study single crystal turbine blade materials – specifically

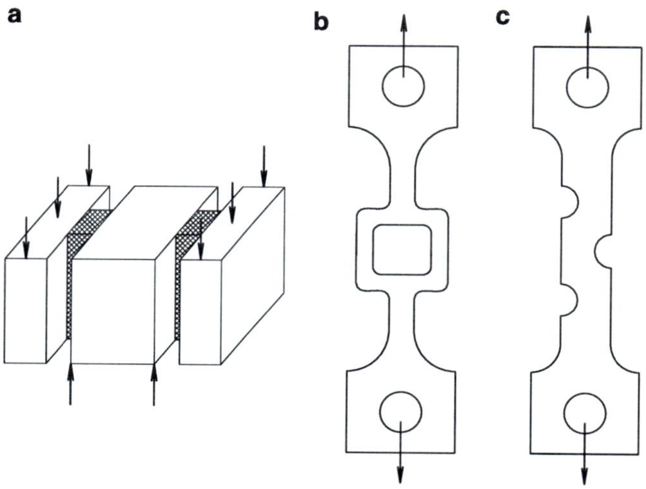

Fig. 8 Specimens that offer assessment of material behavior under special loading conditions; (**a**) double shear creep specimen promoting pure shear in shaded volumes, after Mayer et al. (1995), (**b**) hollow notch specimen, and (***c***) asymmetric notched specimen promoting mixed states of stress, after Touratier et al. (2009)

to examine the deformation and rafting behavior. Imagine the insight into material behavior that can be obtained by merging special test geometry with a full field DIC technique using the actual material structure and orientation! There are clear advances in understanding material behavior that can be made using conventional-sized specimens.

4.3 *Advanced Test Techniques using Subscale Specimens*

Incorporating location-specific properties into the design methodology has several experimental data requirements that are important for validation of the new model-based approach to design. Experimentally, it is important to assess the material behavior at the level of the constituents – at the very fine scale. It is also equally important to roll-up these behaviors into computationally efficient "reduced order" models to enable their use in the design environment. It is this latter function where subscale experimental techniques have the greatest application. By reducing the physical size of the tested volume, the location-specific behavior of a graded microstructure, for example, can be measured. Subscale specimens have a gage volume 20–50X less than a conventional specimen – on the order of 5–20 mm^3. These tests can be conducted in scaled down versions of larger tests using similar test protocol though measurement precision and rig alignment become more important as the size of the gauge length decreases. These tests can also be conducted on loading stages in scanning electron microscopes (SEM) where limitations in size and load capacity usually dictate subscale geometry.

Many techniques have been developed by the nuclear materials industry, as the effects of irradiation on the behavior of materials is complicated and creates a hazardous testing situation where minimizing the amount of irradiated material is seen as a good thing (Lucas 1983; Chuto et al. 2002; Kim et al. 2007; Dai et al. 2006). Tension, fracture, creep, and fatigue tests have all been miniaturized for the nuclear industry and have found widespread use. This is clearly the interrogation of a location-specific property as only the gauge section of the test specimen is irradiated. Generally these subscale techniques have found reasonable agreement for these nuclear industry materials, with properties determined using more conventionally sized specimens.

A significant size effect – when at least one dimension of the specimen is subscale – has been discovered when testing of materials at high temperatures (Cassenti and Staroselsky 2009; Hüttner et al. 2009). These studies are extremely important for the durable design of thin-walled cast single crystal turbine airfoils. Cassenti and Staroselsky (2009) proposed a dislocation-based model to explain the thin-wall debit. By considering the interaction of dislocations with the surface of the sample and the generation of intrusions and extrusions, they were able to demonstrate a cause of the thickness effect on creep. Obviously there could also be a surface oxidation effect as the size of the specimen decreases – as the ratio of the surface area to volume of the gauge section increases. The samples in both of these studies were conventional in length and width but were subscale in thickness (0.2–0.3 mm).

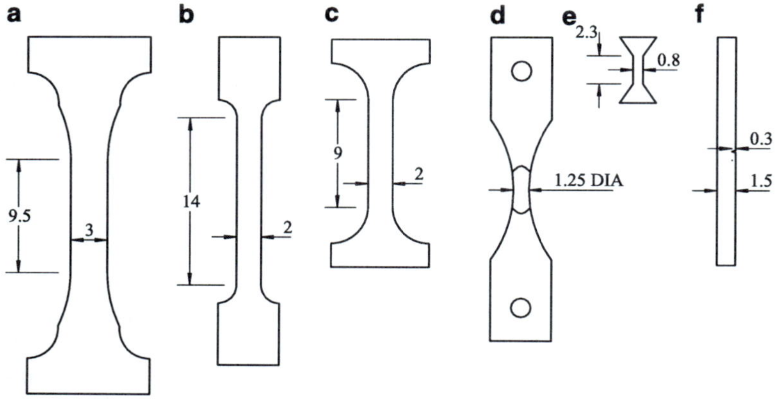

Fig. 9 Selected subscale test specimens (drawn to scale); (**a**) single crystal specimen loaded by two pins at the shoulders suitable for creep and fatigue testing (Tschopp and Rosenberger 2009), (**b**) weld zone specimen loaded by two pins at the shoulders suitable for tension testing (Kim et al. 2009), (**c**) single crystal specimen loaded by shaped grip at shoulders suitable for creep testing (Mälzer et al. 2007), (**d**) mini-sized hourglass-type steel samples (round gage section) suitable for push–pull fatigue testing (Kim et al. 2007), (**e**) weld zone specimen loaded by shaped grip at shoulders suitable for tension testing (Molak et al. 2009), and (**f**) irradiated Charpy test specimen (Chuto et al. 2002) (notch depth is 0.3 mm). All dimensions in millimeter

Several small-scale specimens shown in Fig. 9 have been used to examine different materials and different mechanical properties. Specimen A (Tschopp and Rosenberger 2009) is small to enable the extraction of specimens from the wall of a cast single crystal turbine airfoil to assess the cast-to-size behavior as compared to the typical thick casting that is used to develop data for blade designs. Specimens B (Kim et al. 2009) and E (Molak et al. 2009) are specifically sized so that they can be machined to contain a weld zone or heat-effected zone to assess the location-specific properties. Specimen C (Mälzer et al. 2007) is small due to size limitations in the starting material and to enable testing in the longitudinal and long transverse directions. Specimen D (Kim et al. 2007) assessed the influence of surface finish (machining) on the fatigue behavior of a fusion reactor steel. It is sized so that it could be tested inside a "hot" reactor chamber. Specimen F is a small Charpy specimen for a vanadium reactor alloy. It is small to assess the toughness of the material in an irradiated form – to minimize the radioactive waste from tested specimens.

The gripping and application of load to subscale specimens are more difficult than conventionally sized sampled and require greater precision. The majority of the specimens in Fig. 9 are loaded in tension only through two pins or specially machined "capture" grips that load the shoulder of the specimen. This is usually a robust gripping method and can be accomplished with great precision if the grips and specimens are machined with precision and the rig is well aligned. The specimens of Tschopp and Rosenberger (2009) and Kim et al. (2007) are the only specimens suitable for fatigue loading as they have a large radius going into the gauge section of the specimen. Kim et al. claim that they can conduct tension-compression tests with their specimen (Fig. 9d) without buckling (or backlash?) but do not provide details of the grip arrangement. Usually a pin-loaded specimen is not used in through-zero loading conditions and is not stable in compressive loading; therefore, a more complex loading system must have been used.

Tension, creep, and some fatigue tests require that the strain in the gauge length is measured. This too is more difficult with subscale specimens. Various methods have been used in the samples in Fig. 9. Specimens A (Tschopp and Rosenberger 2009) and B (Kim et al. 2009) use relatively standard alumina rod or clip on extensometers, respectively, having a suitable gauge length for the geometry. Sample C uses an alumina tube-in-tube extensometer that measures the displacement of the high temperature grips. This is not ideal and required an analytical procedure to correct for the displacement (seating) of the specimen in the grips. The fatigue tests on sample D (Kim et al. 2007) were conducted in fully reversed total strain control using a laser diametrical extensometer. Deformation in the tiny weld-zone specimens (E) (Molak et al. 2009) was measured using DIC. It appears that a full field measurement was not obtained (though it could have been), but only the displacement of reference lines was tracked.

Loading and measurement of subscale specimens are also possible in an SEM. Some of the earliest work in this area were conducted by Davidson and Lankford (1986) who measured the displacement fields around small cracks during loading. These aluminum specimens were fatigued outside of the SEM to initiate cracks and then transferred to a loading stage in an SEM for study. They found that strain fields surrounding a small crack are not the same as those surrounding a long crack, indicating that indeed the crack growth kinetics should be different. Chiang (2009) demonstrated very high displacement resolution in a number of materials using a loading stage in an SEM and nanoscale digital speckle photography. For example, it was found that the local strain in lamellar Ti-Al depended on the colony orientation and the orientation of its neighbors.

4.4 Advanced Tests at the Grain Level

In the last 10 years, there has been a substantial increase in test techniques that probe material behaviors at the individual grain level or at the micron-scale (Zupan et al. 2001; Zupan and Hemker 2001; Sharpe et al. 2001; Uchic and Dimiduk 2005). The listed references are only a small subset of the overall "small" test techniques. This section will deal with techniques that can test bulk materials at the fine scale but will not cover the testing of special MEMS materials (Sharpe et al. 2001; Sharpe 2008) that are typically amorphous or polycrystalline silicon. There are also techniques to determine the yield and flow properties of materials using standard nanoindentation tests, but these techniques require analytical treatment of the complex stress state under the indenter. These properties may be influenced by this analytical treatment, and this section will only consider simple, uniaxial states of stress such that

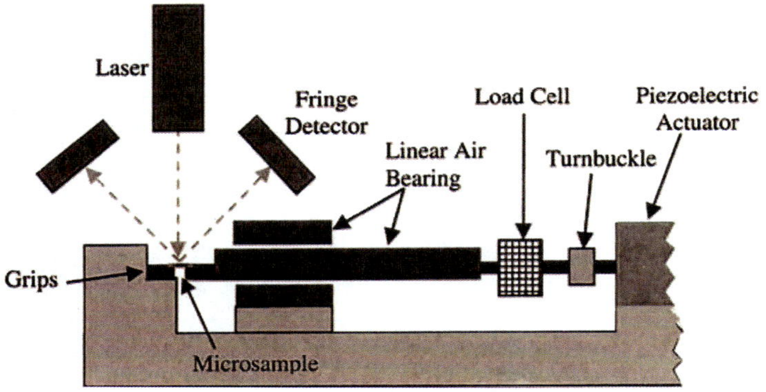

Fig. 10 Schematic of the high temperature microsample test system (Zupan et al. 2001)

the interpretation of the results is more straight forward for the smallest of specimens. The fine-scale tests can be split into techniques that are applicable both inside (Sharpe et al. 2001; Uchic and Dimiduk 2005) and outside the SEM (Zupan et al. 2001; Zupan and Hemker 2001; Sharpe et al. 2001; Uchic et al. 2004; Uchic and Dimiduk 2005). Obviously, the details of the deformation can be more readily observed in situ, but this is not necessary. Some of the earliest techniques were developed by Sharpe and Fowler (1993) who used a tiny dog-bone specimen having an overall length of 3 mm with nominal gauge dimensions of 25–500 μm thick and 200–300 μm wide with a gauge length of 1–1.8 mm. The loading stage made use of a linear air bearing to minimize friction and a piezoelectric actuator for precise displacement control. A schematic of this type of test system is shown in Fig. 10. Feedback via a controller could apply constant load for creep or variable load for fatigue – only a positive stress ratio could be applied. A modified load frame was developed using a voice coil for load application to assess the high cycle fatigue of LIGA Nickel (Aktaa et al. 2005) with success. This technique has been applied to numerous materials at temperatures up to 1, 200°C (Zupan et al. 2001; Zupan and Hemker 2001). Heating is accomplished using direct current methods. Deformation in this case was initially measured via laser interference using two gold lines sputtered on the sample (Sharpe and Fowler 1993). This offered very high resolution at a small gauge length but gave an average measurement of the deformation between the two sputtered lines. Currently, deformation is commonly measured using DIC using optical or SEM imaging (Lewis et al. 2008; Hemker et al. 2008), which can identify local deformation and help to understand the details of the plasticity, etc. Overall, this form of small-scale testing has been successfully applied to a range of materials under a number of different loading conditions (tensile, creep, and fatigue).

Additional test techniques have been developed that probe even smaller volumes of materials such that the amount of deformed material is approaching the deformation scale of the material, i.e., the dislocation cell substructure. This capability is necessary to access how the microconstituents of the material affect the mechanical

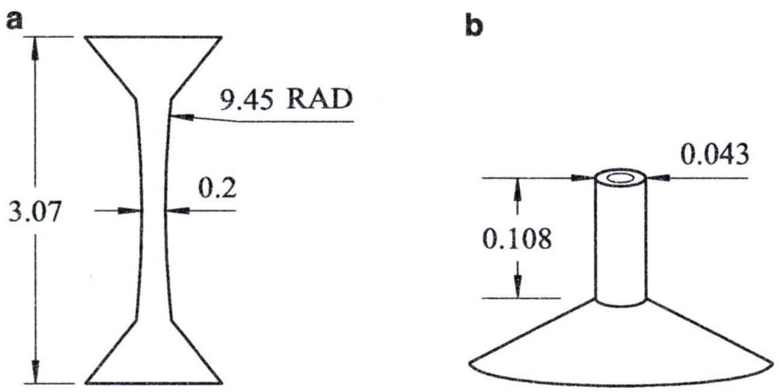

Fig. 11 Scale drawings of small specimens; (**a**) microsample geometry of Sharpe and Fowler (1993) loaded by the tapered ends, and (**b**) the largest micropillar sample of Uchic and Dimiduk (2005) loaded in compression from the top. Dimensions are in millimeter and schematic B is drawn 10× relative to schematic A

behavior. For example, the γ' in nickel-base super alloys has three morphologies (primary, secondary, and tertiary) where the volume fraction and individual size may have an impact on the behavior (Telesman et al. 2008). Figure 11 compares the microscale geometries of Sharpe and Fowler (1993) and Uchic and Dimiduk (2005) (the scales are different). Note the substantially smaller size of these specimens to those in Fig. 9.

So-called single pillar experiments are growing in popularity and have been used to examine the intricacies of deformation at a very small scale for numerous materials (Uchic and Dimiduk 2005; Kiener et al. 2009; Schuster et al. 2007). The technique is best described by Uchic and Dimiduk (2005) and Uchic et al. (2009) wherein a right circular pillar (having a diameter ranging from 0.5 to 43 μm) is machined from a conventionally processed bulk material using focused ion beam (Ga+) milling. The pillar is tested in compression using a conventional nanoindentation system fitted with a flattened diamond tip. Various pillar length to diameter ratios have been used, ranging from 2 to 3. Clearly, column buckling is to be avoided, but higher ratios tend to minimize the stress triaxiality at the ends of the specimen. Loading is applied using a voice coil operating in displacement control. However, for rapid dislocation motion, the feedback is not fast enough to fully prevent the inherent force control of the system. A series of SEM images of a test and the resultant stress strain curve are shown in Fig. 12.

To ensure validity of the results, the angular alignment of the pillar with the indenter is critical. The flat diamond tip is significantly larger than the test specimen, but collinear application of the loading is necessary – even a dust particle can cause eccentricity of loading. There have been questions with regard to the friction between the loading diamond tip and the pillar and the fact that the base of the pillar is the same material as the pillar so that the deformation is exaggerated due to a punching of the pillar into the substrate, but these have been found to be not substantial by Kiener et al. (2009). The effect of lateral constraint was found to be important

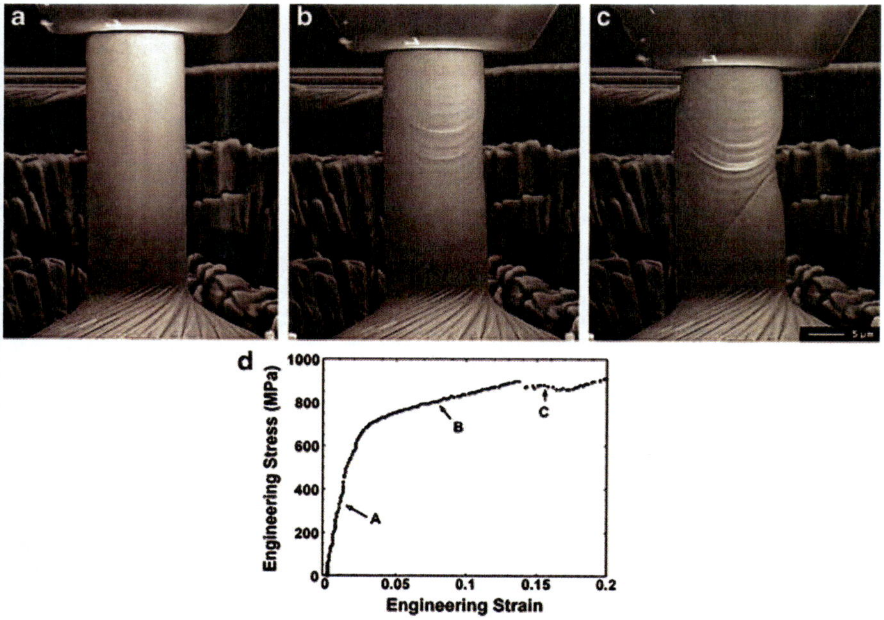

Fig. 12 **(a)–(c)** A series of SEM images collected during a microsample compression test. Strain levels for the three images are: **(a)** 1%, **(b)** 8%, and **(c)** 15%. **(d)** Stress–strain curve from the same test, where the displacement data have been calculated directly from the SEM images. The labels on the stress–strain curve correspond to images **(a)–(c)** (Uchic et al. 2006)

in these tests especially for single crystal specimens oriented for single slip (Shade et al. 2009). This work found that the lateral constraint had a pronounced effect on the strain hardening and other aspects of the stress–strain response. There are also concerns about the surface damage from the Ga+ ions, but the damaged layer was found to have a maximum depth of 50 nm in Cu (Kiener et al. 2007). The depth and severity to the surface damage can also be reduced using a low angle, grazing ion bombardment and a less energetic ion.

The micropillar samples have been the subject of careful analysis both in terms of their experimental suitability to assess material properties (Kiener et al. 2009; Choi et al. 2007; Shade 2008; Shade et al. 2009; Uchic et al. 2009) and to study the fundamentals of plastic deformation at a very fine scale (Uchic et al. 2004; Uchic et al. 2006; Ghosh 2008). A distinct size effect has been discovered in Ni_3Al-1% Ta depending on the diameter of the micropillar, but a 10-μm diameter micropillar of a single crystal nickel-base superalloy matched the bulk material tensile behavior (Uchic et al. 2004). This shows that the results from this scale of testing has to be considered carefully and is likely as small as necessary to fully determine the physical behavior of bulk structural metals at the finest scales.

5 What's Next?

The multiscale test methodologies discussed in this chapter show that there are exciting opportunities at all scales to help bridge the gap between materials simulation and experimental inquiry into the material behavior. The development of location-specific design will require the characterization of material behavior at several different length scales to inform and validate materials modeling. Clearly, the very fine scale of material property characterization is offering interesting insights into the true deformation of materials that will be invaluable for plasticity modeling. These finest scale methods are not in the mainstream by any means and the modeling practitioner needs to evaluate the multiscale results at each step. That is, the deformation behavior at the finest scales is not necessarily acting at the larger scales, so the hierarchical merging of the test scales has to be considered at each step. Improvements in property characterization at the fine scale are also needed. Structures typically fail in tension or due to fatigue loading with cyclic plasticity, yet predominate experimental techniques operate in compression- or tension-only loading. Methods are needed that can simulate the cyclic plastic loading of crystals to assess their cyclic behavior and to promote realistic failure modes. Some of the most extreme environments that materials must survive are not well represented by ambient tests in the vacuum chamber of an SEM. High temperature and/or corrosive environments need to be considered across the scales of testing. Yet, environmental damage typically is diffusion controlled where length is intrinsic to the level of damage. Consider, for example, the material at the tip of a crack growing in a high pressure turbine blade. The stresses here will be very high, and clearly diffusion of the high pressure combustion gasses (and residual oxygen) will be affected by these high stresses at the crack tip. This physical behavior is not considered in continuum crack growth modeling (equation 3) but is a very real phenomenon that controls the propagation of the crack. Techniques at the fine scale need to experimentally investigate these effects to accurately model the phenomena.

The idea of testing at small scales is not purely of academic interest because the volume of material that is stressed at notch and fillet locations is usually quite small, and failure often occurs at a very local level. Many structures, especially high pressure turbine blades, have very fine structural geometries. The development of micro unmanned air vehicles and tiny robotic devices requires that material characterization tests are conducted on a similar scale.

The bulk of the multiscale property characterization is dealing with deformation mechanisms and the material/microstructure that influence the deformation character. This work is, of course, is very important to develop constitutive models of the material's response to loading. However, structural durability is not a direct result of the material's constitutive response. Current design-allowable curves take the form of stress-life (strain-life) for fatigue or LMP for creep rupture that are not direct functions of the deformation character. Or to put it differently, the link between the deformation and damage needs to be built. One needs the constitutive response coupled with the damage accumulation models and a sound lifting strategy to ensure the durability of a component or structure. For example, design for damage-tolerant life will integrate the crack growth curve from an initial crack size (based on intrinsic material "defects" or nondestructive investigation probability of detection limitations) to a critical size for the structure (based on the fracture toughness or limiting crack growth rate of the material) to assess the service life of the structure. These calculations require an assessment of the cyclic stress at the location of interest. This comes from the constitutive response of the material. However, if the grain size of the material is increased, the constitutive response and the crack growth resistance of the material trend in opposite directions. The strength typically is decreased with the increase in grain size, but the crack growth resistance increases. Hence, it is critical to consider and model the damage processes along with the changes in deformation at the various scales. It is also critical to understand the full loading response of the structure including the temperature, environment, loading rates, and time of service or loading.

The goal of linking the behaviors at the various length scales is to ultimately develop model approaches that affect the design engineer. Therefore, the design engineer needs to be an integral member of the integration team to ensure that the model approach is suitable for the product needs. The materials practitioner is required to ensure that the material deforms and fails according to a model approach that matches reality.

Acknowledgments The author gratefully acknowledges Professor S. Ghosh of the Ohio State University and Dr. D.M. Dimiduk of Air Force Research Laboratory for the honor of being asked to submit this chapter. The support of the Air Force Research Laboratory during the preparation of this chapter is gratefully acknowledged.

References

Ahmad E, Manzoor T and Hussain N (2009) Thermomechanical processing in the intercritical region and tensile properties of dual-phase steel. Mater Sci Eng A 508:259–265.

Aktaa J, Reszat JTh, Walter M, Bade K and Hemker KJ (2005) High cycle fatigue and fracture behavior of LIGA Nickel. Scr Mater 52:1217–1221.

Anwander M, Zagar BG, Weiss B, and Weiss H (2000) Noncontacting strain measurements at high temperatures by digital laser speckle technique. Exp Mech 40(1):98–105.

Avril S, Pierron F, Sutton MA and Yan J (2008) Identification of elasto-visco-plastic parameters and characterization of Lüders behavior using digital image correlation and the virtual fields method. Mech Mater 40:729–742.

Brockman RA (2003) Analysis of elastic-plastic deformation in TiAl polycrystals. Int J Plast 19:1749–1772.

Casellas D, Pérez R and Prado JM (2005) Fatigue variability in Al–Si cast alloys. Mater Sci Eng A 398:171–179.

Cashman GT (2010) A review of competing modes in fatigue behavior. Int J Fatigue 32:492–496.

Cassenti B and Staroselsky A (2009) The effect of thickness on the creep response of thin-wall single crystal components. Mater Sci Eng A 508:83–189.

Chiang F-P (2009) Super-resolution digital speckle photography for micro/nano measurements. Opt Lasers Eng 47:274–279.

Choi YS, Uchic MD, Parthasarathy TA and Dimiduk DM (2007) Numerical study on microcompression tests of anisotropic single crystals. Scr Mater 57:849–852.

Chuto T, Satou M, Hasegawa A, Abe K, Nagasaka T and Muroga T (2002) Fabrication using a levitation melting method of V–4Cr–4Ti–Si–Al–Y alloys and their mechanical properties. J Nucl Mater 307:555–559.

Dai Y, Long B, Jia X, Glasbrenner H, Samec K and Groeschel F (2006) Tensile tests and TEM investigations on LiSoR-2 to −4. J Nucl Mater 356:256–263.

Davidson DL and Lankford J (1986) High Resolution Techniques for the Study of Small Cracks. In: Ritchie RO and Lankford J (eds) Small Fatigue Cracks, TMS, Warrendale, PA, 455–470.

Dinda GP, Dasgupta AK and Mazumder J (2009) Laser aided direct metal deposition of Inconel 625 superalloy: Microstructural evolution and thermal stability. Mater Sci Eng A 509:98–104.

Dlouhý A, Kuchaová K and Orlová A (2009) Long-term creep and creep rupture characteristics of TiAl-base intermetallics. Mater Sci Eng A 510:350–355.

Gabb TP, Gayda, J and Miner RV (1986) Orientation and temperature dependence of some mechanical properties of the single-crystal nickel-base superalloy Rene' N4: Part II. Low cycle fatigue. Metall Mater Trans 17(3):497–505.

Gallagher J et al. (2004) Advanced High Cycle Fatigue (HCF) Life Assurance Methodologies. AFRL-ML-WP-TR-2005-4102. Wright-Patterson AFB, Ohio.

Gayda J, Gabb TP and Kantzos PT (2004) The Effect of Dual Microstructure Heat Treatment on an Advanced Nickel-Base Disk Alloy. In: Green KA, et al. (eds) Superalloys 2004, TMS, Warrendale, PA, 323–329.

Ghosh AK (2008) Analysis of the interpretation of yielding and strengthening behavior in small-size samples. Acta Mater 56:2353–2362.

Guitar A, Vigna G and Luppo MI (2009) Microstructure and tensile properties after thermohydrogen processing of Ti–6Al–4V. J Mech Behav Biomed Mater 156–163.

Hemker KJ, Mendis BG and Eberl C (2008) Characterizing the microstructure and mechanical behavior of a two-phase NiCoCrAlY bond coat for thermal barrier systems. Mater Sci Eng A 483:727–730.

Héripré E, Dexet M, Crépin J, Gélébart L, Roos A, Bornert M and Caldemaison D (2007) Coupling between experimental measurements and polycrystal finite element calculations for micromechanical study of metallic materials. Int J Plast 23:1512–1539.

Huebner KH, Dewhirst DL, Smith DE and Byrom TG (2001) The Finite Element Method for Engineers. Wiley-IEEE, New York.

Hüttner R, Gabel J, Glatzel U and Völkl R (2009) First creep results on thin-walled single-crystal superalloys. Mater Sci Eng A 510:307–311.

Igarashi M, Yoshizawa M, Matsuo H, Miyahara O and Iseda A (2009) Long-term creep properties of low C–2.25Cr–1.6W–V–Nb steel (T23/P23) for fossil fired and heat recovery boilers. Mater Sci Eng A 510:104–109.

Janssen JM, Hoefnagels JPM, de Keijser ThH and Geers MGD (2008) Processing induced size effects in plastic yielding upon miniaturization. J Mech Phys Solids 56:2687–2706.

Jha SK, Caton MJ and Larsen JM (2007) A new paradigm of fatigue variability behavior and implications for life prediction. Mater Sci Eng A 468:23–32.

Jha SK, Caton MJ and Larsen JM (2008) Mean vs. Life-limiting fatigue behavior of a nickel-based superalloy. In: Reed RC, et al. (eds) Superalloys, TMS, Warrendale, PA, 565–572.

Jha SK, Larsen JM and Rosenberger AH (2005) The role of competing mechanisms in the fatigue life variability of a nearly fully-lamellar γ-TiAl based alloy. Acta Mater 53:1293–1304.

Johnson DA (2004) An Experimental Investigation of a Limited-ductility Intermetallic. PhD Dissertation, Engineering Sciences, Harvard University.

Kiener D, Motz C and Dehm G (2009) Micro-compression testing: A critical discussion of experimental constraints. Mater Sci Eng A 505:79–87.

Kiener D, Motz C, Rester M, Jenko M and Dehm G (2007) FIB damage of Cu and possible consequences for miniaturized mechanical tests. Mater Sci Eng A 459:262–272.

Kim JW, Lee K, Kim JS and Byun TS (2009) Local mechanical properties of Alloy 82/182 dissimilar weld joint between SA508 Gr.1a and F316 SS at RT and 320 °C. J Nucl Mater 384:212–221.

Kim SW, Tanigawa H, Hirose T, Shiba K and Kohyama A (2007) Effects of surface morphology on fatigue behavior of reduced activation ferritic/martensitic steel. J Nucl Mater 367:568–574.

Lewis AC, van Heerden D, Eberl C, Hemker KJ and Weihs TP (2008) Creep deformation mechanisms in fine-grained niobium. Acta Mater 56:3044–3052.
Liaw PK, Saxena A and Schaefer J (1989) Estimating remaining life of elevated temperature steam pipes – Part 1, material properties. Eng Fract Mech 32(5):675–708.
Lucas GE (1983) The development of small specimen mechanical test techniques. J Nucl Mater 117:327–339.
Lyons JS, Liu J and Sutton MA (1996) High temperature deformation measurements using digital-image correlation. Exp Mech 36(1):1996.
Ma ZY, Pilchak AL, Juhas MC and Williams JC (2008) Microstructural refinement and property enhancement of cast light alloys via friction stir processing. Scr Mater 58:361–366.
Mälzer G, Hayes RW, Mack T and Eggler G (2007) Miniature specimen assessment of creep of the single-crystal superalloy LEK 94 in the 1000 °C temperature range. Metall Mater Trans 38A:314–327.
Mayer C, Eggeler G, Webster GA and Peter G (1995) Double shear creep testing of superalloy single crystals at temperatures above 1000 °C. Mater Sci Eng A 199:121–130.
Molak RM, Paradowski K, Brynk T, Ciupinski L, Pakiela Z and Kurzydlowski KJ (2009) Measurement of mechanical properties in a 316L stainless steel welded joint. Int J Pres Ves Pip 86:43–47.
Morrissey RJ, John R and Porter III WJ (2009) Fatigue variability of a single crystal superalloy at elevated temperature. Int J Fatigue. doi: 10.1016/j.ijfatigue.2009.03.013
National Materials Advisory Board (NMAB) (2008) Committee on Integrated Computational Materials Engineering: Integrated computational materials engineering a transformational discipline for improved competitiveness and national security. National Academies Press.
Paris PC (1962) The Growth of Cracks Due to Variations in Load, PhD Thesis, Lehigh University.
Prevéy PS and Cammett JT (2004) The influence of surface enhancement by low plasticity burnishing on the corrosion fatigue performance of AA7075-T6. Int J Fatigue 26:975–982.
Reuss E (1930) Beruecksichtigung der elastischen Formaenderungen in der Plastizitaetstheorie. Zeits angew Math u Mech 10:266–274.
Schuster BE, Wei Q, Ervin MH, Hruszkewycz SO, Miller MK, Hufnagele, TC and Ramesh KT (2007) Bulk and microscale compressive properties of a Pd-based metallic glass. Scripta Mater 57:517–520.
Shade PA (2008) PhD Dissertation, The Ohio State University.
Shade PA, Wheeler R, Choi YS, Uchic MD, Dimiduk DM and Fraser HL (2009) A combined experimental and simulation study to examine lateral constraint effects on microcompression of single-slip oriented single crystals. Acta Mater 57:4580–4587.
Sharma SK, Jang C and Kang KJ (2009) Effect of thermo-mechanical processing on microstructure and creep properties of the foils of alloy 617. J Nucl Mater 389:420–426.
Sharpe WN and Fowler RO (1993) A Novel Miniature Tension Test Machine. In: Corwin WR, Haggag FM and Server WL (eds) ASTM STP 1204, Small Specimen Test Techniques Applied to Nuclear Reactor Vessel Thermal Annealing and Plant Life Extension, ASTM, Philadelphia, PA, 386–400.
Sharpe WN, Jackson KM, Hemker KJ and Xie Z (2001) Effect of specimen size on Young's modulus and fracture strength of polysilicon. J Microelectromech Syst 10(3):317–326.
Sharpe WN (ed) (2008) Springer Handbook of Experimental Solid Mechanics. Springer, New York.
Sutton MA, Orteu J-J, Schreier HW (2009) Image Correlation for Shape, Motion and Deformation Measurements: Basic Concepts, Theory and Applications. Springer, New York.
Telesman J, Gabb TP, Garg A, Bonacuse P and Gayda J (2008) Effect of Microstructure on Time Dependent Fatigue Crack Growth Behavior in a P/M Turbine Disk Alloy. In: Reed RC, et al. (eds) Superalloys 2008, TMS, Warrendale, PA, 807–816.
Tiwari V, Sutton MA, McNeill SR, Xu S, Deng X, Fourney WL and Bretall D (2009) Application of 3D image correlation for full-field transient plate deformation measurements during blast loading. Int J Impact Eng 36:862–874.

Touratier F, Andrieu E, Poquillon D and Viguiera B (2009) Rafting microstructure during creep of the MC2 nickel-based superalloy at very high temperature. Mater Sci Eng A 510:244–249.
Tschopp MA and Rosenberger AH (2009) Air Force Research Laboratory, Materials and Manufacturing Directorate, AFRL/RXLM. Wright-Patterson AFB, Ohio (unpublished research).
Uchic MD and Dimiduk DM (2005) A methodology to investigate size scale effects in crystalline plasticity using uniaxial compression testing. Mater Sci Eng A 400:268–278.
Uchic MD, Dimiduk DM, Florando JN and Nix WD (2004) Sample dimensions influence strength and crystal plasticity. Science 305:986–989.
Uchic MD, Dimiduk DM, Wheeler R, Shade PA and Fraser HL (2006) Application of microsample testing to study fundamental aspects of plastic flow. Scr Mater 54:759–764.
Uchic MD, Shade PA and Dimiduk DM (2009) Plasticity of micrometer=scale single crystals in compression. Annu Rev Mater Res 39:361–386.
Wallace RM, Annis CG and Sims D (1976) Application of Fracture Mechanics at Elevated Temperature, AFML-TR-76–176 part 2. Wright-Patterson ABF, Ohio.
Woodford DA, Van Steele DR, Amberge K and Stiles D (1992) Creep Strength Evaluation for IN 738 Based on Stress Relaxation. In: Antolovich SD, et al. (eds) Superalloys 1992, TMS, Warredale, PA, 657–664.
Zupan M and Hemker KJ (2001) High temperature microsample tensile testing of γ-TiAl. Mater Sci Eng A 319:810–814.
Zupan M, Hayden MJ, Boehlert CJ and Hemker KJ (2001) Development of high-temperature microsample testing. Exp Mech 41(3):1–6.

Simulation-Assisted Design and Accelerated Insertion of Materials

D.L. McDowell and D. Backman

Abstract Significant advances have been realized in accelerating the insertion of new and improved materials into products within the compressed timeframe of design and prototyping using the emerging computational materials science modeling and systems-based information management and materials design strategies. Recent initiatives in the USA to strengthen the link between materials modeling and simulation, process route, and structure–property relations are discussed, with emphasis on the Accelerated Insertion of Materials (AIM) strategy, tools, and methods. The recent emphasis on Integrated Computational Materials Engineering (ICME), an emergent branch of AIM that is built upon integrating modeling and simulation with product development, is discussed in terms of its common ground with the notion of concurrent design of materials and products – materials design. Materials design includes multiscale modeling of hierarchical materials as an important component, but is much broader in scope. This distinction between materials design and multiscale modeling is considered in some detail, with emphasis on top-down requirements on material structure and performance to meet product requirements. Uncertainty is a ubiquitous aspect of materials design, regardless of whether design decisions are informed by experimental measurements, modeling, and simulation or other heuristics. Some emerging concepts for robust design of materials are briefly described, and challenges for the synthesis of modeling and simulation and materials design are outlined.

Keywords AIM · ICME · Materials design · Modeling and simulation · Multiscale modeling · Top-down design

---

D.L. McDowell (✉)
Georgia Institute of Technology, GWW School of Mechanical Engineering, Atlanta, GA, USA
e-mail: david.mcdowell@me.gatech.edu

S. Ghosh and D. Dimiduk (eds.), *Computational Methods for Microstructure-Property Relationships*, DOI 10.1007/978-1-4419-0643-4_17,

1 Introduction: Systems Engineering and Materials Design

Historically, development and certification of new and improved materials has involved laborious empirical process route studies, extensive mechanical testing and property correlations with process route and resulting microstructure, and time-consuming iteration. The time required for the development of new materials has significantly exceeded the timeframe for new product development, in part due to reliance on empiricism and in part because materials' development has been largely distinct from systems design. In this paradigm, selection of materials has played a dominant role as a key interface of materials development with engineering systems design (Ashby 1999). Traditional materials' development has been highly empirical; dramatic compromise tradeoffs of competing property objectives such as strength and toughness have been widely accepted and taught from textbooks, giving metals the reputation as a mature and heavily constrained/limited class of materials. For example, Ashby materials selection charts (Ashby 1999) enable identification of existing material systems and properties that meet required performance indices for a specified application, which is always an initial step in framing a materials design problem. This is conventionally done by searching databases for properties or characteristics of responses that best meet a set of specified performance indices, often using combinatorial search methods (Shu et al. 2003). Of course, the notion of "properties" is perhaps better reflected by label "responses" since many phenomena of interest cannot be adequately described by single parameter descriptions such as fracture toughness, strength, and ductility. The interface with designers and design systems has led to the representation of materials via simplified sets of properties in systems design, but this is evolving with time as multifunctional performance requirements are becoming more commonplace in product design.

The current revolution in computational multiscale modeling and simulation of structure–property relations of materials is establishing a foundation for new and more efficient ways of enhancing the utility and value of consumer goods by tailoring materials in a manner that is concurrent with the design process. It is no longer necessary to rely solely on properties of existing, already certified materials in the design and product development process. Advances in modeling and simulation can be leveraged to reduce the time for materials development and certification to match the timeframe for prototyping and new product development. This has been a long-standing competitive issue in developing new products. This chapter summarizes the role of computational modeling and simulation, integrated with a systems approach, in fomenting recent advances in materials design and accelerated insertion of materials.

The idea of designing a material with targeted property sets can be traced back to the mid-1980s with the inception of the Steel Research Group (SRG) at Northwestern University (Apelian 2004). Olson's Venn diagram in Fig. 1 (Olson 1997) is foundational to systems-based approaches to materials design. As discussed by Olson (1997), it emphasizes the hierarchy of material structure within the context of process–structure–property–performance relations. It clearly distinguishes the pursuit of top-down, goals/means, inductive systems engineering (design) from

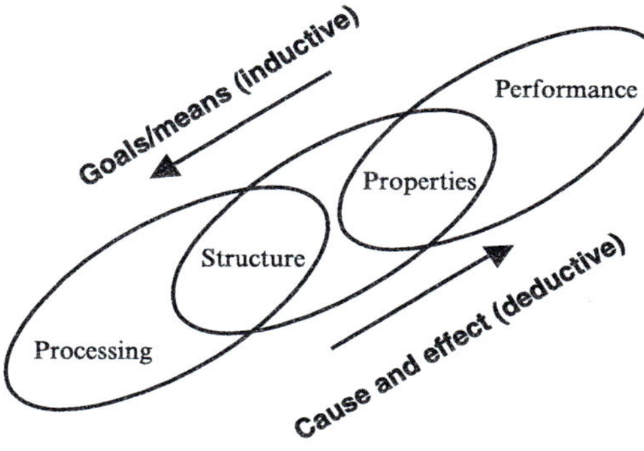

Fig. 1 Olson's linear concept of "Materials by Design" (Olson 1997), with inductive engineering design as a top-down pursuit

bottom-up, cause and effect, deductive, sequential linkages (scientific and engineering analysis). The bottom-up approach has served as the predominant historical model for empirical materials development, with limited top-down guidance from applications. This has traditionally conferred the characteristic of "serendipity" to materials development, with each "hit-or-miss" discovery of a new material followed by a flurry of incorporation of the resulting materials into applications to improve performance. Good examples are fiber-reinforced composite materials that are now commonly used in advanced aircraft and sporting goods. From the metals world, Ni-base superalloys used in aircraft gas turbine engines have been largely developed by empirical means, and have unique characteristics of increasing strength with increasing temperature which delivers outstanding performance in hot section applications. The work of the SRG has demonstrated that even a material as ubiquitous as steel has many variants and has enormous potential for joint improvement of both strength and ductility when subjected to a systematic decomposition of composition, process route, and material subsystems (precipitates, interfaces, grains, dispersoids, etc.) that control the most relevant property sets of interest (Apelian 2004; Olson 2001). This has been accomplished largely by a combined approach of modeling each level of hierarchy with fundamental scientific computing tools, depending on the length scale of each subsystem, with appropriate characterization of the material at each level to calibrate and validate the simulations (McDowell and Olson 2008). For example, interface structure and properties may require first principles or atomistic simulations, whereas effects of morphology of second phases added for strengthening might be amenable to continuum models of various sorts.

In addition to designing materials with targeted property sets, another major implication of concurrent materials and product design is the capability to assign location-specific properties that vary throughout the part in a manner that in some way optimizes performance for a given set of system requirements. This requires an intimate link between systems design and materials synthesis and processing to achieve the necessary heterogeneity of material structure as a function of position. Indeed, part shapes and product functionality can change dramatically in this scenario compared to design with homogeneous material structure and properties, and more weight-efficient designs can be realized.

Materials design relies on process–structure and structure–property relations, which are increasingly amenable to computational modeling and simulation. Prior chapters of this volume have considered advances in digital representation of microstructures, computational structure–property relations, computational methods, and modeling across all relevant material length scales. This chapter adds to this set of topics consideration of virtual design and manufacturing systems of the future that can integrate these advances in computational modeling and simulation with systems design. Systems engineering approaches are often an overlooked aspect that is essential to integrating these tools in a way that addresses top-down product requirements. It is necessary to elaborate on the meaning of systems strategies for materials design, and how this differs from combinatorial searching and bottom-up modeling and simulation. One of the earmarks of a systems strategy is the decomposition of the design problem into a hierarchy of material subsystems and mapping these onto process–structure, structure–property, and property–performance relations, as shown in Fig. 1; these relations may consist of heuristic guidelines, databases, computational models, and/or experimental methods. Management of uncertainty inherent in material process–structure–property–performance relations is of paramount importance. Prospects for continued development are discussed, including challenges and opportunities.

2 Recent Advances in Integrating Materials Simulation and Product Design

A thread of significant, "game-changing" initiatives and R&D programs from the mid-1990s onward that supports this rapidly developing multidiscipline is described in this section. As a precursor, the aforementioned SRG at Northwestern University fleshed out the essential elements of mapping our process–structure–property–performance relations (Apelian 2004; Olson 1997) that are central to systems-based materials design for required property sets. The vision of systems-based materials design and virtual manufacturing was further refined a decade ago for the academic and research communities at a 1998 NSF-sponsored workshop (McDowell and Story 1998) entitled "New Directions in Materials Design Science and Engineering (MDS&E)." This workshop report concluded that a change of culture is necessary in US universities and industries to cultivate and develop the concepts of simulation-based design of materials to support concurrent design of material and products. It also suggested that globalization will naturally drive a specialty (tailored) materials supply/development industry and distributed virtual design and manufacturing. Recommendations included establishing a US roadmap addressing:

1. Development and maintenance of databases to enable distributed materials design
2. Development of principles of systems design that integrate hierarchical materials systems, and

3. Identification of opportunities and deficiencies in science-based modeling, simulation, and characterization "tools" to support concurrent design of materials and products

The DOE-sponsored USCAR program from 1995 to 2000 provided an earlier indication of the feasibility of obtaining substantial improvements by coupling structure–property relations and component level design (Gall et al. 2000, 2001; Fan et al. 2003; McDowell et al. 2003). In this program, concepts for concurrent design of microstructure and associated structure–property relations were applied to achieve an increase in cast automotive vehicle component (suspension arm) fatigue strength with reduction in component weight. Hierarchical computational micromechanics was used to characterize, to first order, cyclic plasticity and fatigue processes associated with cast microstructure attributes ranging from several microns to the order of 1 mm. These simulations involved FE calculations on actual and idealized microstructures and were aimed at a very different goal from typical fatigue analyses, namely understanding the sensitivity of various stages of fatigue crack formation and early growth at hierarchical levels of microstructure (McDowell et al. 2003). The dependence of component fatigue strength on casting porosity and eutectic microstructure was addressed by employing numerical micromechanical simulations using FE methods for a hierarchy of five inclusion types for cast Al-Si alloy A356-T6, spanning the range of length scales relative to the secondary dendrite arm spacing, or dendrite cell size. Resulting predictions of high cycle and low cycle fatigue resistance for a range of types of inclusion conformed to measured variation of experimental fatigue lives. Similar casting porosity-sensitive strength models were developed to couple with process models for casting porosity levels to design component geometry and processing for A356-T6 to reduce weight and gain strength by location-specific structure and properties. Results were validated using full-scale demonstrations.

The DARPA/AFRL Accelerated Insertion of Materials (AIM) program (Apelian 2004; Backman and Dutton 2006) from 2000 to 2005 offered insight into how computational materials science and engineering can be harnessed in the future to assist in developing and certifying materials in a shorter timeframe that more closely matches the product design cycle. AIM was a bold initiative that assembled materials developers, original equipment manufacturers (OEMs), and government and academic researchers in a collaborative, distributed effort to build designer knowledge bases (DKBs) for metallic systems (Ni-base superalloys for gas turbine engine disks) and composite airframe materials. The program was founded upon bringing about three systemic changes: (1) revolutionizing the way designers and materials engineers interact, (2) achieving a leap forward in the application of computational materials science and integration with design engineering tools, and (3) creating an environment in which the design/materials team can learn from and build on previous developments. These three changes enable continual improvement in development productivity, more effective application of materials, and rebalancing of the competition among risk, cost, payoff, and maturity that too often has deterred designers from exploiting new materials.

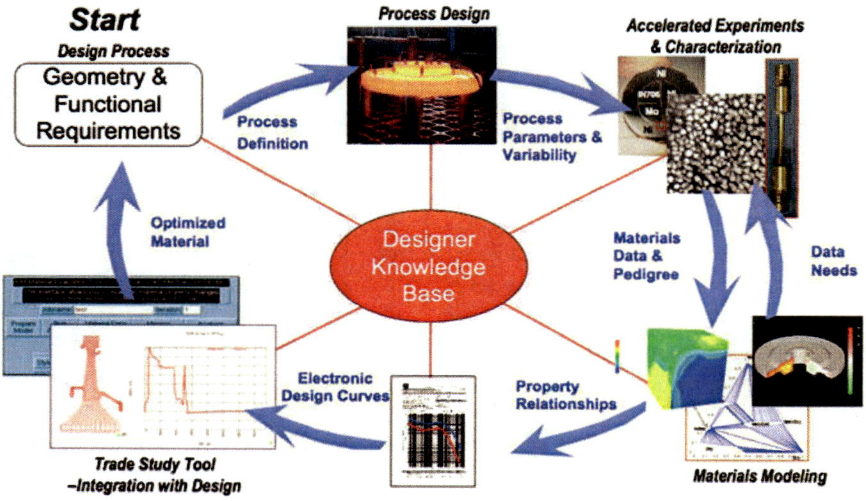

Fig. 2 Schematic illustration of the AIM process flow involving data and modeling as well as integration with the product design process (Backman and Dutton 2006)

The AIM initiative embraced a new process for integrating design and materials engineering and a number of supporting strategies to accelerate acquisition of materials data, as shown in Fig. 2 (Backman and Dutton 2006). The centerpiece of the "AIM system" is the DKB, as schematically shown in Fig. 2. This computational framework provides a means for managing experimental data, executing linked models that consider processing, microstructure, properties, and manufacturability, estimating confidence bounds for system predictions, performing sensitivity studies, and optimizing materials and process designs. The DKB informs product design engineers regarding material performance and feasibility of producing designs, and transfers materials information in digital form to design engineering and analysis systems. The AIM research team established material models, methods for managing uncertainty methods, and use-cases for integration within the DKB:

- *Material models* – Realizing accelerated insertion of new and improved materials via the integrated AIM system relies on models with sufficient fidelity and robustness to confidently predict results of process path, resulting microstructure and associated mechanical properties. The DKB used both available and newly developed physically based models to predict the effects of key process parameters on microstructure and properties. Modeling efforts focused on those that capture the physics governing material-system behavior. Some of the commercially available modeling tools included ThermoCalc™ (thermodynamics), DICTRA™ (kinetics), and DEFORM™ (large strain deformation and heat treat thermal modeling).
- *Uncertainty* – The design of a complex system, such as an aircraft gas turbine engine, must account for various sources of uncertainty inherent in materials

behavior, manufacturing processes, models, etc. The AIM program evaluated data and modeling uncertainty and applied Monte Carlo analysis to estimate the uncertainty produced by variations in processing history, microstructure attributes, measurement errors, and approximations of physically based models.

- *Use cases* – Models, digitization utilities, and an integration framework are necessary ingredients of an AIM system. However, utility and benefits are delivered by identifying important problems whose solution both accelerate development and reduce risk. Use cases that codify an accelerating methodology and describe the steps toward solving such problems are the key to success – they provide direction for AIM tool development, aid development and testing of the DKB, and ensure that elements of the DKB are coordinated to reap an end benefit. These use cases included methodologies to design a heat treatment to better balance properties, identify optimum parameters for a dual heat treatment process, and calculate necessary parameters for thermal processing (heat transfer) using inverse methods.

The AIM metals team at GE established a DKB using iSIGHT™ integration software. DKB development included special computer scripts that facilitate usage by materials engineers without the need for in-depth iSIGHT expertise. The DKB was populated with experimental test data and a series of linked models describing heat treatment, precipitation, evolution of grain structure, and deformation phenomena. Resulting modeling predictions for the fielded Rene'88DT Ni-base superalloy were validated through comparison with microstructure characterization and mechanical property test data stored within the eMatrix database of the DKB, as shown in Figs. 3 and 4. In addition, the team integrated the AIM DKB and methodology with design tools. GE developed a curve generator that combined modeling predictions and property test data to predict design minimum curves for delivery to the computational design environment. In addition, GE developed a Design Trade Study Tool

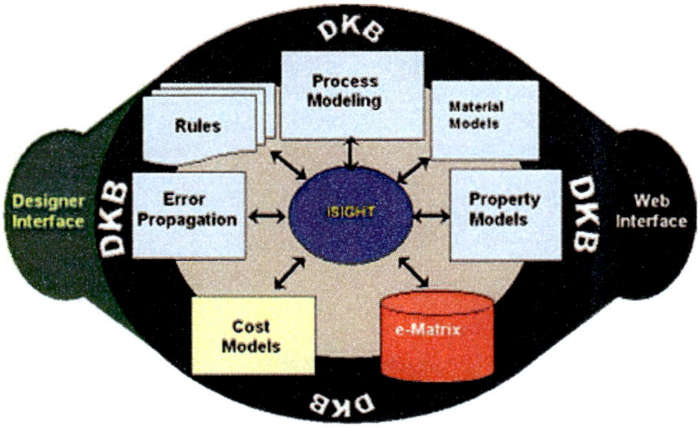

Fig. 3 The AIM designer knowledge base system. It incorporates commercial, proprietary, and emerging computational materials development tools to enable rapid exploration of new materials design space (Backman and Dutton 2006)

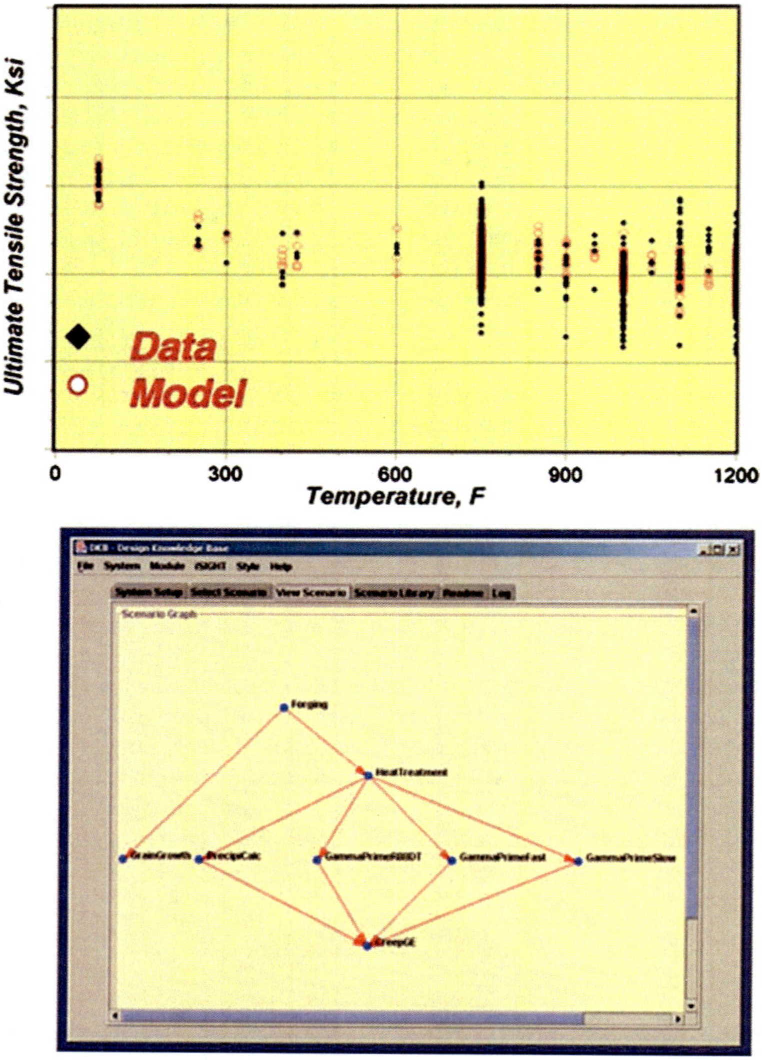

Fig. 4 Property models linked through the DKB to predict the variation due to processing and resultant microstructure, agrees well with experimental data (Backman and Dutton 2006)

that allows the designer and material developer to compute the effect of material changes on the mechanical performance and life of a superalloy disk. Both of these modules are extensible to other materials and engine components.

The GE team demonstrated that AIM can significantly reduce insertion time by eliminating the time lost in iteration and rework loops associated with empirical development activities. By carefully diagramming current workflow processes, the team concluded that development effort for a new disk process could be reduced by 45–70% by inserting AIM analysis as a replacement for experimental trials. In addition, mechanical property modeling can limit both the time and cost of long duration testing, thereby offering further cycle time reductions.

A number of valuable lessons were learned during the course of the DARPA/AFRL AIM program. The program required a significant degree of shared purpose and coordination from team members having diverse backgrounds, perspectives, and responsibilities. While it was not always easy to locate a common, accepted approach leading to useful, pragmatic solutions, AIM brought a realization of the high value of collaboration among the team. This was particularly true for those activities involving design integration where the needs and activities of materials and design engineers are parallel but not necessarily equivalent.

Additionally, establishing useful physically based models for complex, advanced industrial materials is a difficult, open-ended endeavor. Historical experimental designs and test programs do not contain all data necessary to build and calibrate AIM models; furthermore, it is not always obvious a priori which variables and mechanisms are needed to achieve acceptable model fidelity and accuracy. These same limitations retard the development of uncertainty analyses, which depend on sensitivity of solutions and experiments to changes in composition, process route, and resulting material microstructure.

AIM also taught the GE team the power of a systems approach in which sensitivity analysis, designed analytical experiments, and optimization can be combined to deduce solutions that would require untenable levels of experimentation. Clearly, the payoff and reliability of such methods will expand as model fidelity and extrapolative power increases. But useful results can be obtained even with first generation models that were available in the first generation DKB for AIM. As pointed out earlier, use cases provide a compelling approach for motivating and organizing AIM development that leads to the solution of real problems. By establishing specific use cases backward – from analysis output, to required models, to integration of the necessary tools – an AIM systems approach can provide analysis that is used to make better decisions.

Building on these demonstrations, it is clear that a systems approach is needed to integrate modeling and simulation in a targeted, prioritized manner to support design decisions in the process of integrating materials development with product development. Indeed, other recent federal initiatives emphasize the interdisciplinary collaboration of materials modeling and simulation, high performance computing, networking, and information sciences to accelerate the creation of new materials, computing structure and properties using a bottom-up approach. For example, the recent NSF vision for Materials Cyber-Models for Engineering provides a computational materials physics and chemistry perspective (Billinge et al. 2006) in using quantum and molecular modeling tools to explore potentially new materials and compounds, making the link to properties. Such a systems approach that embeds material processing/supply, manufacturing, computational materials science, experimental characterization, and systems engineering and design is similar to the conceptualization of Integrated Computational Materials Engineering (ICME) being pursued by a NAE National Materials Advisory Board study group (Pollock and Allison 2008). ICME is an approach to concurrent design of products and the processes and materials that comprise them. This is achieved by linking materials process–structure and structure–property models at multiple length and time scales with elements of the design system for specific products and applications. The concept of ICME hearkens back to Olson's linear scheme of Fig. 1 (Olson 1997), with the understanding that top-down strategies are essential to supporting inductive, top-down design of materials to meet specific performance requirements.

These apparently diverse and complementary views of concurrent materials and product design, AIM, ICME, Cyberdiscovery, etc., have common ground in terms of linking the overlapping process–structure and structure–property relations in Fig. 1 with product design and development. In all cases, the intent is to enhance the role of modeling and simulation to accelerate materials discovery and development, concurrent with product design and prototyping. Consistent with this intent, use of the term *materials design* in this chapter implies top-down, simulation-supported, decision-based, concurrent design of material hierarchy and product or product family with a ranged set of performance requirements. This view is perhaps most closely aligned with the comprehensive notion of ICME (Pollock and Allison 2008). Some of the key elements of materials design include computational materials science, materials informatics, data mining or combinatorics, multiscale modeling to augment experimental methods, materials selection, experiential learning applied to new materials and products, methods for top-down modeling and quasi-inverse solutions for complex multivariate design spaces, and systems methodologies to support decision-making. The last point is quite important. Modeling and simulation tools provide support for design decisions but do not replace informed human decision-making in concurrent design and development of materials and products. The processes involved are so complex, nonlinear, and stochastic that "human expert in-the-loop" systems are essential for materials design in the foreseeable future. A realistic perspective on the role of modeling and simulation is warranted. In contrast to empirical development of materials, which historically involves meandering through scales and phenomena by "intelligent tweaking" of process route, composition, etc., to meet performance objectives, the goal of systems-based materials design is to explore the extent to which decision support from empirical pathways can be replaced by modeling and simulation. For example, if the number of decisions made in materials design and development based on modeling and simulation is presently less than 10% for a given system, can it be increased to 15% or 30% or even more? This is why we consider materials design to be "simulation-assisted" rather than simulation-based. Informatics elements of decision theory and decision-based design are extremely important. Another key practical question is the extent to which multiple phenomena and objectives can be considered simultaneously rather than sequentially, and the degree to which material life cycle considerations (durability, recycling, long term cost factors, etc.) can be dealt with in the design phase (cradle).

3 Multiscale Materials Modeling and Simulation: Purposes and Utility

The recent report of the NSF *Blue Ribbon Panel on Simulation Based Engineering Science* (Oden et al. 2006) issued broad recommendations regarding the need to more fully integrate modeling and simulation within the curriculum of engineering to tackle a wide range of interdisciplinary and multiscale/multiphysics problems. It would be tempting to equate advances in modeling and simulation per se with materials design or accelerating the insertion of new materials into products. However, phenomena in materials synthesis/processing and structure–property relations are typically far too complex, nonlinear, and path-dependent to be captured by idealized models in a truly predictive sense, even models with rather sophisticated character. Furthermore, the stochastic nature of material structure and properties limits applicability of deterministic approaches.

Essential elements of modeling and simulation include thermodynamics, kinetics, and kinematics; all are built up from the unit process level, but may also be expressed over higher length scales in terms of phenomenological models or higher degree-of-freedom simulations. Since feasible (realizable) structures of materials are established by either absolute or (more commonly) constrained minimization of thermodynamic free energy, thermodynamics is a foundational theme for simulation-assisted materials design. Thermodynamics provides understanding of stable and metastable phases, characterization of lattice and interface structures and energies, defect energies, and driving forces (transition states) for rearrangement of microstructure due to thermally activated processes. As such, it facilitates preliminary design exploration for candidate solutions to coupled material and product design. Kinetics plays an important role as a further step in screening candidate solutions in preliminary design exploration. For example, stability of phases and interfaces at finite temperature in operating environments requires assessment before expensive investment in extensive computation and/or experimental characterization of candidate solutions. Kinematics maps the contributions of various attributes (defects, phases) of microstructure in contributing to overall rearrangement during deformation and failure.

Although design is necessarily a top-down engineering activity, material process–structure and structure–property relations highlighted in Fig. 1 are intrinsically deductive (bottom-up) in nature and are typically computationally intensive. There are several important implications. Due to differences in mechanisms, phenomena, and dominant scales of material hierarchy, as well as differences in time scales of material processing and in-service applications, it is uncommon to conduct modeling in a coupled manner for both process–structure and structure–property relations. Moreover, process–structure and structure–property relations in modeling and simulation are not strictly invertible, owing to:

- Nonlinear, nonequilibrium path-dependent material behavior (computationally expensive and dependent upon initial conditions)

- Dynamic to thermodynamic model transitions in multiscale modeling, with reduction in model degrees of freedom rendering non-uniqueness and conferring a wide range of "suboptimal" solutions to explore
- Approximations made in digital microstructure representation of material structure, including reduction of degrees of freedom
- Dependence of certain properties such as ductility, fracture toughness, and fatigue strength on extreme value attributes of microstructure such as largest grains or phases or other "rare event" failure scenarios that represent a localization problem rather than a homogenization problem
- Microstructure metastability and long-term evolution in service

There are limited examples for which explicit inversion of structure–property relations has been pursued. For example, the groups of Adams (Adams et al. 2004; Adams and Gao 2004; Lyon and Adams 2004) and Kalidindi (Kalidindi et al. 2004, 2005; Knezevic et al. 2008) have tackled the problem of finding crystallographic textures (orientation distribution functions) that satisfy certain requirements on macroscopic anisotropic elastic stiffness of structures. The non-uniqueness in material modeling and simulation casts serious reservations on the viability of naïve single objective optimization for materials design. We will return to this point later when systems design methods are discussed.

3.1 Hierarchical and Concurrent Multiscale Modeling

Recent research trends have focused heavily on multiscale modeling of materials, as outlined in a number of preceding chapters. Traditionally, design has involved selecting materials on the basis of appropriate sets of properties for applications. The emerging paradigm of materials design (Ashby 1999; Apelian 2004; Olson 1997, 2000; McDowell 2001; McDowell et al. 2007) seeks instead to concurrently design the material and product by modifying existing materials or exploring potential new materials using modeling and simulation tools. In this approach, the heterogeneous material microstructure is viewed as a multilevel hierarchy with each level as a subsystem with design degrees of freedom. This notion of hierarchy is primitive in materials design, as phenomena at multiple length and time scales can control behavior. Multiscale modeling typically refers to a means of linking models with different degrees of freedom either within overlapping spatial and temporal domains or in adjacent domains. The aim of materials design is more comprehensive than that of multiscale modeling. Materials design is effectively a multilevel, multiobjective Pareto optimization problem in which ranged sets of solutions are sought that satisfy ranged sets of performance requirements (McDowell et al. 2007; Seepersad 2004; Seepersad et al. 2003, 2004, 2005, 2008; Choi 2005; Choi et al. 2005, 2008a, b; Panchal 2005; Panchal et al. 2005, 2007; Messer et al. 2007). It does not rely on explicit linkage of multiple length scales via numerical or analytical means in a multiscale modeling context unless such linkage has high certainty and is computationally feasible. In fact, it is often preferred to introduce rather elementary modeling concepts at each scale if complex, coupled multiscale models are used for which parameter identification, quantification of uncertainty, and validation of the scheme are difficult. Invariably, multiscale modeling schemes in their own right introduce idealizations or other approximations in terms of initial/boundary conditions or methods of averaging, compounding the uncertainty associated with material constitutive models at each scale.

A number of papers in this volume have addressed the topic of multiscale modeling (spatial homogenization, time scaling, etc.) of inelastic behavior of metallic materials, attempting to bridge descriptions at various scales via either concurrent or hierarchical multiscale modeling. Such methods can add considerable value to materials design if (1) dominant mechanisms and any cross-overs are incorporated, (2) model assumptions and approximations are quantified in terms of measures of uncertainty, and (3) a balanced approach is taken with regard to various aspects of the model (e.g., avoiding undue focus on accuracy of scheme with oversimplified material representations).

For inelastic deformation, environmental effects, and failure processes in metals, dominant mechanisms are associated with phenomena either the atomic scale or over a range of other length scales, as shown in Fig. 5. Models appropriate to each scale considered, from left to right in Fig. 5, include atomistics (classical molecular dynamics), discrete dislocation dynamics, kinetic Monte Carlo methods, statistical mechanics models for distributed defects (discrete and continuous), continuum crystal plasticity, and homogeneous macroscopic plasticity. The minimum length scale associated with each of these windows ranges from interatomic spacing to mean free path for dislocations to grain size to dimensions of components or structures (millimeters). Three comments are pertinent. First, as we transition from left to right, the number of model degrees-of-freedom (DOF) decreases, and complete information characterizing the dynamic state is lost. Statistical thermodynamics is

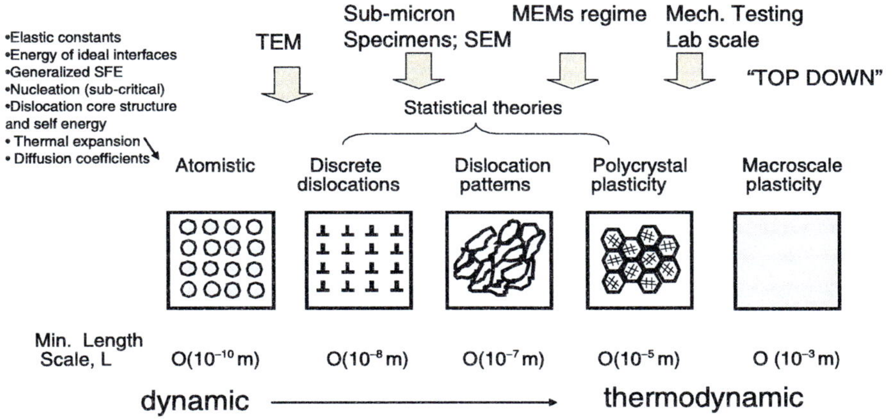

Fig. 5 Minimum length scale and models corresponding to various levels of hierarchy relevant to the inelastic behavior of polycrystalline metals

used in this process, with nonequilibrium thermodynamics treatment of evolving microstructure at larger scales. Second, not every problem requires properties or responses to be estimated at the far right in Fig. 5; for example, modeling of thin films or other material subsystems at small spatial scales might involve only the first few levels. Third, the largest gap in Fig. 5 exists between levels of atomistics and dislocation substructures; there are no well-developed, reliable, and robust methods to bridge this gap.

There are practical constraints on models at all these scales. For quantum calculations, size is limited to the order of 100 atoms and the solution strategy depends on the nature of bonding within the system as well as system configuration. Accounting for free surfaces is problematic. Atomistic simulations also suffer from severe limitations on time step size, but tens of millions of atoms can be considered, offering the potential to bridge into the mesoscale to some extent. Key limitations include uncertainty of the form and parameters of the interatomic potential, particularly for multicomponent systems and in the vicinity of interfaces. With atomistics, it is possible to begin making contact with HRTEM measurements regarding the atomic structure of materials, defect nucleation, and migration, all of which are key outcomes of atomistic simulations. At the mesoscale, microstructure is heterogeneous, but for practical reasons model DOF must be decreased to consider only lower order moments of defect distributions.

Multiscale modeling involves incorporation of information from fine scale models into coarse-grained models over the same spatial domain and over the same time scale, with models executed in either sequential hierarchical (sequential one-way coupled, bottom-up model execution) or concurrent (parallel, two-way coupled model execution) fashion. In some multiscale modeling cases, such as transfer of information from atomistic simulations to continuum discrete dislocation or defect field models, hierarchical models are selected in view of the fact that the time scale of the atomistic simulations cannot match that of the desired application. Concurrent models must exchange information over common temporal and spatial scales.

Hierarchical models often involve statistical description of evolving microstructure at finer scales, based on "handshaking" methods for informing continuum models from high resolution models or experiments. These handshaking methods can range from intuitive formulation of constitutive equations to estimates of model parameters in coarse-grained models based on high resolution simulations (e.g., atomistics or discrete dislocation theory) to development of metamodels or response surface models that reflect material behavior over some parametric range of microstructure and responses. Examples of a combined bottom-up homogenization and handshaking among scales intended to support decision-based design are found in works of McDowell and colleagues (Wang et al. 2006; Shenoy et al. 2007, 2008; McDowell 2008) on microstructure-sensitive multiscale models for cyclic behavior of Ni-base superalloys and α–β Ti alloys. In these models, dislocation density evolution equations are formulated at the scale of either precipitates or homogenized grains, which are then calibrated with experimental elastic stiffness and stress–strain data on single crystals and polycrystals. A key feature of such models is the incorporation of elements of microstructure attributes that are sensitive to process route and affect relevant mechanical properties (precipitate or phase volume fractions, sizes, orientation distribution, etc.) In another hierarchical modeling framework, Hao et al. (2003, 2004) developed a "multiscale fracture simulator" for design of ultra-high strength, high toughness steels that uses information from atomistics in a scaled interface separation relation for interface fracture between metallic matrix and nonmetallic inclusions; additional models are introduced at higher scales for coupled dislocation glide-driven void growth, coalescence and failure by shear localization and macrofracture.

Hierarchical multiscale models may use a set of models for various scales or a self-consistent micromechanics scheme (of either Eshelby–Kröner type or based on a computational scheme) to incorporate fine scale phenomena in coarse-grained constitutive descriptions that invoke the notion of a Representative Volume Element (see chapter titled "Multiscale Characterization and Domain Partitioning for Multiscale Analysis of Heterogeneous Materials"). Yet another class of hierarchical models involves the development and incorporation of scaling laws that describe aggregate dislocation behavior and strengthening effects at higher length and time scales. For example, self-organization of dislocations into periodic low energy substructures (Leffers 1994; Hughes et al. 1997) is consistent with the evolutionary theory of low energy, stress-screened dislocation structures (Kuhlmann-Wilsdorf 1989), and resulting scaling laws can be embedded in continuum crystal plasticity (Butler and McDowell 1998; Horstemeyer and McDowell 1998) or used in other formats to provide design decision-support (see also chapter titled "Challenges Below the Grain Scale and Multiscale Models").

Concurrent multiscale modeling schemes involve two-way (bottom-up and top-down) coupling between models framed at difference scales and are especially useful if relevant fine scale microstructure evolution (such as cracking or slip localization) is distributed at the scale of a structure of interest in a manner that cannot be prescribed a priori. They are often framed across scales treated with continuum models with different degrees of freedom (levels to the right in Fig. 1). For example, Ghosh et al. (2007) introduced a two-way concurrent scheme to model multiscale damage initiation and growth due to microstructure-level damage associated with debonding at fiber–matrix interfaces in composite structures. In their work, a macroscopic continuum damage model was used with explicit high-fidelity micromechanical computation of stress and strain at local hot spots at the microstructure (fiber–matrix) level. An intermediate level of modeling was performed at the scale of a homogenized RVE to monitor the breakdown of coarse-grained continuum laws. In this way, provisions were made to "zoom in" and "zoom out" in terms of fidelity, enabling a broad area search for critical hot spots in components along with microstructure-sensitive simulations of failure processes at these hot spots. Different classes of models were used at each scale, which is an important physical requirement for concurrent multiscale models.

A common shortcoming of such concurrent continuum modeling schemes is the inability to generate sufficiently high stress levels to drive fracture based on atomistically computed criteria, providing impetus to introduce generalized continua approaches that attempt to mimic defect–defect and defect–interface interactions (see chapter titled "Dislocation Mediated Continuum Plasticity: Case Studies on Modeling Scale Dependence, Scale-Invariance, and Directionality of Sharp Yield-Point"). Typically, failure criteria such as cohesive stress or energy are fit to experimental data at higher scales. Effects of fine scale inhomogeneities (dislocations, voids, etc.) on plastic strain localization associated with higher scale inhomogeneities (holes, stress concentrations) have been addressed in recent works that use a second-order scheme for homogenization in terms of deformation gradients at the macroscale, accepting input from lower scale parallel microscale simulations, for example, the works of Kouznetsova et al. (2002, 2004) (see also Vernerey et al. 2007). Recently, a closely related second-order homogenization scheme based on micropolar kinematics has been introduced (Larsson and Diebels 2007) that yields a couple stress measure at the macroscale arising from lower scale heterogeneity.

The aforementioned gap in terms of models that bridge the atomistic and crystal plasticity levels in Fig. 5 is particularly noteworthy. Discrete dislocation dynamics can be coupled with atomistic information at the unit process level (Bulatov 2002), while continuum dislocation field theory with prescribed kinetics of flow requires hierarchical passage of information from fine scale discrete dynamical representations, as in multiscale discrete dislocation plasticity (Bulatov et al. 2001; Zbib et al. 2002; Zbib and de la Rubia 2002). The continuum dislocation kinetics treatment of Arsenlis and Parks (1999, 2002) attempts to track populations of dislocations in a given volume, with recognition of the role of dislocation sources as well as flux into the considered volume; these latter features can be dominant in cases of flow in confined volumes. Some hierarchical models attempt to address long time scales of many practical applications using transition state theory concepts, such as atomistically informed continuum modeling of ensembles of nanocrystalline (NC) Cu grains by Warner et al. (2006) and atomistically informed multiscale self-consistent models of NC materials (Capolungo et al. 2007). Microscopic phase field theory (Wang et al. 2001; Shen and Wang 2003) shows promise as a bridge from atomistic modeling of generalized stacking fault energy to describe partial dislocations in a higher scale continuum theory, when used with transition state theory and appropriate kinetics to model evolving microstructure.

Concurrent atomistic-continuum modeling schemes may use spatial domain decomposition over common temporal scales to address domains in which fine scale models (levels toward left in Fig. 5) are desired in a confined region near a crack tip or a grain boundary interface, for example. In this case, atomistic models can be used to model fracture or dislocation nucleation processes, with an overlapping pad region in which a far field continuum model is provided some form of equivalence or exchange with the atomistic domain (Gumbsch 1995; Rudd and Broughton 1998, 2000; Rafii-Tabar et al. 1998; Weinan and Huang 2001; Shiari et al. 2005; Qu et al. 2005). Filtering of high frequency lattice vibrations is achieved by overlapping the two different descriptions within some region in which finite element nodes coincide with atomic positions. Shilkrot et al. (2002, 2004) have formulated the problem in a way that permits passage of dislocations from the atomistic to continuum domains (only in 2D) and vice versa, the so-called coupled atomistics discrete dislocation (CADD) method (see chapter titled "Emerging Methods for Matching Simulation and Experimental Scales").

Having considered this inexhaustive review of multiscale modeling methods related to linking scales of metal plasticity shown in Fig. 5, we next turn our attention to the top-down exercise of materials design.

3.2 *Materials Design and the Need for Top-Down Methods*

As shown in Fig. 1, materials design requires "top-down" (goals/means), inductive methods to design the process route and resulting hierarchical microstructure to meet a set of specified performance requirements (Olson 1997, 2000). However, the process–structure–property–performance hierarchy shown in Fig. 1 is not the same hierarchy of scales appearing in Fig. 5 (atomistics, dislocations, etc.); process–structure and structure–property relations *each* potentially require application of sets of models that span the hierarchy of length scales in Fig. 5. There are a number of implications of top-down guidance for simulations to support computational materials design (McDowell 2007). First, simulations are typically conducted in bottom-up manner and cannot strictly be inverted from right to left in Fig. 5. Second, it is desirable to use chains of models across scales as far to the right as possible in Fig. 5, dictated by the scale of the application. For example, if system responses of primary relevance are dominated by coarse scale morphology of a polycrystalline material (e.g., attributes of grain size and orientation distributions) and these attributes serve as design variables, then a premium is placed on application of crystal plasticity with homogenized description of the behavior within each grain. In some cases, the analyses should focus on lower, dominant scale(s); an example is quantum engineering of the bond structure at material interfaces (e.g., grain or phase boundaries) to resist defect nucleation or migration, environmental degradation of interfacial strength, etc. Third, the uncertainties associated with process control, forms of models, model parameters, initial conditions, microstructure randomness, and scale transitions are significant and are just as important as other features of the models.

Although sometimes used in the context of materials design (McVeigh et al. 2006), concurrent multiscale models do not always serve the purpose of emphasizing dominant length scales and mechanisms since the uncertainty in the coupling of scales can compound other sources of uncertainty related to material models or structure at each scale. There is considerable uncertainty in any kind of multiscale modeling scheme, including selection of specific scales of hierarchy, approximations made in separating length and time scales in models used, model uncertainty at various scales, approximations made in various scale transition methods, and lack of complete characterization of initial conditions and process history effects. Characterization of sensitivity of responses to microstructure variation is useful for designing materials with properties/responses that are robust with respect to variation of composition, process route, microstructure, and other aspects of integration into products and devices. It is useful in this regard to extend concepts of systems-based robust design introduced by Taguchi (1993) to multilevel concurrent design of materials and products (McDowell et al. 2007; Choi et al. 2008a; Seepersad et al. 2008; McDowell 2007). Ultimately, materials design is a decision-making process that mitigates uncertainty. There are various sources of uncertainty, including (Isukapalli et al. 1998):

- Parameterizable (errors induced by processing, operating conditions, etc.) and unparameterizable (e.g., random microstructure) natural variability.
- Incomplete knowledge of model parameters due to insufficient or inaccurate data.
- Uncertain structure of a model and/or initial conditions due to insufficient knowledge (approximations and simplifications) about a system and variability of its parameters.
- Propagation of natural and model uncertainty through a sequential chain of models.

From a practical perspective, uncertainty of either hierarchical or concurrent multiscale modeling is critical in design. Although the notion of "bottom-up" modeling with seamless scale transitions is scientifically appealing, it often involves a compilation of approximations that may compromise viability for materials design, particularly if used within some kind of optimization scheme. Moreover, the assimilation and bridging of models for phenomena occurring at various scales typically involves numerous experts and organizations; this complicates attempts to seamlessly integrate models across scales. The integration is usually accomplished instead by communication and negotiation among various stakeholders/designers involved. Accordingly, it is necessary to use models in ranges for which they are most appropriate and valid, and to use objective protocols (e.g., utility theory) to assess the value of information that models can deliver in informing design decisions. Moreover, physically based models are often useful to quantify sensitivity to microstructure that enables consideration of trade-offs, regardless of whether model accuracy is sufficient to support predictive design in its own right. Modeling and simulation can provide only partial support for materials design and development for target applications (simulation-assisted design) – experiments and prototyping are indispensible elements in providing decision support. Accordingly, key requirements for models are internal consistency, physically plausibility, microstructure-sensitivity, and ability to calibrate at various levels of hierarchy with measurements. Hence, even first generation models can provide utility in materials design if calibrated and validated within ranges used. Investment in advanced models depends on the value they add to decisions in the design process, which depends on the application.

4 Hierarchical Decision-Making in Materials Design

It should be apparent from the foregoing discussion that Olson's hierarchy in Fig. 1 embeds multiscale modeling as part of the suite of modeling and simulation tools that provide decision-support in design, but materials design and multiscale

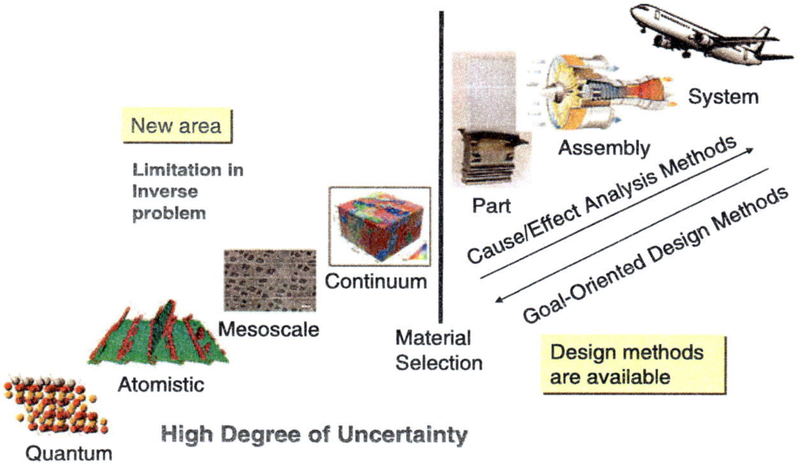

Fig. 6 Hierarchy of levels from atomic scale to system level in concurrent materials and product design. Existing systems design methods focus on levels to the right of the vertical bar, addressing mainly the materials selection problem, only one component in multilevel materials design

modeling are not equivalent pursuits. Figure 6 conceptualizes how well-established methods for design-for-manufacture of parts, sub-assemblies, assemblies, and overall systems can be extended to address the multiple length and time scales that govern process–structure–property–performance relations. As previously discussed, the objective of tailoring the material to specific applications (to the left of the vertical bar in Fig. 6) is distinct from traditional materials selection. The basic challenges revolve around the fact that hierarchical modeling of materials is still in its infancy, and systems-based design methods have not been widely applied to scales of material hierarchy appearing to the left in Fig. 6. Materials design requires methods that use bottom-up modeling and simulation, calibrated and validated by characterization and measurement to the extent possible, facilitating top-down, requirements-driven exploration of length scales shown in Fig. 5. Moreover, the aforementioned sources of uncertainty require more flexibility than is offered by naïve optimization of limited objectives either at individual levels of material hierarchy or at the systems level.

By definition, a multifunctional material is one for which performance dictates multiple property requirements. For example, aircraft gas turbine engine blade materials must meet requirements on conductivity (thermal), oxidation resistance (thermo-chemical), elastic stiffness (mechanical), and high temperature creep and fatigue resistance (thermo-mechanical), among others. Addressing such problems requires multiobjective, instead of single objective, design approaches that provide ranged sets of potentially acceptable solutions for ranged sets of specified performance requirements.

Recent collaborative efforts of one of the authors (DLM) with the Systems Realization Laboratory at Georgia Tech (McDowell et al. 2007; Seepersad 2004;

Seepersad et al. 2003, 2004, 2005, 2008; Choi 2005; Choi et al. 2005, 2008a; Panchal 2005; Panchal et al. 2005, 2007; Messer et al. 2007; McDowell 2007) have cast such materials design problems in the context of robust multilevel, multiobjective decision-based design. They are briefly summarized as follows. Figure 7 fleshes out the linear structure of Fig. 1 as a set of multilevel mappings (*Process–Structure* (*PS*) *relations, Structure–Property* (*SP*) *relations, and Property–Performance* (*PP*) *relations*) (McDowell 2007). These mappings, represented by arrows, can consist of models, experimental measurements, or characterization. Lateral movement at each level of hierarchy is associated with reducing model degrees of freedom (including microstructure representation) in scale transitions. The circled region at upper right in Fig. 7 represents the subdomain of the materials selection problem, which typically pertains to just one or two levels of hierarchy; it involves selection based on tabulated data from models or experiments and may be approached using informatics, e.g., data mining or combinatorics (Ashby 1999; Shu et al. 2003; Billinge et al. 2006). In Fig. 7, each arrow can also be accompanied by a design decision, depending on configuration of the design system.

Figure 8 reinforces the concept that the multilevel mappings in Fig. 7 are distinct from the hierarchy of material length scales in Fig. 5 that affect material responses. It also emphasizes the notion that models at different scales of hierarchy can be calibrated and validated by experimental measurements at each appropriate scale.

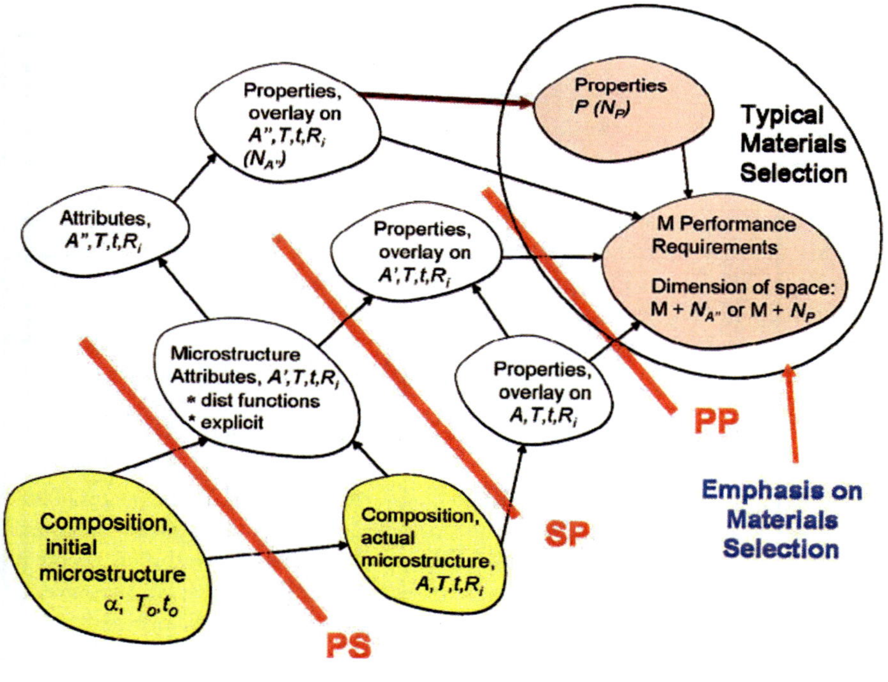

Fig. 7 Hierarchy of mappings in multilevel materials design (McDowell 2007)

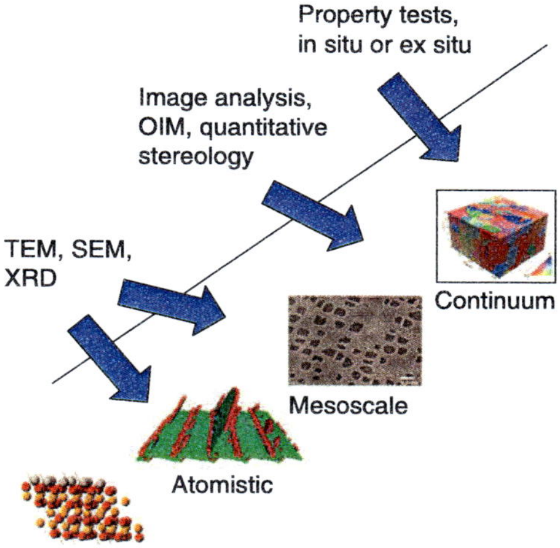

Fig. 8 Hierarchy of models ranging from dynamic (quantum mechanics, atomistics) to thermodynamic (mesoscale statistical continuum and higher length scale continuum models), with insight, calibration, and validation based on time and length scale-dependent measurements

In doing so, the sources and degree of uncertainty in models at each level can be evaluated in a more isolated manner, independent of multiscale modeling assumptions associated with higher scale behavior. In fact, both process–structure and structure–property relations may involve the full range of scales shown in Fig. 8. As in manufacturing process design, the notion of *robust* design (Choi 2005) appears to be central to any reasonable approach in view of pervasive uncertainty. Designs must be robust against variation of initial microstructure, usage factors and history, design goals, and various forms of uncertainty listed in the previous section, including uncertainty in models, tools, and methods used to design. In fact, the configuration of information flow and design decision points is a source of uncertainty in its own right. This includes issues such as the distribution of the design effort, sequencing of simulations and experiments, level of expertise and knowledge of modelers and designers, and other human factors such as degree of interaction and information sharing in the design process. There are important practical implications, namely that robust solutions do not necessarily involve large numbers of iterations (compared to analytical design optimization), are not focused on excessive optimization searches at individual levels, and involve the human being as an interpreter of value of information. Accordingly, ranged sets of solutions, rather than point solutions, are of practical interest. Furthermore, system performance requirements should be specified as ranges rather than single values, and systems performance should be specified rather than property requirements. The language of material properties, prevalent in materials selection, becomes an outmoded lexicon in decision-based concurrent design of materials and products. This is most vividly demonstrated in design of components for which microstructure and properties are functionally graded to meet performance requirements; the material no longer has a monolithic property set. Rather, it is the variation of microstructure that is pertinent to communicating the design, along with component geometry.

A practical approach is to quantify uncertainty to the extent possible and then seek robust solutions that are less sensitive to variation of microstructure and various other sources of uncertainty. Several categories of robust design deal with different sources of uncertainty. Type I robust design, originally proposed by Taguchi (1993), focuses on achieving insensitivity of performance with respect to noise factors – parameters that designers cannot control in a system. Type II robust design relates to insensitivity of a design to variability or uncertainty associated with design variables – parameters that a designer can control in a system. The compromise Decision Support Problem (cDSP) protocol (Mistree et al. 1992) has been introduced as the primary decision support tool, and the Robust Concept Exploration Method (Chen 1995) has been developed to address Types I and II robust design. It is based on goal programming rather than standard linear programming and involves negotiation between multiple designers regarding assignment of goals, constraints, and bounds. In the cDSP, multiple design objectives are set as targets, and the designer can select from among a family of solutions that minimize deviations from these targets in a manner that is subject to a set of constraints (cf. Seepersad 2004; Seepersad et al. 2003, 2004; Chen 1995). For multiple design objectives, robustness establishes preference among candidate solutions, highlighting those that have less sensitivity (i.e., robust) to variation of noise and control parameters. In addition, designs are sought that are robust against variability associated with process route and initial microstructure, forcing functions, cost factors, design goals, and other relevant factors.

Recent collaborative efforts at Georgia Tech have suggested methods to deal with uncertainty associated with microstructure variability and structure and parameters of models (Choi 2005; Choi et al. 2005) as well as chained sequences of models in a multilevel (multiscale) context of material structure and modeling (Panchal 2005; Panchal et al. 2005, 2007). Types I and II robust design has extended in these efforts to include Type III (Choi 2005; Choi et al. 2005), which considers sensitivity to uncertainty embedded within a model (i.e., model parameter/structure uncertainty). Figure 9 clarifies the application of Types I–III robust design; while application of traditional Types I and II robust design methods seek solutions that are insensitive to variations in control or noise parameters, Type III robust design more specifically seeks solutions that minimize the distance between upper and lower uncertainty bounds on the response function(s) of interest associated with material randomness and model structure/parameter uncertainty. These bounds are built up from the statistics obtained from application of models over a parametric range of feasible microstructures and process conditions within relevant regions of design space.

To facilitate top-down exploration for robust design solutions, an iterative approach is essential for bottom-up information flow (simulations, experiments), guided from the top-down by requirements for applications. To this end, Choi et al. (Choi 2005; Choi et al. 2005, 2008a, b) have developed an approach called the Inductive Design Exploration Method (IDEM), illustrated in Fig. 10 (McDowell 2007). IDEM has two major objectives: (1) to guide bottom-up modeling and simulation to explore top-down, requirements-driven design, and (2) to manage uncertainty propagation in model chains. As illustrated in Fig. 10, IDEM requires an

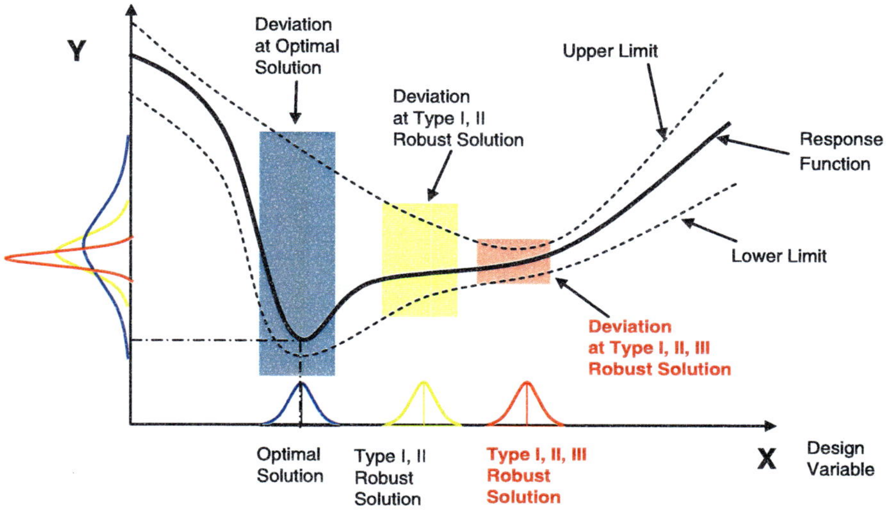

Fig. 9 Illustration of Types I and II robust design solutions relative to optimal solution based on extremum of objective function. Type III robust design minimizes deviation (*dashed line*) of the objective function from the flat region associated with model and microstructure uncertainty (McDowell et al. 2007; Choi 2005)

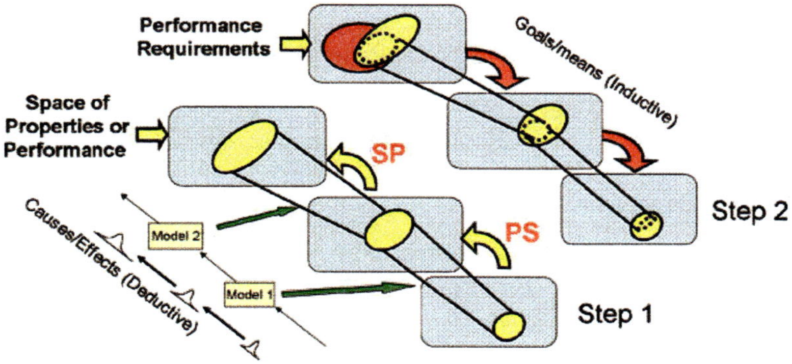

Fig. 10 Schematic of Steps 1 and 2 in IDEM. Step 1 involves bottom-up simulations or experiments, typically conducted in parallel fashion, to map composition into structure and then into properties, with regions in *yellow* showing the feasible ranged sets of points from these mappings. Step 2 involves top-down evaluation of points from the ranged set of specified performance requirements that overlap feasible regions established by bottom-up simulations in Step 1 (Choi 2005; McDowell 2007)

initial configuration of the design process (mappings in Fig. 7). Implementation of IDEM necessitates identifying the connections of inputs and outputs of models, simulations, experiments, databases, and design decision points at various levels of material hierarchy. Moreover, insertion of cDSP decision support "modules" is necessary to achieve a complete graph of information flow. The flow of information and location of decision points (constituting the "design process") is reconfigurable and therefore constitutes an important aspect of the design itself. Model inferences are weighted according to certainty. Quality experimental information can be factored in as desired. Step 1 in Fig. 10 is very important; results from modeling and simulation and available empirical information are used to map potential process–structure and structure–property relations over a sufficiently broad parametric range of compositions, initial structures, and process–structure and structure–property assessments. It involves evaluation of mappings for discrete points at each level of PS, SP, and PP in Figs. 1 and 7; since each of these can be mapped independently without regard to a specific design scenario, it is amenable to massive parallelization of simulations and database mining. In step 2, the results of the step 1 are inverted using concepts of set theory to inductively explore the feasible design spaces of properties, structure, and compositions, working backwards from ranged sets of performance requirements. After step 2, a ranged set of robust solutions can be obtained that factor in model uncertainty by maintaining maximum distance from constraint bounds. Identification of certain subsystems in the material hierarchy with weak coupling to responses of other subsystems is an important step (Choi 2005; Panchal 2005), as these subsystems can be analyzed and designed independently. For example, design of interfaces can be conducted independent of microstructure morphology (grain size, shape, and orientation distribution) in some cases.

Applications of Types I–III robust design methods described above to design of extruded prismatic metals for multifunctional structural and thermal applications (Seepersad 2004; Seepersad et al. 2003, 2004, 2005) and design of multiphase thermite metal–oxide mixtures for target reaction initiation probability under shock compression have been described elsewhere (McDowell et al. 2007; Choi et al. 2005). One challenge is to extend these kinds of concurrent material and product systems design concepts to tailor microstructures that deliver required performance requirements in problems that involve evolution of microstructure, for example:

- Phase morphologies, precipitate/dispersoid distributions, texture, and grain/phase boundary networks in alloy systems for multifunctional performance in terms of strength, ductility, fracture, fatigue, corrosion resistance, shear localization resistance, etc.
- Process path and in-service evolution of microstructure (e.g., plasticity, phase transformation, and diffusion).

Another challenge is to build interfaces that facilitate communication with the design process in terms of high level language regarding "properties" or "responses", at the same time using specific sets of microstructure attributes to represent microstructure with various levels of detail and resolution. The present focus on properties as a medium of communicating design information has historical roots in traditional materials selection. For example, materials development initiatives in industry and government are typically framed in terms of targets on mechanical and thermal properties. An obvious level of the design process shown in Fig. 1 (or Fig. 7) at which to communicate in the "new world" of AIM, ICME, and related approaches for materials design is that of material microstructure, which resides at the interface of process–structure and structure–property relations. This will also be essential to promote horizontal transferability of materials in a range of products with different performance requirements. Moving from a taxonomy based on material properties to microstructure attributes is a transformational step that is yet to be taken, possibly due to the historical separation of materials development (process–structure) by materials suppliers and design requirements (structure–properties and systems design) development at OEMs. As mentioned before, when the design of the material and product is pursued concurrently, it is expected that the material structure will vary through the part; the resulting heterogeneity renders the notion of "properties," an inherently local, homogeneous concept, as an outmoded label for representing the underlying material.

5 Future Prospects: Challenges and Opportunities

Initiatives such as AIM and ICME have been foundational in addressing goals of concurrent design and development of materials and products and providing initial directions for research and development. Systems engineering approaches consistent with these goals are made feasible by the confluence of several fields:

- Computational materials science, micromechanics of materials, and ubiquitous computing
- Advances in high resolution materials characterization and in situ measurements
- Information technology (information theory, databases, digital interfaces, distributed collaboration)
- Decision theory in design (utility theory, goal programming, information economics)

In addition to the change of culture in university design instruction, as well as traditional modes of interaction of materials suppliers and OEMs in industry, proprietary information and licensing issues, etc., there are a number of potential technology barriers that serve as challenges to further development and use of these kinds of methodologies. In materials design problems, one often finds that models are either nonexistent or insufficiently developed to support decision-making. This includes models for both process–structure relations and microstructure–property relations. The ONR/DARPA Dynamic 3-D Digital Structure consortium (Christodoulou 2009) is an example of an initiative that is addressing development of such tools to support materials design, in this case with emphasis on 3D characterization, modeling, and simulation tools and methods. Of course, the issue of utility of 2D versus 3D modeling in design decisions itself is important. One particular need is the coordination of model repositories for rapid availability to design searches. A complicating factor that is rarely addressed is the quantification of uncertainty of model parameters and model structure that is necessary in robust design of materials.

Another very important consideration is that mechanistic models are often the limiting factor in applying decision-based design frameworks; however, guidance is required to decide how to best invest in model development that will maximize payoff or utility in the design process. Not all models are equally important in terms of their role in design, and this assessment depends heavily on the design objectives and requirements. For example, advances in geometric representation of digital polycrystalline microstructures should be accompanied by advances in modeling tools that can account for effects of interfaces and grain boundaries – this reflects the "gulf" of bridging models at atomistic scales and crystal plasticity discussed earlier in relation to Fig. 5. It must be emphasized that as the scales of structure decreases (multilayers, MEMS, etc.), the problem of defect nucleation becomes increasingly important and is still not well characterized. The gap in bridging scales via modeling and simulation is closing each year with advances in discrete dislocation plasticity and statistical theories, but progress in developing predictive methods for dislocation patterning at mesoscales has been slow, in part due to the lack of top-down calibration and in part to the complexity of representing interfaces and dislocation reactions with reduced order models. This is an issue that affects capabilities to model material workhardening, strain localization, and failure initiation. There is still an open issue regarding the critical path of different classes of models in terms of support for materials design. Balanced investment in geometric modeling of microstructures and physically based constitutive models is important. Efforts to identify pertinent scaling relations are very important as they offer potentially more descriptive yet reduced order descriptions. This requires appropriate forums (such as the D-3D Digital Structure consortium) involving all relevant parties (computing and information sciences, materials engineering, mechanics, code developers, materials suppliers, national labs, etc.).

Opportunities for building the infrastructure for materials design (AIM, ICME, etc.) are numerous, including:

- Methods for conducting feasibility studies and early stage robust concept exploration for potential design solutions to assist in framing materials development goals, cost trade-offs, and realistic goals for developing new products with enhanced capabilities.
- Balancing iterations of material process design with structure–property simulations and experiments – managing assets and deciding on nature of interfaces between process–structure and structure–property relations (microstructure-mediated design). Methods for sequential pursuit of design exploration and detail design.
- Linking concurrent material and product design with the information sciences (Broderick et al. 2008), including elements such as knowledge discovery extracted from databases via data mining in interdisciplinary areas such as statistics, materials databases, and results of material modeling to assist in discovery of new materials.
- Parallel processing algorithms for robust concept exploration, moving well beyond combinatorial searching and data mining. Materials design is an ideal candidate for parallelization in the initial design exploration process (cf. IDEM Step 1 in Fig. 10).

Benefits that accrue to pursuing this vision of integrated design of materials and products have transformational implications. These include:

- Quantifying various forms of uncertainty in the design process
- Prioritizing investment in models and computational methods in terms of objective measures of utility in supporting design decisions
- Prioritizing mechanisms and materials science phenomena at various length and time scales to be modeled for a given design problem
- Conducting feasibility studies to establish probable return on investment of candidate new material systems

6 Conclusion

Recent initiatives such as AIM and ICME have been reviewed that use engineering systems approaches for concurrently designing materials and products, accelerating insertion of improved materials in applications. Characteristics of materials design approaches have been outlined, including the distinction of top-down materials design from multiscale modeling. Robust design methods are preferred owing to the prevalence of uncertainty in process control, randomness of microstructure, and nonequilibrium, path-dependent nature of inelastic deformation and associated constitutive models for material behavior, among other sources.

The future of simulation-assisted materials design is promising, particularly in view of ongoing initiatives such as ICME that reinforce its strategic value in industry and create a technology pull for basic research. We envision that strategic planning for materials development programs in the future will draw on this emerging multidiscipline in a way that promotes innovation of products with radical leaps of functionality. This is all made possible by the confluence of engineering science and mechanics, quantitative materials science, materials characterization, materials physics and chemistry, computing and information sciences, and systems engineering. Design curricula and modeling and simulation courses in universities, including materials science departments, will need to respond to this sea change of materials design and development. A premium should be placed on development of effective modes and infrastructure for collaboration of disparate experts and designers involved in the process. For materials design to realize its full potential, collaborative models must address intellectual property issues of data/model sharing and software licensing, including the possibility of pay-per-use licensing. Certainly, standards for certifying validity of tools, as well as specifications of uncertainty, are essential elements of this kind of distributed framework. One can envision that while overall design goals and integration are managed with proprietary limits on information content/flow and system configuration, vendors of material modeling and simulation services could be located anywhere, capitalizing on just-in-time delivery of modeling services from the leading experts within a global context.

Acknowledgments The coauthors are grateful for funding that supported their collaboration in the DARPA AIM program (Dr. Leo Christodoulou, monitor). DLM also wishes to acknowledge support from the DARPA Synthetic Multifunctional Materials Program (Dr. Leo Christodoulou, monitor), the Center for Computational Materials Design (CCMD), a NSF I/UCRC jointly founded by Penn State and Georgia Tech (DLM Co-Director), http://www.ccmd.psu.edu/, as well as support from DARPA/P&W Prognosis (Dr. Leo Christodoulou, DARPA and Dr. Robert Grelotti, P&W, monitors), and ONR D3D tools programs (Dr. Julie Christodoulou, government prime, with Drs. G.B. Olson and H. Jou at QuesTek as monitors). Dr. McDowell especially wishes to thank his many Georgia Tech colleagues (Systems Realization Laboratory faculty Professors F. Mistree and J.K. Allen, and former co-advised graduate students in materials design C.C. Seepersad, H.-J. Choi, and J.H. Panchal) for collaborating to develop first generation robust materials design concepts such as Type III robust design and IDEM reviewed in this chapter.

References

Adams BL, Gao X (2004) 2-point microstructure archetypes for improved elastic properties. J. Comput. Aided Mater. Des. 11(2–3):85–101.

Adams BL, Lyon M, Henrie B (2004) Microstructures by design: linear problems in elastic-plastic design. Int. J. Plast. 20(8–9):1577–1602.

Apelian D (2004) Accelerating technology transition. In: National Research Council Report. National Academies Press, Washington DC.

Arsenlis A, Parks DM (1999) Crystallographic aspects of geometrically-necessary and statistically-stored dislocation density. Acta Mater. 47(5):1597–1611.

Arsenlis A, Parks DM (2002) Modeling the evolution of crystallographic dislocation density in crystal plasticity. J. Mech. Phys. Solids 50:1979–2009.

Ashby MF (1999) Materials Selection in Mechanical Design. 2nd Edition, Butterworth-Heinemann, Oxford, UK.

Backman D, Dutton R (2006) Integrated materials modeling for aerospace components. Symp. on the Role of Computational Methods in Materials Research and Development: Applications of Materials Modeling and Simulation, MS&T '06, Cincinnati, OH, Oct. 18.

Billinge SJE, Rajan K, Sinnot SB (2006) From Cyberinfrastructure to Cyberdiscovery in Materials Science: Enhancing Outcomes in Materials Research, Education and Outreach. Report of NSF-sponsored workshop held in Arlington, VA, Aug. 3–5, http://www.mcc.uiuc.edu/nsf/ciw_2006/. Accessed 8 December 2009.

Broderick S, Suh C, Nowers J, Vogel B, Mallapragada S, Narasimhan B, Rajan K (2008) Informatics for combinatorial materials science. JOM 60(3):56–59.

Bulatov VV (2002) Current developments and trends in dislocation dynamics. J. Comput. Aided Mater. Des. 9(2):133–144.

Bulatov VV, Tang MJ, Zbib HM (2001) Crystal plasticity from dislocation dynamics. MRS Bull. 26(3):191–195.

Butler GC, McDowell DL (1998) Polycrystal constraint and grain subdivision. Int. J. Plast. 14:703–717.

Capolungo L, Spearot DE, Cherkaoui M, McDowell DL, Qu J, Jacob K (2007) Dislocation nucleation from bicrystal interfaces and grain boundary ledges: relationship to nanocrystalline deformation. J. Mech. Phys. Solids 55(11):2300–2327.

Chen W (1995) A robust concept exploration method for configuring complex systems, Ph.D. Dissertation, G.W. Woodruff School of Mechanical Engineering, Georgia Institute of Technology, Atlanta, GA.

Choi H-J (2005) A robust design method for model and propagated uncertainty, Ph.D. Dissertation, G.W. Woodruff School of Mechanical Engineering, Georgia Institute of Technology, Atlanta, GA.

Choi H-J, Austin R, Shepherd K, Allen JK, McDowell DL, Mistree F, Benson DJ (2005) An approach for robust design of reactive powder metal mixtures based on non-deterministic micro-scale shock simulation J. Comput. Aided Mater. Des.12(1):57–85.
Choi H-J, McDowell DL, Allen JK, Mistree F. (2008a) An inductive design exploration method for hierarchical systems design under uncertainty. Eng. Optim. 40(4):287–307.
Choi H-J, McDowell DL, Allen JK, Rosen D, Mistree F (2008b) An inductive design exploration method for robust multiscale materials design. J. Mech. Des. 130(3):031402–1–13.
Christodoulou J (2009) Dynamic 3-dimensional digital structure: a program review. JOM 61(10):21.
Fan J, McDowell DL, Horstemeyer MF, Gall K (2003) Cyclic plasticity at pores and inclusions in cast Al-Si alloys. Eng. Fract. Mech. 70(10):1281–1302.
Gall K, Horstemeyer MF, McDowell DL, Fan J (2000) Finite element analysis of the stress distributions near damaged Si particle clusters in cast Al-Si alloys. Mech. Mater. 32(5):277–301.
Gall K, Horstemeyer MF, Degner BW, McDowell DL, Fan J (2001) On the driving force for fatigue crack formation from inclusions and voids in a cast A356 aluminum alloy. Int. J. Fract. 108:207–233.
Ghosh S, Bai J, Raghavan P (2007) Concurrent multi-level model for damage evolution in microstructurally debonding composites. Mech. Mater. 39(3):241–266.
Gumbsch P (1995) An atomistic study of brittle fracture: toward explicit failure criteria from atomistic modeling. J. Mater. Res. 10:2897–2907.
Hao S, Moran B, Liu WK, Olson GB (2003) A hierarchical multi-physics model for design of high toughness steels. J. Comput. Aided Des. 10:99–142.
Hao S, Liu WK, Moran B, Vernerey F, Olson GB (2004) Multi-scale constitutive model and computational framework for the design of ultra-high strength, high toughness steels. Comput. Methods Appl. Mech. Eng. 193:1865–1908.
Horstemeyer MF, McDowell DL (1998) Modeling effects of dislocation substructure in polycrystal elastoviscoplasticity. Mech. Mater. 27:145–163.
Hughes DA, Liu Q, Chrzan DC, Hansen N (1997) Scaling of microstructural parameters: misorientations of deformation induced boundaries. Acta. Mater. 45(1):105–112.
Isukapalli SS, Roy A, Georgopoulos PG (1998) Stochastic response surface methods (SRSMs) for uncertainty propagation: application to environmental and biological systems. Risk Analysis 18(3):351–363.
Kalidindi SR, Houskamp J, Proust G, Duvvuru H (2005) Microstructure sensitive design with first order homogenization theories and finite element codes. Mater. Sci. Forum 495–497: 23–29.
Kalidindi SR, Houskamp JR, Lyon M, Adams BL (2004) Microstructure sensitive design of an orthotropic plate subjected to tensile load. Int. J. Plast. 20(8–9):1561–1575.
Knezevic M, Kalidindi SR, Mishra RK (2008) Delineation of first-order closures for plastic properties requiring explicit consideration of strain hardening and crystallographic texture evolution. Int. J. Plast. 24(2):327–342.
Kouznetsova V, Geers MGD, Brekelmans WAM (2002) Multi-scale constitutive modelling of heterogeneous materials with a gradient-enhanced computational homogenization scheme. Int. J. Numer. Methods Eng. 54(8):1235–1260.
Kouznetsova VG, Geers MGD, Brekelmans WAM (2004) Multi-scale second-order computational homogenization of multi-phase materials: a nested finite element solution strategy. Comput. Methods Appl. Mech. Eng. 193(48–51):5525–5550.
Kuhlmann-Wilsdorf D (1989) Theory of plastic deformation: properties of low energy dislocation structures Mater. Sci. Eng. A 113:1–41.
Larsson R, Diebels S (2007) A second-order homogenization procedure for multi-scale analysis based on micropolar kinematics. Int. J. Numer. Methods Eng. 69:2485–2512.
Leffers T (1994) Lattice rotations during plastic deformation with grain subdivision. Mater. Sci. Forum 157–162:1815–1820.
Lyon M, Adams BL, (2004) Gradient-based non-linear microstructure design. J. Mech. Phys. Solids 52(11):2569–2586.

McDowell DL (2001) Materials design: a useful research focus for inelastic behavior of structural metals. Special Issue of Theoretical and Applied Fracture Mechanics, Prospects of Mesomechanics in the 21st Century: Current Thinking on Multiscale Mechanics Problems, (eds. G.C. Sih, V.E. Panin) 37:245–259.

McDowell DL (2007) Simulation-assisted materials design for the concurrent design of materials and products. JOM 59(9):21–25.

McDowell DL (2008) Viscoplasticity of heterogeneous metallic materials. Mater. Sci. Eng. R. Rep. 62(3):67–123.

McDowell DL, Olson GB (2008) Concurrent design of hierarchical materials and structures. Sci. Model. Simul. (CMNS) 15(1): 207–240.

McDowell DL, Story TL (1998) New directions in materials design science and engineering, Report of NSF DMR-sponsored workshop held in Atlanta, GA, Oct. 19–21.

McDowell DL, Gall K, Horstemeyer MF, Fan J (2003) Microstructure-based fatigue modeling of cast A356-T6 alloy. Eng. Fract. Mech. 70:49–80.

McDowell DL, Choi H-J, Panchal J, Austin R, Allen JK, Mistree F (2007) Plasticity-related microstructure-property relations for materials design. Key Eng. Mater. 340–341:21–30.

McVeigh C, Vernerey F, Liu WK, Brinson LC (2006) Multiresolution analysis for material design Comput. Methods Appl. Mech. Eng. 195:5053–5076.

Messer M, Panchal JH, Allen JK, McDowell DL, Mistree F (2007) A function-based approach for integrated design of material and product concepts. DETC2007–35743, Proceedings of IDETC/CIE 2007, ASME 2007 International Design Engineering Technical Conferences & Design Automation Conference, Las Vegas, NV, Sept. 4–7.

Mistree F, Hughes OF, Bras BA (1992) The compromise decision support problem and the adaptive linear programming algorithm. Structural Optimization: Status and Promise (ed. M. P. Kamat), AIAA, Washington, DC, 251–290.

Oden JT, Belytschko T, Fish J, Hughes TJR, Johnson C, Keyes D, Laub A, Petzold L, Srolovitz D, Yip S (2006) Simulation-Based Engineering Science: Revolutionizing Engineering Science through Simulation, Report of NSF Blue Ribbon Panel on Simulation-Based Engineering Science. http://www.nsf.gov/pubs/reports/sbes_final_report.pdf. Accessed 8 December 2009.

Olson GB (1997) Computational design of hierarchically structured materials. Science 277(5330):1237–1242.

Olson GB (2000) Designing a new material world. Science 288:993–998.

Olson GB (2001) Brains of steel: mind melding with materials. Int. J. Eng. Educ. 17(4–5):468–471.

Panchal JH (2005) A framework for simulation-based integrated design of multiscale products and design processes, Ph.D. Dissertation, G.W. Woodruff School of Mechanical Engineering, Georgia Institute of Technology, Atlanta, GA.

Panchal JH, Choi H-J, Shepherd J, Allen JK, McDowell DL, Mistree F (2005) A strategy for simulation-based multiscale, multifunctional design of products and design processes. ASME Design Automation Conference, Long Beach, CA. Paper Number: DETC2005–85316.

Panchal JH, Choi H-J, Allen JK, McDowell DL, Mistree F (2007) A systems-based approach for integrated design of materials, products and design process chains. J. Comput. Aided Mater. Des. 14(1):265–293.

Pollock TM, Allison J (2008) Integrated Computational Materials Engineering: A Transformational Discipline for Improved Competitiveness and National Security, Committee on Integrated Computational Materials Engineering, National Materials Advisory Board, National Research Council of the National Academies, ISBN 13:978-0-309-11999-3.

Qu S, Shastry V, Curtin WA, Miller RE (2005) A finite-temperature dynamic coupled atomistic/discrete dislocation method. Model. Simul. Mater. Sci. Eng. 13(7):1101–1118.

Rafii-Tabar H, Hua L, Cross M (1998) A multi-scale atomistic-continuum modeling of crack propagation in a two-dimensional macroscopic plate. J. Phys. Condens. Matter 10(11):2375–2387.

Rudd RE, Broughton JQ (1998) Coarse-grained molecular dynamics and the atomic limit of finite elements. Phys. Rev. B 58(10):R5893–R5896.

Rudd RE, Broughton JQ (2000) Concurrent coupling of length scales in solid state systems. Phys. Status Solidi B 217(1):251–291.

Seepersad CC (2004) A robust topological preliminary design exploration method with materials design applications, Ph.D. Dissertation, G.W. Woodruff School of Mechanical Engineering, Georgia Institute of Technology, Atlanta, GA.

Seepersad CC, Dempsey BM, Allen JK, Mistree F, McDowell DL (2003) Design of multifunctional honeycomb materials. AIAA J. 42(5):1025–1033.

Seepersad CC, Fernandez MG, Panchal JH, Choi H-J, Allen JK, McDowell DL, Mistree F (2004) Foundations for a systems-based approach for materials design. 10th AIAA/ISSMO Multidisciplinary Analysis and Optimization Conference. Albany, NY: AIAA MAO: AIAA-2004-4300.

Seepersad CC, Kumar RS, Allen JK, Mistree F, McDowell DL (2005) Multifunctional design of prismatic cellular materials. J. Comput. Aided Mater. Des. 11(2–3):163–181.

Seepersad CC, Allen JK, McDowell DL, Mistree F (2008) Multifunctional topology design of cellular structures. J. Mech. Des. 130(3):031404–1–13.

Shen C, Wang Y (2003) Modeling dislocation network and dislocation–precipitate interaction at mesoscopic scale using phase field method. Int. J. Multiscale Comput. Eng. 1(1):91–104.

Shenoy MM, Zhang J, McDowell DL (2007) Estimating fatigue sensitivity to polycrystalline Ni-base superalloy microstructures using a computational approach. Fatig. Fract. Eng. Mater. Struct. 30(10):889–904.

Shenoy M, Tjiptowidjojo Y, McDowell DL (2008) Microstructure-sensitive modeling of polycrystalline IN 100. Int. J. Plast. 24:1694–1730.

Shiari B, Miller RE, Curtin WA (2005) Coupled atomistic/discrete dislocation simulations of nanoindentation at finite temperature. ASME J. Eng. Mater. Technol. 127(4):358–368.

Shilkrot LE, Curtin WA, Miller RE (2002) A coupled atomistic/continuum model of defects in solids. J. Mech. Phys. Solids 50:2085–2106.

Shilkrot LE, Miller RE, Curtin WA (2004) Multiscale plasticity modeling: coupled atomistics and discrete dislocation mechanics. J. Mech. Phys. Solids 52:755–787.

Shu C, Rajagopalan A, Ki X, Rajan K (2003) Combinatorial materials design through database science. In: Materials Research Society Symposium – Proceedings, vol 804, Combinatorial and Artificial Intelligence Methods in Materials Science II:333–341.

Taguchi G (1993) Taguchi on Robust Technology Development: Bringing Quality Engineering Upstream; ASME Press, New York.

Vernerey F, Liu WK, Moran B (2007) Multi-scale micromorphic theory for hierarchical materials. J. Mech. Phys. Solids 55(12):2603–2651.

Wang A-J, Kumar RS, Shenoy MM, McDowell DL (2006) Microstructure-based multiscale constitutive modeling of $\gamma - \gamma'$ nickel-base superalloys. Int. J. Multiscale Comput. Eng. 4(5–6):663–692.

Wang YU, Jin YM, Cuitiño AM, Khachaturyan AG (2001) Nanoscale phase field microelasticity theory of dislocations: model and 3D simulations. Acta Mater. 49(10):1847–1857.

Warner DH, Sansoz F, Molinari JF (2006) Atomistic based continuum investigation of plastic deformation in nanocrystalline copper. Int. J. Plast. 22:754–774.

Weinan E, Huang Z (2001) Matching conditions in atomistic-continuum modeling of materials Phys. Rev. Lett. 8713(13):135501.

Zbib HM, de la Rubia TD (2002) A multiscale model of plasticity. Int. J. Plast. 18(9):1133–1163.

Zbib HM, de la Rubia TD, Bulatov V (2002) A multiscale model of plasticity based on discrete dislocation dynamics. ASME J. Eng. Mater. Technol. 124(1):78–87.

Index

S. Ghosh and D. Dimiduk (eds.), *Computational Methods for Microstructure-Property Relationships*, DOI 10.1007/978-1-4419-0643-4,

D

N